Progelhof/Throne

Polymer Engineering Principles

D0223703

SPE Books from Hanser Publishers

Richard C. Progelhof / James L. Throne

Polymer Engineering Principles

Properties, Processes, and
Tests for Design

With 478 Illustrations and 122 Tables

Hanser Publishers, Munich Vienna New York Barcelona

Hanser/Gardner Publications, Inc., Cincinnati

The Authors:
Richard C. Progelhof, Division of Engineering and Engineering Technology, Penn State Erie, The Behrend College, Erie, PA 16563-0101, USA
James L. Throne, Sherwood Technologies, Inc., Hinckley, OH 44233-9676, USA

Distributed in the USA and in Canada by
Hanser/Gardner Publications, Inc.
6600 Clough Pike, Cincinnati, Ohio 45244-4090, USA
Fax: + 1 (513) 527-8950

Distributed in all other countries by
Carl Hanser Verlag
Postfach 86 04 20, 81631 München, Germany
Fax: + 49 (89) 98 48 09

The use of general descriptive names, trademarks, etc., in this publication, even if the former are not especially identified, is not to be taken as a sign that such names, as understood by the Trade Marks and Merchandise Marks Act, may accordingly be used freely by anyone.

While the advice and information in this book are believed to be true and accurate at the date of going to press, neither the authors nor the editors nor the publisher can accept any legal responsibility for any errors or omissions that may be made. The publisher makes no warranty, express or implied, with respect to the material contained herein.

Die Deutsche Bibliothek - CIP-Einheitsaufnahme
Progelhof, Richard C.:
Polymer engineering principles : properties, processes,
and tests for design ; with 122 tables / Richard C. Progelhof ;
James L. Throne. - Munich ; Vienna ; New York ; Barcelona :
Hanser, 1993
    ISBN 3-446-17337-4 brosch.
    ISBN 3-446-15686-0 Pb.
NE: Throne, James L.:

ISBN 1-56990-150-3 (hardcover) Hanser/Gardner Publications, Inc., Cincinnati
ISBN 1-56990-151-1 (paperback) Hanser/Gardner Publications, Inc., Cincinnati

© Carl Hanser Verlag, Munich Vienna New York Barcelona, 1993
Typesetting in the USA by Compset, Inc., Beverly, Massachusetts
Printed and bound in Germany by Passavia Druckerei GmbH, Passau

FOREWORD

The Society of Plastics Engineers is pleased to endorse and sponsor this volume entitled *Polymer Engineering Principles: Properties, Processes, and Tests for Design.* SPE's Technical Volumes Committee believes that this book has the potential to be an SPE best seller because of its far-reaching utility as both a textbook and reference title. The volume includes an attractively concise introduction to polymers, followed by detailed chapters on physical, chemical, and load-bearing properties. The hundreds of figures and tables integrated throughout the text, including later chapters on polymer processing, testing, and design, make this a valuable tool for people in the academic arena as well as practicing engineers.

SPE, through its Technical Volumes Committee, has long sponsored books on various aspects of plastics. Its involvement has ranged from identification of needed volumes and recruitment of authors to peer review and approval and publication of new books.

Technical competence pervades all SPE activities, not only in the publication of books, but also in other areas, such as sponsorship of technical conferences and educational programs. In addition, the Society publishes periodicals, including *Plastics Engineering, Polymer Engineering and Science, Polymer Processing and Rheology, Journal of Vinyl Technology* and *Polymer Composites,* as well as conference proceedings and other publications, all of which are subject to rigorous technical review procedures.

The resource of some 37,000 practicing plastics engineers has made SPE the largest organization of its type worldwide. Further information is available from the Society at 14 Fairfield Drive, Brookfield, Connecticut 06804, U.S.A.

Eugene De Michele
Executive Director
Society of Plastics Engineers

Technical Volumes Committee
Claire Bluestein, Chairperson
Captan Associates, Inc.

PREFACE

Polymers are organic macromolecules. They do not have the inherent strength of traditional materials such as wood, steel or even concrete. They are expensive man-made materials whose physical states can be easily altered by combining more than one polymer, by adding agents to improve processing, by adding fillers or reinforcing fibers, and so on. The resulting polymeric materials can be easily shaped, have excellent chemical resistance, good ductility and volumetric color. On the outside, deliberate attempts to produce useful organic macromolecules can be traced back no more than 150 years. Commercial purely synthetic polymers are truly twentieth century materials.

The engineer or designer needs the properties of a selected material in order to determine its suitability in the required environment. The environment may be entirely mechanical as with the design of a shelf. It may be entirely chemical as in the design of a bleach bottle. Or it may be a combination of environmental effects, such as hydromechanical, chemical and thermal in the case of a brake-fluid hose. The engineer or designer needs to know how plastics differ in performance from one polymer to another, and from that of traditional materials. The designer or engineer usually relies on properties that are obtained from testing routines originally established for traditional materials such as metals. He/she must realize that, unlike traditional material response, polymer behavior changes with changes in intensity and duration of the impressed environmental effect.

The intent of this book is modest. It is designed to lead the reader from an understanding of the basic elements that make up a polymer, through the inherent characteristics of a polymer such as glass transition temperature, melt temperature, molecular weight distribution and degree of crystallinity, through solid and fluid characteristics and basic processing principles, to an understanding of polymer response to testing conditions. The book has six chapters:

1. *Introduction to Polymers*. After a brief history, the nature of a polymer is discussed, beginning with atomic structure, continuing through the concept of a monomer to the building of a large molecule. Combinations of polymers are then discussed, and the nature of the bonding forces that allow polymer toughness and chemical resistance are considered. The comparative polymeric natures of thermoplastic and thermosetting resins are then put into perspective.

2. *The Polymer Solid State*. Several definitions of molecular weight and molecular weight distribution are presented as well as methods for determining them. The crystalline nature of polymers is then discussed. This leads to the molecular nature of

polymer transitions, including glass transition temperature, and then to the transport properties of polymers. And finally, the effect of additives, fillers and reinforcements on the molecular state of polymers is discussed.

3. *Mechanical Properties of Polymers.* Polymers, as a class of materials, exhibit solid-like properties at times and at other times exhibit fluid-like properties. After a brief review of classical material response to applied load, the viscoelastic nature of polymers is considered, under constant load and under dynamic load. The response of polymeric materials to impact or short-term loading and creep or long-term loading is then discussed in the light of viscoelasticity.

4. *Polymer Fluid and Chemical Properties.* Polymers in a liquid state are highly viscoelastic and usually have very high viscosities when compared with simple molecule fluids. As a result, they can behave as elastic liquids and can generate substantial heat while being processed. Polymers are compressible at high processing pressures. The thermodynamics of polymers are discussed in detail. The chapter ends with a discussion of polymer resistance to solvents and chemicals.

5. *Processing.* A brief overview of the major commercial ways of processing polymers into final product shapes is given here. Emphasis is placed on understanding the interaction of the processing environment and the polymer. Extrusion and injection molding are the major commercial thermoplastic processes that are considered in detail. Compression molding is the primary commercial thermosetting resin process that is emphasized. The interrelationship between polymer properties and computer-aided design databases is considered in some detail.

6. *Testing for Design.* This extensive chapter is the major focus of the book, with the preceding chapters needed for full understanding. The chapter focuses on standard test procedures for fluid and solid polymers and how these tests can be used to obtain an understanding of polymer response to real environmental effects. In addition to the standard tensile and flexural testing, friction and wear testing, environmental stress crack testing, weathering, fire retardancy testing and electrical resistance testing are also considered.

These contents are not meant to be new, novel or even evolutionary. The book is not a handbook or a compendium of polymer material properties. For this, we recommend the reader to the following:

D.W. Van Krevelen, *Properties of Polymers,* Elsevier, New York NY (1976).

H. Domininghaus, *Die Kunststoffe und ihre Eigenschaften,* VDI Verlag, Düsseldorf, FRG (1986).

H. Saechtling, *International Plastics Handbook,* Hanser Verlag, Munich (1987).

Modern Plastics Encyclopedia, Published every October by Modern Plastics, 1221 Ave. of the Americas, New York NY 10020.

Encyclopedia of Polymer Science and Engineering, 2nd. Ed., John Wiley & Sons, Inc., 605 Third Ave., New York NY 10158 (1985-1990).

The book is also not meant to be all-inclusive. The rapid growth of the industry has spurred publication of many books that specialize in narrow segments of this material. We do not pretend to summarize other treatises. Instead, wherever appropriate, we refer the reader to them for more information. As an example, in Chapter 5, we present a brief overview of the economically important polymer processes, so that the reader may appreciate how processing constraints influence polymer performance in the final part. There are a dozen or more excellent textbooks that focus specifically on polymer processing and these are cited in the references of that chapter.

We believe that this material will allow the reader to gain a better understanding of the limitations inherent in standard handbook data. With this understanding, the engineer or designer should be better able to select polymer materials that have optimum response to environmental effects. As with any survey text, the key to presenting this material is in knowing what to include. We have focused on polymers that form the major constituent of the final product. The polymer may be "neat" or pure, but it may also be combined with other polymers, adducts such as processing aids, fillers or reinforcing elements. We have chosen not to include the rapidly growing area of composites. It is not at all clear that the mechanical properties of the polymer matrix can be isolated from that of the composite. And despite the intense world-wide research and development efforts in this area, applications are currently restricted to products such as aircraft and aerospace components, where performance dominates material and manufacturing economics.

Many sections in this book owe allegiance to a few classic treatises:

R.D. Deanin, *Polymer Structure, Properties and Applications*, Cahners Books, Boston (1972). A logical progression from the chemical nature of the polymer to its interaction with adducts.

J.M. Dealy, *Rheometers for Molten Plastics*, Van Nostrand Reinhold Co., New York (1982). The best of several books that relate the fundamentals of polymer fluid flow to the current methods of measuring shear viscosity, normal stress difference, elongational viscosity and so on.

M.L. Miller, *The Structure of Polymers*, Reinhold, New York (1966). An early book that relates polymer mechanical response to polymer structure, including an extensive section on crystallinity.

T. Alfrey, Jr., *Mechanical Behavior of High Polymers*, Interscience Publishers, New York (1948). The classic work relating linear viscoelasticity of polymers to their mechanical response to applied load.

S. Turner, *Mechanical Testing of Polymers*, 2nd Ed., George Godwin/PRI, London (1983). A thoughtful discussion of the purpose of testing of polymers, in which the distinction between a material *property* and a material *parameter* is clearly made and reinforced via examples.

One of us (JLT) wishes to thank two years of first-year Polymer Engineering graduate student classes and one year of senior undergraduate Mechanical Engineering students for deciphering, reviewing, correcting, editing and otherwise struggling with the manuscript in a very difficult format.

The influence of the late John L. O'Toole pervades this work. His best written effort was the Annual Design Guide in Modern Plastics Encyclopedia. More importantly, John was a teacher's teacher. He taught us to understand, in general but simple terms, how a polymer responds to controlled environmental forces. He convinced us that temperature must always be considered an important physical parameter. And he taught us that we must always stress that polymer material properties *depend* on the duration and intensity of environmental change. Thus, it is incumbent upon us to reinforce his admonition that the engineer or designer must appreciate that he/she is dealing with material parameters rather than material constants that are typical of traditional materials. We dedicate this book to the memory of our mentor, counselor and friend, John L. O'Toole.

R.C. Progelhof
J.L. Throne

HOW TO USE THIS BOOK
AS A TEXTBOOK

The book naturally progresses from simple differentiation of polymers from more traditional materials such as metals and ceramics, through solid and fluid mechanics of polymers to an analysis of the standard tests that help to define those polymer properties needed to design today's functional products. The text material is not meant to be encyclopedic in scope.

However, by design, the authors have included more material than can be covered in a single course of 40 contact hours or so. We believe that this allows the instructor to tailor the course material to his or her interests without requiring extensive supplemental materials.

The authors recognize that this penalizes the student who must now purchase a book that is more extensive and hence more expensive than one more carefully designed to meet a given course syllabus. However, we believe that the material selected for inclusion is classic in nature and based on well-founded concepts. As a result, the student should have every confidence that the book will remain a good source for information on polymer material properties for many years to come.

There are several ways of adapting the material to a one-semester senior-level undergraduate or introductory graduate-level course. Three scenarios follow:

Scenario One
Emphasis on Solid Polymer Science

Chapter 1   Introduction
Chapter 2   The Polymer Solid State
Chapter 3   Mechanical Properties of Polymers
Chapter 4   Polymer Fluid and Chemical Properties (Sections 4.1 through 4.5)
Chapter 6   Testing for Design (Sections 6.3 through 6.6)

Scenario Two
Emphasis on Fluid and Chemical
Polymer Science

Chapter 1   Introduction
Chapter 2   The Polymer Solid State
Chapter 3   Mechanical Properties of Polymers (Sections 3.1 through 3.7)
Chapter 4   Polymer Fluid and Chemical Properties
Chapter 6   Testing for Design (Sections 6.1 through 6.2, 6.3, 6.7 through 6.10)

Scenario Three
Emphasis on Processing Effects
on Properties

Chapter 1   Introduction
Chapter 2   The Polymer Solid State
Chapter 3   Mechanical Properties of Polymers (Sections 3.1 through 3.7)
Chapter 4   Polymer Fluid and Chemical Properties (Sections 4.1 through 4.5)
Chapter 5   Processing
Chapter 6   Testing for Design (Sections 6.1 through 6.2; Sections 6.6 through 6.7)

At the end of the text, there are *homework problems* for each chapter and each major emphasis within each chapter. *Answers* are given for many problems. In addition, the authors have extensively used open *literature references* to illustrate many of the major points.

To aid in skimming the book, general discussion is in large type and *examples* are given in smaller type. A *glossary* of terms is provided at the end of each chapter and the book contains extensive *subject and author indices*.

CONTENTS

Ques 207

CONVERSION FACTORS

Length:

m	$\times$ 3.28	ft	$\times$ 0.3048	m
μm	$\times$ 10^{-6}	m	$\times$ 10^6	μm
km	$\times$ 1.609	mile	$\times$ 0.622	km
Å	$\times$ 10^{-10}	m	$\times$ 10^{10}	Å
mm	$\times$ 39.37	mils	$\times$ 0.0254	mm

Area:

m^2	$\times$ 10.76	ft^2	$\times$ 0.0929	m^2
cm^2	$\times$ 0.155	in^2	$\times$ 6.452	cm^2
mm^2	$\times$ 1.55 $\times$ 10^{-3}	in^2	$\times$ 645.2	mm^2

Volume:

m^3	$\times$ 35.31	ft^3	$\times$ 0.02832	m^3
m^3	$\times$ 6.102 $\times$ 10^4	in^3	$\times$ 1.639 $\times$ 10^{-5}	in^3
mm^3	$\times$ 6.102 $\times$ 10^{-5}	in^3	$\times$ 1.639 $\times$ 10^4	mm^3
liter	$\times$ 1000	cm^3	$\times$ 0.001	liter
cm^3	$\times$ 29.57	fluid oz	$\times$ 0.0338	cm^3
m^3	$\times$ 264.2	US gallon	$\times$ 3.785 $\times$ 10^{-3}	m^3

Mass:

g	$\times$ 0.0022	lb_m	$\times$ 453.6	g
kg	$\times$ 2.205	lb_m	$\times$ 0.4536	kg
kg	$\times$ 0.001	metric tonne	$\times$ 1000	kg
kg	$\times$ 0.0011	US ton	$\times$ 907.2	kg

Density:

g/cm^3	$\times$ 62.42	lb_m/ft^3	$\times$ 0.016	g/cm^3
g/cm^3	$\times$ 0.03611	lb_m/in^3	$\times$ 27.69	g/cm^3
kg/m^3	$\times$ 0.06242	lb_m/ft^3	$\times$ 16.02	kg/m^3
g/cm^3	$\times$ 0.578	oz/in^3	$\times$ 1.73	g/cm^3
kg/m^3	$\times$ 5.78 $\times$ 10^{-4}	oz/in^3	$\times$ 1.73 $\times$ 10^3	kg/m^3

Force:

N	$\times$ 0.2248	lb_f	$\times$ 4.448	N
kg_f	$\times$ 0.2292	lb_f	$\times$ 4.363	kg_f
kN	$\times$ 0.2248	kip ($10^3\ lb_f$)	$\times$ 4.448	kN
dyne	$\times$ 2.248×10^{-6}	lb_f	$\times$ 2.248×10^{-6}	dyne
dyne	$\times$ 10^6	N	$\times$ 10^5	dyne

Pressure:

Pa	$\times$ 1.45×10^{-4}	lb_f/in^2	$\times$ 6895	Pa
MPa	$\times$ 9.869	atm	$\times$ 1.013×10^5	MPa
Pa	$\times$ 10	dyn/cm^2	$\times$ 0.100	Pa
Pa	$\times$ 7.5×10^{-3}	1 mm Hg	$\times$ 133.3	Pa
Pa	$\times$ 4.012×10^{-3}	1 inch H_2O	$\times$ 248.9	Pa
MPa	$\times$ 10	bar	$\times$ 0.1	MPa
N/mm^2	$\times$ 145	lb_f/in^2	$\times$ 6.895×10^{-3}	N/mm^2

Energy:

J	$\times$ 9.478×10^{-4}	Btu	$\times$ 1055	J
ft lb_f	$\times$ 1.286×10^{-3}	Btu	$\times$ 778	ft lb_f
J	$\times$ 0.2388	cal	$\times$ 4.187	J
J	$\times$ 1×10^7	erg	$\times$ 1×10^{-7}	J
J	$\times$ 2.778×10^{-7}	kW h	$\times$ 3.60×10^6	MJ
J	$\times$ 1	W s	$\times$ 1	kJ
J	$\times$ 0.7375	ft lb_f	$\times$ 1.356	J

Energy, Power, Heat, Fluid Flow Rate:

W	$\times$ 3.413	Btu/h	$\times$ 0.293	W
W	$\times$ 10^7	erg/s	$\times$ 1×10^{-7}	W
W	$\times$ 0.7375	ft lb_f/s	$\times$ 1.356	W
kW	$\times$ 1.34	hp	$\times$ 0.746	kW
liter/min	$\times$ 0.2642	gal/min	$\times$ 3.785	liter/min
liter/min	$\times$ 2.393	ft^3/h	$\times$ 0.4719	liter/min

Heat Flux:

W/m^2	$\times$ 0.317	Btu/h ft^2	$\times$ 3.155	W/m^2
cal/s cm^2	$\times$ 3.687	Btu/h ft^2	$\times$ 0.2712	cal/s cm^2
W/m^2	$\times$ 6.452×10^{-4}	W/in^2	$\times$ 1550	W/m^2

Specific Heat:

J/kg K	$\times$ 2.388×10^{-4}	$Btu/lb_m\ °F$	$\times$ 4187	J/kg K
cal/g °C	$\times$ 1	$Btu/lb_m\ °F$	$\times$ 1	cal/g °C

Thermal Conductivity:

W/m K	$\times$ 0.5777	Btu/h ft °F	$\times$ 1.731	W/m K
W/m K	$\times$ 1.926×10^{-3}	Btu in/s ft^2 °F	$\times$ 519.2	W/m K
W/m K	$\times$ 7.028	Btu in/h ft^2 °F	$\times$ 0.1442	W/m K
W/m K	$\times$ 2.39×10^{-3}	cal/cm s °C	$\times$ 418.4	W/m K

Heat Transfer Coefficient:

W/m^2 K	× 0.1761	Btu/h ft^2 °F	× 5.678	W/m^2 K

Velocity:

km/h	× 0.6205	miles/h	× 1.609	km/h
m/s	× 3.6	km/h	× 0.2778	m/s
m/s	× 39.37	in/s	× 0.0254	m/s
m/s	× 3.281	ft/s	× 0.3048	m/s
m/s	× 1.181 × 10^4	ft/h	× 8.467 × 10^{-5}	m/s

Mass Flow Rate:

kg/s	× 7.937 × 10^3	lb$_m$/h	× 1.26 × 10^{-4}	kg/s
kg/s	× 2.205	lb$_m$/s	× 0.4536	kg/s

Viscosity:

Pa·s	× 10	Poise	× 0.1	Pa·s
Pa·s	× 1000	centipoise	× 0.001	Pa·s
m^2/s	× 10.76	ft^2/s	× 0.0929	m^2/s
Pa·s	× 1.488	lb$_m$/s ft	× 0.672	Pa·s
centipoise	× 1.488 × 10^3	lb$_m$/s ft	× 6.72 × 10^{-4}	centipoise
m^2/s	× 10^6	centistokes	× 1 × 10^{-6}	m^2/s
Pa·s	× 1.45 × 10^{-4}	lb$_f$ s/in^2	× 6.895 × 10^3	Pa·s
Pa·s	× 2.088 × 10^{-2}	lb$_f$ s/ft^2	× 47.88	Pa·s

Stress:

MPa	× 145	lb$_f$/in^2	× 6.895 × 10^{-3}	MPa
MPa	× 0.102	kg$_f$/mm^2	× 9.807	MPa
MPa	× 0.0725	ton$_f$/in^2	× 13.79	MPa
MPa	× 1	MN/m^2	× 1	MPa

Bending Moment:

N m	× 8.85	lb$_f$ in	× 0.113	N m
N m	× 0.7375	lb$_f$ ft	× 1.356	N m
N m/m	× 0.2248	lb$_f$ in/in	× 4.448	N m/m
N m/m	× 1.873 × 10^{-2}	lb$_f$ ft/in	× 53.38	N m/m

Fracture Toughness and Impact Strength:

MPa m$^{\frac{1}{2}}$	× 0.9099	ksi in$^{\frac{1}{2}}$	× 1.099	MPa m$^{\frac{1}{2}}$
J/m	× 0.2248	ft lb$_f$/ft	× 4.448	J/m
J/m	× 0.01874	ft lb$_f$/in	× 53.37	J/m
J/m^2	× 4.757 × 10^{-4}	ft lb$_f$/in^2	× 2102	J/m^2

CHAPTER 1

INTRODUCTION TO POLYMERS

1.1 The Age of Plastics

Anthropologists use many measuring scales to trace the development and evolution of mankind. One common measure is the type of materials used for construction of shelter, weapons, tools, transportation, personal appearance items and utensils. Early man used sharpened stones for cutting edges on weapons and pointed wooden sticks for spears and handles of stone tools. Logs were hollowed into dugout canoes. Natural caves were used for shelter. As man evolved, his technology in materials also evolved. The ability to shape stone into blocks and to bake clay and straw into bricks was learned. Clay was shaped and fired into pottery vessels. Trees were sectioned into beams for shelter construction and boat planking. Wood planks were joined together and carved into furniture, and utensils. Tanners converted animal hides into leather and weavers converted natural fibers such as wool, cotton, and flax into cloth. Along with this development came a need to develop tools of harder and stronger materials to cut and shape more efficiently. In the *bronze age,* copper, zinc and tin were refined and smelted into functional implements such as knives, adzes, swords and shields and decorative implements such as mirrors and jewelry.

The softness of these metals forced the development of even harder metals for cutting, grinding and shaping. Iron eventually replaced the "yellow metals" and the *iron age* was born. As man's metallurgical and material skills improved, an iron alloy, steel, replaced iron, and simple clay was replaced with more refined porcelain. Steel was not only used for cutting and grinding tools, but for all types of devices including machinery, wire and cable, pipe, building beams, ship hulls, swords, cannon and so on. The *steel age* has been with us since about the middle of the nineteenth century. With each evolution in materials of commerce, man has improved his standard of living and has adapted the material for all aspects of daily life.

By the end of the nineteenth century, chemistry was developing as a science. Many new, identifiable chemistry-based materials were being created and commercialized. Many of these new types of substances were long-chained molecules, constructed of large numbers of identical small molecular weight repeating units. These very long molecules, macromolecules, are called *polymers.*

1.2 Introduction to Polymer History

A *polymer* is an identifiable high molecular weight molecule represented by a large number of small repeating molecular units called *mers*. There are many natural polymers, including vegetable and animal fibers such as wool, linen, cotton, and silk. Plant sap latexes include shellac and rubber. Animal proteins such as keratin found in horn, animal hoofs and tortoise shell were probably the first materials to be deliberately manipulated into useful articles such as bowls and cups. Animal horn keratin is probably the first identifiable natural polymer to be the object of polymer synthesis. In the late 1700s, it was found that it could be delaminated into very thin translucent sheets and could be temporarily softened in hot water or steam and shaped under moderate pressure into combs, buttons, and lantern windows. The material was more easily worked and far less expensive than ivory. As the demand grew, the method of fabrication evolved. More rapid ways were sought to delaminate and form the horn. Finally, as the availability of the natural material decreased, substitute materials were sought.*

Shellac was one of the earliest resinous molding materials. Lac is the result of insect secretion of tree sap. Small branches that are encrusted with lac are crushed and washed. Shellac is identified as a collection of polyhydroxy aliphatic carboxylic acids that can be "set" or crosslinked into a three-dimensional structure by adding sulfur and heat. Since it was easily molded under moderate heat and pressure and was translucent and amber in color, it found great acceptance as an ivory substitute. India was and still is the primary source of shellac. As a result, in the United States and Europe in the 1800s, shellac was relatively expensive when compared with horn.

As will be discussed in detail below, there are today two primary classes of polymeric materials: *thermoplastics* and *thermosets*. Thermoplastics have two-dimensional polymeric chains. Thermosets have three-dimensional polymeric chains. Thermoplastics can be thought of as *spaghetti*. When the mass is heated, it can be easily stirred. When the mass is cold, the surface starch and the stiffness in the noodles make stirring very difficult. But, when the mass is reheated, stirring is again easy. Thermoplastics can be reheated and cooled many times without appreciable change in the nature of the polymer.

Thermosetting polymers, or "thermosets", on the other hand, may begin as two-dimensional polymers, but through heat and chemical action, three-dimensional branches are formed. These act to rigidify the polymer. Thermosets can be thought of as *eggs*. When eggs are scrambled and then heated, they retain that scrambled shape regardless of the resulting temperature. Heat and pressure, along with chemicals such as sulfur that promote crosslinking in shellac and rubber, are common to thermoset manufacturing. As a result of their three-dimensional structure, thermoset parts cannot be reground and reprocessed into other parts.

In the early days of polymer synthesis, thermosets and thermoplastics were not so well defined. The early developments in rubber are classic examples of this blurring. Rubber was obtained from a white latex extracted from the bark of certain trees in South America, India and Malaysia. The latex was coagulated and dried by heat. The resulting crepe rubber was tough and ductile but had poor properties such as long-term embrittlement and very low high temperature modulus properties, and a tendency to remain sticky. Mcintosh found that he could temporarily solvate rubber in camphor and so in 1820 developed a waterproof rubber coating for fabric. In 1828, Thomas Hancock added sulfur to rubber to help improve its long-term properties. Since it was necessary to heat the mixture during this step, he called it "Vulcan-ization", after the Greek god of fire. However, it was noted even in the 1820s that not all natural rubbers

---

*As late as 1920, ground horn and dried blood compound was used as a molding material (1). Natural horn adhesives (horsehoof glue) are still used in some applications in woodworking and cabinetry.

yielded identical properties. Charles Goodyear in 1839 carried out very deliberate experiments on the relative roles of sulfur and heat on raw rubber in order to provide a stable, three-dimensional structure with repeatable properties. The first rubber commercialized was caoutchouc or Hevea rubber from the Brazilian Hevea tree. The tree was easy to grow in plantation fashion, had a very high yield, and most importantly, the sap yielded "a linear high-molecular weight [cis configuration] polymer of isoprene, C_5H_8" (2). Hevea could easily be masticated to a liquid paste and would accept additives such as oils as softeners, carbon black and talc fillers, and vulcanizing agents such as sulfur.

In 1843, Dr. George Montgomerie sent samples of gutta percha from Malaya to researchers in England. Research from that time to 1860s on gutta percha and similar rubber compounds showed that although these were isoprenes, they showed trans configuration, contained oxygen, were not as elastomeric, could not be cross-linked as easily, and seemed to have higher viscosity when masticated. As a result, other applications were sought. In the early 1840s, Michael Faraday found that gutta percha had excellent electrical properties and so could make good electrical wire insulation. In 1848, Mr. S. Armstrong insulated telegraph wires for the Morse Telegraph Company with gutta percha. The earliest transatlantic cables were insulated with gutta percha. Furthermore, gutta percha rubber could be vulcanized to a hard rubber. In the 1850s, compression molding techniques were developed to produce hair combs, buttons and photographic picture cases.

During this time, there was significant interest in cellulose. Cellulose is the major constituent of plants, primarily wood. Since it is quite insoluble in most solvents, the purification process is one of dissolving all non-celluloses, such as lignin and other carbohydrates, with caustic soda. It was noticed in 1844 that cellulose from cotton plants was swollen (but not dissolved) with caustic soda. When the alkali soda was washed away, the regenerated cellulose could be fiberized to produce "Mercerized" (named after Mercer, the discoverer of the technique) or low-shrink cotton. In 1846, Prof. Christian Schoenbein, a Basler-Swiss chemist, noted that nitric acid attacked the cellulose, producing a viscous solution that could be cast into films (3). The polymer was called cellulose nitrate. Beginning with cellulose, Alexander Nobel invented dynamite, nitroglycerin absorbed in kieselguhr, in 1866 and Abel improved it a year later by absorbing nitroglycerin in guncotton and potassium nitrate. Also in the 1860s, Alexander Parkes found that camphor softened and solvated the otherwise brittle cellulose nitrate. In 1862, he exhibited combs and cases of "Parkesine" at the Great Exposition of London. His coworker, Daniel Spill, found that camphor-solvated cellulose nitrate was easily molded under steam heat and simple mechanical pressure. He patented his recipes in 1869 (4,5).

During this time, John Wesley Hyatt was also working on camphor-solvated cellulose nitrate. He called it "Celluloid" and founded The Celluloid Company of Albany, NY. He worked with Charles Burroughs of Newark, NJ, to develop ways of processing the material into products such as shirt collars and stays, combs, carriage windows and the first film bases for still and motion photography. He opened the Hyatt Billiard Ball Company of Albany, NY, later called the Albany Billiard Ball Company, to pursue camphor-solvated cellulose nitrate as a substitute for ivory in billiard balls.* For the next 50 years, until his death in 1920, Hyatt continued to develop methods of producing useful products from the materials of the day. The rudiments of blow molding and thermoforming came from Hyatt's experiments with steam forming of Celluloid sheets in a closed mold. The sheets were formed on a Burroughs Hydraulic Planer, the design of which was originally proposed in 1845 for gutta percha. Later Burroughs built processing machinery of his own design for both thermosetting and thermoplastic resins. His equipment became the backbone of the burgeoning plastics machinery industry of the early part of the twentieth century.

By the end of the nineteenth century, the common way of fabricating a solid shape was with compression molding, with the resinous material placed between two heated mold halves, which were then closed under mechanical pressure, usually by manually turning a screw. This technique sufficed for molding of most nineteenth century plastics. In 1870, Smith and Locke invented a machine for transferring soft metals from a closed chamber to a mold cavity. In 1872, Hyatt patented a similar device for heating and transferring a plasticized charge of Celluloid into a mold cavity (7). The working devices were developed by Charles Burroughs. Although ram extrusion had been developed for fabricating lead pipe in the late 1700s, polymer extruders were first used for coating wire with gutta percha in the mid-1840s (8). H. Hewley patented a piston extruder in 1845 for extruding gutta percha pipe. Hyatt cold ram-extruded Celluloid in the 1870s. Most early extruders were designed to operate at cold temperatures on solvated polymers. Screw extrusion was developed for rubber by Royle in 1880 and for rubber and some thermoplastics by Paul Troester in 1892. Undoubtedly Hyatt experimented with his cellulosic variants in screw extruders of similar design by Charles Burroughs.

Until Hyatt, the polymers used were primarily natural. Hyatt can be considered the first to exploit semisynthetic polymers. Baekeland's 1907 discovery of a moldable form for phenol-formaldehyde,** a thermosetting resin, is considered to be the beginning of purely synthetic polymer development (10). Bakelite gained national importance during the 1910s and 1920s when it was used as the ear- and mouth-piece of the first mass-produced telephones. Until the 1920s, thermoset resins dominated the polymer industry. Urea was found to polymerize with formaldehyde to produce resins that could be colored. Until this time, most resins were brown or black. Melamine, a mineral-

---

*It is often thought that Hyatt's 14 April 1868 patent was the replacement of ivory in billiard balls, that it led to a $10,000 prize, and that this was the beginning of the polymer age. However, camphor-solvated nitrocellulose was not the first substitute for ivory in billiard balls. Fiber-reinforced shellac and gutta percha preceded it. Hyatt's USP 76,765 was *not* a patent for camphor-solvated cellulose nitrate; it was a process improvement (pulverization of gum shellac prior to addition of fibrous pulp). Hyatt *is* the father of the polymer age primarily because he recognized business opportunities in polymers and exploited contemporary technologies that were being used to fabricate less synthetic polymers into products that were more expensive or of lower utility than Celluloid. DuBois sums it up correctly: "It isn't the thinker-upper who gets the payoff; it's the man who does something about it" (6).

**In 1872, O. von Baeyer was the first to produce a resin from reacting phenol and formaldehyde. However, he found the resin to be intractable and discarded its development (9).

filled formaldehyde resin, was developed in the 1930s as a tough, readily colored, water-resistant replacement for phenolics and ureas, with diverse applications in dinnerware and electrical insulators, coil bobbins, and appliance plugs.

The history of polymers is not just related to the formation of new polymeric materials. New machine designs were needed to convert them into useful products. The leadership of Charles Burroughs, first with Hyatt, then with the producers of Bakelite, has been chronicled. Eckert and Zeigler marketed a practical plastics injection molding machine in the 1920s. Also during this time, Louis Shaw developed the transfer molding process that allowed manufacture of high precision thermoset parts. In the early 1930s, injection molding of thermoplastic resins had advanced so far that a two-stage screw preplastication scheme, in essence a second-generation polymer processing device, was needed.

It is important to realize that before 1930, very little was known about the molecular architecture of high polymers. In fact, early theories proposed that polymers were really "small molecules held together as colloidal aggregates by unknown forces" (11). The colloidal properties of elasticity, flexibility and high viscosity are now known to be caused by the excessive length of the molecular chain and by chemical crosslinking. By the 1920s, there was sufficient evidence to indicate that these materials were not just colloids. H. Staudinger was the first to show that the viscosity of a polymer-solvent solution increased as the molecular weight of the polymer increased. This supported his hypothesis that polystyrene and natural rubber were linear molecules made of a multitude of small repeat units. Staudinger did not work without precedent however. In 1877, Kekule hypothesized the existence of macromolecules in natural organic polymers and in 1893, Emil Fischer described the structure of cellulose as a long chain of glucose units, becoming the first scientist to identify the small repeat unit or "mer" that makes up a polymer molecule.

In 1927, DuPont made a major commitment to work on linear aliphatic polymers that could be made from "coal, air and water" by establishing a laboratory devoted to synthetic polymer research and employing W.H. Carothers as its head. This unprecedented action would ultimately transform the chemical and dynamite company into a leader in polymer technology. In the next decade, Carothers' group developed neoprene, a purely synthetic rubber, polyamides such as hexamethylene adipate (nylon 66) and polycaprolactam (nylon 6), polyethylene terephthalate (PET), and early versions of polyacrylates. The commercialization of nylon (polyamide 66) in the late 1930s marks the commercial beginning of thermoplastics, for these materials could be remelted and reprocessed, unlike the phenolics and cured rubbers they were replacing. Although nylon is relatively expensive and not considered to be a major commodity polymer today, it is credited with becoming the first polymer to have a household name.

Branched polyethylenes were first developed in the 1930s by Fawcett and Gibson by applying excessive pressure and heat to ethylene gas.* M. Bertholet polymerized styrene with divinyl benzene into an intractable resin in 1866. In the early 1930s, Dow Chemical Corp. developed a feasible process for polymerizing styrene into a useful resin. Vinyl chloride was discovered by H. Regnault in 1838 and was polymerized by

---

*But it required extensive research in selective or stereospecific catalysis by G. Natta and K. Ziegler to develop low-cost methods of producing linear polyethylenes and polypropylene.

E. Baumann in 1872. But as-polymerized polyvinyl chloride (PVC) is nearly intractable. Suitable softening agents (known as plasticizers) were developed in the 1920s and vinyls were commercially processed in the early 1930s. Carbide and Carbon Chemicals, now Union Carbide Corporation or UCC, exhibited sound-recording-quality vinyls* and vinyl floor tiles at the 1933 World Fair.

And so by the 1933–1934 "A Century of Progress" Exposition in Chicago, many major corporations such as General Electric, Boonton Molding, DuPont Chemical, Plax, and Westinghouse displayed articles of commerce and industry, of both thermoplastic and thermoset polymers. Still to come in that decade was the invention of polyurethanes by Otto Bayer in 1937, the commercialization of nylon 6 (polycaprolactam) by I.G. Farben in 1939, the commercialization of unsaturated polyester resins in 1942, the development of silicones by Dow Corning in 1942, the basic patent on epoxy by P. Castan in 1943, and the production of the first fluorocarbon polymers in 1945.

By the end of World War II, thermoplastics were replacing more conventional materials than were thermosets. Epoxies were commercialized in 1947 and ABS (acrylonitrile–butadiene–styrene terpolymer) in 1948. The first fully fluorinated polymer, polytetrafluoroethylene, was commercialized in 1950 and polycarbonate was invented by H. Schnell in 1953. Alan Hay discovered polyphenylene oxide in 1956 and acetal and chlorinated polyether were commercialized in 1959. Polyvinylidene fluoride was commercialized in 1961, phenoxy, polyallomers, and ethylene–propylene rubbers in 1962, ethyl–vinyl acetate, ionomers, and polyimides in 1964.

The improvements in processing after the Second World War are also noteworthy. In the late 1940s, polystyrene bead foam was developed. The first standard mold base was developed in 1943 by DME. This allowed the moldmaker to obtain low-cost standardized parts to construct a mold. Polystyrene foam sheet was first extruded in the 1950s. Injection molding of thermoplastics was revolutionized by W. Willard, who combined the concept of a plasticating extruder screw with the reciprocating drive of a ram injection molding machine to produce the screw injection molding machine. Although blow molding had been used in some fashion from the early Hyatt days, the development of high-density polyethylene in the 1950s forced the development of the modern-day extrusion blow molding process. The polymer film industry was revolutionized by the side feed tubular die that allowed a thin-gage annular film to be blown. In the 1960s, high-speed thin-gage thermoforming equipment was developed. The development of ultra-pure chemical foaming agents allowed thermoplastic structural foam to evolve in the 1960s. In the 1960s, N. Wyeth discovered that polyethylene terephthalate could be biaxially stretched above its glass transition temperature into a container, thus creating a 10 *billion* container per year market. In the late 1960s, the advent of the microprocessor controller for injection molding resulted in better process control and part dimensional tolerance. Invention continues, as seen in the 1970s development

---

*The development of PVC for the sound recording industry represented the beginning of low-cost but intrinsically low noise level and highly wear-resistant recordings and the popularization of slower-speed records (45s, EPs and LPs). It also represented the ultimate demise of shellac recordings, which at one time in the mid-1930s represented more than one-third of the shellac compression and injection molding business (12).

of reaction injection molding of foamed or reinforced polyurethanes for automotive applications and the development of low-temperature forming of crystallizing resins such as PET and PEEK into useful products. Also in the 1970s, several specialty polymers were introduced, including polyamide-imide, polyether sulfone, polyphenyl sulfone, and polybutylene terephthalate. Many blends and alloys were also created. The most significant new development appears to be LCPs or liquid-crystal polymers. These very stiff-chained high-temperature polymers form *in-situ* reinforcing fibers as they cool below their melting temperatures. This allows *organic* reinforcement of the organic polymer, with bond strengths substantially greater than those obtained either by glass-fiber or graphite-fiber reinforcing elements. In addition to the processes discussed above, filament winding, pultrusion, resin transfer molding and sheet and bulk molding of thermosetting resins saw rapid advancement in the 1970s.

And by the 1980s, thermoplastics represented more than 80% (wt) of all polymers produced in the world. The worldwide production of polymers (non-rubbers) in 1972 and 1986 is given in Table 1.1 (13). Projected consumption to 1995 is also shown. It is expected that in the next decade the annual growth of the United States and Europe will lag behind that of emerging nations such as Brazil, South Africa and Korea.

The rapid domination of thermoplastics is due in part to their ease of processing, their reprocessability, their broader range of properties, the ease of compounding adducts (fillers, reinforcements, plasticizers, and other additives) into the neat polymers, their better long-term stability as resins and, in general, their lower sensitivity to common chemicals and water. One factor that must be considered, however, is the rapid development of by-products from thermal and catalytic cracking operations in crude oil refining. In the middle decades of the twentieth century, the rapid acceptance of oil and gas for home heating, replacing coal, and the United States' love affair with the automobile spurred hydrocarbon refining development. The availability of inexpensive hydrocarbons spurred development of the *petrochemical industry,* primarily in the US. The development of ultrapure light hydrocarbon feed stock such as naphtha,* butane, propane and ethane, from both oil refinery operations and from natural gas production, led to thermal cracking to unsaturated low molecular weight hydrocarbons that could be polymerized into thermoplastic polymers.** The production of thermoplastic nylon resin from adipic acid and hexamethylene diamine illustrates the classic transition from coal-based to oil-based petrochemicals during the late 1940s and 1950s. In 1942, phenol was the beginning petrochemical for adipic acid (14). It was obtained from coal tar acids chiefly by sulfonation. By the early 1950s, the source of adipic acid was cyclohexane, either by dehydrogenation of benzene or directly from petroleum, and the source of hexamethylene diamine was butadiene, by dehydrogenation of butane from natural gas.

As we use plastics in load-bearing applications, we need to continually analyze their unique responses to the complex time-dependent stress patterns applied by their environment. Plastics, as a family, can be characterized as *viscoelastic.* That is, at times

---

*Naphtha is a generic term for any light oil having properties intermediate between kerosene and gasoline.
**Owing to tariff restrictions, naphthas cannot be imported to the US for polymer production. Thus, European and African polymer technologies use the less expensive naphtha to produce cheaper thermoplastic polymers having properties that are somewhat different from "equivalent" US species.

Table 1.1. World consumption of plastics, 1972–1995 (13).

Item	1972	1986	1991	1995	Growth/Year Historic	Growth/Year 95/86
Population ($\times 10^6$)	3,860	4,924	5,371	5,716	1.8	1.7
Gross domestic product/capita (GDP, 1982 US$)	2,339	2,779	3,014	3,200	–	–
Gross domestic product, GDP ($\times 10^9$ 1982 US$)	9,027	13,685	16,190	18,290	3.0	3.3
Plastics consumption						
kg/capita	9.3	15.7	17.5	19.1		
kg/$1000 GDP	4.0	5.6	5.8	6.0		
By Region ($\times 10^6$ kg, $\times 10^3$ metric tonnes)						
North America	9,895	22,594	27,550	31,850	5.6	3.9
Latin America	1,415	4,476	5,750	7,055	8.6	5.2
Western Europe	12,995	23,317	27,065	30,200	4.3	2.9
Eastern Europe	4,264	9,802	11,385	12,875	6.1	3.1
Africa/Middle East	434	1,056	1,530	2,045	6.6	7.6
Asia/Oceania	7,064	16,068	20,720	24,975	6.0	5.0
Total World	36,068	77,313	94,000	109,000	5.6	3.9
Net exports	2,298	2,512	1,820	1,600		
Plastics Production ($\times 10^6$ kg, $\times 10^3$ metric tonnes)						
Polyethylene (PE)	10,172	23,181	28,785	34,790	6.1	4.6
Polypropylene (PP)	2,047	8,598	10,995	12,940	10.8	4.6
Polyvinyl Chloride (PVC)	7,638	14,568	17,380	19,700	4.7	3.4
Polystyrene (PS)	4,767	8,380	9,490	10,595	4.1	2.6
Other thermoplastics	4,459	9,472	12,130	14,140	5.5	4.6
Thermosets	9,283	15,626	17,040	18,435	3.8	1.9

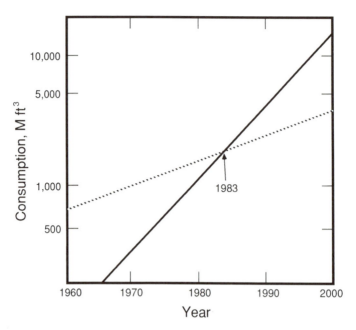

Figure 1.1.   World Volumetric Consumption of Ferrous and Polymeric Materials. Dashed Line: Iron and Steel. Solid Line: Plastics. (40).

they respond in a liquid-like way and at others, as if they are solid-like. From the 1940s on, scientists and engineers—Paul J. Flory, Herman Mark, Turner Alfrey, Jr., Arthur Tobolsky, Robert B. Bird, Bryce Maxwell, John D. Ferry, J. Gordon Williams, Lawrence Nielsen, Arthur Lodge, Fred Billmeyer, Jr., James McKelvey, Ze'ev Tadmor, Peter Vincent and many others—have contributed extensively to the understanding of the relationship between stress, strain, rate of strain and the molecular nature of the polymer. The circle of knowledge necessary to make a particular polymer resin, create the plastic containing that resin, design the part based on a suitable stress analysis and finally process the plastic into the proposed shape has developed into an engineering discipline in the past 30 or 40 years. Thus, complex parts that must meet stringent end use requirements can be designed and manufactured with a modicum of success.

And just as more versatile semisynthetic polymers (Celluloid) replaced natural polymers (vulcanized gutta percha) in the 1860s and more versatile synthetic polymers (Bakelite) replaced semisynthetic ones (Celluloid) near the turn of the century, even more versatile thermoplastic polymers (nylon, polyethylene) replaced thermoset polymers beginning in the 1930s, and certainly by the end of World War II (1940s).*

---

*There is ample evidence that another evolution is underway. Higher performance polymers, particularly those with excellent long-term properties and those that can be reinforced with high performance fibers (such as high modulus glass or graphite) are being considered for the first time as primary elements in load-bearing applications. Thus, the current thermoplastic polymer age may have already given way to the composites age. For a brief discussion on composites, see Chapter 2, Section 2.7.

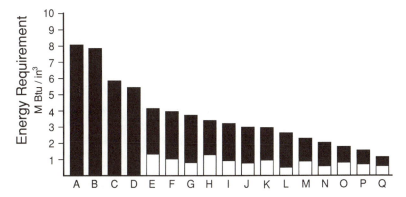

Figure 1.2.   Comparative Materials Energy Requirements. Solid Bar: Fuel. Open Bar: Feedstock. Materials Identification: A: Magnesium; B: Aluminum Die Cast; C: Zinc Die Cast; D: Steel; E: POM (Polyacetal); F: Modified PPO; G: PA-6 (Nylon 6); H: PA-66 (Nylon 66); I: PETP (PET or Polyester); J: PC (Polycarbonate); K: PMMA (Acrylic); L: PP (Polypropylene); M: ABS; N: PVC (Vinyl); O: PS (Polystyrene); P: HDPE; Q: LDPE.

Polymers are materials that can take many forms and shapes. The individual physical properties of a polymer can be tailored by controlling the size and shape of the polymer molecule and the amount and quantity of additives and fillers (from this point on, called "adducts") compounded into the polymer to form the plastic.* As seen in Figure 1.1, the world volumetric consumption of polymers exceeded the volumetric consumption of iron and steel around 1983. By the beginning of the 21st century, it is projected that in the United States, the per capita consumption of *all* materials except cementitious materials such as concrete and baked or fired clays such as adobe, will be 246 liters. Of this, plastics will represent 212 liters, or 86%. On a worldwide basis, the projected per capita consumption will be 19 kg (Table 1.1). Synthetic high molecular weight molecules, macromolecules, or polymers are an essential part of our present and future lifestyle. They are easier to produce than polymers extracted from natural resources. Most important, the energy requirements per unit volume for polymers are significantly lower than those for the metals they frequently replace (Figure 1.2). As our world becomes increasingly energy conscious, we will need to turn more and more to polymers as the principal materials of fabrication. The future growth or potential for this industry is incalculable. It is only *we* who place limitations on this potential.

This book represents a careful examination of the engineering properties of synthetic high molecular weight polymers. The text material considers polymers as they exist today. History is always an important preamble to any topical discussion, since it serves as a guidepost from the past as well as a warning lantern that signals that materials of the future may be substantially different from those in common use today. A well-studied understanding of the underlying relationships between the polymer and its environment certainly can aid in any future transition.

*The distinction between polymer and plastic is detailed below.

1.3 Polymers and Nonpolymers

As noted above, a *polymer* is an identifiable high molecular weight organic molecule whose structure is represented by a large repeating number of low molecular weight units called *mers*. The phrase "high polymer" is often used to describe polymers that have significant physical and mechanical properties. This terminology allows differentiation from low molecular weight organic compounds such as waxes, fatty acids, oils and greases and from certain inorganic ceramics. Although rubbers were some of the earliest materials to be synthesized and although rubbers are true organic high polymers with repeating molecular structures, tradition has considered rubber, both natural and synthetic, in a category separate from polymers. This is due partly to the lack of ridigidy of most rubber structures. Recently, thermoplastic elastomers are meeting the performance characteristics of natural and synthetic rubbers and so the distinction is blurred. Furthermore, the distinction between certain "rubbery" polymers and hard synthetic rubbers is entirely arbitrary.

There is a more clear-cut distinction between a polymer and a plastic. Arnold (15) modified the American Society for Testing and Materials (ASTM) definition, to consider *plastic* as: ". . . a material that contains as an essential ingredient, an organic high molecular weight polymer, is solid and rigid in its finished state, and at some stage in its manufacture or its processing into a finished article, can be shaped by flow."

By inference, a plastic contains at least one polymer, but a polymer is not necessarily a plastic. The polymeric constituent in a plastic is the primary factor distinguishing a plastic from other materials used in design. Very rarely does a designer use a plastic that contains only polymer (viz, 100% polymer). In most instances the polymer is compounded with low molecular weight organic or inorganic substances called *additives*. These additives enhance processing, add color, improve physical properties, prevent degradation and so on (See Table 1.2). Fillers and fibers are added to strengthen the plastic (Table 1.3), to extend its temperature range, or as bulk to reduce the amount of more expensive polymer (Table 1.4). The non-high-polymer materials added to a polymer to produce a plastic are generically referred to as *adducts*.

Table 1.2. Typical small molecule adducts to polymers.

Antioxidants	Heat stabilizers
Antistatic agents	Plasticizers
Colorants and pigments	Processing aids
Crosslinking agents	Emulsifiers
Coupling agents	Lubricants
Fillers	Mold release agents
Bulk	Viscosity depressants
Fibrous	Ultraviolet stabilizers
Flame retardants	Odor suppressors
Foaming agents	

Table 1.3. Typical fibers for reinforcing polymers.

Cellulose fibers	Carbon fibers
α-Cellulose	Asbestos fibers
Pulp preforms	
Cotton flock	Fibrous glass
Jute	Filaments
Sisal	Chopped strand
Rayon	Reinforcing mat
	Glass yarn
Synthetic fibers	Glass ribbon
Polyamide (nylon, PA)	
Polyester (dacron)	Whiskers
Polyacrylonitrile (dynel, orlon, PAN)	
Polyvinyl alcohol (PVOH)	Metallic fibers
Other Fibers	

Table 1.4. Typical fillers for polymers.

Silica products	Metallic oxides
Minerals	Zinc oxide
Sand	Alumina
Quartz	Magnesia
Novaculilte	Titania
Tripoli	Beryllium oxide
Diatomaceous earth	Other inorganic compounds
Synthetic amorphous silica	Barium sulfate
Wet process silica	Silicon carbide
Fumed coloidal silica	Molybdenum disulfide
Silica aerogel	Barium ferrite
Silicates	Metal powders
Minerals	Aluminum
Kaolin (China clay)	Bronze
Mica	Lead
Nepheline silicate	Stainless steel
Talc	Zinc
Wollastonite	
Asbestos	Carbon
Synthetic products	Carbon black
Calcium silicate	Channel black
Aluminum silicate	Furnace black
	Ground petroleum coke
Glass	Pyrolyzed products
Glass flakes	Cellulosic fillers
Hollow glass spheres	Wood flour
Cellular glass nodules	Shell flour
Glass granules	
	Comminuted polymers
Calcium carbonate	
Chalk	
Limestone	
Precipitated calcium carbonate	

Low density polyethylene, used extensively in the thin-film packaging industry, in garbage bags, for example, is one of the few polymers that can be processed without adducts. A comprehensive discussion of the nature and use of adducts is covered in Chapter 2.

To repeat, the essential ingredient of a plastic is an organic high molecular weight polymer molecule with a characteristic repeating molecular structure. The structure of the polymeric constituent is also called "molecular architecture", "configuration", or "atomic confirmation". The discussion that follows focuses on the polymers that are commonly used today in commercial plastics. The polymer architecture, presented below is a summary or overview needed to better understand the relationship between molecular structure and physical and chemical properties of polymers in their plastic state. More complete references are given in the Bibliography below for the interested reader.

1.4 The Molecular State

The *atom* is the building block of a molecule. It is the smallest unit of an element that can exist either alone or in combination with other atoms like it or different from it. During chemical reactions, individual atoms can combine or combinations of atoms may break down, but the individual atoms remain unchanged. Of the more than 120 known elements, most polymers are formed from only eight:

H, hydrogen	F, fluorine
C, carbon	Si, silicon
O, oxygen	S, sulfur
N, nitrogen	Cl, chlorine

There are more than 20 different subatomic particles that form an individual atom. The three most important are the *electron,* with a negative charge, the *proton,* with a positive charge, and the *neutron,* with no charge. In 1913, Niels Bohr visualized the geometric structure of the atom as a small solar system. At the center of the atom is a very small but dense *nucleus* containing protons and neutrons. The total charge of the nucleus is dependent on the number of protons in the nucleus. Orbiting around the nucleus are the negatively charged electrons. Bohr determined that the electrons orbited the nucleus at specific distances from the nucleus just as the planets orbit the sun. The electron orbits are called *electron shells.* The number of electrons in each shell is dependent upon the distance from the nucleus. In Figure 1.3 is the lithium, Li, atom, having 3 protons and 4 neutrons in the nucleus and three electrons orbiting the nucleus in two electron shells. The electrons in the outer shell of any atom are the *valence electrons.* For hydrogen and helium, only two electrons are needed to complete the outer shell. Each of the next eight atoms (lithium, beryllium, boron, carbon, nitrogen, oxygen, fluorine and neon) needs eight electrons to fill its outermost shell to obtain a stable electromagnetic state. The degree to which an atom needs electrons to fill the outer electron shell to achieve stability is called the atom's *valence bond level.* Thus, lithium has one valence electron and seven valence bonds. Helium and neon have completely filled outer shells and so have no valence bonds. This makes them chemically inert. Bohr's concept of specific electron shells at specified radii is considered

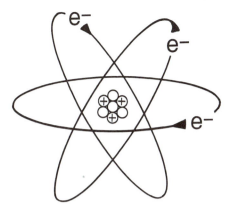

Figure 1.3.   Schematic of the Lithium Atom, with 3 Protons, 4 Neutrons and 3 Electrons.

simplistic today. Electrons are now considered to be focused sites of energy rather than discrete particles. The motion of an electron around a nucleus is now described by a mathematical probability relationship of its being a given distance from the nucleus at any time. The concept of an average electron distance can still be used however to describe atomic interaction.

As noted, each atomic species has a specific number of protons, neutrons and electrons. To distinguish between different atoms, the *atomic number* system is used, based on the number of protons in the nucleus. Thus Li, lithium, has three protons and so its atomic number is *3*. The *periodic table* is the logical classification of atoms by atomic number. As seen in Figure 1.4, each element is listed by its chemical symbol and below this, its atomic number. Lithium, for example, is found in the upper lefthand corner of the periodic table, first column, Group I, second row.

The actual mass of an individual atom is very small. The proton and neutron each has a mass of about 1.7×10^{-24} g. The mass of an electron, at 0.91×10^{-27} g, is considerably smaller than the mass of either the proton or the neutron and so the electron cloud mass is usually ignored when calculating the weight of an atom. Lithium, for example, has 3 protons and 4 neutrons, with an atomic weight of $(3 + 4) \times 1.7 \times 10^{-24}$ g $= 1.19 \times 10^{-23}$ g. Since the mass of any atom is small, an arbitrary reference scale is defined. The most abundant form of carbon, ^{12}C, having 6 protons and 6 neutrons, is the reference element. A single atom of ^{12}C has a defined mass of 12.000 *atomic mass units*, amu. The carbon atom, C, is given in the second row, Group IV column of the periodic table. An enlargement of the carbon box is given as Figure 1.5. The atomic number of carbon is 6. Carbon has 6 protons and 6 neutrons in the nucleus and an equal number of electrons orbiting the nucleus. On the right side of the box is a vertical column containing two numbers, 2 and 4. This is the electron configuration of the atom, with two electrons filling the first electron shell and four in the second electron shell. These four are the valence electrons. Since eight electrons are needed to fill the outer electron shell to achieve electromagnetic stability, the carbon atom has a valence bond level of 4. The number "12.011" below the chemical symbol is the atomic mass of a typical sample of carbon found in nature. If ^{12}C is the reference standard for the mass scale and if the earth was composed entirely of ^{12}C atoms, this number would be exactly "12.000". During the formation of individual atoms, some

IA	IIA	IIIB	IVB	VB	VIB	VIIB	VIII	VIII	VIII	IB	IIB	IIIA	IVA	VA	VIA	VIIA	
								1 H 1.008 (1)									
																	2 He 4.003 (2)
3 Li 6.940 (2 1)	4 Be 9.013 (2 2)											5 B 10.82 (2 3)	6 C 12.010 (2 4)	7 N 14.008 (2 5)	8 O 15.999 (2 6)	9 F 19.00 (2 7)	10 Ne 20.183 (2 8)
11 Na 22.997 (2 8 1)	12 Mg 24.32 (2 8 2)											13 Al 26.98 (2 8 3)	14 Si 28.09 (2 8 4)	15 P 30.975 (2 8 5)	16 S 32.066 (2 8 6)	17 Cl 35.457 (2 8 7)	18 A 39.944 (2 8 8)
19 K 39.100 (2 8 8 1)	20 Ca 40.08 (2 8 8 2)	21 Sc 44.96 (2 8 9 2)	22 Ti 47.90 (2 8 10 2)	23 V 50.95 (2 8 11 2)	24 Cr 52.01 (2 8 13 1)	25 Mn 54.93 (2 8 13 2)	26 Fe 55.85 (2 8 14 2)	27 Co 58.94 (2 8 15 2)	28 Ni 58.69 (2 8 16 2)	29 Cu 63.54 (2 8 18 1)	30 Zn 65.38 (2 8 18 2)	31 Ga 69.72 (2 8 18 3)	32 Ge 72.60 (2 8 18 4)	33 As 74.91 (2 8 18 5)	34 Se 78.96 (2 8 18 6)	35 Br 79.916 (2 8 18 7)	36 Kr 83.80 (2 8 18 8)
37 Rb 85.48 (2 8 18 8 1)	38 Sr 87.63 (2 8 18 8 2)	39 Y 88.92 (2 8 18 9 2)	40 Zr 91.22 (2 8 18 10 2)	41 Nb 92.91 (2 8 18 12 1)	42 Mo 95.95 (2 8 18 13 1)	43 Tc (99) (2 8 18 13 2)	44 Ru 101.7 (2 8 18 15 1)	45 Rh 102.91 (2 8 18 16 1)	46 Pd 106.7 (2 8 18 18)	47 Ag 107.880 (2 8 18 18 1)	48 Cd 112.41 (2 8 18 18 2)	49 In 114.76 (2 8 18 18 3)	50 Sn 118.70 (2 8 18 18 4)	51 Sb 121.76 (2 8 18 18 5)	52 Te 127.61 (2 8 18 18 6)	53 I 126.91 (2 8 18 18 7)	54 Xe 131.3 (2 8 18 18 8)
55 Cs 132.91 (2 8 18 18 8 1)	56 Ba 137.36 (2 8 18 18 8 2)	57-71 See lanthanide series	72 Hf 178.6 (2 8 18 32 10 2)	73 Ta 180.88 (2 8 18 32 11 2)	74 W 183.92 (2 8 18 32 12 2)	75 Re 186.31 (2 8 18 32 13 2)	76 Os 190.2 (2 8 18 32 14 2)	77 Ir 193.1 (2 8 18 32 17)	78 Pt 195.23 (2 8 18 32 17 1)	79 Au 197.2 (2 8 18 32 18 1)	80 Hg 200.61 (2 8 18 32 18 2)	81 Tl 204.39 (2 8 18 32 18 3)	82 Pb 207.21 (2 8 18 32 18 4)	83 Bi 209.00 (2 8 18 32 18 5)	84 Po 210 (2 8 18 32 18 6)	85 At (210) (2 8 18 32 18 7)	86 Rn 222 (2 8 18 32 18 8)
87 Fr (223) (2 8 18 32 18 8 1)	88 Ra 226.05 (2 8 18 32 18 8 2)	89-103 See actinide series															

Lanthanide series (rare-earth elements)	57 La 138.92 (2 8 18 18 9 2)	58 Ce 140.13 (2 8 18 19 9 2)	59 Pr 140.92 (2 8 18 20 9 2)	60 Nd 144.27 (2 8 18 22 8 2)	61 Pm (145) (2 8 18 23 8 2)	62 Sm 150.43 (2 8 18 24 8 2)	63 Eu 152.0 (2 8 18 25 8 2)	64 Gd 156.9 (2 8 18 25 9 2)	65 Tb 159.2 (2 8 18 26 9 2)	66 Dy 162.46 (2 8 18 28 8 2)	67 Ho 164.94 (2 8 18 29 8 2)	68 Er 167.94 (2 8 18 30 8 2)	69 Tm 169.6 (2 8 18 31 8 2)	70 Yb 173.04 (2 8 18 32 8 2)	71 Lu 174.99 (2 8 18 32 9 2)
Actinide series	89 Ac 227 (2 8 18 32 18 9 2)	90 Th 232.12 (2 8 18 32 19 9 2)	91 Pa 231 (2 8 18 32 20 9 2)	92 U 238.07 (2 8 18 32 21 9 2)	93 Np (237) (2 8 18 32 22 9 2)	94 Pu (242) (2 8 18 32 23 9 2)	95 Am (243) (2 8 18 32 24 9 2)	96 Cm (243) (2 8 18 32 25 9 2)	97 Bk (245) (2 8 18 32 26 9 2)	98 Cf (246) (2 8 18 32 27 9 2)	99 Es (254)	100 Fm (253)	101 Md (256)	102 No (253)	103 Lw (257)

Figure 1.4. Periodic Table of the Elements (18).

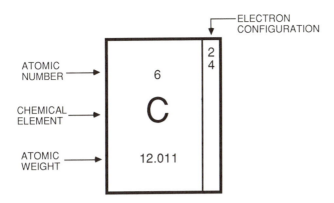

Figure 1.5. The Atomic Description of the Carbon Atom, as Amplified from the Periodic Table of Elements, Figure 1.4.

are given extra neutrons, thus forming heavier carbon atoms. These heavier atoms are called *isotopes*. In the case of carbon, there are sufficient isotopes in nature to allow the sample average amu to be slightly greater than 12.000. The atomic mass of a chemical element in nature is called the element's *atomic weight*.

It was noted that individual atom mass is exceedingly small. Although the amu (atomic mass unit) is useful for comparing relative masses of atoms, it is not convenient for most real problems. In practical problems, the mass of a polymer should be in grams or ounces. This quantity of material contains hundreds of billions of atoms. The *mole* unit has been defined to allow conversion between atomic weight and measurable quantities. The mole is the number of grams of ^{12}C atoms needed to have a mass of exactly 12.000 g, viz, 6.023×10^{23}. The mole is called the *gram atomic weight* and the constant, 6.023×10^{23} atom/mole, is called *Avogadro's number*. This relationship is valid for all atoms. One mole, 6.023×10^{23} atoms, of hydrogen has a mass of 1.008 g, for example. The atomic description of each atom can then be obtained directly from the data in the periodic table. The atomic descriptions for the eight atoms most commonly found in polymers are given in Table 1.5.

Table 1.5. Atomic values for elements frequently found in polymers (16-18).

Element	Chemical symbol	Atomic number	Atomic weight	Valence bonds
Hydrogen	H	1	1.008	1
Carbon	C	6	12.011	4
Nitrogen	N	7	14.0067	3
Oxygen	O	8	15.9994	2
Fluorine	F	9	18.9984	1
Silicon	Si	14	28.086	4
Sulfur	S	16	32.06	2
Chlorine	Cl	17	35.453	1

Atomic Bonding

Each atom with an incomplete outer electron shell is chemically active. It will always attempt to gain a more stable electromagnetic state. This is usually done by transfer of one or more electrons from the outer electron shell of one atom to the outer electron shell of another. This is called *ionic bonding*. In some cases, atoms share one or more electrons from their outer shells. This is called *covalent bonding*. Ionic bonding is the simpler form of atomic bonding. It is based on the balance of forces between positively and negatively charged particles. When an electron is removed from an atom, it will attain a net positive charge. This is the result of now-unequal numbers of positively charged protons and negatively charged electrons. Atoms with net positive or negative charge are *ions*. Postively charged ions are *cations*. Negatively charged ions are *anions*. The energy required to remove an electron from an atom is called the *ionization energy*.

As an example, consider two atoms: sodium, Na, and chlorine, Cl. Sodium is a Group I element with an atomic number of 11. The sodium electrons are in three shells: 2, 8, and 1 electron, respectively. Chlorine on the other hand is a Group VII atom with an atomic number of 17 and electron shells of 2, 8, and 7 electrons, respectively. Thus, chlorine has 7 valence electrons and sodium one. If the sodium atom gives its valence electron to a chlorine atom, each atom will have eight electrons in its outermost electron shell and each will thus be in a stable electromagnetic state. The sodium atom will have 11 protons and 10 electrons and so be cationic or positively charged. The chlorine atom will have 17 protons and 18 electrons and so will be anionic or negatively charged. These two atoms are therefore drawn to each other by the difference in charge, thus forming the stable chemical compound, sodium chloride, common table salt. The *chemical formula* for the compound, NaCl, is a short-hand notation with chemical symbols to represent the composition of the compound.

In covalent bonding, electrons are not transferred from one atom to another. Instead two atoms share electrons. When two atoms mutually share one electron each, the bond is called a *single covalent bond*. Both electrons fill the outer electron shell of each atom and are referred to as a *covalent electron pair*. If two atoms share two covalent electron pairs, the bond is called a *double covalent bond*, and three pairs of electrons form a *triple covalent bond*. Covalent bonds are inherently much stronger than ionic bonds. The relative bonding energies for different covalent and ionic bonds are covered later.

Molecule

When atoms share electrons by covalent bonding in order to allow each atom to attain a stable outer electron shell, the group of atoms is called a *molecule*. A molecule is the smallest chemical unit of a substance that is capable of stable independent existence. Consider the formation of the simple molecule formed by single covalent bonding between atoms of carbon and hydrogen. From Table 1.5, carbon has four valence bonds and hydrogen has one. Carbon needs four electrons to fill the outer electron shell and hydrogen needs one. One electron from the outer electron shell of the carbon shell can form a covalent bond with one of the hydrogen atoms. Four hydrogen electrons must be shared with the carbon in order to achieve a stable electromagnetic state and thus form a stable molecule, CH_4, methane, the main component in natural gas. The methane molecule can be graphically represented in a *structural formula* as:

$$\begin{array}{c} H \\ | \\ H-C-H \\ | \\ H \end{array}$$

Each line between a carbon and a hydrogen atom in the structural formula indicates one pair of shared valence electrons. The Lewis dot diagram can also be used to illustrate electron pairs (16–18). One dot represents an individual electron being shared between the outer electron shells of two atoms. The *electron dot formula* for methane is:

$$\begin{array}{c} H \\ \cdot\cdot \\ H:C:H \\ \cdot\cdot \\ H \end{array}$$

For simplicity, the structural formula is used throughout this book. The actual geometrical configuration of the methane molecule is not two-dimensional as shown here. The four hydrogen atoms are not equally spaced at 90° ($\pi/4$ rad) angles in a plane about the carbon atom. Rather, they are in a very complex arrangement. Considerable research has been conducted to determine the precise atomic orientations in molecules (17–19). The actual atomic orientation can have a significant effect on the property of a polymer if the position of the atoms result in a polarity to the molecule. This is then referred to as a *polar covalent bond*. This is discussed more fully below.

The molecular weight of a molecule is equal to the sum of the molecular weights of the individual atoms forming the molecule. For example, the molecular weight of methane is:

$$CH_4 = 1(C) + 4(H) = 1(12) + 4(1) = 16 \text{ g/g mol}$$

The combination of two carbons in a carbon-hydrogen molecule results in three possible atomic arrangements that satisfy the valence requirements:

$$\begin{array}{c} H \quad H \\ | \quad | \\ H-C-C-H \\ | \quad | \\ H \quad H \end{array} \qquad \begin{array}{c} H \quad H \\ | \quad | \\ C=C \\ | \quad | \\ H \quad H \end{array} \qquad H-C\equiv C-H$$

Ethane (C_2H_6) Ethylene (C_2H_4) Acetylene (C_2H_2)

The most stable form of the three two-carbon hydrocarbon molecules is ethane and the molecule is called *saturated*. Only single covalent bonds exist between carbon atoms. Those molecules with double bonds between the carbon atoms (ethylene) or triple bonds (acetylene) are called *unsaturated* molecules. This terminology is used later to distinguish between an *unsaturated* polyester polymer resin used in sheet molding compounds for boats and bathtubs and *saturated* polyester polymer, polyethylene terephthalate, used to make fibers, as in polyester-cotton, and films, as in photographic film bases.

Ethane is the smallest saturated carbon–hydrogen molecule with a carbon–carbon backbone. Addition of carbon–hydrogen structures to the backbone in a linear, saturated fashion produces:

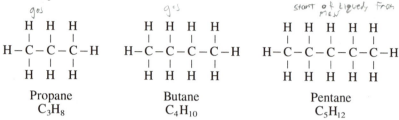

gas   *gas*   *start of liquid from max*

Propane C_3H_8 Butane C_4H_{10} Pentane C_5H_{12}

Pentane, C_5H_{12}, is too heavy to remain a gas at atmospheric pressure and room temperature. It is a liquid. Further additions of sets of one carbon and two hydrogen atoms to the carbon–carbon backbone results in eicosane, $C_{20}H_{42}$, a wax at room temperature. As the molecular weight increases, the ability of the molecule to vaporize into a gas decreases. Heptacontane, $C_{70}H_{142}$, with a molecular weight of 982, decomposes before it vaporizes. The molecule cannot obtain sufficient thermal energy to vaporize before covalent bonds are broken (decomposition).

These molecules are known as linear aliphatic hydrocarbons. The effect of molecular weight (or length of molecule) on transition temperatures is given in Table 1.6. As was noted for methane, all higher molecular weight hydrocarbons do not have simple planar two-dimensional chains but instead have complex orientations (16–18). For simplicity, all high polymers in this text will be graphically represented as linear planar segments.

Other Molecules

To this point, discussion has focused only on carbon and hydrogen atoms in the molecular structure. Many other atomic configurations can exist, of course, as seen in Table 1.8. Consider replacing one of the hydrogen atoms in the unsaturated ethylene chain with an atom that has a similar valance, such as chlorine or fluorine. Double bonded two-carbon molecules with two atoms or molecules covalently bonded to each of the carbons are generically called *vinyls*. As a result, these molecules are called vinyl chloride and vinyl fluoride, respectively:

$$\begin{array}{cc} \text{H} & \text{H} \\ | & | \\ \text{C} & = \text{C} \\ | & | \\ \text{H} & \text{Cl} \end{array} \qquad \begin{array}{cc} \text{H} & \text{H} \\ | & | \\ \text{C} & = \text{C} \\ | & | \\ \text{H} & \text{F} \end{array}$$

Vinyl chloride   Vinyl fluoride

Table 1.6. Homologous series of hydrocarbons and typical properties (20–22).

Hydrocarbon	Chemical formula	Molecular weight	Room temperature density (g/cm^3)	Melting point (°C)	Boiling point (°C)
Methane	CH_4	16	0.00066	-182	-161
Ethane	C_2H_6	30	0.00128	-183	-89
Propane	C_3H_8	44	0.0018	-190	-44
Butane	C_4H_{10}	58	0.0024	-138	-1
Pentane	C_5H_{12}	72	0.628	-130	36
Hexane	C_6H_{14}	86	0.659	-95	69
Heptane	C_7H_{16}	100	0.684	-91	98
Octane	C_8H_{18}	114	0.702	-57	126
Nonane	C_9H_{20}	128	0.719	-54	151
Decane	$C_{10}H_{22}$	142	0.749	-32	174
Undecane	$C_{11}H_{24}$	156	0.750	-26	195
Dodecane	$C_{12}H_{26}$	170	0.751	-12	214
Pentadecane	$C_{15}H_{32}$	212	0.769	-10	271
Eicosane	$C_{20}H_{42}$	282	0.78	37	–
Triacontane	$C_{30}H_{62}$	422	0.78	66	–
Heptacontane	$C_{70}H_{142}$	982	0.79	105	–
Polyethylene	$C_{100}H_{202}$	1,402	0.91	–	
	$C_{1,000}H_{2,002}$	14,002	0.925	–	LMW
	$C_{5,000}H_{10,002}$	70,002	0.94	–	HMW
	$C_{50,000}H_{100,002}$	700,002	0.955	–	
	$C_{500,000}H_{1,000,002}$	7,000,002	0.965	–	UHMW

LMW = Low molecular weight
HMW = High molecular weight
UHMW = Ultra-high molecular weight

The molecular configuration changes when the second, third and even fourth hydrogen atom is replaced with a chlorine or fluorine atom (19):

```
   H   F              H   F              F   F
   |   |              |   |              |   |
   C = C              C = C              C = C
   |   |              |   |              |   |
   H   F              F   F              F   F
```

Vinylidene fluoride   Trifluoroethylene   Tetrafluoroethylene

The complexity of this vinyl molecule is further increased by replacing one of the hydrogen atoms in the ethylene molecule by a group of atoms that have the characteristics of a single atom. Such a group of atoms is called a *radical*. A list of molecular radical groups typically used in high polymers is given in Table 1.7. The simplest molecular radical group is the methyl radical:

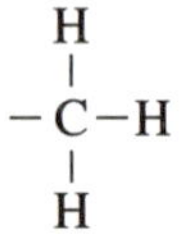

$$\begin{array}{c} H \\ | \\ -C-H \\ | \\ H \end{array}$$

Consider replacing an atomic hydrogen on the ethylene molecule with a methyl radical (here written as $-CH_3$):

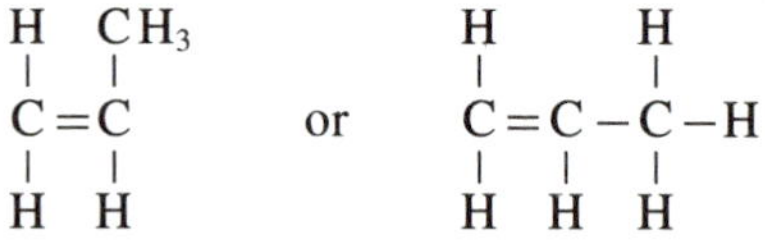

$$\begin{array}{cccc} H \quad CH_3 & & H \qquad H \\ | \quad\; | & & | \qquad\; | \\ C=C & \text{or} & C=C-C-H \\ | \quad | & & | \quad | \quad | \\ H \quad H & & H \quad H \quad H \end{array}$$

Propylene

The molecule propylene is chemically converted to polypropylene. The reader can certainly visualize many other radical group arrangements on the simple ethylenic unsaturated carbon backbone, simply through replacement of one or more of the atomic hydrogens.

Ring Molecules

Consider oxygen in the ethylenic backbone. Oxygen has a valence of 2 and therefore must share its electrons with *two* carbons:

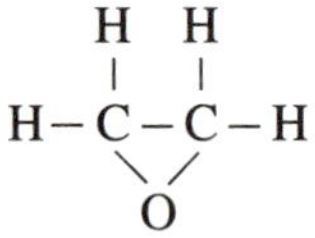

$$\begin{array}{c} H \quad H \\ | \quad\; | \\ H-C-C-H \\ \backslash \;\; / \\ O \end{array}$$

Ethylene oxide

The molecule ethylene oxide forms a saturated ring molecule. This is the simplest ring molecule. Rings of four, five, six, and seven atoms are quite stable. As an example, consider cyclohexane (C_6H_{12}):

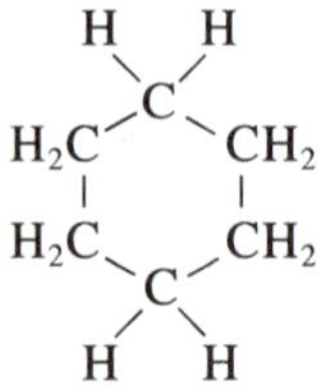

$$\begin{array}{c} H \quad H \\ \backslash \; / \\ C \\ H_2C \quad\;\; CH_2 \\ | \qquad\quad | \\ H_2C \quad\;\; CH_2 \\ \backslash \; / \\ C \\ / \; \backslash \\ H \quad H \end{array}$$

Table 1.7. Polymer radical group structure (23).

monomers on Next p2 25

Radical group name	Formula	Structure
Methyl	— CH_3	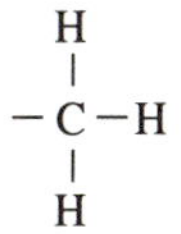 $-\overset{\displaystyle H}{\underset{\displaystyle H}{C}}-H$
Ethyl	— CH_2CH_3	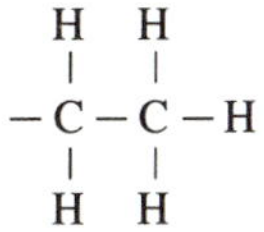 $-\overset{H}{\underset{H}{C}}-\overset{H}{\underset{H}{C}}-H$
Propyl	— $CH_2CH_2CH_3$	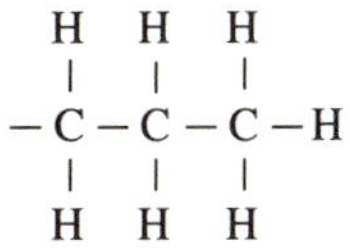 $-\overset{H}{\underset{H}{C}}-\overset{H}{\underset{H}{C}}-\overset{H}{\underset{H}{C}}-H$
Hydroxyl	— OH	$-O-H$
Carboxyl	— COOH	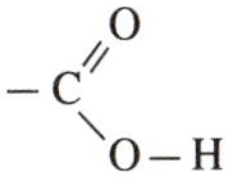 $-C\overset{\displaystyle O}{\underset{\displaystyle O-H}{}}$
Acetyl	— $COCH_3$	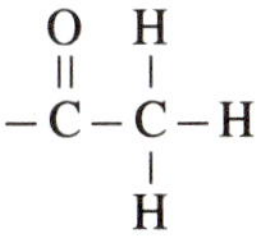 $-\overset{O}{C}-\overset{H}{\underset{H}{C}}-H$
Aldehyde	— CHO	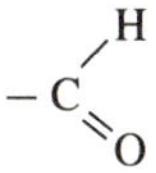 $-C\overset{H}{\underset{O}{}}$
Amino	— NH_2	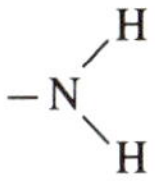 $-N\overset{H}{\underset{H}{}}$

This is a saturated six-ring hydrocarbon. Benzene is a six-ring hydrocarbon, as well, but with three carbon–carbon double bonds:

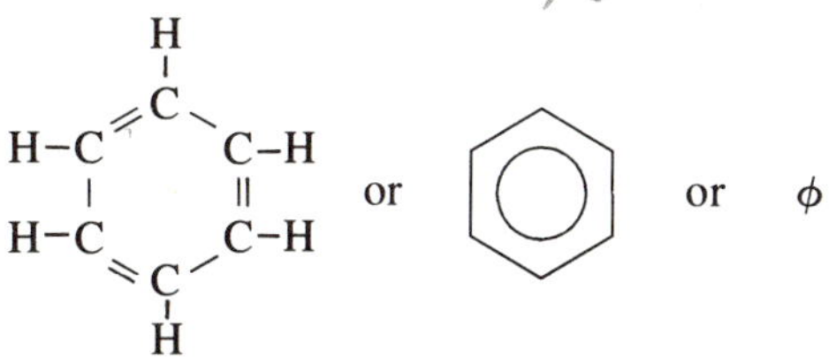

or or φ

Linear molecular chains are called *aliphatic*. Molecular chains that contain benzyl radicals, $-C_6H_5$ groups, are called *aromatic*. The term is derived from the pungent, sharp, or acrid odor exhibited by low molecular weight ring compounds. Note that there are two other short-hand notations used for benzene or the benzyl radical group: the hexagon, with or without the inner circle, and the Greek symbol ϕ. Note that the benzyl radical can be used to substitute for a hydrogen on the ethylenic molecule:

$$
\begin{array}{cc}
\text{H} & \text{H} \\
| & | \\
\text{C} & = \text{C} \\
| & | \\
\text{H} & \phi
\end{array}
$$

The resulting molecule is called vinyl benzene or styrene. This molecule is polymerized into polystyrene.

1.5 Monomer: The Building Block of a Polymer

Note in Table 1.6 that the long chain aliphatic molecules are simply the result of adding, one at a time, identical groups of smaller molecules to the existing carbon–carbon backbone. The group of molecules added each time is called a *mer*. In the case of building a high molecular weight wax from a low molecular weight oil, the group added, one step at a time, is $-CH_3$, the methyl group. In the case of polyethylene, the mer is the ethyl radical, $-CH_2-CH_2-$. This is the mer created from the simplest double-bonded hydrocarbon molecule, ethylene. Ethylene is produced by dehydrogenation of ethane gas:

H-C : 99 Kcal/mol
C-C : 83 kcal/mol   *not as stable*

$$
\begin{array}{cc}
\text{H} & \text{H} \\
| & | \\
\text{H}-\text{C}- & \text{C}-\text{H} \\
| & | \\
\text{H} & \text{H}
\end{array}
\quad
\begin{array}{c}
\text{heat, pressure} \\
\rightarrow \\
\text{catalyst}
\end{array}
\quad
\begin{array}{cc}
\text{H} & \text{H} \\
| & | \\
\text{C} & = \text{C} \\
| & | \\
\text{H} & \text{H}
\end{array}
+ \text{H}_2
$$

Several simple *monomers* based on the ethylene chain by substituting an atom or a radical group for one or more of the hydrogen atoms are listed in Table 1.8.

Not all polymer molecules are formed from a single monomer source. When two or more monomers are used, they are called *comonomers*. For example, low molecular weight unsaturated polyester (MW $\approx$ 1000) is reacted with styrene monomer to make unsaturated polyester resin, a thermoset:

by addition polymerization Type 2

$$
\text{H}\left[\begin{array}{ccccccc}
 & \text{O} & \text{H} & \text{H} & \text{O} & \text{H} & \text{H} \\
 & || & | & | & || & | & | \\
\text{O}-&\text{C}-&\text{C}&=\text{C}-&\text{C}&-\text{O}-\text{C}-&\text{C} \\
 & & & & & | & | \\
 & & & & & \text{H} & \text{H}
\end{array}\right]_x \text{OH}
\quad \text{and} \quad
\begin{array}{cc}
\text{H} & \text{H} \\
| & | \\
\text{C} & = \text{C} \\
| & | \\
\text{H} & \phi
\end{array}
$$

Unsaturated polyester   Styrene

Table 1.8. Monomers based on the ethylenic or vinyl structure.

$$
\begin{array}{cc}
& \text{H} \quad \text{H} \\
& | \quad \; | \\
\text{The basic unit:} & \text{C} = \text{C} \\
& | \quad \; | \\
& \text{H} \quad (\text{A})
\end{array}
$$

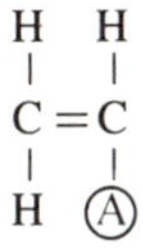

When [A] is:	Monomer is:
—H	Ethylene
—Cl	Vinyl chloride
—F	Vinyl fluoride

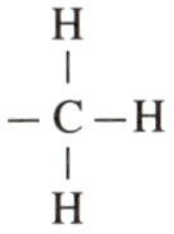

$$
\begin{array}{l}
\quad \text{H} \\
\quad | \\
-\text{C}-\text{H} \\
\quad | \\
\quad \text{H}
\end{array}
$$
Propylene

$$
\begin{array}{l}
\text{O} \qquad \text{H} \\
|| \qquad \; | \\
-\text{C}-\text{O}-\text{C}-\text{H} \\
\qquad \quad | \\
\qquad \quad \text{H}
\end{array}
$$
Methyl acrylate

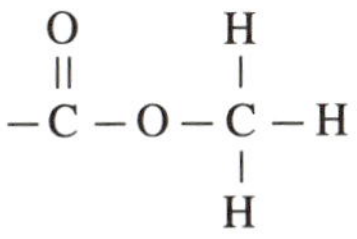

Styrene

condensation

A second example is bisphenol A and phosgene, which when reacted, produce polycarbonate:

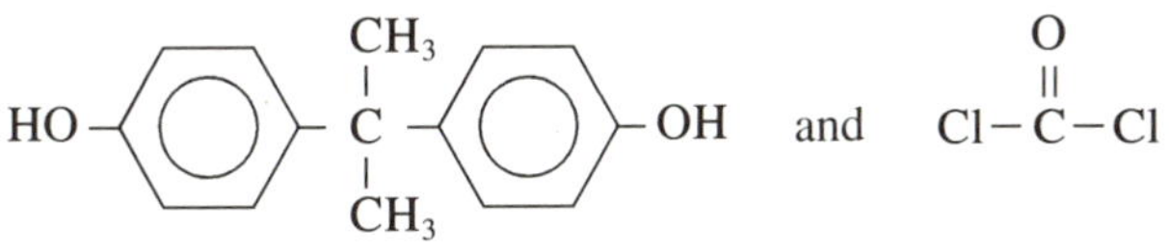

$$
\text{HO} - \!\!\bigcirc\!\! - \underset{\underset{\text{CH}_3}{|}}{\overset{\overset{\text{CH}_3}{|}}{\text{C}}} - \!\!\bigcirc\!\! - \text{OH} \qquad \text{and} \qquad \text{Cl} - \overset{\overset{\text{O}}{||}}{\text{C}} - \text{Cl}
$$

Bisphenol A   Phosgene

= *polycarbonate*

Monomer Functionality

Albright (24) notes that in order to get a high polymer, the monomers or reactants must have functionality of two or more. *Functionality* is either the number of reactive groups on the monomer or the number of chemical bonds that are formed during polymerization. Consider the reaction between a simple alcohol and a simple acid to produce an ester:

$$
\underset{\text{Acid}}{R-\overset{\overset{\displaystyle O}{\|}}{C}-OH} \;+\; \underset{\text{Alcohol}}{HO-R'} \;\rightarrow\; \underset{\text{Ester}}{R-\overset{\overset{\displaystyle O}{\|}}{C}-O-R'} \;+\; \underset{\text{By-product}}{H_2O}
$$

R and R' are any combinations of nonreactive functional groups that make up common acids and alcohols, respectively. The functionality of both the acid and alcohol is one. The functionality of the ester is zero. Albright further shows that a reaction between a difunctional alcohol, also called a "diol", and two molecules of a simple acid yields an ester of zero functionality:

$$
\underset{\text{Acid}}{R-COOH} \;+\; \underset{\text{Diol}}{HO-R'-OH} \;+\; \underset{\text{Acid}}{HOOC-R} \;\rightarrow
$$

$$
\underset{\text{Ester}}{R-O-\overset{\overset{\displaystyle O}{\|}}{C}-R'-\overset{\overset{\displaystyle O}{\|}}{C}-O-R} \;+\; \underset{\text{By-product}}{2\,H_2O}
$$

The reaction between a difunctional acid, called a "dibasic acid", and a difunctional alcohol, a diol, results in a polymer known as a *polyester:*

$$
\underset{\text{Diacid}}{HOOC-R-COOH} \;+\; \underset{\text{Diol (glycol)}}{HO-R'-OH} \;\rightarrow\; \underset{\text{Difunctional ester}}{HOOC-R-\overset{\overset{\displaystyle O}{\|}}{C}-O-R'-OH} \;+\; \underset{\text{By-product}}{H_2O}
$$

The ester so produced is also difunctional and thus capable of continuing the polymerization. The polymer so formed is a linear polymer. Note in the section above that bisphenol A has two functional groups, the hydroxyl groups, $-OH$, and phosgene has two chlorines, Cl. Thus, a high polymer can be formed.

Certain monomers have functionalities greater than two. Consider reaction of a difunctional acid with a trifunctional alcohol, a triol such as glycerine:

$$
\underset{}{HOOC-R-COOH} \;+\; \underset{}{HO-\overset{\overset{\displaystyle OH}{|}}{R'}-OH} \;\rightarrow\; HOOC-R-\overset{\overset{\displaystyle O}{\|}}{C}-O-\overset{\overset{\displaystyle OH}{|}}{R'}-OH
$$

Note that the reactant thus formed remains trifunctional. That is, an acid molecule can react with each of the two hydroxyl groups and an alcohol molecule can react with the carboxyl group ($-COOH$). The polymer thus formed can be three-dimensional or network. Some reactants, such as pentaerythritol (tetramethylol methane) and divinyl benzene, have functionalities of four:

$$CH_2OH$$
$$HOH_2C - C - CH_2OH$$
$$CH_2OH$$

Pentaerythritol

$$CH=CH_2$$

$$CH=CH_2$$

Divinyl benzene

There are dozens of classes of polymers and thousands of variants commercially available today. Most polymers are derived from only a few monomers. Polymers derived from methane, ethylene, propylene and acetylene and the aromatic monomers, benzene, naphthalene, toluene and xylene are given in Appendix A. The chemical structure and trade names of many commercial resins are given later.

1.6 Polymerization:
The Building of a Large Molecule

The process by which monomers and comonomers are chemically linked to form a high molecular weight polymer molecule is called *polymerization*. There are two types of polymerization: addition and condensation. In *condensation polymerization*, reaction results in the formation of a small, usually volatile by-product molecule such as water, acetic acid, or HCl. Reaction is driven to completion by continually removing the by-product molecule. In *addition polymerization*, reaction occurs without a by-product molecule being generated and is usually driven to completion by heat.

Addition Polymerization

Addition reactions can occur in each of several ways. The simplest addition reaction is the *linear addition polymerization* process where a double bond within the monomer molecule is opened, usually by a free radical initiator. In this activated state, the monomer molecule will now react with another unactivated monomer. The resulting dimer now contains an active reacting site and will react with still another unactivated monomer to form a trimer, and so on. The result is a long linear polymer chain. This is the basic process for polymerization of vinyl monomers, the simplest being polyethylene:

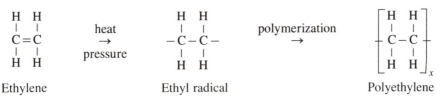

Ethylene   heat → pressure   Ethyl radical   polymerization →   Polyethylene

Pe
Appendix A

The subscript "x" in the polyethylene chemical formula represents the number of mer units that form the polymer molecule; "x" is also called the *degree of polymerization,* DP.*

Theoretically, the linear polymerization of ethylene into a polyethylene molecule should result in a linear chain. However, competing reactions can occur. As a result, instead of obtaining a purely linear polymeric molecule, the actual molecule is primarily linear but with side branches extending from the main chain. The degree and type of side branching along the main chain directly affect the structure of the polymer at room temperature. One of the more obvious differences is in polymer density. Highly branched polymer chains result in lower density, since the branches prevent the main chains from closely packing. The degree of branching can be controlled by proper selection of the polymerization conditions and catalysts. Classically, the polyethylene family is classified by its room temperature densities:

Type of Polyethylene	Density, g/cm^3
Low density (long-chain branched, high pressure) [LDPE]	0.91–0.925
Medium density (short-chain branched, high pressure) [MDPE]	0.925–0.94
High density (linear, low pressure, catalyzed) [HDPE]	0.94–0.965

Note that low density polyethylene is produced by high-pressure polymerization, whereas high density polyethylene is produced by catalysis at low pressure. Recent polymerization development has produced a low density polyethylene with a more linear geometric structure, Figure 1.6. This polymer is *linear low density polyethylene*

---

*Correctly, the *degree of polymerization* must be related to the extent of polymerization, *P*. It can be shown that:

$$F\left(\frac{2}{f}\right)\left(1 - \frac{N}{N_0}\right)$$

where f is the average functionality of the reaction, N is the current moles of polymer molecules and N_0 is the initial number of moles of monomer (or comonomers). Functionality is usually 2. For addition reaction or condensation reaction with a single monomer, such as caprolactam, $DP = N_0/N$. Thus:

$$P = \left(1 - \frac{1}{DP}\right)$$

For condensation reaction with two comonomers of equal reactivity, $DP = N_0/2\,N$. Thus:

$$P = 1 - \frac{1}{2DP}$$

extent of
pulymerization

$P \propto x$

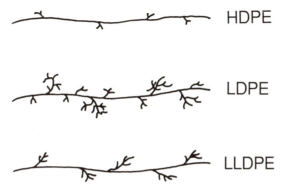

HDPE

LDPE

LLDPE

Figure 1.6.   Schematic Concept of Various Types of Polyethylene, Showing Little Branching for HDPE, Short-Chain Branching for LLDPE, and Extensive Long-Chain Branching for LDPE.

[LLDPE]. The polymer has the lower density of LDPE but the toughness of HDPE. It is apparent that polymerization can result in several types of geometric structures for the polymer being built. When attempting to describe a particular polymer, it will be necessary to know the actual geometric construction or conformation *of the polymer.* In other words, its molecular stereoisomerism is also important.

Example 1.1

Q: Neglecting end groups, estimate the length of a heptacontane molecule.
A: From Table 1.6, the heptacontane molecule is $C_{70}H_{142}$. The geometric configuration of the simple hydrocarbon molecule (Table 1.17) is:

$$\cos \frac{\text{ad }?}{\text{hx}}$$ $$\text{adJ} = \text{hyp cw d}$$

$$\frac{L}{2} = \cdot$$

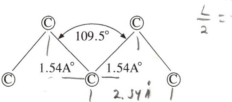

The length between two carbon atoms, L, is given as:

$$L = 2\left\{1.54 \cos\left(\frac{180 - 109.5}{2}\right)\right\} = 2.52 \text{ Å}$$

Mean molecular length, L_T *is:*

$$L_T = L \times \text{number of two-carbon units}$$

$$L_T = 2.52 \times 70/2 = 88.2 \text{ Å}$$

A second type of addition polymerization involves *molecular rearrangement* of atoms between the reacting molecules. An example is the formation of polyurethane:

$$H-O-R-O-H \ + \ O=C=N-R'-N=C=O \ \rightarrow$$

$$\qquad \text{Diol or Glycol} \qquad\qquad\qquad \text{Diisocyanate}$$

$$\left[O-R-O-\overset{\overset{\displaystyle O}{\|}}{C}-\overset{\overset{\displaystyle H}{|}}{N}-R'-\overset{\overset{\displaystyle H}{|}}{N}-\overset{\overset{\displaystyle O}{\|}}{C} \right]_x$$

$$\text{Polyurethane}$$

where R and R′ are specific molecular groups in each monomer. The resulting polymer is a linear polyurethane.

Another example of addition polymerization does not result in a linear polymer. It is the reaction of a low molecular weight unsaturated polyester with styrene monomer. Here the carbon–carbon double bond on the styrene monomer is opened, usually with a peroxide free-radical initiator. The active styrene radical reacts with the carbon–carbon double bond on the backbone of the unsaturated polyester. The chain is extended by addition of a second, third and perhaps fourth styrene monomer to the styrene radical. Eventually after three or four styrene monomer reactions, the active styrene chain encounters another unsaturated polyester carbon–carbon double bond, and reaction is complete. The short chain polystyrene forms a bridge or network between the polyester molecules. The result is a three-dimensional thermoset, theoretically a single molecule the size of the part being fabricated.

A third type of addition polymerization is ring opening or scissoring. Activation of a three- to eight-membered ring containing unsaturation is usually accomplished by heat and a suitable catalytic activator. Consider ethylene oxide, a three-membered ring. Once the monomer is activated by scissoring the carbon–carbon double bond, reaction proceeds in a linear fashion to produce polyethylene oxide:

$$\begin{array}{ccc}
\overset{\displaystyle H}{|} & & \overset{\displaystyle H}{|} \\
C & = & C \\
& \diagdown \ O \diagup &
\end{array} \qquad \overset{\text{heat, catalyst}}{\longrightarrow} \qquad \left[\begin{array}{ccc} \overset{\displaystyle H}{|} & & \overset{\displaystyle H}{|} \\ C & -O- & C \\ \overset{|}{H} & & \overset{|}{H} \end{array} \right]_x$$

$$\text{Ethylene oxide} \qquad\qquad\qquad \text{Polyethylene oxide}$$

Condensation Polymerization

As noted earlier, in *condensation polymerization*, the chemical reaction of comonomers to form the embryonic polymer molecule results in the formation of a small by-product molecule. Since most condensation polymerizations tend to be reversible reactions, this molecule must be removed to allow polymerization to continue. An example of condensation polymerization where water is the by-product is the formation of phenol–formaldehyde (phenolic or Bakelite) from reaction of phenol and formaldehyde:

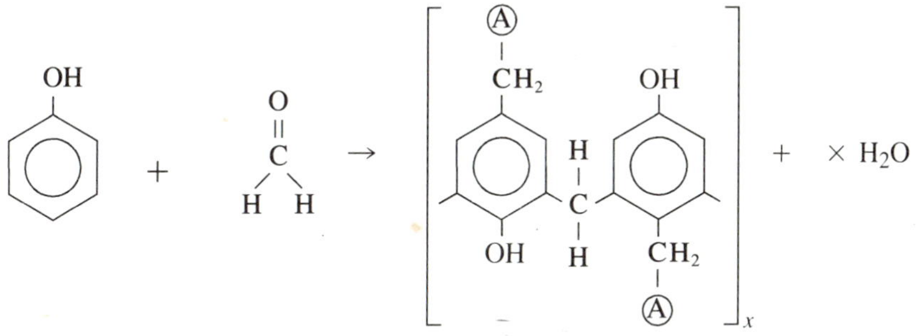

Phenol Formaldehyde Phenol–formaldehyde Water

The vertical lines given by (A) represent covalent bonds attached to an identical molecular structure above and below that shown. The result is a large three-dimensional *network*.

Polycarbonate is produced by condensation polymerization of bisphenol A and phosgene, with HCl being the small by-product molecule:

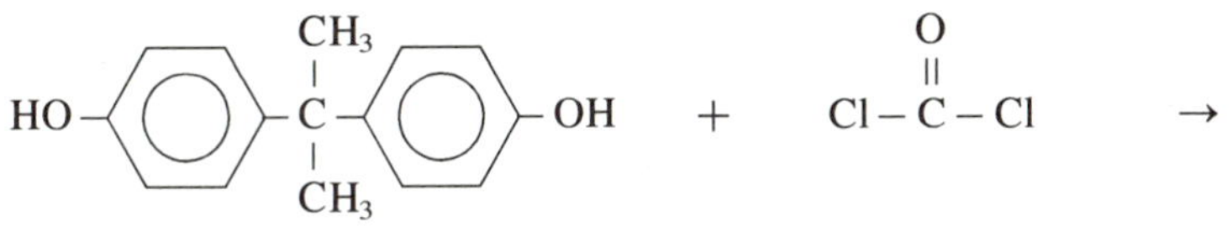

Bisphenol A Phosgene

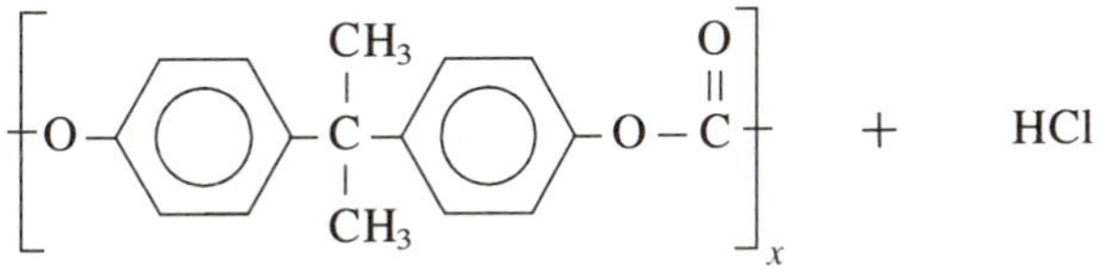

Polycarbonate Hydrogen chloride

Nylon (or polyamide) was one of the earliest thermoplastic condensation polymers. It is formed by reacting a difunctional amine and a difunctional acid. The classic nylon is nylon 66, formed by reacting hexamethylene diamine (6 carbons) with adipic acid (6 carbons), and splitting out water:

$$n\ H_2N\!\left(CH_2-CH_2-CH_2-CH_2-CH_2-CH_2\right)\!NH_2\ +$$

Hexamethylene diamine

$$n\ HOOC\!\left(CH_2-CH_2-CH_2-CH_2\right)\!COOH\ \rightarrow$$

Adipic acid

$$H_2N\!\left[(CH_2)_6-\overset{H}{\underset{|}{N}}-\overset{O}{\overset{\|}{C}}-(CH_2)_4\right]_n\overset{O}{\overset{\|}{C}}-OH\ +\ n\ H_2O$$

"Nylon 66"

The entire family of polyamides is produced by condensation polymerization. Nylon 610, for example, is produced by reacting hexamethylene diamine (6 carbons) with sebacic acid (10 carbons):

$$n \ H_2N + (CH_2)_6 - NH_2 \ + \ n \ HOOC + (CH_2)_8 - COOH \ \rightarrow$$

Hexamethylene diamine   Sebacic acid

$$H_2N \left[(CH_2)_6 - \overset{\overset{H}{|}}{N} - \overset{\overset{O}{||}}{C} - (CH_2)_8 \right]_n \overset{\overset{O}{||}}{C} - OH \ + \ n \ H_2O$$

"Nylon 610"

Nylon 6 or polycaprolactam is formed by reacting ε-amino caproic acid with itself, again removing water in the reaction. It is thought that the first step in the polymerization is the formation of a seven-member ring compound known as ε-caprolactam. Polymerization then occurs via ring opening. Nylon 6 is unique in that the polymer is produced by condensation polymerization from a single monomer:

$$n \ H_2N + (H_2)_5 - COOH \quad \overset{-n H_2O}{\rightarrow} \quad n \ HN \begin{array}{c} CH_2 - CH_2 \\ \diagup \qquad \diagdown \\ \qquad CH_2 \\ \diagdown \qquad \diagup \\ \overset{||}{C} - CH_2 - CH_2 \\ || \\ O \end{array} \rightarrow$$

ε − Amino Caproic acid   ε−Caprolactam

$$H_2N \left[(CH_2)_5 - \overset{\overset{O}{||}}{C} - \overset{\overset{H}{|}}{N} - (CH_2)_5 \right]_n \overset{\overset{O}{||}}{C} - OH$$

"Nylon 6"

Sequential Polymerization

Polymerization can also occur in sequential steps. In *sequential polymerization*, a condensation reaction occurs first between two comonomers to produce a molecule that then reacts with a third molecule via addition polymerization. Since all three constituents have relatively low molecular weights, the materials can be easily mixed and processed. It was noted above that polymerization of unsaturated polyester with styrene monomer was an addition reaction. Unsaturated polyester itself is produced by condensation reaction of ethylene glycol and a mixture of maleic and phthalic acids. Shown here is the reaction to produce the maleate portion of the unsaturated polyester:

Condensation:

$$n\ HO-CH_2-CH_2-OH\ +\ n\ HO-\overset{\overset{\displaystyle O}{\|}}{C}-\underset{\underset{\displaystyle H}{|}}{C}=\underset{\underset{\displaystyle H}{|}}{C}-\overset{\overset{\displaystyle O}{\|}}{C}-OH\ \rightarrow$$

Ethylene glycol   Maleic acid

Addition:

$$n\ H_2O\uparrow+\ \left[O-CH_2-CH_2-O-\overset{\overset{\displaystyle O}{\|}}{C}-CH=CH-\overset{\overset{\displaystyle O}{\|}}{C}-O\right]_n\ +\ H_2C=CH\ \rightarrow$$

Polydiglycol maleate or
Unsaturated polyester

Styrene

Polymer:

$$\left[O-CH_2-O-\overset{\overset{\displaystyle O}{\|}}{C}-\underset{\underset{\underset{\underset{HC}{|}}{CH_2}}{|}}{\overset{\overset{\displaystyle H}{|}}{C}}-CH-\overset{\overset{\displaystyle O}{\|}}{C}-O\right]_x$$

Crosslinking with styrene

The reaction with styrene monomer is an addition reaction, as stated before.

The polymer chemist has many ways in which to form a polymer from common monomers. Both the compositions and geometric configurations of the monomers directly affect the properties of the resulting polymer. As expected, the large number of possible combinations of the basic monomeric building blocks have generated thousands of different polymers. Only a few dozen of these are commercially important, however. Each year, a few new polymers are commercialized. The designer must be aware that new polymeric materials with unique properties are being continually introduced into the market.

Polymer molecules are usually classed either as *linear polymers* or *network* or *three-dimensional polymers*. The classification depends upon the geometric configuration of the covalent bonding between atoms on the backbone of the polymer. The consistency of the atoms along the backbone can be used to further subdivide linear polymers. Several common linear polymers having only carbon–carbon backbones are listed in Table 1.9 (see page 52). Polymers with nitrogen in the backbone are listed in Table 1.10 (see page 58), those with oxygen in the backbone, in Table 1.11 (see page 61) and those with sulfur, in Table 1.12 (see page 66). Three-dimensional polymers are shown in Table 1.13 (see page 67). The distinction between linear and three-dimensional polymers was commented on earlier with the examples of spaghetti as representing linear polymers and scrambled eggs as being like three-dimensional or thermoset polymers.

1.7 Molecular Orientation or Stereoisomerism

To this point, linear polymer molecules have been described as having a basic molecular unit, a mer, repeated n times to form a long, straight molecule. When viewing the covalent linkages between the carbons along the polyethylene polymer molecule, the backbone is symmetrically surrounded by hydrogen atoms. For propylene and many other monomers listed in Table 1.8, the basic molecular element in the polymer molecule is *not* symmetric with respect to the backbone. Each mer has instead a dissimilar atom or radical group replacing one or more of the hydrogen atoms. These groups are called *pendant groups*. For the case of propylene, the molecular appendage is the methyl radical, $-CH_3$. The arrangement of these methyl groups along the linear carbon chain greatly affects the properties of the polypropylene polymer. There are three possible ways of arranging these molecular groups with respect to the carbon–carbon backbone:

> *Atactic*. A random arrangement of the unsymmetrical radical groups or atoms along the backbone plane,
>
> *Isotactic*. An arrangement with all the radical groups or atoms on the same side of the backbone plane, and
>
> *Syndiotactic*. An arrangement with the radical groups or atoms on alternate sides of the backbone plane.

This is called *stereoisomerism*, and the planar representations of these three geometric configurations are seen in Figure 1.7. The actual arrangement of these pendant groups depends strongly on the polymerization process. An excellent example of the significance of molecular orientation on the nature of the polymer is seen in atactic and isotactic polypropylene. Atactic polypropylene is a waxy, tacky polymer with low physical strength and relatively little commercial value, except as an additive in certain hot melt adhesives. In the mid-1950s, G.F. Natta (25) found that certain catalysts from the Group IV to VIII transition elements, particularly titanium trichloride, were *stereospecific*. That is, certain catalyst features were directly related to the molecular conformation of the polymer. These Ziegler–Natta catalysts could be used to produce nearly pure isotactic polypropylene. Catalyst research continues today in an effort to minimize the amount of atactic polypropylene in the more desirable isotactic polypro-

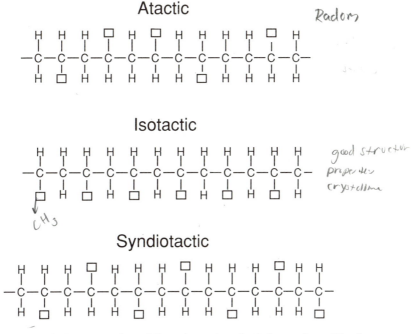

Figure 1.7. Schematic Representation of Stereoisomerism for Polypropylene. The Squares Represent the Pendant Methyl Group, −CH₃ (21).

pylene polymerization. Currently, yields of 94 to 99% isotactic, 1 to 6% atactic are common. Up to 2% atactic can remain in the polymer without appreciably affecting the polymer performance (26). Isotactic polypropylene is a highly crystalline polymer with a melting temperature of about 160°C, is translucent, and has a tensile strength about that of nylon.

1.8 Copolymers and Terpolymers

So far, the polymers discussed have been *homopolymers*. That is, only one monomer or set of comonomers has been used to produce the polymer. As noted, in some cases, polymers of different types can result from the same polymerization. But, it remains that these are homopolymers. *Copolymers* result when more than one monomer or set of comonomers are reacted, either simultaneously or sequentially, to produce a linear polymer. If the resulting polymer has two distinct mer groups, it is called a *copolymer*. If it has three, it is referred to as a *terpolymer*. The reacting mers can form copolymers in several ways:

> *Random.* If A and B mers form a random arrangement along the linear polymer backbone, the polymer is considered to be a random copolymer. One of the earliest

random copolymers is styrene–butadiene rubber (25:75 SBR), developed in the early 1940s to replace natural rubber in automobile and truck tires.

Alternating. If the A and B mers are alternately arranged along the linear polymer backbone, the polymer is known as an alternating or azeotropic copolymer. Such copolymers are rare.

Block. When long blocks of A mers follow long blocks of B mers, and so on, along the polymer backbone, the polymer is a block copolymer. Ethylene oxide and propylene oxide are frequently block-copolymerized to produce surfactants.

Grafted. When the linear backbone has only A mers and long blocks of B mers are attached at regular intervals to functional segments of the A mers, the result is a *graft* copolymer. Usually the A mer has a degree of unsaturation, or functionality. Recall the crosslinking reaction between unsaturated polyester and styrene monomer. The link between unsaturated sites on neighboring polyester backbone is a short polystyrene molecule of 3 or 4 mers. Thus, unsaturated polyester resin represents a graft copolymer. Polystyrene and polyvinyl chloride will also graft to polymethyl methacrylate backbone. In the latter case, the resulting PVC/PMMA copolymer is a tough exterior glazing material.

A graphical representation of these four copolymer categories is given in Figure 1.8.
    The ability to combine polymers to achieve properties of each of the elements is referred to as "tailoring" or "molecular engineering". The nature of the molecular orientation for a given pair or trio of prepolymers depends strongly upon the nature of the individual molecules, their concentration, the size of the blocks or grafts being developed, the sequence of polymerization steps, the relative reactivity of the monomers *and* the polymer chains, the catalyst system, the selectivity of the catalysts, and so on. The objective is to select and control those elements that allow development of the desired copolymer while suppressing those elements that produce unwanted or worthless side reactions.
    There are many, many examples of copolymerization. Consider the classic formation of ABS (acrylonitrile–butadiene–styrene), a tough polymer used in football helmets, computer housings, equipment cabinets and windsurfing boards. The monomers can be copolymerized two at a time. When butadiene and acrylonitrile are randomly copolymerized in the ratio of 3:1, the result is a chemically resistant elastomer called "nitrile rubber". The random copolymerization of styrene and butadiene is also an elastomer, SBR rubber, as mentioned above. If styrene is grafted to polybutadiene such that the butadiene represents only 10 to 25% (wt), the result is an impact polystyrene.* Impact polystyrene is called HIPS or high-impact polystyrene. The copolymer has two phases, with the butadiene rubber as the discrete phase, dispersed in the continuous polystyrene phase. Copolymerization of (25%) acrylonitrile and (75%) styrene monomers results in a random copolymer, SAN, having excellent flexural strength. When

---

*The terminology here is correct. Polybutadiene rubber has unsaturated or double bonds along its backbone. The styrene monomer is grafted to these double bonds and styrene polymerized.

Monomer

A= O

Arcutow

B= ☐

Homopolymer

A. −O−O−O−O−O−O−O−O−O−O−O−O−

B. −☐−☐−☐−☐−☐−☐−☐−☐−☐−☐−☐−☐−

Random Copolymer A-B

O−O−☐−O−☐−☐−O−O−O−☐−O−☐−☐

Alternating Copolymer A-B

−☐−O−☐−O−☐−O−☐−O−☐−O−☐−O−

Block Copolymer A-B

O−O−O−☐−☐−☐−☐−O−O−O−☐−☐−☐−☐−

Grafted Copolymer A-B

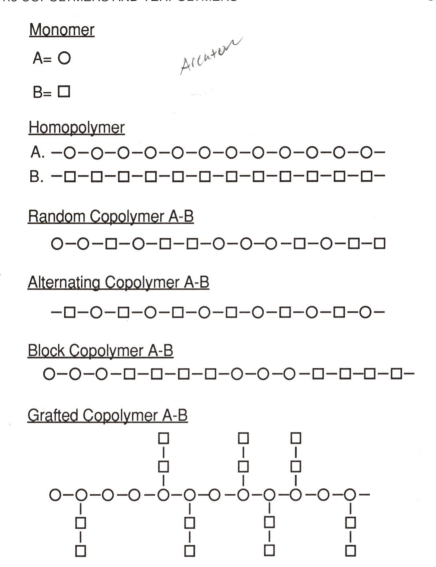

Figure 1.8.   Schematic Examples of Various Types of Copolymer. The Circles Represent Monomer A. The Squares Represent Monomer B.

shorter chains of SAN are grafted to polybutadiene, the result is the terpolymer ABS. Since the butadiene rubber content is about 10%, it becomes the discrete phase, as with high-impact polystyrene, and the SAN becomes the continuous phase.*

---

*Another form of ABS can be produced by melt-compounding two copolymers, SAN and nitrile rubber, to produce a compound having about 25% acrylonitrile, 15% butadiene and 60% styrene. The performances of these two forms of ABS seem to be about the same, thus raising the question of whether the economic expense in forming the terpolymer can be justified.

Example 1.2

Q: For a random linear copolymer, the distribution of monomer sequence lengths of monomer A can be calculated from:

$$N(n) = (1 - x) \, x^{n-1}$$

where x_A is the fraction of polymer A. Find the sequence lengths of polymer A for a mole fraction of copolymer containing 50 mole percent of monomeric material B.

A: Solution:

$$N_A = (1 - 0.5) \, 0.5^{n-1} = 0.5^n$$

Sequence Length (n)	Number Fraction of A Sequences of Length	Cumulative Fraction
1	0.5	0.500
2	0.25	0.750
3	0.125	0.875
4	0.062	0.937
5	0.03125	0.968
10	9.7×10^{-4}	0.998

Several commercial copolymers and terpolymers are listed in Tables 1.14 (see page 70) and 1.15 (see page 72). The chemical equation for each of the copolymers or terpolymers is simply representative of the chemical nature of the mers involved and should not be construed as describing the physical nature of the molecules. This is seen in ABS, where the chemical symbols denote a linear block polymer when it is known that the random SAN copolymer is grafted to the polybutadiene backbone. Very little can be inferred about the physical properties of the copolymers unless the polymer composition and molecular structure are known.

1.9 Polymer Blends and Alloys

A *polymer blend* is a physical combining of two or more polymers without (appreciable) reaction.*

Compounding can show property improvement. For example, if polymer A has a superior physical property, adding a small amount of it to polymer B with a poorer physical property will increase the physical property of B, sometimes substantially.

Blending can reduce the cost of the polymer. Blending a substantial amount of inexpensive polymer B into polymer A may reduce the polymer cost without substantially reducing its overall performance.

---

*The old version of the "ABCs" of polymers is "Alloys, Blends and Compounds". The newer version is "Alloys, Blends and Composites".

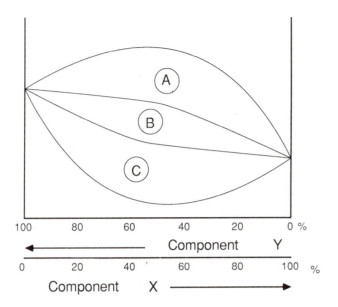

Figure 1.9.   Schematic of the Relative Effects of Blending Polymers, Region A Represents Synergism in Properties. Region B Represents Simple Property Addition. Region C Represents Incompatibility in Properties.

On rare but economically important occasions, blending polymers A and B can result in a synergistic effect. That is the property of the blend can be substantially better than that of either of the two polymers alone. These blends are frequently called *alloys.**

The various types of property response of polymers A and B, when blended, are seen schematically in Figure 1.9. Incompatible blends produce lower than expected property response. Classic incompatible polymers are polyethylene and polystyrene, at all blending levels. A polymer blend shows promise when the properties fall within the additive region. Polymers that are similar in type usually have additive properties. Polyethylene and chlorinated polyethylene show this type of behavior over most of the concentration range. As noted, the rare blend shows synergism. In some cases, synergism can be created by careful formulation of the polymers during polymerization. Consider polyvinyl butyral (PVB). Polyvinyl acetate (PVAc) can be partially reacted to polyvinyl alcohol (PVOH). A mixture of PVAc and PVOH is then reacted with aldehydes to produce polyvinyl butyral (PVB). The final product is usually a blend of the three polymers.

In general, the properties of a blend are usually determined by the *miscibility* of the polymeric constituents. Miscibility implies that a single phase is produced. As noted earlier, HIPS is a two-phase copolymer, simply because butadiene and styrene are not miscible (that is, they are *immiscible*). There are three broad categories for polymer miscibility. Complete miscibility is rare in polymers. The economically most important completely miscible polymer blend is 50% polyphenylene oxide (PPO), 50% high-impact polystyrene (HIPS), known as General Electric Noryl. Polyphenylene oxide homopolymer has excellent thermal stability but is very difficult to process without oxidative and thermal degradation. It is relatively easy to process HIPS but its high-temperature stability is poor. The resulting blend has poorer thermal properties than PPO homopolymer but has substantially improved processability. But the blend has stiffness and strength characteristics that are superior to either PPO or HIPS. The degree of processability can be varied by changing the volume fraction of HIPS in the blend. This allows an entire series of grades of blends to be developed, each having unique thermal stability conditions (27–29).

Completely immiscible polymers represent the second category. When these polymer components are mixed in any proportion, the blend separates into distinct domains or phases, each containing an individual polymer. One polymer acts as the continuous phase and the other is the discrete phase. The classic example of a completely immiscible system is polyethylene and polystyrene. The inherent immiscibility implies very low attraction forces across the interfaces. Since mechanical behavior of the polymer blend is dependent on good adhesion between the phases, these polymer blends tend to have poor mechanical properties. The exceptions are recently developed immiscible blends of inherently fiber-forming polymers, commonly called "liquid crystal" polymers (or LCPs) with engineering resins such as PET and nylon (PA). The LCP fiber formation appears to be occurring on a molecular level with phase separation and attraction forces also occurring on this level. The early results indicate substantial improvement in stiffness and strength, even with an immiscible system.

The most difficult polymer blend category to clearly describe is that of the partially miscible system. The most common system is one in which two completely immiscible polymers are made compatible with a third organic agent, called a *compatibilizer*. The compatibilizer usually increases the interfacial adhesion between the two polymers, resulting in increased mechanical property performance over the uncompatibilized blend. This compatibilization can be introduced in many ways (28). The common ways today are shown in Figure 1.10, where the compatibilizer is a short chain polymer. In one case, the polymer is a block copolymer of each of the two immiscible polymers. When this compatibilizer polymer is compounded in, each block finds the appropriate

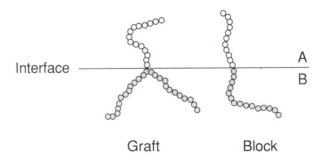

Graft   Block

Figure 1.10.   The Role of a Compatibilizer in a Blend (30).

soluble polymer, thus forming a polymer bridge across the interface. In the second case, the polymer is either grafted or blocked to one main polymer chain but is miscible in the other polymer. The graft then acts as a bridge between the two phases. Both techniques use the polymer as the mechanical bond across the main polymer–polymer interface. The grafting technique gives better uniformity throughout the blend since the probability of a compatibilizer being at the polymer–polymer interface is significantly higher than a random dispersion of a third polymer (the block copolymer) being at the polymer–polymer interface. Some very difficult polymer blends have been achieved by compatibilization, the most significant being "super-tough nylon", made by compounding nylon 66 [PA-66 or poly(hexamethylene adipamide)] with polyethylene, polyethylene oxide and other compatibilizers.

Several commercial polymer blends are given in Table 1.16 (see page 73).

1.10 Molecular Forces and Chemical Bonding
Covalent Bonding

As discussed above, the primary forces holding atoms of the polymer molecule together are covalent bonds, due to electron sharing. The relative distances between atoms and the energy required to break bonds are determined from molecular physics. Values for the interatomic distances and bonding energies of many of the covalent bonds in polymers are listed in Table 1.17 (see page 74). As seen, the strongest bond is for the triple bond $-C\equiv N$ atomic pair at 213 kcal/mol. The carbon–carbon double bond energy is 146 kcal/mol and the carbon–carbon single bond energy is 83 kcal/mol. Note that the oxygen–oxygen single bond energy is the lowest in the table at 35 kcal/mol. This table can be used to determine the type of degradation occurring in individual polymer molecules with increasing temperature or thermal energy.

Example 1.3

Q: Consider two vinyl-group polymers, polyethylene and polyvinyl chloride:

$$\left[\begin{array}{cc} H & H \\ | & | \\ C & - C \\ | & | \\ H & H \end{array}\right]_x \qquad \left[\begin{array}{cc} H & H \\ | & | \\ C & - C \\ | & | \\ H & Cl \end{array}\right]_x$$

Polyethylene   Polyvinyl chloride

Which bond will break first?

A: The accompanying table gives the respective covalent bond energies between the atom pairs found in the two polymers. Also given is the covalent bond that will break first owing to its lowest energy of dissociation in the polymer molecule:

Covalent Bond	Dissociation Energy		Polymer
	kcal/mol	PE	PVC
C – H	99	O	O
C – C	83	X	O
C – Cl	81	—	X

For polyethylene, the carbon–carbon bond will rupture first. The carbon–chlorine bond will rupture first for PVC.

Thus, degradation of polyethylene due to excess thermal energy will result in reduction of polymer molecular weight. For PVC, the first covalent bond to break will be C – Cl. The H – C bond immediately next to it will then break, forming hydrogen chloride gas (HCl). Processors who traditionally mold rigid, unplasticized PVC are well aware that overheating PVC will result in degradation and generation of HCl. If the HCl is retained in the processing equipment at high pressures and temperatures, an explosion can result. A more comprehensive discussion on polymer degradation is covered in Chapter 6.

Secondary Forces

Of equal importance to covalent bonds that hold the polymer molecule together are *secondary forces* between individual linear polymer molecules. These forces directly influence such physical properties as viscosity, surface tension, frictional forces, miscibility, volatility and solubility. Platzer (31) ranks these secondary forces in order of strength as:

1. Ionic bonding,
2. Hydrogen bonding,
3. Dipole interaction,
4. Van der Waals forces.

These are shown in Table 1.18 (see page 75). Consider the weakest force, *van der Waals force*, first. All polymer molecules are attracted to similar molecules by van der Waals forces. Pure hydrocarbon-based polymers such as polyethylene and polypropylene show measurable van der Waals dissociation energies. *Dipole interaction*, sometimes called dipole–dipole attraction, is stronger than van der Waals force, since it involves forces generated by polar groups. These groups are either on the backbone of the polymer, viz, $-C-O-C-$, or one side groups such as $-C\equiv N$ or $-Cl$. Polyacetal is also known simply as "acetal" or polyoxymethylene (POM) in Europe.* POM has a highly polar backbone, viz, $-C-O-C-O-$ and so is highly dipole associated. The polymer has a very high melting temperature (180°C) when compared with similar polyethers and so is used for hot water plumbing fixtures. Polyacrylonitrile is another polymer with very high strength properties that can be directly attributed to dipole associated. Its form:

$$\left[\begin{array}{cc} H & H \\ | & | \\ -C & -C- \\ | & | \\ H & C \\ & ||| \\ & N \end{array} \right]_x$$

polyacrylonitrile (PAN)

Addition polymerization of acrylonitrile

shows very high dipole association with the $-C\equiv N$ radical alternating with hydrogen.

Hydrogen bonding is sometimes considered to be another form of dipole interaction. Hydrogen polarity generates substantially greater forces than those found for other atoms or atomic groups and so hydrogen bonding is always considered separately. The classical hydrogen bonding interactions come about from the following groups:

$$\begin{array}{c} O \quad H \\ || \quad | \\ -C-N- \end{array} \qquad -O-H \qquad -N\begin{array}{c} H \\ \diagup \\ \diagdown \\ H \end{array}$$

Main chain   Pendant groups

These atomic combinations are found in polyamides (nylons), polyurethanes, polyamide-imide, and polyvinyl alcohol. One characteristic of polymers with hydrogen bonding is an affinity with polar molecules such as water, amine-based solvents and plasticizers and to some extent, low molecular weight hydrocarbons.

The strongest secondary forces are *ionic bonds*. These bonds are generated by either monovalent alkali Na^+ that acts only through an association of polar ion parts or by the bivalent Zn^{2+} or Cd^{2+} ions, through true ionic bonding as shown in Table 1.18. True ionic bonding polymers are called *ionomers*. Most ionic polymers are produced by vinyl-group random addition copolymerization with an ionic comonomer. Classic examples are ethylene and sodium methyl methacrylate and vinyl chloride and sodium ethyl acrylate. The random copolymerization results in an amorphous and hence transparent polymer, but the very high intermolecular bond strength acts to provide exceptional flexural stiffness. The "Surlyn" family of Na^+ and Zn^{2+} ionomers marketed by duPont are the most successful ionomers, but ionic polymers are frequently used as adhesively tenacious paint primers against metallic polar surfaces.

---

*Polyacetal is polymerized from formaldehyde. Polyether is the generic family to which it belongs. It was one of the earliest polymers produced and was the polymer first studied by Staudinger in the 1930s as representative of a "polymer" and not a colloidal gel.

Note that secondary forces are substantially smaller than covalent bonding forces. As a result, when thermal energy and/or differential stresses are applied to a polymer exhibiting substantial secondary bonding forces, these bonds yield first, rather than the breaking of covalent bonds. The materials thus are free to distort or flow under shear. These bonds do not yield sequentially as with a zipper. Instead they act in groups. Thus, polymers with very strong secondary forces such as hydrogen bonding usually have very high viscosities and are more difficult to melt process than polymers with weak secondary forces such as van der Waals forces. Owing to the high concentration of highly polar fluorine atoms, polytetrafluoroethylene has very strong secondary bonds. Simple shapes can be melt processed from PTFE by high pressure ram extrusion, but complex shapes are usually machined from block or rod or sinter-molded from powder. The difficulty in processing PTFE is of course contrasted to the ease of processing linear polyethylene, having only van der Waals forces. When the applied forces or thermal stresses are removed, the secondary forces reform to provide improved polymeric strength in the desired product.

1.11 The High Polymer: Redefinition

As noted, the term *high polymer* has been loosely defined. Polymers used in plastics are required to hold a specific shape and to support a finite load. Therefore, the most obvious definition is based on a physical property, such as tensile strength. Furthermore, the strength of polymer molecules acting in concert depends primarily on the atomic arrangement within the polymer molecule itself and then whether the molecular structure is a three-dimensional matrix bound by covalent bonds or a series of linear molecules held together by secondary forces. Theoretically, every part made from a polymer system that forms a three-dimensional matrix is a single molecule of super-high molecular weight. In reality, of course, the system is an entangled network of very high molecular weight polymers, considered to be an *interpenetrating,* crosslinked network.

The term *high polymer* is thus usually restricted to linear or uncrosslinked polymers. Since different types and strengths of secondary bonding forces occur between different types of molecules, each linear or uncrosslinked resin family will have its own criteria. Each polymer family exhibits critical degree of polymerization, DP_c. That represents the chain length or number of monomer units that make up the linear polymer molecule where noticeable mechanical strength such as tensile strength, occurs. Typically, polyamides have a critical value of $DP_c > 40$, cellulosics, $DP_c > 60$ and PVCs, $DP_c > 100$. This variation in minimum chain length is a strong function of the magnitude of the secondary forces between individual polymer molecules. As noted, polyamides exhibit very strong hydrogen bonds, whereas vinyls have only the much weaker van der Waals effect. Therefore, vinyls need a longer chain length than the polyamides to reach the same level of intermolecular forces. Above these minimum chain lengths, mechanical properties increase rapidly, then approach asymptotic values. The approach to an asymptotic value occurs for polyamides at $DP_c > 150$, cellulosics, $DP_c > 250$ and PVCs, $DP_c > 400$. This effect is seen schematically in Figure 1.11 for polyamide

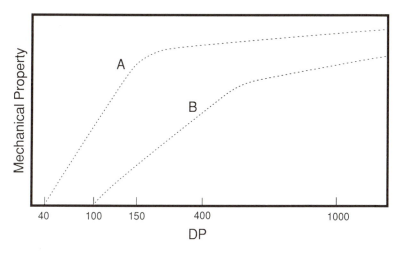

Figure 1.11.   Schematic Example of the Effect of the Degree of Polymerization (DP) on the Mechanical Strength for Polyamide (PA), A, and PVC, B (32).

and PVC. It is apparent that physical properties still increase with increasing molecular weight above these points, but at substantially reduced rates. This is seen for elastic modulus (Figure 1.12) and tensile strength (Figure 1.13). Typically, polymers used in plastics have a minimum degree of polymerization of about 600. Therefore, properties are rather independent of molecular weight variation in most polymers. These polymers can therefore be classed as *high polymers*.

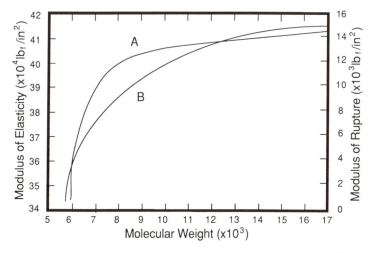

Figure 1.12.   The Effect of Molecular Weight on the Tensile Moduli of PVC/PVAc (Polyvinyl Acetate) Copolymer. Curve A: Modulus of Elasticity. Curve B: Modulus of Rupture (33).

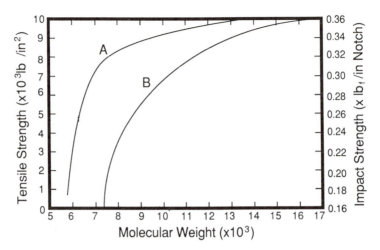

Figure 1.13. Effect of Molecular Weight on Tensile Strength and Impact Strength of PVC/PVAc Copolymer. A: Tensile Strength. B: Notched Izod Impact Strength (34).

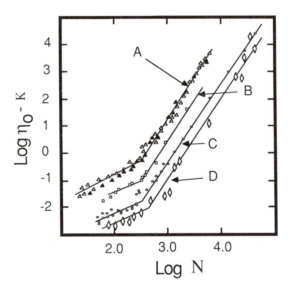

Figure 1.14. The Effect of Molecular Weight (Degree of Polymerization) on the Viscosity of Several Polymers. Curve A and Triangles: Polystyrene. Solid Triangles: Anionic. Open Triangles: Free Radical. Curve B and Open Squares: Polyvinyl Acetate. Curve C and Open Circles: Polyisobutylene. Curve D and Open Diamonds: Polydimethyl Siloxane (35).

Another physical property that manifests itself in a similar fashion is melt viscosity. As seen in Figure 1.14 for several polymers, the melt viscosity is strongly dependent upon the degree of polymerization. Note that there is a substantial change in the dependency of melt viscosity with degree of polymerization at about DP = 1000. This effect is thought to be related to increased molecular entanglements and increased secondary forces above this level of polymerization. The inflection point in the viscosity curve is frequently used to define the minimum degree of polymerization for a high polymer.

1.12 Thermoplastic and Thermosetting Polymers: A Summary

As noted in the first section of this chapter, polymers are commonly classified as *thermoplastic* or *thermosetting*, depending upon the response of the polymer to cyclic heating. The difference can be traced back to the nature of the monomer, the polymerization process, the polymeric molecular structure and the bonding systems within the polymer. Typically, *thermoplastic* polymers are linear polymers formed by addition polymerization. The bonding occurring along the polymer backbone is covalent. The bonding between polymer molecules is via secondary forces. As the temperature of the polymer is increased, the bonding energy of secondary forces between individual molecules decreases and the thermal energy of the linear polymer molecule increases. When a stress is applied to the polymer, the individual polymer molecules have the tendency to slide past one another. This is a gross effect, called melt flow. Upon cooling, the thermal energy of the individual molecules decrease, the secondary forces increase and secondary bonds are reestablished. Therefore, so long as the polymer temperature is below its degradation temperature, the response of the polymer of applied cyclic temperature is considered reversible.

A schematic diagram of the steps involved in manufacturing a part from a thermoplastic molding polymer is shown in Figure 1.15. A difunctional monomer is polymerized into a linear high polymer. The polymer is compounded with additives to form a *thermoplastic* molding compound in the form of small pellets. These pellets are subsequently remelted or resoftened in the processing machinery into a homogeneous melt and shaped by flow into the desired geometry. The polymer is cooled to a solid or rigid state and removed. Theoretically, the molecular weight of the polymer in the final product is equal to the molecular weight of the polymer formed in the polymerization process.

Practically however, processing usually involves thermal and shear events that can result in some degradation. The extent of degradation depends upon the shear and temperature history of the particular process(es) and the inherent thermal stability of the polymer molecule. It is given that no matter how carefully the manufacturer processes a thermoplastic, reground and recycled polymer resin will have reduced properties when compared with the original resin. This is one of the reasons resin suppliers specify upper limits on the quantity of recycled polymer that should be mixed with

Thermoplastic

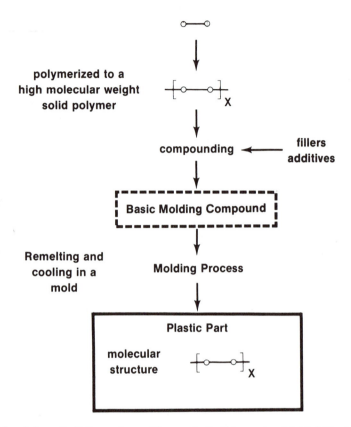

Figure 1.15.   Schematic Pathway from a Thermoplastic Polymer to a Molded Plastic Part.

virgin resin. And as will be discussed in detail in Chapter 6, when a plastic part fails, the design engineer must determine whether the failure was the result of poor design, improper processing and/or bad molding compound. When dealing with a thermoplastic, it is apparent that adequate quality control of incoming resin can minimize the potential of bad molding compounds entering the manufacturing process. A determination of the effect of processing can be made by examining certain aspects of the polymer molecular structure before and after processing. Thus, the understanding of the relationship between polymer manufacture and processing enables the design engineer to quickly assess potential problem areas.

A *thermosetting polymer* is one in which the final produce ideally consists of a single large three-dimensional matrix and practically consists of a few such interpenetrating matrices. The formation of the three-dimensional matrix or network is completed in the final molding or processing step but can be initiated in earlier, intermediate steps. The *branching networks* are called *crosslinks* and the resultant polymer is a network, three-dimensional matrix, or crosslinked polymer. There are many ways to produce a thermosetting polymer. Note that when thermal energy is added to a crosslinked polymer, there are now weak secondary forces. The only molecular forces are covalent bonds. Thus, a thermosetting polymer can only fail by covalent bond-breaking in either the main chain or in the crosslinks, viz, degradation. Crosslinking polymers can be heated, softened and made to flow under shear stress only, before being fully crosslinked. In general, thermosetting polymers are hard, strong, rigid and have good heat resistance.

In the schematic diagram of the manufacturing steps from thermosetting molding compound to finished part (Figure 1.16), two monomers having functionalities of at least 2, with one having functionality greater than 2, are polymerized into a low molecular weight linear polymer. This polymer, a *crosslinking agent* (in this example), a third monomer (styrene, in this example), and other adducts (fillers, compatibilizers, fire retardants) are then compounded to produce a *thermosetting molding compound*. This compound is usually supplied to the part manufacturer in the form of large pellets, briquettes, sheets or logs. The compound may then be preheated and mechanically or hydraulically forced into the desired shape in a heated mold. The heat activates the

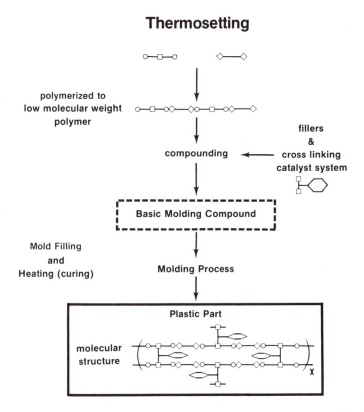

Figure 1.16.   Schematic Pathway from a Thermosetting Polymer to a Molded Plastic Part.

cross-linking agent, causing the low molecular weight linear polymer to form branching networks from the polymer backbone. The result is a networked, three-dimensional thermosetting polymer part.

The problems associated with thermosetting compound manufacture are significantly different from those with thermoplastic resin manufacture. Of most importance, the molding compound is a chemically active compound. The rate of reaction is strongly dependent upon temperature. In many cases, particularly for critical parts, the molding compound is shipped and stored under refrigeration. Even with these extreme conditions, some chain extension reaction takes place. As a result, "shelf life" of a thermosetting resin is important. This relates to the degree of chain extension that can be tolerated for a particular molding condition. This tolerance must include part volume, flow length, molding temperatures and so on. The quality control (QC) scheme for incoming thermosetting molding compounds must relate to the degree of chain extension. Standard QC tests involve forms of flowability, such as spiral flow and length-to-diameter flow for bars of different thicknesses.

For compression, transfer and injection molding, polymerization occurs in the mold. The molecular structure of the product will be completely different from that of the initial molding compound. For many types of thermosets, incomplete chain extension is the result of improper molding. As will be discussed in detail in Chapter 5, temperature and time are significant molding parameters for thermosets. When molding is conducted at the proper mold and material temperature but the molding cycle is short, the center of the molded part will be insufficiently reacted or "cured". The result is similar for adequate cycle time but insufficient molding temperature. Insufficient cure can usually be detected by hardness, as discussed in Chapter 6. Frequently, "post-curing" or holding the molded part in a heated environment, usually hot air, is sufficient to finish the curing process. Typical molding and curing times and temperatures are provided by the molding compound supplier.

In some cases, linear polymers that are normally thermoplastic can be made to form crosslinked structures. This is done by chemically attacking active sites on the polymer backbone either by a small molecule free radical (such as a peroxide) or by exposure to γ or β radiation. The number of crosslinking elements per 1000 linear bonds along the polymer backbone is known as the *netting index* (42). A thermosetting resin such as phenolic has a netting index of 500 or so. Polyurethane has a netting index of about 250 and unsaturated polyester resin has a netting index of about 100 to 200. Crosslinked polyethylene may have a netting index of about 1 to 10. Crosslinked thermoplastics may soften dramatically at transition temperatures such as crystalline melting temperatures but do not flow. These polymers can be penetrated by solvents and swelled but are never completely soluble. The solubility of solvent in polymer increases with decreasing netting index, as might be expected.

At the beginning of this chapter, thermoplastic polymers were likened to spaghetti. The analogy can be extended a bit more at this point. The individual polymer molecules can now be likened to the individual strands of noodles. The secondary bonds between polymer molecules are likened to starch on the surface of the noodle strand. When the noodles are heated, the starch effect decreases, in a fashion similar to relative reduction in secondary bond strength to polymer interchain motion. The ease of stirring warmed noodles is similar to the ease of shearing linear polymer molecules. And as an additional extension of the analogy discussed in detail below, low molecular weight lubricants are frequently added to polymeric materials to allow ease of flow during processing. The "spaghetti" analog adds oil or butter to the warmed noodles.

Earlier in the chapter, thermosetting polymers were likened to scrambled eggs, in that once the polymer was heated and molded, it could not be reprocessed, much like scrambled eggs cannot be unscrambled. The reacted or *cured* thermosetting polymer is a three-dimensional atomic network of covalent bonds. This structure can be modeled with *tinker toys*. Each disk represents an atom and each stick a covalent bond holding the atoms together. If a force is applied to one point of the tinker toy structure, it is transmitted throughout the structure. For elastomeric thermosetting polymers, the sticks in the model are replaced with springs having temperature-dependent spring constants. Regardless of the nature of the model bonds, the basic response of a thermosetting polymer to applied load can be visualized with a simple three-dimensional stick-and-ball model.

These simplistic analogs are used throughout the text that follows as elementary ways of generally distinguishing between the *expected response* of a thermosetting polymeric structure and that of a thermoplastic polymeric structure to an *externally applied load*. As will be seen, simple variations on these analogs will allow a general understanding of certain phenomena that occur during processing as well as a grasp of the effect of temperature, time and environment on the molded part.

Table 1.9. Structure, common trade names and suppliers for linear polymers with a *carbon* backbone (20,22,36–39).

The basic unit:

$$\left[\begin{matrix} H & Ⓐ \\ | & | \\ C & C \\ | & | \\ H & H \end{matrix}\right]$$

Ⓐ	Polymer	Trade name (Resin Supplier;[1] Country[2]
—H	Polyethylene (PE)	AC-Polyethylene (Allied-Signal, US)
		Alathon (DuPont, US)
		Alkathene (BXL, GB)
		Bakelite (UCC, US)
		Baylon (Bayer AG, DE)
		Bralen (Chemapol, CS)
		Carlona (Shell, GB)
		Chemplex (Chemplex Co., US)
		Dow PE (Dow Chemical, US)
		Dowlex (Dow Chemical, US)
		Dylan (Sinclair, US)
		Eltex (Solvay, BE)
		Eraclene (Anic, IT)
		Ertineno (Union ERT, ES)
		Escorene (Esso, GB)
		Fentene (Montepolimeri, IT)
		Finathene (Fina, BE)
		Fortiflex (Soltex Polymer, US)
		Hi-Fax (Hercules, Inc., US)
		Hipten (Hemijska Ind., YU)
		Hiplex (Hemijska Ind., YU),
		Hi-Zex (Mitsui, JP)
		Hostalen (Hoechst AG, DE)

Lacqtene (ATochem, FR)
Lortex (CdF Chimie RT, FR)
Lupolen (BASF AG, DE)
Marlex (Phillips Petrol., US)
Marlex TF 130 (Phillips Petroleum, US)
Microthene (USI Chemicals, US)
Mirathen (VEB Leuna, DD)
Mirason (Mitsui Polychemicals, JP)
Moplen (Montepolimeri, IT)
Natene (Rhône-Poulenc, FR)
Novapol LL (Novacor Chemicals, CA)
Novatec (Mitsubishi, JP)
Okiten (INA-OKI, YU)
Paxon (Allied Fibers, US)
Petrothene (USI Chemicals, US)
Poly-Eth (Gulf Oil, US)
Rexlon (Nippon Petrochemicals, JP)
Rigidex (BP Chemicals, US)
Ropol (Ploiestipetrochemical, RO)
Rotothene (Rototron Corp., US)
Rumiten (Rumianca, IT)
Sclair (Du Pont Canada, CA)
Scolefin (AHB Chemie Export, DD)

Table 1.9. *Continued.*

Ⓐ	Polymer	Trade name (Resin Supplier,[1] Country[2])	
—H	Polyethylene (PE) (*continued*)	Sholex (Showa Denko, JP)	Super Dylan (Arco Chemical, US)
		Staflene (Nisseki Plastic Chem., JP)	Tenite (Eastman Chemical, US)
		Stamylan (DSM, NL)	Tipolen (Tiszai Vegyi Kombinat, HU)
		Stamylex (DSM, NL)	Ultzex (Mitsui Petrochemical, JP)
		Sumikathene (Sumitono Chemical, JP)	Vestolen (Hüls AG, DE)
		Suntec (Asahi Chemical, HP)	Yukalon (Mitsubishi Petrochemical, JP)
—CH₃	Polypropylene (PP)	Carlona P (Shell, GB)	Noblene (Mitsubishi Petrochemical, JP)
		Daplen (Chemie Linz, AT)	Novolen (BASF AG, DE)
		Eltex P (Solvay, FR)	Poly-Pro (Ube Industries, JP)
		Escorene (Exxon, US)	Pro-fax (Hercules Inc., US)
		Fortilene (Soltex, US)	Propathene (ICI, GB)
		Hostalen PP (Hoeschst AG, DE)	Rexene (El Paso Olefins, US)
		Kastilene (Anic, IT)	Sho-Allomer (Showa Denko, JP)
		Lacqtène P (Atochem, FR)	Stamylan P (DSM, NL)
		Markex (Phillips, US)	Tatren (Chemapol, CS)
		Moplen (Montepolimeri, IT)	Tenite (Eastman Chemical, US)
		Napryl (Rhône-Poulenc, FR)	Vestolen P (Hüls AG, DE)
—Cl	Polyvinyl chloride (PVC)	Benvic (Solvay, BE)	Halvic (Halvic Kunststoffwerke, AT)
		Bovil (OHIS, YU)	Hipnil (Hemijska Industria, YU)
		Breon (BP Chemical, GB)	Hispavic (Hispavic Industrial, ES)
		Carina (Shell Intern., GB)	Hostalit (Hoechst AG, DE)
		Corvic (ICI Ltd., GB)	Juvinil (Jugovinil, YU)
		Dalvin (Diamond Shamrock, US)	Kohinor (Pantasote Inc., US)
		Ekavyl (PCUK, FR)	Lonzavyl (Lonza AG, CH)
		Geon (BF Goodrich, US)	Marvinol (Uniroyal, US)

Table 1.9. *Continued.*

ⓐ	Polymer	Trade name (Resin Supplier;[1] Country[2])	
—Cl	Polyvinyl chloride (PVC) *(continued)*	Mipolam (Dynamit Nobel AG, DE)	Trosiplast (Dynamit Nobel AG, DE)
		Nipeon (Nippon Zeon Co., JP)	Trovidur (Dynamit Nobel AG, DE)
		Nipolit (Chisso Corp., JP)	Varlan (DSM, NL)
		Norvinyl (Norsk Hydro AS, NO)	Vestolit (Hüls AG, DE)
		Ongrovil (Barsodi Veggi Komb, HU)	Vinidur (BASF AG, DE)
		Pevikon (Kema Nobel AG, SE)	Vinika (Mitsubishi Monstanto Chem., JP)
		Pliovic (Goodyear, US)	Vinnol (Wacker Chemie, DE)
		Ravinil (Anic, IT)	Vinoflex (BASF AG, DE)
		Rosevil (Chem. Kombinat Borzesti, RO)	Vinychlon (Mitsui Toatsu Chem., JP)
		Scon (Norsk Hydro Polymers Ltd.;, GB)	Vipla (Montepolimeri, IT)
		Shell Polyvinylchloride (Shell Plastics, GB)	Viplast (Montepolimeri, IT)
		Sicron (Montepolimeri, IT)	Vygen (General Tire & Rubber, US)
		Solvic (Deutsche Solvay Werke, DE)	Welvic (ICI Ltd, GB)
⬡	Polystyrene (PS)	Afcolene (Rhône-Poulenc, FR)	Hostyren (Hoechst AG, DE)
		Arralene (Arrahona, ES)	Lastirol (LATI, IT)
		Carinex (Shell, US)	Lustrex (BP Int., GB)
		Diarex (Mitsubishi Monsanto, JP)	Polystyrol (BASF AG, DE)
		Dylene (ARCO Polymers, US)	Restirolo (S.I.R., IT)
		Edistir (Montepolimeri, IT)	Styrolux (BASF AG, DE)
		Esbrite (Sumitomo Chemical, JP)	Styron (Dow Chemical, US)
		Lacqrène (Atochem, FR)	Toporex (Mitsui Toatsu, US)
		Gédex (CdF Chimie, FR)	Vestyron (Hüls AG, DE)
— CH₂— CH₃	Polybutylene (PB)	Shell PB (Shell, US)	
—CN	Polyacrylonitrile (PAN)	Barex (Vistron/Sohio/BP Chemical, US)	Cycopac (Borg-Warner/GE, US)

Table 1.9. *Continued.*

Ⓐ	Polymer	Trade name (Resin Supplier,[1] Country[2])
$-CH_2-CH-CH_3$ with CH_3	4-Methyl pentene-1 (TPX)	TPX (Mitsui Petrochem., JP)
$-F$	Polyvinyl fluoride (PVF)	Tedlar Folie (Du Pont Chemical, US)
$-OH$	Polyvinyl alcohol (PVOH)	Alcotex (Revertex, GB) Elvanol (Du Pont Chemical, US) Mowiol (Hoechst AG, DE) Polyviol (Wacker-Chemie, DE) Rhodoviol (Rhône-Poulenc, FR) Vinavol (Hoechst AG, DE)
(carbazole structure) N—	Polyvinyl carbazol (PVK)	Luvican (BASF AG, DE)
$-O-\underset{\parallel}{\underset{O}{C}}-CH_3$	Polyvinyl acetate (PVAc)	
$-O-CH_3$	Polyvinyl ether (PVE)	
$-\underset{\parallel}{\underset{O}{C}}-O-CH_3$	Polymethyl acrylate (PMA)	

Table 1.9. *Continued.*

The basic unit:

$$\left[\begin{array}{c} H \quad Ⓐ \\ | \quad | \\ -C-C- \\ | \quad | \\ H \quad Ⓑ \end{array} \right]$$

Ⓐ	Ⓑ	Polymer	Trade name (Resin Supplier,[1] Country[2])
—Cl	—Cl	Polyvinylidene chloride (PVDC)	Difan (BASF AG, DE) Ixan (Solvay, BE) Saran (Dow Chemical, US) Vilit (Huls AG, DE) Viclan (ICI Ltd, GB)
—F	—F	Polyvinylidene fluoride (PVDF)	Dyflor (Dynamit Nobel AG, DE) Foraflon (PCUK, FR) Kureha (Kureha Chem. Ind., JP) Kynar (Pennwalt Corp., US) Solef (Solvay, BE) Vidar (SWK Trostberg AG, DE)
—CH₃	⌬ (phenyl)	α-Methyl styrene (αMS)	
—CH₃	—C—O—CH₃	Polymethyl methacrylate (PMMA)	Altulite (Altulor, Groupe CdF, FR) Biodrak (A. Drakopoulos, GR) Casorcryl (Casolith B.K., NL) Degalan (Degussa, DE) Deglas (Degussa, DE) Delpet (Asahi Chemicals, JP) Dewoglas (Degussa, DE) Diakon (ICI, GB) Lacrilex (LATI, IT) Lucite (Du Pont, US) Oroglas (Röhm & Haas, US) Perspex (ICI Ltd., GB) Piacryl (VEB Piesteritz, DD) Plexiglas (Rohm GmbH, DE) Resartglas (Resart-Ihm, DE) Sadur (AMMA) (Resart-Ihm, DE) Shinkolithe (Mitsubishi Rayon, JP) Sumipex (Sumitomo Chemical, JP) Swedcast (Swedlow, Inc., US) Umaplex (Synthesia, CS) Vedril (Vedril SpA, IT)

Table 1.9. *Continued.*

The basic unit:

$$\begin{bmatrix} A & & C \\ & | & \\ C & - & C \\ & | & \\ B & & D \end{bmatrix}$$

A	B	C	D	Polymer	Trade name (Resin Supplier,[1] Country[2])
F	F	F	F	Polytetrafluoroethylene (PTFE)	Fluon (ICI, GB) Halon (Ausimont, US) Hostaflon TF (Hoechst AG, DE) Neoflon (Daikin Ind., JP) Polyflon (Daikin Kogyo, JP) Soreflon (PCUK, FR) Teflon (Du Pont, US)
F	F	F	Cl	Polychlorotrifluoroethylene (PCTFE)	Kel-F (3M, US) Voltalef (PCUK, FR)
F	F	F	$-\overset{\displaystyle F}{\underset{\displaystyle F}{C}}-F$	Polyhexafluoropropylene (PHFP)	

[1]Note: Owing to the great volatility in ownership of resin producers, this list may contain names of companies that no longer make the specific resin.

[2]Country Key: AT, Austria; BE, Belgium; CA, Canada; CH, Switzerland; CS, Czechoslovakia; DD, DE, Germany; DK, Denmark; ES, Spain; FI, Finland; FR, France; GB, Great Britain; GR, Greece; HU, Hungary; IT, Italy; JP, Japan; MI, NL, Netherlands; ND, Norway; RO, Romania; US, United States; and YU, Yugoslavia.

Table 1.10. Linear polymers with *nitrogen* in the backbone structure, common trade names and suppliers for (20, 22, 36–39).

Polymer	Trade name (Supplier,[1] Country[2])	
$\left[\,\text{H}-\text{N}-\left(\begin{smallmatrix}\text{H}\\ \text{C}\\ \text{H}\end{smallmatrix}\right)_{5}-\overset{\text{O}}{\underset{}{\text{C}}}\,\right]$ Nylon 6 (PA 6)	Akulon (Akzo, NL) Amilan (Toney Ind., JP) Capron (Allied-Signal. US) Durethan B (Bayer AG, DE) Fabenyl (Ems Chemie, CH) Fosta Nylon (American Hoechst, US) Grilon (Emser Werke AG, CH) Latamid (L.A.T.I., IT) Maranyl F (ICI, GB) Miramid (VEB Leuna, DD)	Nivionplast (Anicfibre, IT) Orgamide (Atochem, FR) Plaskon (Allied-Signal, US) Renyl (Montedison, IT) Silon (Silon-Werk, CS) Sniamid (Technopolimeri, IT) Technyl C (Rhône-Poulenc, FR) Torayca (Toray, JP) Ultramid B (BASF AG, DE) Zytel (Du Pont, US)
$\left[\,\text{H}-\text{N}-\left(\begin{smallmatrix}\text{H}\\ \text{C}\\ \text{H}\end{smallmatrix}\right)_{10}-\overset{\text{O}}{\underset{}{\text{C}}}\,\right]$ Nylon 11 (PA 11)	Rilsan B (Atochem, FR)	
$\left[\,\text{H}-\text{N}-\left(\begin{smallmatrix}\text{H}\\ \text{C}\\ \text{H}\end{smallmatrix}\right)_{11}-\overset{\text{O}}{\underset{}{\text{C}}}\,\right]$ Nylon 12 (PA 12)	Rilsan A (Atochem, FR) Grilamid (Emser-Werke AG, CH) Vestamid (Hüls AG, DE)	

Table 1.10. *Continued.*

Polymer	Trade name (Supplier,[1] Country[2])
Nylon 66 (PA 66) $-\left[-N-(CH_2)_6-N-C(=O)-(CH_2)_4-C(=O)-\right]-$ (H on N)	Akulon (Akzo, NL) Durethan A (Bayer AG, DE) Leona (Asahi Kasei Chem., JP) Maranyl A (ICI, GB) Sniamid (Technopolimeri, IT) Minlon (Du Pont, US) Technyl A (Rhône-Poulenc, FR) Toray (Toyo Rayon, JP) Ultramid A (BASF AG, DE) Verton (ICI, GB) Zytel E (Du Pont, US)
Nylon 610 (PA 610) $-\left[-N-(CH_2)_6-N-C(=O)-(CH_2)_8-C(=O)-\right]-$	Maranyl B (ICI, GB) Technyl D (Rhône-Poulenc, FR) Toray (Toyo Rayon, JP) Ultramid S (BASF AG, DE) Zytel (Du Pont, US)
Nylon 612 (PA 612) $-\left[-N-(CH_2)_6-N-C(=O)-(CH_2)_{10}-C(=O)-\right]-$	Zytel (Du Pont, US)

Table 1.10. *Continued.*

Polymer	Trade name (Supplier,[1] Country[2])
Polyamide-imide (PAI)	Torlon (Amoco, US)
Polyurethane, linear (PUR)	Caprolan (Elastogran/BASF, DE) Cytor (American Cyanamid, US) Desmopan (Bayer AG, DE) Elastollan (Elastogran/BASF, DE) Estane (BF Goodrich, US) Fabeltan (Ems Chemie AG, CH) O-Thane (Quinn & Co., US) Pellethane (Upjohn, US) Texin (Mobay Chemical Corp., US)

[1]Note: Owing to the great volatility in ownership of resin producers, this list may contain names of companies that no longer make the specific resin.

[2]Country Key: AT, Austria; BE, Belgium; CA, Canada; CH, Switzerland; CS, Czechoslovakia; DD; DE, Germany; DK, Denmark; ES, Spain; FI, Finland; FR, France; GB, Great Britain; GR, Greece; HU, Hungary; IT, Italy; JP, Japan; MI; NL, Netherlands; ND, Norway; RO, Romania; US, United States; and YU, Yugoslavia.

Table 1.11. Structure, common trade names and suppliers for linear polymers with *oxygen* in the backbone (20,22,36–39).

Polymer	Trade name (Supplier,[1] Country[2])
Polycarbonate (PC)	Jupilon (Mitsubishi Chem., JP) Lexan (GE Plastics, US) Makrolon (Bayer AG, DE) Merlon (Mobay Chem., US) Novarex (Mitsubishi Chem., JP) Orgalan (Atochem, FR) Panlite (Teijin Ind., JP) Sinvet (Anic, IT)
Polyacetal (POM)	Delrin (Du Pont, US) Tenac (Asahi Chem. Ind., JP)
Phenoxy	Phenoxy (Amoco Perf., US)
Phenylene oxide (PPO)	

From Bisphenol A & phosgen condensation Reaction

Table 1.11. *Continued.*

Polymer	Trade name (Supplier,[1] Country[2])
Polyethylene oxide (PEOX)	Polyox (Union Carbide, US)
Polyethylene terephthalate (PET)	Arnite (Akzo Plastics, NL) Cleartuf (General Tire, US) Crastin (Ciba Geigy, CH) FR-PET (Teijin, JP) Hostadur (Hoechst AG, DE) Hostaphan (Hoechst AG, DE) Melinar (ICI, GB) Petlon (Mobay Chem. Corp., US) Rynite (Du Pont, US) Tenite (Eastman Chemical Ind., US) Ultradur (BASF AG, DE) Vestadur (Hüls AG, DE)
Polybutylene terephthalate (PBT)	Arnite (Akzo Plastics, NL) Celanex (Celanese, US) Crastin (Ciba Geigy, CH) Hostadur (Hoechst AG, DE) Miranoven (VEB Leuna-Werke, DD) Novadur (Mitsubishi Chemical, JP) Orgater (Atochem, FR) Pibiter (Montepolimeri, IT) Pocan (Bayer AG, DE) Shinko Lac (Mitsubishi Rayon Co., JP) Tufpet (Toyobo Corp., JP) Ultradur (BASF AG, DE) Valox (GE Plastics, US) Vestodur (Hüls AG, DE)

Table 1.11. *Continued.*

Polymer	Trade name (Supplier,[1] Country[2])
Polyether ether ketone (PEEK)	Victrex (ICI, GB)
Polyether ketone (PEK)	Hostatek (Hoechst AG, DE) Ultrapek (BASF AG, DE) Victrex (ICI, GB)

The cellulosic basic unit:

Table 1.11. *Continued.*

Ⓐ	Ⓑ	Polymer	Trade Name (Supplier,[1] Country[2])
—H	—H	Cellulose	
$-C(=O)-CH_3$	$-C(=O)-CH_3$	Cellulose acetate (CA)	Ampol (American Polymers, US) Cellidor CP (Bayer AG, DE) Cellit (Bayer AG, DE) Tenite (Eastman Chemical Inc., US)
$-C(=O)-CH_2-CH_3$	$-C(=O)-CH_2-CH_3$	Cellulose propionate (CP)	Ampol (American Polymers, US) Cellidor B (Bayer AG, DE) Flo-Tyrate (Seitetsu Kagaku Co., JP) Tenite (Eastman Chemical Inc., US) Tenex (Teijin, JP)
$-C(=O)-CH_3$	$-C(=O)-CH_2-CH_2-CH_3$	Cellulose acetate butyrate (CAB)	Ampec (American Polymers, US) Ethocel (Dow, US)
$-CH_2-CH_3$	$-CH_2-CH_3$	Ethyl cellulose (EC)	
$-NO_2$	$-NO_2$	Cellulose nitrate (CN)	Celluloid (Celanese, US)
$-C(=O)-CH_3$	$-C(=O)-CH_2-CH_3$	Cellulose acetate propionate (CAP)	

Table 1.11. *Continued.*

Polymer	Trade name (Resin Supplier)
Polyether–imide (PEI)	Ultem (GE Plastics, US)
Polyester–imide	Allobec (Dr. Beck & Co., DE) Cellatherm (Reichhold Chemie, DE) Imidex (GE Plastics, US) Isomid (Schenectady Chem., US) Teritherm (P.D. George Co., US)
Polyimide 2080	Polyimide 2080 (Upjohn, US)

[1]Note: Owing to the great volatility in ownership of resin producers, this list may contain names of companies that no longer make the specific resin.

[2]Country Key: AT, Austria; BE , Belgium; CA, Canada; CH, Switzerland; CS, Czechoslovakia; DD; DE, Germany; DK, Denmark; ES, Spain; FI, Finland; FR, France; GB, Great Britain; GR, Greece; HU, Hungary; IT, Italy; JP, Japan; MI; NL, Netherlands; ND, Norway; RO, Romania; US, United States; and YU, Yugoslavia.

Table 1.12. Structure, common trade names and suppliers for linear polymers with *sulfur* in the backbone (20,22,36–39)

Polymer	Trade name (Supplier,[1] Country[2])
Polyphenylene sulfide (PPS)	Ryton (Phillips Chemicals, US) Experimental PPS (Bayer AG, DE)
Polysulfone (PSU, PSO$_2$)	Udel (Amoco Perf., US) Ultrason S (BASF AG, DE)
Polyether sulfone (PES, PESO$_2$)	Victrex (ICI, GB) Ultrason (BASF AG, DE)
Polyarylether sulfone (PAES, PAESO$_2$)	Radel (Amoco Perf., US) Astrel (Carborundum, US)

[1]Note: Owing to the great volatility in ownership of resin producers, this list may contain names of companies that no longer make the specific resin.

[2]Country Key: AT, Austria; BE, Belgium; CA, Canada; CH, Switzerland; CS, Czechoslovakia; DD; DE, Germany; DK, Denmark; ES, Spain; FI, Finland; FR, France; GB, Great Britain; GR, Greece; HU, Hungary; IT, Italy; JP, Japan; MI; NL, Netherlands; ND, Norway; RO, Romania; US, United States; and YU, Yugoslavia.

Table 1.13. Structure, common trade names and suppliers for polymers with network geometries.

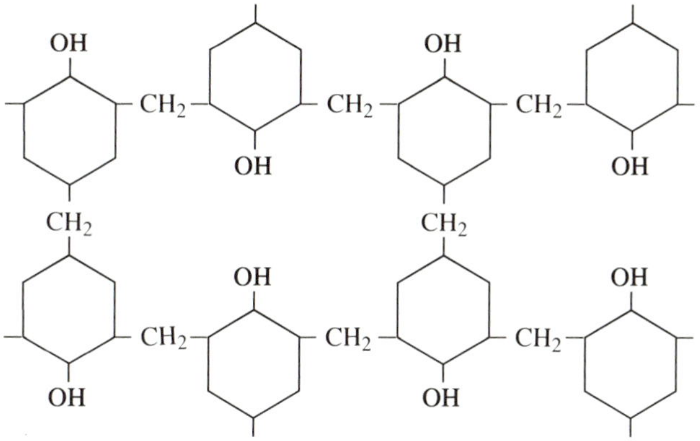

Phenol-formaldehyde (PF)

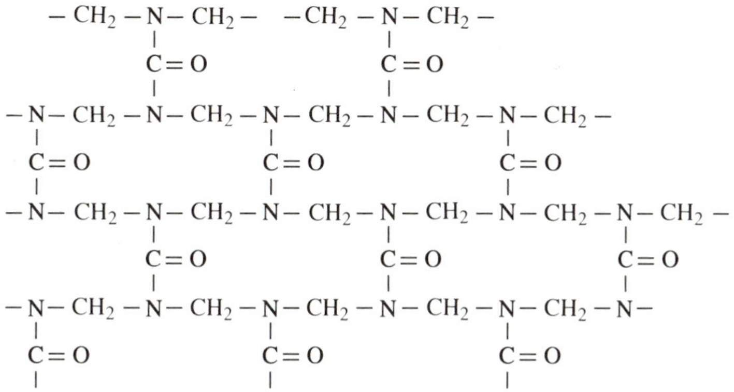

$$- CH_2 - N - CH_2 - \quad - CH_2 - N - CH_2 -$$
$$C=O \qquad\qquad C=O$$
$$- N - CH_2 - N - CH_2 - N - CH_2 - N - CH_2 - N - CH_2 -$$
$$C=O \qquad\qquad C=O \qquad\qquad C=O$$
$$- N - CH_2 - N - CH_2 - N - CH_2 - N - CH_2 - N - CH_2 - N - CH_2 -$$
$$C=O \qquad\qquad C=O \qquad\qquad C=O$$
$$- N - CH_2 - N - CH_2 - N - CH_2 - N - CH_2 - N - CH_2 - N -$$
$$C=O \qquad\qquad C=O \qquad\qquad C=O$$

Urea formaldehyde (UF)

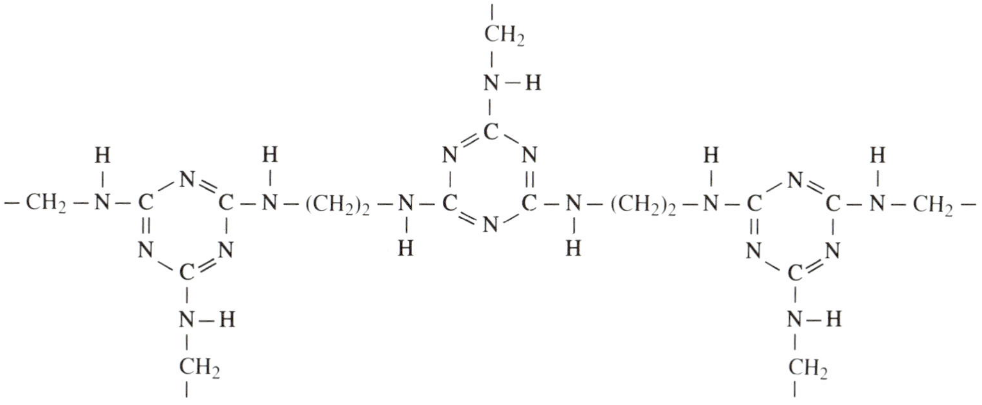

Melamine-formaldehyde (MF)

Approximate chemical formulae

Table 1.13. *Continued.*

Polymer type	Trade name (Resin Supplier,[1] Country[2])
Alkyd	Durez (Durez, US)
	Glaskyd (American Cyanamid, US)
	Plaskon (Plaskon, US)
	Plenco (Plastics Eng. Co., US), (Plumb Chemicals, US)
Allyl (DAP)	Cosmic (Prolam, US)
	Durez (Durez, US)
	Plaskon (Plaskon, US)
	Rx (Rogers Corp., US)
Epoxy	Hysol (Hysol/Morton Salt, US)
	Hyflow (Hysol/Morton Salt, US)
	Plaskon (Plaskon, US), (Fiberite, US), (Furane, US)
Melamine	Cymel (American Cyanamid, US)
	Perstorp (Perstorp AG, DE)
	Plenco (Plastics Engineering, US)
Phenolic	Durez (Durez, US)
	Genal (GE Plastics, US)
	Fiberite (Fiberite, US)
	RCI (Reichhold, US)
	Resinoid (Resinoid Eng., US)
	Rx (Rogers Corp., US)
	Valite (Valite Corp., US)
Polyester, thermoset (UPE)	Fiberite (Fiberite, US)
	Plenco (Plastics Engineering, US)
	Polylite (Reichhold, US)
	Premi-Glass (Premix, US), (Glastic, US), (Haysite, US), (Plumb Chemicals, US)
Polyimide, thermoset	Kinel (Rhone-Poulenc, FR), (Monsanto, US)
Polyurethane, thermoset (PUR) (polymer, isocyanates, polyols, foam systems)	Baydur (Mobay, US)
	Dermathane (Upjohn, US)
	Mondur (Mobay, US)
	Multrano (Mobay, US)
	Niax (Union Carbide, US)
	PAPI (Upjohn, US)
	Pluracol (BASF Wyandotte, US)
	Polyol (Union Carbide, US)
	Rubicast (Rubicon, DE)

Table 1.13. *Continued.*

Polymer type	Trade name (Resin Supplier,[1] Country[2])
Polyurethane, thermoset (PUR) (*continued*)	Rubinate (Rubicon, DE)
	Varanol (Dow Chemical, US)
Urea	Plaskon (Plaskon, US)
	Beetle (American Cyanamid, US)
	Skanopal (Perstorp AG, DE), (Budd Chemical, US)

[1]Note: Owing to the great volatility in ownership of resin producers, this list may contain names of companies that no longer make the specific resin.

[2]Country Key: AT, Austria; BE, Belgium; CA, Canada; CH, Switzerland; CS, Czechoslovakia; DD; DE, Germany; DK, Denmark; ES, Spain; FI, Finland; FR, France; GB, Great Britain; GR, Greece; HU, Hungary; IT, Italy; JP, Japan; MI; NL, Netherlands ND, Norway; RO, Romania; US, United States; and YU, Yugoslavia.

Table 1.14. Structure, common trade names and suppliers
for linear copolymers (20,22,36–39)

Polymer	Trade name (Resin Supplier,[1] Country[2])
Styrene–acrylonitrile (SAN)	Afcolène (Rhône-Poulenc, FR) Kostil (Montepolimeri, IT) Lacqrene (ATC Chemie, FR) Litac (Mitsui Toatsu Chem., JP) Luran (BASF AG, DE) Lustran (Monsanto Co., US) Restil (S.I.R., IT) Sanrex (Mitsubishi Monsanto, JP) Tyril (Dow Chemical, US) Vestyron (Hüls AG, DE)
Styrene–butadiene (SBR)	Diarex (Mitsubishi Monsanto, JP) Daki Polistren (INA, YU) Dylene (Arco Polymers, US) Fosta Tuf-Flex (Hoechst, DE) K-Resin (Phillips Chemicals, US) Luran (BASF AG, DE) Parastyren (Paraisten Kalkki Oy, FI) Plaper (Mitsubishi Monsanto, JP) Polystyrol (BASF AG, DE) Rhodopas (Rhône-Poulenc, FR) Styrolux (BASF AG, DE) Styropol (Carl Gordon Ind., US) Vestyron (Hüls AG, DE)
Ethylene vinyl acetate (EVA)	Acraldon (Bayer AG, DE) Baylon (Bayer AG, DE) Evaclene (Anic, IT) Evaflex (Mitsui, JP) Evatane (ICI, GB) Evatate (Sumitomo Chemical, JP) Levapren (Bayer AG, DE) Lupolen (BASF AG, DE) Miravithen (VEB Leuna, DD) Soablen (Nippon Synthetic, JP) Soarlex (Nippon Synthetic, JP) Ultrathene (USI Chemicals, US) Wacke VAE (Wacker-Chemie, DE)

Table 1.14. *Continued.*

Polymer	Trade name (Resin Supplier,[1] Country[2])
$\begin{bmatrix} \begin{matrix} H & H \\ \| & \| \\ C & - & C \\ \| & \| \\ H & H \end{matrix} \end{bmatrix}_x \begin{bmatrix} \begin{matrix} H & H \\ \| & \| \\ C & - & C \\ \| & \| \\ H & OH \end{matrix} \end{bmatrix}_y$ Ethylene vinyl alcohol (EVOH)	Clarene (Deutsche Solvay-Werke, DE) Lavasint (Bayer AG, DE)
$\begin{bmatrix} \begin{matrix} H \\ \| \\ C - O \\ \| \\ H \end{matrix} \end{bmatrix}_x \begin{bmatrix} \begin{matrix} H & H \\ \| & \| \\ C & - & C \\ \| & \| \\ H & H \end{matrix} \end{bmatrix}_y$ Acetal (POM) (y = o) Acetal Copolymer (y $\neq$ o)	Celcon (Celanese, US) Delrin (Du Pont, US) Duracon (Daicel-Polyplastics, JP) Hostaform (Hoechst AGe, DE) Kematal (Celanese, US) Ultraform (BASF AG, DE)
$\begin{bmatrix} \begin{matrix} F & F \\ \| & \| \\ C & - & C \\ \| & \| \\ F & F \end{matrix} \end{bmatrix}_x \begin{bmatrix} \begin{matrix} F & F \\ \| & \| \\ C & - & C \\ \| & \| \\ F & FCF \\ & \| \\ & F \end{matrix} \end{bmatrix}_y$ Tetrafluoroethylene–hexafluoropropylene, Fluorinated ethylene–propylene (FEP)	Neoflon (Daikin, JP) Teflon (Du Pont, US)
$\begin{bmatrix} \begin{matrix} H & H \\ \| & \| \\ C & - & C \\ \| & \| \\ H & H \end{matrix} \end{bmatrix}_x \begin{bmatrix} \begin{matrix} F & F \\ \| & \| \\ C & - & C \\ \| & \| \\ F & Cl \end{matrix} \end{bmatrix}_y$ Ethylene–chlorotrifluoroethylene (ECTFE)	Halar (Ausimont, US)

[1]Note: Owing to the great volatility in ownership of resin producers, this list may contain names of companies that no longer make the specific resin.

[2]Country Key: AT, Austria; BE , Belgium; CA, Canada; CH, Switzerland; CS, Czechoslovakia; DD; DE, Germany; DK, Denmark; ES, Spain; FI, Finland; FR, France; GB, Great Britain; GR, Greece; HU, Hungary; IT, Italy; JP, Japan; MI; NL, Netherlands; ND, Norway; RO, Romania; US, United States; and YU, Yugoslavia.

Table 1.15. Structure, common trade names and suppliers for linear terpolymers (20,22, 36–39).

Polymer
Acrylonitrile–butadiene–styrene (ABS)

Trade name (Resin Supplier,[1] Country[2])
Cycolac (Borg-Warner/GE Plastics, US)
Cycopac (Borg-Warner/GE Plastics, US)
Kralastic (Uniroyal, US)
Lastiflex (LATI, IT)
Lustran (Monsanto, US)
Lustropak (Monsanto, US)
Novodur (Bayer AG, DE)
Restiran (S.I.R., IT)
Ronfalin (DSM, NE)
Royalite (Uniroyal, US)
Shinko Lac (Mitsubishi Rayon, JP)
Sternite (Chemical Products, GB)
Sunloid (Tsutsunaka Plastic Ind., JP)
Terluran (BASF AG, DE)
Toyolac (Toray Ind., JP)
Urtal (Montepolimeri, IT)

[1]Note: Owing to the great volatility in ownership of resin producers, this list may contain names of companies that no longer make the specific resin.

[2]Country Key: AT, Austria; BE, Belgium; CA, Canada; CH, Switzerland; CS, Czechoslovakia; DD; DE, Germany DK, Denmark; ES, Spain; FI, Finland; FR, France; GB, Great Britain; GR, Greece; HU, Hungary; IT, Italy; JP, Japan; MI; NL, Netherlands; ND, Norway; RO, Romania; US, United States; and YU, Yugoslavia.

Table 1.16. Common trade names and suppliers for commercial linear polymer blends (20,22, 36–39).

Polymer	Trade name (Supplier,[1] Country[2])
PVC/PMMA	DKE 450 (Du Pont, US)
	Kydex (Rohm & Haas, US)
	Polydene (A. Schulman, DE)
ABS/PVC	Abson (Abtec, US)
	Cycovin KAB (Borg-Warner/GE, US)
	Cycoly (Borg-Warner/GE, US)
	Kralastic FVM (Uniroyal, US)
	Polyman 509 (A. Schulman, DE)
	Ryulex (Dainippon, JP)
ABS/PC	Bayblend (Bayer AG, DE)
	Cyclolac (Borg-Warner/GE, US)
	Cycoly (Borg-Warner/GE, US)
	Moldex A (Anic, IT)
ABS/TPU	Cycoly (Borg-Warner/GE, US)
	Estane (BF Goodrich, US)
	Pellethane (Upjohn, US)
PPO/PS (mPPO)	Noryl (GE Plastics, US)
	Xyron (Asahi Dow, JP)
POM/TPU	Delrin 100 ST (Du Pont, US)
	Hostaform C (Hoechst AG, DE)
PC/PBT	Makroblend (Bayer AG, DE)
	Xenoy (GE Plastics, US)
PSO$_2$/ABS	Mindel (Amoco Perf., US)
PSO$_2$/SAN	Ucardel (Amoco Perf., US)
Chlorinated polyether/PVC	Hostalit (Hoechst AG, DE)

[1]Note: Owing to the great volatility in ownership of resin producers, this list may contain names of companies that no longer make the specific resin.

[2]Country Key: AT, Austria; BE, Belgium; CA, Canada; CH, Switzerland; CS, Czechoslovakia; DD; DE, Germany; DK, Denmark; ES, Spain; FI, Finland; FR, France; GB, Great Britain; GR, Greece; HU, Hungary; IT, Italy; JP, Japan; MI; NL, Netherlands; ND, Norway; RO, Romania; US, United States; YU, Yugoslavia.

Table 1.17. Covalent bonding parameters (19, 28, 29, 40, 41).

Covalent bond	Bond distance (Å)	Dissociation energy (kcal/mol)	Bond angles for carbon, oxygen and silicon elements
C ≡ N	1.15	213	
C ≡ C	1.20	194	
C = O	1.21	174	
C = N	1.27	147	109.5° (C with H₂)
C = C	1.34	146	
C = F¹	1.32–1.39	103–123	
O – H	0.96	111	
C – H	1.10	99	108° (C–O, H₂)
N – H	1.01	93	
Si – O	1.64	88	110° (C–O)
C – O	1.46	86	
C – C	1.54	83	
C – Cl	1.77	81	142° (Si, (CH₃)₂)
S – H	1.34	81	
C – N	1.47	73	110° (Si–O)
C – Si	1.87	69	
C – S	1.81	62	
O – O	1.32	35	

¹The bond length decreases and the dissociation energy increases as additional fluorine atoms are substituted on the same carbon atom.

Table 1.18. Secondary bonding forces (26).

Type	Structure		Dissociation Energy kcal/mole
Ionic Bonding	100% ionized	50% ionized	10–20
Hydrogen Bonding			3–7
Dipole Interaction			1.5–3
Van der Waals			0.5–2

References

1. J.H. DuBois, *Plastics History USA*, Cahners Books, Boston (1972), p. 6.
2. B. Golding, *Polymers and Resins: Their Chemistry and Chemical Engineering*, Van Nostrand, New York (1959), p. 159.
3. J. Leeming, *Rayon*, Chemical Publishing Co., Cleveland (1950), p. 9.
4. D. Spill, USP 97,454 (1869).
5. D. Spill, USP 101,175 (1869).
6. J.H. DuBois, *Plastics History USA*, Cahners Books, Boston (1972), p. 39.
7. J.H. DuBois, *Plastics History USA*, Cahners Books, Boston (1972), p. 216.
8. J.H. DuBois, *Plastics History USA*, Cahners Books, Boston (1972), p. 312.
9. B. Golding, *Polymers and Resins: Their Chemistry and Chemical Engineering*, Von Nostrand, New York (1959), p. 243.
10. J.H. DuBois, *Plastics History USA*, Cahners Books, Boston (1972), p. 78.
11. B. Golding, *Polymers and Resins: Their Chemistry and Chemical Engineering*, Von Nostrand, New York (1959), p. 1.
12. J.H. DuBois, *Plastics History USA*, Cahners Books, Boston (1972), p. 267.
13. Anon., Freedonia Group, Inc., Cleveland OH, 1987.
14. J. Wakeman and N. Weil, Ind. Eng. Chem., *34* (1942), p. 1387.
15. L.K. Arnold, *Introduction to Plastics*, Iowa State University Press, Ames IA (1968), p. 3.
16. H. Metcalf, J. Williams and J. Castka, *Modern Chemistry*, Holt, Rinehart & Winston, New York (1978).
17. L.S. Wasserman, *Chemistry*, Wadsworth Publishing, Belmont CA (1974).
18. W.A. Nevill, *General Chemistry*, McGraw-Hill Book Co., New York (1967), p. 45.
19. F.W. Billmeyer, Jr., *Textbook of Polymer Science*, Interscience, New York (1966).
20. R.C. Progelhof, Plast. Eng., *4*:10 (1981), p. 17.
21. S. Rosen, *Fundamental Principles of Polymeric Materials for Practicing Engineers*, Barnes & Noble, New York (1971).
22. I. Rubin, *Injection Molding*, John Wiley & Sons, New York (1973).
23. P. Grafton, *A Brief Discussion of the Commonly Used Plastics and Their Origin*, Boonton Molding Co., Boonton NJ (1973).
24. L.F. Albright, *Chemistry and Engineering of High Polymers*, Purdue University (1987), p. 1–7.
25. G.F. Natta, Sci. Amer., *205*:2 (1961), p. 33.
26. L.F. Albright, *Chemistry and Engineering of High Polymers*, Purdue University (1987), p. 11–12.
27. O. Olabisi, L. M. Robeson, and M. T. Shaw, *Polymer–Polymer Miscibility*, Academic Press, New York (1979).
28. D. Paul and J. Barlow, J. Macromol. Sci., *C18*:1 (1980), p. 109.
29. M. Shen and H. Kawai, AIChE J., *24*:1 (1979), p. 1.
30. D. Paul and S. Newman, Eds., *Polymer Blends*, John Wiley & Sons, New York (1978).
31. N. Platzer, Ind. Eng. Chem., *61*:5 (1969), p. 10.
32. R.D. Deanin, *Polymer Structure, Properties and Applications*, Cahners Books, Boston (1972), Chapter 3.
33. R.D. Deanin, *Polymer Structure, Properties and Applications*, Cahners Books, Boston (1972), p. 68.
34. R.D. Deanin, *Polymer Structure, Properties and Applications*, Cahners Books, Boston (1972), p. 77.
35. T.G. Fox and V.R. Allen, J. Chem. Phys., *41* (1964), p. 344.

36. M. Chien and R.A. Weiss, Polym. Eng. Sci., 28 (1988), p. 6.
37. R. Juran, Ed., Mod. Plast. Encyclopedia, 64:10A (1988).
38. H. Saechtling, *International Plastics Handbook*, Hanser, Munich (1987).
39. H. Domininghaus, *Die Kunstoffe und ihre Eigenschafen*, VDI Verlag, Dusseldorf, FRG (1986).
40. Anon., *The Need for Plastics Education*, SPE/SPI Education Committee Report, 1985.
41. R. Seymour, *Introduction to Polymer Chemistry*, McGraw-Hill, New York (1971).
42. H. Bowen, et al., Spec. Pub. 11, Chemical Soc. London (1958).

Glossary

Addition Reaction   Polymerization by opening a double bond.
Additive   A low molecular weight substance added to a polymer in small quantity to enhance specific polymer properties. Can be organic or inorganic. See *Adduct*.
Adduct   An additive to a polymer. Can be processing aid, colorant, stabilizer, filler, reinforcement, and so on.
Aliphatic   Polymer that contains no ring structure. —— Linear chains
Alloy   A physical mixture of two entities, in a precise ratio, especially where some synergism is manifest.
Aromatic   Polymer that contains ring structure such as benzyl group.
Atactic   Random arrangement of unsymmetrical radical groups.
Backbone   The main feature of a polymer chain.
Blend   A physical mixture of two entities.
Branches   Side chains that extend from the polymer backbone.
Compatibilizer   Organic agent that increases interfacial adhesion in incompatible blends.
Condensation Reaction   Polymerization by splitting out one or more small molecules.
Copolymer   A polymer made of two monomeric units.
Covalent Bond   Sharing of electrons between atoms.
Crosslinking   The development of molecular bridges between polymer chains.
Curing   For thermosetting polymers, the establishment of the three-dimensional structure. In thermoplastics, it means "cooling to a rigid structure", without crosslinking.
Diacid   A pure acid having two reactive carboxyl groups.
Diamine   A pure amine having two reactive amino groups.
Diol   A pure alcohol having two reactive hydroxyl groups.
Functionality   The number of reactive groups on a monomer OR the number of chemical bonds formed during polymerization.
Homopolymer   A polymer made of identical monomeric units.
Hydrogen Bonding   A loose attraction between polar groups on neighboring polymer chains.
Isotactic   Ordered arrangement of radical groups on the same side of the polymer backbone.
Mer   Small molecular unit in a polymer structure.
Miscibility   Characteristics of polymer blends that result in a single phase.
Monomer   A pure reactive chemical that is the building block of a polymer.
Pendant Group   A side group on a polymer chain.
Plastic   A commercial product that contains at least one polymer.
Polymer   A high molecular weight molecule.
Saturated   Single covalent bonds between carbon atoms on polymer backbone.
Sequential Reaction   One type of polymerization, followed by another.

Stereoisomers Polymers that have the same chemical composition but have pendant groups that occupy differing positions along the backbone.

Syndiotactic Ordered, alternating arrangement of radical groups along the polymer backbone.

Terpolymer A polymer made of three monomeric units.

Thermoplastic Polymer with two-dimensional, linear chain.

Thermoset Polymer with three-dimensional, crosslinked chain.

Unsaturated Double or triple covalent bonds between carbon atoms on polymer backbone.

Vinyl The generic ABC = CDE monomer, where A, B, D, and E can be atoms or other molecular moieties. C is the carbon atom.

Viscoelastic Materials that exhibit both fluid-like and solid-like properties.

CHAPTER 2

THE POLYMER SOLID STATE

Recd
119 · 133
157 · 169

2.1 Introduction

The extreme lengths of polymeric chains are the primary cause of the non-simple interaction of these materials with environmental changes. For more traditional materials such as steel and water, phase transitions from liquid to solid occur at clearly defined temperatures. Only under extreme processing conditions can these transition temperatures be altered. The nature of the traditional material structure is usually quite well-defined, predictable and for the most part controllable. For polymers, many concepts of material behavior must be reinterpreted. This section concentrates on those elements that affect the structure of the polymer solid state.

2.2 Polymer Molecular Weight

As noted in Section 1.4, the molecular weight of any material is the weight, in grams, for every *mole* of that species. Thus, the molecular weight of carbon is 12, since 6.023×10^{23} atoms of carbon weigh exactly 12 g. Monomers that are reacted to form a polymer are usually simple molecules with well-defined molecular weight. For example, ethylene, $H_2C = CH_2$, has a molecular weight of 28. If the ethylene molecule were magnified 10^8 times, it would be approximately 20 mm (0.74 inch) long. Polymerization is the act of controlling the reaction between a growing polymer chain and activated monomers. Even though the overall reaction is controlled by controlling initiators, catalysts, monomer concentration, evolution of reactive by-products, temperature, and pressure, polymerization is still a series of random events occurring on a microscopic scale. Even the most linear polymer chain grows in a stochastic fashion. As a result, on the whole, polymer chains are of different lengths at any time in the polymerization process. And at the termination of the process, the number of monomers used to produce each polymer chain will vary substantially from chain to chain. If the process were truly random, the distribution of the length of polymer molecules would be similar to a standard frequency or Gaussian distribution curve with the number of chains of a given length plotted against that length (Figure 2.1). If the molecular weight of the *repeat unit,* or that portion of the monomer that participates in the polymer backbone, is known, the molecular weight of a polymer of a given chain length is known. For example, the molecular weight of the ethylene repeat unit, $- CH_2 - CH_2 -$, is 28. A polyethylene polymer having about 10,000 repeat units has a molecular weight of about 280,000.* Thus, the polymer chain length can be plotted against molecular weight, as seen in Figure 2.2. Note that if magnified 10^8 times, this polyethylene polymer molecule, if fully elongated, would be 200 m (656 ft) long.

---

*The correct molecular weight of a given polymer molecule is obtained by multiplying the molecular weight of the repeat unit by the number of repeat units, *then adding in* the molecular weights of the *end groups*. Correctly, the end group on the polyethylene molecule is ($- CH_3$). Since there are two, the molecular weight of a polyethylene polymer with 10,000 repeat units and two end groups is 280,030. However, since the exact number of repeat units is never known, the contribution of the end groups is usually ignored.

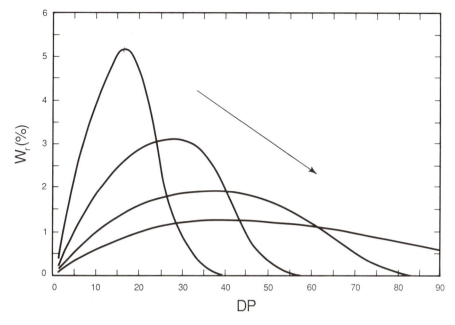

Figure 2.1. Molecular Weight Distributions from a Batchwise Reaction. Arrow Shows Effect of Increasing Monomer Concentration (1).

Very few polymerization reactions yield pure linear polymers. Instead, complex side reactions occur. To some extent, these effects are controlled by controlling catalyst and reactor processing conditions. But only in exceptional cases can they be eliminated. As a result, most polymer *molecular weight distribution* plots are typically non-Gaussian, as seen in Figure 2.2. Curve A represents a very narrow molecular weight distribution polymer, with a maximum occurring at about 20,000 and a molecular weight range of 3000 to 130,000 or so. Curve B has a maximum of about 70,000, and a molecular weight range from about 300 to more than 8,000,000. The shape of the molecular weight distribution curve and the locations of the various maxima are very important, since most polymer physical properties can be related in some way to this curve. The methods of measuring molecular weight and molecular weight distributions are discussed in Section 2.3.

Statistical Analysis Techniques

Because of the distribution in chain lengths of any polymer, statistical analysis is needed for characterization. Consider some terms that are basic to statistical analysis. The *arithmetic mean* or sample mean is the sum of all values, x, divided by the total number of values, n:

(2.1)
$$\text{Mean} = \sum_{1}^{n} \frac{x_i}{n}$$

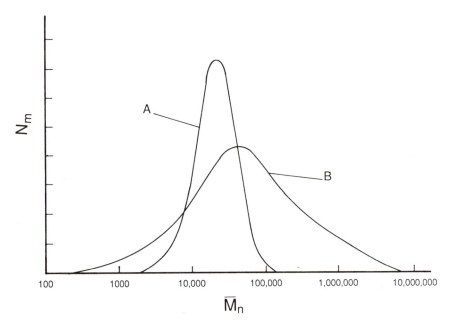

Figure 2.2.   Typical Molecular Weight Distributions. A: M_n = 20,000, PI = 1.65. B: M_n = 70,000, PI = 28 (2).

The *median* is the value at the mid-point in the grouping of values. If the sample has an odd number of values, the median is the one in the middle. If the sample has an even number, the median is the average of the two values closest to the mid-point. The *mode* is the value that occurs most frequently. This is usually obtained by making a histogram. Each of these averages describes a particular shape to the statistical curve. Similarly, a set of averages for molecular weight distribution can be used to describe the nature of the polymer. The type of statistical average used for polymers depends to a large degree on the method used to obtain the distribution.

Polymer Molecular Weight Distribution

The simplest averaging technique to describe the molecular weight distribution of a polymer sample is *number average molecular weight*. It is akin to the arithmetic mean, in that it is a sum of the number of chains of a given length, times the molecular weight of the chain, divided by the total number of chains:

(2.2) $$M_n = \frac{\sum_1^\infty n_i \, MW_i}{\sum_1^\infty n_i}$$ Sensitive to short molecules

Note that the weight of the ith chain is given as:

$$(2.3) \qquad\qquad w_i = n_i\, MW_i$$

Typically, *number average molecular weights* are determined by end group analysis, boiling point elevation, freezing point depression, osmotic pressure, and vapor pressure osmometry (3).

The *weight average molecular weight* is the sum of the weights of the various chain lengths, times the molecular weight of the chain, divided by the total weight of all the chains:

$$(2.4) \qquad\qquad M_w = \frac{\sum\limits_{1}^{\infty} w_i\, MW_i}{\sum\limits_{1}^{\infty} w_i} \qquad \text{sensitive to long molecule}$$

This can be written in terms of the numbers of chains of a given length, as:

$$(2.5) \qquad\qquad M_w = \frac{\sum\limits_{1}^{\infty} n_i\, MW_i^2}{\sum\limits_{1}^{\infty} n_i\, MW_i}$$

Light scattering is one way of directly measuring weight average molecular weight.

The *Z-average molecular weight* is given as:

$$(2.6) \qquad\qquad M_z = \frac{\sum\limits_{1}^{\infty} n_i\, MW_i^3}{\sum\limits_{1}^{\infty} n_i\, MW_i^2} \qquad \text{sensitive to very high MW}$$

Ultracentrifugal techniques are used to characterize Z-average molecular weights (1).

Note the general form for these molecular weights:

$$(2.7) \qquad\qquad M_a = \frac{\sum\limits_{1}^{\infty} n_i\, MW_i^{(a+1)}}{\sum\limits_{1}^{\infty} n_i\, MW_i^a}$$

The shape of any curve is determined by moments. The higher the moment, the greater is the influence of the higher values on the curve. Thus, M_n, the number average molecular weight, is quite sensitive to shorter molecules. M_w is more sensitive to the longer molecules and M_z is quite sensitive to the very high molecular weight chains.

As a result, for any given molecular weight distribution, the value for M_z will be higher than that for M_w, and the value for M_n will be lower than that for M_w.

When characterizing a specific polymer molecular weight distribution, it is common to form the ratio of weight average to number average molecular weights. This is called the *polydispersity index*, PI, or *dispersity index*, DI:

$$(2.8) \qquad PI = \frac{M_w}{M_n} = \frac{\sum_1^\infty n_i \, MW_i^2 \cdot \sum_1^\infty n_i}{\left[\sum_1^\infty n_i \, MW_i\right]^2}$$

The classical statistically random polymer has a polydispersity index of 2. Most polymers have values somewhat larger than 2. Values of 3 to 8 are not uncommon, with special purpose polymers having values as large as 30.

There is another empirical technique to describe molecular weight distribution. This is based on the effect individual polymer molecules in very dilute solution have on the viscosity of a solvating fluid. When molecular weights are determined by solution viscosity means, the results are presented as:

$$(2.9) \qquad [\eta] = K \, M^a$$

where K is a material-specific constant and a is a correlation coefficient with values between 0.5 and 1.0. If the solution viscosity is used to calculate a molecular weight distribution, this expression is used:

$$(2.10) \qquad M_v = \left[\frac{\sum_1^\infty n_i \, MW_i^{(a+1)}}{\sum_1^\infty n_i \, MW_i}\right]^{\left(\frac{1}{a}\right)}$$

M_v is called the *solution average molecular weight*. Note that if $a = 1$, M_v is identical to M_w. Usually the value of M_v is slightly less than that for M_w and substantially greater than that for M_n.

Application of MWD to Sample Distributions

Most of the methods of measuring molecular weight yield either number average or weight average values. Several prototypical distributions are presented in this section to illustrate the significance and utility of these and other averaging techniques. In Figure 2.3 and in Table 2.1 are given six possible molecular weight distributions for a

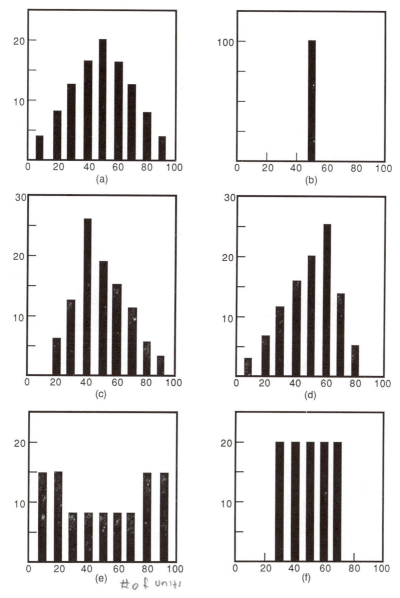

Figure 2.3. Sample Molecular Weight Distributions. a: Symmetrical. b: Monodisperse.
c: Asymmetrical, Skewed Left. d: Asymmetrical, Skewed Right. e: Bimodal. f: Uniform.

Table 2.1. Number of monomer units in each sample shown in Figure 2.3.

Sample	Number of monomer units								
	10	20	30	40	50	60	70	80	90
a	4	8	12	16	20	16	12	8	4
b					100				
c		6	13	26	19	15	11	7	3
d	3	6	11	16	20	25	14	5	
e	15	15	8	8	8	8	8	15	15
f			20	20	20	20	20		

polymer having 5000 repeat units.* Therefore, all polymers have the same number average molecular weight, by design. The ordinate (x-axis) in Figure 2.3 is the number of monomer units. The abscissa (y-axis) is the number of polymer molecules having the specific number of monomer units.

Example 2.1

Q: Calculate the M_n, M_w, M_z, and polydispersity index for the symmetrical distribution in Figure 2.3.

A: Based on the monomer units per molecule, n_x, and the molecular weight of the mer unit, $M_x = S_x\,k$, the number average molecular weight, the weight average molecular weight, and the Z-average molecular weight equations are rewritten as:

$$M_n = \frac{\sum n_i \, M w_i}{\sum n_i}$$

$$M_n = \frac{\sum_1^m n_x S_x\,k}{\sum_1^m n_x} \quad \text{— total mer units for all samples}$$

$$M_w = \frac{\sum_1^m n_x S_x^2\,k}{\sum_1^m n_x S_x}$$

$$M_z = \frac{\sum_1^m n_x S_x^3\,k}{\sum_1^m n_x S_x^2}$$

---

*For this illustration only, the polymer molecules are assumed to form in increments of 10 repeat units. In reality, the actual distributions would form a continuous distribution of single increments of the repeat unit.

The working numbers are given in Table 2.2. The three statistical averages are:

$$\frac{\sum (MW\ unit)\cdot(\#\ of\ units\ /molecule)}{S_x} \quad n_x$$

$$M_n = \frac{5000\ k}{100} = 50\ k$$

total mer units for all molecules in sample

$$M_w = \frac{290,000\ k}{5000} = 58\ k$$

$$M_z = \frac{18,500,000\ k}{290,000} = 63.6\ k$$

$$PI = \frac{M_w}{M_n} = \frac{58}{50} = 1.16$$

In Table 2.3 are given the values for the three moments of each distribution type. Note that sample b of Figure 2.3 is monodisperse. That is, all the molecules have the same chain length. Thus, the polydispersity index, PI, is 1. The polymer having a distribution that is skewed to the low molecular weight range, sample c of Figure 2.3, shows lower M_w and M_z values than those for the polymer with a distribution skewed toward higher molecular weights, sample d of Figure 2.3. Note however that the PIs are about equal. Thus, the PI value does not give a good indication of the absolute values of the smallest and largest polymer molecules present in the sample, nor does it describe the shape of the molecular weight distribution curve. The Z-average reflects the influence of the very high molecular weights in the bimodal distribution of sample e in Figure 2.3.

It is important to note that these statistical averaging techniques have limited utility in comparing materials of similar molecular weight distributions. If the numerical values of these averages are substantially different for two polymers, it can be concluded that the polymers are different. If the values are essentially the same, however, the

MW of mer unit for diff. unit

# of units/molecule polymer

Table 2.2 Computations for Example 2.1, symmetrical distribution.

S_x	n_x	$n_x S_x$	$n_x S_x^2$	$n_x S_x^3$
10	4	40	400	4,000
20	8	160	3,200	64,000
30	12	320	10,800	324,000
40	16	640	25,000	1,024,000
50	20	1,000	50,000	2,500,000
60	16	960	57,600	3,456,000
70	12	840	58,800	4,116,000
80	8	640	51,200	4,096,000
90	4	360	32,400	2,916,000
	100 $\sum n_x$	5,000[1]	290,000	18,500,000

[1]Total number of mer units.

Table 2.3. Statistical moments for each sample shown in Figure 2.3.

Sample	Number of molecules	ΣnS	M_n/K	M_w/K	PI	Z/K	Shape of distribution
a	100	5,000	50	58	1.16	63.8	Symmetrical
b	100	5,000	50	50	1.00	50	Monodisperse
c	100	5,000	50	56	1.12	61.4	Asymmetrical (right)
d	100	5,000	50	55.6	1.11	59.7	Asymmetrical (left)
e	100	5,000	50	66.6	1.33	74.9	Bimodal
f	100	5,000	50	54	1.08	57.4	Uniform

polymers may not necessarily be the same. The practical use of these averages is in characterizing the *relative* magnitudes of the distributions. For example, in Figure 2.2, the number average molecular weight distribution of sample A is 20,000 and that for B is 70,000. Polymer B, therefore, should behave as if it is a higher molecular weight polymer, *even though* it has a substantially broader molecular weight distribution.

2.3 Methods for Measuring Molecular Weight

Since polymer molecular weight is intrinsically tied to polymer performance and since polymer molecular weight can be found only as a distribution of values, much effort has been expended in finding easy, rapid ways of determining representative values. As noted, test results can be strongly influenced either by very low molecular weight elements or very high ones. As a result, test methods tend to produce biased values. And so frequently more than one test is run to characterize a polymer. Many tests are outlined in Table 2.4 (4).

Gel permeation chromatography, or GPC, is based on the rate at which different size molecules diffuse through a packed bed of absorbing gel beads. A fixed amount of polymer in suitable solvent solution is injected into a GPC column. Smaller molecules have greater solubility in the gel beads and so diffuse through the bed at a slower rate than larger molecules. This means that the first polymer issuing from the column has the highest molecular weight. By monitoring the solution concentration as a function of time and comparing the results with a standard calibration curve for that type of polymer in that type of column, the molecular weight distribution can be obtained. GPC now includes high pressure and liquid chromatographic techniques (5)

Molecular-level electron microscopy (MLEM) involves viewing a thin film of polymer under a very high resolution electron microscope. For some polymers, the diameter of the individual polymer coils and the density of the solid polymer can be determined. This technique is quite sophisticated, involving computer-aided counting techniques to accurately measure the molecular weight distribution. Most of the techniques identified in Table 2.4 are used to present the engineer and designer with a true picture of the nature of the polymer with which they are dealing.

Another technique that is used extensively is *solution viscosity* (6). Here, the molecular weight distribution is not measured directly. Instead the polymer is dissolved in an appropriate solvent and the polymer-containing solvent viscosity is measured. This viscosity is then related to the molecular weight distribution of the polymer. Viscosity, as noted in Chapter 4, is the property of a material to resist shearing or elongational stress. When a high molecular weight material is dissolved in a suitable solvent, the viscosity of the resulting solution (polymer + solvent) is greater than that of the solvent alone. The magnitude of the viscosity increase depends upon the concentration, length, geometry and flexibility of the polymer molecule as well as the interactions, if any, between neighboring polymer molecules and between polymers and solvents. In 1920, Einstein theoretically predicted that the relative viscosity, that is, the ratio of solution viscosity to solvent viscosity, would be linearly dependent upon the concentration of rigid noninteracting spheres:

(2.11) $$\eta_{rel} = \eta/\eta_s = 1 + 2.5\,\phi$$

where ϕ is the volume fraction of spheres in suspension in the solvent, η is the viscosity of the solution (or suspension) and η_s is the solvent viscosity. Even though polymer molecules are not rigid spheres, in very dilute solution they can be considered as near-spherical coils or rigid rods. As a result, the *relative viscosity* of a dilute solution of polymer in solvent can be written as:

(2.12) $$\eta_{rel} = 1 + K\,c$$

where c is the polymer concentration, usually in g/ml or g/dl (dl = deciliter, 0.1 l), and K is a constant for a specific polymer–solvent combination. The *specific viscosity* of a polymer is simply the relative viscosity minus one:

(2.13) $$\eta_{sp} = K\,c$$

The specific viscosity represents the fractional increase in viscosity due to the polymers in solution.

If polymers were truly noninteracting spheres or spherical elements, a *reduced viscosity*, being the ratio of the specific viscosity and concentration, should yield the material-specific constant K:

(2.14) $$\eta_{red} = \frac{\eta_{sp}}{c} = K$$

For a large number of polymer–solvent systems, the reduced viscosity is usually dependent upon concentration:

(2.15) $$\eta_{red} = K\,f(c)$$

The concentration dependency at very dilute concentrations is usually linear:

(2.16) $$\eta_{red,c\ small} = K\,(1 + ac)$$

Table 2.4.　Limitations of methods for measuring molecular weight (4).

Method	Type of result[1]	Best molecular weight range	Time required		Apparatus		Supporting measurements[3]	Usual limitations
			Measurement	Calculation	Source[2]	Cost		
Gas density	M_n	Gases	1 h	Short	Lab/Comm	Low	—	Gas samples
Vapor density	M_n	Up to 200	Short	Short	Lab/Comm	Low	—	Sample bp > 200°C
Cryoscopic	M_n	Up to 10^4	Short	Short	Lab/Comm	Low/med	Calibration	Insensitive at high molecular weight
Ebulliometric	M_n	Up to 10^4	Short	Short	Commercial	Medium	Calibration	Insensitive at high molecular weight
Vapor pressure, thermoelectric	M_n	Up to 10^4	Short	Short	Commercial	High	Calibration	Insensitive at high molecular weight
Isopiestic	M_n	Up to 10^3	Days	Short	Laboratory	Low	Calibration	Insensitive at high molecular weight
Osmotic pressure	M_n	2×10^4 to 5×10^5	Short to 18 h	Short	Comm/Lab	Med/High/Low	—	Membrane porosity
Light-scattering dissymmetry	M_w	10^4 to 10^5	$1\ h^4$	$Short^4$	Commercial	High	dn	Single solvent, clarify
Angular extrapolation	M_w	5×10^4 to 10^7	$2\ h^4$	$4\ h^4$	Commercial	High	dn	Single solvent, clarify
Sedimentation, velocity	M_z	10^4 to 10^7	$2\ h^4$	$8\ h^4$	Commercial	Very high	dn, V, ρ	Nearly ideal solvent

Method	M[1]	Range	Time	Time	Comm[2]		dn, V, ρ[3]	Requirements
Sedimentation, equilibrium	M_w, M_z	500 and up	Days to weeks	8 h	Commercial	Very high	dn, V, ρ	Nearly ideal solvent
Sedimentation, archibald	M_w	500 and up	1 h	8 h	Commercial	Very high	dn, V, ρ	–
Solution viscosity	M_v	10^3 to 10^7	Short	Short	Commercial	Low	Calibration	–
End-group analysis	M_n	Up to 2×10^4	1 to 4 h	Short	Laboratory	Low	–	Condensation polymers
X-ray diffraction	M	Any crystal	1 to 8 h	Varies	Commercial	Very high	–	High crystallinity
Surface pressure	M_n	2×10^3 to 10^5	1 h	Short	Laboratory	Medium	–	Surface-active sample and clean surfaces
Mass spectrometer	M	Up to 700	1 to 4 h	Short	Commercial	Very high	Calibration	Volatile sample at $\leq 200°C$

[1] M = true molecular weight
M_n = number-average molecular weight
M_w = weight-average molecular weight
M_v = viscosity-average molecular weight
M_z = molecular weight distribution
[2] Lab = apparatus may be constructed or modified in the laboratory
Comm = commercially available
[3] Calibration = calibration with a suitable standard
dn = determination of the refractive increment
V = determination of partial specific volume
ρ = determination of density
[4] Includes measurements of concentration series

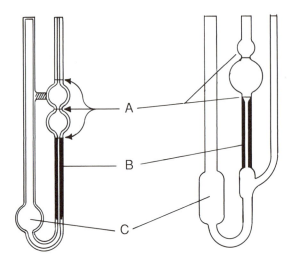

Figure 2.4. Capillary Viscometers Used to Measure Polymer–Solvent Viscosities. Left: Ostwald. Right: Ubbelode. A: Timing Lines. B: Capillary. C: Reservoir.

As a result, the experimental procedure involves dissolving a polymer in a suitable solvent at a low concentration (say, c_1), measuring its reduced viscosity (η_{red}), then cutting the concentration in half, say ($c_1/2$). The reduced viscosity at this concentration is measured, and the process repeated until a linear relationship is formed. Extrapolation to infinite dilution ($c = 0$) yields the material-specific constant K. The extrapolated viscosity obtained is called the *limiting viscosity number* or more traditionally, *intrinsic viscosity,* [η].* Experimentally (and with some theoretical justification), intrinsic viscosity can be seen to be directly related to molecular weight:

(2.17) $$[\eta] = k\,M^a$$

where k and a are polymer–solvent specific constants and M is the viscosity average molecular weight distribution.

There are several techniques for measuring solution viscosity. Polymer is charged to a simple glass capillary viscometer such as shown in Figure 2.4. The device is placed in a constant temperature bath, usually at 30°C. The viscometer is inverted and a fixed amount of fluid is allowed to flow under gravity and its own head through the capillary. The time required for this is directly related to fluid viscosity by:

(2.18) $$\eta = \rho\left[at + \frac{b}{t}\right]$$

---

*Note that intrinsic viscosity, specific viscosity and inherent viscosity terms are not true engineering or material-specific viscosities. They are viscosity ratios. As a result, they are *dimensionless*. Viscosity is *not* dimensionless, being a ratio of applied stress to resulting rate of deformation.

where ρ is the solution density and a and b are viscometer constants. For long times ($t > 100$ s) and very fine-diameter capillaries, the equation can be approximated as:

(2.19)
$$\eta \approx \rho \, a \, t$$

The efflux time for the solvent can be varied by using viscometers of different capillary diameters. It decreases with increasing capillary diameter. The efflux time for the solvent should be greater than 100 s as well so that the nonlinear time term containing b can be ignored. The ratio of the solution viscosity to the solvent viscosity can then be written as:

(2.20)
$$\frac{\eta}{\eta_s} = \frac{\rho t}{\rho_s t_s}$$

If the polymer concentration is very low, the density of the solvent and that of the solution are essentially equal. Thus:

(2.21)
$$\frac{\eta}{\eta_s} = \frac{t}{t_s}$$

In order to obtain intrinsic viscosity for a polymer of an unknown molecular weight, a calibration curve relating intrinsic viscosity to molecular weight for that polymer is needed. This is usually obtained by measuring the intrinsic viscosity of a monodisperse polymer (viz, a polymer with PI = 1). A typical calibration curve for polyisobutylene in cyclohexane is seen in Figure 2.5. These data are fit with:

(2.22)
$$[\eta] = k \, M^a$$

where $k = 27.6 \times 10^{-5}$ and $a = 0.69$. The general form of this equation is called the Mark-Houwink equation. A comprehensive tabulation of k and a for many polymer–solvent combinations is given in Table 2.5 (6).

Example 2.2

Q: To illustrate the concept of solution viscosity, consider the capillary viscometry data for polyisobutylene in cyclohexane at 30°C in Table 2.6. Using the appropriate values for polymer–solvent constants k and a in Table 2.5, determine the intrinsic viscosity and the solution viscosity average molecular weight.

A: The reduction of data for a typical run, say number 2, is:

$$c = 0.238 \text{ g/dl}$$

$$t = 87.97 \text{ s}$$

$$t_s = 76.83 \text{ s}$$

$$\eta_s = 0.83 \text{ Pa s}$$

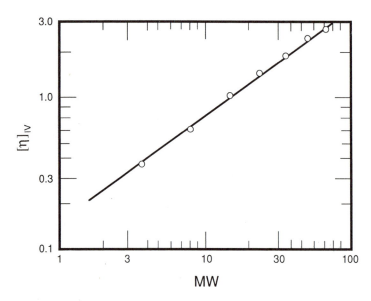

Figure 2.5.   Molecular Weight–Dependent Intrinsic Viscosity for Polyisobutylene in Cyclo-hexane at 30°C. $[\eta]_{IV}$ is Intrinsic Viscosity. Molecular Weight Units $\times 10^4$ (7).

Table 2.5.   Intrinsic viscosity constants, k and a, for various polymer-solvent systems (7).

Polymer	Solvent	Temperature (°C)	$k \times 10^5$ (dl/g)	a
Polyethylene				
(LDPE)	Decalin	135	62	0.70
(HDPE)		70	38.7	0.738
Polyisobutylene	Cyclohexane	30	27.6	0.69
Polypropylene (PP)				
Atactic	Decalin	135	11.0	0.80
Isotactic		135	10.0	0.80
Polymethyl methacrylate				
(PMMA)	Acetone	30	5.83	0.72
	Toluene	25	7.1	0.73
Polystyrene (PS)	Benzene	25	9.18	0.743
	Cyclohexane	35	76.0	0.5
	Toluene	30	11.0	0.725
Nylon 6 (PA-6)	m-Cresol	25	180.0	0.654
Nylon 66 (PA-66)	90% HCOOH	25	51.6	0.687
Polyacrylonitrile				
(PAN)	Dimethyl formamide	25	23.3	0.75
Polyvinyl chloride				
(PVC)	Tetrahydrofuran (THF)	20	3.63	0.92

Table 2.6. Experimental capillary viscosity data for polyisobutylene in cyclohexane at 30°C (8).

Run	Polyisobutylene in cyclohexane (g/dl)	Time (s)	Viscosity η_{rel}	η_{sp}	η_{red}
1	0	76.83	—	—	—
2	0.238	87.97	1.14	0.145	0.609
3	0.305	91.43	1.19	0.190	0.623
4	0.437	98.34	1.28	0.280	0.641
5	0.728	114.94	1.58	0.496	0.681
6	0.896	125.31	1.63	0.631	0.704
7	1.000	132.09	1.72	0.719	0.719

Relative viscosity:

$$\eta_{rel} = \frac{\eta}{\eta_s} = \frac{t}{t_s} = \frac{87.97\ s}{76.83\ s} = 1.145$$

Solution viscosity:

$$\eta_s = \eta_{rel,s} = 1.145 \times 0.83\ Pa\ s = 0.95\ Pa\ s$$

Specific viscosity:

$$\eta_{sp} = \eta_{rel} - 1 = 1.145\text{-}1 = 0.145$$

Reduced viscosity:

$$\eta_{red} = \frac{\eta_{sp}}{c} = \frac{0.145}{(0.238\ g/dl)} = 0.609\ dl/g$$

The above results and those for the other five runs are given in Table 2.6. The intrinsic viscosity is obtained by plotting the reduced viscosity values against concentration, as shown in Figure 2.6, then extrapolating to zero concentration. An intrinsic viscosity value, $[\eta] = 0.575$, is obtained.

The solution viscosity average molecular weight is obtained from:

$$M_v = \left[\frac{[\eta]}{K}\right]^{1/a} = \left[\frac{0.575}{27.6 \times 10^{-5}}\right]^{1/0.69} = 64,500\ g/g\ mol.$$

At intrinsic viscosity values above 5, substantial error may be introduced owing to the relatively high solution shear rates and velocity gradients in the capillary tube. Correction factors can be used under these conditions (9). It is common practice to measure a solution viscosity only at one *specific* polymer concentration (typically 1.0 g/dl). These values are used for comparative purposes only, since they include a molecular

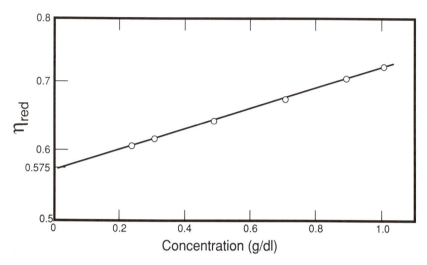

Figure 2.6.   Concentration-Dependent Reduced Viscosity for Polyisobutylene in Cyclohexane at 30°C.

interaction effect. For the same polymer–solvent system, increasing viscosity *at this single point* can be directly related to increasing molecular weight. This technique is frequently used for quality control during production when comparing batch-to-batch reactor output and for processes that require careful control of polymer feedstock molecular weight.

2.4 Molecular Order
of Thermoplastics

A thermoplastic polymer is in a fluid or "molten" state when the thermal energy of the individual polymer molecule is substantially greater than the secondary bonding forces that act to hold adjacent polymer molecules together. In this state, the polymer chain segments are free to rotate and reptate,* pendant groups are free to rotate and vibrate, and the entire chain can move relatively easily past neighboring chains under applied stress. Thermoplastic polymers are frequently characterized as "linear" or one-dimensional in the limit. As noted in Section 2.2, polyethylene is a linear molecule, with a molecular weight in its high density form, of about 280,000. If this polymer is fully extended and magnified 10^8 times, it would be about 5 mm (0.2 in) in diameter and 200 m (656 ft) in length. Its effective length-to-diameter ratio would be 37,000. For all intents, chain segments near one end should be unaffected by actions on chain segments near the other. However, polymeric chains are rarely fully extended. Instead they reside in coiled configurations, with extensive intersections and entanglements of

---

*Reptation is snake-like motion.

neighboring chains. Actual polymer chain configurations and probable end-to-end dimensions of polymers are given in many polymer chemistry texts (10,11). At very high temperatures, the disorder on a molecular scale is nearly complete, with polymer reptation, entanglement and disentanglement occurring at a high rate. Furthermore, the high level of thermal energy results in a relatively large volume occupied by the active molecule, viz, a high specific volume (point 1 in Figure 2.7). As the polymer temperature is decreased, the degree of agitation decreases, resulting in a reduction in the amount of space occupied by each molecule. The specific volume of the polymer decreases, as seen as point 2 in Figure 2.7. As the polymer further cools, the attractive forces between neighboring molecules increase very rapidly, until at some point there is an abrupt compaction or compression of the polymer molecules into a tight, orderly structure. This is seen as a rapid drop in the specific volume of the polymer, curve A. For nonpolymeric materials, this phase change is known as solidification or freezing and the temperature at which this occurs is called the freezing temperature. For polymers, this phase change is called *crystallization*. Note in Figure 2.7, that the specific volume does not abruptly drop to a new value. Crystallization begins at T_c, the crystallization onset temperature, but the approach to a steady specific volume change continues to occur as the temperature continues to fall. Polymer crystallization is a time-dependent phenomenon that will be discussed in detail in the next section.

Note that not all polymers crystallize (as indicated by curve B in Figure 2.7). Polymers that do not exhibit this tendency to form a denser, more ordered structure are called *amorphous* polymers. Some polymers may have a very slight amount of crys-

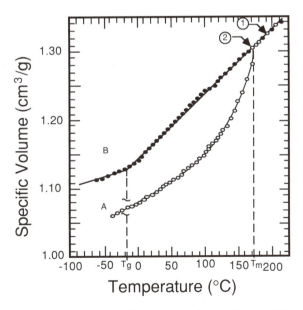

Figure 2.7. Temperature-Dependent Specific Volume of Amorphous and Crystalline Polypropylene. A and Open Circles: Crystalline Polymer. B and Closed Circles: Amorphous, Atactic Polymer (12).

tallinity as such as polyvinyl chloride (PVC) with about 5 to 20% crystallinity. But the properties of these polymers are more like those of traditional amorphous polymers than those of crystalline ones and so are called amorphous. The freezing or melting phenomenon is a *first-order thermodynamic transition* and is discussed in detail in Chapter 3. Not all polymers exhibit a second-order thermodynamic transition known as the *glass transition*. Thermosets usually do not and highly oriented crystalline thermoplastic polymers may not. Although this will be discussed in more detail in Chapter 3, it is observed as a change in the *slope* of the specific volume–temperature curve, as seen at about $-18°C$ for atactic polypropylene in Figure 2.7.

As seen in Figure 2.7, the specific volume change for an amorphous polymer cooling from the melt state to room conditions is substantially less than that for a crystalline polymer. The specific volume changes can be directly related to individual part shrinkage in the molding process (see Chapter 6). Injection molders and moldmakers know by experience that the shrinkage of a crystalline plastic may be two to three times that of an amorphous one.

The level to which a polymer crystallizes depends strongly upon the molecular nature of the polymer and the time–temperature profile imposed on the polymer by its environment. Not all polymers that *can* crystallize *do* crystallize. The degree and uniformity of the crystalline structure in a molded part directly affect its performance and dimensional stability. Before the interaction of the process with the degree of crystallinity in a molded part can be discussed, the nature of the crystalline structure and the way in which this structure is formed must be understood. For amorphous polymers and those crystalline ones with substantial amounts of amorphous regions, the frozen-in alignment or orientation of the polymer molecules is a direct result of the molding or flow process. This can cause significant directional property variations. It can be beneficial, but it can be detrimental as well.

In order to understand polymer crystallinity, the nature of the polymer must be discussed. Since only a very few polymers are highly crystalline, the role of the unordered or amorphous regions must also be considered in some detail.

Crystal Structure

Since crystalline polymers play economically important roles in many load-bearing applications, the nature of the polymer crystal itself is considered first. The precise nature and causes of polymer crystallinity are under intensive study (13–21). Early studies focused on crystallizing polymers from dilute solutions. From these studies, an understanding of the nature of the single polymer crystal has developed. This understanding has been translated to the structure of crystals formed during near-isothermal quiescent crystallization from polymer melt. And this knowledge is being extended to crystallization in non-isothermal shear fields. This last area typifies conditions in modern processing practice of injection molding, extrusion and blow molding, as examples (Chapter 5).

A popular conceptual model for the structure of a polymer single crystal or *crystallite* is the *regular chain-folding re-entry model,* as seen in Figure 2.8. The crystallite is typically of platelet form, several hundred thousand Å (several ten thousand nm) on a

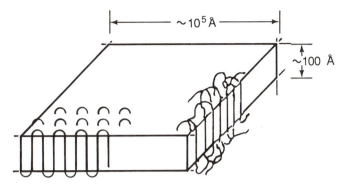

Figure 2.8.   Schematic Representation of the Relationship of Single Polymer Molecules Forming a Single Polymer Crystal (22).

side and 100 to 500 Å (10 to 50 nm) in thickness. X-ray analysis of crystallites shows
that polymer chains are aligned perpendicular to the large flat faces of the crystallite.
Since the length of an individual polymer molecule is substantially greater than the
thickness of the crystallite, the individual polymer molecules fold back and forth on
themselves in a tight configuration. The fold length for re-entry is less than 10 Å
(1 nm). The regular re-entry concept results in a very tightly packed crystallite and a
high density, low specific volume configuration. If the polymer backbone is somewhat
stiff, if there is a finite amount of short-chain branching, or if the crystallization occurs
in less-than-ideal conditions, the chain folding is not regular. A *switchboard re-entry*
model is proposed for non-ideal chain folding.

If the crystallite model is a valid model for bulk crystallization from the melt, it is
easy to see why certain polymers tend to crystallize and others do not. Polymers that
tend to crystallize have flexible backbones so that the molecule can easily pack onto
the crystallite surface (Figure 2.9). And polymers that have regular ordered atomic
structures allow for high packing levels on the surface. High-density polyethylene with
its simple carbon backbone and regularly spaced hydrogen atoms has a flexible backbone and a very regular atomic structure and so crystallizes to a very high level.
Polypropylene has a flexible backbone, but the regularly spaced methyl pendant group
(on every other carbon) must be in the proper configuration to allow backbone rotation
to accommodate it. If the methyl group is not always on the same side of the backbone,
it interferes with packing and prevents crystallization. This type of polypropylene is
called *atactic* and it is amorphous. If the methyl group is always on the same side of
the backbone, polypropylene readily crystallizes. This type of polypropylene is called
isotactic. See Figure 1.7 for schematics of these types. The conformation of polymer
groups along a backbone is called *stereoisomerism*. The commercial success of polypropylene is due entirely to the development of catalysts that selectively produce stereo-
isomeric isotactic polypropylene. Bisphenol A–polycarbonate has no pendant groups
or side chains and the backbone is geometrically symmetric. But the polymer backbone
is very stiff owing to the two phenyl groups in the backbone. This restricts folding and
as a result commercial polycarbonate is non-crystalline or amorphous. Polystyrene is

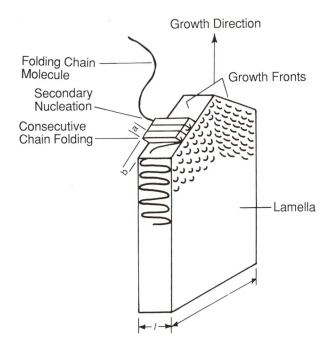

Figure 2.9.   Schematic Representation of the Mechanism of Single Polymer Molecules Migrating to the Surface of a Growing Single Polymer Crystal (23).

characterized by a regular pendant phenyl group (on every other backbone carbon) with random conformation along the backbone (atacticity). This polymer is characterized by substantial backbone stiffness (electron pairing between the backbone carbons and those in the phenyl ring) and by a very large pendant group that interferes with folding. As a result, atactic polystyrene is amorphous.

Thus, the polymer chemical structure presented in Section 1.6 for homopolymers and Section 1.8 for copolymers can be used to screen those polymers that *might have* a tendency to crystallize. Of course, at this point, discussion has focused on the order in an isolated single crystal or crystallite. This analysis is now extended to bulk crystallization.

The Bulk Crystallization Process

The crystallite discussed in the last section is formed by polymer molecules being drawn from a low viscosity solvent into a tight, highly ordered structure by intermolecular forces. There are two competing rate-dependent processes at work during crystal growth. The polymer molecule in the disordered region ahead of the growing front must be drawn to the growing front by secondary forces that are sufficiently strong to overcome viscous and entanglement forces in the disordered region. This is called *molecular mobility.* And a specific amount of energy must be expended in bending and aligning the molecule into the proper configuration at the surface of the growing front. This is called *surface nucleation*. At high isothermal temperature, the rate of molecular mobility is high but the rate of surface nucleation is low. As a result, crystallization rate is low. At low isothermal temperature, the opposite holds, and again, crystallization rate is low. At some intermediate isothermal temperature, crystalline growth rate is maximum. Although this is apparent for isolated crystallites, it also holds for crystallization from the melt.

In practical polymer processing, crystals do not grow in near-isothermal isolated conditions. Crystallization is usually initiated at an "energy well". If no well-defined site is isolated, crystallization is due to random fluctuations in molecular alignment leading to localized order. Crystallization is said to be due to *homogeneous nucleation*. *Heterogeneous nucleation* begins from insoluble impurities such as catalyst residue, colorants and/or fillers, or unmelted crystallites in the melt. Heterogeneous nucleation dominates crystallization in nearly all commercial molding processes.

Crystallite growth from a nucleus in an isothermal polymer melt is substantially different from single crystallite growth from a dilute solution. Usually many crystallites begin to grow from a single nucleus. The *lamellae* preferentially grow outward from the point of initiation. As growth continues the ribbon-like lamellae can twist, twin (form two or more branches), or bisect (Figure 2.10). Very low molecular weight molecules, stereo-irregular molecules and some isotactic molecules remain in an unordered fashion between the growing lamellae. Some crystallizing molecules can have both ends in growing lamellae with the middle portion in the unordered or amorphous material between and around the lamellae (Figure 2.11). The twisting behavior of the many ribbon-like lamellae growing from a single nucleus results in the formation of a nearly spherical network of crystallites (Figure 2.12). This crystal growth pattern is called *spherulitic crystal growth* and the network of lamellae is a *spherulite*. Radial growth of an individual lamellae continues until the growing front depletes the region of available highly mobile molecules and the secondary forces are insufficient to attract the less mobile ones. The growth rate of the spherulite depends on the size of the

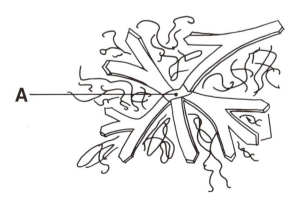

Figure 2.10. Schematic Representation of Bulk Crystal Nucleation and Growth, Showing Twinning. Point A: Point of Nucleation (24).

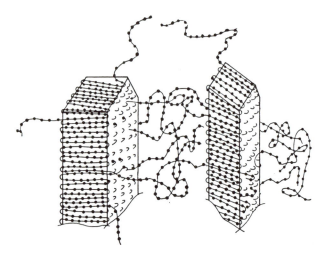

Figure 2.11.   Schematic Representation of Molecular Structure Between Lamellae (25).

polymer molecule, its geometry, the mobility of the polymer chain in the amorphous melt, the energy of the polymer molecule and the rate of surface nucleation. Consider the simplistic example of nucleation and spherulitic growth in a cubic array (Figure 2.13). As crystallization proceeds, all spheres nucleate and grow at essentially similar rates. This is called *primary crystallization*. Growth at this rate continues until the spheres impinge. Continued growth occurs either with distortion of the spheres or by nucleation and growth of crystallites in the regions between the spheres. This is called

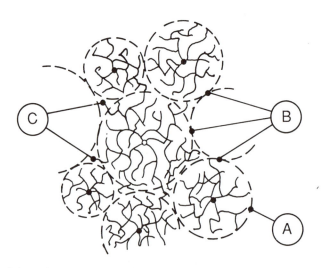

Figure 2.12.   Schematic Representation of Interacting Spherulitic Polymer Crystal Structure. A: Spherulite. B: Impingement Region. C: Interspherulite Region (26).

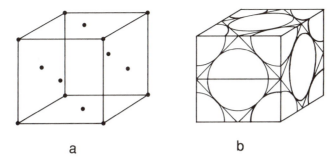

a   b

Figure 2.13.   Model of Bulk Spherulitic Growth. a: Initiation of Crystallization, Time = t_1. b: Impingement of Spherulites, Time = t_2 (26). (See Figure 2.16.)

secondary crystallization and typically occurs at a substantially reduced rate. It must be understood that the growth of spherulites in an isothermal medium is not instantaneous. The time required to reach a final level of crystallization depends on the nature of the polymer molecules and the melt temperature.

Differential Scanning Calorimetry

The *differential scanning calorimeter* or DSC has become an essential tool in measuring and understanding thermal transitions in polymers. Although the device is relatively expensive ($25,000 or more), it is easy to operate and quite rugged. A known weight of polymer, typically 3 to 10 mg, is placed into a special container and that is placed into a test cell. The heating or cooling rate is selected, usually 10 or 20°C/min, and the initial and final temperatures preset. The instrument measures the amount of energy that must be added to or removed from the sample as a function of time in order to maintain the proscribed heating or cooling rate. A typical DSC heating curve is seen in Figure 2.14. The differential amount of energy needed to maintain the rate is plotted as the abscissa or *y*-axis. Either time or temperature is plotted as the ordinate or *x*-axis. These two parameters are related since the heating/cooling rate is predetermined. Points B and C represent phase changes or first-order thermodynamic phase transitions.* These transitions are identified by discontinuities in the curve. The polymer at point B has enough molecular mobility that it continues to crystallize. At point C, the formed crystallites now begin to melt. Point A represents a second-order thermodynamic phase transition known as the glass transition (see Section 4.6). This transition is identified by a discontinuity in the *slope* of the curve.

It must be remembered that because of the very nature of the polymer macromolecule, transitions are *not* as instantaneous as they are with simple molecules. In order to obtain useful thermodynamic data, the DSC heating/cooling rate must be sufficiently

*Point B is called the "cold crystallization" condition and can occur only if the polymer was first quenched from the melt at a cooling rate sufficient to prevent crystallization.

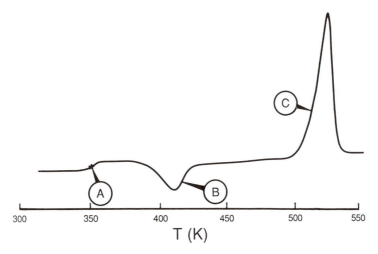

Figure 2.14. Differential Scanning Calorimetry (DSC) Heating Curve for Amorphous Poly-ethylene Terephthalate Film, Quenched from Melt. A: Glass Transition Region. B: Cold Crys-tallization Region. C: Crystalline Melting Region (27).

slow to allow for *near*-isothermal transitions. Because of the inherent transient nature of the DSC test, absolute isothermal conditions are usually not feasible. This point will be emphasized shortly. The DSC heating curve for the polymer that was heated to above its melting temperature in Figure 2.14 and slowly cooled is shown as Figure 2.15. The polymer has been heated to the melt temperature, this is, some temperature substantially (50°C or more) above the crystallization temperature, and held at this

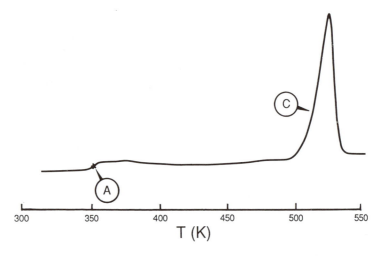

Figure 2.15. Differential Scanning Calorimetry (DSC) Reheating Curve for Polyethylene Ter-ephthalate Film, Crystallized from the Melt. Regions as Given in Figure 2.14 (27).

temperature for several minutes to allow for complete melting of all crystallites. The cooling curve again shows points C and A, the crystallization and glass transition points, but of course point B is absent.

Note the importance of these data. If a crystallizable polymer is not given sufficient time to crystallize during a cooling process such as that experienced in injection molding, extrusion, blow molding or blown film processing, the DSC heating curve will show substantial "cold" crystallization prior to the melting temperature range. If sufficient time *has* been provided for crystallization, there will be little cold crystallization evident in the DSC heating curve. Thus, the DSC is used extensively as a diagnostic tool when dealing with process-related quality control of crystallizable polymers.

Recall that the crystallization or melting process is a phase change not unlike freezing or melting of water. For simple molecules, freezing results in 100% crystallization and a significant liberation of energy per unit mass, called the *latent heat of fusion*. Similarly, heat is liberated when polymers transform from the disordered amorphous state to the thermodynamically ordered crystalline one. This energy per unit mass is called the *energy of crystallization*. On the DSC plot (Figure 2.16), the onset of crystallization is seen as a transition in the curve, time t_1. As will be seen below, the rate of crystallization is highly time-dependent. As a result, the DSC curve drops precipitously. The spherulites grow at a very high rate until impingement. At that point (time t_2), the crystallization rate drops, the amount of energy liberated per unit time slows rapidly, and as a result the DSC curve begins to rise rapidly. When crystallization is complete, the heat removal rate returns to a near-constant value, time t_3. Thus, the first portion of the crystallization curve represents spherulitic growth, the bottom of the curve represents spherulite impingement, and the last portion represents secondary growth. The relative size of spherulites is related to $(t_2 - t_1)$. The addition of a nucleating agent

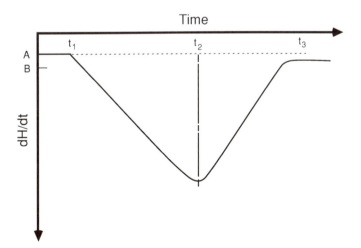

Figure 2.16. The Crystallization Portion of the DSC Curve, for Slow Cooling from the Melt. t_1: Beginning of Crystallization. t_2: Spherulitic Impingement. t_3: Completion of Crystallization (28).

Table 2.7. Latent heat of fusion, ΔH_f (cal/g), (100% crystallinity) for several polymers (29–31).

Polymer	Heat of fusion, ΔH_f (cal/g)
Polyethylene (PE)	68.4
Polypropylene (PP)	45
Polyethylene terephthalate (PET)	30
Nylon 6 (PA-6)	45.3
Nylon 66 (PA-66)	46.6

results in more spherulites per unit volume, smaller spherulites at impingement and a smaller value of $(t_2 - t_1)$.

Since the DSC measures differential energy, the area under the curve between crystallization onset and completion of secondary crystallization is a measure of the energy of crystallization per unit mass of material. If the amount of energy needed to crystallize the polymer to 100% is known, the degree of crystallinity in the solid polymer can be immediately determined. Latent heats of fusion (to 100% crystallinity) for many crystallizable polymers are given in Table 2.7.

If the heating and cooling rates are infinitesimally slow, or the rate of crystallization is infinitely fast, the crystallizing temperature and melt temperature should be about the same.* As the rates of heating and cooling increase relative to the crystallization rate, these values diverge. This is seen in the melting and crystallizing DSC curves at only 2°C/min for some polyolefins are shown in Figure 2.17. The rate of crystallization is intrinsically related to the mobility of the polymer. Thus, for a homologous series of polymers having differing molecular weights, degrees of branching and even different end groups, DSC curves should be different. This is seen for polyethylenes of differing densities in Figure 2.18. The DSC can be used to determine the amount of each type of polymer, as seen in Figure 2.19, where a commercial polyethylene shows a mixture of high density and low density polyethylenes. This is shown as *two* crystallization peaks in each trace. Note that the area under the peak at about 108°C in the 3014 sample (Curve A) is larger than that in the 4104 sample (Curve B), indicating that 3014 contains a greater fraction of the lower melting LDPE.

It was noted above that the DSC heating/cooling rate must be compatible with the polymer crystallization rate. If not, spurious results are obtained. Figure 2.20 represents a series of DSC plots of increasing heating rate from room temperature. Note that not only does the cold crystallization temperature appear to increase with increasing heating rate, but the energy of crystallization is increasing as well. This means that the degree of crystallinity is increasing. Conversely, if the cooling rate is fast enough, the polymer is quenched and essentially no crystallization is observed. The polymer remains amorphous.

---

*Polymer freezing points are usually depressed and melting points elevated by low molecular weight polymers, adducts, and such. As a practical result, these values are rarely exactly equal.

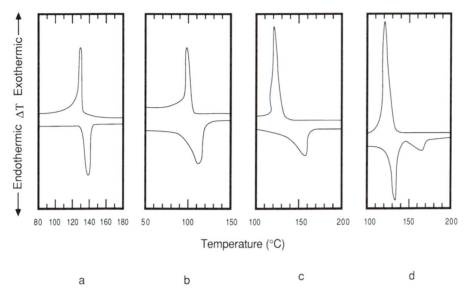

Figure 2.17. DSC Heating and Cooling Curves for Four Polyolefin Polymers. a: HDPE, Marlex 50. b: LDPE, DYNH. c: Isotactic Polypropylene. d: HDPE + Isotactic Polypropylene (32).

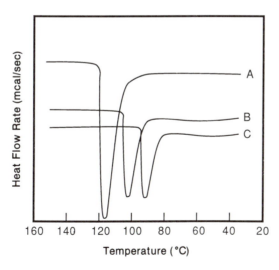

Figure 2.18. Comparison of the Crystallization Portions of DSC Cooling Curves for Three Polyethylene Types. a: High Density PE. b: Medium Density PE. c: Low Density PE. Cooling Rate: 10°C/min (33).

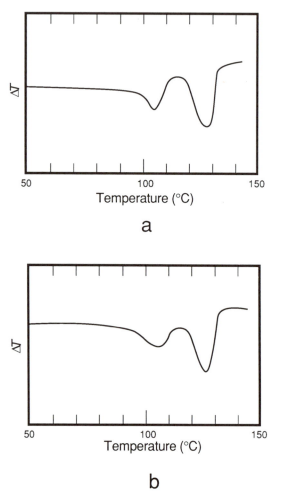

Figure 2.19. DSC Comparison of Two High-Density/Low-Density Polyethylene Blends. a: Dylan 3014. b: Dylan 4014 (34).

The crystalline morphology is also affected by cooling rate. As seen in Figure 2.21 for five polymers, increasing cooling rate results in decrease in the diameter of the spherulites. This is because homogeneous or spontaneous nucleation is preferred at lower temperatures. Homogeneous nucleation always produces many more spherulites per unit volume than does heterogeneous nucleation. And if the polymer temperature drops rapidly enough, the mobility of the chains traveling to the growing sites decreases rapidly, further limiting spherulitic size.

The DSC is also used as an analytical tool in determining the ways in which polymers behave when blended or copolymerized. DSC curves for different mole fraction co-polymers of nylon 66 and 610 (PA-66 and PA-610) are shown in Figure 2.22. As the mole fraction of PA-610 increases, the melt temperature decreases, the temperature

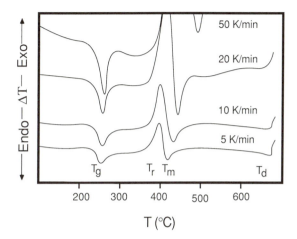

Figure 2.20.   Differential Scanning Calorimetry Scan of Poly(oxy-2,6-dimethyl (-1,4-phenyl-ene)), Showing the Effect of Heating Rate on Transition Temperatures. T_g, Glass Transition Temperature. T_r, Onset of Crystallization or Cold Crystallization Temperature. T_m, Melt Temperature. T_d, Onset of Decomposition (35).

range increases and the area under the peak, the energy of crystallization, decreases. The same is true as the mole fraction of PA-66 is increased, starting with pure PA-610. At 50 and 60 (mol)% PA-610, two distinct crystallization peaks are found. This indicates formation of a block copolymer. If the melt temperatures are plotted against copolymer composition, a phase diagram results, with a copolymer eutectic point occurring at about a 30:70 (mol)% PA-66:PA-610 composition (36).

X-Ray Techniques

The crystal structure in metals is easily characterized by transmission X-ray analysis. Metal structures have been categorized by the way in which the principal crystal planes intersect, e.g., cubic or rhombohedral, and by the way in which individual crystals pack, e.g., body-centered cubic or face-centered cubic. Pure polymer crystallites can also be characterized in this manner (37). However, the nature of crystal growth in spherulites is random isotropic and so standard metallographic techniques fail. Furthermore, polymers are rarely fully crystalline. And the intensity of the crystal plane orientation is greatly diminished by small amounts of amorphous material. Anisotropy as in fiber-stretching will yield preferred crystal plane orientation and therefore useful X-ray patterns. Two types of X-ray analysis are used together to examine the crystal structure of polymers: wide-angle X-ray scattering (WAXS) and small-angle X-ray scattering (SAXS).

Consider the polymer folded-chain lamellae to be aligned as playing cards in a deck (Figure 2.23). The spacing between these cards is d. As the angle of incident monochromatic light of wavelength, λ, to the plane of a lamella changes, the length of the

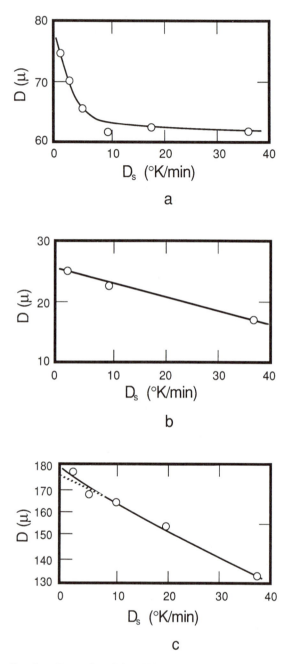

Figure 2.21.   Cooling-Rate Dependent Spherulitic Diameters for Several Polymers. a: Polyethylene. b: Polyethylene Terephthalate. c: Nylon 6 (PA-6).

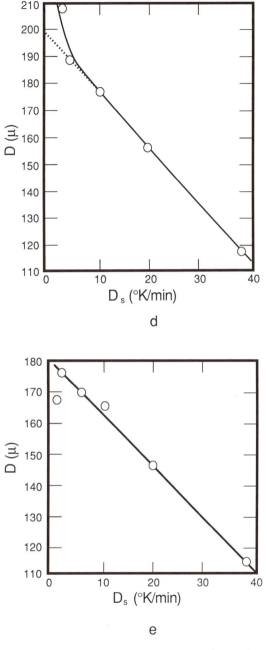

Figure 2.21.   (*cont.*) d: Nylon 66 (PA-66). e: Polypropylene (38).

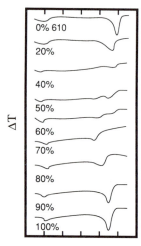

Temperature (°C)

Figure 2.22.   Comparison of DSC Heating Curves for Blends of Nylon 66 and Nylon 610 (PA-66 and PA-610) (39).

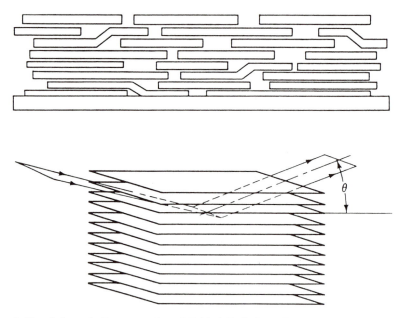

Figure 2.23.   Schematic Representation of Folded-Chain Lamellae as Akin to Playing Cards, for X-Ray Diffraction Analysis (40).

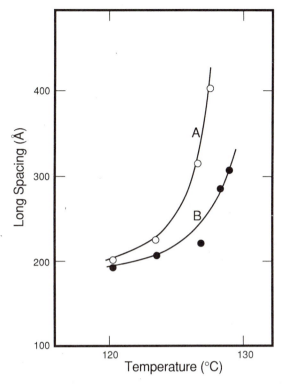

Figure 2.24.   Effect of Annealing on Lamellae Thickness for Two Polymers having Different Molecular Weight Values. M_v is Viscosity-Average Molecular Weight. A: 52,000. B: 520,000 (41).

path of the rays scattered by the adjacent lamellae changes in accordance with Bragg's law of optics:

$$(2.23) \qquad\qquad \sin \theta = \frac{n\lambda}{2d}$$

where n is a **diffraction order**. At very low angles of beam incidence, $\sin \theta \approx \theta$. As an example, if $\lambda = 1.54$ Å (0.154 nm)* and $d = 100$ Å (10 nm) (42), the first diffraction order ($n = 1$) occurs at 0.0077 rad or 0.44°. X-ray spectroscopy is the desired means of measuring lamella thickness, or long spacing, at this very small angle. SAXS is used primarily to monitor changes in the way in which the crystallites are formed. Increasing the temperature for maximum crystallization rate results in increased long spacing and thus thicker lamellae. The effects of molecular weight and molecular weight distribution on the annealing characteristics of polymers are also observed (Figure 2.24).

*0.154 nm is the CuK radiation wavelength used for most commercial X-ray measurements.

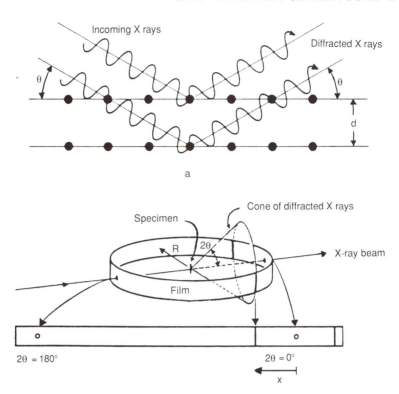

Figure 2.25. X-Ray Crystallography for Polymers. X-Ray Beam, Passes Through Cylindrical Film, and Specimen. Cone of Diffracted X-Rays Forms Intense Bands, as Crescents on Film (43).

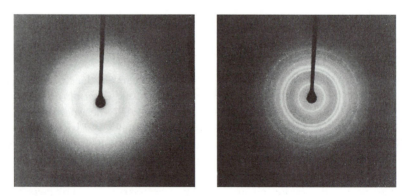

Figure 2.26. X-Ray Diffraction Patterns for Amorphous Atactic Polystyrene (a) and Oriented, Semicrystalline Isotactic Polystyrene (b). Sharp, Intense Bands in (b) Show Extent of Crystallinity (44).

X-ray diffraction is used primarily to examine the crystallographic nature of a polymer. Crystallography is beyond the scope of this introductory work. Suffice it to say that semicrystalline polymers, as with many crystalline materials such as metals and pure ceramics, exhibit structural order. Polyethylene, for example, has an *orthorhombic unit cell.* As the X-ray beam impinges a collection of crystallites in random alignment, the beam is reflected into cones, as seen in Figure 2.25. Amorphous polymers have no order and so the resulting X-ray image is a series of low intensity halos (Figure 2.26). Crystalline polymers show very distinct rings. The radial position of these rings can be related directly to the orientation in the crystallites. Highly drawn fibers and films show molecular axes that lie in the direction of elongation. The X-ray defraction normal to the direction of elongation shows a series of discrete reflections (Figure 2.27).

As is apparent, SAXS and wide-angle X-ray diffraction (WAXD, although sometimes called wide-angle X-ray scattering (WAXS) in deference to the symbolic notation, SAXS), are frequently used together to characterize the nature of the polymer crystallite formed during processing. For example, the characteristic X-ray patterns as a function of elongation for uniaxially drawn polyethylene are seen in Figure 2.28. Note for WAXS that the halos form point-shaped images as the draw continues. For SAXS, the halo resolves early into nonsymmetry, with points changing in plane between about 230% and 310% elongation, as the crystallites are deformed out of their initial orientation. From these and other analytical tools, it can be surmised that during fiber (and biaxial film) orientation, the following phenomena occur in a somewhat sequential manner (45):

• Lamellae rigidly slip past one another. Since the lamellae lying in the draw direction cannot do so, the spherulite distorts. Strain is accommodated by interlamellar amorphous material.
• Tie chains extend and deformation occurs by the slip-tilting of the lamellae.

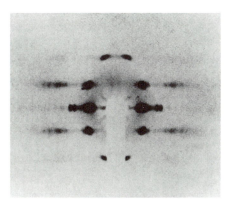

Figure 2.27.   X-Ray Diffraction Pattern of Uniaxially Oriented, Semicrystalline Polypropylene, Showing High Degree of Crystal Orientation in the Draw Direction (46).

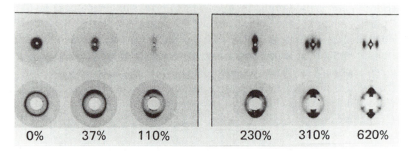

Figure 2.28.   The Combination of SAXS (Top Photos) and WAXS (Bottom Photos) Illustrate Degree of Uniaxial Orientation in HDPE (47).

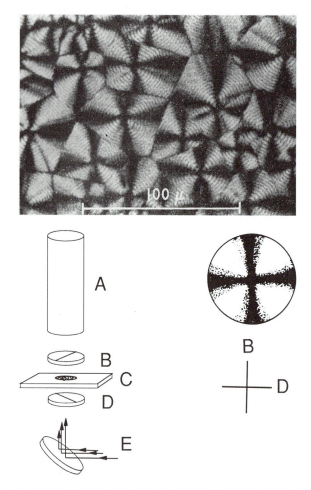

Figure 2.29.   The Nature of Spherulitic Growth, Seen as "Maltese Crosses", as Photographed Through Crossed Polarizers. A: Optical Column, B: Analyzer, C: Specimen, D: Polarizer, E: Mirror (48).

- Within any spherulite, parallel lamellae are shared relative to one another. Slip occurs, with tie chains holding lamellae together.
- Finally, lamellae segments are now aligned with the tensile axis.

Micrography

The *optical microscope* with a temperature-controlled hot stage is still the most popular way of observing crystal growth. The useful upper limit of optical magnification is about $500\times$. Above that, the depth of field is so limited that interpretation is difficult. Of course, at this level of magnification, only spherulitic growth can be observed. Individual lamellae are too small. Some measure of the nature of the spherulite can be observed by viewing through crossed polarizing lenses. The twisting lamellae refract the light to produce "Maltese cross" patterns on the growing spherulites (Figure 2.29). The optical microscope can be used to gain information about the rate of polymer crystallization.

The *scanning electron microscope* (SEM) is used for examination at magnifications in excess of $500\times$. Unfortunately, the typical electron beam used in SEM cannot be directed against organic materials, as it vaporizes them at point of contact. As a result, the solid polymer must be protected first by a sputter-coated monomolecular layer of nickel or gold. This means that *in-situ* observations of the growing spherulites cannot be made. Instead, a rather tedious series of experiments must be carried out, with the polymer quenched at various times throughout its crystallization. And care must be taken to ensure that the nature of the "time-frozen" spherulite is being observed and not artifacts of the experiment or sample preparation. SEM is probably better utilized as a way of observing the processing effects during compounding of fillers and fibers or blending potentially incompatible polymers.

Birefringence

Transparent polymers have three principal indices of refraction, n_1, n_2, and n_3. If $n_1 = n_2 = n_3$, the material is isotropic. The difference between any two pairs of indices, such as $\Delta n_{12} = n_1 - n_2$, is called *birefringence*. Amorphous polymers become birefringent when strained. In crystalline polymers, chain alignment causes birefringence. The total birefringence is:

(2.24)
$$\Delta n_T = x\,\Delta n_c + (1-x)\,\Delta n_a$$

where x is the fraction of crystallinity and Δn_c and Δn_a are birefringence values for the crystalline and amorphous portions, respectively. As seen in Figure 2.30, the amorphous portion of the polypropylene fiber continues to orient, or increase in anisotropy or birefringence, during drawing. Usually birefringence is measured with light polarized at $45°$ ($\pi/4$ rad) to the primary direction of orientation. The order of intereference is obtained by counting the number of color bands per unit dimension. Thick samples or high refractive index (above air) and highly anisotropic samples yield strongest birefringence. Polystyrene exhibits an exceptional degree of birefringence, with strain

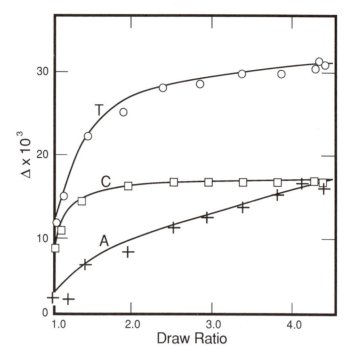

Figure 2.30.   Polypropylene Fiber Birefringence as a Function of Draw Ratio. A: Amorphous Contribution. C: Crystalline Contribution. T: Total Contribution (= A + C) (49).

effects easily seen in collimated white light. Birefringence is an important tool in accurately determining the degree of residual stress in parts molded of transparent polymers.

Dilatometry

One of the most accurate ways of determining the rate of isothermal bulk crystallization is by dilatometry. A *dilatometer* is an analytical instrument for measuring small changes in specific volume of a material. For isothermal crystallization measurements, the dilatometer bulb containing the polymer sample is first heated to a temperature far above the polymer melting point (at least 50°C). After thermal equilibrium (10 min or so), the dilatometer bulb is quenched in a second isothermal bath. The height of the mercury column atop the sample is then measured with time. The change in sample volume is directly related to the polymer specific volume and thus the degree of crystallinity of the polymer. The specific volume change for isothermal crystallization of polyethylene is shown in Figure 2.31. Note that crystallization does not begin immediately upon quenching. The time prior to onset of crystallization is called the *crystallization induction time* and is highly temperature sensitive as noted in Table 2.8. Furthermore the rate of crystallization as given by the slope of the specific volume-

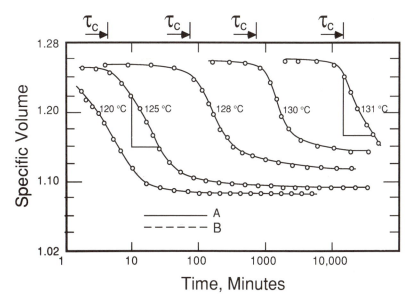

Figure 2.31.   Effect of Isothermal Crystallization on Specific Volume of Polyethylene. Crystallization Begins at Time, τ_c. A and Solid Horizontal Line: Unit Crystal Density at 131°C. B and Dashed Horizontal Line: Unit Crystal Density at 120°C (50).

time curve, is also temperature dependent. And finally, note that this polymer does not achieve 100% crystallinity as shown by the theoretical specific volumes of the ideal polyethylene crystal at 120 and 131°C. In fact, the asymptotic (ultimate or infinite-time) crystallinity is dependent upon the isothermal crystallization temperature.

Practical Morphology

In practical molding practice, the isothermal condition is almost never encountered.*
Typically, a relatively thick molten polymer layer contacts a cold metallic surface. The polymer at the interface is quenched far below its equilibrium crystallization temperature but the polymer at some distance into the polymer layer is cooled quite slowly. As a result, the morphology of the polymer at the surface can be substantially different from that in its interior. For polyoxymethylene (POM) and high-density polyethylene (HDPE), the lamellar structure at the interface is oriented perpendicular to both the surface and direction of flow (Figure 2.32). Just inside this layer is a "transcrystalline" layer, with the lamellae forming twisted ribbons, with orientation still perpendicular

---

*An exception to this can occur during blown film forming, where, for a short time, the rate of heat liberated by crystallization is balanced, for the most part, by the rate of heat removal to the polymer surroundings (see Section 5.2).

Table 2.8. Isothermal crystallization induction time,
τ_c, for high-density polyethylene (51).

Temperature, T_o (°C)	Induction time, τ_c (min)
120	—
125	2.4
128	45
130	660
131	900

to the surface. Toward the center of the polymer layer, the cooling rate is slow enough to allow spherulitic growth. The thickness of these various crystallite layers depends upon the relative magnitudes of the crystallization rate and the cooling rate. For relatively slowly crystallizing polymers such as polyethylene terephthalate (PET), the material at the polymer-metal interface is quenched so rapidly that essentially no crystallization occurs. Thus thick sections of PET can have *four* distinct types of morphology.

It is apparent that if the rate of cooling from the melt state is sufficiently fast relative to the rate of crystallization, the polymer at room temperature can have negligible crystallinity *even though it has a strong tendency to crystallize*. PET is a classic example of a polymer that is produced as an amorphous film by rapid quenching. Polyvinylidene chloride (PVDC or "Saran"), polyphenylene sulfide (PPS), and polyetherether ketone (PEEK) are others. Rapid quenching of a polymer with a high rate of crystallization such as HDPE does not result in a totally amorphous film. Instead it results in a film with substantially reduced haze. Haze is caused by the presence of spherulites typically 400 to 10,000 Å (40 to 1000 nm) in size. Particles of this size interfere with the transmission of light. Quenching inhibits spherulitic growth to this level, thus minimizing haze. "Shrink wrap" is another application of controlled crystallization. When film in which crystallization has been inhibited is heated, crystallization can occur. The result is a decrease in specific volume or reduction in area. The film shrinks onto whatever it is wrapped around.*

Rapidly crystallizing polymers do not always form spherulitic structures during cooling. If a polymer is allowed to crystallize in a very high shear field, the crystallites align in extended chain fashion. The classic example of high shear crystallization is uniaxial fiber formation. As anticipated, the degree of crystal orientation is a strong function of the draw ratio (essentially the degree of reduction of the fiber cross sectional area). This is quantified with X-ray analysis (Figure 2.27). In most bulk molding processes such as injection molding, blow molding, and extrusion, the shear field is not as intense as in fiber forming. As a result, the resulting crystallite structure is a

---

*Not all shrinkage is due to crystallization, however. Frequently film is highly biaxially oriented and this orientation is "set" or held in place by rapid cooling. When the film is reheated, the polymer is annealed and relaxation of residual stress causes a reduction in area.

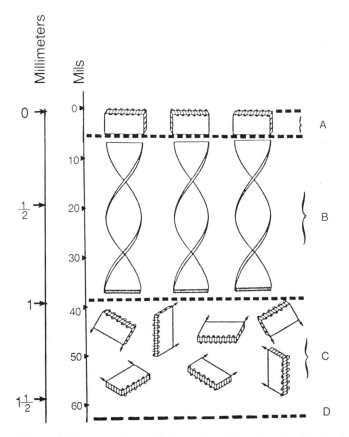

Figure 2.32.   Schematic Representation of Polymer Morphology in a Rapidly Cooled Crystalline Polymer-Injection-Molded Acetal (POM). T_{melt} = 205°C. T_{mold} = 30°C. A: Skin Region. B: Transcrystalline Region. C: Spherulitic Core. D: Sample Centerline (52).

combination of spherulitic growth and extended chain crystallization. The structure is called "shishkebab," as seen in Figure 2.33.

With certain polymers, spherulitic growth is not always the preferred crystalline state even at very slow cooling rates. Some lamellae tend to form very high aspect-ratio structures that easily twin or split into more than one growing front. The resulting structure is called "dendritic" and the structure resembles an evergreen tree. Dendrites usually grow very rapidly, even though the rate of crystallization as measured by dilatometric means is relatively slow. As a result, the dendrites rapidly surround and immobilize large segments of unordered molecules. Increasing cooling rate results in increasing amounts of amorphous materials trapped within the dendritic structure (Figure 2.34). Polymers with dendritic structures tend to have properties that are relatively insensitive to cooling rate.

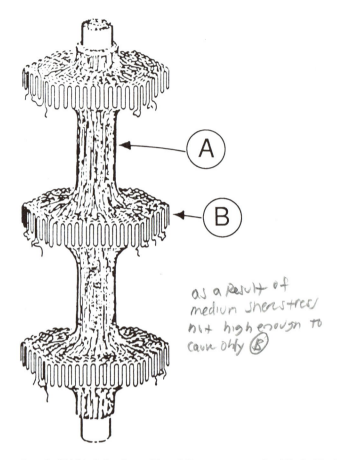

*as a Result of
medium shearstress
not high enough to
cause only Ⓑ*

Figure 2.33. Schematic of Shishkebab Crystalline Microstructure. A: Fibril Nucleus.
B: Folded Chain Lamella (53).

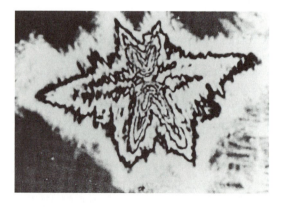

Figure 2.34. Typical Dendritic Crystalline Structure Formed by Rapid Crystallization of Polyethylene from Stirred Xylene Solution (70°C) (54).

Crystallization Kinetics

The specific volume data obtained from isothermal dilatometric measurements during crystallization can be replotted as the ratio of the instant volume change, ΔV, to the volume change measured if *isothermal* crystallization is allowed to continue "forever", ΔV_∞. An example of this for polypropylyene is given as Figure 2.35. If the data are replotted as $-\ln(1 - \Delta V/\Delta V_\infty)$ against the time (Figure 2.36), a straight line results for early times. The equation that results is given as:

$$(2.25) \qquad -\ln\left(\frac{1 - \Delta V}{\Delta V_\infty}\right) = A\, t^n$$

If $\phi = \Delta V$ and $\phi_\infty = \Delta V_\infty$, this equation can be rewritten as:

$$(2.26) \qquad \frac{\phi}{\phi_\infty} = 1 - \exp(-A\, t^n)$$

Avrami obtained this expression by careful theoretical analysis of the nature of homogeneous and heterogeneous nucleation and the geometry of the growing crystal (57–60). The rate constant A includes nucleation parameters and mechanistic nature of growth of the polymer. The exponent n is related to the geometry of the growing crystal. The Avrami constants are usually experimentally determined for a given polymer. Some typical values are given in Table 2.9. Although theoretical values for n are

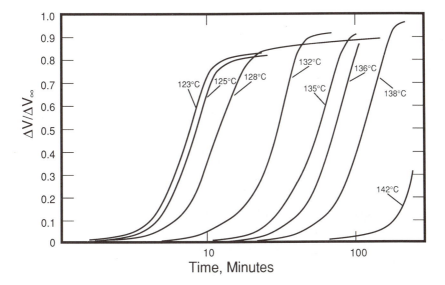

Figure 2.35.   Effect of Temperature on Specific Volume for Isothermal Crystallization of Polypropylene (59).

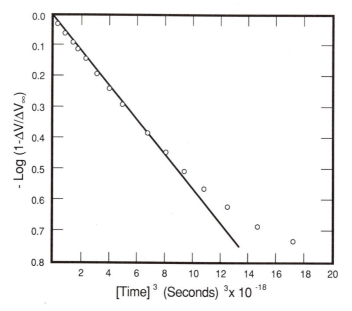

Figure 2.36.   An Avrami-Type Replot of the Specific Volume Change Data of Figure 2.35. Isothermal Temperature is 128°C. Linearity is Shown For Time³ and Short Times (60).

always integers, experimental values almost never are, owing to the mixture of types of crystals that are growing simultaneously.

Example 2.3

Q: Plot the Avrami time-dependency of ϕ/ϕ_∞ for $A = 10$, $n = 3$.
A: The equation is:

$$\frac{\phi}{\phi_\infty} = 1 - \exp[-10\,\tau^3]$$

It is best plotted on semilogarithmic paper, as seen in Figure 2.37.

Refer again to Figures 2.35 and 2.36. If the mechanism of crystallization remained constant, that is, A and n were constant, the slope of the curve would remain constant until the curve intersected the $\Delta V/\Delta V_\infty = 1$ line. That does not happen. The time where the slope of the curve changes correlates well with the time where growing spherulites impinge and secondary crystallization begins. Typically, the slope changes from a value of $n = 3$ to 4 to a value of $n = 1$ or so. Note that the change in the value for n also theoretically indicates that the nature of crystal growth is changing from three-dimensional spherulitic to one-dimensional rod-like (Table 2.9). The Avrami equation continues to predict volumetric change even with this adjusted slope. These data support the interpretation of DSC crystallization data, Figure 2.16, given in Section 2.4.

Table 2.9.   Values for Avrami constant, A, and exponent, n, for various types of polymer crystal nucleation and growth mechanisms (61).

Shape of entity	Avrami constant, A		Avrami exponent, n	
	Instantaneous	Sporadic	Instantaneous	Sporadic
Rod	$k_0 k_1 A^*$	$(1/2)kk_1 A^*$	1	2
Disk	$k_0 k_2^2 \pi d$	$(1/3)kk_2^2 \pi d$	2	3
Sphere	$(4/3)k_0 k_3^2 \pi$	$(4/3)kk_3^3 \pi$	3	4

A^* = Rod cross sectional area
d = Disk diameter
k = Linear increase in nuclei with time, growth rate proportional to nuclei density
k_0 = Constant concentration of nuclei, growth rate proportional to nuclei density
k_1 = Volumetric proportionality with growth rate, rods
k_2 = Volumetric proportionality with growth rate, disks
k_3 = Volumetric proportionality with growth rate, spheres

Example 2.4

Q: Repeat Example 2.3, with the Avrami exponent becoming $n = 1$ when $\phi/\phi_\infty = 0.6$.
A: The results are shown on Figure 2.37, together with the results of Example 2.3.

The Avrami constant A is highly temperature sensitive, as seen for chlorinated polyether in Figure 2.38. Recall that the value of A is a measure of the rate of crystallization. Large values of A represent rapid crystallization. For chlorinated polyether, A

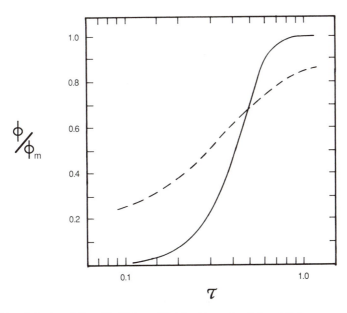

Figure 2.37.   A Reduced Specific Volume Plot for Examples 2.3 and 2.4.

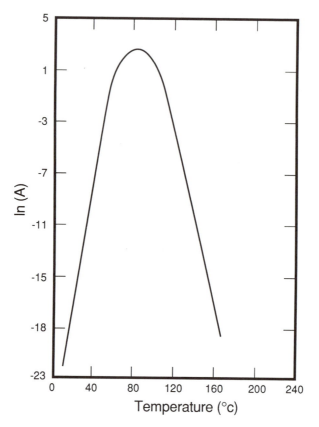

Figure 2.38.   Experimental Avrami Constant, A, for Chlorinated Polyether (62).

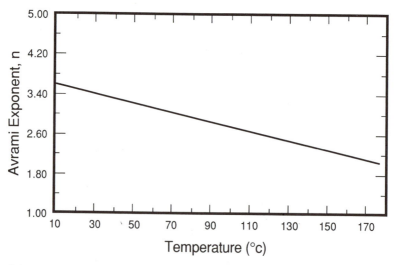

Figure 2.39.   Experimental Avrami Exponent, *n*, for Chlorinated Polyether (62).

is negligible above 150°C and below 40°C. The maximum value of A occurs at about 92°C. All crystallizing polymers show similar temperature dependencies for A.

The exponent n appears to be monotonic with temperature (Figure 2.39), but keep in mind that crystallization time is raised to the power of this exponent. A small change in the value of n can result in substantial changes in the rate of crystallization. It was noted earlier that polymers do not begin measurable crystallization the instant the polymer temperature is lowered into the crystallizing zone. The delay or *induction time* is attributed to molecular ordering processes that are highly temperature-dependent. As seen in Figure 2.40 for chlorinated polyether, the induction time shows a dramatic minimum at 92°C, a value between the glass transition temperature and the melt temperature. This minimum nearly always coincides with maximum value for the Avrami constant A.

Example 2.5

Q: The "quench time" for a crystallizable polymer is the time below which the cooled polymer remains amorphous. From Figure 2.40, determine the quench time for chlorinated polyether.

A: From Figure 2.40, the induction time can be empirically curve-fit with:

$$\ln \tau = 9.51 - 0.252\,T + 0.001346\,T^2$$

where τ is the induction time in min and T is the polymer temperature, °C. The temperature at which the induction time is minimum can be found by differentiating this expression and setting the derivative to zero. This yields $T = 93.6°C$. The minimum induction time at this temperature is:

$$\ln \tau_{T=93.6} = -2.28$$

$$\tau = 0.1017 \text{ min} = 6.1 \text{ s}$$

Frequently, the isothermal *half-time* of crystallization for crystalline polymers is reported. This is simply the time required for the value of the differential specific volume of the polymer to reach 50% of its value at infinite time:

(2.27)
$$\Delta V = 0.5\,\Delta V_\infty \text{ or}$$

(2.28)
$$t_{1/2} = \left(\frac{\ln 2}{A}\right)^{1/n}$$

Note that this time should always be measured from crystallization onset. The isothermal half-time has a minimum value exactly at the temperature where the crystallization rate is maximum and the induction time is minimum (63). The half-time value is Gaussian-symmetric about the minimum value:

(2.29)
$$t_{1/2}(T) = t_{1/2,min}\,\exp\left[\frac{4\,\ln 2\,(T - T_{max})^2}{D^2}\right]$$

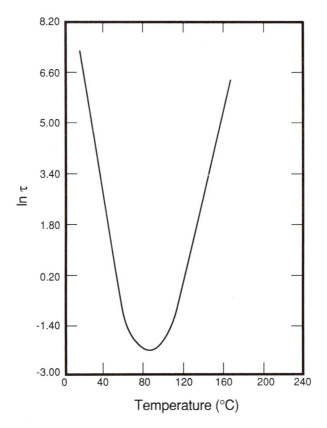

Figure 2.40. Experimental Isothermal Induction Time, τ, as a Function of Temperature for Chlorinated Polyether (62).

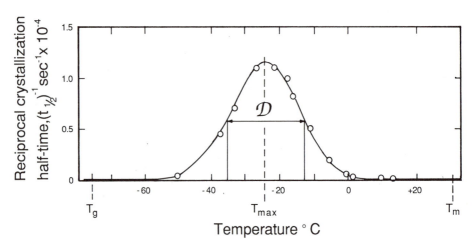

Figure 2.41. Experimental Verification of Gaussian-Type Temperature-Dependent Crystallization Half-Time. Material: Isotropic Natural Rubber (64).

Table 2.10. Temperatures for maximum crystallization rate and values for D, half-width of the rate curve for several polymers (65).

Polymer	Temperature (°C)			$t_{1/2}$, min (s)	K_{max}, (s^{-1})	D (°C)	G (°C/s)
	T_m	T_g	T_{max}				
Natural rubber	30	−75	−24	5000	1.4×10^{-4}	23	0.0034
Polypropylene (isotactic)	180	−20	65	1.25	0.55	60	35.0
Polyethylene terephthalate	267	67	190	42.0	0.0162	64	1.1
Nylon 6	228	45	146	5.0	0.14	46	6.8
Nylon 66	264	45	150	0.42	1.64	80	139
Polystyrene (isotactic)	240	100	170	185	3.7×10^{-3}	40	0.16
Polyethylene succinate,							
$M = 820$	83		40	360	1.9×10^{-3}	60	0.12
$M = 4415$	106.5		30	2490	2.8×10^{-4}	35	0.01

T_{max} is the temperature at maximum crystallization rate ($t_{1/2,min}$) and D is the half-width of the rate curve (Figure 2.41). Typical values for $t_{1/2,min}$, T_{max} and D are given in Table 2.10. The *effective* crystallization temperature ranges for many materials are only 20 to 40°C ($\pm$ 10 to 20° of T_{max}), even though crystallization could occur across the entire temperature range of T_g to T_m (as much as 200°C).

Another measure of the isothermal rate of crystallization is the polymer half-time value at 30°C *below* its melt temperature. As seen in Table 2.11, polyethylene has a very high rate whereas polyethylene terephthalate crystallizes very slowly at this temperature.

The *hot stage microscope* can also be used to observe spherulitic growth rates. The effects of changes in polymer characteristics such as molecular weight, or the nature of the nucleant such as the addition of a highly active inorganic nucleant, can be measured via hot stage microscopy. The effect of molecular weight on the growth rate

Table 2.11. The isothermal rates of crystallization for several polymers at temperatures 30°C *below* their reported melt temperatures (66).

Polymer	Crystallization rate (µm/min)
Polyethylene	5000
Polyhexamethylene adipamide (PA 66)	1200
Polyoxymethylene (POM)	400
Polycaprolactam (PA 6)	150
Polytrifluorochloroethylene (PTFE)	30
Isotactic polypropylene (PP)	20
Polyethylene terephthalate (PET)	10
Isotactic polystyrene (PS)	0.25
Polyvinyl chloride (PVC)	0.01

of one polymer is seen in Figure 2.42. As expected, the longer the polymer molecule becomes, the more resistance it offers to crystallizing forces and the slower the rate of crystallization becomes. Maximum crystallite growth rates for several polymers are listed in Table 2.12.

Keep in mind that even though the product made from a crystalline polymer is used at temperatures far below the maximum crystallite growth rate temperature, crystallization can continue. So long as the polymer molecule is mobile, crystallization can proceed, albeit at an extremely slow rate. Polymers apparently cannot crystallize at temperatures below their glass transition temperatures. But polyethylenes and polypropylene, as examples, have glass transition temperatures far below room temperature. Continuing crystallization means continuing decrease in specific volume. Shrinkage and warpage are directly related to local differential changes in specific volume. In many instances, crystallization is deliberately continued by reheating a fixtured molded crystalline polymer part to a temperature above its glass transition temperature (and usually to its maximum crystallite growth rate temperature, Table 2.12) and holding it there for several minutes to several hours. This is called *annealing*.

Example 2.6

Q: Assume a part is molded from nylon 6 (PA-6) resin. The part has a maximum wall thickness of 6 mm. In order to anneal this part in a forced-air convection oven, what temperature should be used and how long should it be annealed?

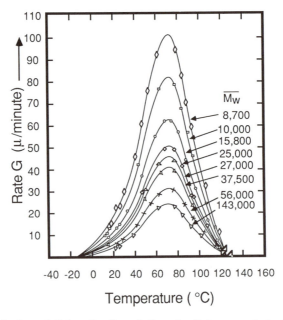

Figure 2.42.   Isothermal Spherulite Growth Rate for Polytetramethyl-p-Phenylene Siloxane (67).

Table 2.12. Maximum crystallite growth rates for several polymers (68).

Polymer	Average mol. wt.	T_m (°C)	T_g (°K)	$T_R{}^1$ (°C)	G_{max} (μm/min)	T_R/T_m
Isotactic polystyrene	190,000	523	–	347	6.0	0.66
	1,380,000	523	–	347	1.5	0.66
	Not given	523	–	341	0.03	0.65
	185,000	523	–	343	0.2	0.66
Polyoxymethylene	40,000	453	203	361	400	0.80
Nylon 66	17,200	538	333	413	—	0.77
Nylon 6	24,700	500	300	411	120	0.82
Polytetramethyl-p-Phenylene siloxane	1,400,000	423	258	338	21	0.80
Nylon 56	Not given	539	318	458	200	0.85
Nylon 96	Not given	519	318	452	130	0.87
1-Polypropylene oxide	10,300	348	—	285	50	0.82
Polythylene succinate	1,500	370	—	328	41	0.89
	2,700	376	—	328	23	0.87

[1]Temperature of maximum growth rate

A: From Table 2.12, the maximum crystalline growth rate for nylon 6 occurs at 411°K or 138°C. For a small forced convection oven, the minimum annealing time can be obtained from an empirical relationship based upon the annealing time of one hour for a 3 mm thick section. The relationship, based upon the transient temperature response of a plate, is proportional to thickness squared, i.e.:

$$t = (\text{maximum part wall thickness}/3 \text{ mm})^2 \text{ (in h)}$$

For the 6 mm part, the annealing time is estimated to be 4 h.

Ziabicki (69) shows that if a polymer is cooled from T_m to T_g, a measure of the degree of *nonisothermal crystallinity* can be obtained by integration:

$$(2.30) \qquad \int_{T_g}^{T_m} \frac{dT}{[t_{1/2}(T)]} \approx 1.064 \frac{D}{t_{1/2, \, max}} = G$$

G is the "Ziabicki kinetic crystallizability parameter" of a polymer (Table 2.10). When comparing the nonisothermal crystallization potentials of two polymers cooling at the same rate, the one with the larger value of G will tend to crystallize to a higher level (ϕ/ϕ_∞).

Much of the above discussion assumes that crystallization occurs isothermally. In polymer processing, this can occur only when the rate of heat removal exactly balances the heat of crystallization. There is some evidence that this is at least partially correct for certain operating conditions in fiber and blown film extrusion (70,71). A compar-

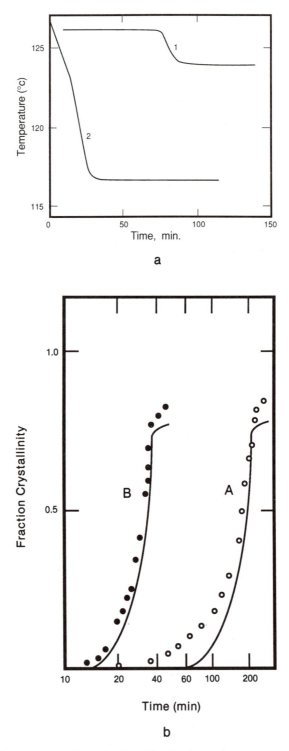

Figure 2.43. Non-Isothermal Crystallization Kinetics for HDPE. a: Cooling Curves for Two Quenching Profiles (1 and 2). b: Time-Dependent Crystallinity for the Two Profiles. Solid Lines Represent Non-Isothermal Ziabicki Kinetic Model (72).

ison of experimentally determined nonisothermal crystallization levels with the iso-thermal Avrami model shows reasonable agreement (Figure 2.43). Thus, as a first approximation, the simple differential form for the Avrami equation can be used for nonisothermal crystallization kinetics.

Molecular Order: A Summary

The engineer and designer must keep in mind that a part injection molded from a crystalline polymer is in essence a molecular composite. That is, the molecular struc-ture of the surface of the part is different from that at the part centerline, sometimes decidedly so. The nature of the structure is dependent on the type of plastic and the molding or fabricating conditions. The conformational nature of the polymer can change with its method of manufacture due for example to catalyst concentration and type and/or nature of the solvent, and with the nature of the adducts such as additive packages for specialized applications such as ultraviolet protection or fire retardancy, filler characteristics and coupling agents. Furthermore the molding conditions can change (cavity filling rate, mold cooling rate) even during each process cycle. This implies that the complex molecular structure may change dramatically throughout the part. There are many verifications of this in the polymer literature (73,74) for unfilled or "neat" and filled and/or reinforced crystalline polymers.

Thus, even though the molded part is a complex composite of macroscopic materials including polymer, adducts, fillers, reinforcements and so on, it is also a complex composition on a molecular level. The latter is most vexing, since only pathological tests reveal the nature of the composite. And many of these tests require extensive effort, sometimes calling for exotic equipment such as electron microscopes and X-ray diffraction devices. Usually the DSC and the mechanical spectrometer can be used to deduce the general nature of the molecular structure, such as the degree of crystallinity or role of the adducts on phase transition, but not the specific nature such as rate of crystallization or type of crystallinity.

2.5 Polymeric Transitions

In Section 2.4, it was noted that the crystallization/melting phenomenon of polymers is accompanied by a near-step change in specific volume at a given temperature. This phase change is called a first-order thermodynamic transition and is easily observed and measured using differential scanning calorimetry (DSC). It was also noted that DSC is used to detect second-order thermodynamic transitions such as glass transition temperature, T_g. In this section, the typical polymeric transitions are reviewed in detail.

It is well known that all matter exists in one of three states: gaseous, liquid, or solid. The relationship between states or phases and temperature and pressure is usually dis-played as a *phase diagram*. The phase diagram for water, shown in Figure 2.44, is typical for most simple molecules. Note that at atmospheric pressure, water at $-10°C$, point 1, exists as a solid. As the solid is heated, its molecular activity increases until at 0°C, point 2, it melts. The solid–liquid *phase transition* requires a fixed amount of energy per unit mass. This energy is called the *heat of fusion* for simple molecules. When the liquid water temperature reaches 100°C, point 3, another phase transition

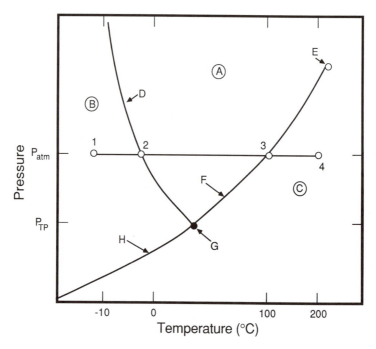

Figure 2.44.   Phase Diagram for Water. A: Liquid Phase. B: Solid Phase. C: Vapor Phase. D: Fusion Line. E: Critical Point. F: Vaporization Line. G: Triple Point. H: Sublimation Line.

occurs, this time from liquid to vapor or gas. The energy associated with this phase change is called the *heat of vaporization*. With continued heating, point 4, the water vapor temperature continues to increase, with no subsequent phase change. Another characteristic of simple molecules is the *triple point,* below which the material changes from solid directly to vapor, and the process is called sublimation. And another is the *critical point,* above which liquid and vapor phases are indistinct, with no heat of vaporization.

Polymer molecules and in fact most long chain oligomeric molecules have only solid and liquid phases. As the temperature of a polymeric liquid increases, it begins to decompose, degrade or depolymerize. The polymeric vapor state does not exist. The extreme lengths of polymer chains act to inhibit thermodynamic transitions. Furthermore, shorter polymer chains are more mobile than longer ones and so react to forces in a much more rapid fashion. These elements act to broaden phase transition regions. A typical isobaric or constant pressure phase diagram for a narrow molecular weight linear amorphous polymer is given as Figure 2.45. At very low molecular weight, M_1, the polymer acts like a simple material. It is essentially solid below T_1 and liquid above T_1. A vapor phase could theoretically exist at very high temperatures if the polymer did not degrade or decompose. As the molecular weight is increased, the definitions of solid and liquid begin to blur. The *glass transition* represents the point above which short-range chain mobility can occur. By short range, it is meant that chain segments of up to 50 repeat units can act in concert to rotate, vibrate and reptate. A more complete discussion is given below. The polymer above the glass transition is rubbery

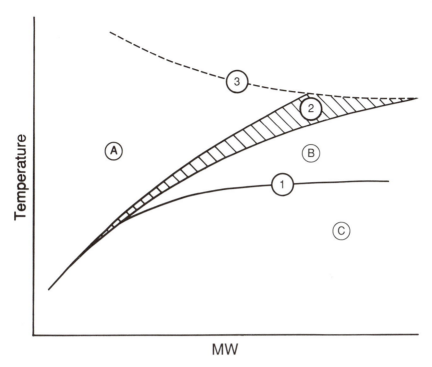

Figure 2.45. Phase Diagram for a Simple Amorphous Linear Polymer at Atmospheric Pressure. Region A: Viscous Liquid. Region B: Rubbery Phase. Region C: Glassy Phase. 1: Glass Transition Line. 2: Diffuse Transition Zone. 3: Thermal Decomposition Line (75).

solid and below, it is brittle solid. As the polymer temperature is increased above the glass transition temperature, T_g, the polymer becomes less and less solid and more and more fluid. There is some evidence that an amorphous polymer becomes a true liquid (albeit an elastic liquid) at a poorly defined transition known as the *softening point*. The softening point is probably the point at which the stiffest elements of the polymer chain structure or the highest molecular weight chains begin to rotate, vibrate and reptate.

If the polymer is crystalline, the crystalline melting point must be included in the phase diagram (Figure 2.46). Most crystalline polymers are in reality only semicrystalline. Polypropylene is at best 40% crystalline and polyethylene terephthalate is at best 60% crystalline, whereas polytetrafluoroethylene (PTFE) is 90 to 95% crystalline. As the temperature of a solid crystalline polymer is increased, the non-crystalline portion undergoes the glassy solid–rubbery solid glass transition. The crystalline portion of the polymer undergoes no transition. So the polymer retains considerable mechanical strength and in fact frequently gains in toughness.* At a high level of crystallinity, the

---

*Toughness is usually defined as the area under the stress–strain curve. The initial slope of the stress–strain curve is sometimes called the "Modulus of Toughness". This is discussed in Section 4.9.

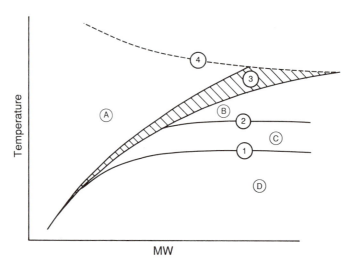

Figure 2.46.   Phase Diagram for a Simple Semicrystalline Linear Polymer at Atmospheric Pressure. Region A: Viscous Liquid. Region B: Rubbery Phase. Region C: Rubbery–Crystalline or Leathery Phase. Region D: Rigid–Crystalline Phase. 1: Glass Transition Line. 2: Crystalline Melting Line. 3: Diffuse Transition Zone. 4: Thermal Decomposition Line (75).

effect of amorphous glass transition on mechanical properties may be negligible. The melting temperature signals the relatively abrupt destruction of the crystalline phase. This usually results in a rubbery amorphous gel with relatively little mechanical strength. Continued heating results in a relatively gradual shift to an elastic liquid.

The Glass Transition Temperature

As noted, the second-order thermodynamic transition known as the glass transition is marked by a change in the polymer character from a brittle solid to a rubbery state. Typically the modulus or stiffness of the polymer drops by a factor of a thousand or more.

Classical room temperature *glassy* polymers are:

Polystyrene (T_g = 100°C)

Polymethyl methacrylate (T_g = 105°C)

Amorphous polyethylene terephthalate (T_g = 70°C)

Those that are *rubbery* at room temperature include:

Silicone rubber (T_g = −123°C)

Polyisobutylene (T_g = −75°C)

Polyisoprene (natural rubber) (T_g = −73°C)

Polymers that are *rubbery–crystalline* at room temperature include:

High-density polyethylene ($T_g = -100°C$, $T_m = 125°C$)

Isotactic polypropylene ($T_g = -10°C$, $T_m = 165°C$)

Polymers that are *brittle–crystalline* at room temperature include:

Crystalline PET ($T_g = 70°C$, $T_m = 265°C$)

Polyphenylene sulfide ($T_g = 88°C$, $T_m = 290°C$)

Glass transition and crystalline melting temperatures for many other polymers are given in Tables 2.13 and 2.14.

Table 2.13. Glass transition temperatures and melt temperatures for several amorphous homopolymers, copolymers and blends (9,11,15,18,75).

Polymer	T_g (°C)	T_m^1 (°C)
Homopolymers		
Cellulosics		
Cellulose acetate (CA)	50	230
Cellulose acetate butyrate (CAB)	0	140
Cellulose acetate propionate (CAP)	20	190
Polymethyl methacrylate (PMMA)	90 to 105	160
Polyamide-imide (PAI)	275	–
Polyaryl ester	190	–
Polyaryl ether	160	–
Polycarbonate (PC)	150	220
Polystyrene (PS)	100	240
Polyurethane (PUR)	–20	–
Polyvinyl chloride (PVC)	87 to 135	212
Polysulfones		
Polysulfone	200 to 150	–
Polyaryl sulfone	205	–
Polyether sulfone	230	–
Polyphenyl sulfone	220 to 230	–
Copolymers and blends		
Acrylonitrile-butadiene-styrene (ABS)	80 to 125	190
Polynitrile	104	317
Polyphenylene oxide (modified) (mPPO)	104 to 120	–
Styrene-acrylonitrile (SAN)	115 to 125	–
Vinyls		
Vinyl chloride-Vinyl acetate	75 to 105	–
Vinyl chloride-ethylene	10 to 27	
Vinyl chloride-propylene	35 to 50	

[1]Melting temperature of pure crystal only. Polymer is considered to be amorphous.

Table 2.14. Glass transition temperatures and melt temperatures for several crystalline tendency homopolymers, copolymers and blends (9,11,15,18,75).

Polymer	T_g (°C)	T_m (°C)
Homopolymers		
polyacetal, polyoxymethylene (POM)	−50	181
Fluoropolymers		
Polytetrafluoroethylene (PTFE)	−33, 126	327
Polychlorotrifluoroethylene (PCTFE)	45	220
Ionomer	−45	90 to 96
Nylons		
PA-6	50	220
PA-66	50	265
PA-610	40	227
PA-612	~30	217
PA-11	46	194
PA-12	37	179
Polyacrylonitrile (PAN)	104,130	313
Polybutylene (PB)	−25	128
Polyesters		
Polybutylene terephthalate (PBT)	~50	240
Polyethylene terephthalate (PET)	69	267
Polyethylene (PE)	−125, −33	110 to 140
Polyethylene oxide (PEOX)	−56	66
Polyphenylene sulfide (PPS)	88	290
Polypropylene (PP)	−10, −20	176
Polyvinylidene chloride (PVDC)	−17	198
Polyvinylidene fluoride (PVDF)	−35	156
Copolymers and blends		
Polyacetal–polyethylene	~40	175
Polyethylene vinyl acetate (EVA)	−42	65 to 90
Fluoropolymers		
FEP	11	275
ECTFE	~0	245
ETFE	~20	270

As seen in Figure 2.7, the glass transition temperature for atactic polypropylene is −18°C. Since it is an amorphous polymer, there is no further phase change with increasing temperature. The glass transition temperature for isotactic polypropylene homopolymer is also about −18°C.* Below T_g, both stereoisomers are brittle solid. Above T_g, the atactic polymer is rubbery and the isotactic is rubbery–crystalline. Above T_m, the isotactic polymer is a highly elastic liquid. One process, thermoforming, depends upon the deformability of the isotactic polypropylene polymer just below

---

*Actually a range from −20 to −10°C is more appropriate.

T_m, in its rubbery–crystalline region. In a similar fashion, stretch blow molding of slightly crystalline polyethylene terephthalate takes place just above its T_g. If the stretch-blown product is annealed at about 140°C (very near its maximum crystallizing rate temperature), the final product is said to be "heat set" and capable of withstanding sterilizing conditions of about 100°C or so without distortion.

The glass transition was described as being the point where short-range chain mobility occurred. Rosen (76) categorizes polymer molecular motions as:

1. Translational motion (sliding, reptation) of entire molecules. This is called flow or permanent deformation.
2. Cooperative wriggling and jumping of segments of molecules approximately 40 to 50 carbon atoms in length. This permits chain flexing and uncoiling. High temperature creep and second stage creep are also thought to be examples of this type of motion.
3. Motion of a few (5 to 10) atoms along the main chain or rotation and vibration of side chains and pendant groups.
4. Vibrations of atoms about equilibrium positions. This is similar to crystal lattice vibration except that there is no precise location of the atoms in amorphous polymers and only an imperfect one in crystalline polymers.

Below the glass transition temperature, the only polymer motions are types 3 and 4. Just above, type 2 becomes important and in the liquid state, type 1 dominates. Note that the very imperfect nature of the polymer structure allows for some void around the polymer chains. This void or *free volume* is about 2.5% for all thermoplastics at T_g. As the polymer is heated, the free volume increases and the increased space allows for more molecular motion. If an external stress is applied to the polymer in this state, the molecular motion will be such as to relieve the stress. This is the onset of the rubbery phase of the polymer.

The most important factor that determines the glass transition temperature of a polymer is backbone flexibility. As seen in Table 2.15, polymers with flexible backbones have low glass transition temperature values and those with rigid backbones have high values. Steric hindrance caused by the location or size of a pendant group can also lead to increased glass transition temperature, Table 2.16. Polyethylene has no pendant group and its glass transition temperature is about −120°C. The large phenyl group on polystyrene causes substantial steric hindrance and a glass transition temperature of about 100°C obtains. Molecular asymmetry yields increased T_g, as seen by comparing values for symmetric polyvinylidene chloride and unsymmetric polyvinyl chloride in Table 2.17. Increasing polarity yields increased T_g as well, as shown in Table 2.18. This is seen by comparing the T_g of −10°C for nonpolar polypropylene with a value of 103°C for the very polar polyacrylonitrile. Additives and low molecular weight polymers can dramatically reduce the glass transition temperature of a polymer. The classic example of this is plasticized polyvinyl chloride (PVC). The glass transition temperature for unplasticized or rigid PVC (RPVC) is 87°C. At room temperature, RPVC is a brittle, hard solid. Small amounts of compatible low molecular weight additives such as dioctyl phthalate (DOP) or diisooctyl phthalate (DIOP) serve to "lubricate" the rigid molecules and to allow greater chain mobility (Figure 2.47). This

Table 2.15. The effect of backbone rigidity on glass transition temperatures of linear homopolymers (77,78).

Polymer[1]	Structure	T_g (°C)
Polyethylene	$\begin{bmatrix} \overset{\displaystyle H}{\underset{\displaystyle H}{\overset{\mid}{\underset{\mid}{C}}}} - \overset{\displaystyle H}{\underset{\displaystyle H}{\overset{\mid}{\underset{\mid}{C}}}} \end{bmatrix}$	−120
Polybutadiene	$\begin{bmatrix} \overset{H}{\underset{H}{\overset{\mid}{\underset{\mid}{C}}}} - \overset{H}{\overset{\mid}{C}} = \overset{H}{\overset{\mid}{C}} - \overset{H}{\underset{H}{\overset{\mid}{\underset{\mid}{C}}}} \end{bmatrix}$	−90
Polyoxymethylene	$\begin{bmatrix} \overset{H}{\underset{H}{\overset{\mid}{\underset{\mid}{C}}}} - O \end{bmatrix}$	−50[2] −85[3]
Polyvinyl formal	(structure)	105
Polycarbonate	(structure)	150

[1]Polymers listed in order of increasing backbone rigidity.
[2] Crystal segment.
[3]Amorphous segment.

Table 2.16. The effect of steric hindrance on glass transition temperatures of linear vinyl-type homopolymers (77,78).

Polymer	Structure	T_g (°C)
Polyethylene	$\left[\begin{array}{cc} H & H \\ | & | \\ -C & -C- \\ | & | \\ H & H \end{array}\right]$	−120
Polypropylene	$\left[\begin{array}{cc} H & H \\ | & | \\ -C & -C- \\ | & | \\ H & H-C-H \\ & | \\ & H \end{array}\right]$	−20
Polystyrene	$\left[\begin{array}{cc} H & H \\ | & | \\ -C & -C- \\ | & | \\ H & \bigcirc \end{array}\right]$	100

Table 2.17. The effect of side-group symmetry on glass transition temperatures of vinyl-type linear homopolymers (77,78).

Polymer	Structure	T_g (°C)
Polyvinyl chloride	$\left[\begin{array}{cc} H & H \\ | & | \\ -C & -C- \\ | & | \\ H & Cl \end{array}\right]$	87
Polyvinylidene chloride	$\left[\begin{array}{cc} H & Cl \\ | & | \\ -C & -C- \\ | & | \\ H & Cl \end{array}\right]$	−17

Table 2.18. The effect of polar side groups on glass transition temperatures of linear homopolymers (77,78).

Polymer	Polarity	Structure	T_g (°C)
Polypropylene	Non-polar	$\left[\begin{array}{cc} H & H \\ | & | \\ C & - & C \\ | & | \\ H & H-C-H \\ & | \\ & H \end{array}\right]$	−10
Polyvinyl chloride	Slightly	$\left[\begin{array}{cc} H & H \\ | & | \\ C & - & C \\ | & | \\ H & Cl \end{array}\right]$	87
Polyacrylonitrile	Very	$\left[\begin{array}{cc} H & H \\ | & | \\ C & - & C \\ | & | \\ H & C\equiv N \end{array}\right]$	103

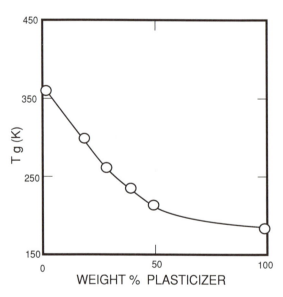

Figure 2.47. Effect of DOP Plasticizer on Glass Transition Temperature of Polyvinyl Chloride (PVC) (79).

lowers glass transition temperature value to $-20°C$ or more. Thus "flexible vinyl" or plasticized polyvinyl chloride (FPVC) remains in the rubbery phase even at $-10°C$. •

In Figures 2.45 and 2.46, it was noted that the glass transition temperature increases with increasing chain length. The molecular-weight dependency of the T_g of polystyrene is seen in Figure 2.48. This can be written as:

$$(2.31) \qquad T_g = T_{g,\infty} - \frac{K}{M_n}$$

where $T_{g,\infty}$ is the glass transition value extrapolated to infinite molecular weight, K is a material-specific constant and M_n is the number average molecular weight of the polymer. K for the polystyrene in Figure 2.48 is 2×10^5 °C g/g mol. For most commercial polymers, the degree of polymerization is in excess of 500 or so. As a result, the value for the measured glass transition temperature is about equal to that at infinite molecular weight.

The glass transition temperature is also related to the extent of free volume in the polymer. The amount of free volume decreases with increased applied pressure. Thus, the glass transition temperature for any polymer increases with increasing pressure. This is seen in Figure 2.49 for polystyrene. For this polymer, there is a linear relationship:

$$(2.32) \qquad T_g = 90°C + A\,P$$

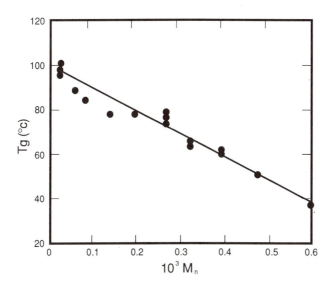

Figure 2.48. Number-Average Molecular Weight Effect on Glass Transition Temperature for Polystyrene. T_g in °C. (80).

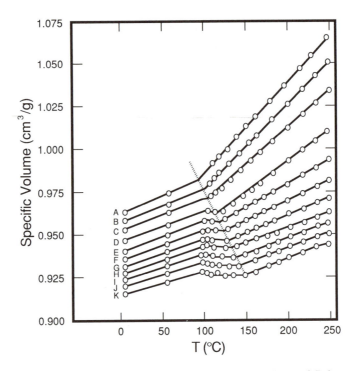

Figure 2.49. Pressure- and Temperature-Dependent Specific Volume of Polystyrene. A: 0 kg/cm^2 = Pressure. B: 200. C: 400. D: 600. E: 800. F: 1000. G: 1200. H: 1400. I: 1600. J: 1800. K: 2000 kg/cm^2 (81).

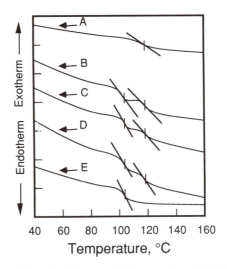

Figure 2.50. The Effect of Blending Polymethyl Methacrylate (PMMA) and Polystyrene (PS) on Glass Transition Temperatures, as Noted by DSC. A: 100% PMMA. B: 70% PMMA/30% PS. C: 60% PMMA/40% PS. D: 50% PMMA/50% PS. E: 100% PS.

where A $= 0.033$ (P in lb_f/in^2) or $= 1.595$ (P in MPa). At 40 MPa (5865 lb_f/in^2) applied pressure, $T_g = 146°C$, an increase of 65°C over room temperature values. Not all polymers exhibit this great an increase however.

The glass transition temperatures of homopolymers are preserved during blending or compounding if the polymers are immiscible. DSC can be used for identification, with all species exhibiting their respective changes in the slopes of the time-dependent energy curves, Figure 2.50. If the blend is miscible, as with polystyrene (PS) and polyphenylene oxide (PPO), the blend exhibits a single T_g, Figure 2.51.

For copolymers, the molecular structure is changed. This can affect the overall mobility of the polymer at any temperature and this in turn can change the location of the glass transition temperatures. In a long block copolymer, glass transition temperatures for both polymers may be evident but may be slightly displaced toward one another. In a very short block copolymer, only one T_g may be observed. In a random copolymer, only one glass transition occurs. There are two empirical relationships for determining the relative location of the random and short block copolymer transitions between the two homopolymer transitions:

(2.33)   Volume Fraction: $T_g = \phi_1 T_{g,1} + \phi_2 T_{g,2}$

(2.34)   Weight Fraction: $\dfrac{1}{T_g} = \dfrac{W_1}{T_{g,1}} + \dfrac{W_2}{T_{g,2}}$

where ϕ_i and W_i are the respective volume and weight fractions of the ith polymer. T_g is given in absolute temperature.

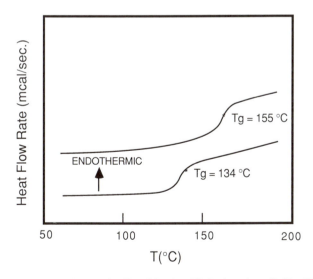

Figure 2.51.   Heating DSC Curves for Two Blends of Polyphenylene Oxide (PPO) and Polystyrene (PS). Note the Single T_g, Indicating a Thermodynamically Compatible Blend (82).

Example 2.7

Q: Consider two polymers of *identical* density. One is rubbery at room temperature with $T_g = -20°C$. The other is brittle solid with $T_g = 50°C$. Determine the volume fraction of minor component when the glass transition temperature of the copolymer is 27°C (about room temperature).

A: If the volume fraction model is valid, the copolymer glass transition temperature is 27°C (room temperature) when the volume fraction of the first is 0.33. If the weight fraction model is valid, room temperature glass transition occurs when the weight fraction of the first is 0.277. Since the densities of the two are equal, room temperature glass transition of a real copolymer probably occurs at a volume or weight fraction of between 0.28 and 0.33 of the first polymer.

Actual random copolymerization of polymers having this volume fraction range should be carried out to determine which model is appropriate.

When a polymer is crosslinked, its molecular motion is severely restricted. Thermosetting polymers are classified into two broad categories. The first are those resins that exhibit a glass transition, such as epoxide, urea, alkyd thermosetting polyesters and diallyl phthalates. However, the glass transition temperatures of these thermosets can

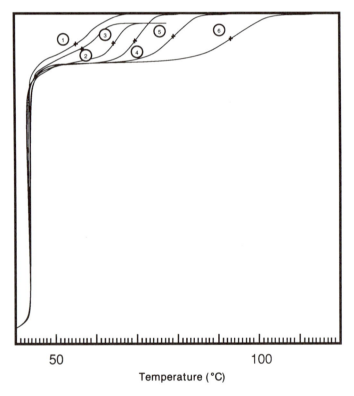

Temperature (°C)

Figure 2.52. The Effect of Curing Time at 140°C on Transitions in Epoxy Molding Compounds, from DSC Heating Curves. 1: Initial (Uncured). 2: Heating to 140°C and Immediately Quenching. 3: 1 Min Curing at 140°C. 4: 3 Min. 5: 7 Min. 6: 15 Min at 140°C (83).

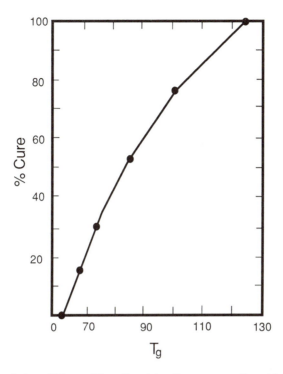

Figure 2.53.   Correlation of Epoxy Glass Transition Temperature, T_g, with Degree of Cure,
From Figure 2.52 (84).

be changed substantially by the composition of the polymer. Phenolic (PF), melamine
and silicone thermosetting polymers do not exhibit glass transition temperatures. These
resins are covalently bonded in such a manner that the molecule will degrade before it
reaches a region of substantial molecular mobility such as the glass transition temperature range.

Since many thermoset molding compounds can contain significant amounts of low
molecular weight materials, these compounds will exhibit a glass transition temperature
that is dependent on the degree of cure. The glass transition temperature increases with
increasing level of cure, as seen in Figures 2.52 and 2.53. Here, the effect of thermal
cure on the glass transition temperature of an epoxide is obtained by DSC. Again, the
DSC is an excellent analytical device for determining the degree of cure a thermosetting
resin has obtained during conventional processing and subsequent annealing.

The Practical Significance of
Polymer Transitions

As noted, a polymeric transition represents a change in the physical characteristics of
the polymer. These changes are very important when considering a polymer for a given
product application. As noted, polymers exist in either amorphous or crystalline states.

The designer must consider the nature of the state and the possible tendency for certain polymers, when amorphous, to crystallize during the lifetime of the product.

First, consider the amorphous polymer. The most significant transition from a design viewpoint is the glass transition temperature. As noted, above T_g, amorphous polymers are rubbery or flexible. Since amorphous polymers do not melt, increasing temperature above T_g implies increasing polymer flexibility until the polymer is fluid. Below the glass transition temperature, the polymer is rigid or hard. As noted, T_g is dependent upon the molecular weight of the polymer. Since polymers have relatively broad molecular weight distributions, the glass transition temperature is usually not a single temperature, unlike the freezing point or boiling point of simple molecules such as water. Instead, the glass transition temperature has a range in values of 5°C to as much as 20°C. Traditionally, the glass transition temperature listed in the literature is the mean value between the onset and completion of the transition. The onset of glass transition is the temperature where a portion of the polymer molecular structure loses rigidity. In certain part designs where temperature-dependent tolerance is important, the lowest value of T_g is needed. For example, if a part is under load, permanent deformation can occur even at relatively low stresses. A good example is PVC. Its listed T_g is 95°C but the upper use design temperature can be as low as 70°C. The listed T_g of PMMA is 105°C, 10 degrees higher than that for PVC, but PMMA can be designed to handle substantially higher temperatures owing to its narrower molecular weight distribution.

For polymers that can crystallize, there are (at least) two transition temperatures. As noted, the glass transition temperature has the same implication in terms of design as it does for amorphous polymers. The onset of the crystalline melting temperature represents the upper use temperature of a crystalline polymer. As noted, from a practical viewpoint, a crystalline polymer never achieves 100% crystallinity. For example, PE crystallizes to about 70 to 90%. Nylon 66 (PA-66) crystallizes to about 50% and PET to about 60%. Again, the portion of the polymer that does not crystallize remains in an amorphous state. However, usually the crystalline portion of the polymer is sufficiently stiff to allow the polymer to withstand applied stresses above T_g.

As seen in Figure 2.46, a polymer that can crystallize can be in one of three states. It can be a crystalline-glassy polymer, a crystalline–rubbery polymer, or a completely rubbery polymer (above T_m, the melt temperature). Typically, these polymers are used in either of the first two states. A product designed of a polymer in the first state will be hard and brittle. Nylon 66 at room temperature is a crystalline–glassy polymer. Its T_g is 50°C.*

A product designed of a polymer in the second state will be hard but ductile. For example, polyethylene at room temperature is a crystalline-rubbery polymer. Its T_g is about -110°C. Care must be taken in designing very near T_g. For example, products of PP homopolymer are hard–ductile from room temperature to 150°C (near the melting temperature). But near its -10°C T_g, PP becomes hard–brittle and its impact strength

---

*Note that if a part fabricated of nylon 6 is designed to operate at temperatures above 50°C, the polymer will be in the second, rigid-rubbery state and will perform substantially differently than at room temperature.

becomes very low. Driving over a PP garden hose left in the driveway on a winter night will reinforce this point.

Another factor that is sometimes overlooked when designing with polymers that crystallize is that certain polymers crystallize very slowly. Thus, during processing, the polymer may not have achieved an equilibrium crystallization level. When the product is used at an elevated temperature, crystallization may continue. Since the density of the polymer in a crystalline state is greater than when it is in an amorphous state, the product dimensions may decrease and it may lose critical tolerance. Under certain conditions, the designer may need to anneal the product, that is, hold it at an elevated temperature for some time, in order to stabilize its dimensions.

One of the cardinal rules every designer must follow is to select and test the polymer of choice in the use temperature range called for in the part specifications.

2.6 Transport Properties of Polymers

Transport mechanics usually detail momentum, energy and mass transport. The primary physical property in momentum transport is viscosity, with ancillary properties of viscoelasticity and elasticity also important. The elements of momentum transport are detailed in Chapter 4. In energy transport, thermal conductivity, specific heat and density or specific gravity are primary physical properties. Thermal diffusivity, an ancillary property is also important. In mass transport, diffusivity is important. Solubility, a thermodynamic property, is also important. Many of the elements of diffusivity and solubility are related to the interaction of polymers with a chemical environment and so are also detailed in Chapter 4. In this section, those properties that influence energy transport will be reviewed.

Heat Capacity

As will be discussed in Chapter 3, heat capacity is a thermodynamic property. The isobaric heat capacity of any material is the rate of change of enthalpy with temperature at constant pressure:

$$(2.35) \qquad\qquad c_p = \left(\frac{\partial H}{\partial T}\right)_P$$

Any material that shows a second-order transition at a specific temperature also shows a *discontinuity in slope* in its enthalpy curve. A material exhibiting a first-order transition at a specific temperature also shows a *discontinuity in function* in its heat capacity value. For polymers, these discontinuities are not so apparent, since first- and second-order transitions usually occur over small but finite temperature ranges. This is seen in schematic in Figure 2.54. An *effective* specific heat during crystallization is obtained

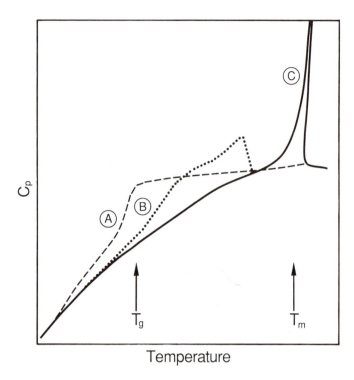

Temperature

Figure 2.54.   Schematic Representation of Temperature-Dependent Heat Capacity for Amorphous and Crystalline Polymers. T_g is Glass Transition Temperature. T_m is Crystalline Melting Temperature. A: Amorphous Polymer. B: Semicrystalline Polymer. C: Crystalline Polymer (85).

by dividing the area under the enthalpy-temperature curve by the temperature difference determined from the beginning and ending temperatures of the phase change:

(2.36)
$$c_{p,cryst} = \int_{T_1}^{T_2} \frac{(dH/dT)\ dT}{(T_2 - T_1)}$$

For polypropylene, for example, the effective heat capacity during crystallization is 1.8 cal/g°C (1.8 Btu/lb$_m$°F) between T = 155 and 165°C. Usually, the specific heat of the liquid and solid polymer phases at temperatures far from phase changes are nearly independent of temperature. The specific heat of the rubbery amorphous polymer phase is also relatively independent of temperature, although it may decrease in value somewhat as the temperature approaches T_g. The rubbery phase heat capacity value of a polymer is usually lower than that in the liquid phase. For semicrystalline polymers, the measured specific heat value $c_{p,t}$, is usually a simple linear sum of the values for the crystalline and amorphous phases:

(2.37) $c_{p,t} = X\ c_{p,c} + (1 - X)\ c_{p,a}$

where X is the fraction crystallized, $c_{p,c}$ is the heat capacity for the crystalline phase and $c_{p,a}$ is that of the amorphous phase. Below the glass transition temperature, the heat capacities of most polymers decrease linearly with temperature.

Thermal Conductivity

Conduction heat transfer is the mechanism of energy transfer from a warm portion of a solid body to its cooler portion. Heat flow per unit area perpendicular to the direction of transfer is called *heat flux, q* (cal/cm^2 s or Btu/ft^2 h). The heat flux is proportional to the thermal gradient, dT/dx. The proportionality parameter is the *thermal conductivity.* *

(2.38)
$$q = -k\frac{dT}{dx}$$

For most applications, thermal conductivity can be considered essentially equal or isotropic in all directions and so a single value, a scalar, suffices. For uniaxially and nonuniformly biaxially stretched fibers and films, the assumption of a single value for thermal conductivity is technically incorrect.

For highly structured materials such as metals, thermal energy is transferred by electrons. In all polymers, the mechanism is by mobile chain segment interference or collision. As a result, values of thermal conductivity are not affected by second-order phase transitions such as glass transitions. Thermal conductivity values for amorphous polymers are only weakly dependent on temperature from the glassy region to the fluid state. It is thought that the primary mechanism for energy transfer is polymer backbone motion, with secondary energy transfer across secondary bonds. Thus, higher molecular weight molecules should have higher thermal conductivities, to a limit. This is seen in Figure 2.55. And thermal conductivity increases with increasing orientation as seen in Figure 2.56.

The presence of a highly ordered crystallite structure increases the ability of the polymer to transmit thermal energy. This is seen for several polyethylenes in Figure 2.57. Crystalline polymers also show a gradual decrease in thermal conductivity values as the material temperature approaches the melt temperature. At the melt temperature, there is an abrupt drop in value, Figure 2.58. This is attributed to the greatly increased free volume and more randomly ordered state in the polymer liquid.

Thermal conductivity values for both amorphous and crystalline polymers appear to increase gradually with increasing applied hydraulic pressure, but the change in value is slight and monotonic, so long as no morphological transitions occur.

Measurement of thermal conductivity of polymers from below glass transition to far into the fluid phase is very difficult. Traditionally, thermal conductivity is determined

---

*Thermal flux can be written for each of the principle axes as can thermal gradient. As a result, thermal conductivity is correctly a second-order tensor. The correct expression is:
$$q_i = -k_{ij}\, dT/dx_j$$

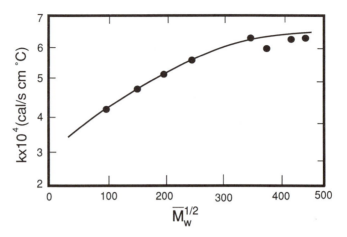

Figure 2.55.   Molecular Weight Dependency of Thermal Conductivity for Linear Polyethylene Near Its Crystallization Temperature, 140°C (86).

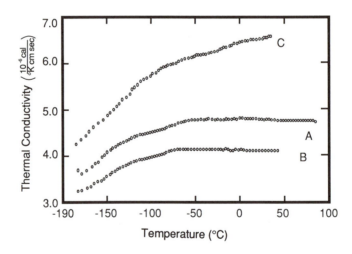

Figure 2.56.   Temperature Dependence of Thermal Conductivity for Polymethyl Methacrylate (PMMA) Film. A: Unstretched. B: Perpendicular to the Direction of 375% Stretch. C: Parallel to the Direction of 375% Stretch (87).

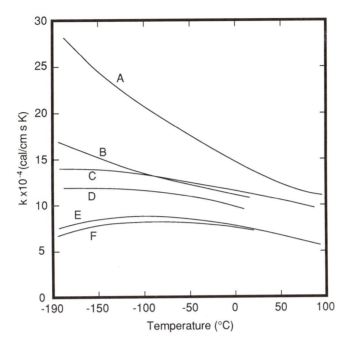

Figure 2.57. Effect of Polyethylene Density and Degree of Crystallinity on Temperature-Dependent Thermal Conductivity. A: specific gravity = 0.982 g/cm³. B: = 0.962. C: = 0.961. D: = 0.951. E: = 0.923. F: = 0.918 g/cm³ (88).

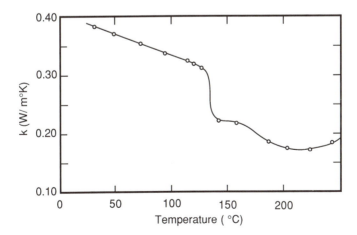

Figure 2.58. Temperature Dependency of Thermal Conductivity for High Density Polyethylene (89).

by placing a planar sheet of material of known thickness between plates of known but different temperatures, then measuring the rate of heat transfer from the warmer plate to the cooler one. For metals, the differential temperature needs to be but a degree or so, since the material thermal conductivity is very high. Owing to the very low thermal conductivity values of polymers, the thermal gradient must be substantial (tens of degrees) in order to measure the steady-state heat transfer rate. This introduces a major source of error. At long times, edge effects (viz, heat loss from the edges of the specimen to its surroundings) can become important. Edge heaters are used to compensate for this effect. In the fluid or liquid phase, long times allow the establishment of buoyancy currents caused by the temperature difference. This too is a source of error. Further, since the time needed to stabilize at steady state can be quite large, determining the temperature-dependent thermal conductivity curve can be quite laborious. More rapid ways, such as energy-pulsing a heated probe and measuring the transient decay of the energy at fixed distances from the source, yield accurate data but commercial devices are not available.

Thermal Diffusivity

Thermal diffusivity is a combined thermophysical property of materials. It is obtained from:

$$(2.39) \qquad \alpha = \frac{k}{\rho \, c_p}$$

where α is the thermal diffusivity (cm^2/s or ft^2/h), k is thermal conductivity, ρ is density, and c_p is specific heat (the last three in proper units). This property is obtained naturally from time-dependent or unsteady state conduction heat transfer. Typically, it is obtained from individual values of the three primary properties. Direct measurement yields values that are far less error-prone (76). Thermal diffusivity values are quite temperature *insensitive*, regardless of whether they are for amorphous or semicrystalline polymers (Figure 2.59).

Example 2.8

Q: The surface of a very thick block of polycarbonate (PC) is raised from 100°C to 120°C. Determine the temperature 0.25 cm below the surface after 30 s.

A: The mathematical relationship for transient heat conduction into a semi-infinite solid is (90):

$$\frac{\partial T}{\partial t} = \alpha \, \frac{\partial^2 T}{\partial x^2},$$

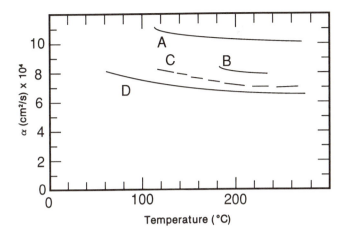

Figure 2.59.   Temperature Dependency of Thermal Diffusivity for Several Polymers. A: Poly-carbonate (PC). B: Acetal (POM). C: Butyrate (PVB). D: Polymethyl Methacrylate (PMMA) (91).

where α is the thermal diffusivity. The thermal diffusivity of PC at 100°C is obtained from Figure 2.59; $\alpha = 11 \times 10^{-4}$ cm²/s. To obtain the temperature, a dimensionless penetration parameter must be calculated:

$$\frac{x}{2} \sqrt{(\alpha t)} = 0.688$$

The dimensionless temperature at $x = 0.25$ cm is then obtained from Figure 2.60 (89):

$$\frac{[T(x,t) - T_s]}{[T_o - T_s]} = 0.62$$

$$T_{x = 1/4 \text{ cm, } t = 30 \text{ s}} = 108°C$$

Coefficient of Thermal Expansion

The final dimensions of a plastic part are dependent upon variables such as temperature, moisture content, curing conditions, loss of plasticizer or solvent, release of molded-in stresses, and phase changes. These are discussed in further detail in Chapter 6. The effect of temperature on polymer dimensions is quantified as the *thermal expansion coefficient*, κ. κ is defined as:

(2.40)
$$\kappa = \frac{1}{L} \frac{\Delta L}{\Delta T}$$

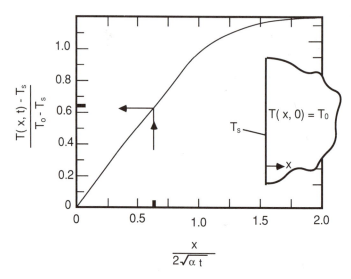

Figure 2.60.   Transient Heat Conduction into a Semi-Infinite Slab Subject to a Step Change in Surface Temperature at Time, $t = 0$ (92).

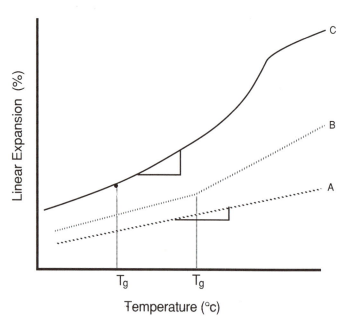

Figure 2.61.   Representative Temperature-Dependent Linear Expansion Coefficients for Three Types of Materials. A: Metal. B: Amorphous Polymer. C: Crystalline Polymer. T_g is Glass Transition Temperature.

and has the units of m/m°C or reciprocal temperature, °C^{-1}. The thermal coefficient of expansion is the slope of the temperature-dependent expansion curve. Metals typically have κ values in the range of 10 to 30 μm/m°C. Amorphous polymer values range from 50 to 150 μm/m°C. Crystalline polymer values range from 100 to 250 μm/m°C. These are seen in schematic in Figure 2.61. The value of thermal expansion coefficient depends strongly on molecular orientation and can change substantially in value as the polymer passes through transitions.

2.7 The Nature of Polymer Additives, Fillers and Reinforcements

Frequently, polymers require *adducts* to be used for specific applications. Some additives such as antioxidants and heat stabilizers, are required in specific resins such as polypropylene to prevent discoloration and mechanical property loss. These are typically special purpose, very high purity chemicals such as butylhydroxytoluene (BHT) or sodium stearate and are usually added at very low levels (about 0.1% (wt)). Organic ultraviolet stabilizers such as benzophenone and antistatic agents are also fine chemicals that are added at relatively low dosage. Crosslinking agents, such as peroxides for unsaturated polyester resins and polyethylenes, and coupling agents for reinforced polymers or for certain rubber-toughened polymers, are added at low dosage levels as well (up to 1.0% (wt)). Chemical foaming agents are also added at relatively low dosage (up to 1.0% (wt)). Colorants are of two types. Dyes and organic pigments are usually used as tints and are added to 1.0% (wt) or so. Typical chemicals are phthalocyanine green dye. Inorganic pigments are typically added to 1.0% (wt). Titanium dioxide (TiO_2) is an inorganic oxide opacifier. Ferric oxides are used as reddish-brown colorants and chromium oxides are dark green. Carbon black is added to 0.3% (wt) or so as a colorant, opacifier and an ultraviolet screen. Adducts of higher molecular weight (waxes, as an example) or of higher dosage include processing aids, such as emulsifiers, lubricants, viscosity depressants, mold release agents, and plasticizers. Dosage can reach 10.0% (wt).

Flame retardants can be either inorganic or organic. Inorganic flame retardants usually inhibit combustion through partial decomposition into water or CO_2. Typical inorganic flame retardants are alumina trihydrate (ATH) and antimony trioxide. Inorganic flame retardants are usually most effective in dosages in excess of 20% (wt). At high dosage levels, they can dramatically change mechanical, thermal and flow behavior of the host polymer. Under these conditions, they should be considered fillers. Organic flame retardants are special-purpose waxes, oligomers or low molecular weight polymers that contain high fractions of bromine, chlorine or phosphorus. Upon decomposition, halogen gases scavenge oxygen at the polymer–gas interface to reduce the combustion potential.

In some cases, these adducts can affect the morphological character of the polymer. For example, for crystalline polymers, heterogeneous nucleation occurs on nonpolymer substances. In pristine (unfilled/unreinforced or "neat") polymers, catalyst residue can trigger heterogeneous nucleation. In nearly all commercial crystalline polymers, adducts can initiate nucleation. In polyethylene for example, haze, usually caused by spherulites large enough to interfere with visible light transmission, can be reduced by deliberately adding small amounts of certain low molecular weight adducts such as stearic or oleic acid. In some cases, small amounts of inorganics are deliberately added to control crystallization levels. Talc will change the crystallization kinetics of nylon 66 (PA-66) and polyethylene terephthalate (PET). Accidental nucleants can be introduced as colorants or pigments. In these cases, parts molded under identical conditions but with different colorant or adduct packages can have decidedly different overall dimensions and part-to-part tolerances.

Specific types of adduct can cause more substantial effects than others. For example, dyes are low molecular weight organics. At typical dosage of 0.5 to 5.0% (wt), they do not appreciably alter the physical properties of polymers. Pigments, on the other hand, are usually very fine inorganic powders having particle sizes of 0.1 to 1.0 μm. As seen in Figure 2.62 even at 2% (wt) ferric oxide or carbon black,* the thermal conductivity of polystyrene–acrylonitrile copolymer (SAN) is noticeably decreased.

Plasticizers

Plasticizers are compatible low molecular weight organics that are added to reduce the melt viscosity and solid modulus of an otherwise rigid polymer. The classic polymer–plasticizer system is polyvinyl chloride and an octyl phthalate. It is well known that polyvinyl chloride is a brittle, nearly amorphous polymer with a very high viscosity at flow conditions. Owing to this and its propensity to degrade at normal shear conditions, it is nearly impossible to melt process pure polyvinyl chloride. As a result, softeners or plasticizers such as dioctyl phthalate (DOP) and diisooctyl phthalate (DIOP) are added to 10 to 30% (wt). The glass transition temperature of the system shifts to lower temperatures (Figure 2.47) and the temperature range broadens substantially with increasing plasticizer concentration. As expected, hardness, tear strength and modulus at room conditions decrease and ultimate elongation increases with increasing plasticizer concentration. Other properties such as thermal conductivity (Figure 2.63) change in relatively complex and at times unpredictable manners. The compatibility of plasticizers and other low molecular weight adducts with polymers is discussed in greater detail in Chapter 4.

Fillers

Hundreds of materials have been added to polymers as *fillers*. Many are tabulated in Appendix B. Fillers are usually considered to have shapes that are nonfibrous, nearly spherical, compact or platy. They are usually inexpensive inorganics. Usually, fillers

---

*Carbon black is considered a pigment here.

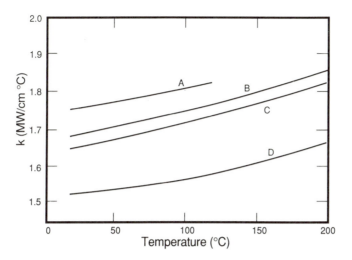

Figure 2.62.   Effect of Pigments on Temperature-Dependent Thermal Conductivity of Amorphous Styrene-Acrylonitrile (SAN) Polymer. A: Natural. B: Chocolate Pigment, 2.08% (wt). C: Blue, 2.57% (wt) D: White, 7.19% (wt). (93).

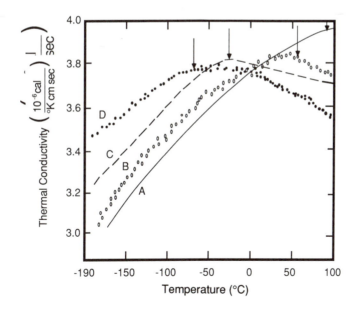

Figure 2.63.   Effect of Plasticizer Concentration on Temperature-Dependent Thermal Conductivity of Plasticized Polyvinyl Chloride (FPVC). Arrows Point to Measured Glass Transition Temperature, T_g. A: No Plasticizer. B: 10% (wt) DOP. C: 20% (wt). D: 40% (wt) (94).

Table 2.19. The US consumption of fillers in polymers (95).

Material	Consumption ($\times$ 10³ metric tonnes)			
	1975	1980	1990	2000
Alumina trihydrate	50	200	800	1,600
Asbestos	180	350	800	1,700
Carbonates	700	1,500	3,500	9,000
Cellulosic types	40	90	300	500
Glass	5	15	50	200
Silicas	25	80	300	500
Silicates	6	15	50	100
Talc minerals	40	200	900	1,800
Miscellaneous	14	50	200	500
Total	1,060	2,500	6,900	15,000

are added to improve polymer stiffness and continuous use temperature, albeit at significant decrease in elongation at break values.* As seen in Table 2.19, 5600 Mlb (2.5 MT) of fillers were used in plastics in 1980: 60% was carbonate-based, usually as calcium carbonate. Typical filler dimensions are 0.1 to 100 μm with most commercial fillers having dimensions of 1 to 10 μm. This means that filler dimensions are typically 100 to 1000 times the thickness of lamellae (at 0.01 μm) and about 0.01 to 1 times the diameter of crystalline spherulites (at 10 to 100 μm). Fillers act as heterogeneous nucleation sites, thus producing products having very fine spherulitic structures. At the same time, fillers act to inhibit ultimate spherulitic growth (X_∞). Thus filled polymers may have a lower level of crystallinity. The lower limit on filler loading is typically 20% (wt) or so. Below that, the slight improvement in desired properties does not warrant the expense of compounding. The upper limit is governed by the ability to compound the filler into the polymer. At low shear rates, the compound viscosity increases very rapidly with filler loading. Although newer compounding techniques are being developed to handle very high loading, the practical upper limit for many traditional inorganic fillers such as talc, calcium carbonate, titanium dioxide, and clay, in many polymers is about 60% (wt).

Properties of Filled Polymers

Since the filler density is high, the *volume* occupied by the filler is low even at high loadings. The density of the filled polymer can be obtained from:

$$\frac{1}{\rho} = \frac{1 - X}{\rho_p} + \frac{X}{\rho_f}$$

(2.41)

---

*It is a fallacy that fillers are added to reduce cost. Usually the compounding cost and coupling agent materials cost offset any savings in polymer materials cost.

Table 2.20. The effect of the shape of the filler particle and the type of packing on the maximum packing fraction, P, for uniformly sized particles (96).

Particle shape	Type of packing	P, Maximum packing fraction
Spheres	Hexagonal close packing	0.7405
Spheres	Face-centered cubic	0.7405
Spheres	Body-centered cubic	0.60
Spheres	Simple cubic	0.5236
Spheres	Random close packing	0.637
Spheres	Random loose packing	0.601
Fibers	Parallel hexagonal packing	0.907
Fibers	Parallel cubic packing	0.785
Fibers	Parallel random packing	0.82
Fibers	Random orientation	0.52

where ρ is the density of the filled polymer, ρ_p is that of the unfilled polymer, ρ_f is that of the filler and X is the weight fraction of filler.

Other than plasticizers and inorganic flame retardants, most adducts do not appreciably change the mechanical, thermal or rheological (flow) characteristics of polymers. Fillers and fibers do. The effects due to filler level are usually normalized to the *maximum packing fraction*, P. The maximum packing fractions for simple shapes of uniform sizes can be determined theoretically (Table 2.20). For example, the volume occupied by perfect spheres of diameter D spaced on simple cubic array is $\pi D^3/6$. If all spheres touch, the cubic space around each sphere is D^3. Therefore the maximum packing fraction for this example is $P = \pi/6 = 0.524$. This, of course, is not the highest possible value of P for equal-sized spheres. The equilateral pyramid yields a hexagonal close packed structure having $P = 0.74$. The actual maximum packing density is usually between $P = 0.60$ (random loose packing) and $P = 0.64$ (random close packing) (97). $P = 0.625$ is considered a good value to use (98). Most filler particle sizes are distributed over a finite range of diameters. The particle size distribution can be obtained by colony counter or sieving (99). The simplest particle size distribution is bimodal with both groups having uniform diameter particles, D_1 and D_2. Figure 2.64 shows the relative bulk volume as a function of bimodal sphere composition. As small spheres are added to the large sphere system, the packing fraction increases, since the small spheres fill in the interstices between the large spheres and no large spheres need to be removed. Similarly, when large spheres are added to the small sphere system, some small spheres must be removed, but the void fraction around these spheres is replaced with solid material (a portion of the large sphere). The packing fraction again increases. As seen, the solid line represents a diameter ratio, large sphere to small sphere, R, of infinity. The intercept is at 28% small spheres ($R = \infty$) and $P = 0.85$. Reciprocal packing fractions for intermediate bimodal ratios are shown in Figure 2.65. Packing fractions for other configurations are given in Table 2.20.

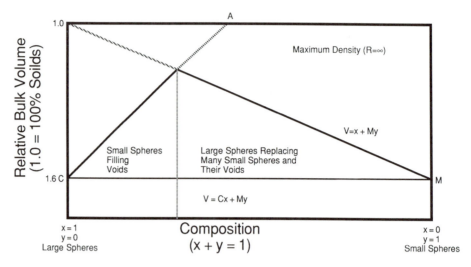

Figure 2.64.   Schematic of Packing Concept for Theoretical Packing of a Two-Sphere System (100).

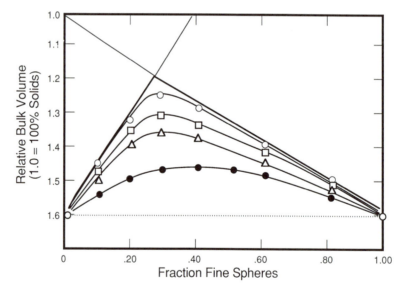

Figure 2.65.   Sphere–Sphere Packing with R, Radius Ratio, as Parameter. Solid Line: $R = \infty$. Open Circles: $R = 16.5$. Open Squares: $R = 6.5$. Triangles: R = 4.8. Solid Circles: R = 3.4. Dotted Line: $R = 1$ (101).

Typically, for continuously graded fillers, the particle size distribution can be obtained by sieving or other means. The maximum packing fraction can be obtained by employing the following scheme (102):

1. Choose a screening protocol that allows size differentials of $\sqrt{2}$ from the minimum particle size to the maximum.
2. Determine void volume for each screen size by oil absorption technique such as ASTM D281. (An average value is probably sufficient for this.)
3. Calculate $K = d_{max}/d_{min}$ where d_{max} is the average maximum equivalent spherical diameter and d_{min} is minimum one. From Figure 2.66, determine the number of component sizes for the measured void fraction.
4. Prepare a table for the number of sizes differing by $\sqrt{2}$ over the range selected.
5. Calculate a factor r to relate one size to the one above it (this number to be slightly greater than 1):

(2.42) $$r = V^{n/m}$$

where V is the fractional void volume, n is the number of component sizes *minus one*, m is the number of size intervals, in steps from the smallest to the largest diameter.

6. Multiply each size by r beginning with 1 for the finest size.

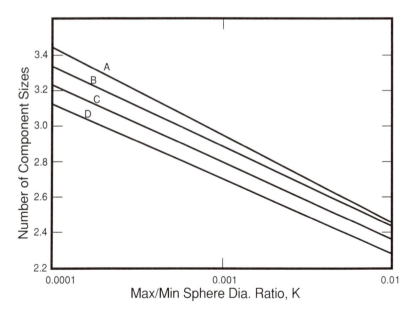

Figure 2.66. Relationship between Sphere Ratio and Component Size for Several Volume Fractions. A: V = 60% Voids. B: 50%. C: 40%. D: 30% (103).

7. Calculate the fractional void in a continuous distribution from:

(2.43)
$$V_v = \frac{r^{\log d} - r^{\log d_{min}}}{r^{\log d_{max}} - r^{\log d_{min}}}$$

where d is the size interval, $\sqrt{2} = 1.414$.
8. Calculate the maximum packing fraction, $P = 1 - V_v$.

Example 2.9 illustrates this method for a very narrow particle size distribution (1.0 μm to 16.0 μm).

Example 2.9

Q: Consider a continuously graded filler having a particle size range of 1.0 to 16 μm. Determine the packing fraction if the volume fraction by oil absorption is 40%.

A: The K ratio $= d_{max}/d_{min} = 16.1/K = 0.0625$. The value of the component sizes, n, in Figure 2.66 is extrapolated to be 2.0. The sizes are given in Table 2.21.

The r-factor, that relates one size to the one above it is given as:

$$r = v^{n/m} = 2.768$$

This number is used to obtain the percent individual volume in the second column of the table. Here $n = 1.0$, $m = 9$.

The fractional void is then given as:

$$V_v = \frac{[2.768^{\log 1.414} - 2.768^{\log 1.0}]}{[2.768^{\log 16} - 2.768^{\log 1.0}],} = 0.0688$$

The maximum packing fraction, $P = 1 - V_v = 93.13\%$.

Table 2.21.

Size × $\sqrt{2}$	Volume of Each Size × r	% Individual Volume
1.00	1.00	0.0185
1.41	2.77	0.0514
2.00	7.67	0.1423
2.83	21.24	0.3938
4.00	58.70	1.088
5.66	162.5	3.012
8.00	449.7	8.338
11.3	1244.7	23.09
16.0	3445.7	63.88
	5393.7	100.0

Mechanical Properties

Fillers affect the stress–strain behavior of the polymer matrix. The elastic modulus or filled polymer stiffness is probably most influenced. Nielsen (104) proposes a relationship between the moduli of the filled material, the polymer and the filler:

$$(2.44) \qquad \frac{M}{M_p} = \frac{1 + A B \phi}{1 - B \psi \phi}$$

where

$$A = k_E - 1$$

$$B = \frac{M_f/M_p - 1}{M_f/M_p + A}$$

$$\psi = 1 + (1 - P) \frac{\phi}{P^2}$$

k_E is the Einstein coefficient, given in Table 2.22, ϕ is the filler volume fraction, P is the maximum packing fraction, M is the modulus of the filled system, M_p is that of the polymer and M_f is that of the filler. The Einstein coefficient, k_E, is a measure of the shape of the filler particle. $k_E = 2.5$ for regular shapes (spheres) and becomes large for fibrous or acicular particles. The yield stress of a filled polymer, σ_y, can be related to that of the unfilled polymer by:

$$(2.45) \qquad \sigma_y = \sigma_{y.p} [1 - (\phi/P)^{2/3}]$$

Ultimate elongation of filled polymers usually decreases with increasing filler loading, but elongation at yield is usually unaffected. One measure is:

$$(2.46) \qquad \varepsilon = \varepsilon_p [1 - (\phi/P)^{1/3}]$$

where ε is the elongation of the filled polymer and ε_p is that of the unfilled or neat polymer. As is expected, stretching tends to debond* the filler particles from the surrounding matrix. The microcavitation that occurs leads to stress concentration and ultimately premature failure. Improved tensile characteristics of filled polymers are achieved by adding coupling agents that inhibit debonding.

Similarly, creep of filled systems can be related to creep of the unfilled polymer:

$$(2.47) \qquad \varepsilon M(t) = \varepsilon_p M_p(t)$$

where the subscript "p" refers to the unfilled polymer property. $M(t)$ can be estimated from the Nielsen expression above.

---

*This is also called polymer dewetting.

Table 2.22. Values of the Einstein coefficient, k_E, for various types of fillers (96).

Filler type	k_E[1]
Spheres, one size, maximum packing	2.50
Spheres, random close packing	2.50
Spheres, random loose packing	2.50
Rods or ellipsoids, random packing[2]	
Aspect ratio $= 2$	2.58
$= 4$	3.08
$= 6$	3.80
$= 10$	5.93
Mixed sizes, irregular shapes, minimum surface area	4.00[5]
Mixed sizes, plates and flakes	$5+$[5]
Agglomerates of spheres	$2.5/\phi_a^{3.5}$
Agglomerates generally	$k_e/\phi_p^{4.6}$

[1]Correction factors for the mechanical case of Poisson's ratio, μ_1, of the matrix:

μ_1	Factor
0.50	1.00
0.40	0.90
0.35	0.87
0.30	0.84
0.20	0.80

[2]At high rates of shear and low V_f for the rheological case, these shapes tend to orient and to reduce k_E.

[3]ϕ_a is the volume fraction of the agglomerate which is spherical.

[4]ϕ_p is the volume fraction of particles in the agglomerate, and k_E is the appropriate value for the particle shape.

[5]These values may be approximated from oil absorption data or more precisely determined by intrinsic viscosity.

[6]These values are generally less than the maximum packing fraction for the particular particle shape.

Thermal Properties

Even though the absolute values of modulus and strength are substantially influenced by filler type and concentration (Figure 2.67), the general nature of these temperature-dependent properties is unchanged. Filler can affect *perceived* physical strength, however. A classic example is deflection temperature under load (DTUL). Since this test is detailed in Chapter 6, it will not be discussed here, except to note that DTUL is a measure of the temperature at which the polymer distorts under a specific load. If the main feature of a filler is to increase the overall modulus of the polymer, the temperature where deformation under the specific load is reached will increase. *But* the general shape of the temperature-dependent strength does not change. It is more significant to compare loss and storage moduli of filled and unfilled polymers (Figure 2.67). Note

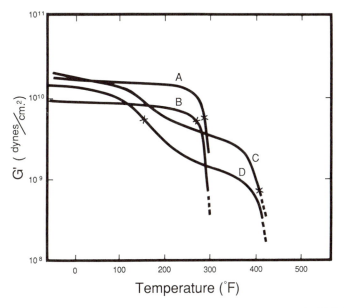

Figure 2.67.   Dynamic Elastic Moduli of Filled and Unfilled Polymers. A: 20% (wt) Glass-Reinforced Polycarbonate (PC), DTUL = 295°F. B: Unfilled PC, DTUL = 275°F. C: 30% (wt) Glass-Reinforced Nylon-6 (PA-6), Dry, DTUL = 420°F. D: Unfilled PA-6, Dry, DTUL = 120°F. DTUL = Deflection Temperature Under Load. DTULs Denoted by x (105).

that the filler increases the rigidity of the plastic but not the upper temperature capability of the plastic even though the DTUL test* indicates increased upper temperature limits.

The filler can contribute significantly to the thermal conductivity. Nielsen's expression applies here also:

$$\frac{k}{k_p} = \frac{1 + A\,B\,\phi}{1 - B\,\psi\,\phi}$$

(2.48)

$$\text{where } A = k_E - 1$$

$$B = \frac{k_f/k_p - 1}{k_f/k_p + A}$$

$$\psi = 1 + \frac{(1 - P)\,\phi}{P^2}$$

k_E and P can be obtained from Tables 2.20 and 2.22: k, k_p and k_f are the thermal conductivities of the filled polymer, the unfilled polymer and the filler. Most inorganic

*This particular effect is real and is discussed in detail in the discussion of the DTUL test in Chapter 6.

fillers can be considered semiconductors. That is, they have relatively low thermal conductivity values. As expected, metal fillers have the greatest effect on thermal conductivity.

Example 2.10

Q: Estimate the room temperature thermal conductivity of nylon 6 (PA-6) polymer containing 30% (vol) calcium carbonate. Assume the filler particles are spherical and of uniform size.

A: The thermal conductivities of the two materials are:

$$k_p = k_{Nylon\ 6} = 0.21\ \text{W/m K} \qquad \text{(Appendix C)}$$
$$k_f = k_{CaCO_3} = 2.34\ \text{W/m K} \qquad \text{(Appendix D)}$$

$$A = k_E - 1 = 0.5$$

$$\phi = 0.3$$

$$P = 0.601$$

$$\psi = 1 + 0.3 \times \frac{(1 - 0.601)}{(0.601)^2} = 1.33$$

$$B = \frac{[2.34/0.21 - 1]}{[2.34/0.21 + A]} = 0.871$$

$$\frac{k}{k_p} = \frac{[1 + 0.5 \times 0.871 \times 0.3]}{[1 - 0.871 \times 0.3 \times 1.33]} = 1.73$$

The thermal conductivity value of the filled resin is 73% greater than that of the neat resin, or:

$$k = 1.73 \times 0.21 = 0.363\ \text{W/m K}$$

Specific heat can be obtained from a simple law of mixtures applied to the *volume* of the filled polymer:

(2.49)
$$c_p' = c_{p.p}\ \rho_p\ (1 - \phi) + c_{p.f}\ \rho_f\ \phi$$

where c_p' is the volumetric specific heat (cal/°C cm^3). This can be combined with an expression for the density of the filled polymer to obtain the specific heat of the filled polymer. Typically, the *volumetric* specific heat of a filled polymer is slightly greater than that for the neat polymer. The filled polymer *thermal diffusivity*, $\alpha' = k/\rho\ c_p'$, is also slightly greater than that of the neat polymer, since k usually increases more rapidly with increasing filler loading than does c_p'.

The linear coefficient of thermal expansion can be significantly reduced by the addition of a filler, Figure 2.68. Most of the commercial fillers in Appendix B exhibit thermal expansion coefficients of 0.3 to 8.0 μm/m °C. Thus, mixtures of fillers and polymers show substantially reduced values. The degree of reduction is a function of the filler type and the volume fraction. This has great importance when the filler content of a resin is changed in an injection molded part, as an example.

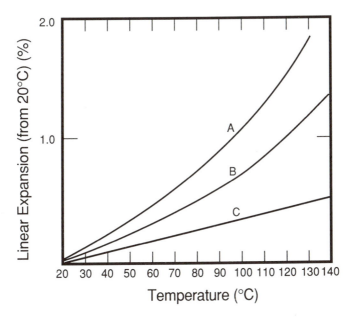

Figure 2.68.   Effect of Filler Loading on Temperature-Dependent Linear Expansion Coefficient for Polypropylene. Expansion Measured from 20°C. A: Unfilled Polypropylene (PP). B: PP + 20% (wt) Talc. C: PP + 20% (wt) Glass Fiber (106).

The permittivity or dielectric constant is given as $e = \Delta/E$, where Δ is the electrical flux density and E is the field strength. The permittivity for a filled system can be obtained directly from the thermal conductivity equation by substituting e for k everywhere. Thus, the permittivity increases with increasing filler loading.

Filled Polymer Summary

Table 2.23 best summarizes the relative influence of fillers on general mechanical properties of the filled polymer. Similarly, Table 2.24 gives the nature of the filler needed to achieve specific property improvement without substantial sacrifices in other properties.

Fibers

Fillers are usually considered to be compact materials. Fibers on the other hand are long when compared with their diameter. That is, the *aspect ratio, L/D* is greater than 10. Fiber shape is often called acicular or elongated. There are two classes of fibers: short fibers and continuous fibers. Fiber strengths are compared by dividing the individual fiber property by the resin density to yield a *specific* property. Specific tensile strengths and moduli for many high performance fibers are shown in Figure 2.69.

Table 2.23. Effect of various types of fillers on the composite properties
of filled resins (107).

Physical property of composite	Maximum improvement by filler		Filler			Relative filler effect on matrix type[5]	
	At	When V_f/P_f is	Bond matrix[2]	Type[3]	Size range[4]	Rigid	Flexible
Modulus, M[1]	P_f	High	Max	Solid, max M	Broad	Least	Most
Tensile strength	V_f/P_f	Low	Max	Fiber	Narrow	Least	Most
Flexural strength	V_f/P_f	Medium	Max	Fiber	Narrow	Least	Most
Elongation	V_f/P_f	Low	Min	Sphere	Narrow, fine	Least	Most
Tear strength	V_f/P_f	Low	Max	Fiber	Narrow	—	—
Impact strength	V_f/P_f	Low	Min	Fiber	Narrow, fine	Most	Least
Compressive strength	P_f	High	Max	Solid, max M	Broad	Least	Most
Creep	P_f	High	Max	Fiber	Broad	Least	Most
Hardness	P_f	High	Max	Solid, high mohs	Broad	Least	Most
Coefficient of friction	P_f	High	Max	Solid	Broad	Least	Most
Abrasion resistance	P_f	High	Max	Solid, high mohs	Broad	Most	Least
Density	P_f	High	—	Sphere, low density	Broad	—	—

[1]Young's, shear or bulk.
[2]Surface modification of filler may be necessary to maximize bonding.
[3]Tables 2-1 and 2-16 of Ref. 114 give details.
[4]Broad size range produces highest P_f.
[5]At a testing temperature below T_g of matrix for rigid, and above T_g for flexible.

Nearly all reinforcing fibers are inorganic. Organic fibers of aromatic polyamide or "aramid" (trade name: Nomex or Kevlar) and ultraoriented polyethylene terephthalate (trade name: Compet) are exceptions. There are many naturally occurring inorganic fibers, including asbestos and wollastonite (calcium silicate). There are frequently several types of fibers within a given fiber category. For example, there are six types of asbestos fiber, including chrysotile and crocidolite. The former is used with thermosets in brake pads and with PVC in floor tile. The latter is used as reinforcing in most thermoplastics. Some fibers are fabricated from naturally occurring ores such as mineral wool (fibers obtained from blast furnace slag at about 75% (wt) calcium silicate), franklinite (an anhydrous crystalline form of calcium sulfate) and dawsonite (an anhydrous fiber form of sodium-aluminum carbonate).

Glass fibers are formed by melt spinning and drawing refined silicates. Glass fibers can be produced either in continuous strand or in chopped or milled form. Fiber diam-

Table 2.24. Some guidelines for an effective selection of fillers for development of specific properties of filled resins (108).

Improvement	Matrix type[1]	Without loss of	Matrix should be[2]	Best filler type[3]	Concentration[4]
Thermoplastics					
Modulus	High T_g	Impact	Ductile	Fine, narrow range	Low
Modulus	High T_g	Tensile	Ductile, tough	Fine, fiber or flake, max bonding	Low/moderate
Modulus	High T_g	Flexural	Tough	High P_f, max bonding	Moderate/high
Abrasion resistance	High T_g	Compressive strength	Ductile, tough	High hardness, high P_f	High
Lower density	High T_g	Compressive strength	Hard, tough	Glass microballoons	Moderate/high
Modulus	Low T_g	Low viscosity	Low melt viscosity	High P_f, wide range	Low
Tensile	Low T_g	Impact	High elongation type	Fine, narrow range, moderate bonding	Low
Creep	Low T_g	Tensile	Moderate elongation type	Fine, wide range, max bonding	Low/moderate
Tear resistance	Low T_g	Impact	Moderate/high elongation type	Fine, narrow range, moderate bonding	Low
Thermosets					
Modulus	High T_g	Impact	Ductile	Fine, wide range	Low/moderate
Modulus	High T_g	Tensile	Moderate ductile	Fiber or flake	Moderate
Flexural	High T_g	–	Moderate ductile	Fiber or flake	Moderate
Compressive strength	High T_g	Impact	Moderate ductile	Fine or flake	Moderate
Tear resistance	Low T_g	Abrasion resistance	Tough	Fine, hard, max bonding	Low/moderate

[1]Versus temperature of testing.
[2]Matrix may require modification to obtain indicated property.
[3]Note that best choice may not be the most efficient filler type.
[4]According to relative filler volume, $V_f P_f$, where high is in excess of 0.5, moderate is 0.25 to 0.5 and low is below 0.25.

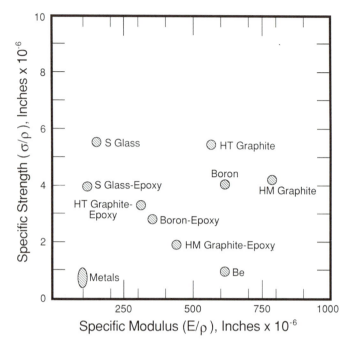

Figure 2.69. Comparison of Specific Strength and Specific Modulus for Many Reinforcing Fibers. Specific Property is the Property Divided by Specific Gravity.

ters are given in "denier", or the weight in grams of 9000 m of single filament fiber.* Most glass reinforcing fibers have equivalent diameters of 10 to 20 μm (1.6 to 6.5 denier). Commercial fiber glass is available as fiber bundles of 204 filaments and can be provided as twisted yarn, roving (or tow), chopped strands, woven roving, uniaxial and biaxial woven fabric, tape, and mixtures of these such as chopped strand/woven roving mat for reaction injection molding (see Chapter 5). As seen in Table 2.25, the tensile modulus for glass is about 10 Mlb_f/in^2 [69 GPa], the tensile strength is about 400 klb_f/in^2 [2.8 GPa], and the specific gravity is about 2.5 (g/cm^3). The ultimate elongation is about 1.5%. E glass is the common fiber used to reinforce polymers. S glass is used for high performance applications.

    Aromatic polyamides are prepared from aromatic diamines and chlorides of aromatic (phthalic) diacids. A typical structure is:

$$+NH-O-NH-CO-O-CO+_x$$

Aramids are known for their high specific strength and ultimate elongation when compared with glass. The individual equivalent fiber diameter is about 12 μm and its

---

*The international and SI yarn standard is Tex, the weight in grams of 1000 meters.

Table 2.25. Specific chemical and physical characteristics for various glass fiber types (109).

Description, definition and characterization	A-glass Typical soda–lime–silica glass, limited for reinforcement due to poor resistance to water	C-glass Chemical glass – possesses improved durability, making it preferred composition for applications requiring corrosion resistance
Chemical properties		
chemical composition (%)		
SiO_2	72.0	65.0
Al_2O_3	0.6	4.0
Fe_2O_3	–	0.2
CaO	10.0	14.0
MgO	2.5	3.0
B_2O_3	–	6.0
Na_2O	15.2	8.0
K_2O	–	–
ZrO_2	–	–
SO_3	0.7	0.1
F_2	–	–
Chemical resistance–14μm fiber		
% Weight loss, 1 h boil-in		
H_2O	11.1	0.13
1.0N H_2SO_4	6.2	0.10
0.1N NaOH	15.0	2.28
Mechanical properties		
Specific gravity	2.50	2.49
Virgin tensile strength ($\times 10^6$ lb$_f$/in^2)	0.35	0.40
Tensile modulus elasticity ($\times 10^6$ lb$_f$/in^2)	9.80	10.0
Thermal properties		
Softening point, °F	1,300	1,380
Coefficient of thermal expansion, ($\times 10^{-7}$ in/in/°F)	90	40
Optical properties		
Index of refraction	1.512	1.541
Electrical properties		
Dielectric constant, 72°F, 10^6 Hz	6.90	6.24
Loss tangent, 72°F, 10^6 Hz	0.0085	0.0052

density is about 1.45 g/cm^3. Its tensile strength is about 400 klb$_f$/in^2 [2.8 GPa], its modulus is about 12 to 19 Mlb$_f$/in^2 [83 to 131 GPa] and its dry ultimate elongation is 2 to 4% (depending upon the fiber processing condition). Aramid applications are restricted to continuous use temperatures below about 360°F (180°C). The fibers are produced in the same general structure types (yarns, roving, chopped strands, woven structures, fabrics, tapes) as glass and graphite fibers. The raw material cost is high

Table 2.25. *Continued.*

D-Glass	E-Glass	S-Glass	YM-31A
Glass with improved dielectric strength and low density, developed for improved electrical performance	Borosilicate type, used for major share of all reinforcement applications	Glass with high tensile strength and modulus, developed for aerospace applications	Special composition developed for high modulus to impart greater rigidity to reinforced structures
74.5	54.0	65.0	53.7
0.3	14.0	25.0	–
Trace	0.2	–	0.5
0.5	17.5	–	12.9
–	4.5	10.0	9.0
22.0	8.0	–	$(Li_2O-3.0)$
1.0	0.6	–	$(BeO - 8.0)$
1.5		–	$(TiO_2 - 8.0)$
$(Li_2O-0.5)$	–	–	2.0
–	–	–	$(CeO - 3.0)$
–	0.1	–	–
–	1.7	–	–
–	48.2	–	–
–	9.7	–	–
2.16	2.54	2.49	2.89
0.35	0.50	0.665	0.50
7.50	10.5	12.5	15.9
1,420	1,555	1,580	N.A. – Vitrifies
17	28	16	–
1.47	1.547	1.523	1.635
3.56	5.80	4.53	–
0.0005	0.001	0.002	–

N.A. = Not applicable.

and the process is quite expensive. As a result, these fibers are used primarily in high performance applications. Table 2.26 compares the physical properties of high performance epoxy reinforced with E-glass and Aramid "49".

Carbon and graphite fibers are made by oxidizing rayon or polyacrylonitrile (PAN), carbonizing the fibers at about 3600°F (2000°C) and then graphitizing at about 5000°F (2750°C) in an inert atmosphere of nitrogen or helium. Typical filament equivalent diameters are 1 to 10 μm. Graphite fiber tensile modulus can be as high as 50 to 60 Mlb_f/in^2 [345 to 415 GPa] and tensile strength as high as 350 to 450 klb_f/in^2 [2.4 to

Table 2.26.   A comparison of physical properties of epoxy-fiber composites.
[1/8-in panels with 181 style fabric. Epoxy resin matrix cured at 350°F
in 75 lbf/in² autoclave.] (110).

Property[1]	E-Glass	Kevlar 49
Laminate density (lb/in³)	0.065	0.048
Tensile strength ($\times$ 10³ lbf/in²)		
Room temperature, dry	63	75
Room temperature, wet	55	73
300°F (150°C), 0.5 h	50	50
Compressive strength ($\times$ 10³ lbf/in²)		
Room temperature, dry	62	25
300°F (150°C), 0.5 h	61	15
Modulus (tension and compression) ($\times$ 10⁶ lbf/in²)		
Room temperature, dry	3.6	4.5
Room temperature, wet	3.2	4.0
300°F (150°C), 0.5 h	3.4	4.5

[1]Specimens were 1/8-inch panels with |8| style fabric. Epoxy resin matrix cured at 350°F in 75 lbf/in² autoclave.

3.1 GPa] (but not usually at the same time). Ultimate elongation usually does not exceed 1%. The fiber density is 1.7 to 1.9 g/cm³, depending upon the method of manufacture. Graphite fibers are produced as fiber bundles and bundle forms similar to glass fibers. Chopped graphite fibers are used to short-fiber reinforce thermoplastics and thermosets in the same way as chopped glass fibers. This is seen for nylon 66 (PA-66) in Table 2.27 (108). The higher performance level of graphite fiber is tempered by its much higher unit cost, increased difficulty in compounding the fibers into the polymer, and increased problems in coupling the fibers to the polymer. *Coupling agents* are chemicals that are added in small amounts to the surface of the fiber to insure intimate bonding between the polymer and the fiber. Bonding can be particularly vexing, since many of the common coupling agents cannot be used at the high processing temperatures of many high performance polymers.

Polymeric materials that contain reinforcing fibers are commonly called *composites*. Filled and foamed polymers are sometimes called composites, as well. Polymeric materials that contain continuous fibers placed in a specific pattern are commonly called *laminates*. However, these terms are frequently misused and should therefore be considered general until properly defined for the specific material. Below, fiber-reinforced polymers are called composites, regardless of whether the fibers are short or continuous.

Note that fiber orientation affects the modulus. The in-fiber direction is called the longitudinal direction. The cross-fiber direction is called the transverse direction. If fibers are essentially infinite in length ($L/D > 100$), the longitudinal modulus can be obtained from:

(2.50)
$$E_L = E_p (1 - \phi) + E_f \phi$$

Table 2.27. Properties of graphite-and graphite-glass fiber-reinforced thermoplastics (111).

Property	ASTM method	Unreinforced resin	Nylon 66, reinforcing fiber (%)				
			20 Carbon	30 Carbon	40 Carbon	20 Carbon 20 Glass	40 Glass
Specific gravity	D792	1.14	1.23	1.28	1.34	1.40	1.46
Water absorption, 24 h, %	D570	1.6	0.6	0.5	0.4	0.5	0.6
Equilibrium after continuous immersion		8.0	2.7	2.4	2.1	–	3.0
Mold shrinkage, 1/8-in Average section (10^{-3} in/in)	D955	15.0	2.0–3.0	1.5–2.5	1.5–2.5	2.5–3.5	3.5–4.0
Tensile strength (lb$_f$/in^2)	D638	11,800	28,000	35,000	40,000	34,000	31,000
Tensile elongation (%)	D638	10	3–4	3–4	3–4	3–4	2–3
Flexural strength (lb$_f$/in^2)	D790	15,000	42,000	51,000	60,000	49,000	42,000
Flexural modulus ($\times 10^6$ lb$_f$/in^2)	D790	0.4	2.4	2.9	3.4	2.8	1.6
Shear strength (lb$_f$/in^2)	D732	9,600	12,000	13,000	14,000	13,000	12,000
Impact strength, Izod (ft-lb/in)	D256						
Notched, 1/4-inch		0.9	1.1	1.5	1.6	1.8	2.6
Unnotched, 1/4-inch		–	8.0	12.0	13.0	16.0	19.0
Heat distortion temperature at 264 lb$_f$/in^2 (°F)	D648	150	495	495	500	500	500
Coefficient of thermal expansion ($\times 10^{-5}$ in/in°F)	D696	4.5	1.4	1.05	0.80	1.15	1.4
Thermal conductivity (Btu-in/h ft^2 °F)	C177	1.7	5.5	7.0	8.5	6.4	3.6
Surface resistivity (Ω/sq)	–	10^{15}	25–30	3–5	1–3	–	10^{15}
Flammability	UL S94	–	–	–	–	–	–

Table 2.27. *Continued.*

Property	ASTM method	Polysulfone			Polyester		
		Unreinforced	30% Glass	30% Carbon	Unreinforced	30% Glass	30% Carbon
Specific gravity	D792	1.24	1.45	1.37	1.32	1.52	1.47
Water absorption, 24 h (%)	D570	0.20	0.20	0.15	0.08	0.06	0.04
Equilibrium after continuous immersion		0.60	0.58	0.38	—	0.45	0.23
Mold shrinkage, 1/8-in Average section (10^{-3} in/in)	D955	7–8	2–3	2–3	17–23	3–4	1–2
Tensile strength (lb_f/in^2)	D638	10,200	18,000	19,000	8,000	19,500	20,000
Tensile elongation (%)	D638	50–100	3–4	2–3	10	3–4	2–3
Flexural strength (lb_f/in^2)	D790	15,400	24,000	25,500	13,000	28,000	29,000
Flexural modulus ($\times 10^6$ lb_f/in^2)	D790	0.39	1.20	2.05	0.35	1.35	2.00
Shear strength (lb_f/in^2)	D732	9,000	9,500	7,000	8,000	8,000	—
Impact strength, Izod (ft-lb/in)	D256						
Notched, 1/4-inch		1.2	1.8	1.1	0.3	1.6	1.2
Unnotched, 1/4-inch		60.0	14.0	4–5	25	9–10	4–5
Heat distortion temperature at 264 lb_f/in^2 (°F)	D648	345	365	365	155	430	430
Coefficient of thermal expansion ($\times 10^{-5}$ in/in°F)	D696	3.1	1.4	0.7	5.3	1.2	0.5
Thermal conductivity (Btu-in/h ft² °F)	C177	1.8	2.2	5.5	1.1	3.2	6.5
Surface resistivity (Ω/sq)		10^{16}	10^{16}	1–3	10^{16}	5×10^{15}	2–4
Flammability	UL S94	VE-O	VE-O	VE-O	—	—	—

Table 2.27. *Continued.*

Property	Polyphenylene sulfide			Ethylene tetrafluoroethylene			Vinylidene fluoride	
	Unreinforced	30% Glass	30% Carbon	Unreinforced	30% Glass	30% Carbon	Unreinforced	20% Carbon
Specific gravity	1.34	1.56	1.45	1.70	1.89	1.73	1.76	1.77
Water absorption, 24 h (%)	0.20	0.04	0.04	0.02	0.018	0.015	0.04	0.03
Equilibrium after continuous immersion	–	–	0.10	–	–	0.24	–	–
Mold shrinkage 1/8-in Average section (10^{-3} in/in)	10	2	1	15–20	2–3	1.5–2.5	–	2.5–3.5
Tensile strength (lb_f/in^2)	10,800	20,000	27,000	6,500	14,000	15,000	5,500	12,300
Tensile elongation (%)	3–4	3–4	2–3	150	4–5	2–3	100	3–4
Flexural strength (lb_f/in^2)	20,000	29,000	34,000	10,500	19,000	20,000	7,000	17,200
Flexural modulus ($\times 10^6$ lb_f/in^2)	0.60	1.60	2.45	0.20	1.05	1.65	0.20	1.10
Shear strength (lb_f/in^2)	–	–	–	6,000	7,000	7,000	6,000	7,500
Impact strength, Izod (ft-lb/in) Notched, 1/4-inch	0.3	1.4	1.1	NB	7–8	4–5	2.3	2.6
Unnotched, 1/4-inch	3–4	8–9	5–6	NB	17–18	10	–	7–8
Heat distortion temperature at 264 lb_f/in^2 (°F)	278	500	500	165	460	465	190	295
Coefficient of thermal expansion ($\times 10^{-5}$ in/in°F)	3.0	1.3	0.6	4.2	1.7	0.8	8.5	3.0
Thermal conductivity (Btu-in/h ft² °F)	2.0	2.8	5.2	1.65	2.8	5.6	2.1	3.0
Surface resistivity (Ω/sq)	10^{16}	10^{16}	1–3	5×10^{14}	10^{15}	3–4	5×10^{14}	3–5
Flammability	VE-O	VE-O	VE-O	VE-O	VE-O	VE-O	VE-O	VE-O

NB, did not break.

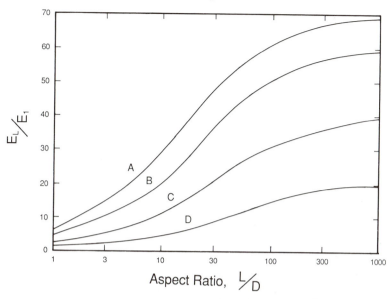

Figure 2.70.   The Effect of Discontinuous Fiber Aspect Ratio on Relative Longitudinal Young's Modulus. Fiber-to-Resin Matrix Modulus Ratio, $E_2/E_1 = 100$. Curve A: Fiber Vol Fraction, $\phi_2 = 0.7$. B: $\phi_2 = 0.6$. C: $\phi_2 = 0.4$. D: $\phi_2 = 0.2$ (112).

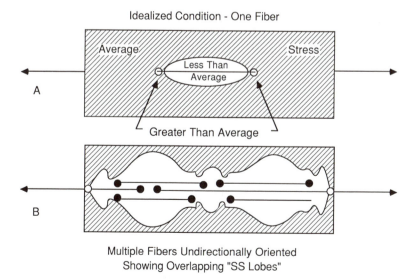

Figure 2.71.   Schematic of Stress Distribution Around a Single and Overlapping Multiple Fibers. Shaded Area: Average Stress. Open Area: Less Than Average Stress (113).

where E_L, E_p, and E_f are the moduli of the composite in the longitudinal direction, polymer matrix, and fiber; ϕ is the volume fraction of fiber in the composite. In the transverse direction, the fiber appears as if it is a filler. Thus, the Nielsen equation applies, with appropriate selection for the maximum packing fraction. For short fibers ($L/D < 100$ and typically 25 to 50), the equation above overestimates the longitudinal modulus. On the other hand, direct application of the Nielsen equation for compact fillers underestimates the modulus. This is seen in Figure 2.70 for fiber-to-polymer modulus ratio of 100. For fibers that are oriented randomly in a plane, the modulus can be obtained from (113):

$$(2.51) \qquad E_r = 0.375\, E_L + 0.625\, E_T$$

E_r, E_L and E_T are the two-dimensional, longitudinal and tangential moduli, calculated from appropriate uniaxial fiber equations. The loss in strength when fibers become short is due primarily to increasing frequency of stress concentration at the fiber tips. For a single fiber (Figure 2.71), stress is concentrated only at the fiber tip. The schematic representation is known as a "stress/strain lobe" or simply a SS lobe. For multiple fibers (Figure 2.71), the stress concentration is not uniform along the fiber bundle. For high modulus matrix polymers, the SS lobe will be short and wide. Thus short fibers at low concentrations can be used. For most polymers, the SS lobe will be long and narrow, as seen in Figure 2.72. Thus, fibers with high L/Ds, high moduli, and

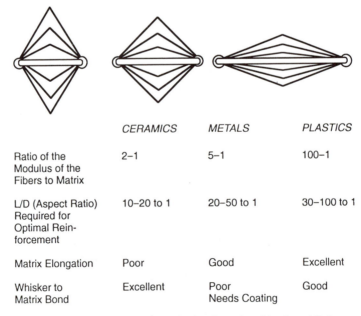

	CERAMICS	METALS	PLASTICS
Ratio of the Modulus of the Fibers to Matrix	2–1	5–1	100–1
L/D (Aspect Ratio) Required for Optimal Reinforcement	10–20 to 1	20–50 to 1	30–100 to 1
Matrix Elongation	Poor	Good	Excellent
Whisker to Matrix Bond	Excellent	Poor Needs Coating	Good

Figure 2.72.   A Comparison of the "SS" Lobe for Ceramics, Metals and Polymers (115).

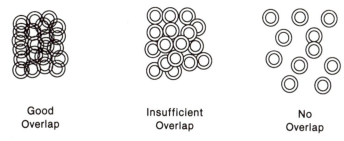

| Good
Overlap | Insufficient
Overlap | No
Overlap |

Figure 2.73. A Schematic of the Effect of Fiber Population on the Distribution of Stresses in Composite Structures (116).

high concentrations are needed to insure sufficient overlapping of the SS lobe (Figure 2.73) to prevent early failure.

Composite technology is one of the most rapidly growing new areas in polymers. Additional information about the mechanical nature of fiber-reinforced polymers can be found in (117,118).

References

1. J.L. Throne, *Plastics Process Engineering*, Marcel Dekker, New York (1979), p. 106.
2. R.M. Ogorkiewicz, *Thermoplastics*, John Wiley & Sons, New York (1970), p. 17.
3. W.J. Freeman, "Characterization of Polymers", in *Encyclopedia of Polymer Science and Engineering*, Vol. 3, 2nd Ed., John Wiley & Sons, New York (1985), p. 290.
4. G.L. Beyer, "Molecular Weight Determination", in N.M. Bikales, Ed., *Characterization of Polymers*, Wiley-Interscience, New York (1971), p. 58.
5. B.G. Belenkii and L.Z. Vilenchik, *Modern Liquid Chromatography of Macromolecules*, Elsevier, New York (1983).
6. M. Kurata, M. Iwama and K. Kamada, in "Viscosity-Molecular Weight Relationships", in J. Brandrup and E.H. Immergut, Eds., *Polymer Handbook*, John Wiley & Sons, New York (1964), pp. 1–4.
7. P.J. Flory, *Principles of Polymer Chemistry*, Cornell University Press, Ithaca, NY (1953), p. 311.
8. P.J. Flory, *Principles of Polymer Chemistry*, Cornell University Press, Ithaca NY (1953), p. 311.
9. F.W. Billmeyer, Jr., *Textbook of Polymer Science*, Wiley-Interscience, New York (1966), p. 17.
10. F.W. Billmeyer, Jr., *Textbook of Polymer Science*, Wiley-Interscience, New York (1966), Chapter 3.
11. P.J. Flory, *Principles of Polymer Chemistry*, Cornell University Press, Ithaca, NY (1953), Chapter 3.
12. P.H. Geil, "Polymer Morphology", in H.S. Kaufman and J.J. Falcetta, Eds., *Introduction to Polymer Science and Technology*, Wiley-Interscience, New York (1977), p. 244.
13. P.H. Geil, "Polymer Morphology", in H.S. Kaufman and J.J. Falcetta, Eds., *Introduction to Polymer Science and Technology*, Wiley-Interscience, New York (1977), pp. 208–238.

14. D. Williams, *Polymer Science and Engineering*, Prentice-Hall, New York (1971).

15. J.M. Schultz, *Polymer Materials Science*, Prentice-Hall, New York (1974).

16. Z. Tadmor and C. Gogos, *Principles of Polymer Processing*, Wiley-Interscience, New York (1979).

17. M. Kohan, *Nylon Plastics*, Wiley-Interscience, New York (1973).

18. A. Tobolsky and J. Mark, *Polymer Science and Materials*, Wiley-Interscience, New York (1971).

19. H.-G. Elias, *Macromolecules: Structure and Properties*, Vol. 1, 2nd Ed., Plenum, New York (1984).

20. H. Tadokoro, *Structure of Crystalline Polymers*, Wiley-Interscience, New York (1979).

21. D.C. Bassett, *Principles of Polymer Morphology*, Cambridge University Press, Cambridge (1981).

22. S.L. Rosen, *Fundamental Principles of Polymer Materials*, Wiley-Interscience, New York (1982), p. 39.

23. D.G. Bright, "Quantitative Studies of Polymer Crystallization Under Non-Isothermal Conditions", PhD Dissertation, Georgia Institute of Technology (1975), p. 32.

24. D.G. Bright, "Quantitative Studies of Polymer Crystallization Under Non-Isothermal Conditions", PhD Dissertation, Georgia Institute of Technology (1975), p. 14.

25. I.I. Rubin, *Injection Molding*, John Wiley & Sons, New York (1972), p. 191.

26. D.G. Bright, "Quantitative Studies of Polymer Crystallization Under Non-Isothermal Conditions", PhD Dissertation, Georgia Institute of Technology (1975), p. 14.

27. W.P. Brennan, "Characterization and Quality Control of Engineering Thermoplastics by Thermal Analysis", Thermal Analysis Application Study 22, Perkin-Elmer, Norwalk, CT (1974), p. 10.

28. D.G. Bright, "Quantitative Studies of Polymer Crystallization Under Non-Isothermal Conditions", PhD Dissertation, Georgia Institute of Technology (1975), p. 30.

29. F.W. Billmeyer, Jr., *Textbook of Polymer Science*, Wiley-Interscience, New York (1966), p. 161.

30. F. Rodriguez, *Principles of Polymer Systems*, McGraw-Hill, New York (1970), p. 37.

31. M.L. Miller, *The Structure of Polymers*, Reinhold, New York (1966), p. 529.

32. B. Ke, "Differential Thermal Analysis", in B. Ke, Ed., *Newer Methods of Polymer Characterization*, Polymer Review Series No. 6, Wiley-Interscience (1964), p. 36.

33. W.P. Brennan, "Characterization of Polyethylene Films by Differential Scanning Calorimetry", Thermal Analysis Application Study 24, Perkin-Elmer, Norwalk, CT (1978), p. 10.

34. B. Ke, "Differential Thermal Analysis", in B. Ke, Ed., *Newer Methods of Polymer Characterization*, Polymer Review Series No. 6, Wiley-Interscience (1964), p. 37.

35. H.-G. Elias, *Macromolecules: Structure and Properties*, Vol. 1, 2nd Ed., Plenum, New York (1984), p. 382.

36. B. Ke, "Differential Thermal Analysis", in N.M. Bikales, Ed., *Characterization of Polymers*, Wiley-Interscience, New York (1971), p. 200.

37. H.-G. Elias, *Macromolecules: Structure and Properties*, Vol. 1, 2nd Ed., Plenum, New York (1984), Figure 4.7, pp. 100–101.

38. D.G. Bright, "Quantitative Studies of Polymer Crystallization Under Non-Isothermal Conditions", PhD Dissertation, Georgia Institute of Technology (1975), pp. 58–63.

39. B. Ke and W.A. Sisko, J. Polym. Sci., *50* (1961), p. 87.

40. J.M. Schultz, *Polymer Materials Science*, Prentice-Hall, New York (1974), p. 61.

41. M. Takayanagi and F. Nagatashi, Mem. Fac. Eng., Kyushu Univ., *4* (1965), p. 33.

42. J.M. Schultz, *Polymer Materials Science*, Prentice-Hall, New York (1974), Chapter 2.

43. D.P. Askeland, *The Science and Engineering of Materials*, PWS Engineering, Boston (1984), p. 59.

44. H.-G. Elias, *Macromolecules: Structure and Properties*, Vol. 1, 2nd Ed., Plenum, New York (1984), p. 156.

45. J.M. Schultz, *Polymer Materials Science*, Prentice-Hall, New York (1974). p. 499.

46. H.-G. Elias, *Macromolecules: Structure and Properties*, Vol. 1, 2nd Ed., Plenum, New York (1984), p. 159.

47. H.-G. Elias, *Macromolecules: Structure and Properties*, Vol. 1, 2nd Ed., Plenum, New York (1984). p. 193.

48. J.M. Schultz, *Polymer Materials Science*, Prentice-Hall, New York (1974), pp. 156–162.

49. R.J. Samuels, J. Polym. Sci., *20C* (1967), p. 253.

50. T. Huseby and S. Matsuoka, Mod. Plast., *41*:11 (1964), p. 117.

51. T. Huseby and S. Matsuoka, Mod. Plast., *41*:11 (1964), p. 117.

52. E.S. Clark, Appl. Polym. Symp., *20* (1973), p. 325.

53. A.J. Penning, J.M.M.A. van der Mark, and A.M. Kiel, Kolloid Z., *237* (1970), p. 336.

54. B. Wunderlich, *Macromolecular Physics*, Academic Press, New York (1973); see also, H.-G. Elias, *Macromolecules: Structure and Properties*, Vol. 1, 2nd Ed., Plenum, New York (1984), p. 182.

55. M. Avrami, J. Chem. Phys., *7* (1939), p. 1103.

56. M. Avrami, J. Chem. Phys., *8* (1940), p. 212.

57. M. Avrami, J. Chem. Phys., *9* (1941), p. 17.

58. L. Mandelkern, *Crystallization of Polymers*, McGraw-Hill, New York (1964).

59. L. Marker, P.M. Hay, G.P. Tilby, R.M. Early and O.J. Sweeting, J. Polym. Sci., *38* (1959), p. 36.

60. L. Marker, P.M. Hay, G.P. Tilby, R.M. Early and O.J. Sweeting, J. Polym. Sci., *38* (1959), p. 33.

61. H.-G. Elias, *Macromolecules: Structure and Properties*, Vol. 1, 2nd Ed., Plenum, New York (1984), Table 10.2, p. 395.

62. W. Sifleet, "Unsteady Heat Transfer in a Crystallizing Polymer", MS ChE Thesis, Ohio University, Athens (1970).

63. A. Ziabicki, *Fundamentals of Fibre Formation: The Science of Fibre Spinning and Drawing*, John Wiley & Sons, New York (1976), p. 105.

64. A. Ziabicki, *Fundamentals of Fibre Formation: The Science of Fibre Spinning and Drawing*, John Wiley & Sons, New York (1976), p. 112.

65. A. Ziabicki, *Fundamentals of Fibre Formation: The Science of Fibre Spinning and Drawing*, John Wiley & Sons, New York (1976), p. 113.

66. H.-G. Elias, *Macromolecules: Structure and Properties*, Vol. 1, 2nd Ed., Plenum, New York (1984), Table 10.3, p. 395.

67. K. Steiner, K.J. Lucas and K. Ueberreiter, Kolloid Z., *214* (1966), p. 23.

68. J.M. Schultz, *Polymer Materials Science*, Prentice-Hall, New York (1974), p. 187.

69. A. Ziabicki, *Fundamentals of Fibre Formation: The Science of Fibre Spinning and Drawing*, John Wiley & Sons, New York (1976), p. 112.

70. T. Kanai and J.L. White, Polym. Eng. Sci., *24* (1984), p. 1185.

71. K. Nakamura, T. Watanabe, K. Katayama and T. Amano, J. Appl. Polym. Sci., *18* (1974), p. 616.

72. A. Ziabicki, *Fundamentals of Fibre Formation: The Science of Fibre Spinning and Drawing*, John Wiley & Sons, New York (1976), p. 115.

73. J.L. Throne, *Plastics Process Engineering,* Marcel Dekker, New York (1979), pp. 558–564.

74. J.A. Brydson, *Flow Properties of Polymer Melts,* Van Nostrand Reinhold, New York (1970), pp. 95–97.

75. D.W. Van Krevelen, *Properties of Polymers: Their Estimation and Correlation with Chemical Structure,* Elsevier Scientific, New York (1976), p. 22.

76. S.L. Rosen, *Fundamental Principles of Polymer Materials,* Wiley-Interscience, New York (1982), pp. 67–72.

77. C.D. Armeniades and E. Baer, "Transitions and Relaxations in Polymers", in H.S. Kaufman and J.J. Falcetta, Eds., *Introduction to Polymer Science and Technology,* Wiley-Interscience, New York (1977), p. 274.

78. L. Nielsen, *Mechanical Properties of Polymers,* Reinhold Publishing Co., New York (1958), Chapter 2.

79. W.P. Brennan, "T_g and Plasticizers", Thermal Analysis Applications Study 11, Perkin-Elmer, Norwalk, CT (1973), p. 5.

80. T.G. Fox and P.K. Flory, J. Polym. Sci., *14* (1954), p. 315.

81. K.H. Von Hellwege, W. Knappe and P. Lehman, Kolloid Z.: Polymere, *183* (1962), p. 110.

82. W. Brennan, "Characterization and Quality Control of Engineering Thermoplastics by Thermal Analysis", Perkin-Elmer Thermal Analysis Applications Study 22 (1977), p. 13.

83. A.P. Gay, "Establishing a Correlation Between the Degree of Cure and the Glass Transition Temperature of Epoxy Resins", Perkin-Elmer Thermal Analysis Applications Study 2 (1972), p. 8.

84. A.P. Gay, "Establishing a Correlation Between the Degree of Cure and the Glass Transition Temperature of Epoxy Resins", Perkin-Elmer Thermal Analysis Applications Study 2 (1972), p. 9.

85. R.D. Deanin, *Polymer Structure, Properties and Applications,* Cahners, Boston (1972), p. 246.

86. D. Hansen and C. Ho, J. Polym. Sci. A, *3* (1965), p. 665.

87. K. Eiermann and K.H. Hellwege, J. Polym. Sci., *57* (1962), p. 102.

88. K. Eiermann, J. Polym. .Sci. C, *6* (1968), p. 157.

89. D. Hands and F. Horsfall, J. Phys. E, *8* (1975), p. 689.

90. F. Kreith and W.Z. Black, *Basic Heat Transfer,* Harper and Row, New York (1980), Chapter 4.

91. R.H. Shoulberg, J. Appl. Polym. Sci., *7* (1963), p. 1601.

92. F. Kreith and W.Z. Black, *Basic Heat Transfer,* Harper and Row (1980), p. 144.

93. W.M. Underwood and J.R. Taylor, Polym. Eng., Sci., *18* (1978), p. 561.

94. K. Eiermann and K.H. Hellwege, J. Polym. Sci., *57* (1962), p. 101.

95. J.L. Milewski and H.S. Katz, "Introduction", in J.L. Milewski and J.S. Katz, Eds., *Handbook of Fillers and Reinforcements for Plastics,* Van Nostrand Reinhold, New York (1978), p. 5.

96. L.E. Nielsen, *Mechanical Properties of Polymers and Composites,* Vol. 2, Marcel Dekker, New York (1974), p. 382.

97. L.E. Nielsen, *Mechanical Properties of Polymers and Composites,* Vol. 2, Marcel Dekker, New York (1974), p. 382.

98. J.V. Milewski, "Packing Concepts in the Utilization of Filler and Reinforcement Combinations", in H.S. Katz and J.V. Milewski, Eds., *Handbook of Fillers and Reinforcements for Plastics,* Van Nostrand Reinhold, New York (1978), p. 68.

99. C. Orr and D. Dalla Valle, *Fine Particle Measurements, Size Surface and Pore Volume*, Macmillan, New York (1960), Chapter 2.

100. J.V. Milewski, "Packing Concepts in the Utilization of Filler and Reinforcement Combinations", in H.S. Katz and J.V. Milewski, Eds., *Handbook of Fillers and Reinforcements for Plastics*, Van Nostrand Reinhold, New York (1978), p. 69.

101. J.V. Milewski, "Packing Concepts in the Utilization of Filler and Reinforcement Combinations", in H.S. Katz and J.V. Milewski, Eds., *Handbook of Fillers and Reinforcements for Plastics*, Van Nostrand Reinhold, New York (1978), p. 70.

102. T.H. Ferrigno, "Principles of Filler Selection and Use", in H.S. Katz and J.V. Milewski, Eds., *Handbook of Fillers and Reinforcements for Plastics*, Von Nostrand Reinhold, New York (1978), pp. 20–21.

103. T.H. Ferrigno, "Principles of Filler Selection and Use", in H.S. Katz and J.V. Milewski, Eds., *Handbook of Fillers and Reinforcements for Plastics*, Van Nostrand Reinhold, New York (1978), p. 19.

104. L.E. Nielsen, *Mechanical Properties of Polymers and Composites*, Vol. 2, Marcel Dekker, New York (1974), p. 460.

105. J. O'Toole, in *Modern Plastics Encyclopedia*, 63:10A (1986), p. 500.

106. R.M. Ogorkiewicz, *Thermoplastics*, John Wiley & Sons, New York (1970), p. 141.

107. T.H. Ferrigno, "Principles of Filler Selection and Use", in H.S. Katz and J.V. Milewski, Eds., *Handbook of Fillers and Reinforcements for Plastics*, Van Nostrand Reinhold, New York (1978), p. 30.

108. T.H. Ferrigno, "Principles of Filler Selection and Use", in H.S. Katz and J.V. Milewski, Eds., *Handbook of Fillers and Reinforcements for Plastics*, Van Nostrand Reinhold, New York (1978), p. 55.

109. J.G. Mohr, "Fiber Glass", in H.S. Katz and J.V. Milewski, Eds., *Handbook of Fillers and Reinforcements for Plastics*, Van Nostrand Reinhold, New York (1978), p. 481.

110. D.L.G. Sturgeon and R.I. Lacy, "High Modulus Organic Fibers", in H.S. Katz and J.V. Milewski, Eds., *Handbook of Fillers and Reinforcements for Plastics*, Van Nostrand Reinhold, New York (1978), p. 539.

111. H.S. Katz, "Carbon–Graphite Filaments", in H.S. Katz and J.V. Milewski, Eds., *Handbook of Fillers and Reinforcements for Plastics*, Van Nostrand Reinhold, New York (1978), pp. 573–575.

112. L.E. Nielsen, *Mechanical Properties of Polymers and Composites*, Vol. 2, Marcel Dekker, New York (1978), p. 460.

113. L.E. Nielsen, *Mechanical Properties of Polymers and Composites*, Vol. 2, Marcel Dekker, New York (1974), p. 462.

114. J.V. Milewski and H.S. Katz, "Whiskers", in H.S. Katz and J.V. Milewski, Eds., *Handbook of Fillers and Reinforcements for Plastics*, Van Nostrand Reinhold, New York (1978), p. 455.

115. J.V. Milewski and H.S. Katz, "Whiskers", in H.S. Katz and J.V. Milewski, Eds., *Handbook of Fillers and Reinforcements for Plastics*, Van Nostrand Reinhold, New York (1978), p. 456.

116. J.V. Milewski and H.S. Katz, "Whiskers", in H.S. Katz and J.V. Milewski, Eds., *Handbook of Fillers and Reinforcements for Plastics*, Van Nostrand Reinhold, New York (1978), p. 457.

117. M. Grayson, *Encyclopedia of Composite Materials and Components*, Wiley-Interscience, New York (1983).

118. "Composites", *ASM International Engineered Materials Handbook*, Vol. 1, American Society for Metals, Metals Park, OH (1988).

Glossary

Amorphous   Non-crystalline.

Annealing   Holding a polymer at an elevated temperature for an extended period of time.

Aspect Ratio   Length-to-diameter ratio.

Biaxial Orientation   Two-dimensional stretching of polymer film or sheet.

Birefringence   Difference in indices of refraction, measured separately along two principal directions.

Composite   Polymer containing a high concentration of reinforcing fibers.

Crystalline   Having a degree of order or regularity on a molecular level.

Crystalline Tendency   Characteristic of a polymer with the tendency to form a crystalline structure.

Crystallite   Ordered polymer structure.

Crystallography   The study of the structure of crystalline materials.

Dendritic   Rapid crystal growth that results in Christmas-treelike crystalline morphology.

Diffusivity   In heat transfer, a measure of the movement of energy through a material. In mass transfer, a measure of the movement of molecules through one another.

Dilatometry   Measurement of the change in volume.

DSC   Differential scanning calorimetry.

End Group   The functional structure on each end of a polymer chain.

Extended Chain   A crystalline structure where the polymer chains lie parallel to one another, in bundle fashion, for substantial distances.

First-Order Transition   Change in the value of specific volume curve at a given temperature. Denotes melting or freezing.

Folded Chain   Flexible polymer chains form lamellae by folding back upon themselves.

Free Volume   The measured difference between the theoretical molecular packing density of a polymer and its actual density. The space is submolecular in dimension.

Glass Transition Temperature   The temperature range where substantial backbone mobility, of 20 to 50 repeat units, is achieved. Below this range, the polymer is glassy–brittle. Above, it is rubbery–ductile.

Glassy Polymer   A polymer below its glass transition temperature, or alternatively, one below its brittle temperature.

Half-Time   The time required for a polymer to achieve half its maximum specific volume change due to crystallization at a given temperature.

Heat Capacity   Rate of change of enthalpy per unit temperature change.

Heat Flux   Heat flow per unit area.

Heterogeneous Nucleation   Crystal growth initiating from insoluble impurities in the polymer.

Homogeneous Nucleation   Crystal growth initiating solely from aggregation of polymer chains.

Induction Time   The time prior to measured beginning of isothermal crystallization.

Intrinsic Viscosity   Extrapolated limit of reduced viscosity, as polymer concentration goes to zero.

Kinetics   The study of the time- or rate-dependency of a system. For example, for polymers, the study of the rate of crystallization.

Lamella   Ribbon-like structure of a growing crystallite.

Laminate   Polymer containing reinforcing elements in very specific patterns.

Lubricant   Additive that increases polymer processing range. See *Plasticizer.*

Modulus of Toughness   Initial slope of an isothermal stress-strain curve.

Morphology   Study of the structure of matter.

Molecular Weight Distribution   The distribution of molecular weights of polymer chains.

Nucleant   Additive that induces phase change activity.

Nucleation   The mechanism for initial formation of a growing crystallite.

Permittivity   Dielectric constant.

Plasticizer   Additive that increases polymer processing range and reduces its mechanical strengths.

Primary Crystallization   Unimpeded growth of polymer crystals.

Reduced Viscosity   Relative viscosity, divided by polymer concentration.

Relative Viscosity   Ratio of the viscosity of a polymer dissolved in a solvent to that of the solvent alone.

Repeat Unit   That portion of the monomer that occurs repeatedly as an identifiable unit in the polymer chain.

Reptation   Snake-like motion.

Rubbery Polymer   A polymer above its glass transition temperature.

SAXS   Small-angle X-ray scattering.

Secondary Crystallization   Crystal growth after polymer crystals have begun to impinge.

Second-Order Transition   Change in the slope of specific volume curve at a specific temperature. Denotes glass transition temperature.

SEM   Scanning electron microscope.

Specific Heat   Rate of change of enthalpy with temperature.

Specific Viscosity   Relative viscosity minus one.

Spherulite   A polymer crystallite that is grown quiescently as a sphere.

SS Lobe   Stress–strain lobe; stress concentration around a fiber tip in a matrix under strain.

Thermal Conductivity   Property of a substance that characterizes the steady-state transfer of thermal energy.

Thermal Diffusivity   Property of a substance that characterizes the transient transfer of thermal energy.

Thermal Expansion Coefficient   Slope of the temperature-dependent dimension.

Toughness   Area under an isothermal stress–strain curve.

Uniaxial Orientation   One-dimensional stretching of polymer film or sheet.

WAXS   Wide-angle X-ray scattering.

CHAPTER 3

MECHANICAL PROPERTIES
OF POLYMERS

Reed

190 - 201
202 - 212
257 · 266
267 - 272
273 - 276

3.1 Introduction to Classical Materials

All materials have traditionally been classified as either *viscous fluids* or *elastic solids*. Water, air, oils and greases are viscous fluids. Ferrous metals, such as steel and iron, non-ferrous metals, such as aluminum and zinc, and non-metals such as concrete and wood are considered to be elastic solids. This classification has worked well in most traditional designs. Comprehensive design technologies have been developed over the years for predicting the responses of elastic solids or viscous fluids to applied forces. Design limits have also been developed under which these different materials react as *classical materials*. Plastics do not usually respond to applied forces in manners similar to either of the classical material types. This is true whether the plastics are solid and carrying a load or are liquids and are being shaped by flow. Polymeric responses to applied loads may include responses of both viscous liquids and elastic solids simultaneously. A polymer that exhibits varying degrees of these combined responses is called a *viscoelastic material*.

The designer, engineer or technologist should recognize that there can be a very broad spectrum of material responses to applied forces. The purely viscous fluid and the elastic solid are the two extremes of the practical spectrum of material responses*. Under certain environmental conditions, some classical materials exhibit viscoelastic characteristics. The most notable example is the load response of metals at high temperatures. An elastic material elongates a fixed distance under a given load. If the load remains constant, the material dimensions remain unchanged. Metals are considered to be classical materials. However, the metal blades of a jet engine operating at very high temperatures and high centrifugal forces can elongate with time. This time-dependent elongation, blade creep, is a characteristic response of viscoelastic materials, not elastic ones.

---

*A fluid having no viscosity is called an *inviscid fluid*. It is an ideal case of fluid flow, with some application in boundary layer theory. A fluid having an infinite elastic modulus is called an *ideal solid*. It is the ideal case of elastic solids. These represent the *true* extremes in material response, but have no practical application in polymers.

This chapter begins with a consideration of the response of classical materials to applied load or deformation. Then responses of polymeric materials to similar loading conditions are compared. It will be apparent from the examples that polymers are some of the few design materials that exhibit viscoelastic behavior over the full range of design and manufacturing conditions. To characterize viscoelastic response to applied load (that is, the interaction between applied stress and material strain and rate of strain), combinations of mechanical elements such as springs and viscous dashpots are used. These elements serve only for illustration and are not quantitative analogs to real polymers. The responses of real polymers are considerably more complex than the responses of mechanical analogs.

Note that ultimate properties and ultimate design strengths are not considered in this chapter. These quantities are best determined from experiments designed to measure the specific viscoelastic response of a specific polymer to a specific loading sequence for the product lifetime. These data are then used to formulate specific failure criteria. This subject forms the basis for Chapter 6.

3.2 Stress

Stress is defined as the *intensity of force* (1), that is, the force per unit area perpendicular or normal to the force. Consider a bar with a uniform cross-sectional area, Figure 3.1. A force S is applied axially to the bar. Axial forces that tend to elongate the bar are called *tensile* forces. Those that tend to contract the bar are *compressive* forces. By sign convention, tensile forces are positive and compressive forces are negative. When the force is applied uniformly across the bar, the stress is called a *normal stress*, denoted as σ. The normal stress in Figure 3.1 is given as:

(3.1) $\sigma = S/A$

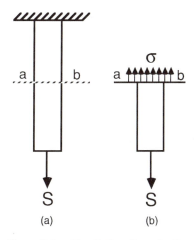

Figure 3.1.   Tensile Loading of a Bar.

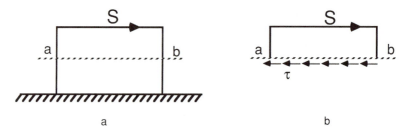

Figure 3.2.   Shear Loading of a Block.

where A is the cross-sectional area normal to the applied force. If the force, S, is applied parallel to the *surface* of the bar, the resulting stress is called a *shear stress*, Figure 3.2, and is denoted as τ. The shear stress is given as:

(3.2) $$\tau = S/A_s$$

where A_s is the surface area parallel to the applied force. In many practical applications, the applied load may not be purely normal or parallel to a given cross-section. The resultant stress, s, is then resolved into normal stress and shear stress components, Figure 3.3. Note that the concept of combined stress also holds for other types of applied load, such as torsion with compression or torsion under tension.

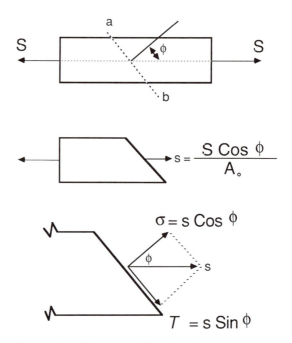

Figure 3.3.   Schematic of Combined Stresses on a Bar.

Most standard mechanical property tests apply simple stress, such as simple tension, simple compression or simple shear, to the specimen. The specimen is designed to produce measurable strain only in the direction of applied stress. If the practical application of these test results involves a combined stress mode, a more thorough understanding of the complexity of combined stress is essential for proper interpretation. The concept of combined stress and the resolution of stress into the three-dimensional principal components is detailed in standard texts on strength of materials and stress analysis.

3.3 Strain

The application of stress to any object usually results in measurable deformation. The application of a simple stress results in a deformation in the direction of stress application. Consider tensile stress and strain. The local deformation per unit length of the body is the material strain, given as ε:

(3.3)
$$\varepsilon = \frac{\Delta L}{L}$$

If the applied force results in a normal stress, the resulting deformation is called a normal strain. If the applied force results in a shear stress, the resulting deformation is a shear strain. Consider the applied tensile stress in Figure 3.4. At some distance from

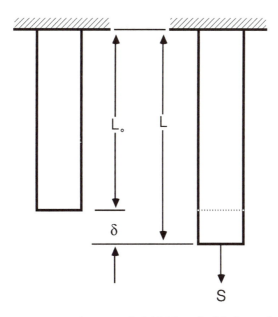

Figure 3.4.   Axial Deformation of a Bar. L_0 Is Initial Length, L Is Instant Length, δ Is Increase in Length owing to Load. S.

the point of application of the force, the differential strain at every point along the bar axis will be uniform. There are two ways of defining strain. The first is in terms of the initial length, L_o, and the second is in terms of the time-dependent length, L_t. The first is referred to as *engineering strain*. If the deformation, δ, is small when compared with the original bar length, L_o, the strain is given as:

(3.4)
$$\varepsilon = \frac{L - L_o}{L_o} = \frac{\delta}{L_o}$$

and is called the engineering strain. *True strain* is given by summing all the differential strains over the bar length:

(3.5)
$$\varepsilon = \int_{L_o}^{L} \frac{dL}{L} = \ln \left(\frac{L}{L_o} \right)$$

Similarly, there are two ways of defining applied stress. The first is given in terms of the initial cross-sectional area, A_o, and the second is in terms of the time-dependent cross-sectional area, A_t. The combinations yield four ways of defining stress-strain behavior, where F_t is the time-dependent tensile force:

Strain	Stress		
$\varepsilon = \Delta L/L_t$	$\sigma = F_t/A_t$	\} True Stress	(3.6)
$\varepsilon = \Delta L/L_o$	$\sigma = F_t/A_t$		(3.7)
$\varepsilon = \Delta L/L_t$	$\sigma = F_t/A_o$	\} Engineering Stress	(3.8)
$\varepsilon = \Delta L/L_o$	$\sigma = F_t/A_o$		(3.9)

For materials that have relatively low design limits on strain (brittle metals such as steel and cast iron, and ceramics), the engineering strain is a good approximation to the true strain. For highly ductile materials such as soft metals (gold, copper) and most plastics, the approximation is poor at high strain levels. For example, at an engineering strain level of 20%, the error in approximating true strain is about 10%.

The rate of *deformation* when the applied force is normal to the cross-sectional area of the specimen is the rate of change of strain with time, and is denoted as $\dot{\varepsilon}$:

(3.10)
$$\dot{\varepsilon} = \lim_{\Delta t \to 0} \frac{\Delta \varepsilon}{\Delta t}$$

When the applied force is in shear, the effect is angular deformation, Figure 3.5. The shearing strain, τ, is given as:

(3.11)
$$\gamma = \frac{a}{b} = \tan \theta$$

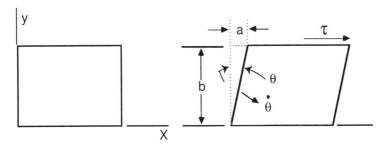

Figure 3.5.   Shear Deformation of a Block. τ Is the Applied Shear Stress, θ Is the Instant Deformation Angle. $\dot{\theta}$ Is the Instant Rate of Change of the Deformation Angle.

and has the units of radians. For very low strain levels, this can be approximated as:

(3.12) $$\gamma \approx \theta$$

The rate of shearing strain or rate of shear deformation, $\dot{\gamma}$, is given as:

(3.13) $$\dot{\gamma} = \frac{\lim}{\Delta t \to 0} \frac{\Delta \theta}{\Delta t} = \frac{d\theta}{dt}$$

The rate of shear deformation, deformation rate, strain rate, or rate-of-strain, has the units of rad/s or simply, s^{-1}, and can be considered to be an instantaneous angular velocity.

3.4 Constitutive Equation

The material strain response to deforming forces is called a "constitutive equation of state". Applied stress can be divided into components in three principal directions and each component can be divided into components normal and parallel, in shear, to each component direction. The action of applied stress is time-dependent deformation in each of the component directions. The defining relationship is a second-order tensor constitutive equation of state between stress, strain and rate-of-strain. If the extent of deformation is relatively small, the coordinate system can remain fixed in space. Stationary coordinate systems are used primarily for elastic deformation, solid mechanical plastic deformation, linear viscoelasticity and certain very restricted forms of elastic liquid deformation, such as creep. If substantial deformation results, the stationary coordinate system becomes cumbersome. A coordinate system that moves with the deforming element is desired and the constitutive equation of motion is said to be in "convected" coordinates. Viscous and viscoelastic fluid flow are best described with convected coordinates. Other viscous flow parameters are discussed in detail in Chapter 4.

The simplest models for solids and liquids are those whose constitutive relationships depend only on displacement, for solids, and rate of displacement, for liquids. These simple models form the basis for understanding the response of classical materials to applied stresses.

The Purely Elastic Material

A material that exhibits a linear relationship between stress and strain is called a *Hookean solid.* A Hookean material elongates a fixed amount when a fixed load is applied. Doubling the load results in doubling the elongation. If the load remains constant, the elongation will remain unchanged, regardless of the time under load. Materials that are considered purely Hookean in normal applications do exhibit non-Hookean behavior under some conditions. The jet engine blade creep discussed earlier is an example of time-dependent behavior and not a characteristic response of a purely Hookean material.

The Hookean proportionality between stress and strain is *modulus*. In tension, the modulus is Young's modulus or the modulus of elasticity, E. In compression, it is the compressive modulus. In shear, it is the shear modulus or the modulus of rigidity, G:

(3.14)
(3.15)

$$\text{Tensile:} \quad \sigma = E\,\varepsilon$$
$$\text{Shear:} \quad \tau = G\,\gamma$$

Hooks Law

A Hookean material is purely elastic in that it responds instantaneously to changes in applied load and always returns to a predetermined deformation state, Figure 3.6. Since Hookean solid deformation is a function only of the applied stress, the representation of stress–strain for various rates of strain results in a single line with the modulus

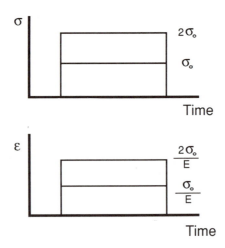

Figure 3.6.   Schematic of the Response of a Hookean Solid, (Lower Graph) to a Step Change in Applied Load, (Upper Graph). σ_0 Is the Applied Stress, $\varepsilon = \sigma_0/E$ Is the Resulting Strain. E Is the Young's Modulus.

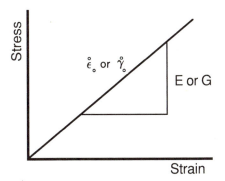

Figure 3.7.   Strain Rate Response of a Hookean or Linear Elastic Member to Applied Stress. $\dot{\varepsilon}_o$ for Tensile Loading and $\dot{\gamma}_o$ for Shear Loading Follow Along the Stress–Strain Line Having a Slope E, Young's Modulus, for Tensile Loading and G, Shear Modulus, for Shear Loading.

value as the slope, Figure 3.7. The mechanical analog for a Hookean material is a linear elastic spring, Figure 3.8.

Example 3.1

Q: The structure shown in Figure 3.9 is subjected to an axial load of 8000 lb$_f$. a: Determine the stresses in sections 1, 2 and 3. b: What is the displacement of the right surface if the material has a modulus of 1×10^6 lb/$_f$/in^2?

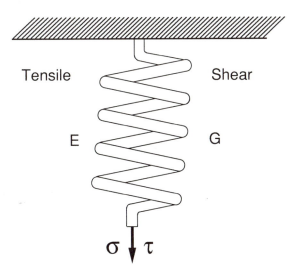

Figure 3.8.   Mechanical Analog of a Hookean or Linear Elastic Element. The Applied Load Is σ in Tension and τ in Shear. The Spring Constant Is E, Young's Modulus, in Tension and G, Shear Modulus in Shear.

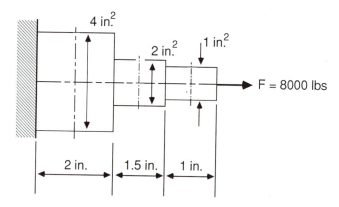

Figure 3.9.   Load-Bearing Member (Example 3.1).

A: a: The stresses are:

$$\sigma_1 = P/A = 8000 \text{ lb}_f/4 \text{ in}^2 = 2000 \text{ lb}_f/\text{in}^2$$

$$\sigma_2 = 4000 \text{ lb}_f/\text{in}^2$$

$$\sigma_3 = 8000 \text{ lb}_f/\text{in}^2$$

b: The displacement is:

$$\Delta = \sum_1^3 \Delta_i$$

where $\Delta_i = \varepsilon_i L_i = \sigma_i \dfrac{L_i}{E_i}$.

$$\Delta = \left[\frac{2000 \times 2}{1 \times 10^6}\right] + \left[\frac{4000 \times 1.5}{1 \times 10^6}\right] + \left[\frac{8000 \times 1}{1 \times 10^6}\right]$$

$$= [4 + 6 + 8] \times 10^{-3} \text{ in} = 0.018 \text{ in}$$

[Note that the shortest section, number 3, exhibits the greatest deformation due to its high stress.]

Poisson's Ratio

When a tensile load is applied to a bar, Figures 3.4 and 3.10, the bar changes in lateral, non-axial dimensions, x and z, as well as in axial dimension, y. The ratios of the lateral strains to the axial strain are given as:

(3.16) $-\varepsilon_x/\varepsilon_y$ and $-\varepsilon_z/\varepsilon_y$

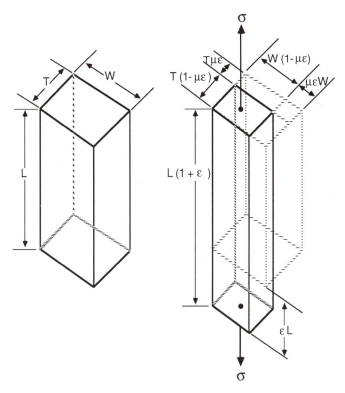

Figure 3.10.   Schematic of the Meaning of Poisson's Ratio. The Lateral Strain on a Bar Owing to Longitudinal Stresses. Bar Is Hookean or Linear Elastic Material. σ Is Applied Tensile Stress, ε Is Measured Elongation in Applied Stress Direction. $\mu\varepsilon$ Is the Measured Contraction in the Non-Stressed Directions.

These ratios are constant and equal and are called Poisson's ratio, μ. A Hookean material is isotropic, with elastic properties equal in all directions. The value for Poisson's ratio for a Hookean material is 0.25. Typical Poisson's ratio values for other materials are given in Table 3.1.

The generalized constitutive equation for an isotropic Hookean material is called the generalized Hooke's Law and is:

(3.17)

$$\sigma_x = \left[\frac{E}{(1 + \mu)(1 - 2\mu)}\right][(1 - \mu)\varepsilon_x + \mu\,\varepsilon_y + \mu\,\varepsilon_z]$$

$$\sigma_y = \left[\frac{E}{(1 + \mu)(1 - 2\mu)}\right][\mu\,\varepsilon_x + (1 - \mu)\varepsilon_y + \mu\,\varepsilon_z]$$

$$\sigma_z = \left[\frac{E}{(1 + \mu)(1 - 2\mu)}\right][\mu\,\varepsilon_x + \mu\,\varepsilon_y + (1 - \mu)\varepsilon_z]$$

where x, y, and z are any three mutually perpendicular directions.

Table 3.1. Poisson's ratios for common and polymeric materials. (1, 2–6).

Material	Poisson's Ratio, v
Aluminum	0.34
Aluminum alloy	0.32
Copper	0.35
Iron	0.28
Carbon steel	0.28
Magnesium	0.33
Titanium	0.34
Concrete	0.08–0.12
Cork	0.00
Rubber	0.50
Cellulose acetate (CA)	0.44
Cellulose acetate butyrate (CAB)	0.46
Polycarbonate (PC)	0.38
Polyether sulfone (PES)	0.40
High density polyethylene (HDPE)	0.38
Ultra-high molecular weight polyethylene (UHMWPE)	0.46
Polymethyl methacrylate (PMMA)	0.35
Polypropylene (PP)	0.33
Polystyrene (PS)	0.32
Low density polystyrene foam	0.03

Bulk Modulus

When any solid is stressed, its volume changes. Consider a bar in simple tension, Figure 3.10. The unstressed bar volume, V_{oj} is simply LTW, length × thickness × width. The volume of the stressed bar is:

$$(3.18) \qquad V = LTW(1 + \varepsilon)(1 - \mu\varepsilon)(1 - \mu\varepsilon)$$

The volumetric strain is the ratio of the change in volume to the initial volume:

$$(3.19) \qquad \Delta V/V_o = \varepsilon(1 - 2\mu) + \varepsilon^2\mu(\mu - 2) + \varepsilon^3\mu^2$$

For small strain, the last two terms on the right can be neglected. Bulk modulus, K, is defined as the ratio of an applied hydrostatic pressure to the resultant volume strain:

$$(3.20) \qquad K = \frac{P}{\Delta V/V_o}$$

Relationships Between Elastic Constants

The four elastic constants, E, G, μ, and K, are used to describe any solid material. For isotropic Hookean solids, these constants are related:

(3.21)
$$G = \frac{E}{2(1 + \mu)}$$

(3.22)
$$K = \frac{E}{3(1 - 2\mu)}$$

(3.23)
$$\mu = \frac{3K - 2G}{6K + 2G}$$

Thus, only two elastic constants need to be determined experimentally. The other two are obtained from these expressions. Chapter 6 details various tests to measure the basic mechanical properties of a plastic. The significance of any test conducted on a material depends on the ability of the tester to correctly interpret the results. A basic understanding of the relationship between an applied load or loads, stress distribution, and deformation and rate of deformation, strain and strain-rate distribution, is necessary.

Newtonian Fluid

Many fluids show a linear relationship between the applied stress and rate of deformation.* The proportionality constant is called *viscosity,* and these fluids are called *Newtonian fluids*:

(3.24)   Tensile: $\sigma = \eta_e \dot{\varepsilon}$

(3.25)   Shear: $\tau = \eta \dot{\gamma}$

For fluids under axial tension, the proportionality constant is the Trouton elongational viscosity, η_e. For fluids in shear, the proportionality is the Newtonian shear viscosity, η. For simple uniaxial stretching, the Trouton elongational viscosity is directly proportional to the Newtonian viscosity. The viscosity ratio is three:

(3.26) $\eta_e = 3\eta$

In biaxial extension, the viscosity ratio is 6 and in pure shear extensional viscosity, the viscosity ratio is 4 (7). Note that for Newtonian fluids, applied shear results in

---

*A more extensive discussion of the viscous nature of polymeric fluids is given in Chapter 4. This section is included here to establish the concept of the dashpot as a mechanical analog to Newtonian viscosity.

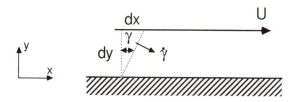

Figure 3.11. Schematic of Simple Shear Flow. U Is Velocity of Parallel Plate at Distance dy from Fixed Plate. τ Is the Measured Deformation of the Fluid Element. $\dot{\gamma}$ Is the Instantaneous Rate of Deformation of the Fluid Element.

deformation or strain that increases linearly with time. The physical significance of shear rate, $\dot{\gamma}$, can be demonstrated by considering the shearing force on a very thin layer of fluid being applied by a plate moving at constant velocity U, Figure 3.11. The time rate of change in deformation is given as:

$$(3.27) \qquad \frac{dU}{dy} = \frac{d(dx/dt)}{dy} = \frac{d(dx/dy)}{dt} = d\gamma/dt = \dot{\gamma}$$

The units of shear rate, s^{-1}, are commonly called "reciprocal seconds".

Example 3.2

Q: A purely viscous fluid is continuously sheared between two parallel plates, $\delta = 0.001$ m apart. The top plate moves at $U = 0.1$ m/s, and the bottom plate is stationary. If the stress exerted on the top plate is 2000 Pa, what is the shear rate and the viscosity of the fluid?

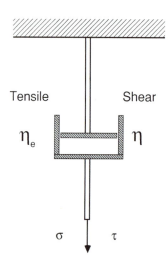

Tensile   Shear

η_e η

σ τ

Figure 3.12. Mechanical Analog of a Newtonian Fluid. σ Is Applied Tensile Load and τ Is Applied Shear Load. The Fluid Resistance Is Shown as a Dashpot Having an Elongational Viscosity, η_e, in Tension and Shear Viscosity, η, in Shear.

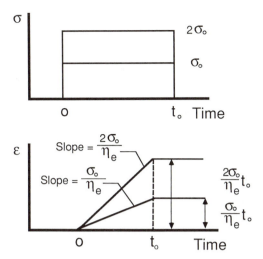

Figure 3.13.   Schematic of Response of a Newtonian Fluid, (Lower Graph) to a Step Change in Applied Tensile Load, σ_0, (Upper Graph). The Rate of Deformation, $\dot{\varepsilon}$, Is the Slope of the ε-Time Curve, $\dot{\varepsilon} = \sigma_0/\eta_e$, Where η_e Is the Elongational Viscosity. Note Permanent Deformation, $\varepsilon = \sigma_0 t_0/\eta_e$, at Time t_0 When the Applied Tensile Load Is Removed.

A: The shear rate is:

$$\dot{\gamma} = \frac{U}{\delta} = \frac{0.1 \text{ m/s}}{0.001 \text{ m}} = 100 \text{ s}^{-1}$$

The fluid viscosity is:

$$\eta = \frac{\tau}{\dot{\gamma}} = \frac{2000 \text{ Pa}}{100 \text{ s}^{-1}} = 20 \text{ Pa s}$$

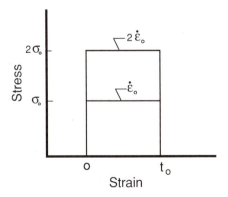

Figure 3.14.   Strain Rate Response of a Newtonian Fluid to Applied Tensile Load, σ_0.

The *mechanical analog* of a Newtonian fluid is a linear damping element or dashpot, Figure 3.12. The response of a viscous dashpot to applied force is a linear time-dependent displacement. The rate of displacement is independent of the initial position of the dashpot. Upon removal of the applied force, the rate of displacement of the dashpot instantaneously drops to zero and the dashpot position remains fixed. The deformational response of a Newtonian fluid, and its mechanical equivalent, to a constant applied stress is seen in Figure 3.13. The Newtonian fluid strain response to applied stress is seen in Figure 3.14. Note that if the stress is doubled, the rate of deformation or strain rate is doubled.

3.5 RESPONSE OF CLASSICAL MATERIALS TO CYCLIC DEFORMATION

Probably the clearest difference between Hookean and Newtonian material responses to applied load is seen when the load is applied cyclically. Many natural phenomena apply loads in one form of periodic fashion or another. Structural members such as bridge segments and airplane wings are exposed to cyclic loading. Automobile tires and motor housings also undergo periodic loading. Keep in mind that Hookean solids and Newtonian viscous liquids represent two extremes of material response to load. Most real materials in general and plastic materials in particular respond to cyclic loads in manners that include elements of both. The analysis given is for shear loading. Similar results can be obtained for tensile or compression loading.

Consider the following cyclic shear deformation:

$$(3.28) \qquad \gamma = \gamma_0 \sin(\omega t)$$

where γ_0 is the amplitude of the deformation and ω is the frequency, Figure 3.15. The strain rate, $\dot{\gamma}$, is the time rate of change of the deformation:

$$(3.29) \qquad \dot{\gamma} = \frac{d\gamma}{dt} = \gamma_0 \, \omega \cos(\omega t)$$

The stress developed in a Hookean solid is proportional to the imposed strain:

$$(3.30) \qquad \tau = G \gamma = G \gamma_0 \sin(\omega t)$$

The stress in a Newtonian fluid is proportional to the strain rate, $\dot{\gamma}$:

$$(3.31) \qquad \tau = \eta\dot{\gamma} = \eta \gamma_0 \, \omega \cos(\omega t)$$

These responses are shown in Figure 3.15. Note that the Hookean deformation is in phase with the applied stress, whereas the Newtonian deformation is out of phase with

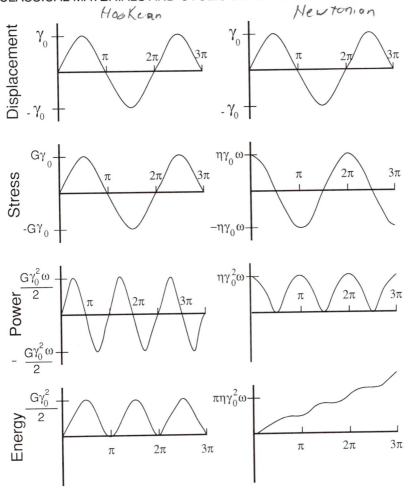

Figure 3.15.   Comparison of the Responses of a Hookean, Linear Elastic Solid, (Left Column) and a Newtonian, Purely Viscous Fluid, (Right Column) to Applied Sinusoidal Tensile Load. The Top Row Represents the Displacements to the Applied Load, Second Row. The Elastic Element Is in Phase With the Applied Load. The Viscous Element Is Out of Phase With the Applied Load. The Power Curve, Third Row, Shows Fully Recoverable Energy for an Elastic Element and Completely Unrecoverable Energy for a Viscous Element. The Elastic Element Is a Conservative Energy Element, Fourth Row, Whereas the Viscous Element Is a Completely Dissipative Element.

the applied stress by exactly $\pi/2$ rad or $90°$. The relationships between in-phase and out-of-phase material responses to an applied forcing function are used in Chapter 4 in the quantitative description of a viscoelastic material response to applied load.

   Another way of looking at these responses is to consider power or energy consumption. Power is the rate at which work is being performed. Work is the product of the component of a force in the direction of movement times the displacement in that direction:

$$(3.32) \qquad W = F_s S$$

If the component of force varies with position, work is given in integral form:

$$(3.33) \qquad W = \int F_s \, dS$$

For any material, work can be determined in terms of tensile or shear forces:

$$(3.34) \qquad \text{Tensile:} \quad W = \int (A\sigma)(L \, d\varepsilon)$$

$$(3.35) \qquad \text{Shear:} \quad W = \int (A\tau)(H \, d\tau)$$

where the first term in parentheses is the force and the second is the displacement. The work per unit volume can be written as:

$$(3.36) \qquad \text{Tensile:} \quad w = W/V = \int \sigma \, d\varepsilon$$

$$(3.37) \qquad \text{Shear:} \quad w = W/V = \int \tau \, d\gamma$$

Power, then, is the rate at which work is being done:

$$(3.38) \qquad \text{Tensile:} \quad P = dw/dt = \sigma \, d\varepsilon/dt$$

$$(3.39) \qquad \text{Shear:} \quad P = dw/dt = \tau \, d\gamma/dt$$

For a Hookean solid, power is:

$$(3.40) \qquad P = G \, \gamma_0^2 \, \omega \, \sin(\omega \, t) \, \cos(\omega \, t)$$

and for a Newtonian fluid:

$$(3.41) \qquad P = \eta \, \gamma_0^2 \, \omega^2 \, \cos^2(\omega \, t)$$

As seen in Figure 3.15, the instantaneous power for a Hookean solid is cyclic and is symmetric about zero. A Hookean solid conserves power, producing as much energy in the contraction portion of the periodic cycle as it dissipates in the expansion portion. The Newtonian fluid power consumption is substantially different. The Newtonian fluid dissipates power in both portions of the power cycle. Another way of looking at these simple concepts is to consider the total energy required to deform the body at any instant in time. The work done per unit volume is just the integral of power as a function of time to the present:

$$(3.42) \qquad W = \int P \, dt$$

For the Hookean solid:

(3.43)
$$W = \frac{G \, \gamma_0}{2} \sin^2(\omega \, t)$$

and for the Newtonian fluid:

(3.44)
$$W = \frac{\eta \, \gamma_0^2 \, \omega}{2} \left[\omega \, t + 1/2 \, \frac{\sin(2 \, \omega \, t)}{2} \right]$$

The total energy for each of these materials is seen in the fourth set of curves in Figure 3.15. It is apparent that a Hookean solid is a conservative element in that the total net work necessary to deform it always becomes equal to zero for each half-cycle of deformation. Consider the spring analog. The spring is stretched during the first quarter of the cycle. The work necessary to stretch the spring is stored in the spring as potential energy. In the second quarter of the cycle, the spring is allowed to return to its original position. During this portion of the cycle, the spring does work on the agent applying the force, in turn reducing its potential energy to zero. This process is repeated during the second half of the cycle, with the spring being compressed rather than extended. The net result is that at the end of each cycle, the spring is in the equilibrium position having expended zero work.

Example 3.3

Q: If a fluid is being deformed in a manner such that the power input to the fluid is constant at 1×10^3 W/m^3 for $\tau = 30$ s, what is the total work done on the fluid?
A: Work is given by:

$$W = \int P \, dt = P \, \tau$$
$$W = 1 \times 10^3 \ W/m^3 \times 30 \ s = 30 \times 10^3 \ W \ s/m^3.$$

On the other hand, the Newtonian fluid is a completely dissipative element. After each complete cycle, the net work on the material has increased by an amount $[\pi \eta \, \gamma_0^2 \, \omega]$ over the work level of the preceding cycle. For the mechanical model, a complete cycle on a dashpot represents first extending the dashpot a given distance, then compressing the dashpot an equal distance beyond the original position, then extending it again to the original position. There is no restoring force in the dashpot. Moving the dashpot from the extended position to the original position and then an equal distance to the compressed position requires constant application of force. Moving the dashpot through a complete cycle, then, requires continual application of energy. The work done on the dashpot is converted to thermal energy, which is dissipated to the dashpot surroundings. Thus the cyclical work done on a Newtonian fluid is called *viscous dissipation* energy. The rate of dissipation per unit volume is a function of the fluid viscosity, forcing frequency and the square of the amount of displacement.

To summarize simple material response to cyclic loads, Hookean solids represent true energy-conserving, in-phase, *storage* elements. Newtonian fluids represent true energy-dissipating, out-of-phase, *dissipative* elements. Polymers in general have both elastic solid and viscous fluid characteristics. The dissipative nature of polymer fluids may result in significant internal heating of an otherwise solid plastic under cyclic fatigue loading. On the other hand, good dissipative characteristics are desired when solid polymers are bonded by ultrasonic means. This discussion is considered in greater detail in Chapter 6, on fatigue characteristics of polymers.

3.6 Deformational Behavior of Polymers

Polymers as a class are viscoelastic. That is, to some degree, all polymers have both solid and fluid characteristics in response to applied stress. To some degree, polymers at temperatures substantially (50°C or more) below their glass transition temperatures usually behave more like true solids than fluids. Similarly, polymers at temperatures substantially (50°C or more) above their melt or softening temperatures usually behave more like true fluids than solids. Between these extreme temperatures, the response of polymers to applied loads depends on many factors, including residual strains, time of load, loading rate, and physical properties such as thermal conductivity.

Fluid State

To better understand the general concepts of polymer response under load, consider first a polymer as a fluid in shear. As noted, the Newtonian or classical fluid response to applied load is a fixed deformation rate, with the proportionality constant being Newtonian viscosity, a shear-rate-independent material property. Many polymers exhibit shear rate curves that are not linear with applied stress, Figure 3.16. The slopes of the stress–strain rate curves, when replotted, show viscosities that decrease with increasing strain rates, shear rates, or rates of deformation. Shear-thinning is seen for several polymers in Figure 3.17. Polymers with non-constant viscosities are called non-Newtonian fluids, and can be represented in a most simple fashion as:

$$(3.45) \qquad\qquad \tau = \eta(\dot\gamma)\,\dot\gamma$$

or

$$(3.46) \qquad\qquad \tau = K\,\dot\gamma^n$$

The second expression is called the *power-law* viscosity relationship. K and n are experimentally determined coefficients, sometimes called Ostwald–deWaale power-law coefficients. Note that most polymer viscosity–shear rate curves are not linear on log–log coordinates over the entire scale length. Some polymers, such as PA-6 (nylon 6) and PMMA, exhibit constant viscosity regions at low shear rates. Thus, for many

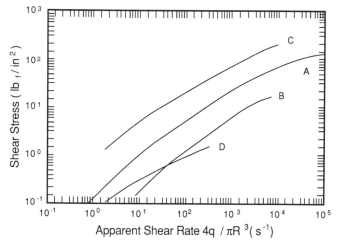

Figure 3.16.   Stress–Rate-of-Strain Curves for Several Polymers. (8). A: PMMA at 250°C. B: PA-6 (Nylon 6) at 288°C. C: PE at 190°C. D: HIPS at 230°C.

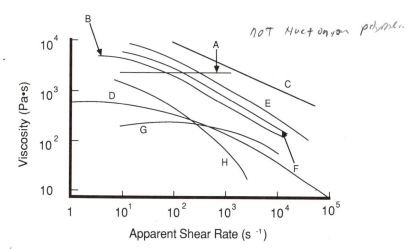

Figure 3.17.   Representative Shear Viscosities of Some Commercial Polymers at Their Processing Temperatures. (9). Note That Actual Responses Can Vary Substantially With Different Types of Each Resin.

Curve	Polymer	Temperature (°C)
A	PC	288
B	LDPE	235
C	GP-PS	230
D	PMMA	250
E	PMMA	200
F	HDPE	232
G	PA-6 (nylon 6)	288
H	PP	230

polymers, the simple power-law model usually cannot be used over the entire shear field. This is discussed in greater detail in Chapter 4.

Solid State

The simple Hookean solid is characterized by a constant, time-independent deformation under load. Recall that Hookean solids should show instantaneous deformation step changes to new strain levels. In contrast, polymers can exhibit time-dependent deformations, Figure 3.18. For cellulose acetate, some instantaneous strain is apparent, but deformation then continues, initially at relatively rapid rate, then at a much slower constant, measurable rate. In this latter section, then, the rate of deformation is constant at constant applied load, as with a Newtonian fluid. At low stress levels, the value for the constant rate of deformation is low, as is the case with a Newtonian fluid. At the highest stress level, materials usually fracture before the Newtonian plateau can be reached. This is called *creep rupture* and will be studied in detail in Chapter 4. Ma-

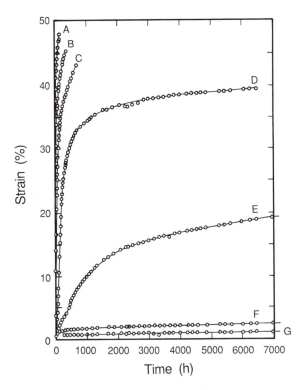

Figure 3.18.   Tensile Strain of Cellulose Acetate at 25°C. (10). Parameter, Applied Stress. Curve A: 2695 lb_f/in^2. B: 2505. C: 2305. D: 2008. E: 1690. F: 1320. G: 1018. Specimen Did Not Fracture at 7000 Hours at Loads of 1690 lb_f/in^2 or Less.

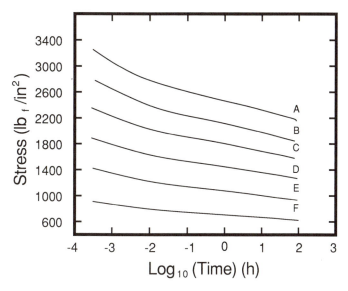

Figure 3.19.   Stress Relaxation of Unfilled PTFE in Compression in the Gasketometer (Chapter 6). Annular Gasket, 2 in OD × 1.65 in ID × 0.080 in Thickness. (11). Curve A: Initial Stress = 3500 lb_f/in^2. B: 3000. C: 2500. D: 2000. E: 1500. F: 1000 lb_f/in^2.

terials that continue to deform under constant applied load are said to *creep**. The materials engineer must therefore be aware that polymers will *always* continue to creep or deform under load and that creep rupture depends on stress level, time and temperature. Again, these points are detailed in Chapter 4.

The actual shape of the time-dependent stress–strain curve depends on the geometry of the polymer molecule, the degree of crosslinking, polymer molecular weight, molecular weight distribution, additives, fillers, crystallinity, and other factors.

Another characteristic response of a viscoelastic solid to applied load is compressive stress behavior under constant compressive deformation. For a Hookean material under constant compressive strain, the resulting stress would be constant, as given by $\sigma = E \varepsilon_0$, where $\varepsilon_0 = \delta/L_0$. As long as the strain is applied, the stress is constant. When the strain is removed, the stress returns to zero and the material recovers to its original dimension. In Figure 3.19, an unfilled PTFE O-ring is compressed to various levels of strain and the induced stress measured with time. As is apparent, stress decays with time. This can occur only if the polymer flows while under compression. If, after an extended period of time, the compression load is removed (the stress is allowed to return to zero), the polymer will be found to be permanently deformed. The practical

---

*Many ductile non-polymeric materials creep under load, including lead, copper, gold, and montmorillonite clay. These materials are sometimes called "plastic" materials.

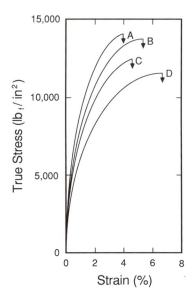

Figure 3.20.   Effect of the Rate of Tensile Testing on the Stress–Strain Behavior of PMMA, $M_n = 3.16 \times 10^6$, 77°F. (12). Curve A: 0.02 in/min. B: 0.08. C: 0.32. D: 1.28 in/min.

results of this phenomenon, called *stress relaxation,* is extrusion of gaskets under high compression and loosening of container caps, nuts, and other threaded parts with time.

Recall that Hookean solids always respond instantaneously to applied load, regardless of the rate at which the load is applied. As seen in Figure 3.20 for PMMA, the slope of the stresss–strain curve, the elastic modulus, increases with increasing strain rate. Furthermore, the ultimate tensile strength, stress level at the time of failure, also increases with increasing strain rate. And the shape of the stress–strain curve can change dramatically with small changes in temperature, unlike Hookean solid behavior. Many polymers exhibit similar non-Hookean responses.

Linear Viscoelastic Analysis

From the discussion above, it is apparent that polymers can attain some of the characteristics of solids and some of the characteristics of fluids. In many melt pumping processes, fluid characteristics dominate. In most mechanical short-term testing schemes, solid characteristics dominate. Note, however, that the application may be controlled by the minor elements. Melt spinning, for example, depends on the elastic nature of the polymer liquid. Creep is a manifestation of the viscous nature of the polymer solid. And so on. As noted in Section 3.4, the nature of material response to applied forces is usually described by a constitutive equation of state. The simplest equations of state are the Hookean equation for a solid and the Newtonian equation for a fluid. Each has only one measurable constant. The simplest equation for a non-Newtonian fluid is the viscous-only power-law equation, with two measurable constants. Hundreds of equations containing many constants for both melt and solid phases of polymers have been proposed over the past century. Many equations consider the polymer to be a moleculeless continuum of strain field response to an applied strain tensor. Others consider the polymer to be a series of springs and dashpots, with temporary connections representing molecule entanglements or temporary (hydrogen) bonds or (Van der Waals) forces and permanent connections representing crosslinks, crystallites, or stiff chain segments. And others consider molecular chains to be spring-connected dumbbells that move past one another in reptile-like fashion. Certainly, as the model becomes more elaborate, the ability to describe polymer behavior over a wide range of processing and application conditions becomes greater. At the same time, as the model becomes more exotic, the likelihood increases that the required constants cannot be determined without elaborate experiments. Unfortunately, there are very few practical examples of applications of well-defined, elegant constitutive equations of state with full complements of values for required constants to predict real processing conditions.

Mechanical analogs can be used to illustrate certain viscoelastic material behavior, but extension to prediction of polymer processing and application response should be avoided, if possible. However, from a purely engineering context, these models can be used to illustrate in a gross fashion how a polymer might react to specific imposed boundary conditions. Polymers are simply not made of springs and dashpots in series, parallel, or any other combination. They are a complex arrangement of partially coiled macromolecules of differing lengths, with impinging electron clouds, at times with bulky side groups, moving against other molecules and an assortment of small molecule additives, either by applied shear or simply Brownian motion. In short, no mechanical analog can adequately describe even the simplest polymer system. But analogs can be used to demonstrate the interrelationship between the elastic and viscous response of a polymer system to applied forces.

The analysis presented in this section is limited to *linear viscoelasticity*. In reality, no polymer is linear viscoelastic in either the solid or fluid phase. Thus, the concern should be whether a given polymer response can be approximated by a simple linear viscoelastic model *at the particular conditions imposed*. If so, then a better understanding of polymer behavior obtains. The results of tests described in detail in Chapters 4 and 6 will allow the designer to make an intelligent decision about the applicability of a given model to a specific design condition. By linear, it is meant that stresses are low enough to cause only a small deformation or deformation rate so that the coordinates of any deformed element can be fixed in space and so that stress–strain–rate-of-strain can be written with relatively few arbitrary constants. A more general theory allows not only for movement and distortion of the coordinate system but also for a nonlinear relationship between applied stress and rate of deformation*. For the purpose of this discussion, one-dimensional stress–strain–rate-of-strain models are presented.

---

*Recall that a linear relationship results in a material *constant,* called Newtonian viscosity. The nonlinear relationship should allow for the shear-rate dependency of viscosity observed for most practical polymers. This is not meant to imply that a simple linear viscoelastic model cannot be devised that includes power-law behavior, for example. A true general viscoelastic theory should yield a nonlinear relationship as a natural feature of the theory, not as an added artifact.

Maxwell Element

In an attempt to characterize the long-term response of asphaltum to applied load, James Maxwell in 1867 proposed an analog model consisting of a series combination of a linear elastic element, a spring, and a linear viscous element, a dashpot (13). This model, now referred to as the *Maxwell element,* is shown as Figure 3.21, annotated for both tensile and shear loads. The applied stress is equal on both the spring and dashpot. In tensile notation:

$$(3.47) \qquad \sigma = \sigma_{spring} = \sigma_{dashpot}$$

The total displacement is the sum of the displacements of each component:

$$(3.48) \qquad \varepsilon_T = \varepsilon_{spring} + \varepsilon_{dashpot}$$

Recall the tensile constitutive equations for each component:

$$(3.49) \qquad \text{Spring:} \quad \sigma = E\,\varepsilon$$

$$(3.50) \qquad \text{Dashpot:} \quad \sigma = \eta_e\,\dot{\varepsilon} \qquad \text{tr ajton}$$

To determine the time-dependent deformation under applied load, it is necessary to differentiate the total displacement equation with respect to time and then substitute the appropriate terms for each component. This yields a first-order ordinary differential equation:

$$(3.51) \qquad \text{Tensile:} \quad \dot{\sigma} + \left(\frac{E}{\eta_e}\right)\sigma = E\dot{\varepsilon}_T$$

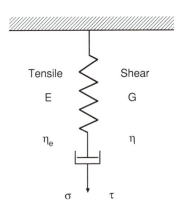

Figure 3.21.   Schematic of Maxwell Element in Tension and Shear Loading. σ Is Tensile Stress. τ Is Shear Stress. η_e Is Elongational Viscosity. η Is Shear Viscosity. E Is Elastic Modulus. G Is Shear Modulus.

A similar expression for shear can be obtained by differentiating the appropriate shear displacement functions for each of the components. The result is:

$$(3.52) \qquad\qquad \text{Shear:} \qquad \dot{\tau} + \left(\frac{G}{\eta}\right)\tau = G\dot{\gamma}_T$$

where $\dot{\gamma}_T$ is the total shear rate of deformation. One characteristic of any first-order differential equation of this type is the time constant, λ. This time constant provides a measure of the rate of response of the system to a step input. At the time equal to one time constant, the system has achieved $[1 - 1/e]$ or 63.2% of the total response. At the time equal to two time constants, the response has reached $[1 - 1/e^2]$ or 86.5% of the total response, and so on. For the Maxwell element:

$$(3.53) \qquad\qquad \text{Tensile:} \qquad \lambda = \frac{\eta_e}{E}$$

$$(3.54) \qquad\qquad \text{Shear:} \qquad \lambda = \frac{\eta}{G}$$

The physical significance of the time constant is discussed shortly. The solution of the differential equation depends on the initial conditions imposed on the system.

Uniform Loading: "Creep". Consider the response of a Maxwell element to a uniform tensile load from time zero to time t_0, at which time the load is instantaneously removed. This condition represents a creep loading condition. Under a uniform applied load, the tensile stress, σ, does not change with time. Therefore, $\sigma = \sigma_0$ and $\dot{\sigma} = 0$. The solution to the differential equation is:

$$(3.55) \qquad\qquad \varepsilon = \overset{\text{creep}}{\left(\frac{\sigma_0}{\eta_e}\right) t} + \overset{\text{elastic}}{\frac{\sigma_0}{E}}$$

This response is shown in Figure 3.22. Consider the overall response in terms of the responses of the individual components. Upon application of the load, the spring immediately extends to an equilibrium value σ_0/E. Under uniform load, the dashpot extends at a constant velocity, σ_0/η_e. The position of the dashpot at any time is equal to the velocity times time $(\sigma_0/\eta_e) \times t$. Since the displacement of the Maxwell element is the sum of the displacements of the spring and dashpot, the above equation obtains. Note that when the load is removed at time t_0, the Maxwell element does not return to its initial position. Instead, the spring recovers a distance σ_0/E but the dashpot displacement remains at the position $(\sigma_0/\eta_e) t_0$. Since a nonzero, permanent displacement is a measure of a fluid, the Maxwell element response is likened to a viscoelastic fluid (elastic because some deformation occurs instantaneously upon loading and because some deformation is recovered instantaneously upon unloading).

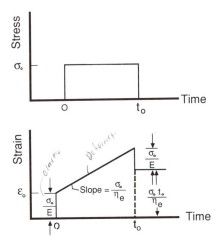

Figure 3.22. Response of the Maxwell Element, Figure 3.21, to Constant Applied Tensile Stress, σ_0. σ_0/E Is the Instantaneous Elongational Strain Response, $= \varepsilon_0$. The Slope of the Time-Dependent Strain Is σ_0/η_e. The Permanent Offset at Time, t_0, Is $\sigma_0 t_0/\eta_e$.

Uniform Deformation: "Stress Relaxation".   Consider the Maxwell element response to application of a constant, initial strain or displacement. This yields time-dependent stress relaxation. Consider a polymer segment, such as a gasket shown in Figure 3.23, compressed to a strain level ε_0. The initial compressive stress is:

$$(3.56) \qquad\qquad \sigma_0 = \varepsilon_0 E$$

Under constant displacement (strain), $\varepsilon = \varepsilon_0$, and $\dot{\varepsilon} = 0$. The Maxwell element differential equation is:

$$(3.57) \qquad\qquad \dot{\sigma} + \left(\frac{E}{\eta_e}\right)\sigma = 0$$

Integration yields:

$$(3.58) \qquad . \qquad \sigma = \sigma_0 \exp\left[\frac{-Et}{\eta_e}\right] = \sigma_0 \exp\left[\frac{-t}{\lambda}\right]$$

This is shown in Figure 3.24. The time constant, λ, represents the time required for the stress to decay to $1/e$ or 36.8% of its initial value. It is the intercept of the tangent line drawn to the stress–time curve at zero time. Note that for long times, the stress asymptotically approaches zero. This relaxation phenomenon explains why gaskets and bottle caps must be "retorqued" periodically.

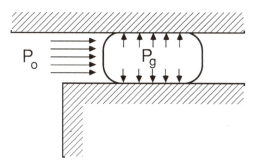

Figure 3.23.   The Geometry for a Gasket. P_g Is the Pressure Exerted by the Gasket Material Against the Sealing Surfaces. P_0 Is the Differential Pressure Applied Against the Gasket, Causing It to Extrude.

Uniform Strain Rate.   Uniform stress and constant strain represent ideals for a step change initial conditions. Consider the Maxwell element response to uniform strain rate, $\dot{\varepsilon} = \dot{\varepsilon}_0$. This represents the actual conditions applied to a polymer specimen in a simple tensile test. The differential equation can be written as:

$$(3.59) \qquad \dot{\sigma} + \left(\frac{E}{\eta_e}\right)\sigma = E\dot{\varepsilon}_0$$

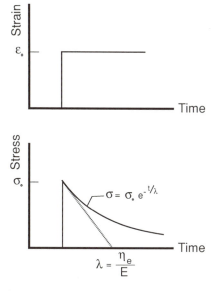

Figure 3.24.   Response of the Maxwell Element of Figure 3.21 to Constant Strain, ε_0. This Is Characteristic of the Gasketometer, Figure 3.23. The Stress Decay Is Given as $\sigma = \sigma_0\, e^{(-t/\lambda)}$ where λ Is the Time Constant for Stress Relaxation, $\lambda = \eta_e/E$.

If the initial stress is zero, $\sigma(0) = 0$, the solution is:

$$(3.60) \qquad \sigma = \dot{\varepsilon}_0\, \eta_e \left(1 - \exp\left[\frac{-E\varepsilon}{\dot{\varepsilon}_0\, \eta_e}\right]\right)$$

This equation is graphed in Figure 3.25. Consider a very important application of this analysis. Recall that Young's modulus for any material is experimentally determined in simple tensile testing by measuring the slope of the stress–strain curve at zero strain. For purely elastic or Hookean materials, the slope is independent of the rate of strain. In other words, the rate at which the grips are moved apart does not affect the modulus of an elastic material. For viscoelastic materials, this is patently not the case. The very simple Maxwell element qualitatively illustrates the way in which many polymers respond to the simple tensile test. Consider two limiting conditions of this response. When $\eta_e\,\dot{\varepsilon}_0 = 0$, $\sigma = 0$. If the exponential term is expanded in infinite series, the condition, $\eta_e\,\dot{\varepsilon}_0 = \infty$ yields $\sigma = E\varepsilon_0$. Thus, the apparent modulus of a viscoelastic material as modeled by the Maxwell element changes with the rate at which the grips are separated. Of course, the same is true in part design, where the apparent modulus is a function of the rate of environmental loading. Note that the apparent modulus increases with increasing loading rate. That is, the model material appears stiffer with increasing loading rate. This result is contrary to what is expected in the conventional strength-of-materials analysis.

To summarize, the Maxwell element is a mechanical analog that may model a simple linear viscoelastic fluid. Its response qualitatively describes the phe-

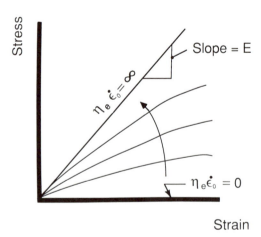

Figure 3.25.   Response of the Maxwell Element of Figure 3.21 to Constant Strain Rate, $\dot{\varepsilon}_0$. The Stress–Strain Response Slope Is $\eta_e\dot{\varepsilon}$. The Slope When $\eta_e\dot{\varepsilon} = \infty$ Is E, the Young's Modulus.

nomenon of stress relaxation and apparent changing modulus in tensile testing of polymers. It does not adequately describe static loading response or creep of polymers however.

Voigt–Kelvin Element

If the simple spring and dashpot model is rearranged with the components in parallel, the unit is the Voigt–Kelvin element, Figure 3.26. The schematic shows notation for the element in tensile and shear modes. If both components deform uniformly under load:

(3.61)
$$\varepsilon = \varepsilon_{\text{dashpot}} = \varepsilon_{\text{spring}}$$

The total stress applied to the element is the sum of the stress on each component:

(3.62)
$$\sigma_T = \sigma_{\text{dashpot}} + \sigma_{\text{spring}}$$

Through appropriate substitution, the first order differential constitutive equations for tensile and shear load for the Voigt–Kelvin element are:

(3.63) Tensile: $\dot{\varepsilon} + \left(\dfrac{E}{\eta_e}\right)\varepsilon = \dfrac{\sigma}{\eta_e}$

(3.64) Shear: $\dot{\gamma} + \left(\dfrac{G}{\eta}\right)\gamma = \dfrac{\gamma}{\eta}$

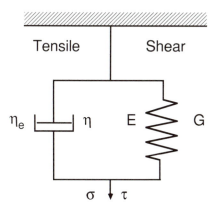

Figure 3.26. Schematic of the Voigt–Kelvin Mechanical Analog in Tensile and Shear Loading. σ Is Tensile Load. τ Is Shear Load. η_e Is Elongational Viscosity. η_e Is Shear Viscosity. E Is Young's or Elastic Modulus. G Is Shear Modulus.

The time constant for the Voigt–Kelvin element under tension is $\lambda_T = \eta_e/E$ and under shear is $\lambda_s = \eta/G$. Consider deformation under constant tensile load, $\sigma = \sigma_0$. The time-dependent deformation is given as:

$$(3.65) \qquad \varepsilon = \frac{\sigma_0}{E}\left(1 - \exp\left[\frac{-Et}{\eta_e}\right]\right)$$

Note that instantaneous deformation under load, Figure 3.27, is zero. At long times, deformation approaches a fixed value, $\varepsilon_\infty = \sigma_0/E$. If the stress is removed at time t_0, the time-dependent deformation is:

$$(3.66) \qquad \varepsilon = \left\{\frac{\sigma_0}{E}\left(1 - \exp\left[\frac{-t_0}{\lambda}\right]\right)\right\} \exp\left[\frac{-(t-t_0)}{\lambda}\right]$$

The terms in braces { } represent the strain at t_0. Note that the Voigt–Kelvin element always returns to its initial undeformed condition. Although this element exhibits a decreasing creep rate after initial loading, it does not exhibit the initial step deformation seen in real polymeric materials, Figure 3.18. Since this element eventually returns to an undeformed condition after stress has been removed, this element represents a viscoelastic solid.

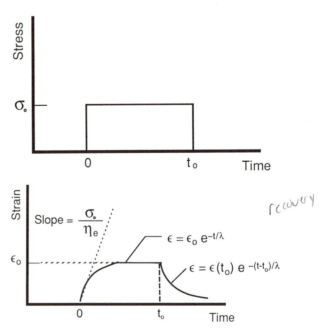

Figure 3.27.   Response of the Voigt–Kelvin Element of Figure 3.26 to Constant Stress, σ_0. The Initial Time-Dependent Strain Has a Slope of σ_0/η_e. The Strain, ε, Is an Exponential Function of Time and λ, $\varepsilon = \varepsilon_0 \exp(-t/\lambda)$, Where λ Is the Time Constant.

Note that the Voigt–Kelvin element response can be analyzed in terms of the responses of the in-parallel components. When a load is first applied, the dashpot resists immediate deformation. The force causes the dashpot to elongate at an instantaneous velocity of σ_o/η_e. As time continues, the rate of elongation of the dashpot decreases, with the load being increasingly borne by the spring. Eventually the dashpot has elongated to such a degree that the entire load is borne by the spring. The linear spring under load σ_o has an ultimate deformation of σ_o/E. When the load is removed from the element, the restoring force is provided by the spring, which forces the dashpot back to its initial position. The time required to attain the initial position is a function of the ratio of dashpot resistance, η_e, to restoring force of the spring, $E\varepsilon_o$.

The Voigt–Kelvin element does not describe the phenomenon of stress relaxation. The application of a finite strain at zero time requires the dashpot to elongate instantaneously. This requires the dashpot to have an infinite resistance. To overcome an infinite resistance requires an infinite stress, which is an unnatural situation.

Four-Parameter Element

It is apparent that the Maxwell and Voigt–Kelvin elements possess characteristics that allow modeling of certain stress–strain–rate-of-strain situations but not others. Consider then a series of combination of the elements, the four-parameter element. This element has been called a "crude, qualitative description" of the viscoelastic response of a linear amorphous polymer. Figure 3.28 shows only the tensile aspects of the element. A similar four-parameter element can be constructed for shear. Consider tensile deformation of the element.

$$(3.67) \qquad \varepsilon_T = \varepsilon_M + \varepsilon_{VK}$$

where ε_M is the deformation of the Maxwell element and ε_{VK} is that of the Voigt–Kelvin element. Recall the constitutive equations for these elements:

$$(3.68) \qquad \text{Maxwell:} \qquad \dot{\sigma} + \left(\frac{E_1}{\eta_{e_1}}\right)\sigma = E\dot{\varepsilon}_M$$

$$(3.69) \qquad \text{Voigt–Kelvin:} \qquad \dot{\varepsilon}_{VK} + \left(\frac{E_2}{\eta_{e_2}}\right)\varepsilon_{VK} = \frac{\sigma}{\eta_e}$$

These equations are combined to eliminate ε_M and ε_{VK} to obtain a single second-order linear differential constitutive equation for the four-parameter element:

$$(3.70) \qquad \frac{d^2\sigma}{dt^2} + \left\{\frac{E_1}{\eta_{e_1}} + \frac{E_1}{\eta_{e_2}} + \frac{E_2}{\eta_{e_2}}\right\}\frac{d\sigma}{dt} + \left(\frac{E_1 E_2}{\eta_{e_1}\eta_{e_2}}\right)\sigma = E_1\frac{d^2\varepsilon}{dt^2} + \left(\frac{E_1 E_2}{\eta_{e_2}}\right)\frac{d\varepsilon}{dt}$$

As with the simpler mechanical analogs, the four-parameter element response to constant stress, strain rate and instantaneous fixed strain resulting in stress relaxation can

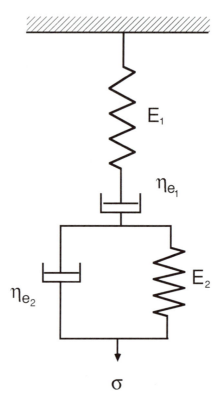

Figure 3.28.   Schematic of the Four-Parameter Element, Incorporating Maxwell and Voigt–Kelvin Elements in Series. Notation for Tensile Loading Only. E_1 and E_2 Are the Tensile Moduli of the Maxwell and Voigt–Kelvin Elements, Respectively. η_{e1} and η_{e2} Are the Elongational Viscosities of the Maxwell and Voigt–Kelvin Elements, Respectively. σ Is the Applied Stress.

be ascertained by solving the above equation with appropriate boundary conditions. Consider the instantaneous application of a fixed tensile load as illustration. Solutions to the other cases are given in Table 3.2. At $\sigma = \sigma_0$, the time-dependent stress terms are zero. The resulting equation is:

$$(3.71) \qquad \frac{d^2\varepsilon}{dt^2} + \left(\frac{E_2}{\eta_{e_2}}\right)\frac{d\varepsilon}{dt} = \left(\frac{E_2}{\eta_{e_1}\eta_{e_2}}\right)\sigma_0$$

Two initial conditions are required for solution. At zero time, the initial deformation is given as:

$$(3.72) \qquad \varepsilon(0) = \frac{\sigma_0}{E_1}$$

Table 3.2. Linear viscoelastic stress-strain models. (14).

Maxwell	Voigt–Kelvin	Four-parameter

Maxwell

$\sigma = \sigma_{\text{spring}} = \sigma_{\text{dashpot}}$

$\varepsilon_T = \varepsilon_{\text{spring}} + \varepsilon_{\text{dashpot}}$

$\dot{\sigma} + \left(\dfrac{E}{\eta_e}\right)\sigma = E\,\dot{\varepsilon}_T$

Time constant

$\lambda = \dfrac{\eta_e}{E}$

Uniform load (Creep)

$\dfrac{d\sigma}{dt} = \dot{\sigma} = 0$

$\varepsilon_M = \left(\dfrac{\sigma_o}{\eta_e}\right)t + \left(\dfrac{\sigma_o}{E}\right)$

Voigt–Kelvin

$\varepsilon = \varepsilon_{\text{spring}} = \varepsilon_{\text{dashpot}}$

$\sigma_T = \sigma_{\text{spring}} + \sigma_{\text{dashpot}}$

$\dot{\varepsilon} + \left(\dfrac{E}{\eta_e}\right)\varepsilon = \dfrac{\sigma}{\eta_e}$

$\lambda = \dfrac{\eta_e}{E}$

$\dot{\sigma} = 0$

$\varepsilon_{VK} = \left(\dfrac{\sigma_o}{E}\right)(1 - e^{-t/\lambda})$

Four-parameter

$\varepsilon_T = \varepsilon_M + \varepsilon_{VK}$

$\ddot{\sigma} + \left(\dfrac{E_1}{\eta_{o2}} + \dfrac{E_2}{\eta_{o2}} + \dfrac{E_1}{\eta_{o1}}\right)\dot{\sigma} - \left(\dfrac{E_1\,E_2}{\eta_{o1}\,\eta_{o2}}\right)\sigma = E_1\,\ddot{\varepsilon} + \left(\dfrac{E_1\,E_2}{\eta_{o1}\,\eta_{o2}}\right)\dot{\varepsilon}$

$\dot{\sigma} = 0$

$\varepsilon_{FP} = \varepsilon_M + \varepsilon_{VK}$

Table 3.2. *Continued.*

Stress relaxation

$$\frac{d\varepsilon}{dt} = \dot{\varepsilon} = 0 \qquad\qquad \dot{\varepsilon} = 0 \qquad\qquad \dot{\varepsilon} = 0$$

$$\sigma = \sigma_o\, e^{-t/\lambda} \qquad\qquad \sigma = E\,\varepsilon_o \qquad\qquad \sigma = A\, e^{-\alpha t} + B\, e^{-\beta t}$$

where:

$$2\alpha = \left[\frac{\eta_{e_1}^2}{E_1\eta_{e_2}} + \frac{E_2}{\eta_{e_1}} + \frac{E_1}{\eta_{e_1}}\right] + \sqrt{\left[\frac{\eta_{e_1}^2}{E_1\eta_{e_2}} + \frac{E_2}{\eta_{e_1}} + \frac{E_1}{\eta_{e_1}}\right]^2 - \left(4\,\frac{E_1 E_2}{\eta_{e_1}\eta_{e_2}}\right)}$$

$$2\beta = \left[\frac{\eta_{e_1}^2}{E_1\eta_{e_2}} + \frac{E_2}{\eta_{e_1}} + \frac{E_1}{\eta_{e_1}}\right] - \sqrt{\left[\frac{\eta_{e_1}^2}{E_1\eta_{e_2}} + \frac{E_2}{\eta_{e_1}} + \frac{E_1}{\eta_{e_1}}\right]^2 - \left(4\,\frac{E_1 E_2}{\eta_{e_1}\eta_{e_2}}\right)}$$

$$A = \frac{[\varepsilon E_1^2(1/\eta_{e_1} + 1/\eta_{e_2}) - \beta E_1]}{\alpha - \beta}$$

$$B = E_1\varepsilon - A$$

Uniform strain rate

$$\dot\varepsilon = \dot\varepsilon_o$$

$$(\varepsilon = \dot\varepsilon_o t)$$

$$\dot\varepsilon = \dot\varepsilon_o$$

$$\sigma = \eta_{e_2}\dot\varepsilon_o + \left[\frac{E_1\dot\varepsilon_o - \eta_{e_2}\beta}{\beta - \alpha}\right]e^{-\alpha\varepsilon/\dot\varepsilon_o}$$
$$+ \left[\frac{\eta_{e_2}\alpha - E_1\dot\varepsilon_o}{\beta - \alpha}\right]e^{-\beta\varepsilon/\dot\varepsilon_o}$$

Stress removal at $t = t_o$

$$\varepsilon = \left(\frac{\sigma_o}{E}\right)[1 - e^{-t_o/\lambda}]e^{-(t-t_o)/\lambda}$$

$$\varepsilon = \frac{\sigma_o t_o}{\eta_e}$$
(permanent set)

$$\varepsilon = \left(\frac{\sigma_o}{\eta_{e_1}}\right) + \left(\frac{\sigma_o}{E_2}\right)[1 - e^{-E_2 t/\eta_{e_2}}] \times$$
$$[e^{-t-t_o E_2/\eta_{e2}}]^{-1}$$

Creep modulus, E_c $\left(E_c = \frac{\sigma_o}{\varepsilon}\right)$

$$E_c = \left(\frac{t}{\eta_e} + \frac{1}{E}\right)^{-1}$$
(stress relaxation)

$$E_c = \frac{E}{E_c[1 - e^{-tE/\eta_e}]}$$
(creep)

$$E_c = \left[\frac{1}{E_1} + \frac{t}{\eta_{e_1}} + \left(\frac{1}{E_1}\right)(1 - e^{-E_2 t/\eta_{e2}})\right]^{-1}$$
(viscoelastic – impact)

This is the instantaneous deformation of the Maxwell spring. The second condition is not obvious. Assume that at zero time, the rate of deformation of the four-parameters element is equal to the sum of the rates of deformation of each of the two simpler elements:

$$(3.73) \qquad \dot{\varepsilon} = \frac{\sigma_0}{\eta_{e_1}} + \frac{\sigma_0}{\eta_{e_2}}$$

where η_{e_1} is the tensile viscosity of the Maxwell dashpot and η_{e_2} is that for the Voigt-Kelvin dashpot. With these two boundary conditions, the four-parameter element response is given as:

$$(3.74) \qquad \varepsilon = \left\{ \frac{\sigma_0}{E_1} + \frac{\sigma_0 t}{\eta_{e_1}} \right\} + \left\{ \frac{\sigma_0}{E_2} \left(1 - \exp\left[\frac{-E_2 t}{\eta_{e_2}} \right] \right) \right\}$$

The term in the first set of braces { } is the Maxwell element response, and that in the second is the Voigt–Kelvin element response. The response is shown in Figure 3.29. Note that the initial response is identical to that of the Maxwell element, with instantaneous deformation or *elastic strain* equal to σ_0/E_1. The Voigt–Kelvin element initially has zero deformation. Furthermore, the Voigt–Kelvin element responds with an exponential time-dependent deformation with a maximum value of σ_0/E_2. The addition of these *elastic* components results in the early portion of the time-dependent strain curve for the four-parameter element being labeled as the region of *retarded elastic strain*. When the Voigt–Kelvin element has achieved maximum elongation, the Maxwell dashpot continues to extend. For long times, the element is exhibiting *viscous flow*, with the slope of the stress–time curve given simply as σ_0/η_{e2}. Thus, the four-parameter element can be thought of as a simple sum response of *three* elements: a Voigt–Kelvin retarded elastic solid, a Hookean elastic element, and a Newtonian viscous fluid.

Instantaneous load removal, at time t_0, yields additional information about the four-parameter element. The deformation is described mathematically as:

$$(3.75) \qquad \varepsilon = \left\{ \frac{\sigma_0}{E_2} \left(1 - \exp\left[\frac{-E_2\,t_0}{\eta_{e_2}} \right] \right) \exp\left[-(t - t_0) \frac{E_2}{\eta_{e_2}} \right] \right\} + \frac{\sigma_0\, t_0}{\eta_{e_1}}$$

The first term in the braces { } represents the Voigt–Kelvin element deformation at $t = t_0$. The second represents the Maxwell element deformation at $t = t_0$. Note in Figure 3.29 that the Voigt–Kelvin element causes the retarded elastic strain recovery, whereas the Maxwell spring causes an instantaneous strain recovery and the Maxwell dashpot results in the permanent deformation or "set". The general shape of this curve qualitatively represents the creep data of cellulose acetate, Figure 3.18, and many other polymers.

In summary, the four-parameter element qualitatively accounts for many of the actual observed responses for polymers under long-term constant load. It includes instanta-

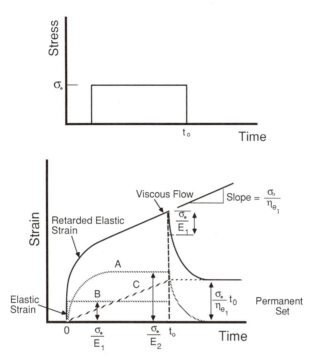

Figure 3.29.   Response of the Four-Parameter Element of Figure 3.28 to a Constant Stress, σ_o. Line A Represents the Voigt–Kelvin Segment Response. Line B Represents the Response of the Maxwell Element Spring. Line C Represents the Response of the Maxwell Element Dashpot. The Solid Line Represents the Response of the Combined Elements. Note the Four Sections to the Response. The Initial Response Is Pure Elastic Strain. The Response Following Is Retarded Elastic Strain. The Primary Result of the Voigt–Kelvin Element Pure Viscous Flow Follows, the Result of the Maxwell Element Dashpot. Finally, After Load Removal, the Permanent Set, the Result of the Maxwell Element Dashpot.

neous elastic strain, retarded elastic strain, viscous flow, instantaneous elastic strain recovery on unloading, retarded strain recovery and permanent deformation or set.

Apparent Modulus of Four-Parameter Element.   When the four-parameter element is put in constant stress, σ_0, the strain is seen to be directly proportional to stress:

$$(3.76) \qquad \varepsilon = \sigma_0 \left[\frac{1}{E_1} + \frac{t}{\eta_{e_1}} + \left(\frac{1}{E_2} \right) \left(1 - \exp \left[\frac{-E_2\, t}{\eta_{e_2}} \right] \right) \right]$$

This can be written as:

$$(3.77) \qquad\qquad\qquad\qquad \varepsilon = \frac{\sigma_0}{E_{\text{app}}(t)}$$

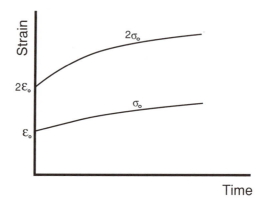

Figure 3.30.   Time-Dependent Strain for Two Applied Stresses for the Four-Parameter Element.

where $E_{app}(t)$ is the time-dependent apparent creep modulus, given as the reciprocal of terms in the large brackets in the equation above. Note that doubling the applied stress doubles the strain at any time, Figure 3.30. Furthermore, the apparent creep modulus is independent of applied stress and decreases with increasing time, as seen in Figure 3.31. This model response can be compared with tensile creep modulus data at different stress levels for 30% GR-PA-6 (glass-reinforced nylon 6) and 25% GR-PTFE (glass-reinforced PTFE) in Figure 3.32. GR-PA-6 exhibits a stress-independent creep modulus and so can be modeled by the four-element linear viscoelastic element. GR-PTFE has a stress-dependent creep modulus and so this linear viscoelastic model is not applicable. Creep data for many other materials can be found in the appropriate data section in the annual Modern Plastics Encyclopedia (16).

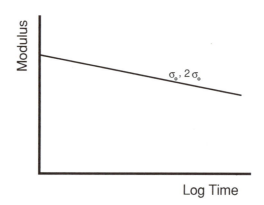

Figure 3.31.   The Apparent Creep Modulus of a Four-Parameter Element, Showing No Effect of Applied Strain Level.

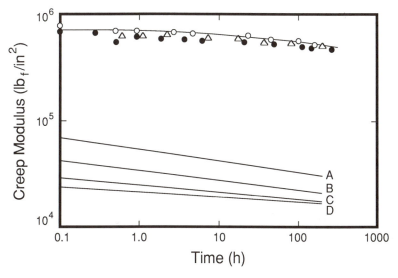

Figure 3.32.   Creep Moduli of Two Glass Fiber-Reinforced Thermoplastics. The Top Curve and Data Points Are 30%(wt) Glass Fiber-Reinforced Nylon 6 (PA-6) in Tension (Dry). Open Circles: 8000 lb_f/in^2. Closed Circles: 6000. Triangles: 4000. Bottom Set of Curves are 25%(wt) Glass Fiber-Reinforced PTFE in Compression. Curve A: 1000 lb_f/in^2. B: 2000. C: 3000. D: 4000 lb_f/in^2. (15).

It can be concluded that the four-parameter element response to creep loading in many cases adequately describes real polymer response. This simple model helps visualization of the complex time-dependent nature of polymeric materials, and points up the risk and uncertainty in applying short-time test data to long-time polymer performance.

The engineer must remember, however, that the four-parameter element is a linear viscoelastic model based on only two polymeric frequency responses, λ_1 and λ_2. In actuality, all polymers exhibit a spectrum of responses, some being short-time responses and some being long-time responses. This is discussed in more detail below.

Model-to-Molecule Analogy

Many researchers have tried to relate the various components of the four-parameter element with individual molecular response mechanisms in polymers. Rosen (17) describes these for a polymer in shear as:

a. Dashpot 1 [the Maxwell dashpot] represents molecular slippage. This slip of polymer molecules past one another is responsible for flow. The value of η_1 alone (molecular friction in slip) governs the equilibrium flow of the material.

b. Spring 1 [the Maxwell spring] represents the elastic straining of bond angles and lengths. All bonds in polymer chains have equilibrium angles and lengths. The value of G_1 characterizes the resistance to deformation for these equilibrium values. Since these deformations involve interatomic bonding, they occur essentially instantaneously from a macroscopic point of view. This type of elasticity is known thermodynamically as energy elasticity, because the straining of bond angles and lengths increases the material's internal energy.

c. Dashpot 2 [the Voigt–Kelvin dashpot] represents the resistance of the polymer chains to uncoiling and coiling, caused by temporary mechanical entanglements of the chains and molecular friction during these processes. Since coiling and uncoiling require cooperative motion of many chain segments, they cannot occur instantaneously and hence account for retarded elasticity.

d. Spring 2 [the Voigt–Kelvin spring] represents the restoring force brought about by the thermal agitation of the chain segments, which tends to return chains oriented by a stress to their most random or highest entropy configuration.

Since each of the components in the four-parameter element can be heuristically related to a general molecular characteristic of polymers, the effect of changing molecular weight and of crosslinking can be analyzed in terms of the elemental components as well. Consider the effect of increasing molecular weight of a linear polymer. The molecular resistance to flow is characterized by dashpot 1. Increasing the length of an individual molecule implies increasing the effective magnitude of the secondary forces between individual molecules. Increased bonding forces are realized by increased resistance in dashpot 1, or increased elongational viscosity, η_e. Shear viscosity, η, is proportional to weight average molecular weight, M_w, to the 3.4 power. As noted in Figure 3.29, an increasing elongational viscosity of the Maxwell dashpot, η_{e1}, results in a decrease in the slope of the viscous flow portion of the stress–time curve. Furthermore, an increasing viscosity causes a decrease in the magnitude of the permanent set or deformation, $\sigma_o t_o / \eta_{e1}$. This is seen in Figure 3.33.

Crosslinking is the covalent linking of linear molecules. As the extent of crosslinking increases (viz, the number of crosslinks per 1000 backbone carbon atoms or netting index increases), the effective polymer molecular weight increases. This results in a dramatic reduction in the degree of molecular mobility. Even at very light crosslinking, as with peroxide-crosslinked HDPE where there is no more than one link per thousand backbone carbon atoms, molecular slippage is almost totally inhibited. Thus, the effective viscosity for the Maxwell component dashpot approaches infinity. But, light crosslinking does not prohibit coiling and uncoiling of molecular chain segments. As a result, retarded elastic strain, as modeled by the Voigt–Kelvin components, is still observed. Since the resulting structure does not exhibit a permanent set, it is now considered to be a viscoelastic solid rather than a liquid. Thus, crosslinked polyolefins, particularly LDPE and HDPE, yield classic viscoelastic solid behavior under sustained load.

Since the mobility of chain segments is severely restricted by crosslinking, an increasing level of crosslinking should result in restriction in coiling and uncoiling tendencies. Thus, the restoring forces within the molecule should dominate. For the four-parameter element, the Voigt–Kelvin viscosity, η_{e2}, should increase dramatically. As

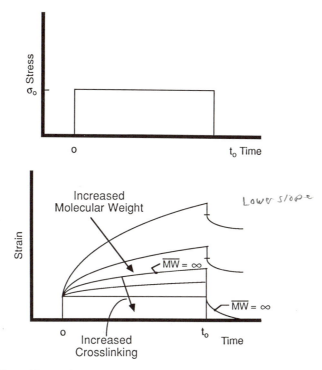

Figure 3.33. Effect of Increasing Molecular Weight and Crosslinking on the Response of a Four-Parameter Element to Constant Applied Stress, σ_0. (18).

a result, the overall stiffness and relative elasticity of the polymer should increase. The resulting time-dependent strain is seen in Figure 3.33*.

Of course, this type of analysis is not restricted to crosslinking. Any thermodynamic change can cause shifts in the molecular viscous-to-elastic response. For example, increasing crystallinity implies decreasing chain mobility, in general, and so polymer response to applied load would resemble the curves for increasing crosslinking. Similarly, polymers below their glass transition temperatures have relatively little viscous mobility, whereas above T_g, the elastic contribution is dramatically reduced. Keep in

---

*Note from this discussion that the creep resistance of a polymer can be improved substantially by crosslinking. Consider fabrication of a polyethylene container. As discussed in detail in Chapter 5, two common methods of fabricating containers are blow molding and rotational molding. The polyethylene used in blow molding has a much higher molecular weight than that used in rotational molding. Therefore, blow-molded containers have inherently greater creep resistance than rotationally molded containers, when the resin is uncrosslinked polyethylene. However, crosslinkable polyethylenes have been developed specifically for rotational molding. Although these materials are more expensive than uncrosslinked blow-molding grade polyethylenes, their greater creep resistance frequently offsets their additional cost.

mind, however, that the four-parameter element is one of the simplest models that can be used to mirror many types of polymer response. Many polymers cannot be modeled with the simple model.

A Three-Parameter Element

Oldroyd (19) examined elastic and viscous characteristics of bitumen, emulsions and suspensions of Newtonian fluids in other fluids and proposed a differential constitutive equation of state relating shear stress and rate of deformation:

$$(3.78) \qquad \tau + \lambda_1 \dot{\tau} = \mu_o (\dot{\gamma} + \lambda_2 \ddot{\gamma})$$

where $\ddot{\gamma}$ is the time rate of change of the shear rate and $\dot{\tau}$ is the time rate of change of the applied shear stress. μ_o, λ_1 and λ_2 are three material parameters. μ_o is the viscosity at very low shear rates, called the zero-shear viscosity. λ_1 is the relaxation time, representing stress decay when strain is set to zero. λ_2 is the relaxation time, representing stress decay when strain is set to zero. λ_2 is the retardation time, representing the strain decay when the applied shearing stress is set to zero. Consider the three-parameter element, Figure 3.34. This is the four-parameter element without the Maxwell elastic element. The strain on the Voigt–Kelvin element is $\gamma_1(t)$ and that on the Maxwell dashpot is $\gamma_2(t)$. The Voigt–Kelvin element response is:

$$(3.79) \qquad \tau = G\tau + \eta \dot{\gamma}$$

In terms of μ_o, λ_1 and λ_2, this is written as:

$$(3.80) \qquad \tau(t) = [\mu_o/(\lambda_1 - \lambda_2)] \gamma_1 + [\mu_o \lambda_2/(\lambda_1 - \lambda_2)] \dot{\gamma}_2$$

The Maxwell dashpot element is also written as:

$$(3.81) \qquad \tau(t) = \mu_o \dot{\gamma}_2$$

Since the total strain is $\tau(t) = \gamma_1(t) + \tau_2(t)$:

$$(3.82) \qquad \gamma(t) = \left[\frac{(1 + \lambda_2 d)}{M_o(\lambda_1 - \lambda_2)} + \frac{d}{\mu_o} \right] \tau(t)$$

where d is the operator, d/dt. When the Oldroyd equation is compared with this one, it is apparent that the Maxwell dashpot viscosity, η_1 equals μ_o, the Voigt-Kelvin viscosity, η_2 equals $\mu_o\lambda_1/(\lambda_1 - \lambda_2)$, and the Voigt–Kelvin elastic modulus, G equals $\mu_o/(\lambda_1 - \lambda_2)$. Thus, a four-parameter element, reconstructed as a three-parameter one, can be used to describe linear voscoelastic fluid flow as well. Note that the absence of the Maxwell spring element implies negligible elastic resistance to deformation on the interatomic bond level. This is expected for fluids.

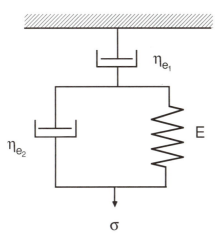

Figure 3.34.   Schematic of Three-Parameter Element. σ Is Applied Tensile Load. E Is Young's Modulus for Voigt–Kelvin Element. η_{e1} Is Elongational Viscosity for Voigt–Kelvin Element. η_{e2} Is Elongational Viscosity for Element Exhibiting No Initial Elastic Strain.

Method of Stress Superposition

In many practical applications, the applied stress on a plastic part is not constant. Rather, different levels of stress are applied and removed from the part at different times. Consider the stress history shown in Figure 3.35. Linear systems such as the four-parameter linear viscoelastic element obey the Boltzmann superposition principle. That is, partial deformation effects may be added together to obtain the resultant deformation. In Figure 3.35, the initial uniform stress σ_0 at time t_0 initially deforms the body σ_0/E_1. The deformation increases with time until the stress is increased at time t_1 to the value σ_1. The deformational curve can be viewed by the Boltzmann principle as a sum of the deformation caused by the original stress level σ_0 applied at t_0 plus a second stress $(\sigma_1 - \sigma_0)$ applied at time t_1.

At time t_2, all stress is removed. The resulting deformation is the sum of the recoveries of the two stress components, σ_0 and $(\sigma_1 - \sigma_0)$. Then at time t_3, another stress, σ_3, is applied. The response to the system of this stress is added to the continued relaxation strain of the two preceding histories. Mathematically the deformation of the element is the sum of all events preceding the time in question. The total creep strain, $\varepsilon(t)$ is given as:

$$(3.83) \qquad \varepsilon(t) = \frac{\sigma_0}{E(t)} + \frac{\sigma_1 - \sigma_0}{E(t - t_1)} + \ldots + \frac{\sigma_n - \sigma_{n-1}}{E(t - t_n)}$$

The reciprocal of modulus is compliance. The compliance in tension is $D = 1/E$, and in shear is $J = 1/G$. For the tensile strain equation above, in compliance terms:

$$(3.84) \quad \varepsilon(t) = \sigma_0 D(t) + (\sigma_1 - \sigma_0) D(t - t_1) + \ldots + (\sigma_n - \sigma_{n-1}) D(t - t_n)$$

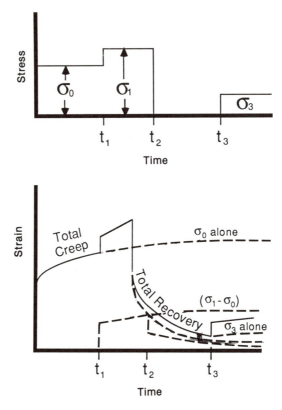

Figure 3.35.   Example of the Method of Superposition Applied to Four-Parameter Element. σ_o Is Initial Tensile Load. Load Is Increased to σ_1 at Time t_1. Load Is Removed at Time t_2. Another Load, σ_3, Is Applied at Time t_3.

For a finite number of loading and unloading steps:

$$(3.85) \qquad \varepsilon(t) = \sum_{i=1}^{n} \Delta \sigma_i D(t - t_i)$$

For an infinite number:

$$(3.86) \qquad \varepsilon(t) = \int_{-\infty}^{t} \left[\frac{\partial \sigma(\tau)}{\partial \tau} \right] D(t - \tau) \, d\tau$$

or

$$(3.87) \qquad \varepsilon(t) = \int_{-\infty}^{t} \left[\frac{d \varepsilon(U)}{dU} \right] \left[\frac{dU}{E(t - U)} \right]$$

The complete stress history on the polymer contributes to the observed deformation at time t. Therefore integration is from $(-\infty)$ to the current time, t.

Relaxation Spectrum

Real polymers do not exhibit single relaxation times as modeled by the linear viscoelastic elements. Molecular weight distribution and a statistical distribution and time-dependency in the number of entanglements should produce a spectrum of relaxation times. A system of simple linear viscoelastic elements can be used to illustrate more complex polymer response to imposed loading. It was shown, for example, in Section 3.6 for Figure 3.23 that the simple two-parameter Maxwell element qualitatively describes stress relaxation under constant deformation, ε_o. The individual element response is:

$$(3.88) \qquad \sigma(t) = \varepsilon_o E \exp\left[\frac{-t}{\lambda}\right]$$

where $\lambda = \eta_e/E$. Consider a system of n Maxwell elements in parallel, Figure 3.36. This is the Maxwell–Weichert element. Each Maxwell element is deformed to a value $\varepsilon_i = \varepsilon_o$. The total stress on the Maxwell–Weichert element is given as:

$$(3.89) \qquad \sigma_T = \sum_{i=1}^{n} \sigma_i$$

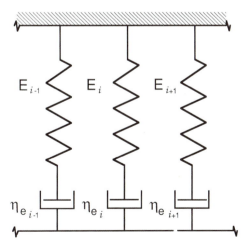

Figure 3.36.   Schematic of the Parallel Maxwell Element Model Known as the Maxwell–Weichert Element. η_{el} Is the Elongational Viscosity of the ith Element. E_i Is the Young's Modulus of the ith Element.

or upon substitution:

$$(3.90) \qquad \sigma(t) = \varepsilon_o \sum_{i=1}^{n} E_i \exp\left[\frac{-t}{\lambda_i}\right]$$

where:

$$(3.91) \qquad \lambda_i = \frac{\eta_{ei}}{E_i}$$

The apparent modulus is given as:

$$(3.92) \qquad E(t) = \frac{\sigma(t)}{\sigma_o} = \sum_{i=1}^{n} E_i \exp\left[\frac{-t}{\lambda_i}\right]$$

For an infinite number of Maxwell elements:

$$(3.93) \qquad E(t) = \int_{0}^{\infty} E(\lambda) \exp\left[\frac{-t}{\lambda}\right] d\lambda$$

If one of the Maxwell elements has a dashpot with an infinite viscosity, the limiting modulus for that element will always be finite:

$$(3.94) \qquad \sigma(t) = \sigma_o E_1 + \sum_{i=2}^{n} \varepsilon_o E_i \exp\left[\frac{-t}{\lambda_i}\right]$$

For long times, the summation term approaches zero. Thus $\sigma(t)$ approaches $\sigma_o E_1$ and $E(t)$ approaches E_1.

Example 3.4

Q: a. Predict the time-dependent apparent modulus of the dual Maxwell–Weichert element of Figure 3.37, with:

$$E_1 = 5 \times 10^5 \text{ lb}_f/\text{in}^2 \text{ and } \lambda_1 = 6 \text{ s}$$

$$E_2 = 7 \times 10^2 \text{ lb}_f/\text{in}^2 \text{ and } \lambda_2 = 1000 \text{ s}$$

b. Determine the deformation of 25 s if a stress of 1000 lb_f/in^2 is applied.

A: a: The apparent modulus of the two-element Maxwell–Weichert model is:

$$E(t) = E_1 \exp\left[\frac{-t}{\lambda_1}\right] + E_2 \exp\left[\frac{-t}{\lambda_2}\right]$$

$$= 5 \times 10^5 \exp[-0.1666\ t] + 700 \exp[-0.001\ t]$$

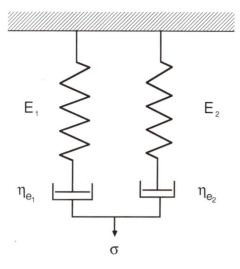

Figure 3.37. Schematic for a Dual Maxwell–Weichert Element. η Is the Applied Tensile Load. η_{e1} and η_{e2} Are the Elongational Viscosities. E_1 and E_2 Are the Young's Moduli of the Two Elements.

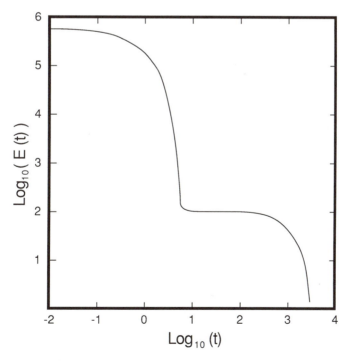

Figure 3.38. Time-Dependent Creep Modulus of the Dual Maxwell–Weichert Element of Figure 3.37 in Response to Constant Applied Tensile Load.

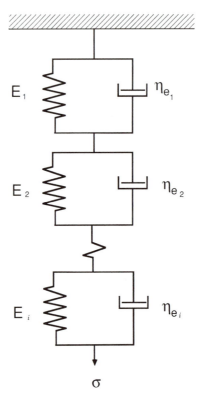

Figure 3.39.   Schematic of the Generalized Voigt–Kelvin Mechanical Analog in Tension. E_i Is the Young's Modulus of the ith Element. η_{ei} Is the Elongational Viscosity of the ith Element. σ Is the Tensile Load.

The time-dependent modulus is plotted in Figure 3.38.

    b. The extent of deformation at 25 s is:

$$E(25) = 3374 + 683 = 4057 \ \text{lb}_f/\text{in}^2$$

$$\varepsilon(25) = \frac{\sigma}{E(25)} = \frac{1000}{4057} = 0.247 \text{ or } 24.7\%$$

The Voigt–Kelvin element can also be generalized. A series of Voigt–Kelvin elements, Figure 3.39, can be used to illustrate creep for real polymers. The response of a single Voigt–Kelvin element to a constant stress is:

(3.95)
$$\varepsilon(t) = \sigma_0 \, D_1 \left(1 - \exp\left[\frac{-E}{\lambda_1} \right] \right)$$

where $D_1 = 1/E_1$, the compliance, and $\lambda_1 = \eta_e'/E_1$. The response for a finite series of elements is:

$$(3.96) \qquad \varepsilon(t) = \sigma_0 \sum_{i=1}^{n} D_i \left(1 - \exp \left[\frac{-t}{\lambda_i} \right] \right)$$

The compliance is given as:

$$(3.97) \qquad D(t) = \frac{\varepsilon(t)}{\sigma_0} = \sum_{i=1}^{n} D_i \left(1 - \exp \left[\frac{-t}{\lambda_i} \right] \right)$$

For an infinite number of Voigt–Kelvin elements in series, the summation is replaced with an integral:

$$(3.98) \qquad D(t) = \int_0^\infty D(\lambda) \left(1 - \exp \left[\frac{-t}{\lambda} \right] \right) d\lambda$$

Dynamic Response

To this point, mechanical analogs have been used to illustrate linear viscoelastic behavior under constant stress or strain. Many engineering applications require materials to perform under cyclic or periodic loads. The characteristics of the material undergoing cyclic deformation can be readily measured in several relatively simple but elegant ways (see Section 6.3 on Dynamic Mechanical Analysis). Consider the plastic response to a cyclical shear strain displacement:

$$(3.99) \qquad \gamma = \gamma_0 \sin(\omega t)$$

The responses of a single spring and dashpot to this forcing function are shown in Figure 3.40. It was noted in Section 3.5 that the spring represents a Hookean solid and its response is always in phase with the forcing function. The dashpot represents a Newtonian fluid and its response is always $\pi/2$ rad or 90° out of phase with the forcing function. A viscoelastic material contains both elements and so its response should be intermediate to these extremes, Figure 3.41. The shear stress is resolved into two components, one in phase with the displacement and the other $\pi/2$ rad or 90° out of phase.

$$(3.100) \qquad \tau^* = \tau' + i.\tau''$$

$$(3.101) \qquad \gamma^* = \gamma'$$

The complex shear modulus G^* is then given as:

$$(3.102) \qquad G^* = \tau^*/\gamma^* = (\tau'/\gamma') + i.(\tau''/\gamma')$$

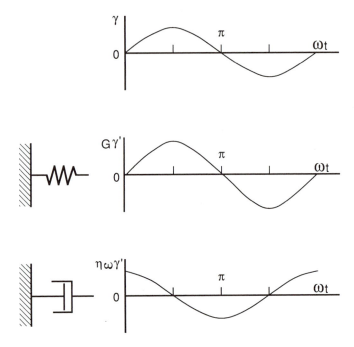

Figure 3.40. Responses of the Classical Elements to Applied Sinusoidal Strain, Top Curve. The Second Curve Shows the Response of a Simple Spring, a Hookean Material Analog. The Curve Is in Phase with the Applied Strain. The Third Curve Shows the Response of a Simple Dashpot, a Newtonian Fluid Analog. The Curve Is $\pi/2$ Radians out of Phase with the Applied Strain.

The in-phase component is called the real response. The out-of-phase portion is the imaginary response. Thus $G' = (\tau''/\gamma')$ is the in-phase portion of the modulus and is called the *storage modulus*. It represents the fraction of energy that is elastically recoverable. $G'' = (\tau''/\gamma')$ is the out-of-phase portion of the modulus and is called the *loss modulus*. It represents the fraction of inputted energy that is dissipated through viscous flow. And finally, $G^* = G' + i\,G''$.

The ratio of loss modulus to storage modulus is *tan* δ. This is called the loss tangent:

$$(3.103) \qquad \tan\delta = \tau''/\tau' = G''/G'$$

where δ is the phase angle, Figure 3.41. Similarly, the compliance of a material, J^*, is given as:

$$(3.104) \qquad J^* = 1/G^* = \gamma^*/\tau^*$$

and the loss and storage compliances, respectively, are:

$$(3.105) \qquad J'' = \frac{-G''}{G'^2 + G''^2}$$

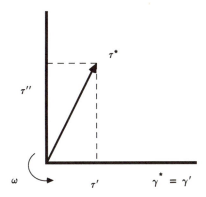

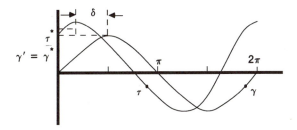

Figure 3.41.   Response of a Simple Viscoelastic Element to Applied Sinusoidal Shear Strain. γ Is the Deformation in Response to the Applied Shear, τ. δ Is the Phase Angle.

(3.106)
$$J' = \frac{G'}{G'^2 + G''^2}$$

and the complex shear *compliance* is:

(3.107)
$$J^* = J' - iJ''$$

The response of a polymer under static cyclic load is shown in schematic in Figure 3.42. At high loading frequencies, a typical polymer appears glassy. That is, the polymer has a very high value for storage modulus and nearly no value for loss modulus. At very low cyclic frequency, the polymer is rubber-like, having a relatively low value for storage modulus and again nearly no value for loss modulus. In both extremes, the polymer storage modulus appears as independent of frequency. At intermediate frequencies, however, the polymer appears as a viscoelastic solid with a frequency-dependent storage modulus. The loss modulus on the other hand shows a maximum in the curve. This maximum usually occurs in the region where the frequency dependency of the storage modulus is the maximum. The value of the loss tangent, tan δ, also exhibits a maximum in nearly the same region. At times, compliance is used rather

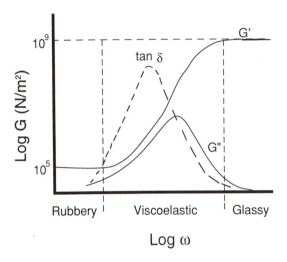

Figure 3.42. The Response of a Real Polymer to Applied Sinusoidal Strain. G Is Modulus, ω Is the Sinusoidal Frequency. When the Polymer Is Rubbery, Both the Storage and Loss Portions of the Modulus, G' and G'', Are Small. Tan δ, the Phase Shift Parameter Is Small. When the Polymer Is Glassy (Below Glass Transition Temperature), the Storage Modulus, G' Is Large, but the Loss Modulus, G''', Is Small. The Phase Shift Angle Is Small. In the Viscoelastic Region, G' Is Decreasing with Temperature and G'' Reaches a Maximum Value. tan δ Becomes Very Large in This Region.

than complex modulus. The equivalent frequency-dependent compliance curves are given in Figure 3.43.

Recall that the simple two-element Maxwell element was in general agreement with the real polymer stress relaxation behavior and that the simple two-element Voigt–Kelvin element agreed with creep. Consider these elements under cyclic load.

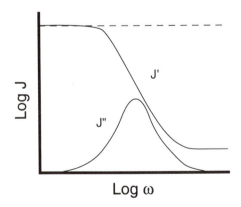

Figure 3.43. The Compliance of a Polymer as a Function of Applied Sinusoidal Strain. J' Is the Storage Portion of the Compliance, J'' Is the Loss Portion of the Compliance.

Maxwell Element Under Cyclic Load

The Maxwell element in shear is given as:

$$(3.108) \qquad \dot{\tau} + \left(\frac{G}{\eta}\right) \tau = G\dot{\gamma}$$

Since the dynamic input is given as:

$$(3.109) \qquad \gamma(t) = \gamma_0 \sin(\omega t)$$

the solution of the equation for stress yields:

$$(3.110) \quad \tau(t) = C_1 \exp\left[\frac{-t}{\lambda}\right] + \left[\frac{G \gamma_0 \omega \lambda}{1 + (\omega \lambda)^2}\right] [\omega \lambda \sin(\omega t) + \cos(\omega t)]$$

where $\lambda = \eta/G$, the time constant. The constant C_1 is an arbitrary value dependent on the initial stress condition of the system at zero time. For long time, steady state, the first term becomes negligible. The steady-state response of a Maxwell element to cyclic loading is then given as:

$$(3.111) \qquad \tau = \left[\frac{G.\gamma \omega \lambda}{1 + (\omega \lambda)^2}\right] [\omega \lambda \sin(\omega t) + \cos(\omega t)]$$

The real component of the applied strain is γ_0. Thus, the elastic stress component in phase with the strain is τ':

$$(3.112) \qquad \tau' = \left[\frac{G \gamma_0 (\omega \lambda)^2}{1 + (\omega \lambda)^2}\right]$$

and the viscous stress component is τ'':

$$(3.113) \qquad \tau'' = \left[\frac{G \gamma_0 \omega \lambda}{1 + (\omega \lambda)^2}\right]$$

The loss tangent, tan δ, is:

$$(3.114) \qquad \tan \delta = \frac{\tau''}{\tau'} = \frac{1}{\omega \lambda}$$

The storage and loss moduli are, respectively:

$$(3.115) \qquad G'' = \frac{G (\omega \lambda)^2}{1 + (\omega \lambda)^2}$$

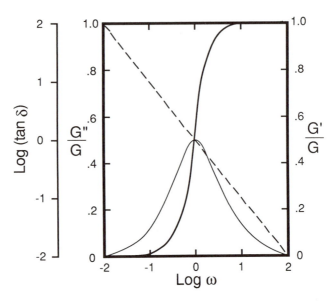

Figure 3.44.   Response of the Maxwell Element to Sinusoidal Strain. The Dashed Line Is tan δ, the Phase Shift Angle. The Heavy Solid Line Is Reduced Storage Modulus, G'/G. The Light Solid Line Is Reduced Loss Modulus, G''/G.

(3.116) $$G'' = \frac{G\,\omega\,\lambda}{1 + (\omega\,\lambda)^2}$$

These are shown graphically in Figure 3.44. Note that the Maxwell model storage and loss moduli mimic the experimental data of Figure 3.42, but that the Maxwell model loss tangent does not show a maximum.

The Voigt–Kelvin Element Under Cyclic Load

The Voigt–Kelvin element can be analyzed in a similar fashion. The element in shear is given as:

(3.117) $$\tau = G\,\gamma + \eta\,\dot\gamma$$

The dynamic input is:

(3.118) $$\gamma(t) = \gamma_0\,\sin(\omega\,t)$$

and the steady-state response of the Voigt–Kelvin element to cyclic loading is:

(3.119) $$\tau = G\,\gamma_0\,\sin(\omega\,t) + \eta\,\omega\,\gamma_0\,\cos(\omega\,t)$$

The in-phase or elastic stress component, τ', the out-of-phase, or viscous stress component, τ'', the loss tangent, $\tan \delta$, and the loss, storage, and complex moduli are, respectively:

(3.120) $$\tau' = G \gamma_0$$

(3.121) $$\tau'' = \eta \omega \gamma_0$$

(3.122) $$\tan \delta = \frac{\tau''}{\tau'} = \frac{\eta \omega}{G}$$

(3.123) $$G'' = \eta \omega$$

(3.124) $$G' = G$$

(3.125) $$G^* = [G^2 + (\eta \omega)^2]^{1/2}$$

As with the Maxwell element, there is adequate modeling of the loss and storage compliance terms but the frequency-dependent loss tangent is unrealistic.

Four-Parameter Element Under Cyclic Load

The four-parameter element provides yet another possible model for polymer response to cyclic load (see Figure 3.45). As noted above, the constitutive equation is second-order in its response to load. If a sinusoidal *stress* input:

(3.126) $$\tau = \tau_0 \cos(\omega t)$$

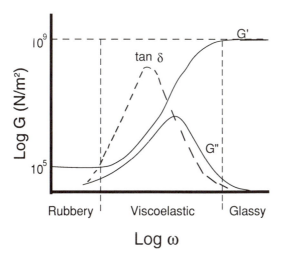

Figure 3.45.   The Response of a Real Polymer to Applied Sinusoidal Strain. Symbols as in Figure 3.42.

is applied to the elements at steady state, where the exponentially transient terms containing arbitrary constants become negligible, the time-dependent *strain* is given as:

$$(3.127) \qquad \gamma = \tau_0 \left[\frac{\cos(\omega\,t)}{E_1} + \frac{\sin(\omega\,t - \delta)}{(\omega^2\,\eta_2^2 + E_2^2)^{1/2}} + \frac{\sin(\omega\,t)}{\omega\,\eta_1} \right]$$

The first term inside the bracket represents the Maxwell spring and is in phase with the stress. The third term represents the Maxwell dashpot and so is out of phase by $\pi/2$ rad with the stress. The second term represents the contribution of Voigt–Kelvin element. δ is the phase angle. The four-parameter element loss tangent is given as:

$$(3.128) \qquad \tan\delta = \frac{-E_2}{\omega\,\eta_2}$$

When ω is small, δ approaches $(-)\,\pi/2$ rad and so this element responds as if it were in phase, an elastic element. When ω is large, δ approaches 0 and this element responds as if it were $\pi/2$ rad out-of-phase, a viscous element. Thus the four-parameter element shows good qualitative agreement with actual material response. However, once again, the loss tangent does not show the maximum so typical of real polymeric materials.

More Complex Models

It was noted earlier that sets of elements such as parallel Maxwell elements or Voigt–Kelvin elements in series can be used to model real polymer responses to time-independent static loads or deflections. Similarly, combinations of these elements can be used to model polymer responses to cyclic loads and strains. The responses of Maxwell–Weichert parallel elements and Voigt–Kelvin elements in series are given in Table 3.3. Complex models can more accurately reflect real polymer responses, but the number of parameters to be experimentally determined increases in direct proportion to the number of elements employed. Unfortunately, none of these groups of elements yield loss tangent curves that show maxima.

The Maxwell–Weichert element is written in terms of the complex viscoelastic modulus as:

$$(3.129) \qquad G(t) = \sum_{i=1}^{n} G_i \exp\left[\frac{-t}{\lambda_i}\right]$$

and the complex compliance $J(t)$ is written for the series of Voigt–Kelvin elements as:

$$(3.130) \qquad J(t) = \sum_{i=1}^{n} J_i \left(1 - \exp\left[\frac{-t}{\lambda_i}\right] \right)$$

The complex compliance for the Maxwell–Weichert element and the complex modulus for the series of Voigt–Kelvin elements cannot be written simply.

Table 3.3. Responses of generalized Maxwell (Maxwell–Weichert) and Voigt–Kelvin elements to creep, stress relaxation, and sinusoidal inputs.

Experiment	Maxwell–Weichert model	Voigt–Kelvin model
Creep	—	$D = \sum\limits_{i=1}^{n} D_i\,(1 - e^{-t/\lambda_i})$
Stress relaxation	$E = \sum\limits_{i=1}^{n} E_i\,e^{-t/\lambda_i}$	
Sinusoidal input	$E' = \sum\limits_{i=1}^{n} \dfrac{E_i\,(\omega\,\lambda_i)^2}{1 + (\omega\,\lambda_i)^2}$	$D' = \sum\limits_{i=1}^{n} \dfrac{D_i}{(1 + (\omega\lambda_i)^2}$
	$E'' = \sum\limits_{i=1}^{n} \dfrac{E_i\,(\omega\,\lambda_i)}{1 + (\omega\,\lambda_i)^2}$	$D'' = \sum\limits_{i=1}^{n} \dfrac{D_i}{1 + (\omega\lambda_i)^2}$

The continuous relaxation spectrum of the Maxwell–Weichert element is given as:

$$(3.131) \qquad G(t) = G_e + \int_{-\infty}^{\infty} H \exp\left[\frac{-t}{\lambda}\right] d\ln\lambda$$

where $H\,d\ln\lambda$ represents the contribution to rigidity for relaxation times between $\ln\lambda$ and $\ln + d\ln\lambda$. This equation is considered to be the definition for H (20). For steady state of or uncrosslinked polymers, H should approach zero. For crosslinked polymers, H should drop to a relatively low level, but should approach a constant, non-zero value. For short times, the mechanical behavior should be that of a perfectly elastic solid and so H should approach zero here too. However, even at very short times or high frequencies, there is an appreciable amount of viscous dissipation in real systems. As with the analysis on the simple Maxwell element, H is small in the glassy region, rises rapidly, then drops steeply into the rubbery zone wehre it either plateaus or falls to very low values. For high molecular weight, uncrosslinked polymers, a plateau zone is experimentally seen in Figure 3.46 between the maximum and near-zero zone. Ferry defines the eight types of polymers in Figure 3.46 as:

1. Dilute polymer solution (PS, MW = 0.86 M, 0.015 g/ml Aroclor 1248).
2. Amorphous, low molecular weight polymer (PVAc, MW = 10.5 k).
3. Amorphous, high molecular weight polymer (PS, MW = 0.6 M).
4. Amorphous, high molecular weight polymer with long side chains (poly-*n*-octyl methacrylate, MW = 3.62 M).
5. Amorphous, high molecular weight polymer below its glass transition temperature (PMMA).

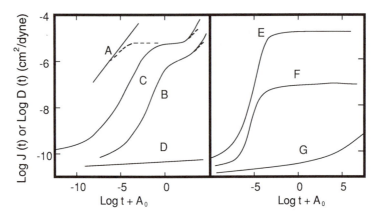

Figure 3.46.   Creep Compliance Spectra for Several Classical Polymeric Types. (21). J Is Creep Compliance. Uncrosslinked Polymers on Left, Crosslinked Polymers on Right. A_0 Is Shift Factor, in [ ] Below:

A: Amorphous Low-MW Polymer $[A_0 = -5]$.
B: Amorphous High-MW Polymer $[-6]$.
C: Amorphous High-MW Polymer with Long Side Groups $[-9]$.
D: Amorphous High-MW Polymer Below T_g $[0]$.
E: Lightly Crosslinked Amorphous Polymer $[-7]$.
F: Dilute Crosslinked Gel $[-1]$.
G: Highly Crystalline Polymer $[2]$.

6.   Lightly crosslinked amorphous polymer (Hevea rubber).
7.   Dilute crosslinked gel (10% PVC in solvent, styrene-butadiene random copolymer, vulcanized with dicumyl peroxide).
8.   Highly crystalline polymer (HDPE).

Note that if H is known for Maxwell–Weichert elements, the elements of the complex modulus $G^*(t)$ can also be determined:

(3.132)
$$G' = G_e + \int_{-\infty}^{\infty} \left[\frac{H.\omega^2 \lambda^2}{1 + (\omega \lambda)^2} \right] d \ln \lambda$$

(3.133)
$$G'' = \int_{-\infty}^{\infty} \left[\frac{H \omega \lambda}{1 + (\omega \lambda^2)} \right] d \ln \lambda$$

The series Voigt–Kelvin model can be written in terms of a continuous retardation spectrum, L, as:

(3.134)
$$J(t) = J_g + \int_{-\infty}^{\infty} L \left(1 - \exp \left[\frac{-t}{\lambda} \right] \right) d \ln \lambda + \frac{t}{\eta}$$

Again J_g represents the deformation due to non-infinite stress. The retardation spectrum, L, is seen for seven characteristic polymers in Figure 3.47. The storage and loss compliance terms can be obtained from $J^*(t)$ in terms of L as follows:

$$(3.135) \qquad J' = J_e + \int_{-\infty}^{\infty} \left[\frac{L}{1 + (\omega \lambda)^2} \right] d \ln \lambda$$

$$(3.136) \qquad J'' = \int_{-\infty}^{\infty} \left[\frac{L \omega \lambda}{1 + (\omega \lambda)^2} \right] d \ln \lambda + \frac{1}{\omega \eta}$$

It is apparent that the continuous forms for these simple mechanical analogs can be used as guides to understanding the rather complex stress–strain behavior of some polymers. These analog forms can become quite complex. For example, for the continuous Maxwell–Weichert analog, the relationship between stress and rate of deformation is given as:

$$(3.137) \qquad \tau = \dot{\gamma} \int_{-\infty}^{\infty} \lambda H \left(1 - \exp \left[\frac{-t}{\lambda} \right] \right) d \ln \gamma + G_e^* \dot{\gamma} t$$

A stress–strain (τ, γ) plot produces a curve at constant strain rate. The slope of the curve yields the time-dependent continuous-element modulus, $G(t)$. Similar analysis with the series Voigt–Kelvin continuous element analog for constant rate of stress

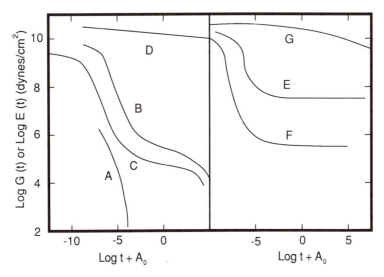

Figure 3.47.   Stress Relaxation Moduli for Several Classical Polymeric Types (22). Notation Identical to That in Figure 3.46.

yields a stress–strain curve, the slope of which is the continous-element creep compliance.

Keep in mind that regardless of the complexity of these models, they remain only mechanical analogs of real polymer performance under periodic loads. Polymers are not comprised of Hookean springs and Newtonian dashpots, regardless of how elegantly the elements are hooked together. Polymer performance should never be "force-fit" into an analog interpretation, regardless of the sophistication of the model. Even the most exotic assemblage of springs and dashpots cannot adequately explain relatively simple isothermal steady-state polymer behavior such as stirring rod climbing or shear banding. The complexities of such phenomena as high-frequency fatigue failure and nonisothermal fiber spinning are certainly beyond the best efforts of "patching up" these simple analogs. Accurate, intelligent observation is probably more important to the understanding of polymer stress–strain–rate-of-strain behavior than blind application of exotic multi-parameter equations of state.

3.7 Time-Temperature Superposition

The effect of temperature on the response of a polymer above its glass transition temperature can be as significant as the effect of frequency on storage modulus, Figure 3.48, and of time on modulus, Figure 3.49. Consider the simple Maxwell element. From molecular theory (23), the time constant, λ, is proportional to the square of the root-mean-square end-to-end length of the polymer module, $a^2 Z$, proportional to a measure of intermolecular friction, f, and inversely proportional to temperature, T:

$$(3.138) \qquad \lambda(T) = \frac{k \, a^2 \, Z f}{T}$$

The complex modulus, G^*, is temperature-dependent, as well:

$$(3.139) \qquad G^*(t;T) = k' \, T \exp\left[\frac{-t}{\lambda(T)}\right]$$

and the real and imaginary modulus components are given as:

$$(3.140) \qquad G'(\omega;T) = G^*(t;T) \frac{(\omega \, \lambda)^2}{1 + (\omega \, \lambda)^2}$$

$$(3.141) \qquad G''(\omega;T) = G^*(t;T) \frac{\omega \, \lambda}{1 + (\omega \, \lambda)^2}$$

The temperature effect on the time constant can be lumped into a single parameter, a_T, called a shift factor, as:

$$(3.142) \qquad \frac{\lambda(T)}{\lambda(T_o)} = a_T = \frac{[a^2 f]_{T_o} \, T}{[a^2 f]_T \, T_o}$$

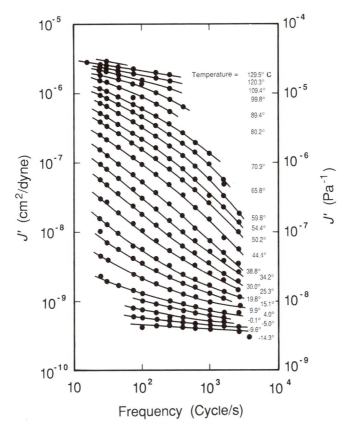

Figure 3.48.   Storage Compliance, J', as a Function of Frequency for Poly-*n*-Octyl Methacrylate in Temperature Range Between Glassy and Rubbery Regions (24).

T_0 is an arbitrary temperature that can be selected in the middle of the temperature range of interest*. The proper relationship between a_T and temperature is obtained by plotting $G^*(t;T) = G^*(t) T_0/T$ against t, $G'(\omega;T) = G' T_0/T$ and $G''(\omega;T) = G'' T_0/T$ against ω, and so on. These curves are then shifted until they overlap, Figure 3.50. The amount of shift is then plotted against temperature difference, $(T-T_0)$. As seen in Figure 3.51, $J'(\omega;T)$, the storage compliance data of Figure 3.48 for poly-*n*-octyl methacrylate is reduced to a single curve. a_T as a function of temperature is given in Figure 3.52. The polycarbonate torsional creep data of Figure 3.49 are reduced to two curves of differing molecular weights in Figure 3.53. The a_T shift factor for the PC is given as Figure 3.54. The functional dependence of a_T on temperature is:

$$(3.143) \qquad \log_{10} a_T = \frac{-C_{1o} (T-T_0)}{C_{2o} + T - T_0}$$

---

*Correctly, the shift factor should include the ratio of densities at the subject and reference temperatures. For clarity, this ratio is assumed to be unity in all subsequent equations.

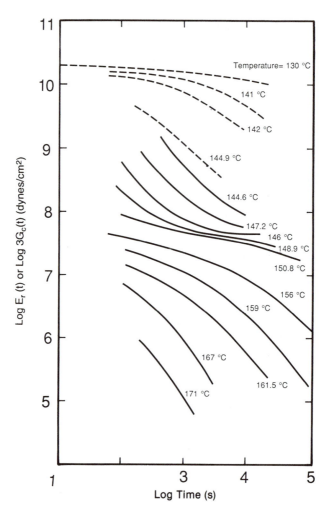

Figure 3.49.   Effect of Temperature on Torsional Creep Modulus, G_c, or Stress Relaxation Modulus, E_r, for Polycarbonate, $M_w = 40,000$ (25). Dashed Lines: Creep Data. Solid Lines: Stress Relaxation Data.

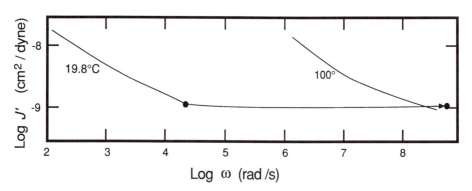

Figure 3.50.   Example of Shifting Storage Compliance Data, J', at 19.8°C to 100°C Loss Compliance Data Curve. (26).

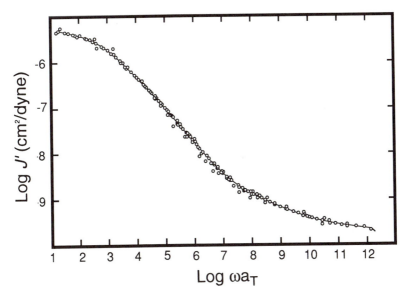

Figure 3.51. Master Curve for Storage Compliance, J', of Poly-*n*-Octyl Methacrylate Data Given in Figure 3.48. (27). a_T Is the Shift Factor, Given in Figure 3.52.

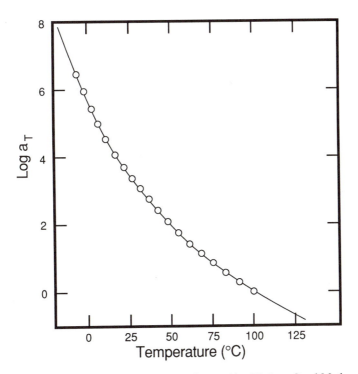

Figure 3.52. Shift Factor, a_T, for Storage Compliance, J', of Poly-*n*-Octyl Methacrylate Data of Figure 3.48. (28).

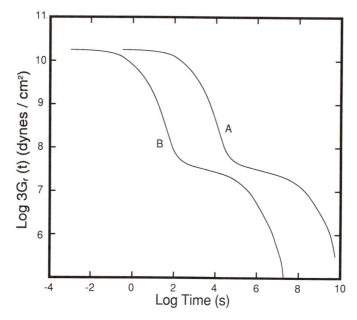

Figure 3.53.   Master Curve for Stress Relaxation Modulus, G_r, for Polycarbonate. Curve A: M_w = 90,000. Curve B: M_w = 40,000, Figure 3.49. (25).

Experimentally, the most convenient reference temperature is the glass transition temperature, T_g. The above equation is frequently written as:

(3.144)
$$\log_{10} a_T = \frac{-C_{1g}\,(T - T_g)}{C_{2g} + T - T_g}$$

Table 3.4.   WLF coefficients for temperature superposition.

Polymer	T_g (°K)	C_{1g}	C_{2g} (°C)
Polystyrene (PS)	373	14.5	50.4
Polyisobutylene (PIB)	202	16.6	104.4
Polyvinyl acetate (PVAc)	305	15.6	46.8
Polyurethane (PUR, Elastomer)	238	15.6	32.6
Natural rubber (Hevea)	200	16.7	53.6
Polymethyl acrylate (PMA)	276	18.1	45.0
Polyethyl methacrylate (PEMA)	335	17.6	65.6
Poly-n-octyl methacrylate	253	16.1	107.3
Polycarbonate (PC)	423	16.14	56.0
"Universal constants"	–	17.44	51.6

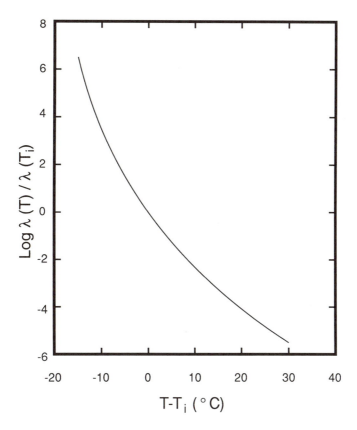

Figure 3.54. Shift Factor, a_T, for Polycarbonate of Figure 3.53 (29). Data Points for Both M_w Materials Fall On Solid Line. Solid Line: $\log_{10} a_T = 16.14\ (T - T_i) / (56 + T - T_i)$.

where C_{1g} and C_{2g} are called the Williams–Landel–Ferry or *WLF* coefficients. This expression is considered relatively reliable for rubbery-state amorphous polymers over the temperature range of $T = T_g$ to $T = T_g + 100°C$. WLF coefficients for several polymers are given in Table 3.4. For polymers that are not given in the table, homologous polymer values can be used. As a first approximation, $C_{1g} = 17.44$ and $C_{2g} = 51.6$ can be considered as "universal constants" (29).

Example 3.5

Q: If it takes an amorphous polymer with $T_g = 35°C$ 1 hour to reach 10^9 dyn/cm^2 modulus at 75°C, how long does it take it to reach the same modulus at 50°C?

A: Since the plastic is unknown, the universal constants from Table 3.4 are used:

$$\log_{10} a_T = \frac{-C_{1g}\ (T - T_g)}{C_{2g} + T - T_g}$$

where $C_{1g} = 17.44$ and $C_{2g} = 51.55$.

$$\text{At } 75°C, \log_{10} a_T = \frac{-17.44 \times 40}{51.55 + 40} = -7.62$$

or

$$a_{75} = 2.4 \times 10^{-8}$$

$$\text{At } 50°C, \log_{10} a_T = \frac{-17.44 \times 15}{51.55 + 15} = -3.93$$

or

$$a_{50} = 1.173 \times 10^{-4}$$

$$\frac{t_{75}}{t_{50}} = \frac{2.4 \times 10^{-8}}{1.173 \times 10^{-4}} = 2.046 \times 10^{-4}$$

$$t_{50} = \frac{1 \text{ h}}{2.046 \times 10^{-4}} = 4890 \text{ hours.}$$

Note that the WLF empirical function, a_T, can be used to shift shear viscosity as well:

$$(3.145) \qquad \eta(T) = \frac{\eta_0 \, a_T \, T}{T_0}$$

where η_0 is determined at a reference temperature, T_0, and both viscosities are determined at the same shear rate. Again the temperature must be less than $T_g + 100°C$ for good correlation.

Thermodynamically, it can be shown that for an amorphous polymer above its glass transition temperature, the WLF equation is a measure of the difference in apparent energy of activation for viscoelastic relaxation and the Arrhenius-dependent energy of activation for viscous-only behavior. The viscoelastic activation energy, ΔH_a, is written as:

$$(3.146) \qquad \Delta H_a = \frac{R \, d \ln a_T}{d(1/T)}$$

or upon differentiation of the WLF equation:

$$(3.147) \qquad \Delta H_a = \frac{2.303 \, R \, C_{1g} \, C_{2g} \, T^2}{(C_{2g} + T - T_g)^2}$$

At high temperatures, the WLF equation reduces to the Arrhenius form, with $\Delta H_a = 2.303\ R\ C_{1g}\ C_{2g}$. The viscous activation energy is given as:

$$(3.148) \qquad \Delta H_\eta = \Delta H_a - R\ T + a_T\ R\ T^2$$

where R is the universal gas constant. For polymers below the glass transition temperature, the WLF empirical equation is not applicable. Instead this equation has been proposed:

$$(3.149) \qquad \log_{10} a_T = \frac{\Delta H_a}{R}\left(\frac{1}{T} - \frac{1}{T_0}\right)$$

This relationship holds for highly crystalline polymers below their melt temperatures, as well, since the crystalline structure acts to minimize molecular motions that lead to intermolecular friction, the hypothesis on which the WLF model is based.

Note that the simple Maxwell element used here to demonstrate the WLF arithmetic can be replaced with the Voigt–Kelvin element for stress relaxation, the four-parameters element for more complex stress–strain responses, and in fact the more complex parallel Maxwell–Weichert element, the series Voigt–Kelvin element, and even the very complex tensor-based convected stress–strain constitutive equations of state. In each, the approach to the empirical WLF superposition model is the same, and for nearly all cases, the superposition model is independent of the nature of the constitutive equation.

3.8 Polymeric Yield and Ductility

One common method of measuring the strength of a polymer is by uniaxial tensile stress. The specific tensile test used for most plastics is ASTM D638. This test is an adaptation from a standard used to evaluate metal response to uniaxial tensile load; see Section 6.3. As noted above, the applied force per unit cross-sectional area of the specimen is the stress. *Engineering stress* is the force per unit area of the unstressed specimen. *True stress* is the force per unit area of the stressed specimen. Since Poisson's ratio for most polymers is less than 0.5, the value for ideal materials, the area of the specimen decreases with increasing axial strain. The deformation or strain in a specimen is usually reported as engineering strain:

$$(3.150) \qquad \frac{\varepsilon}{L_0} = \frac{\delta}{L_0}$$

where δ is the specimen elongation between appropriate calibration marks, rather than as true strain:

$$(3.151) \qquad \frac{\varepsilon}{L} = \ln\left(\frac{L}{L_0}\right)$$

As an example, load-elongation data for mild steel are used to calculate both types of strains in Figure 3.55. The shapes of these stress–strain curves are typical for many ductile materials. These curves exhibit several distinctive characteristics. The initial portion of the curve is linear; that is, the strain is directly proportional to stress. The material in this region is Hookean. Point A is the *proportional limit* or the maximum stress for which the material remains Hookean. Of course, the slope of the stress–strain curve at this point is simply Young's modulus or the elastic modulus.

As the stress is increased, the curve begins to bend slightly. Point B is called the *elastic limit*. Typically, this is the point beyond which the sample cannot recover to zero strain when the stress is removed. As the stress is increased, the *yield point* C is reached. The stress at this point is called the *yield stress*. At this point, the material has exceeded its elastic limit, complete recovery is impossible, and incipient plastic drawing begins. For many materials, a slight increase in stress at this point results in a large increase in true strain. In some materials, this is seen as a dramatic drop in the *engineering strain*. As the load is increased, *drawing* or *necking* continues. The ma-

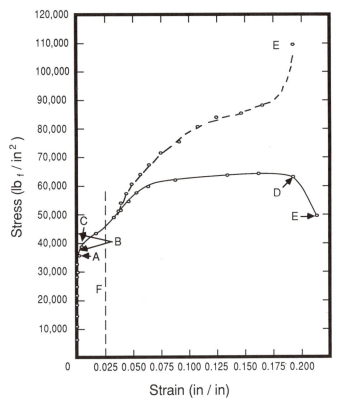

Figure 3.55.   Stress–Strain Diagram for Mild Steel. Solid Line: Engineering Stress–Strain Curve. Dashed Line: True Stress–Strain Curve. Point A: Proportional Limit. B: Elastic Limit. C: Yield Point. D: Ultimate Strength. E: Breaking Stress. F: 0.2% Offset Line.

terial resistance to applied load becomes greater and the engineering stress–strain curve again increases. The highest point on the engineering stress–strain curve, D, is the *ultimate strength* of a material. If the load is continued beyond this point, the material ruptures. The stress at this point is called the *breaking* or *rupture strength*.

The area under the engineering stress–strain curve is a measure of the *toughness* of the material. Toughness is the total work needed to break a specimen, with units of MP (−) m^3 or lb$_f$ (−) in. Toughness can be written as:

$$(3.152) \qquad \text{toughness} = \int_0^\infty F \, d\,\delta$$

When this is written as:

$$(3.153) \qquad \frac{\text{toughness}}{\text{volume}} = \int_0^\delta \left(\frac{F}{A_o}\right) d\left(\frac{\delta}{L_o}\right)$$

$$= \int_0^\delta \left(\frac{\sigma}{A_o}\right) d\left(\frac{\delta}{L_o}\right)$$

the modulus of toughness is obtained.

Not all metals exhibit the so-called *ductile* stress–strain diagram as seen in Figure 3.55. Metals such as cast iron exhibit no appreciable yielding before rupture and so are called *brittle*. Other metals such as copper or lead exhibit very low proportional limits (point A), and very high strain levels prior to rupture. Aluminum on the other hand exhibits a stress–strain curve similar to that of a ductile metal but without a marked yield point. Schematics of stress–strain diagrams of typical metals are shown in Figure 3.56.

When stress–strain curves are not linear, the modulus of the material can be measured in several different ways. There are four different techniques described in ASTM E11 and shown in Figure 3.57. The most common methods used for polymers are the initial tangent modulus method and the secant modulus method.

For metals, mechanical property tensile tests and the methods of data reduction have been developed and refined over decades. Similarly, tests for shear modulus, bulk modulus, Poisson's ratio, impact strength, and so on have been perfected for metals. These testing procedures have been modified for mechanical determination of solid polymer properties. The validity of these tests for polymers must meet one fundamental criterion: Does the test include all the variables that directly or indirectly influence polymer response?

As an example, mechanical tensile tests were primarily developed to determine the Hookean limits of metals. Polymers are viscoelastic. As a result, those elements of the test that are insignificant for metals such as rate of deformation, material temperature and environmental preconditioning, and the nature of yielding, are critical to the interpretation of polymeric material response. Thus, each test must be evaluated in light of

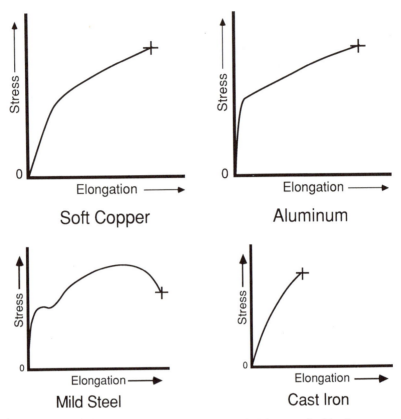

Figure 3.56.   Schematics of Stress–Strain Diagrams for Metals.

the particular characteristics of the polymer under test. As seen in Chapter 6, mechanical property tests for polymers must be categorized in terms of the relationship between the time of test and the material time constant, viz, ultrashort term testing (impact), short term testing (tensile test), long time testing (creep), and cyclic loading. These are discussed separately below.

Yield in Polymers

As noted earlier, polymers typically behave as solids below the glass transition temperature, or as fluids such as crystalline material response above the melt point. However, in the past 30 years, the study of polymeric behavior in the yield region has intensified. Part of this interest stems from the observation that many practical polymer processing schemes, such as rolling, thermoforming, stretch blow molding, fiber forming and tubular blown film forming, depend on nonisothermal elongation of softened polymers. Practical processing elongation under load probably occurs most efficaciously in or near the yield region of the polymer. Furthermore, polymer engineers are

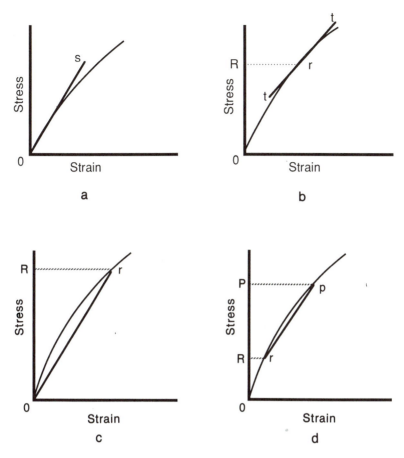

Figure 3.57.   Four Accepted Methods of Determining Elastic Modulus from Stress–Strain Diagram a: Initial Tangent Modulus [*s* Is Slope of Curve]. b: Tangent Modulus at Stress R [*t* Is Tangent to Curve]. c: Secant Modulus Between Origin and Stress *R*. d: Chord Modulus Between Stresses *R* and *P*. (30).

now aware that significant molecular orientation or reorientation occurs in the yield zone. This affords a convenient study of morphological characteristics of crystalline polymers using X-ray scattering techniques and of amorphous polymers using stress birefringence. Furthermore, the study of precise necking phenomena such as slip or shear bands or kink bands suggests that polymer deformation processes proceed in manners similar to ductile crystalline metals. Early on, it was thought that polymer yielding was simply a localized softening caused by internal friction of disentanglement. Although molecular theories have demonstrated that other mechanisms dominate, the concept of localized softening *cannot* be entirely discounted. Local increases in temperature and the time-dependent nature of the polymer frequently confound general interpretation of polymer behavior in the yield region. Even if the effect of temperature can be compensated or accounted for, the reduction in the rate of change of the engineering stress–strain curve with strain is real, intrinsic to ductile polymers, and is sometimes called *strain softening* (31).

Under certain conditions, most (possibly all) polymers can be shown to display engineering stress–strain curves similar to that shown in Figure 3.58. Many polymers have intrinsic or "natural" draw ratios in the yield region. That is, under yielding stress, the cross section of the neck that forms has a very specific relation to the initial specimen cross section. Experimentally, this is seen as a neck region of uniform dimension, with continuing extension resulting in material being drawn from the regions at both ends of the stable neck region. Other polymers have no natural draw ratios.

The specific cause of initiation of the neck is not always known. For example, if the specimen cross section is differentially lower at a particular location but the sample is homogeneous otherwise, necking occurs there. If free volume is slightly higher or if the material has some inhomogeneity in the region, necking occurs. Or, if the local yield stress of the material is low at a specific location, the yield point is reached there before it is reached elsewhere. If the polymer strain-hardens upon drawing to a given amount, a *natural draw ratio,* the point of yielding is transferred elsewhere along the sample. The most logical points are on either side of the localized neck. One obvious effect of incipient localized deformation is shear banding. In metals, stress is relieved by forming localized highly elongated dislocation bands, known as Luders bands. In polymers, they are called yield bands, shear bands or kink bands (32,33). For semicrystalline polymers, these bands are seen as microscopic highly oriented fiber bundles at approximately 30° ($\pi/6$ rad) to the draw axis. Schultz (34) describes necking as stress-induced destruction of crystalline units.

Traditional classifications can be used to describe polymeric performance under load, as shown in Figure 3.59:

Soft, Weak. These are polymers that show low modulus, low (or no) yield point and low enlongation at break.

Soft, Tough. These materials have low modulus, low yield point but very high elongation at break.

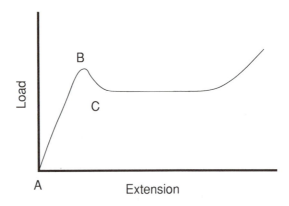

Figure 3.58.   Schematic of Engineering Stress–Strain Curve for a Cold-Drawn Polymer. (35). B: Yield Point. C: Necking or Cold Drawing.

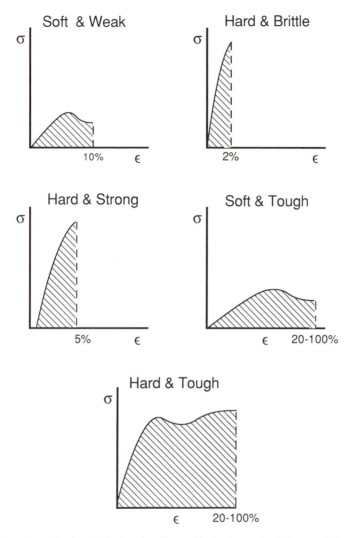

Figure 3.59.   Classification of Engineering Stress–Strain Curves for Polymers (36). σ, Applied
Stress. ε, Resulting Strain.

Hard, Brittle. These polymers have high modulus, no yield point and very low
elongation at break. (These are typical Hookean materials.)

Hard, Tough. These polymers have high modulus, high yield strength and relatively
high elongation at break.

Hard, Strong. These materials have high modulus, high yield strength and relatively
low elongation at break.

Note that as with more traditional materials, polymers that are brittle in one stress–
strain form can be considered to be ductile in another, as with PS in Figure 3.60.

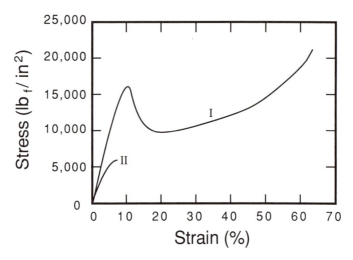

Figure 3.60. Engineering Stress–Strain Curve for General Purpose Polystyrene at 77°F. (37). Curve I: Compression. II: Tension.

Consider a true stress–strain curve for a polymer that shows a substantial yield region, Figure 3.61. This polymer is considered to be one that *cold-draws* or has a stable or natural neck. The line AD represents the true stress at yield. The line AE represents the true stress at the point where strain-hardening occurs. If R is the draw ratio (being $\varepsilon - 1$), then the nature of a polymer can be explained in terms of the relationship between the rate of change of the stress with R and the ratio of stress to

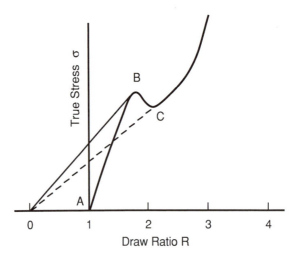

Figure 3.61. Schematic of True Stress–Strain Curve for a Cold-Drawing Polymer. (38). σ, Stress. R, Draw Ratio, $\varepsilon + 1$.

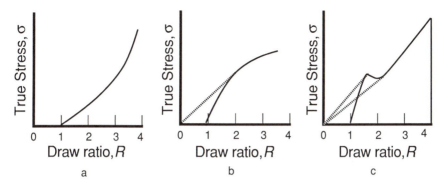

Figure 3.62.   Schematics of Three Types of True Stress–Strain Curves for Polymers. (39). σ, True Stress. R, Draw Ratio, $\varepsilon + 1$. Curve a: $d\sigma/dR > \sigma/R$. b: $d\sigma/dR = \sigma/R$ at One Point. c: $d\sigma/dR = \sigma/R$ at Two Points.

R, that is, $d\sigma/dR$ versus σ/R, Figure 3.62. If $d\sigma/dR > \sigma/R$, the polymer does not neck. If $d\sigma/dR = \sigma/R$ at one point, the polymer necks at that point. The neck gets progressively thinner until fracture occurs. If $d\sigma/dR = \sigma/R$ at two points, stable necking and cold-drawing occur. Yield always involves a change in the slope of the true stress–strain curve.

Some general observations about polymer yielding are in order. The general shapes of tensile stress–strain curves for ductile amorphous and many semicrystalline polymers are very similar, Figure 3.63. The *yield stresses* and tensile moduli of all ductile polymers increase with decreasing temperature. The *yield strain* on the other hand is nearly independent of temperature at 5 to 15% (40). The yield stress for amorphous

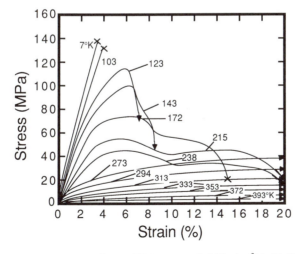

Figure 3.63.   Stress–Strain Curves for HDPE (sp.gr. = 0.964 g/cm^3) in Helium at Strain Rate, $\dot{\varepsilon} = 0.02$ min^{-1}. (41).

polymers approaches zero as the temperature decreases to the glass transition temperature. This is attributed to decreased chain mobility and increased intermolecular forces at T_g.

Semicrystalline polymers show several characteristic yield regions. Typically the crystalline portion of semicrystalline polymers is substantially stiffer than the amorphous portion. When the amorphous portion is glassy, the yield strength decreases with increasing crystallinity. When the amorphous portion is rubbery and the crystalline portion is quite stiff, the yield strength increases with increasing crystallinity. And when the amorphous portion is rubbery and the crystalline portion is soft, as at high temperature but below T_m, the melting temperature, the yield strength is for the most part independent of the degree of crystallinity.

3.9 The Nature of Fracture

When any polymer exceeds a molecular cohesive strength level, it fails. Alfrey (14) points out that the molecular process of breaking cannot be mathematically described, since "before the break, we have one piece of material; after it we have two or more". Ultimate strengths of polymers can be calculated from bond strengths and intermolecular forces. Typically, failure occurs at loading levels orders of magnitude below theoretical molecular polymer strengths, principally because levels of fracture are nearly always determined by flaws, imperfections, voids, and molecular irregularities. Certainly, one should expect that not all entanglements can be undone during typical times of applied stress. Thus, molecular backbone rupture should occur to some degree, depending of course on the severity of the applied stress. As seen in Figure 3.64,

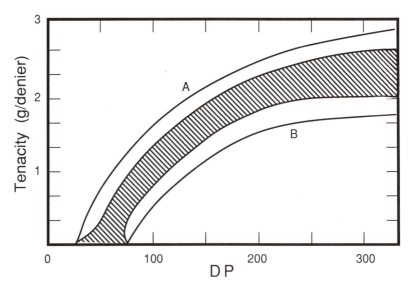

Figure 3.64.   Tenacity (Fiber Tensile Strength) as a Function of Degree of Polymerization, DP. (42). Curve A: Polyesters, Polyamides, Other Condensation Polymers. Curve B: Polyolefins, Other Hydrocarbon Polymers. Shaded Area: Most Other Polymers.

increasing molecular weight results in increasing tensile strength, but the relationship appears asymptotic. Owing to their high degree of hydrogen bonding, polyesters and polyamides show intrinsically higher ultimate tensile strengths than olefins of a similar molecular weight.

Again, Alfrey notes that "the structural features that determine breaking strength are strongly affected by fortuitous and uncontrollable small differences in conditions of preparation". This is particularly true if the polymer does not yield prior to fracture (*brittle fracture*). Yielding can be considered phenomenologically for ductile materials, but even then the final failure mechanism (*ductile failure*) is usually described empirically.

Vincent (43) identifies five types of fracture, Figure 3.65:

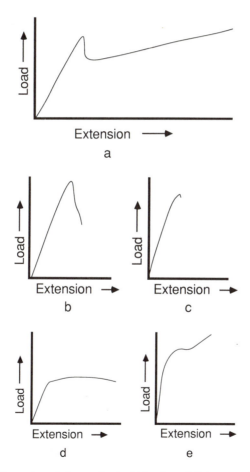

Figure 3.65.   Five Examples of Stress–Strain Behavior of Polymers. (44). Curve a: Cold-Drawing Polymer, Such as Polycarbonate (PC). b: Neck-Rupturing Polymer, Such as Unplasticized PVC (RPVC). c: Slight Yielding Without Neck Formation, Such as Polyacetal (Polyoxymethylene), POM) Copolymer. d: Yielding Without Neck Formation, Such as Plasticized PVC (FPVC). e: Polymer with Uniform Extension, Such as PTFE.

Uniform Extension. The load increases until fracture occurs.

Cold Drawing. Substantial elongation and load decrease occur beyond the yield point of the material. Substantial overall stabilized necking occurs prior to failure.

Necking Rupture. Localized necking occurs and rupture follows without stabilization of the neck.

Necking Rupture with Very Thin Necking Stabilization. This is a variation of necking rupture, with substantial engineering elongation occurring, albeit at rapidly decreasing load.

Brittle Fracture. The load rises monotonically to fracture.

In metals, characteristic fracture patterns are used to determine the source and nature of the fracture (45). Typically, under continuous load, the source of fracture is seen optically as a shiny, featureless circular or parabolic "mirror" zone, Figure 3.66. The zone is terminated by a series of parabolic rings called "ribs", "beach markings" or "arrest lines", with interstitial radial lines sometimes called "hackles" that point back toward the mirror zone. The ring spacing is an imperfect function of the local strain rate. In polymers, similar patterns are seen. Interpretation is difficult since the polymer may respond to the strain rate in either a ductile, yielding fashion or a brittle fashion. Typically, with very low strain rates or with a highly ductile or high molecular weight polymer, the arrest lines are missing or difficult to identify and the mirror zone size

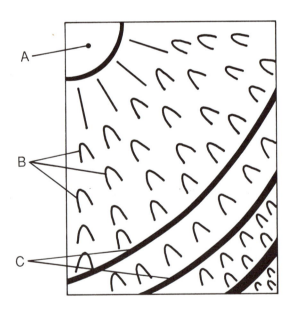

Figure 3.66.   Schematic of Mirror Zone Fracture Patterns. (47). A: Mirror Zone, Initiation Zone of Fracture. B: Chevrons or New Fracture Points, Pointing Toward Initiation Zone. C: Beach Markings or Arrest Lines.

increases substantially. The visual radial lines are seen to be a myriad of microscopic sites of local necking or cold drawing. In crystalline polymers, drawing seems to occur predominantly at the edges of spherulites. In periodic low strain rate loading, the arrest lines are usually quite distinct, and the fracture surface architecture has substantially fewer features than in the material between the lines (46).

High Speed Loading

The theoretical strength of a polymer is usually the strength of the backbone carbon–carbon bond. The value depends on the method of calculation, but is thought to be about 2 to 4 $\times$ 10^{11} dyn/cm^2 (2.9 to 5.8 $\times$ 10^6 lb$_f$/in^2, or 20 to 40 GPa). Typical tensile strengths for polymers are 10^7 to 10^9 dyn/cm^2 (1450 to 145 $\times$ 10^3 lb$_f$/in^2, or 1 to 100 MPa). These values are about 0.01 to 1% of theoretical. Certainly strength is reduced partly because the polymer chain is rarely, if ever in bulk, fully oriented prior to fracture. Instead, initial fracture occurs at some defect or irregularity in the structure, then spreads across the surface (nearly) normal to the direction of applied load. In 1921, Griffith (48) proposed a relationship between σ_c, the applied stress, Y, the modulus of a brittle material, γ, the energy required to continue a crack, that is, the fracture surface energy, and c, the crack length:

(3.154)
$$\sigma_c = \alpha \sqrt{\frac{Y\gamma}{c}}$$

α is a proportionality constant. It was originally proposed that γ was of the same order of magnitude as or identical to surface energy. However, for brittle polymers, the value of τ is about 1000 times that of surface energy and about 250 times that of carbon–carbon bond fracture. The very high energy is attributed to molecular-level yielding at the crack tip even for very brittle polymers. Thus, there are several similarities between necking rupture and brittle fracture. For example, it appears that yielding is achieved in brittle fracture as well as necking fracture. A measure of orientation and a significant amount of voids are seen in brittle fracture surfaces. The latter is thought to occur because locally yielding material is restrained by surrounding material in both necking rupture and brittle fracture. This localized restraint results in microscopic cavitation and voids. Thus, brittle fracture, as espoused by Griffith, is a limiting case of necking rupture. However, the yielding zone is obviously orders of magnitude less for brittle materials than for necking materials.

Of course, a material can be brittle or ductile, depending on several factors, such as strain rate, temperature, molecular weight, crosslinking, fillers and additives, and glass transition temperature. Three principles are proposed for understanding these factors:

If the environmental conditions change to increase the stress at a given strain, the probability of fracture at that strain increases.

If the level of stress concentration increases at a given strain level, the probability of fracture increases.

Table 3.5.   Effect of molecular weight on environmental stress–crack resistance for high-density polyethylene (HDPE). (49).

Melt index[a] (g/10 min)	Intrinsic viscosity[b]	Environmental stress–cracking time, F_{20}, h[c]
0.08	1.35	> 500
0.22	1.17	> 500
0.35	1.04	> 500
0.60	1.10	> 500
1.30	0.98	> 500
7.8	0.94	< 0.5
16	0.895	< 0.5

[a]ASTM D1238, condition *E*. melt index is an inverse function of HDPE molecular weight.
[b]In xylene at 85°C.
[c]ASTM D1693.

If elongational stability increases, the probability of fracture decreases.

Many factors influencing fracture are described in terms of these three principles in Table 3.5. Some of these are discussed below.

Temperature

The yield stress and strength at fracture of a brittle material always decrease with temperature. As seen in Figure 3.67 for PVC, the intersection of the brittle strength and the yield stress is the *brittle point* or temperature. Brittle strengths are less temperature-sensitive than yield stresses. Table 3.6 lists brittle temperatures for several polymers. Note that the brittle temperature is not equal to the glass transition temperature of the polymer. In many cases, it is substantially lower than the glass transition temperature. For inherently brittle polymers such as PS and PMMA, however, the two nearly coincide.

Strain Rate

As noted above, as the strain rate increases, polymers tend to become more brittle with the most apparent result being an increase in yield stress, as seen in Figure 3.68. This rate is for the most part linear with logarithmic extensional speed:

$$(3.155) \qquad \sigma_Y = a \ln \dot{\varepsilon} + b$$

The energy to break drops rapidly with increasing extensional speed, Figure 3.69. The exceptions to the linearity of Figures 3.68 and 3.69 are due to changes in the mechanism of elongation. PP, for example, deforms to fracture by cold drawing at moderate-to-low elongational rates. At high rates, it necks to failure. Typically, as the

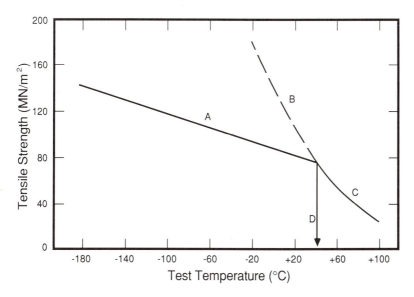

Figure 3.67. The Effect of Temperature on the Ductile-Brittle Transition Region of Unplasticized PVC (RPVC). (50). Curve A: Brittle Strength. B: Yield Stress, Estimated. C: Yield Stress. D: Brittle Point.

rate of deformation increases, the polymer mode of failure moves from cold-drawing failure to necking rupture to brittle failure. Increasing strain rate usually indicates decreased level of fracture toughness. And yield stress increases with increasing strain rate.

At low strain rates, as in standard tensile testing, the heat generated by elongational work is for the most part dissipated to the environment. And so the experiments are near-isothermal. At high strain rates, the energy cannot be dissipated prior to material failure. The result is a localized increase in temperature in the necking region and *adiabatic* strain response to applied stress. Local temperature increases of 50° to more than 300°C have been measured during high speed elongation. The effective yield stress decreases with increasing temperature and if there is insufficient orientation hardening,

Table 3.6. Brittle temperatures of several polymers. (51, 52). Unnotched charpy impact specimens.

Polymer	Brittle temperature (°C)
Polystyrene (PS)	100
Polymethyl methacrylate (PMMA)	80
Poly-4-methyl pentene−1	45
Polypropylene (PP)	10
Unplasticized polyvinyl chloride (RPVC)	−50
Polyisobutylene (PIB)	−63
High density polyethylene (HDPE, <1 melt index)	−180

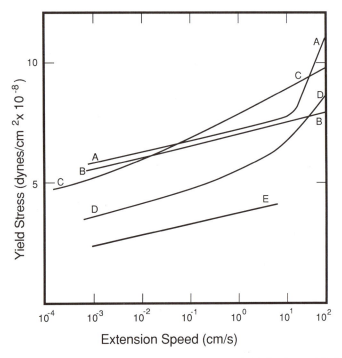

Figure 3.68. Strain-Rate-Dependent Tensile Yield Stress at 20°C for Several Polymers. (53). A: Amorphous Polyethylene Terephthalate (PET). B: Polyacetal Polyoxymethylene (POM) Co-polymer. C: Unplasticized PVC (RPVC). D: Nylon 66 (PA-66), Ambient Humidity. E: Poly-4-Methyl-Pentene.

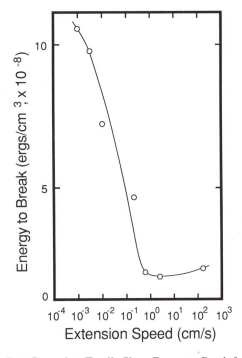

Figure 3.69. Strain-Rate-Dependent Tensile Shear Energy to Break for HDPE at 20°C. (54).

cold drawing is prevented. Polymers with high degrees of entanglement or crystallinity and large temperature coefficients for the yield stresses should show the greatest "high speed embrittlement" or reduction in fracture toughness.

Flexural brittle strengths are about 1.5 to 2 times those of tensile brittle strengths. One hypothesis is that in tension the source of fracture (defect or irregularity) can be anywhere in the cross-sectional normal to the applied load. In flexure, the defect must be relatively close to the maximum fiber stress region in order to substantially influence the material strength.

Impact

The extreme in high speed extension is *impact*. The polymeric material is subjected to an impulse load of very short duration of micro- to milliseconds. The load can be either uniaxial or biaxial. In biaxial impulse loading, the load can be a point source or distributed. Laboratory simulation of impact data and interpretation are very difficult. Several impact testers are described in Section 6.4. Most are adaptations of testers designed for less-ductile materials such as steels or ceramics. To get a specimen to fracture at the point of impact rather than in the grips or with brittle materials, in several places, a notch is molded or cut in the specimen. The notch is a stress concentrator and so fracture occurs there. As with high speed elongation, Section 6.4, the material ahead of the fracture tip yields. The zone of plastic deformation can extend far ahead of the fracture tip with highly ductile polymers such as polyolefins. With brittle polymers such as PS and PMMA, it can be restricted to a very small zone. Of course, the initiation of fracture is due entirely to the deliberate notch. As the notch radius is reduced, the stress concentration at the notch or fracture tip increases:

(3.156)
$$k = 1 + 2 \sqrt{c/r}$$

where k is the elastic stress concentration factor, c is the depth of the notch and r is the notch tip radius. For *every* material, the amount of energy required to initiate a crack is substantially greater than that needed to propagate it. As a result, with very sharp notches, the measured impact strength is related to crack propagation energy. For blunt notches, the impact strength is a sum of crack propagation and crack initiation energies. And for unnotched impact, the strength should be primarily the energy to initiate a crack, Figure 3.70. Many materials show little "notch sensitivity". That is, they show little change in impact strength with notch radius. Multiphase materials such as rubber-modified PS and nylon, show relatively little decrease in impact strength with decreased notch radius. Filled ductile materials such as talc-filled PP or HDPE show substantial decreases in impact strength with increasing filler level. If the crack initiation energy is substantially greater than crack propagation energy, the polymer impact energy becomes thickness-dependent. Typically, impact strength drops with increasing material thickness, Figure 3.71.

Crack initiation energy is related to yield strength. Increasing temperature lowers yield strength and crack propagation energy. Therefore if the energy required to initiate a crack dominates, a rapid increase in impact strength with temperature is seen as the polymer becomes increasingly ductile, Figure 3.72. Note that for rigid PVC, crack

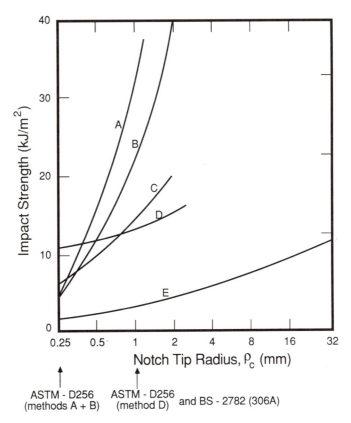

Figure 3.70. Representative Examples of Notch-Sensitive and Notch-Insensitive Polymers, Shown as Charpy Impact Strength. (55). Curve A: PVC. B: Nylon (PA). C: Polyacetal (POM). D: ABS. E: PMMA.

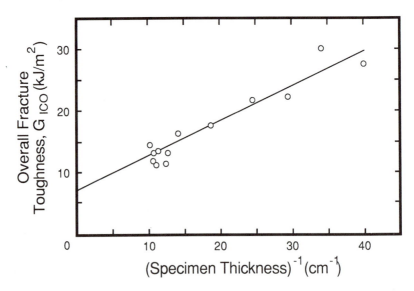

Figure 3.71. Overall Fracture Toughness as a Linear Function of Reciprocal Specimen Thickness for Lexan (tm) Polycarbonate (PC) at $-20°C$. (56).

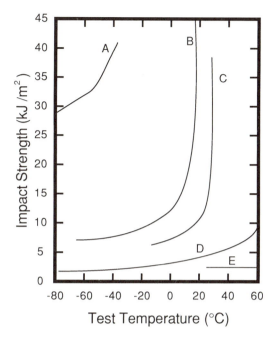

Figure 3.72.   Temperature-Dependent Impact Strength for PVC. (57). Curve A: Unnotched. B: Notch Radius, 2 mm. C: 1 mm. D: 0.25 mm. E: $\approx$ 10 μm Radius.

propagation energy is quite low. As a result, sharp notches dominate impact strength regardless of material temperature. A comparison of overall impact behavior and tensile yield stress is given in Table 3.7.

Long-Term Behavior

Short-term behavior of polymers cannot be successfully extrapolated to long-term behavior. Although the mechanical models of Section 3.6 can be used to describe the time-dependent shape of the stress–strain curve, they cannot predict the conditions of long-term failure. There are two general categories of long-term behavior—response to continued load and response to period load. The former is creep, the latter is fatigue.

Creep Behavior

Creep is usually considered as polymer response to long-term fixed loads. Practically, the load can be applied uniaxially or biaxially and the polymer can be in tension, torsion, shear, or flexure, or it can respond in a mixed-mode fashion. Uniaxial tensile creep is the preferred experimental mode although biaxial creep is used to evaluate materials for pipe applications. Usually the polymer response is given as time-dependent strain under constant stress, Figure 3.73. Additional information about material response is obtained by cross-plotting. Ductile polymers respond in slightly different

Table 3.7. Approximate relationships between polymer behavior under impact and tensile yield stress, dyn/cm^2 (58).

Class of behavior in impact	Polymer	Approximate tensile yield stress ($\times$ 10^8 dyn/cm^2)
Not brittle even when severely notched	Rubbers	< 1
	Low-density polyethylene	1
	Polytetrafluoroethylene	1.5
	Poly-1-butene	2
	Poly-1-pentene	< 1
	Poly-n-hexyl methacrylate	< 1
Not brittle when unnotched. Brittle when severely notched	High-density polyethylene	3
	Polypropylene	3.5
	Chlorinated polyether	4
	Polyvinyl chloride	6
	Polyoxymethylene	7
	Polyethylene terephthalate	6
	Polycarbonate (Bisphenol A)	6.5
	Polyphenylene oxide	7.5
	Polydiphenyl ether sulfone	7.5
	Nylon 66 (Dry)	9
Brittle even when unnotched	Poly-4-methyl-1-pentene	3
	Polystyrene	7 to 10
	Polymethyl methacrylate	11
	Polycyclohexyl methacrylate	Too brittle
	Polyacrylonitrile	Too brittle
	Poly-N-vinylcarbazole	Too brittle

fashion from brittle materials. When a uniaxial tensile load is applied smoothly and in relatively rapid fashion (full load in about 1 s), instantaneous, Hookean elongation occurs. Continued load application results in continued elongation at an ever-decreasing rate until rupture occurs. The time-dependent strain curves for typical brittle and ductile polymers are shown in Figure 3.74. The rupture envelope is shown as a dashed line. Both types of polymer exhibit initial Hookean response to applied load. Both types show linear increases in strain with logarithmic time. But ductile polymers show a relatively rapid increase in strain rate prior to failure. Brittle polymers give no warning of impending rupture.

Catastrophic failure condition is probably not the appropriate failure criterion for design. Before the rupture condition is reached, certain other fracture indicators are seen. At relatively low strain levels, microscopic *crazes* can be initiated. An incipient craze is usually depicted as a local increase in void volume or the formation of microvoids having dimensions greater than the wavelength of visible light. This results in a local change in index of refraction or a white cloudy appearance in a transparent or translucent part. It is sometimes called *stress whitening*. Examples of this are the onset of the debonding between a polymer and a reinforcing glass fiber and the localized

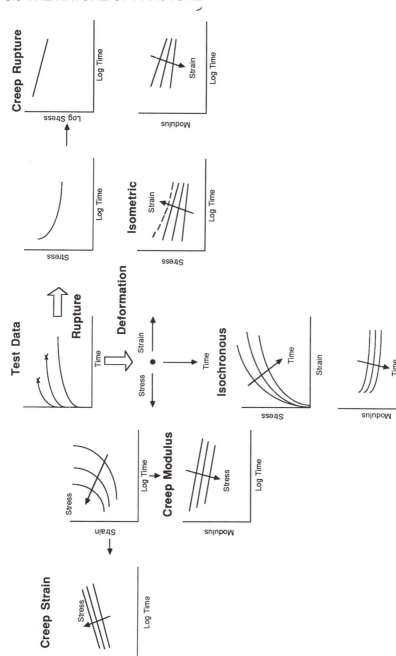

Figure 3.73. Graphical Representation of Creep Data.

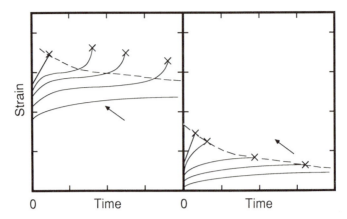

Figure 3.74.   Creep Rupture for Ductile (Left Graph) and Non-Ductile (Right Graph) Polymers.
(58). Arrows Show Increasing Applied Stress.

debonding around rubber particles in impact-modified polystyrenes and ABS. Continued time under load or increased load leads to formation of crazes. The microvoid opens to produce two microscopic surfaces with polymeric material in the form of fibrils between. The fibrils are predominantly aligned in the direction of the applied load. The fibrils provide some coherence and load-bearing capability. Even though the polymer is weakened, it has not failed.

Crazing is the failure condition for many polymers. PS and PMMA are classical polymers that craze prior to failing. At increasing temperatures, crazing results in

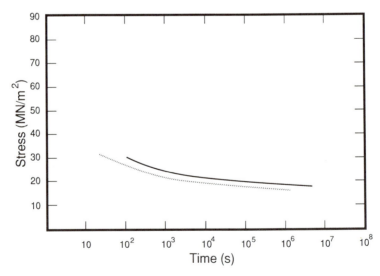

Figure 3.75.   Onset of Stress Whitening and Necking Under Long-Term Load for Polypropylene Homopolymer. (59). Solid Line: Necking. Broken Line: Stress Whitening.

significantly lower values for elongation at failure. If the polymer is brittle, crazing is followed by *stress cracking*. The fibrils are unable to sustain the applied load, and microcracks become visible and highly unstable. Fracture or creep rupture follows. If the polymer is ductile, it will neck prior to failure. At constant stress, the strain at necking failure is substantially greater than either whitening or crazing. Necking failure stress decreases with increasing temperature. Schematic time-dependent design curves are seen in Figures 3.75–3.78 for ductile and brittle polymers.

Figure 3.73 can be replotted in one of three ways, holding one of the three parameters—stress, strain, or time—constant in subsequent graphs. Consider plots of constant stress, the set of graphs on the left. The first graph is a semi-log plot of time-dependent strain. Typically, the curves are nonlinear. Replotting on log–log scale usually yields a plot with a series of parallel linear lines. This is a *creep strain* plot. Note that the ratio of stress to strain is modulus. As a result, the initial data can be replotted as time-dependent *creep modulus* in the graph below. The majority of creep design data published in the United States is reported in this fashion.

Consider time to be constant. The first graph below the center is an *isochronous creep* plot. The majority of creep design data published in Europe is given in this fashion. The data can be replotted in terms of a strain-dependent creep modulus with time as the parameter.

If strain is held constant, the time-dependent stress graph is called *isometric creep*. Again, isometric creep data are found extensively in European publications. And again, replotting the data yields an isometric modulus graph.

If tensile creep data are plotted on a linear time scale, the short-time performance is magnified. As a result, the data show the relative effects of retarded elastic strain,

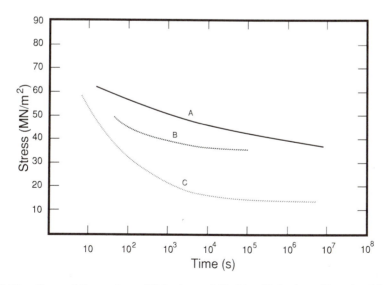

Figure 3.76.   Onset of Craze, Stress Whitening and Necking Under Long-Term Load for Unplasticized PVC (RPVC). (60). A: Ductile Failure or Necking. B: Stress Whitening. C: Craze Initiation.

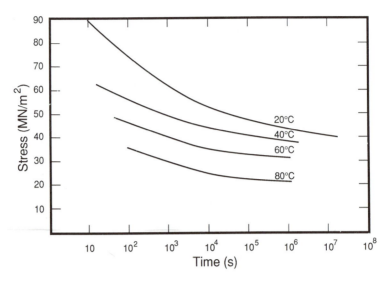

Figure 3.77. Temperature-Dependent Onset of Necking Under Long-Term Load for Polymethyl Methacrylate (PMMA). (61).

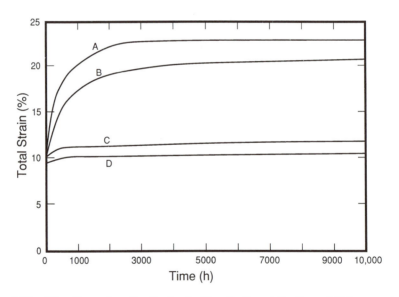

Figure 3.78. Time-Dependent Tensile Strain (Tensile Creep) for Several Polymers at 72°F, 3000 lb$_f$/in^2. (62). Curve A: Polyacetal (POM). B: Heat-Resistant ABS. C: Polycarbonate (PC). D: Polysulfone.

Figure 3.78. If the data are plotted on long–time scale, the effect of an elastic or viscous flow behavior is observed. Creep rupture strength can be more readily determined with this type of graph, Figure 3.79. Typically creep data should never be extrapolated more than one decade. The decade beginning at 10 hours lasts only 90 hours. The decade beginning at 1000 hours lasts 375 days or more than one year. And the decade beginning at 10,000 hours lasts more than 10 years. In many civil engineering applications, performance specifications are written to 120 years.

The actual creep rupture curve for a particular polymer depends on its molecular weight, molecular weight distribution, fillers, additives and any residual stress owing to processing. However, several interesting and significant conclusions can be drawn from creep rupture curves. Recall that the ultimate tensile stress of gray cast iron is independent of time at 20,000 lb$_f$/in^2 (138 MPa). If a part is fabricated of this material and loaded, and does not fail immediately, it most likely will never fail because of tensile rupture. The stress level at rupture decreases with time for all polymers. Consider PS, Figure 3.79, at 4600 lb$_f$/in^2 (32 MPa) tensile load. The specimen will statistically fail in creep in 1000 hours (point A). Any design using polymer under load must determine the design life of the product to be able to determine the ultimate allowable stress on that material. It is a common and risky practice to obtain the ulti-

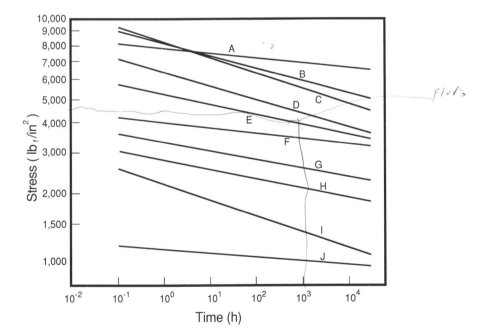

Figure 3.79. Time-Dependent Creep Rupture Strength for Several Polymers at 20°C. (63). Curve A: Polycarbonate (PC). B: SAN. C: Polyacetal (POM). D: General Purpose Polystyrene. E: ABS. F: Impact Polystyrene. G: Polypropylene (PP). H: Cellulose Acetate Butyrate (CAB). I: High-Density Polyethylene (HDPE). J: Low-Density Polyethylene (LDPE).

mate stress of a polymer given in resin supplier data sheets, then apply a safety factor to the value to determine a design stress. Typically, all ultimate stress values listed in data sheets are obtained from short-term tensile tests. As an example, the PS in Figure 3.79 has a data sheet ultimate stress of 8500 lb$_f$/in^2 (58.6 MPa). If a safety factor of 2 is used to determine the design stress (4250 lb$_f$/in^2 or 29.3 MPa), the part will fail in 7000 hours or less than one year. The designer must learn that for polymers, ultimate creep stresses are highly time-dependent. Additional considerations are made in Section 6.5.

Fatigue

When materials are subjected to periodic or cyclic loads in excess of their elastic limit, they can fail by *fatigue*. The ultimate fatigue stress is usually plotted against the number of cycles to failure as an "*S–N*" curve, Figure 3.80. For metals, the fatigue stress value is usually lower than the time-independent creep stress at the same loading. It decreases with cycles to failure and frequently plateaus at 20% to 30% of the creep stress. The plateau is called the "fatigue limit" of the material. As long as the specimen can dissipate the elongational heat generated, the fatigue stress is independent of cyclic frequency.

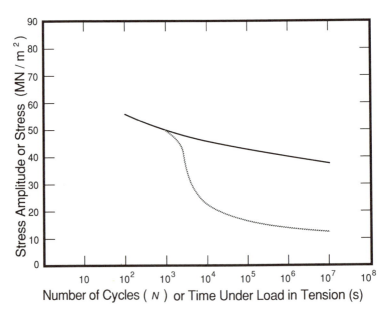

Figure 3.80.   Comparison of Time-Dependent and Cycle-Dependent Ultimate Stress for Unplasticized PVC (RPVC) at 20°C, 65% RH. (64). Solid Line: Ductile Failure Curve Under Continuous Load. Broken Line: Fracture Under Cyclic Load, Showing Ductile-Brittle Transition to Brittle Fracture.

In polymers, *fatigue strength* depends on several additional parameters. As noted above, creep stress is time-dependent. Thus at low cyclic frequency, the fatigue stress would drop with the number of cycles owing to creep. And polymers have high internal friction and low thermal conductivity. As a result, most do not easily dissipate elongational heat. Rapid local temperature rise can lead to fatigue failure at a very low number of cycles if the cyclic frequency is high. It is always tempting to simulate long-term fatigue by stressing the specimen at high frequency to reduce testing time. For example, at 1 Hz it takes 278 hours to achieve 10^6 cycles. At 60 Hz, it takes only 4.6 hours to reach the same testing level. Accelerated testing can lead to major errors in interpreting the data. As noted, the ultimate fatigue failure stress is not a design criterion for polymers.

As with creep loading, cyclic loading produces stress whitening, crazing, and cracking prior to either brittle or ductile necking failure. Unlike creep loading failure, interpretation of these early failure modes is confounded by local heating at high cyclic frequency. A schematic incorporating many of these elements is given as Figure 3.81. If the loading can be considered isothermal, the S–N curve may show a dramatic ductile–brittle fracture transition at cyclic times that are far less than what might be observed at continuous loading. It is thought that the yield zone ahead of the fracture tip is formed and maintained as a viscoelastic or time-dependent element. If the time

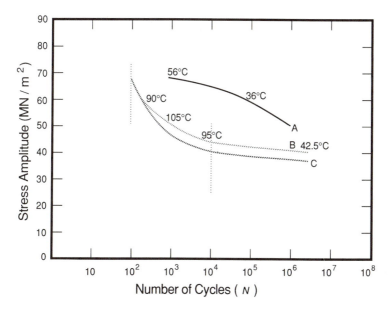

Figure 3.81.   Effect of Heat Removal From Polymer During Cyclic Fatigue. (65). Flexural Frequency: 2.5 Hz. Temperatures Refer to Surface Measurements. Line A: Forced Air Cooling. Line B: Natural Cooling. Line C: No Cooling. Vertical Line Demarks Region of Thermal Softening Failure (Left) and Region of Crack Propagation Failure (Right). Note: Forced Air Cooling (Line A) Failure Is by Crack Propagation.

response for the polymer is substantially longer than the time constant of the applied load, the yield zone is restricted, there is insufficient orientation hardening, and local necking follows. Thus the material appears brittle. Unfortunately, this hypothesis is difficult to verify in every failure case owing to temperature rise in the stressed zone. Other factors include deliberate or accidental stress concentration, environmental factors such as an aggressive medium, degree of crystallinity and broad molecular weight distribution.

Strain Recovery and Stress Relaxation

When a material is stressed below its elastic yield limit, it recovers completely when the stress is removed. When the stress level exceeds the elastic limit, strain recovery is not complete. For some materials with very low elastic limit, such as lead and copper, recovery is nil. The material is said to have dead-bend characteristics. Polymers behave in similar manners, except that because polymers are viscoelastic, the degree of recovery is usually time-dependent. Of course, no failure occurs after the load is removed. But it has been shown that at very low stress levels, creep compliance and recovery compliance are identical for a given level of stress, Figure 3.82. It is thought that this is because polymer response to isothermal application of energy is unique. At higher stress levels, it appears that the viscoelasticity is nonlinear and that the elastic recovery is greater than the elastic portion of polymer response during loading.

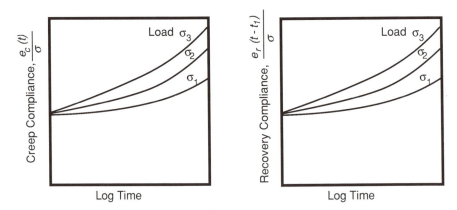

Figure 3.82.   Schematic of Creep Compliance and Recovery Compliance for Non-Linear Viscoelastic Material at Three Loading Levels, σ_1, σ_2, and σ_3. (66). $e_c(t)/\sigma$, Creep Compliance. $e_r(t - t_1)/\sigma$, Recovery Compliance. This Model Follows Boltzmann Superposition Principle, Showing That Creep and Recovery Curves Are Identical, for This Case of Non-Linear Viscoelasticity.

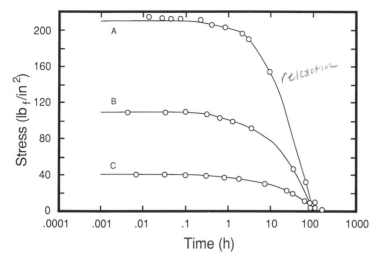

Figure 3.83.   Stress Relaxation Data for Gum Rubber at 100°C. (67). Material: Hevea Gum Stock. Solid Lines: Kinetic Theory of Elasticity. Open Circles: Experimental Data. Curve A: 100% Elongation. B: 50%. C: 20%.

   On the other hand, if the strain level on the polymer is held constant, the applied stress decays. The shape of the typical *isometric relaxation* curve is substantially different from that for the classic *isometric creep* curves [compare Figure 3.83 and Figure 3.73]. Typically, the stress is sustained until the polymer fails in a near-catastrophic fashion. The failed polymer always appears brittle in the fracture zone, indicating a relatively rapid ductile–brittle transition akin to that seen in cyclic fatigue. This implies permanent carbon–carbon bond failure in the final stages of relaxation failure rather than the reversible breaking and remaking of such secondary bonds as Van der Waals or hydrogen bonds, more typical of creep and early stress relaxation.

Failure Envelope

The gist of this section can be summarized by reviewing a typical stress–strain failure envelope, Figure 3.84. The solid boundary encloses the rupture points. Increasing strain rate and decreasing temperature result first in an increased ultimate strain, then decreased ultimate strain. The dashed lines can be used to illustrate the interrelationship between creep behavior and stress relaxation. Consider a polymer at point D under creep at constant stress. As time progresses, the polymer behavior moves at constant stress to increased strain and failure at point F. On the other hand, consider a polymer at point D under stress relaxation at constant strain. The polymer behavior moves at constant strain to decreased stress and failure at point E.

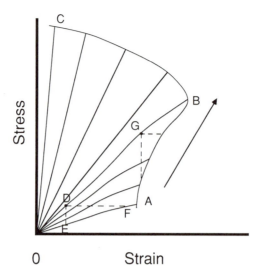

Figure 3.84.   Schematic of Rupture Envelope. (2). Arrow Denotes Increasing Strain Rate or Decreasing Temperature. Envelope Connects Rupture Points. Dashed Lines Illustrate Stress Relaxation and Creep Under Varying Conditions.

References

1. S. Timoshenko and G.H. MacCullough, *Elements of Strength of Materials,* 3rd Ed., Van Nostrand, New York (1949).
2. I.M. Ward, *Mechanical Properties of Solid Polymers,* John Wiley & Sons, New York (1971), p. 370.
3. B. Sandor, *Strength of Materials,* Prentice-Hall, Englewood Cliffs, NJ (1978).
4. I. Granet, *Modern Materials Science,* Reston Publishing Co., Reson, VA (1980).
5. A. Glanvill, *Plastics Engineers Data Book,* Industrial Press, New York (1971), p. 175.
6. A.R. Luxmore and D.R.J. Owen, "Syntactic Foams", in N.C. Hilyard, Ed., *Mechanics of Cellular Plastics,* Applied Science, London (1982), Chapter 9.
7. J.M. Dealy, *Rheometers for Molten Plastics,* Van Nostrand Reinhold, New York (1984), pp. 32–41.
8. R.S. Westover, "Processing Properties", in E.C. Bernhardt, Ed., *Processing of Thermoplastic Materials,* Reinhold, New York (1958), Chapter 3.
9. A.B. Glanvill, *Plastics Engineers Data Book,* Industrial Press, New York (1971), p. 165.
10. W.N. Findlay, Modern Plastics *19* (1942), p. 73.
11. M.N. Riddell, G.A. Toelcke and J.L. O'Toole, SPE Tech. Pap., *16* (1970), p. 219.
12. J.K. Knowles and A.G.H. Dietz, Trans. Amer. Soc. Mech. Eng., *77* (1955), p. 178.
13. J.C. Maxwell, Phil. Trans. Roy. Soc., *A157* (1867), p. 49.
14. T. Alfrey, Jr., *Mechanical Behavior of High Polymers,* Interscience, New York (1948), Chapter 8.
15. M. Riddell and R. Lamonte, Plast. Design Forum, *26*:3 (1976), p. 43.
16. Anon., "Creep Charts", in Modern Plastics Encyclopedia, *64*:10A (1987), available from MPE, 1221 Avenue of the Americas, New York, NY 10020.
17. S.L. Rosen, *Fundamental Principles of Polymer Materials,* Wiley-Interscience, New York (1982), p. 245.

18. S.L. Rosen, *Fundamental Principles of Polymer Materials*, Wiley-Interscience, New York (1982), p. 247.
19. J.G. Oldroyd, Proc. Roy. Soc., *A218* (1953), p. 122.
20. J.D. Ferry, *Viscoelastic Properties of Polymers*, John Wiley & Sons, New York (1961), p. 45.
21. J.D. Ferry, *Viscoelastic Properties of Polymers*, John Wiley & Sons, New York (1961), p. 46.
22. J.D. Ferry, *Viscoelastic Properties of Polymers*, John Wiley & Sons, New York (1961), p. 29.
23. J.D. Ferry, *Viscoelastic Properties of Polymers*, John Wiley & Sons, New York (1961), p. 154.
24. J.D. Ferry, *Viscoelastic Properties of Polymers*, 3rd Ed., John Wiley & Sons, New York (1960), p. 265.
25. J. Mercier, J.J. Aklonis, M. Litt, and A.V. Tobolsky, Appl. Polym. Sci., *9* (1965), p. 447.
26. J.D. Ferry, *Viscoelastic Properties of Polymers*, 3rd Ed., John Wiley & Sons, New York (1960), p. 267.
27. J.D. Ferry, *Viscoelastic Properties of Polymers*, 3rd Ed., John Wiley & Sons, New York (1960), p. 268.
28. J.D. Ferry, *Viscoelastic Properties of Polymers*, 3rd Ed., John Wiley & Sons, New York (1960), p. 269.
29. J.D. Ferry, *Viscoelastic Properties of Polymers*, John Wiley & Sons, New York (1961), p. 219.
30. ASTM E11, *Annual Book of ASTM Standards*, Philadelphia, PA (1988).
31. I.M. Ward, *Mechanical Properties of Solid Polymers*, John Wiley & Sons, New York (1971), p. 295.
32. M. Horio, Int. Polym. Proc., *1* (1986), p. 4.
33. K.J. Pascoe, "Fracture Mechanics", in W. Brostow and R.D. Corneliussen, Eds., *Failure of Plastics*, Hanser, Munich (1985).
34. J.M. Schultz, *Polymer Materials Science*, Prentice-Hall, Englewood Cliffs, NJ (1974), p. 202.
35. I.M. Ward, *Mechanical Properties of Solid Polymers*, John Wiley & Sons, New York (1971), p. 271.
36. T. Alfrey, Jr., *Mechanical Behavior of High Polymers*, Interscience, New York (1948), p. 490.
37. T. Alfrey, Jr., *Mechanical Behavior of High Polymers*, Interscience, New York (1948), p. 517.
38. I.M. Ward, *Mechanical Properties of Solid Polymers*, John Wiley & Sons, New York (1971), p. 276.
39. I.M. Ward, *Mechanical Properties of Solid Polymers*, John Wiley & Sons, New York (1971), p. 277.
40. A. Moet, "Fatigue Failure", in W. Brostow and R.D. Corneliussen, Eds., *Failure of Plastics*, Hanser, Munich (1985).
41. N. Brown, "Yield Behavior of Polymers", in W. Brostow and R.D. Corneliussen, Eds., *Failure of Plastics*, Hanser, Munich (1985), p. 104.
42. K.H. Mark, Paper Trade J., *113*:3 (1941), p. 34.
43. P.I. Vincent, "Fracture", in N.M. Bikales, Ed., *Mechanical Properties of Polymers*, John Wiley & Sons, New York (1971).
44. P.I. Vincent, "Short-Term Strength and Impact Behaviour", in R.M. Ogorkiewicz, Ed., *Thermoplastics: Properties and Design*, John Wiley & Sons, New York (1974), p. 70.

45. A.J. Kinloch and R.J. Young, *Fracture Behaviour of Polymers*, Applied Science, London (1983), Chapter 3.

46. R.W. Hertzberg and J.A. Manson, *Fatigue of Engineering Plastics*, Academic Press, New York (1980), p. 37.

47. I.M. Ward, *Mechanical Properties of Solid Polymers*, 2nd Ed., John Wiley & Sons, New York (1983), p. 359.

48. A.A. Griffith, Phil. Trans. Roy. Soc., London, Ser. A, *221* (1921), p. 163.

49. P.I. Vincent, "Fracture", in N.M. Bikales, Ed., *Mechanical Properties of Polymers*, John Wiley & Sons, New York (1971), p. 82.

50. P.I. Vincent, "Short-Term Strength and Impact Behaviour", in R.M. Ogorkiewicz, Ed., *Thermoplastics: Properties and Design*, John Wiley & Sons, New York (1974), p. 71.

51. P.I. Vincent, "Short-Term Strength and Impact Behaviour", in R.M. Ogorkiewicz, Ed., *Thermoplastics: Properties and Design*, John Wiley & Sons, New York (1974), p. 74.

52. R.D. Deanin, *Polymer Structure, Properties and Applications*, Cahners Books, Boston (1972), pp. 92–93.

53. P.I. Vincent, "Fracture", in N.M. Bikales, Ed., *Mechanical Properties of Polymers*, John Wiley & Sons, New York (1971), p. 137.

54. P.I. Vincent, "Short-Term Strength and Impact Behaviour", in R.M. Ogorkiewicz, Ed., *Thermoplastics: Properties and Design*, John Wiley & Sons, New York (1974), p. 77.

55. A.J. Kinloch and R.J. Young, *Fracture Behaviour of Polymers*, Applied Science, New York (1983), p. 190.

56. R.A.W. Fraser and I.M. Ward, Polymer, *19* (1978), p. 220.

57. P.I. Vincent, *Impact Tests and Service Performance of Thermoplastics*, Plastics Institute, London (1971).

58. K.G. Smack and J. O'Toole, SPE RETEC, New Haven, CT (1967).

59. K.V. Gotham, "Long-Term Durability", in R.M. Ogorkiewicz, Ed., *Thermoplastics: Properties and Design*, John Wiley & Sons, New York (1974), p. 54.

60. K.V. Gotham, "Long-Term Durability", in R.M. Ogorkiewicz, Ed., *Thermoplastics: Properties and Design*, John Wiley & Sons, New York (1974), p. 55.

61. K.V. Gotham, "Long-Term Disability", in R.M. Ogorkiewicz, Ed., *Thermoplastics: Properties and Design*, John Wiley & Sons, New York (1974), p. 56.

62. Anon., *UDEL Polysulfone: Design Engineering Data*, Union Carbide Corporation, Report F-44689 (1975).

63. R. Kahl, "Principles of Plastics Material Selection", Center for Prof. Adv., East Brunswick, NJ (1979).

64. K.V. Gotham, "Long-Term Durability", in R.M. Ogorkiewicz, Ed., *Thermoplastics: Properties and Design*, John Wiley & Sons, New York (1964), p. 61.

65. K.V. Gotham, "Long-Term Durability", in R.M. Ogorkiewicz, Ed., *Thermoplastics: Properties and Design*, John Wiley & Sons, New York (1974), p. 64.

66. I.M. Ward, *Mechanical Properties of Solid Polymers*, John Wiley & Sons, New York (1971), p. 213.

67. T. Alfrey, Jr., *Mechanical Behavior of High Polymers*, Interscience, New York (1948), p. 329.

Glossary

Apparent Modulus   Time-dependent ratio of stress to strain.

Brittle Point   The temperature at which strength and yield stress intersect. Also called the *Brittle Temperature*.

Cold Draw   Localized stable elongation or necking.

Compliance   Reciprocal complex modulus, $J^* = 1/G^*$.

Constitutive Equation   A set of coordinate-invariant equations relating applied stress to resulting time-dependent material deformation.

Craze   Localized increase in void volume or microvoid formation.

Creep   Long-term material flow or deformation under constant load.

Elastic Limit   Maximum strain, where full elastic recovery is possible.

Fatigue Limit   Cycle-dependent fatigue stress plateau.

Fibril   Microscopic bundles of oriented polymer between crack surfaces.

Hookean Solid   A solid that exhibits deformation in direct proportion to applied stress.

Ideal Solid   A solid that exhibits no deformation, regardless of the magnitude of applied stress.

Imaginary Response   Out-of-phase, dissipative or energy loss response.

Inviscid Fluid   A fluid that exhibits no resistance to applied stress, regardless of its magnitude.

Linear Viscoelasticity   A condition where all material elements are time-independent.

Loss Element   Element that is out of phase with applied stress. An energy-dissipating element.

Modulus   Measured initial slope of uniaxial stress–strain curve.

Newtonian Fluid   A fluid that exhibits rate of deformation or strain rate in direct proportion to applied stress.

Notch Sensitivity   A change in impact strength with notch radius.

Power Law Relationship   A stress–strain rate relationship, wherein viscosity is proportional to strain rate to a power.

Proportional Limit   Maximum strain at which material remains Hookean.

Real Response   In-phase or energy storage response.

Rupture Strength   Material strength at maximum elongation.

Shift Factor   Term used to superimpose time and temperature effects.

Spectrum   Series of time-dependent responses.

Storage Element   Element in phase with the applied stress. An energy-restoring element.

Stress Relaxation   Time-dependent reduction in applied load for material under constant deformation.

Superposition   Technique for shifting responses to a single curve.

Tan δ   Ratio of loss and storage moduli.

Trouton Viscosity   Proportionality between applied tensile stress and elongational deformation rate.

Ultimate Strength   Maximum material strength.

Viscoelastic   Material exhibiting both elastic and viscous properties.

Viscous-Only Material   Material that exhibits no time-dependent material properties.

Whitening   Formation of microcracks. See *Craze*.

WLF Equation   The Williams–Landel–Ferry empirical equation relating shift factor to actual temperature and reference temperature.

Yield Point   Point beyond which complete elastic recovery is impossible and plastic drawing begins.

Yield Zone   Highly stressed region ahead of an advancing fracture.

CHAPTER 4

POLYMER FLUID AND CHEMICAL PROPERTIES

no Reading

4.1 Introduction

As noted in the preceding chapters, polymers are extremely long molecules. In nearly every polymerization process, the molecules become entangled as they form. Crystalline-tendency polymers maintain some degree of order as they cool from the melt state, but amorphous polymers are chiefly characterized by random molecular entanglements. Even in quite dilute solutions in shear, the molecules form temporary entanglements. When the concentration of polymer chains increases, the resistance to shearing forces also increases, since polymeric chains must disentangle to move in a differential fashion. Polymer melts exhibit the highest level of entanglement and so have very high viscosities. As might be anticipated, the rate of disentanglement and the resistance to sliding of polymer chain upon polymer chain is *not* directly proportional to the applied stress. This means that polymeric fluids do not behave as small molecule fluids, that is, as Newtonian fluids.

The study of the response of polymeric fluids to applied stress is called *rheology*. As noted below, small molecule Newtonian fluids at constant temperature require but one material *constant* to describe their response to applied shear—*viscosity*. The complex nature of polymers requires that material *functions* be used to describe the response to the simplest applied shear field. This text does not cover the development of the more complete models, known as *constitutive equations of state* for fluid flow, for these material functions. The interested reader can find additional material in References 1–5. However, a basic understanding of polymer fluid response to applied stress is needed for several reasons.

In general, polymers have very high viscosities. Polymer fluid flow is therefore always laminar. The intrinsically high viscosity implies that robust pumping equipment is needed to shape polymer melts into useful products. While polymeric processing temperatures are not as high as temperatures for processing molten metal or glass, the robustness of the equipment implies relatively high capital cost and high energy costs. Furthermore, high forces and pressures needed in processing can cause molecular damage. That is, disentanglement can occur by chain scission as well as differential chain movement. Reduction in molecular weight means a reduction in the desired design properties in the commercial part. Since a finite portion of the polymer processed is *not* a portion of the finished product, economics dictate that this portion must be reprocessed. Reprocessing means grinding, remelting and reprocessing at elevated temperatures, pressures and shear rates. This reprocessing also reduces the average molecular chain length and thus the physical properties of the polymer and resulting end product. This chapter begins with a brief review of the fundamentals of polymer rheology.

Very frequently, polymers are mixed with smaller molecules, such as plasticizers, volatile liquids and solvents. And as products, polymers contact smaller molecules such as gases and aggressive chemicals. The way in which a polymer reacts to the small molecule environment will dictate whether it dissolves and so can be solvent welded or it doesn't and so can serve as a container. The appropriate approach to polymer-small molecule interaction is through *solution thermodynamics*. As with rheology, however, this subject is better treated in detail elsewhere. Some useful references for additional study are given at the proper time. The small molecule–polymer interaction is briefly reviewed to illustrate the proper terminology when considering, for example, adhesion, plastication, water vapor transmission, permeability of gases such as oxygen and carbon dioxide, and chemical attack. In Chapter 6, some standard tests using these terms are discussed in detail.

4.2 The Nature of Flow

Polymers are viscoelastic materials. That is, depending on the applied stress, they can behave as viscous fluids, as elastic solids, or as materials having characteristics similar to both these general material types. Chapters 5 and 6 examine polymer response to applied load. This chapter emphasizes the fluid nature of polymers.

A simple fluid is one that does not support applied shear stress but instead deforms at a predictable rate. Furthermore, a simple fluid does not return to its original undeformed state once the applied shear stress has been removed. Complex fluids such as polymers do not always behave as simple fluids under load. To understand the family of fluids, it is important to consider the spectrum of material response to applied load. One scenario progresses from *ideal or rigid solid* to *ideal or inviscid fluid*:

A *rigid solid* shows *no* deformation under load, regardless of the nature or severity of the applied load.

A *Hookean solid* shows linear deformation under applied load. The proportionality is called the elastic modulus. A Hookean solid shows no creep, no yielding, no hysteresis,* no plasticity, and recovery is complete upon removal of the applied load (Figure 4.1).

An *elastoplastic solid* is a material having Hookean behavior to a given level of deformation, at which point the material yields. Yielding produces continuous deformation under constant load (Figure 4.2).

A time-dependent or *elastoviscous solid* shows a yielding behavior that depends on the rate at which the stress is applied. At high loading rates, the material may behave as if it has a high modulus and low yielding elongation. At low loading rates, the

---

*A material is said to exhibit *hysteresis* when its response to increasing load is different from its response to decreasing load.

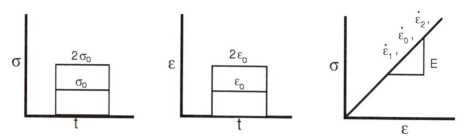

Figure 4.1.   Response of a Hookean Solid, ε_0, to Applied Stress, σ_0. E is Elastic Modulus, $\dot{\varepsilon}_0$, $\dot{\varepsilon}_1$, $\dot{\varepsilon}_2$ Represent the Relative Effects of Strain Rate on the Stress–Strain Curve.

material may behave as if it has a low modulus or no modulus at all. When the applied stress is removed, the material shows substantial hysteresis. Given sufficient time, however, the material recovers its initial dimensions (Figure 4.3).

A *viscoelastic liquid* exhibits time dependency in a manner similar to the elasto-viscous solid, except that it retains a permanent deformation once the applied load has been removed. This "offset" or "permanent set" is a measure of the *fluid* nature of the material. This material is thus called a *liquid*.

The term "elastic liquid" is usually reserved for a specific class of viscoelastic liquids. Significant theoretical efforts have been placed on explaining the strain response of certain materials to applied stresses. The stress–strain relationships are called "constitutive equations of state" (6). For these selected materials, these relationships must meet very specific phenomenological conditions. The complex

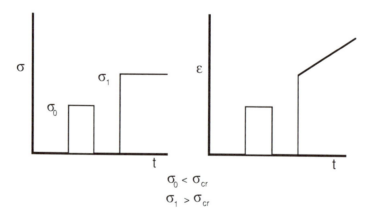

Figure 4.2.   Time-Dependent Response of an Elastoplastic Solid to Applied Stress, σ. σ_{cr} Represents the Critical Stress Level At Which Elastic Limit is Exceeded and Plastic Deformation Begins. Note that $\sigma_0 < \sigma_{cr}$ Whereas $\sigma_1 > \sigma_{cr}$.

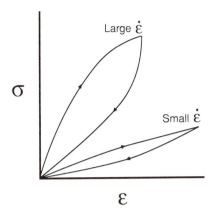

Figure 4.3. Hysteretic Response of an Elastoviscous Solid to Applied Stress. Note That the Elastoviscous Solid Appears More Rigid for Greater Strain Rate; $\dot{\varepsilon}_{large}$, Than for Lesser Strain Rate, $\dot{\varepsilon}_{small}$. Furthermore, the Hysteresis Effect, the Area Under the Stress–Strain Envelope, is Greater for $\dot{\varepsilon}_{large}$.

arithmetic allows for identification of several hierarchical levels of materials. An elastic liquid that meets the lowest hierarchy is called a "simple fluid".*

The concept of a viscous-only, or simple *non-Newtonian fluid* is used most frequently to describe polymeric fluids. There are many models for non-Newtonian fluids. All include the polymer response to applied strain in terms of strain and strain rate.** The simplest, the Ostwald–deWaale model, involves only two material parameters. Others use three or more parameters.

A *Newtonian fluid* is one in which there is a direct correspondence between the applied stress and the fluid rate-of-strain or deformation (Figure 4.4). The proportionality is *viscosity,* a material constant.

When the viscosity of a fluid is zero, the fluid is said to be an *inviscid fluid.*

In short, polymeric materials in a softened or melt state can be considered as viscous liquids, viscoelastic liquids, elastic liquids, or elastoviscous solids. Brydson (7) notes that "it is not normally crucial to have to define the boundaries between such behaviour

---

*It should be noted that the arithmetic describing a simple fluid is anything but simple. As will be seen below, the equation describing stress, strain and rate-of-strain for liquids is a second-order tensor with coordinates that must deform with the fluid element. Another type of elastic liquid is a "second-order fluid". Here the material must be Newtonian at low shear rates and the normal stress differences must be proportional to the square of the shear rate. Owing to the complex arithmetic needed to define these types of fluids, they are not discussed in detail in this introduction to polymer rheology.
**The terms "strain rate", "rate-of-strain" and "rate-of-deformation" are used interchangeably in this book and in most technical literature on polymeric fluids.

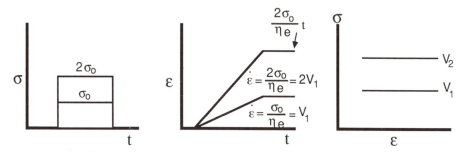

Figure 4.4.   Response of a Newtonian Fluid to Applied Stress, σ_0. $\dot{\varepsilon} = \dot{\sigma}_0/\eta_e$, Where η_e is the Elongational Newtonian Viscosity. The Fluid Remains Deformed When the Applied Stress is Removed.

[sic] particularly as the mathematical theories may be applied to all four types of materials (viscous, elastoviscous, viscoelastic and elastic)''.

There are two major categories of viscoelastic liquids. Linear viscoelastic fluids exhibit no strain rate dependency of material functions such as modulus or viscosity. *Linear viscoelasticity* is usually restricted to very small deformation levels. For most polymeric materials, material functions are strain rate- or amplitude-dependent, particularly at the large strain levels experienced in polymer processing. Thus, although linear viscoelasticity is described in detail in Chapters 3 and 6, it is only a simplification of the actual behavior of real polymers.

Hookean and Newtonian Material Relationships

The proportionality between Hookean stress and resulting strain is modulus. The modulus is time-independent and is therefore a material constant. This can be written as:

$$(4.1) \qquad\qquad \sigma = E\,\varepsilon$$

where σ is the applied stress, ε is the resulting strain and E is the modulus.

The proportionality between Newtonian stress and rate of strain is viscosity. Newtonian viscosity is time- and stress-rate independent and thus is a material constant. The linear relationship for a Newtonian fluid can be written as:

$$(4.2) \qquad\qquad \sigma = \eta_e\,\dot{\varepsilon}$$

where σ is the applied stress, $\dot{\varepsilon}$ is the rate-of-strain, and η_e is the elongational Newtonian viscosity. In shear, this relationship is usually written as:

$$(4.3) \qquad\qquad \tau = \eta\,\dot{\gamma}$$

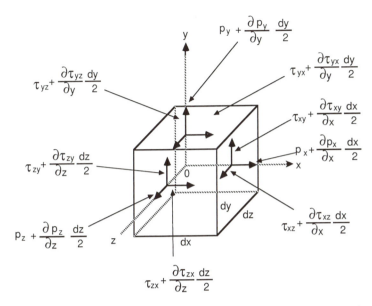

Figure 4.5.   Shear and Normal Stress Components on a Fluid Element. In General, τ_{ij} Represents the Resulting Stress in the i-Direction Owing to Applied Shearing Force in the j-Direction. p_k is the Resulting Stress in the k-Direction Owing to Normal Forces Perpendicular to the k-P plane.

where τ is the applied shear stress and $\dot\gamma$ is the resulting rate of deformation. η is now the Newtonian shear viscosity. To better understand the concept of a polymeric fluid, consider stress distribution on a small element, Figure 4.5. At any time, there are nine components of stress acting on the element. These stresses are described by a second-order tensor, in subscript notation, as σ_{ij}. The subscript i represents the coordinate of the element against which the stress is applied and the subscript j represents the co-ordinate direction in which the stress is applied. Thus, σ_{12} represents the stress acting in the direction x_2 but applied on the elemental plane perpendicular to direction x_1. The stress matrix is symmetric, that is, $\sigma_{ij} = \sigma_{ji}$. By convention, σ_{ii} is referred to as a "normal stress" and σ_{ij} $(i \neq j)$ is called a "shear stress". Alone, normal stresses have no rheological significance (8). Instead, normal stress differences, such as $[\sigma_{11} - \sigma_{22}]$ and $[\sigma_{11} - \sigma_{33}]$, are observed.

Even very dilute solutions of polymers in solvents exhibit normal stress differences. In Figure 4.6, 1.5% (wt) polyacrylamide (Separan AP 30) in a 50/50 mixture of water and glycerin exhibits a first normal stress difference dependency on shear rate that mirrors low-density polyethylene melt (10). The normal stress difference for the Newtonian 50/50 mixture of water and glycerin is zero.

Shear stresses can be written in terms of viscous stresses, (τ_{ij}) and hydrostatic pressure (or head, π) as:

(4.4) $$\sigma_{ij} = \tau_{ij} + \pi\,\delta_{ij}$$

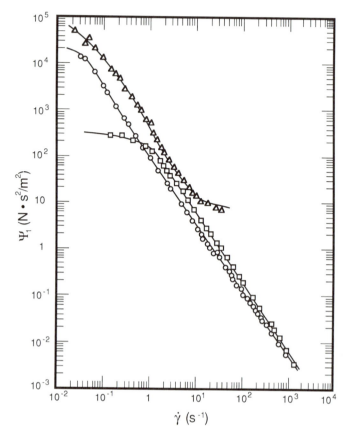

Figure 4.6.   Example of Primary Normal Stress Difference, ψ_1, Dependency on Shear Rate, $\dot{\gamma}$, for Two Polymer Solutions and a Soap Solution (9). Circles: 1.5% (wt) Polyacrylamide in Water/Glycerin Mixture. Triangles: 2% (wt) Polyisobutylene in Primol. Squares: 7% (wt) Aluminum Laurate in Decalin and m-Cresol.

where δ_{ij} is the Kronecker delta, $= 1$ when $i = j$, and $= 0$ when $i \neq j$.

Recall that fluid flow implies deformation rate in the element. Thus, the viscous stress in any direction (ij) can be written in terms of the shear rates $(\mathrm{d}v_i/\mathrm{d}x_j)$ as:

$$(4.5) \qquad \tau_{ij} = \eta \left(\frac{\partial v_j}{\partial x_i} + \frac{\partial v_i}{\partial x_j} \right)$$

The sum of shear rates is called the "tensor rate of deformation". As with shear stress, it has nine terms, six of which are independent. Note that this equation is a rheological constitutive equation of state. It contains the applied forces, the material response and a proportionality term, in this case η, a viscosity, which here is a constant. Thus, this expression is a *Newtonian constitutive equation of state*.

Viscometric Flow

Simple "viscometric" flow is flow in which the fluid behaves as if it is in steady simple shear. In this flow field, three functions describe the material response to applied load: viscosity and first and second normal stress differences. For non-Newtonian fluids, all three are shear rate- and temperature-dependent:

$$(4.6) \qquad \eta(\dot{\gamma};T), \; N_1(\dot{\gamma};T), \; N_2(\dot{\gamma};T)$$

where $N_1 = \tau_{11} - \tau_{22}$ and $N_2 = \tau_{22} - \tau_{33}$. Note that for a Newtonian fluid, $\tau_{11} = \tau_{22} = \tau_{33} = 0$. Thus the normal stress differences for Newtonian fluids are always zero. And as noted earlier, for Newtonian fluids, $\eta = \eta \; (T \text{ only})$.

As noted below, when the applied stresses are small, the resulting extent of deformation is small. Thus, the polymer can be treated as a linear viscoelastic fluid. Further, in the limit as the shear rate, $\dot{\gamma}$, goes to zero, the rate of change of viscosity with shear rate, that is, $d\eta/d\gamma$, must also go to zero. In other words, as the shear rate goes to zero, the fluid viscosity must approach a constant, that is, the Newtonian state.

Power-Law Fluids

In many cases, polymer fluid flow can be described as power-law non-Newtonian in both viscosity and the first normal stress difference:

$$(4.7) \qquad \eta = k \, \dot{\gamma}^{n-1}$$

$$(4.8) \qquad N_1 = c \, \dot{\gamma}^{m-1}$$

where k, c, n, and m are arbitrary constants. When $n = m = 1$, the equation reverts to the Newtonian forms. For polymers, $n < 1$ and $m < 3$. Thus, for the viscosity equation, it is apparent that as $\dot{\gamma} \to 0$, $\eta \to \infty$. This violates one of the material tenets, that the viscosity must be finite at limiting applied load. Nevertheless, the power-law constitutive equation, sometimes called the Ostwald–deWaale model, is one of the most widely used equations for polymeric flow. Values of the Ostwald–deWaale constants (k,n) are given later in Table 4.24 for several polymers.

There have been many models designed to "correct" the inconsistencies in the two-parameter (k,n) Ostwald–deWaale model. Some of these are given in Table 4.1. One model that yields the Newtonian viscosity limit as the shear rate goes to zero is the Ellis model:

$$(4.9) \qquad \frac{1}{\eta} = \left(\frac{1}{\eta_o}\right)\left[1 + \left(\frac{\tau}{\tau_{1/2}}\right)^{(1-n)/n}\right]$$

Table 4.1.   Some viscosity models.

Model	Equation
One-Constant (Newtonian)	$\tau = \eta \, \dot{\gamma}$
Two-Constant (Power-Law)	$\tau = k \, (\dot{\gamma})^n$
Three-Constant (Curve-Fit)	$\log \eta = E_o + E_1 \log \dot{\gamma} + E_2 \log^2 (\dot{\gamma})$
(Eyring–Powell)	$\tau = \eta \, \dot{\gamma} + (1/B) \, \sinh^{-1} (\dot{\gamma}/A)$
(Ellis)	$\eta_0/\eta(\tau) = 1 + (\tau/\tau_{1/2})^{(1 - n)/n}$
Four-Constant (Carreau)	$\dfrac{\eta(\dot{\gamma}) - \eta_\infty}{\eta_0 - \eta_\infty} = [1 + (\lambda \dot{\gamma})^2]^{n - 1/2}$

As τ approaches zero, $\eta = \eta_0$. As τ becomes large, the second term in the bracket is much greater than unity, and so the viscosity dependency on shear stress becomes power-law. $\tau_{1/2}$ is the shear stress at the point where the viscosity has fallen to $\eta_0/2$. In essence then, the Ellis model involves three material parameters (η_0, n, and $\tau_{1/2}$). The transition from low-shear Newtonian viscosity to power-law-dependent viscosity is called the lower critical shear stress, $\dot{\gamma}_{cr,L}$.

To mirror the viscous-only response of most polymers, a four-constant model is recommended. There is ample evidence that polymers attain a measure of Newtonianness at both low and high shear rates (Figure 4.7). The Ellis model accounts for low-shear Newtonian effects only. Since nearly all processing occurs below the upper critical shear rate, $\dot{\gamma}_{cr,U}$, this added complexity does not appear necessary in all modeling. Furthermore, since for many processes, both shear and energy transfer occur at the polymer-metal interface at shear rates above the lower critical shear rate, $\dot{\gamma}_{cr,L}$, the majority of theoretical analysis use only the power-law model. Table 4.24 gives Ellis constants for several commercial polymers.

As noted, for most polymers, n < 1 in the power-law equation. This implies that the polymer viscosity decreases with increasing shear rate. This material is formally known as a "pseudoplastic fluid". The stress-rate-of-strain relationship of this fluid is seen in Figure 4.8. Certain paints, inks and polyvinyl chloride pastes show an *increase* in viscosity with increasing shear rate. A material having this behavior is called a "dilatant fluid". Dilatancy is seen also in Figure 4.8.

Certain materials such as slurries and suspensions and highly filled polymeric fluids tend to behave as elastoplastics. The ideal case is a "Bingham fluid". Below a given yield point, the material behaves as a classical Hookean elastic. Above, it flows as if it is a Newtonian fluid. The ideal Bingham plastic and Newtonian responses to applied shear load are compared in Figure 4.9. Talc-filled polypropylene behaves as a Bingham fluid at high filler loadings. At low applied loads, the rates of deformation for real polymers may tend toward the elastic condition.

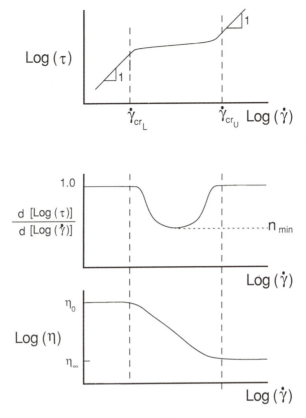

Figure 4.7.   One Interpretation of the Molecular Nature of Polymer Molecules at Various Levels of Shear Rate. At Low Shear Rate, Molecules Reside in a Highly Entangled, Random State. This Yields a Zero-Shear, Newtonian Viscosity. At Very High Shear Rate, Molecules Reside in Linear, Untangled Bundles. This Yields an Infinite-Shear, Newtonian Viscosity. Between these Molecular States is a State of Partial Alignment, Resulting in a Non-Newtonian Viscosity. $\dot{\gamma}_{crL}$ and $\dot{\gamma}_{crU}$ Represent the Shear Stresses at Which the Transition from Newtonian to Non-Newtonian Behavior is Observed. The Power-Law Index, n, is Seen to Have a Minimum Value in the Region between the Two Newtonian Extremes.

The power-law fluid model and its variants assume that the polymer viscosity is shear-rate- and temperature-dependent, but time-independent. Not all polymeric materials have time-independent viscosities. The most obvious example is the viscosity of a reacting polymer, such as unsaturated polyester resin (body putty or boat resin), epoxy (adhesive), or oxygen-activated silicone (caulk). Other materials have a hierarchical structure that is broken down (or built) by shear. These materials are called thixotropic (or, obversely, rheopectic). Paints are common thixotropic materials.

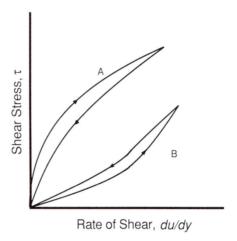

Figure 4.8. Pseudoplastic Fluids Exhibit Higher Effective Viscosities When Shear Rate Increases Than When it Decreases, Curves A. Dilatant Fluids Exhibit Higher Effective Viscosities When the Shear Rate Decreases Than When it Increases, Curves B.

Time-dependent material flow conditions can be written in simple shear stress terms as:

$$(4.10) \qquad\qquad \tau = \eta(\dot{\gamma}, t; T)\, \dot{\gamma}$$

As mentioned, the time-dependent shear viscosity of a reactive fluid is a classic example, as seen in Figure 4.10.

Example 4.1

Q: Determine the power law index exponent for PMMA (acrylic) at 200°C from Figure 4.11 at shear rates greater than 500 s^{-1}.

A: Since $\eta = k\, \dot{\gamma}^{n-1}$ can be written as:

$$\log \eta = (n - 1) \log \dot{\gamma} + \log k$$
$$y = mx + b$$

the slope of the log–log viscosity–shear rate curve is $n-1$. Select two arbitrary points on the curve:

$$\eta_1 = 0.1 \ \mathrm{lb_f/in^2} \ \text{at} \ \dot{\gamma}_1 = 750 \ \mathrm{s^{-1}}$$

$$\eta_2 = 0.01 \ \mathrm{lb_f/in^2} \ \text{at} \ \dot{\gamma}_2 = 20{,}000 \ \mathrm{s^{-1}}$$

$$n - 1 = \frac{\log 0.1 - \log 0.01}{\log 750 - \log 20{,}000}, \ \text{or}$$

$$n = 0.30$$

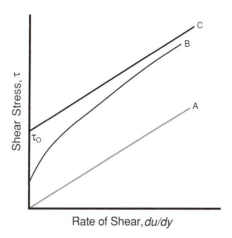

Figure 4.9.   A Comparison of Newtonian, Non-Newtonian and Bingham Plastic Fluids. The
Slope of the Newtonian Stress–Strain Rate Curve A, the Viscosity, is Constant. The Slope of
the Non-Newtonian Stress–Strain Rate Curve B, the Viscosity, Decreases with Increasing Shear
Stress, Indicating Shear-Thinning Behavior. The Bingham Plastic Requires That the Applied
Shear Stress Exceed a Critical Value, τ_0, Before Flow Begins. If the Shear Stress is Less
Than the Critical Value, the Bingham Plastic Behaves as a Hookean Solid. If the Shear Stress
is Greater Than the Critical Value, the Bingham Plastic Behaves as a Newtonian Fluid,
Curve C.

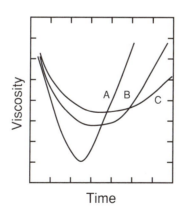

Figure 4.10.   Schematic Representation of a Reactive Liquid, Where Crosslinking Reactions
Cause the Apparent Viscosity to Increase with Time. These Curves are Typical of Polyurethane,
Unsaturated Polyester Resin, and Epoxy Viscosities, Beginning with Resin at Room Tempera-
ture. A: Mold Temperature = T_1. B: Mold Temperature = $T_2 < T_1$. C: Mold Temperature =
$T_3 < T_2$. Note That Increasing Temperature Causes the Initial Viscosity to Decrease Rapidly,
but Increases the Rate of Reaction. This Causes the Viscosity to Increase Rapidly at Earlier
Times.

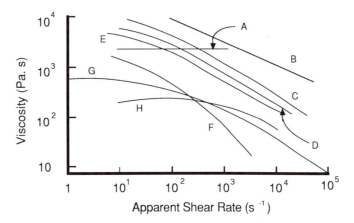

Figure 4.11. Shear Rate-Dependent Viscosities for Several Commercial Polymers at Typical Processing Temperatures (11). Curve A: Polycarbonate (PC) (288°C). B: General Purpose PS (230°C). C: PMMA (200°C). D: HDPE (232°C). E: LDPE (235°C). F: PP (230°C). G: PMMA (250°C). H: Nylon 6 (PA-6) (288°C).

Rheological Property Definition

Understand that a *rheological property* is a measurable *material function*. Dealy (12) points out that, for polymeric fluids, "no single rheological property gives a complete rheological characterization of the material." As noted, for steady-state, isothermal viscometric flows, three properties are sought—viscosity and the first and second normal stress differences. Temperature and time are also important parameters. However, Dealy further points out that *viscosity* is a function of only one variable—shear rate. As a result, it can be represented by a single curve, that curve being defined, of course, for a given flow condition, time and temperature. The complexity of the polymer rheological response to applied stress obviates development of reliable scaling laws. That is, tests done in one type of rheological device may not be directly related to those done in other devices, *even though* the geometries are quite similar. This is discussed further in Chapter 6.

   The shear-rate-dependent viscosities of several polymer melts are given in Figure 4.11. Note the wide variation in viscosities for these commercial materials. Furthermore, certain polymers such as polystyrene and polyamide-imide show nearly no "Newtonian zero-shear plateau", whereas others such as polyethylene and nylon 66 (PA-66) show a marked region. Some polymers show substantial shear-thinning at high shear rates whereas others such as polycarbonate show near-Newtonian behavior. The power-law viscosity model is used extensively for polymer flow calculations. The power-law constants, n and k, and the range of power-law validity for several polymers are given in Table 4.24 (104). Normal stress difference data are harder to obtain, primarily because accurate measuring methods have been developed relatively recently. Nevertheless, normal stress differences for several polymers are given in Figure 4.12.

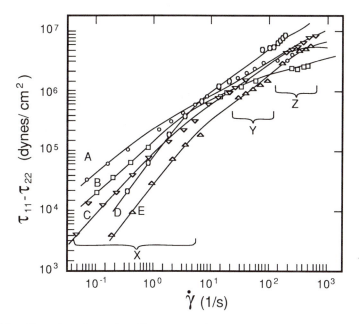

Figure 4.12. Shear Rate-Dependent First Normal Stress Difference for Several Polymers at 200°C (13). $\tau_{11} - \tau_{22}$, Normal Stress Difference. $\dot{\gamma}$, Shear Rate. Curve A: HDPE. Curve B: PS. Curve C: LDPE. Curve D: PMMA. Curve E: PP. Zone X: Rheogoniometer Data. Zone Y: Slit Die Data. Zone Z: Capillary Die Data.

Extensional Flows

In addition to simple viscometric flows, some of which are discussed below, extensional or stretching flow is important. Extensional flow is sometimes referred to as uniaxial extension or simple extension. In uniaxial extension (Figure 4.13), the $[\tau_{22} - \tau_{33}]$ normal stress difference, is zero. Thus, only the viscosity, now called the elongational viscosity, and the first normal stress difference, $[\tau_{11} - \tau_{22}]$, are important. If L is the instantaneous length of the filament, then a measure of strain is given as:

$$(4.11) \qquad d\varepsilon = \frac{dL}{L}$$

If V is the velocity at the end of the filament, then:

$$(4.12) \qquad \dot{\varepsilon} = \frac{V}{L} = \frac{d \ln L}{dt}$$

If the sample is straining at the same rate everywhere along its length, then the local velocity of any element is given as:

$$(4.13) \qquad \boxed{v_1 = \dot{\varepsilon}\, x_1}$$

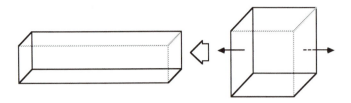

Figure 4.13.   Extensional Flow of a Polymer Element — Schematic.

Since the polymer can be considered incompressible, the changes in the off-axis dimensions (x_2 and x_3) are given as:

(4.14)
$$v_2 = \frac{-\dot\varepsilon\, x_2}{2}$$

(4.15)
$$v_3 = \frac{-\dot\varepsilon\, x_3}{2}$$

In cylindrical coordinates:

(4.16)
$$v_1 = \dot\varepsilon\, z$$

(4.17)
$$v_2 = \frac{-\dot\varepsilon\, r}{2}$$

(4.18)
$$v_3 = 0$$

Since the flow is axisymmetric:

(4.19)
$$\tau_{22} = \tau_{33} = \tau_{rr}$$

where τ_{rr} is the normal stress difference in the radial direction. As a result, there is but one material function, η_e, the extensional viscosity:

(4.20)
$$\eta_e(\dot\varepsilon) = \frac{[\tau_{11} - \tau_{22}]}{\dot\varepsilon} = \frac{[\tau_{zz} - \tau_{rr}]}{\dot\varepsilon}$$

η_e is the extensional viscosity, also known as the Trouton viscosity.*

---

*For the special case of a Newtonian fluid, it can be shown that $\eta_e = 3\,\eta$, where η is the Newtonian viscosity, a material constant. Thus, η_e for Newtonian fluids is also a material constant.

4.3 Steady Flow Through a Capillary Tube

The classic viscometric flow problem involves flow of a fluid through a tube having an infinite length-to-diameter ratio (Figure 4.14). From a force balance:

$$(4.21) \qquad F_1 = P \pi r^2$$

$$(4.22) \qquad F_2 = -\left[P + \left(\frac{\partial P}{\partial z} \right) dz \right] \pi r^2$$

where P is the applied pressure, r is the local radius, z is the local position, and $(\partial P/\partial z)$ is the rate of change of pressure with position. The drag on the surface of the element is given as:

$$(4.23) \qquad F_3 = 2 \pi r \, dz \, \tau$$

where τ is the shear stress. The force balance $(F_1 + F_2 + F_3 = 0)$ yields:

$$(4.24) \qquad \tau = \left(\frac{r}{2} \right) \left(\frac{\partial P}{\partial z} \right)$$

If $(\partial P/\partial z)$ is independent of z, then this equation can be written as:

$$(4.25) \qquad \tau = \frac{r \, \Delta P}{2L}$$

The shear stress on the material thus varies from zero at the center line $(r = 0)$ to maximum at the capillary wall $(r = R)$. For the latter:

$$(4.26) \qquad \tau_w = \frac{R \, \Delta P}{2L}$$

Note that this is valid, regardless of the viscoelastic nature of the fluid. The flow in a pipe is characterized by the Reynolds number, a nondimensional term, Re. For a non-Newtonian fluid flowing in a circular tube, this term is defined by (14):

$$(4.27) \qquad \boxed{ \text{Re} = \frac{8 \, \rho \, \bar{v}^2}{\tau_w} }$$

where ρ is the fluid density, $\bar{v}$ is the average velocity, and τ_w is the shear stress at the tube wall, in consistent units. Below a value of about Re = 2000, the fluid motion remains layered. This is called laminar flow.

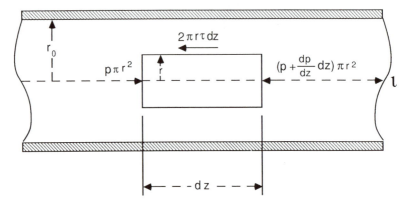

Figure 4.14.   Force Balance of a Fluid Element in Cylindrical Tube Flow.

Example 4.2

Q: Estimate the Reynolds number for 0.9 g/cm^3 density molten polymer flowing at 27.5 g/min ($\dot{m}$), in a 0.1-inch diameter pipe 10 inches long if the pressure drop is 40 lb$_f$/in^2.

A: The shear stress at the wall is given as:

$$\tau_w = \frac{R\,\Delta P}{2L}$$

$$= \frac{0.05 \text{ in} \times 40 \text{ lb}_f/\text{in}^2}{(2 \times 10 \text{ in})} = 0.1 \text{ lb}_f/\text{in}^2$$

$$= \frac{0.1 \text{ lb}_f/\text{in}^2 \times 6985 \text{ (N/m}^2)}{(\text{lb}_f/\text{in}^2)}$$

$$= 689.5 \text{ N/m}^2 = 6895 \text{ dyn/cm}^2$$

The average velocity, $\bar{v}$, is given as:

$$\bar{v} = \dot{m}/\rho\,A$$

$$= \frac{27.5 \text{ g/min} \times (1/60) \text{ min/s}}{0.9 \text{ g/cm}^3 \times \pi \times (0.5 \times 62.4)^2} = 10 \text{ cm/s} = 3.96 \text{ in/s}$$

The Reynolds number is:

$$\text{Re} = \frac{8 \times 0.9 \text{ g/cm}^3 \,(10 \text{ cm/s})^2}{6895 \text{ dyn/cm}^2} = 0.104$$

The fluid is in deep laminar flow (Re < 2000).

Now consider the shear rate at every point within the flow element. If δr is the thickness of a shell of fluid (see Figure 4.14 above), the volume flow rate of the shell is:

$$(4.28) \qquad Q_s = 2 \pi r \, \delta r \, v_z,$$

where v_z is the linear flow rate. If this volume is integrated between $r = 0$ and $r = R$, the volumetric flow rate, Q, is obtained:

$$(4.29) \qquad Q = 2 \pi \int_0^R r v_z \, dr.$$

Now $v_r = v_r(r)$, so the above equation can be integrated by parts to yield:

$$(4.30) \qquad Q = 2\pi \left[\frac{r^2}{2} v_z \Big|_0^R - \int_0^R \frac{r^2}{2} \, dv_z \right]$$

When $r = R$, $v_z = 0$, so:

$$(4.31) \qquad Q = -\pi \int_0^R r^2 \, dv_z = -\pi \int_0^R r^2 \left(\frac{\partial v_z}{\partial r} \right) dr$$

When $dr = (R/\tau_w) \, d\tau$, the flow rate, Q, can be written in terms of the shear stress, τ, as:

$$(4.32) \qquad Q = -\left(\frac{\pi R^3}{\tau_w^3} \right) \int_0^{\tau_w} \dot{\gamma} \, \tau^2 \, d\tau$$

If both sides are differentiated with respect to τ_w (a measurable constant):

$$(4.33) \qquad \left(\frac{1}{\pi R^3} \right) \left[\tau_w^3 \frac{dQ}{d\tau_w} + 3\tau_w^2 Q \right] = -\dot{\gamma}_w \tau_w^2$$

But $\tau_w = (R/2) \, (dP/dx) = R \, \Delta P/2L$. Thus:

$$(4.34) \qquad -\dot{\gamma}_w = \left(\frac{1}{\pi R^3} \right) \left[3Q + \Delta P \left(\frac{dQ}{d \, \Delta P} \right) \right] = \left(\frac{4Q}{\pi R^3} \right) \left[\frac{3}{4} + \frac{1}{4} \frac{d \ln Q}{d \ln \Delta P} \right]$$

This equation relates the measurable quantities of radius, R, pressure drop, ΔP and volumetric flow rate, Q, to shear rate at the wall of the capillary. It is referred to as the Rabinowitsch, Rabinowitsch–Mooney, or Weissenberg–Rabinowitsch–Mooney equation. Brydson (15) notes that for a time-independent fluid:

$$(4.35) \qquad \qquad \tau_w = f(\dot\gamma)_w \text{ [only]}$$

Therefore:

$$(4.36) \qquad \tau_w = f\left[\left(\frac{3}{4}\right) \left(\frac{4Q}{\pi R^3}\right) + \left(\frac{\tau_w}{4}\right) \left(\frac{d(4Q/\pi R^3)}{d\tau_w}\right) \right]$$

Thus, if τ_w is plotted against $(4Q/\pi R^3)$, a unique curve results. $(4Q/\pi R^3)$ is called the *apparent wall shear rate*, $\dot\gamma_{w,a}$, then:

$$(4.37) \qquad \qquad \tau_w = f(\dot\gamma_{w,a})$$

Thus, the Rabinowitsch expression can be written as:

$$(4.38) \qquad -\dot\gamma_w = \left(\frac{3}{4}\right)(\dot\gamma_{w,a}) + \left(\frac{1}{4}\right)(\dot\gamma_{w,a})\left[\frac{d \ln (\dot\gamma_{w,a})}{d \ln \tau_w}\right]$$

If a local "power-law index", n', is defined as:

$$(4.39) \qquad \qquad n' = \frac{d \ln (R \; \Delta P/2L)}{d \ln (4Q/\pi R^3)}$$

then:

$$(4.40) \qquad \qquad -\dot\gamma_{w,a} = \left[\frac{3n' + 1}{4n'}\right]\left(\frac{4Q}{\pi R^3}\right)$$

This is the Metzner form of the Rabinowitsch–Mooney equation. If n' is locally constant, then the equation can be integrated to yield:

$$(4.41) \qquad \qquad \tau_w = \frac{R \; \Delta P}{2L} = k'\left(\frac{4Q}{\pi R^3}\right)^{n'}$$

k' is called the consistency index and n' is the flow behavior index.

Note that if n' is independent of shear rate, as with a power-law fluid, the following identities are obtained:

(4.42) $$n' = n,$$

(4.43) $$k' = k\left(\frac{3n + 1}{4n}\right)^n$$

Example 4.3

Q: Nylon 6, (PA-6) Capron 8200 flows isothermally at 503°K through a 0.5-inch ID by 18-inch long pipe. If the flow rate is 2.0 in³/s, what is the pressure drop? For Capron 8200, $100 < \dot{\gamma} < 2000$, $k = 1950$ N s^n/m² and $n = 0.66$.

A: The apparent wall shear rate, $\dot{\gamma}_{w,a}$ is given as:

$$\dot{\gamma}_{w,a} = 4Q/\pi R^3$$

$$= 4 \times 2/(0.5/2)^3 \, \pi = 163 \text{ s}^{-1}$$

The shear rate, $\dot{\gamma}_w$, is given as:

$$\dot{\gamma}_w = \left[\frac{3n + 1}{4n}\right] \dot{\gamma}_{w,a}$$

$$= \left[\frac{3 \times 0.66 + 1}{4 \times 0.66}\right] \times 163 = 184 \text{ s}^{-1}$$

The shear stress at the wall is given as:

$$\tau_w = k \, (\dot{\gamma}_w)^n$$

$$= 1950 \times (184)^{0.66} = 60.9 \text{ kPa}$$

$$= 60900 \, / \, (6895 \text{ Pa/lb}_f/\text{in}^2) = 8.84 \text{ lb}_f/\text{in}^2$$

The pressure drop is then:

$$\Delta P = \frac{2L \, \tau_w}{R} = 2 \times 18 \times 8.84/(0.5/2)$$

$$= 1272 \text{ lb}_f/\text{in}^2.$$

It can also be shown that the radius-dependent velocity profile is given as:

(4.44) $$v_z(r) = \left(\frac{nR}{n + 1}\right)\left[-\frac{R\Delta P}{2kL}\right]^{1/n} \times \left[1 - \left(\frac{r}{R}\right)^{\frac{n+1}{n}}\right]$$

Note that for n approaching 0, the velocity profile approaches plug flow. That is, $v_z(r) \neq f(r)$. When $n = 1$, the velocity profile is traditionally laminar parabolic. For

$n = 1$, $k = \mu$, the Newtonian viscosity, and the pressure drop-flow rate equation becomes:

$$(4.45) \qquad \frac{R\,\Delta P}{2L} = \mu \,\frac{4Q}{\pi R^3}$$

or rearranging:

$$(4.46) \qquad Q = \frac{\pi R^4\,\Delta P}{8\,\mu L}$$

This equation is called the Hagen–Poiseuille* equation.

Example 4.4

Q: Construct a relationship between the shear stress at the wall and the apparent shear rate where the tube radius is not a factor.

A: The working equations are $\tau_w = R\,\Delta P/2L$ and $\dot{\gamma}_{w,a} = 4Q/\pi R^3$. Solve the first for R:

$$R = \frac{2L\tau_w}{\Delta P}$$

Substitute into the second expression:

$$\dot{\gamma}_{w,a} = \frac{Q\Delta P^3}{2\pi L^3 \tau_w^3}$$

Solve for τ_w:

$$\tau_w = (Q/2\pi)^{1/3}\,(\Delta P/L)\,\dot{\gamma}_{w,a}^{-1/3}$$

For a given flow rate–pressure drop situation:

$$\tau_w = C\dot{\gamma}_{w,a}^{-1/3}$$

This is plotted in Figure 4.15, where $\ln \tau_w = \ln C - (1/3)\ln \dot{\gamma}_{w,a}$

Several useful relationships for laminar flow in a circular tube are given in Table 4.2.

---

*Poiseuille was a nineteenth-century fluid mechanician after whom the now-obsolete unit for viscosity, the "poise," was named. There is some controversy over the English or American pronunciation of Poiseuille's name. According to Bird et al. (16), the correct pronunciation is 'Pwah-zoo-ya.'' Other pronunciations include "Paw-sell" and "Poy-sell." Bird is probably correct. But since the accepted unit for viscosity is now Pa · s, it is apparent that all old rheologists have lost their "poise."

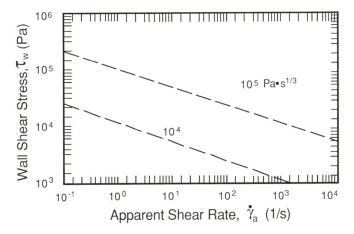

Figure 4.15. Comparison of Wall Shear Stress, τ_w, and Apparent Shear Rate at the Wall, $\dot{\gamma}_{w.a}$, for Flow in a Cylindrical Tube.

Table 4.2. Relationships for polymer flow in a circular pipe: power-law and Newtonian.

Term	Power-Law	Newtonian
Velocity, v	$V^*\left[1 - \left(\dfrac{r}{R}\right)^{\frac{(n+1)}{n}}\right]$	$V^*\left[1 - \left(\dfrac{r}{R}\right)^2\right]$
Volume flow rate, Q	$\dfrac{n+1}{3n+1}\, V^*\,\pi R^2$ or $\dfrac{-n}{3n+1}\,\pi R^3\left[\dfrac{R}{2K}\dfrac{dP}{dz}\right]^{1/n}$	$\dfrac{1}{2}\, V^*\,\pi R^2$ or $-\dfrac{\pi}{8}\dfrac{R^4}{\mu}\dfrac{dP}{dz}$
Average velocity, $\overline{V}$	$\dfrac{n+1}{3n+1}\, V^*$	$\dfrac{1}{2}\, V^*$
Wall shear rate, $\dot{\gamma}_\omega$	$-\dfrac{4Q}{\pi R^3}\left[\dfrac{3n+1}{4n}\right]$ or $\dfrac{4Q}{\pi R^3}\left[\dfrac{3}{4}+\left(\dfrac{d\ln Q}{d\ln(dP/dz)}\right)\right]$	$-\dfrac{4Q}{\pi R^3}$
Shear stress, τ	$\tau_w\,\dfrac{r}{R}$	$\tau_w\,\dfrac{r}{R}$

Note: V^* = Centerline Velocity.

The Rabinowitsch equation can be used to predict pressure drop-flow rate relationships for any time-independent fluid, as long as a unique relationship between *apparent shear rate at the wall* and *shear stress at the wall* is known. The slope of such a relationship at a given shear rate defines n and from this can be obtained k. As noted in Chapter 6, pressure drop–flow rate data for flow through a capillary tube must be corrected for "end effects", that is, non-viscometric entrance and exit losses. As a result, the Metzner–Weissenberg–Rabinowitsch–Mooney equation does not have wide acceptance for prediction of flow of general time-independent fluids through capillaries.

4.4 Time Dependency and Anomalous Elastic Liquid Effects

Early on, it was observed that long chain polymeric materials exhibited unusual time-dependent behavior. Weissenberg (17) noted that when a rod was rotated in an otherwise-quiescent polymeric liquid, the liquid did not form a vortex, as with Newtonian fluids. Instead, it climbed the rotating rod. Lodge (18) demonstrated that when the stress applied to a deformed polymeric element is abruptly released, the material rapidly retracts toward an undeformed state. This indicates the elastic nature of polymers. Kapoor (19) demonstrated that when steady flow of an elastic fluid in a capillary tube is stopped, the fluid "recoils" or flows back toward the pressure source.

The tubeless siphon (20) is a phenomenon unique to elastic or time-dependent liquids. Consider a siphon: a tube draws liquid from a reservoir up and over the edge of the reservoir. For a Newtonian fluid, when the tube is raised above the level of the reservoir, the siphon effect stops. For an elastic liquid, the flow continues. Another effect is called the "Uebler effect" (21). For Newtonian fluids, when a bubble is introduced in the throat of a flow contraction, it accelerates to the speed of the fluid. "Bubble tracing" is a common way of studying flow patterns in Newtonian fluids. For elastic or time-dependent liquids, a large bubble (on the order of 10 to 15% of the small diameter of the contraction) will simply stop at the entrance to the contraction.

Pure Newtonian fluids in laminar flow contract as they issue from a capillary tube. The degree of contraction is due entirely to the rearrangement in the velocity profile from parabolic to plug. The extent of contraction is 13%. That is, the extrudate is 13% smaller in diameter than the capillary tube inside diameter. Extrudate swell* is another example of the complex nature of a polymer in the fluid state (22). When an elastic polymer issues from a circular tube or die, the fluid stream diameter continues to increase with distance from the die until an equilibrium diameter is achieved (Figure 4.16).

Extrudate swell is directly associated with the elastic or storage component of the fluid. As the fluid is forced through the die under pressure, it undergoes a high degree of orientation. Molecules that are naturally coiled or entangled when at rest are pref-

---

*Extrudate swell is sometimes called jet swell or die swell. The term "jet swell" is not as clear as extrudate swell. "Die swell" is patently incorrect.

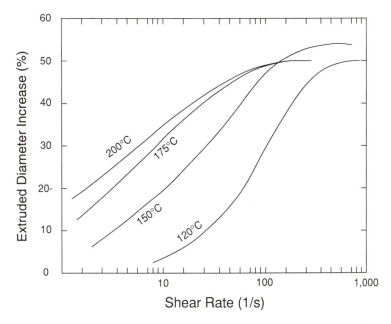

Figure 4.16. Temperature-Dependent Extrude Swell of Low-Density Polyethylene.

erentially oriented in the flow direction. When the restraining forces are removed to the die exit, the stored energy in the polymer enables the polymer chains to recoil and reentangle, thereby increasing isotropy and increasing extrudate diameter. Since polymers are envisaged as combinations of elastic and viscous components, recovery is time dependent, and the extent and rate of recovery depend strongly on temperature, molecular weight distribution, extent of shear, the length of time shear was applied, and the nature of the polymer. For certain very elastic liquids, sometimes called "memory fluids", the extrudate diameter can be substantially larger than the capillary tube diameter. For example, polymethyl methacrylate dissolved in dimethyl phthalate exhibits a fourfold extrudate swell (23). For large tube length-to-diameter ratios, the ratio of extrudate diameter to tube diameter can be predicted (24) by:

$$(4.47) \qquad D_e/D = 0.1 + \left[1 + \frac{(\tau_{11} - \tau_{22})}{2\tau_{21,w}^2} \right]^{1/6}$$

Here $(\tau_{11} - \tau_{22})$ is the first normal stress difference and τ_{21} is the shear stress, both determined at the capillary wall. The elastic effects of a polymer melt can be related to its tensile and shear strain recovery. As shown in Figure 4.17, the extrudate swell ratio, B_{SR}, for a constant diameter capillary tube decreases with increasing capillary L/D ratio. At zero length, the polymer fluid does not experience capillary tube-induced shear. Hence, all polymeric recovery is the result of tensile stress. On the other hand, the extrudate swell ratio approaches a limiting value, $B_{Sr,\infty}$, as the value of L/D be-

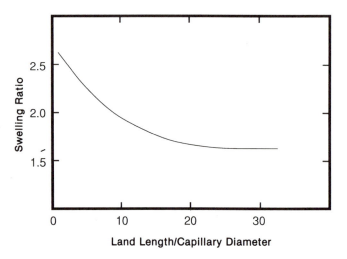

Figure 4.17.   *L/D*-Dependent Capillary Extrudate Swell, Uniform Diameter Capillary (26).

comes large. This value is the result of shear-induced strain. Any tensile stress imposed at the die entry has dissipated by the time the fluid reaches the long *L/D* die exit. These two extremes represent the limiting cases for extrudate swell (25). These limiting conditions are considered below.

Zero-Length Capillary

Consider the annular fluid element in Figure 4.18. The fluid element undergoes a change in shape in the radial direction, given by:

$$(4.48) \qquad \varepsilon_R = \ln(1 + \varepsilon) = \ln\left[1 + \frac{dr' - dr}{dr}\right]$$

or:

$$(4.49) \qquad \exp(\varepsilon_R) = 1 + \frac{dr' - dr}{dr}$$

This can be written as:

$$(4.50) \qquad \exp(\varepsilon_R)\, dr = dr'$$

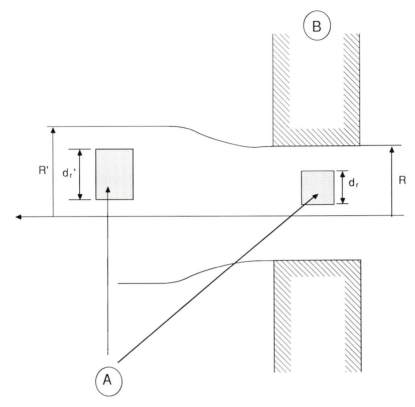

Figure 4.18.   Polymer Stream Emerging from a Short L/D Die (27). A: Annular Fluid Element. B: Short L/D Die.

If the swelling ratio is defined as:

(4.51) $(B_{ER})^2$ = area of swollen extrudate/area of capillary

then:

(4.52) $$(B_{ER})^2 = \int_0^R 2\,\pi\,r dr' \;/\; \int_0^R 2\,\pi\,r dr$$

(4.53) $$= \exp(\varepsilon_R)$$

or:

(4.54) $$B_{ER} = [\exp(\varepsilon_R)]^{1/2}$$

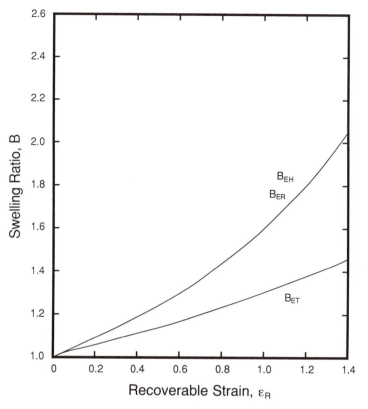

Figure 4.19.   Effect of Tensile Recoverable Strain on Swelling Ratio for Short *L/D* Capillary and Slot Dies (28). ε_R: Recoverable Strain. *B*: Swelling Ratio.

For a short rectangular slit, *T* in width and *H* in thickness, the swelling ratios are:

(4.55) $$B_{ET} = [\exp(\varepsilon_R)]^{1/4}$$

and:

(4.56) $$B_{EH} = [\exp(\varepsilon_R)]^{1/2}$$

These are shown in Figure 4.19.

Large L/D Capillary

For flows in large *L/D* capillaries, or for flows characterized by a residence time in the capillary that is long when compared with the largest polymer relaxation time, the swelling ratio at the die exit is the result of shear-imposed strain. The annular fluid element in Figure 4.20 has a shear strain, γ_r, at the die exit:

(4.57)
$$\gamma_r = \tan \alpha = \frac{ed}{ae}$$

The areal ratio between the distorted inside and the recovered outside annular elements is:

$$\frac{\text{area of outside annulus}}{\text{initial area of annulus}} = \frac{2\pi r dr^1}{2\pi r dr} = \frac{ad}{ae}$$

(4.58)
$$= \frac{(ae^2 + ed^2)^{1/2}}{ae}$$

$$= (1 + \left(\frac{ed}{ae}\right)^2)^{1/2}$$

$$= (1 + \gamma_r^2)^{1/2}$$

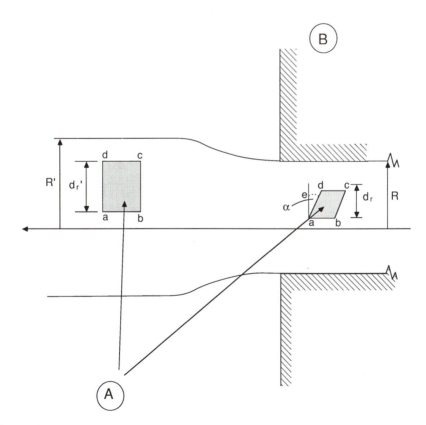

Figure 4.20.   Polymer Stream Emerging from a Long L/D Die (29). *A*: Annular Fluid Element. *B*: Long L/D Die.

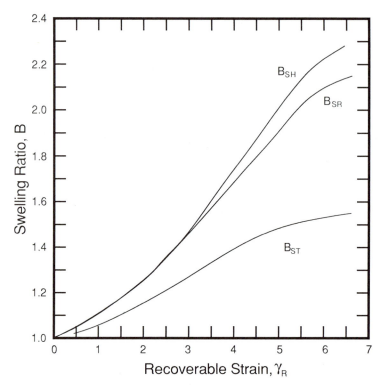

Figure 4.21. Effect of Tensile Recoverable Strain on Swelling Ratio for Long *L/D* Capillary and Slot Dies (30). $\dot{\gamma}_R$: Recoverable Strain. *B*: Swelling Ratio.

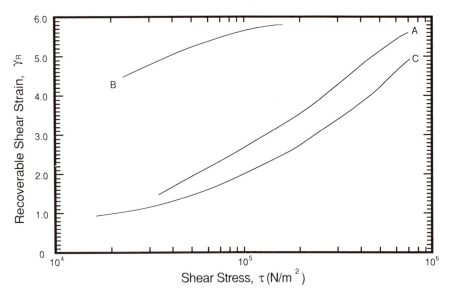

Figure 4.22. Shear-Rate Dependent Recoverable Shear Strain for Three Olefins (31). Curve A: LDPE$_1$ at 170°C. B: LDPE$_2$ at 170°C. C: Ethylene–Propylene Copolymer at 230°C.

Now the swelling ratio is given as:

$$(4.59) \qquad B_{SR}^2 = \text{outside/inside} = \frac{\displaystyle\int_0^R (1 + \gamma_r^2)^{1/2} 2\pi r dr}{\displaystyle\int_0^R 2\pi r dr}$$

Or if the shear strain is proportional to radius:

$$(4.60) \qquad \gamma_r = (r/R)\, \gamma_R$$

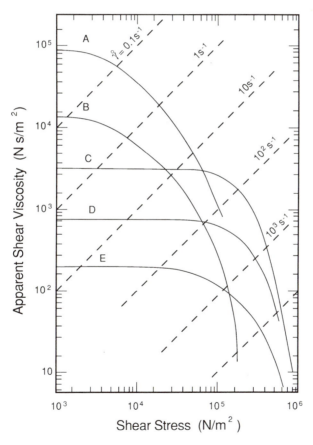

Figure 4.23.   Shear-Rate Dependent Viscosities for Several Polymers (32). Dashed Lines: Shear Rate. A: LDPE₁ at 170°C. B: Ethylene–Propylene Copolymer at 230°C. C: PMMA at 230°C. D: POM (Acetal Copolymer) at 200°C. E: PA-66 (Nylon 66) at 285°C.

then:

(4.61)
$$B_{SR} = \left[\frac{2}{3} \gamma_R \left[\left(1 + \frac{1}{\gamma_R^2} \right)^{3/2} - \frac{1}{\gamma_R^3} \right] \right]^{1/2}$$

For rectangular channels of width T and thickness H:

(4.62) $$B_{ST} = \left[\frac{1}{2} (1 + \gamma_R^2)^{1/2} + \frac{1}{2\gamma_R} \ln \left[\gamma_R + (1 + \gamma_R^2)^{1/2} \right] \right]^{1/3}$$

(4.63) $B_{SH} = B_{ST}^2$

The limiting swelling ratios for tube and slot flow are shown in Figure 4.21.

To use this analysis, the amount of recoverable strain as a function of volumetric flow rate must be known for a given resin. The amount of recoverable strain is shown in Figure 4.22 as a function of shear rate for long L/D capillaries for some polyolefins. The corresponding shear-dependent viscosity data are shown in Figure 4.23.

Probably the effect that has caused the most controversy in elastic liquid behavior is the "pressure hole effect". Simply put, when fluids flow over a depression in the wall of a container, the flow streamlines tend to dip into and out of the "hole". Pressure transducer holes are frequently placed in pipelines to measure pressure drop of polymeric materials. For elastic liquids, the non-zero first normal stress difference can cause abnormally low pressures to be measured at such holes. With proper instrumentation, the "pressure hole error" so measured can be used to measure the first normal stress difference (33). Other "elastic liquid anomalies" are described in Bird et al. (34). A thorough examination of the characteristics of polymer fluids that cause this type of non-Newtonian behavior is clearly beyond the scope of this work (1–5). Suffice it to say that elastic, conservative, restorative forces are present even in the flowable state of many polymers.

4.5 External Effects on Viscosity

There are many elements that can change the stress–rate-of-strain behavior of a polymer in shear. Some of these are temperature, pressure, level and nature of adducts, fillers and reinforcements and molecular weight distribution. Here the temperature effect is considered. Chapter 6 briefly treats the effect of pressure.

Temperature Effect on Viscosity

As noted throughout this text, temperature is an important parameter for material property change in polymers. Temperature effects are important in two regards. First, a given polymer can be processed over a wide temperature range. It is certainly the case that as the polymer temperature increases, its viscosity decreases. If the applied force is constant, resistance to flow decreases and thus flow rate increases. The Arrhenius equation is frequently used to describe the temperature dependency of polymers:

(4.64)
$$\eta = A \exp(E/RT)$$

where η is a defined shear viscosity, E is a viscous energy of activation for the polymer, R is the gas constant, T is absolute temperature, and A is a pre-exponential coefficient. Note that this equation is a two-parameter equation. For most polymers, a plot of $\ln \eta$ vs $1/T$ yields a reasonably straight line over 50 to 150°C temperature range. This equation is difficult to use when analytical integration is involved. As a result, another exponential equation is sometimes preferred:

(4.65)
$$\eta = \alpha \exp(-\beta T)$$

where α and β are empirical coefficients. These coefficients are determined by plotting $\ln \eta$ vs. T (absolute). McKelvey (35) points out that the term β^{-1} "represents the number of degrees that the temperature must be raised at constant shear rate in order to decrease the viscosity by the factor (1/e)". He shows that if η_1 is the viscosity at T_1 and η_2 is at T_2, then:

(4.66)
$$\eta_1 = \eta_2 \exp[\beta(T_2 - T_1)]$$

If $(T_2 - T_1) = 1/\beta$, then $\eta_1 = \eta_2/e$. McKelvey points out that β is really determined at constant shear rate. As seen in Table 4.3, the "1/e" reduction in viscosity for polymethyl methacrylate requires only an 18°C increase in temperature whereas an 85°C temperature increase is needed for low density polyethylene. Of course, this assumes a constant shear rate. It can be shown using the Arrhenius equation that there are two energies of activation, one at constant shear rate and the other at constant shear stress. The relationship is given as:

(4.67)
$$\frac{E\tau}{E\dot{\gamma}} = 1 - \dot{\gamma}\left(\frac{\partial \eta}{\partial \tau}\right)_T$$

Table 4.3. Temperature dependence of viscosity for several commercial polymers (36).

Material	Shear rate (s^{-1})	$1/b_{\dot{\gamma}}$ (°C)	Trade Name
Polymethyl methacrylate	100	24	Lucite 140
Polymethyl methacrylate	27	18	Plexiglas V100
Cellulose acetate	100	32	Tenite acetate 036-H2
Nylon 6	100	60	Plaskon Nylon 8206
Nylon 66	100	56	Zytel 101 NC10
Polyethylene	100	85	Bakelite DYNH
Polyethylene	100	70	Alathon-10
Polystyrene	100	73	Styron 475
Polyvinyl chloride, rigid	40	51	Geon 8750
Polyvinyl chloride, flexible	100	40	Opalon 71329

This comes about because:

(4.68)
$$\left(\frac{\partial \eta}{\partial T}\right)_\tau = -\eta \frac{E_\tau}{RT^2}$$

(4.69)
$$\left(\frac{\partial \eta}{\partial T}\right)_{\dot{\gamma}} = -\eta \frac{E_{\dot{\gamma}}}{RT^2}$$

Activation energies for low-density polyethylene are given in Table 4.4. Consider these simple examples.

Example 4.5

Q: Determine the temperature increase needed to reduce the constant-stress viscosity of polyethylene by 50%. The shear stress of PE at 120°C is 1000 Pa = 10,000 dyn/cm².

A: From Table 4.4 at $\tau = 1000$ Pa:

$$E_\tau = 15{,}000 \text{ cal/g mol}$$

$$\eta_2/\eta_1 = \exp[E_\tau/RT_2]/\exp[E_\tau/RT_1]$$

$$\ln[\eta_2/\eta_1] = (E_\tau/R)[1/T_2 - 1/T_1]$$

$$\ln(1/2) = (15{,}000/1.987)[1/T_2 - 1/(120 + 273)]$$

$$1/T_2 = 0.002453 °K^{-1}$$

$$T_2 = 407.7 °K = 134.7 °C$$

$$\Delta T = 134.7 - 120 = 14.7 °C$$

Table 4.4. Activation energies for polyethylene. Temperature range: 108–230°C. $\dot{\gamma}$ is shear rate. τ is shear stress. $E_{\dot{\gamma}}$ is shear-rate dependent energy of activation. E_τ is shear-stress dependent energy of activation (37).

Shear rate $\dot{\gamma}$, (s⁻¹)	Shear-rate-dependent energy of activation, $E_{\dot{\gamma}}$ (kcal/mol)	Shear stress τ, (dyn/cm²)	Shear-stress-dependent energy of activation, E_τ, (kcal/mol)
0	12.8	0	12.8
10^{-1}	11.4	10^4	15.0
10^0	10.3	10^5	17.8
10^1	8.5	10^6	19.0
10^2	7.2		
10^3	6.1		

Example 4.6

Q: Consider the polymer in Example 4.5. What would the temperature increase be if the process occurred at constant shear rate, $\dot\gamma = 100 \text{ s}^{-1}$?

A: From Table 4.4, the energy of activation is:

$$E_{\dot\gamma} = 7200 \text{ cal/g mol}$$

Now

$$1/T_2 = 1/(120 + 273) + \ln(0.5) \times 1.987/7200$$

$$T_2 = 424.9°K = 151.9°C$$

$$\Delta T = 151.9 - 120 = 31.9°C$$

Note that if the polymer is a power-law (Ostwald–deWaale) fluid:

$$(4.70) \qquad\qquad E_\tau = E_{\dot\gamma}/n$$

Viscous Dissipation

When a Hookean solid is deformed under load and the load is then removed, the deformation path of recovery follows exactly the initial strain path. The material is said to have no stress–strain hysteresis. When a non-Hookean material is strained and allowed to recover, the outbound and inbound paths may not coincide. The unrecovered work of deformation is converted to internal energy and ultimately released as heat. When polymers are stressed, the molecular disentanglements are the primary cause for the great resistance to deformation, viz, viscosity. This leads to substantial unrecoverable work of deformation and heat buildup. This is called "viscous dissipation". One measure of viscous dissipation is given by:

$$(4.71) \qquad\qquad \phi = \tau\dot\gamma \; [= N/s \; m^2 = W/m^3]$$

Example 4.7

Q: Estimate the magnitudes of viscous dissipation for polystyrene flowing through a gate and cavity of an injection mold. The following conditions are relevant:

$$\text{Gate: } \dot\gamma = 100{,}000 \text{ s}^{-1}, \tau = 10{,}000 \text{ Pa}$$

$$\text{Cavity: } \dot\gamma = 1{,}000 \text{ s}^{-1}, \tau = 10 \text{ Pa}$$

A: The solution is:

$$\phi = \dot{\gamma}\,\tau$$

$$\text{Gate: } \phi = 10^9 \text{ W/m}^3$$

$$\text{Cavity: } \phi = 10^4 \text{ W/m}^3$$

Observation: Viscous dissipation through gates is always substantially greater than that through cavities.

Consider the capillary flow example seen earlier. For the arithmetic of Example 4.3, the material was assumed to be isothermal. Consider now the effect of viscous dissipation for a power-law fluid (38,39). The two equations that describe non-isothermal power-law flow are:

$$(4.72) \qquad 0 = -\frac{dP}{dz} + \frac{1}{r}\frac{d}{dr}\left[\eta r \frac{dv_z}{dr}\right]$$

$$(4.73) \qquad \rho c_p v_z \frac{\partial T}{\partial z} = \frac{k}{r}\frac{\partial}{\partial r}\left[r\frac{\partial T}{\partial r}\right] + \eta \left(\frac{dv_z}{dr}\right)^2$$

The second equation, the energy equation, is one-way coupled to the momentum equation, the first equation. That is, the momentum equation can be solved first, and the solution of this equation, for $v_z(r)$, substituted into the energy equation. If $\eta = K(-dv_z/dr)^{n-1}$ (where now K is the power-law coefficient since k is reserved for thermal conductivity in the energy equation), then:

$$(4.74) \qquad v_z = v_{max}[1 - (r/R)^{(n+1)/n}]$$

where $v_{max} = [\Delta PR/2KL]^{1/n}[Rn/(n+1)]$. This expression is now substituted into the energy equation for $v_z(r)$, and the expression integrated. The expression cannot be solved directly analytically. A dimensionless temperature rise, Θ ($\chi = r/R$, $\zeta = kz/\rho c_p v_{max}R^2$), is formulated:

$$(4.75) \qquad \Theta = \frac{k(T - T_0)(1/n + 3)^3}{4KR[v_{max}(1/n + 1)/R]^{n+1}}$$

The dimensionless radial- and axial-dependent temperature for $n = 0.5$ and two boundary conditions, constant wall temperature and zero wall heat flux, are shown in Figures 4.24 and 4.25. Note that the maximum temperature buildup usually occurs at the exit of the capillary tube. This fixes $\zeta = \zeta_{max}$. With this, the dimensionless local temperature, Θ, can be determined from Figure 4.24 or 4.25, depending on the wall condition. Thus:

$$(4.76) \qquad (T - T_0)_{max} = \frac{4KR^2(\tau_w/k)^{(n+1)/n}}{k(1/n + 3)^2}\theta_{max}$$

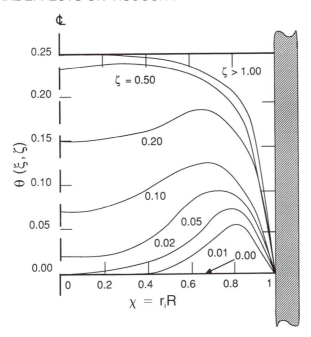

Figure 4.24. The Effect of Viscous Heating on Temperature Profiles in Tube Flow—Constant Wall Temperature. Power-Law Index, $n = 0.5$ (40). Dimensionless Temperature Rise, $\theta = k (T - T_o) (1/n + 3)^2 /[4mR^2 (v_{max}/R)^{(n+1)} (1/n + 1)^{(n+1)}]$. Dimensionless Radius, $\chi = r/R$. Dimensionless Distance Down the Tube, $\zeta = kz/\rho c_p v_{max} R^2$.

If the material parameters (ρ, c_p, k, K, n) are known and the processing conditions (wall condition, wall shear stress) are known, then the local maximum temperature can be found.

Example 4.8

Q: Estimate the temperature rise of a polyethylene melt that enters a circular tube die of diameter 0.08 cm and length 1.5 cm at 460°K at a wall shear stress of 100,000 N/m². The following polymer property data are relevant:

$$\rho c_p = 1.8 \times 10^6 \text{ J/m}^3\text{°K}$$

$$k = 4.184 \times 10^{-2} \text{ W/m°K}$$

$$n = 0.5$$

$$K = 6.9 \times 10^3 \text{ W s}^{1/2}\text{/m}^2$$

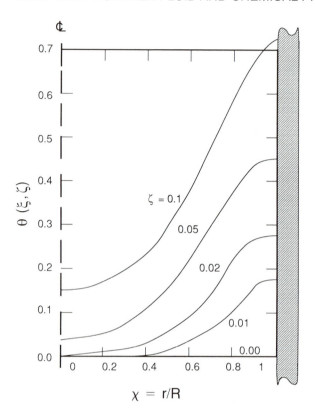

Figure 4.25. The Effect of Viscous Heating on Temperature Profiles in Tube Flow-Zero Wall Heat Flux (40). Coordinates Same as Figure 4.24.

A: The dimensionless distance at the tube exit is given by:

$$\zeta = kz \, (1 \, + \, 1/n) \, K^{1/n} \, / \, \rho c_p \tau_w^{1/n} \, R^3$$
$$= (4.184 \times 10^{-2}) \, (0.015) \, (3) \, (6.9 \times 10^3)^2 \, / \, (1.8 \times 10^6) \, (1 \times 10^5)^2 \, (0.0004)^3$$
$$= 0.0778$$

From Figure 4.25, the maximum dimensionless temperature rise, $\Theta = 0.62$. The actual temperature rise is obtained from:

$$\Delta T_{max} = 4KR^2 \, (\tau_w/K)^{(1 + 1/n)} \, \Theta \, / \, k \, (1/n \, + \, 3)^2$$
$$= 4 \, (6.9 \times 10^3) \, (0.0004)^2 \, (1 \times 10^5/6.9 \times 10^3)^3 \, (0.62)/$$
$$(4.184 \times 10^{-2}) \, (5)^2$$
$$\Delta T_{max} = 8°C.$$

To help with calculation of this important factor, a nomograph Figure 4.26 (41), can be used. To use this nomograph, the capillary wall shear stress and shear rate must be

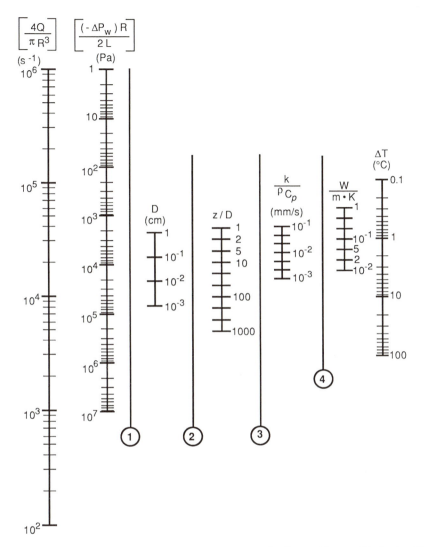

Figure 4.26.   Nomograph for Estimation of Temperature Rise Due to Viscous Dissipation (42).

known. These form the values in the two left-hand scales. The diameter and length of the capillary form the values for the next two scales. The thermal diffusivity is used on either of the next two scales, depending on the units. Thermal diffusivity, $\alpha = k/\rho\, c_p$, and thermal conductivity values can be obtained from Appendix C. To use this nomograph, find first the shear stress and shear rate points. Extend the line to vertical [1]. Using this intersection and the capillary diameter, extend the line to vertical [2]. Use this line and the capillary length to extend the line to vertical [3]. Then use this line and the thermal diffusivity value to extend the line to the ΔT vertical. This establishes the average or bulk temperature rise for the conditions.

4.6 Polymer Melt Thermodynamics

In many processes, many polymers can be considered to be incompressible liquids. However, as applied pressures and temperatures increase, certain polymers exhibit substantial levels of compressibility. This section highlights the response of pure polymers to applied pressures and temperatures.

Pressure-Volume-Temperature Relationships

Polymers and simple fluids respond to applied forces in different ways. At times, polymers appear to behave like solids under load, while at other times, they appear to behave like fluids. Earlier, polymers were characterized as elastic liquids or viscoelastic solids. Their complex responses to applied loads were detailed as stress–strain–rate-of-strain constitutive equations of state. Polymers also do not behave as simple materials in a thermodynamic sense. Pressure, temperature, and specific volume for all materials are related through a *thermodynamic constitutive equation of state*:

(4.77) $$v = v(P,T)$$

Table 4.5. Universal gas constant values.

Temperature scale	Pressure units	Volume units	Weight units	Energy units	R, gas constant
Kelvin (°K)	–	–	g mol	calories	1.9872
Kelvin (°K)	–	–	g mol	joules (absolute)	8.3144
Kelvin (°K)	–	–	g mol	joules (international)	8.3130
Kelvin (°K)	atm	cm³	g mol	atm cm²	82.057
Kelvin (°K)	atm	liter	g mol	atm liters	0.08205
Kelvin (°K)	mm Hg	liter	g mol	mm Hg liter	62.361
Kelvin (°K)	bar	liter	g mol	bar liters	0.08314
Kelvin (°K)	kg/cm²	liter	g mol	kg/cm² liter	0.08478
Kelvin (°K)	atm	ft³	lb mol	atm ft³	1.314
Kelvin (°K)	mm Hg	ft³	lb mol	mm Hg ft³	998.9
Kelvin (°K)	–	–	lb mol	chu or pcu	1.9872
Rankine (°R)	–	–	lb mol	Btu	1.9872
Rankine (°R)	–	–	lb mol	hp h	0.0007805
Rankine (°R)	–	–	lb mol	kW h	0.0005819
Rankine (°R)	atm	ft³	lb mol	atm ft³	0.7302
Rankine (°R)	in Hg	ft³	lb mol	in Hg ft³	21.85
Rankine (°R)	mm Hg	ft³	lb mol	mm Hg ft³	555.0
Rankine (°R)	lb_f/in² absolute	ft³	lb mol	lb ft³/in²	10.73
Rankine (°R)	lb_f/in² absolute	ft³	lb mol	ft lb	1545.0

where v is molar specific volume, P is applied hydrostatic pressure and T is absolute temperature. The simplest thermodynamic equation of state is that of an ideal gas:

$$(4.78) \qquad\qquad Pv = RT$$

where R is the universal gas constant (Table 4.5). The ideal equation of state assumes that the molecules are perfect elastic bodies with point masses. At moderate pressures and very high temperatures, monomolecular gases such as helium approach ideal gas behavior. Typically, however, most gases are non-ideal although quite compressible, even relatively close to their critical points. On the other hand, solids are relatively incompressible. That is, their specific volume is temperature dependent but nearly pressure independent. For most simple fluids, a step change in pressure or temperature usually results in an instantaneous change in specific volume.

One approach to a non-ideal "gas-type" thermodynamic equation of state is to include the presence of the molecules, in terms of their mass and less-than-ideal energy interaction at the moment of impact. The most popular model is the Van der Waals equation of state:

$$(4.79) \qquad\qquad (P \; a/v^2) \; (v - b) = RT$$

where "a/v^2" represents the effect of intermolecular forces between individual molecules and "b" represents the volume occupied by the molecules themselves. This two-parameter model allows for qualitative P-v-T prediction for many gases.

Equilibrium Specific Volume for Polymers

As noted, the polymer thermodynamic P-v-T equation of state depends on the viscoelastic response of the polymer to changes in process conditions. If the process rate of change with time is relatively long compared with the polymer response time, the polymer specific volume will be in equilibrium. Thus, characteristic P-v-T curves can be obtained (Figures 4.27 and 4.28). Note the abrupt change in the slope of the specific volume curve at about 110°C and 0.1 MPa for PS, an amorphous polymer. This is the glass transition temperature region. Note further that this transition temperature region is pressure-dependent, increasing to about 200°C at 300 MPa. LDPE shows an abrupt change in the specific volume curve slope as well, with the transition occurring at about 125°C at 0.1 MPa and about 180°C at 200 MPa. This is the region of crystallization of polyethylene. Pressure-dependent transitions were discussed in detail in Chapter 2.

Polymer Thermodynamic
Equation of State

Polymers, as elastic liquids, are frequently subjected to very high hydrostatic pressures and temperatures during processing. 100 MPa pressure is not unusual in injection molding, for example, and pressures of 500 MPa are achieved in certain compression mold-

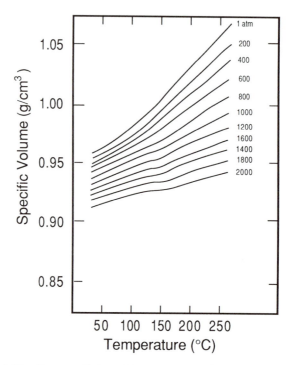

Figure 4.27. Pressure–Volume–Temperature Diagram for Polystyrene (43).

ing and forging operations. At these extreme conditions, most polymers are compressible. Spencer and Gilmore (45) proposed a modified Van der Waals equation to describe the P-v-T relationship for polymers:

$$(4.80) \qquad (P + \pi)(v - \omega) = R'T$$

where $R' = R/M^*$, and M^* is an empirical molar interaction coefficient. π is similar to the Van der Waals intermolecular force term and ω is similar to the molecular volume. However, since molecular interpretation of these coefficients is questionable at times, the Spencer–Gilmore equation should be considered only as an empirical *three-parameter equation of state*. Spencer–Gilmore coefficients for several polymers are given in Table 4.6. Note that the units are given in traditional mixed units. For certain polymers such as polyethylene and polystyrene, M^* equals the molecular weight of the polymeric mer.

Example 4.9

Q: Estimate the specific volume of polystyrene at 150°C and 4000 lb_f/in^2. The density of polystyrene at room temperature is 1.05 g/cm^3. Determine the increase in free volume at elevated temperature and pressure. Compare the results from Figure 4.28 and the Spencer–Gilmore equation.

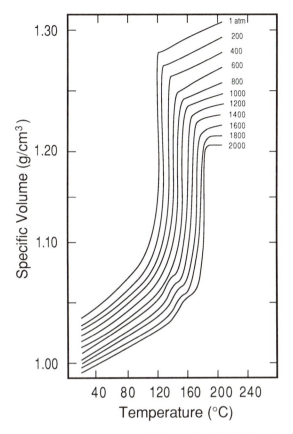

Figure 4.28.   Pressure–Volume–Temperature Diagram for LDPE (Hostalen GF 5250) (44).

A: From Figure 4.28 for $P = 4000/14.7 = 272$ atm, $v = 0.988$ cm^3/g. The density is $1/v = 1.012$ g/cm^3. Thus the free volume has increased $[1.05/1.012 - 1] \times 100 = 3.7\%$. From the Spencer–Gilmore equation:

$$v = \frac{R'T}{(P + \pi)} + \omega,$$

$$R' = 82.06/M^* = 82.06/104$$

$$\omega = 0.822 \text{ cm}^3/\text{g}$$

$$\pi = 1840 \text{ atm}$$

Therefore:

$$v = 82.06 (150 + 273)/(272 + 1840) (104) + 0.822 = 0.980 \text{ cm}^3/\text{g}$$

The increased free volume calculated by the Spencer–Gilmore equation is given as:

$$[1.05/(1/0.980) - 1] \times 100 = 2.9\%.$$

Table 4.6. Parameters for the Spencer–Gilmore thermodynamic equation of state for several polymers (45, 46, 47).

Polymer	Polymer characteristic Amorphous	Polymer characteristic Crystalline	M^* (g/mol)	π (atm)	ω_i (cm³/g)	Molecular weight of Mer (g/mol)	T_m (°C)
Polystyrene	X		104	1840	0.822	104	
Polymethyl methacrylate	X		100	2130	0.734	100	
Ethyl cellulose	X		60.5	2370	0.720	219	
Cellulose acetate butyrate	X		54.5	2810	0.688	275	
Polycarbonate	X		56.1	3135	0.669	254	
Polypropylene		X	41.0	1600	0.620	42	176
Polybutylene terephthalate		X	113.2	2239	0.712	–	–
Nylon 610		X	111	10,768	0.9064	282	–
Polyethylene (0.92 g/cm³)		X	28.1	3240	0.875	28	120
Polyethylene terephthalate		X	37.0	4275	0.574	–	–

$R = 82.06$ cm³ atm/g mol °K = 8.314 J/kg mol °K

The P-v-T relationship should be valid regardless of the thermodynamic state of a material. As noted in Chapter 2, the glass transition temperature can be seen experimentally as a change in specific volume, thermal conductivity, volume expansivity and/or specific heat of a polymer at a specific temperature. These changes are called second-order thermodynamic transitions. The Spencer–Gilmore equation does not predict second-order transitions. Crystallization is a time-dependent change in volume at a specific temperature. Since it is a phase change, it is a first-order thermodynamic transition. The Spencer–Gilmore equation cannot be used here, either. Despite these serious limitations, this relatively simple equation does qualitatively describe other thermodynamic behavior. For example, consider the differential form for P-v-T:

$$(4.81) \qquad dv = \left(\frac{\partial v}{\partial T}\right)_P dT + \left(\frac{\partial v}{\partial P}\right)_T dP$$

where $(\partial v/\partial X)_Y$ represents the volume rate of change with respect to X with Y held constant. This equation can be written as:

$$(4.82) \qquad dv/v = \kappa \, dT - \beta \, dP$$

where $\kappa = (\partial \ln v/\partial T)_P$, the volume expansivity or coefficient of thermal expansion, with units of $\deg^{-1}$, and $\beta = -(\partial \ln v/\partial P)_T$, the isothermal or bulk compressibility, with units of pressure^{-1}. Through thermodynamic manipulation:

$$(4.83) \qquad (\partial P/\partial T)_v = \kappa / \beta$$

For the Spencer–Gilmore equation:

$$(4.84) \qquad \kappa = \frac{R/v}{M*(P + \pi)}$$

$$(4.85) \qquad \beta = \frac{RT/v}{M*(P + \pi)^2}$$

and the ratio is:

$$(4.86) \qquad \kappa/\beta = (P + \pi)/T$$

For PS, as calculated from the Spencer–Gilmore coefficients, the pressure and temperature effect on coefficient of thermal expansion, κ, is shown in Figure 4.29A and the effect on bulk compressibility, β, is shown in Figure 4.29B.

When heat is added to a polymer, its temperature increases. The magnitude of increase is directly related to polymer thermodynamic properties such as enthalpy, internal energy, specific heat, and energy of crystallization and, of course, the manner in which the heat is being added from the environment. The enthalpy of a polymer, or heat content per unit mass, is referenced to some base state and is a measure of the amount of energy required to raise the polymer energy level to a proscribed condition. Enthalpic tables for many polymers are given in Appendix E. The total amount of energy at constant pressure required change the temperature of a polymer is given as:

$$(4.87) \qquad Q = m (H_2 - H_1)$$

where H_1 is the enthalpy at temperature T_1, H_2 is its enthalpy at temperature T_2 and m is the mass of the polymer. Typical curves at atmospheric pressure for amorphous polystyrene (PS) and crystalline low-density polyethylene (LDPE) are shown as Figure 4.30.

From the definition of enthalpy, the specific heats of a material can be obtained arithmetically:

$$(4.88) \qquad dH = dQ + v\, dP$$

where dH is the change in enthalpy from a base state. dQ is the change in internal energy from a base state. The specific heat at constant pressure is given as:

$$(4.89) \qquad c_p = (\partial H/\partial T)_P = (\partial Q/\partial T)_P$$

The specific heat at constant volume, c_v, is given as the rate of change of internal energy with temperature at constant volume. Usually constant pressure specific heat is used to calculate the effect of energy input on polymer temperature. Thus, wherever the term "specific heat" is used without qualification regarding pressure or volume, constant pressure specific heat, c_p, should be assumed. Typical temperature-dependent

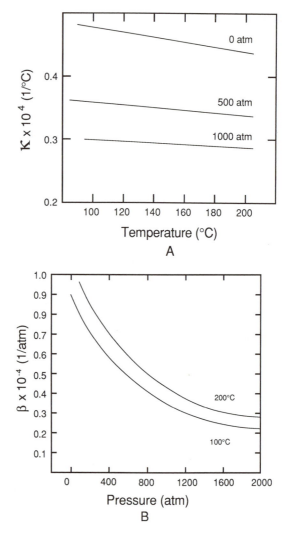

Figure 4.29. Coefficient of Thermal Expansion, κ, and Coefficient of Bulk Compressibility, β, for Polystyrene as Predicted by the Spencer–Gilmore Equation of State. A: Thermal Expansion, κ. B: Bulk Compressibility; β.

specific heat curves for several polymers including amorphous PS and crystalline LDPE are shown in Figures 4.31 and 4.32, respectively.

The change in enthalpy at constant pressure can be found if the temperature dependence of the specific heat is known:

(4.90)
$$(H_2 - H_1)_P = \int_{T_1}^{T_2} c_p \, dT$$

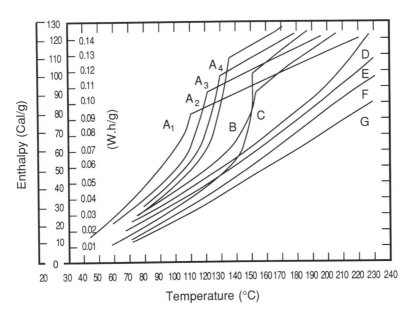

Figure 4.30. Temperature-Dependent Enthalpies of Several Commercial Polymers at Atmospheric Pressure (48). Curves A_1–A_4: PE, Increasing Density. B: PP. C: Polyacetal (POM). D: Nylon (PA). E: PS. F: PC. G: Rigid PVC (RPVC).

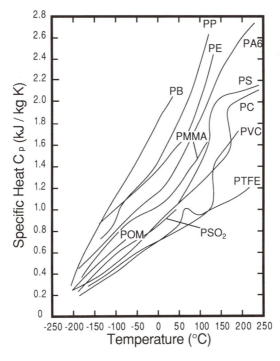

Figure 4.31. Typical Atmospheric Pressure Temperature-Dependent Specific Heats of Several Polymers. (49).

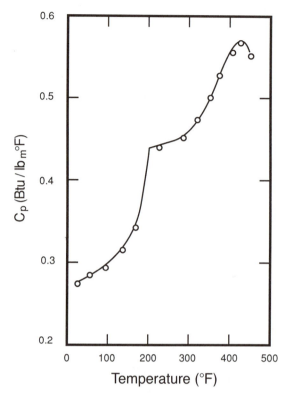

Figure 4.32. Typical Atmospheric Pressure, Temperature-Dependent Specific Heat of a Semi-crystalline Polymer, in This Case, LDPE, MW = 300,000, sp. gr. = 0.9213 (50). Discontinuity at about 200°F is Crystalline Melt Temperature.

Typically, specific heat measurements can be determined from differential scanning calorimetry (DSC), see Chapter 2. If the data can be fit with a piecewise linear curve between T_1 and T_2, as:

(4.91) $c_p = a + b T$

then:

$$(4.92) \qquad \Delta H_P = \int_{T_1}^{T_2} (a + bT) \, dT = (T_2 - T_1) \, [a + (T_1 + T_2) \, b/2]$$

Example 4.10

Q: What is the change in enthalpy of high-impact polystyrene (HIPS) when heated from 100°C to 200°C at atmospheric pressure?

A: From Reference (50):

$$c_p = -200 + 5\,T\,[\text{J/kg}^\circ\text{K}]$$
$$\Delta H_P = (473 - 373)\,[-200 + (5/2)\,(473 + 373)] = 1.915 \times 10^5 \text{ J/kg}$$

The specific heat of a polymer containing particulates or reinforcing elements can be obtained from a sum of the mass-weighted specific heats of the components. Thus for two components, polymer and filler:

(4.93) $$c_{p,\text{mix}} = (m_p/m_{\text{total}})\,c_{p,p} + (m_f/m_{\text{total}})\,c_{p,f}$$

where m_p and m_f are the masses of polymer and filler, $m_{\text{total}} = m_p + m_f$, and $c_{p,p}$ and $c_{p,f}$ are the specific heats of polymer and filler. For a multicomponent system:

(4.94) $$c_p = \sum_i^n (m_i\,/\,m_{\text{total}})\,c_{p,i}$$

In terms of volume fractions, where ϕ_i is the volume fraction of the ith component:

(4.95) $$c_p = \sum (\rho_i\,\phi_i\,c_{p,i})/\sum \rho_i\,\phi_i$$

Note that an experimental measurement of the constant-volume change of pressure with temperature yields a way of obtaining the constant π, the first Spencer–Gilmore constant. Integration of isothermal enthalpy with regard to pressure yields:

(4.96) $$(H - H_o)_T = \int_{P_o}^{P} \left[v - T\left(\frac{\partial v}{\partial T}\right)_P \right] dP$$

Substituting β and the definition for v into this equation, the integration yields:

(4.97) $$(H - H_o)_T = \omega\,(P - P_o)$$

Thus, an isothermal change in enthalpy with pressure yields a value for ω, the second Spencer–Gilmore constant. The third, M^*, can then be obtained from any value of v at P and T. Griskey and co-workers constructed standard enthalpic and entropic tables for many common polymers. These are given in Appendix E. Example 4.11 illustrates how these can be used to determine many thermodynamic properties of these polymers.

Example 4.11

Q: This question deals with the thermodynamics of polypropylene (PP) and uses the thermodynamic table for PP in Appendix E. There are six questions.

Q1: Determine the amount of energy needed to heat PP from 80°F and 14.7 lb_f/in^2, absolute (1 atm) to 400°F and 9000 lb_f/in^2.

Q2: From the question above, determine the fraction of energy required to pressurize PP from 1 atm to 9000 lb_f/in^2 at 400°F.

Q3: Determine the heat capacity of PP at 420° and 14.7 lb_f/in^2, absolute.

Q4: Determine the heat capacity of PP at 320°F and 14.7 lb_f/in^2, absolute.

Q5: Comment on the differences in the two values of Q3 and Q4.

Q6: Determine the minimum horsepower required to compress 1 lb of 400°F PP in a compression molding press from 14.7 lb_f/in^2, absolute to 9000 lb_f/in^2 in 1 s. 1 HP = 1.413 Btu/s.

A: Again, there are six answers.

A1: $H_{400,9000}$ = 276.99 Btu/lb

$H_{80,14.7}$ = 22.4 Btu/lb

ΔH = 254.59 Btu/lb.

A2: Energy needed to compress 400°F PP isothermally = 276.99 − 254.1

= 22.89 Btu/lb

Since the total energy required is 254.59 Btu/lb, the fraction of energy required to compress is:

$$22.89/254.59 = 0.09$$

or 9% of the total.

A3:

$$c_p = \Delta H/\Delta T = (H_{440,14.7} - H_{400,14.7})/\Delta T$$
$$c_{p,420,14.7} = (278.4 - 254.1)/(440 - 400)$$
$$= 0.61 \text{ Btu/lb °F}$$

A4:

$$c_{p,320,14.7} = (216.9* - 146.1)/(360 - 280)$$
$$= 0.885 \text{ Btu/lb °F.}$$

A5: The melting temperature of PP is approximately 320°F. Therefore the enthalpy value at 280°F does not include the latent heat of melting, whereas that at 360°F does. The literature value for the heat capacity of PP away from transition temperatures is usually reported as 0.6 Btu/lb °F.

A6: [See also Q2/A2]

$$\Delta H = H_{9000} - H_{14.7}$$

$$= 276.99 - 254.10$$

$$= 22.89 \text{ Btu/lb}$$

$$\text{Energy to compress} = 22.89 \times 1 \text{ lb/1 s}$$

$$= 22.89 \text{ (Btu/s) } / 1.413$$

$$= 16.2 \text{ HP}$$

---

*This value interpolated from 400 and 320F data.

One additional relationship is worth considering. The total change of entropy, S, is given in terms of volume change and specific heat as:

$$(4.98) \qquad T \, dS = c_p \, dT - T \, (\partial V / \partial T)_P \, dP$$

In terms of volume expansivity, κ:

$$(4.99) \qquad T \, dS = c_p \, dT - T \, v \, \kappa \, dP$$

Consider an isothermal system. Now $dT = 0$. Furthermore, at near-equilibrium conditions, $dQ \approx T \, dS$. The resulting equation can be written as:

$$(4.100) \qquad Q = -T \int v \, \kappa \, dP \approx -T \, \kappa \int v \, dP$$

For polymers, $v = v(P)$ is a rather weak function. Therefore, as an approximation, $v \approx v_{avg}$. Then:

$$(4.101) \qquad Q \approx -T \, \kappa \, v_{avg} \, P$$

This is a measure of the amount of heat liberated during isothermal compression of a polymer.

For an isentropic process, where no heat is added or removed, $dS = 0$. So:

$$(4.102) \qquad c_p \, dT = T \, v \, \kappa \, dP$$

For a small change in pressure, the change in temperature is seen to be directly proportional to the volume expansivity and inversely proportional to the specific heat of the polymer. Of course, this must occur at temperatures away from any first or second-order thermodynamic changes.

Certainly, more elegant thermodynamic equations of state have been proposed. Most require the addition of at least one additional constant (51). As a result, at least one additional experiment is needed to obtain a working P-v-T relationship for any given polymer.

Non-Equilibrium Specific Volume for Polymers

Most non-polymeric materials respond to changes in pressure or temperature by instantaneous specific volume change. On rare occasions, non-equilibrium conditions can exist for simple materials. For example, steam issuing at high speed from a nozzle can be momentarily supersaturated at the nozzle mouth. Under many processing conditions, the viscoelastic nature of a polymer can inhibit thermodynamic transitions, particularly when temperatures and pressures are changing rapidly. For example, if the response time of the polymer is much greater than the rate of change in processing conditions, the specific volume will continue to change for some time after the processing conditions are stabilized. Retarded dimensional change is sometimes called the *dimensional stability* of the polymer. One measure of non-equilibrium thermodynamic change in specific volume is volumetric shrinkage, defined as:

$$(4.103) \qquad S_v = (v_m - v_f)/v_m = 1 - v_f/v_m$$

where v_f and v_m are the specific volumes of the polymer at use condition and at the processing condition, respectively. For injection molding, the processing condition is that in the mold cavity when the gate freezes (see Chapter 5). One way of calculating linear shrinkage is to assume volumetric shrinkage is isotropic. Thus:

$$(4.104) \qquad S_L = 1 - L_f/L_m = 1 - (v_f/v_m)^{1/3} = 1 - (1 - S_v)^{1/3}$$

If the cube root is expanded in series form, the linear shrinkage can be approximated as:

$$(4.105) \qquad S_L = (1/3) S_f + \text{higher order terms}$$

Of course, shrinkage values obtained using specific volume changes are only approximate since geometric considerations, molecular orientation, anisotropy, non-equilibrium cooling and other effects are neglected.

Thermodynamics of Rapid Cooling

The rapid cooling of a polymer results in a thermodynamic non-equilibrium state. The equations of state presented above cannot be used to estimate the specific volume of a polymer. Consider the quenching of an amorphous polymer initially in the softened or fluid state.* If the polymer is slowly cooled at constant pressure, its specific volume would follow curve A–B in Figure 4.33. Given the proper coefficients, the specific volume could be predicted from the Spencer–Gilmore equation or an equivalent equation of state. On the other hand, if the polymer is quenched in a water bath or on chill rolls, there is insufficient time for molecular relaxation before the molecule mobility is inhibited. The resulting specific volume curve is shown as A–C in Figure 4.33. As noted above, the volume change from C to B can occur at use conditions and is a measure of the dimensional stability of the polymer and the product made of it. Note that the specific volume change from A to C can be considered to be a result of an instantaneous constant pressure process. Consider first the isobaric cooling process. Write the specific volume as:

$$(4.106) \qquad v = v(T, t)$$

---

*This analysis follows that in McKelvey (53).

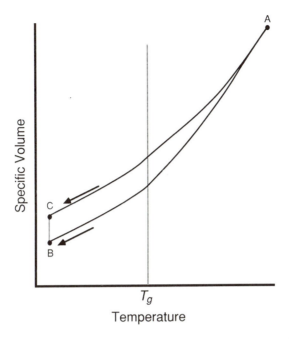

Figure 4.33. Schematic Volume–Temperature Curves of an Amorphous Polymer Cooled Through its Glass Transition Temperature (52). The Rate of Cooling can Affect the Instantaneous Specific Volume of the Polymer at Room Temperature.

where T is temperature and t is time. Now the differential form is:

$$(4.107) \qquad dv = \left(\frac{\partial v}{\partial t}\right)_T dt + \left(\frac{\partial v}{\partial T}\right)_t dT$$

This can be written in terms of dv/dT:

$$(4.108) \qquad \frac{dv}{dT} = \left(\frac{\partial v}{\partial T}\right)_t + \left(\frac{\partial v}{\partial t}\right)_T \left(\frac{1}{r}\right)$$

where $r = dT/dt$, the rate of cooling. Experimentally, softened PS is subjected to a step change in temperature (54). An instantaneous change in specific volume results, followed by a gradual change as a function of time. This response is shown in Figure 4.34. The initial change is shown as χ, and the time-dependent change is shown as:

$$(4.109) \qquad (\partial v/\partial t)_T = \gamma(v_e - v)$$

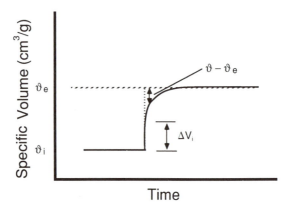

Figure 4.34.   Schematic of Time-Dependent Change in Specific Volume of an Amorphous Polymer Such as PS to a Step Change in Environmental Temperature (55).

where γ is the proportionality constant, called an isothermal rate constant, and v_e is the specific volume at equilibrium. The Spencer–Gilmore equation of state can be written for constant pressure as:

$$(4.110) \qquad\qquad v = aT + b$$

With appropriate differentiation and substitution:

$$(4.111) \qquad\qquad dv/dT = \chi + (\gamma/r)(aT + b - v)$$

Note that under intense quenching, r approaches $-\infty$. Therefore the second term on the right approaches zero and:

$$(4.112) \qquad\qquad dv/dT = \chi$$

That is, the final specific volume change simply equals the initial instantaneous change. Note from Figure 4.34 that:

$$(4.113) \qquad\qquad v_A - v_C = \chi(T_A - T_C)$$

and:

$$(4.114) \qquad\qquad v_A - v_B = a(T_A - T_B)$$

Therefore:

$$(4.115) \qquad\qquad v_C = v_B + (a - \chi)(T_A - T_B)$$

The first term on the right is the equilibrium volume. The second represents the volume change upon relaxation. Thus, the maximum amount of post-mold shrinkage or dimensional change owing to a step change in temperature for an amorphous polymer is seen to be a function of the isothermal rate constant, γ, and the degree of quench.

To obtain a general solution to the equation above, a temperature-dependent expression for γ is needed. Experimentally, it has been shown that:

(4.116) $$\gamma = K \exp(-C/T)$$

where K has the units of time^{-1} and C has the units of temperature. Although a closed analytical solution is not available, numerical results indicate that substantial deviation from equilibrium specific volume occurs at relatively low cooling rates (Figure 4.35).

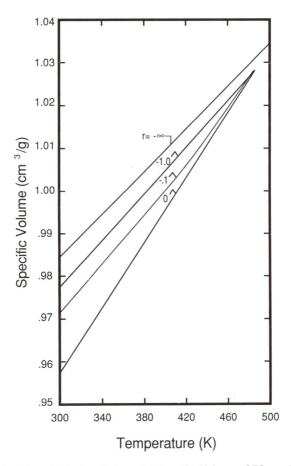

Figure 4.35.   The Effect of Cooling Rate on the Specific Volume of PS as a Function of Final Temperature. r is the Cooling Rate. $K = 18.3$ h^{-1}. $C = 6731°$K. $\chi = 2.5 \times 10^4$ cm^3/g $°$K. $b = 0.845$ cm^3/g. $a = 3.75 \times 10^{-4}$ cm^3/g $°$K. $T_o = 500°$K. $v_e = 1.0325$ cm^3/g.

Note that molded parts are frequently annealed to minimize residual stresses that occur during the molding process and are frozen in by rapid cooling. If the polymer specific volume is substantially far from equilibrium, substantial volumetric changes can also occur during annealing. This effect is extremely important in the quality control of injection molded parts, regardless of whether the polymer is amorphous or crystalline. It is also important if the polymer has not reached crystalline equilibrium and the product is to be used at temperatures that approach "cold crystallization temperature", that is, the temperature at which the rate of crystallization becomes significant. And it is also important in extrusion, where dimensions measured at the cutoff saw or the take-up rolls can change during storage of the product.

Isothermal Expansion

The volume change C to B in Figure 4.33 is isothermal expansion. The volume change as a function of time is obtained by integrating:

(4.117)
$$(\partial v/\partial t)_T = \gamma(v_e - v)$$

This yields:

(4.118) $\quad (v_e - v)/(v_e - v_i) = (v_B - v)/(v_B - v_C) = \exp[-\gamma t]$

As is apparent, γ is a time constant for a first-order response to a step change in conditions. As noted, γ is usually experimentally determined. If the exponential relationship holds, the approach to equilibrium specific volume is exponentially dependent on a coefficient that is exponentially dependent on temperature. Thus, even under no-load conditions, dimensions of molded products can continue to change even after long periods of time. In this case, the cause is simply the result of the time-dependent characteristics of polymers in a thermodynamic sense.

Example 4.12

Q: Consider the dimensional stability of polystyrene sheet extruded at 500°K and quenched in a water bath at 300°K. Determine the specific volume immediately after quench. Then determine the specific volume as a function of time. The initial specific volume is given from:

$$v_e = 3.75 \times 10^{-4} T + 0.845$$

A: The equilibrium specific volumes at 300°K and 500°K are:

$$v_{e,500} = 1.0325 \text{ cm}^3/\text{g}$$
$$v_{e,300} = 0.9575 \text{ cm}^3/\text{g}$$

Assume that the rate of temperature drop during quench is very rapid. The specific volume immediately after the quench can be read from Figure 4.35 or calculated from:

$$v_C = v_B + (a - \chi)(T_A - T_B)$$
$$= 0.9575 + (3.75 - 2.5) \times 10^{-4} (500 - 300)$$
$$= 0.9825 \text{ cm}^3/\text{g}$$

The rate of contraction between points A and C can be described as:

$$(v_B - v)/(v_B - v_C) = \exp[-\gamma t]$$

where the rate constant, γ, is given as:

$$\gamma = 1.8 \times 10^8 \exp[-6731/T]$$
$$\gamma_{300} = 0.033 \text{ h}^{-1}$$

Thus the time-dependent specific volume is given as:

$$v = 0.9575 + 0.025 \exp[-0.033 t]$$

If the sheet dimensions are measured upon leaving the water bath, the linear shrinkage is approximated by:

$$S_L \approx (1/3)(1 - v/v_0)$$
$$\approx (1/3)[1 - (0.975 + 0.025 \exp[-0.033 t])/0.9825]$$

To complete 96% of the overall volume change, $v_C - v$, requires about 60 h.

For crystalline and semicrystalline polymers, the effect of time- and temperature-dependent crystallization must be included. Consider, in words, the mechanisms involved during rapid cooling. Crystallization is usually thermodynamically exothermic. This means that as a polymer crystallizes, heat is liberated. If the rate of crystallization is rapid enough, the rate of cooling is reduced, stopped, or in some cases, reversed. Thus, r in Example 4.12 is no longer a constant. The classical model for crystallization kinetics is based on volumetric change during crystallization and is called the Avrami model (see Chapter 2). However, the rate constants for this model are usually obtained from isothermal experiments. Polymers that have very high crystallization rates can be quenched very rapidly to produce products such as film or sheet with relatively little crystallinity. In the amorphous regions, these polymers do experience the time-dependent specific volume changes described above, however. If the temperature of such a polymer is raised, crystallization begins or continues and a significant volume change takes place, usually with substantial changes in other properties, such as haze, toughness and tear resistance.

Since polymers apparently are rarely in true thermodynamic equilibrium, the time-dependent nonisothermal thermodynamic analysis of a crystallizing polymer is usually not approached using the simple models considered above. Instead, statistical non-equilibrium thermodynamics seems to yield a more appropriate approach. This subject is beyond the intent of this work, however (56–58).

Energy of Crystallization

Note in Figure 4.30 that for an amorphous polymer such as PS, the enthalpy curve exhibits an experimentally measurable discontinuity in slope as polymer temperature increases from room condition to fluid processing conditions. This discontinuity occurs at the polymer glass transition temperature. The specific heat at constant pressure, being the slope of the enthalpy–temperature curve, shows a discontinuity at the glass transition temperature. Thermodynamically, this is known as a *second-order transition*. Amorphous polymers exhibit no phase change. On the other hand, a crystalline material such as LDPE (Figure 4.30) exhibits a distinct discontinuity in the temperature-dependent enthalpy at its melt temperature. The specific heat curve shows a distinct singularity at the melt temperature, since the rate of change of enthalpy at that temperature is infinite. The melt temperature condition is thermodynamically a *first-order transition*. Crystalline polymers also show specific heat discontinuities at their glass transition temperatures.*

As detailed in Chapter 2, the differential scanning calorimeter (DSC) or its variants can be used to determine accurately the region of glass transition, the crystalline melting temperature, and the amount of energy expended during each phase transition. Typically, the device measures time-dependent material temperature while heating or cooling it at a fixed time rate of change of temperature, in °C/min. This is then compared with the response of a standard to determine specific heat and enthalpic changes. The 10°C/min heating rate DSC curves for three different polyethylenes are shown in Figure 4.36. The regions where the curves peak represent the temperature ranges of most rapid melting. Similarly, these materials were cooled at 10°C/min, and these DSC cooling curves are seen in Figure 4.36. Again the regions in which the curves peak out represent the locations of greatest rate of crystallization. Note for example that MDPE crystallites melt fastest at about 125°C but reform at about 120°C. This point is discussed in detail in Chapter 2. Typically, the temperature of maximum crystallizing rate, during cooling, is below the melting temperature even though the areas under the curves, being the energy of crystallization, ΔH_c, are nearly identical.** As a rule of thumb, the crystallization temperature is about 0.9 times the melt temperature, in absolute units.

---

*Note that the glass transition temperature for LDPE is about $-100°C$, far below the lowest temperature shown in Figure 4.32.

**Crystallization is a rate-dependent phenomenon. If the cooling or heating rates are of the same order of magnitude as the crystallization or melting rate, *kinetic* equilibrium will not be achieved. As a result, heating and cooling can yield *slightly* different areas under the time-dependent thermal curves. For further details, see the discussion of differential scanning calorimetry in Chapter 2.

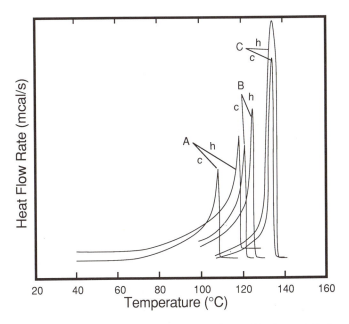

Figure 4.36.   Comparison of Differential Scanning Calorimetry Heating and Coding Curves for
Polyethylenes of Different Densities (59). Curves A: Low-Density PE. B: Medium-Density PE.
C: High-Density PE. All Samples Weighed 7.1 mg. Range: 10 mcal/s. Scan Rate: 10°C/min. h:
Heating. c: Cooling.

The morphology of crystallization is discussed in detail in Chapter 2. The energy
necessary to change the highly ordered crystalline state to a near-random "fluid" state
is called the latent heat of melting for non-polymer materials. For polymers, it is the
energy of melting. Conversely the energy liberated during reordering from the random
state to the crystalline state is called the latent heat of fusion for non-polymer materials
and, for polymers, it is the energy of crystallization. DSC measurements show that the
energy of melting and that of crystallization are nearly identical. The theoretical energy
of crystallization can be calculated (see Table 2.7). The value obtained by measuring
the area under the curve then represents a measure of the degree of crystallization. For
example, as shown in Figure 4.37, the area is given as 54.5 cal/g for HDPE. The
energy of crystallization to 100% is calculated to be 68.4 cal/g. Therefore the initial
sample of HDPE is 79.7% crystalline. If the liquid HDPE is then cooled at essentially
the same cooling rate, the area under the curve will yield a value of 54.5 cal/g, thus
returning the sample to its original 79.7% crystallinity. As noted in Chapter 2, the DSC
heating or cooling rate is, for the most part, governed by laboratory practice. Ideally,
temperature should be changed as slowly as possible to allow for the crystallization
kinetic changes to take place in near-isothermal conditions. Practically, it is necessary
to examine as many samples as possible in a given time frame. The typical DSC scan
range is 5 to 40°C/min. Note that accurate analytical testing requires repeated scans at
slower and slower scan rates until the peak temperatures (for transition) and areas under

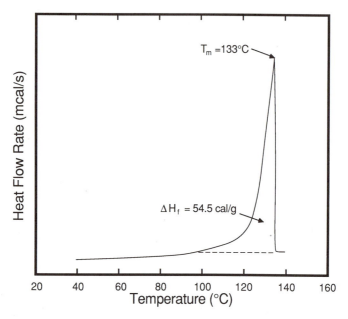

Figure 4.37.   Differential Scanning Calorimetry of the Crystallization Region of High-Density Polyethylene (HDPE) (59). Sample Weight: 7.1 mg. Range: 10 mcal/s. Scan rate: 10°C/min. Peak Temperature: T_m = 133°C. Area Under DSC Curve, ΔH_f = 54.5 cal/g. Heat of Fusion for HDPE at 100% Crystallinity = 68.4 cal/g. Thus Percent Crystallinity is 100 × 54.5/68.4 = 79.7%.

the curves do not change from scan to scan. From a practical processing viewpoint, even at the highest scan rates, DSCs cannot heat or cool fast enough to duplicate primary processing effects as shear heating or quench cooling. The DSC can be used to simulate certain types of annealing and reheating for secondary processing, however.

4.7 Polymer Solution Thermodynamics

The section that follows includes a brief summary of polymer solution thermodynamics. This terminology is needed to ensure a better understanding of the relationship between a polymer and smaller molecules such as gases and liquids in solubility and permeation. For more details in polymer solution thermodynamics, the reader is referred to References 60 to 62.

Free Volume

As noted, polymers do not or cannot pack tightly, owing to steric hindrance, stiffness of the polymer backbone, or repulsion of elements along the backbone. The theoretical densities of certain polymers are given in Table 4.7, along with typical processing

Table 4.7. Crystalline and amorphous densities for several polymers. Crystalline densities calculated from assumed crystal structure.

Polymer	Crystal system	Density (g/cm³)	
		Crystalline	Amorphous
Hydrocarbons			
Polyethylene	Ortho	1.00	0.852
		1.014	
	Pseudo-Mono	0.965	
Polypropylene	Mono	0.936	0.85
		0.937	
Polystyrene	Rho	1.11	1.04 to 1.065
Poly-α-butylene	Rho	0.95	0.87
1,2 Polybutadiene, syndiotactic	Ortho	0.963	
1,2 Polybutadiene, isotactic	Rho	0.96	
Poly-3-methyl butene-1	Mono	0.93	
Poly-4-methyl pentene-1	Tet	0.813	
Polar Vinyls			
Polyvinyl chloride	Ortho (assumed)	1.44	< 1.39
Polyvinylidene chloride	Mono	1.94	1.66
	Mono	1.959	
Polytetrafluoroethylene			
< 20°C	Pseudo-hex	2.40	
> 20°C	Hex	2.36	
Polychlorotrifluoroethylene	Hex	2.10	2.08
Polyvinyl alcohol	Mono	1.35	1.291
Polymethyl methacrylate, isotactic	Pseudo-ortho, tri	1.23	1.22
Polymethyl methacrylate, syndiotactic	—	—	1.19
Polyvinylidene bromide	Mono	3.065	
Others			
Polyethylene terephthalate	Tri	1.455	1.335
Polyoxymethylene	Hex	1.506	1.25
Polypropylene oxide	Ortho	1.096	
Polycarbonate, bisphenol A	Ortho	1.30	
Nylon 66 (α)	Tri	1.24	1.09
		1.220	1.069
Nylon 66 (β)	Tri	1.25	
Nylon 6	Mono	1.23	
	Mono	1.21	

Ortho = Orthorhombic　　Mono = Monoclinic
Rho = Rhombohedral　　Tet = Tetragonal
Hex = Hexagonal　　Tri = Triclinic

Table 4.8.   Properties of nylon 6, 12 and 66 (PA 6, 12, and 66). Dry and at 50% relative humidity. (63).

Property	Nylon 66		Extracted Nylon 6		Nylon 12	
	Dry	50% RH	Dry	50% RH	Dry	50% RH
Tensile Stress (lb$_f$/in^2) at:						
$-40°F$ ($-40\%C$)	15,700	14,900	17,400	–	–	–
73°F (23°C)	12,000	11,200	11,800	10,000	8,000	7,600
170°F (77°C)	9,000	5,900	9,900	8,500	–	–
Yield stress (lb$_f$/in^2) at:						
$-40°F$ ($-40°C$)	15,700	14,900	17,000	–	11,500	–
73°F (23°C)	12,000	8,500	11,800	6,400	7,500	5,900
170°F (77°C)	6,500	5,900	5,900	4,000	3,500	–
Elongation at break (percent) at:						
$-40°F$ ($-40°C$)	20	20	5	–	–	–
73°F (23°C)	60	300	200	300	250	250
170°F (77°C)	340	350	310	325	–	–
Elongation at yield (percent) at:						
$-40°F$ ($-40°C$)	4	–	–	–	–	–
73°F (23°C)	5	25	–	–	10	20
170°F (77°C)	30	30	–	–	–	–

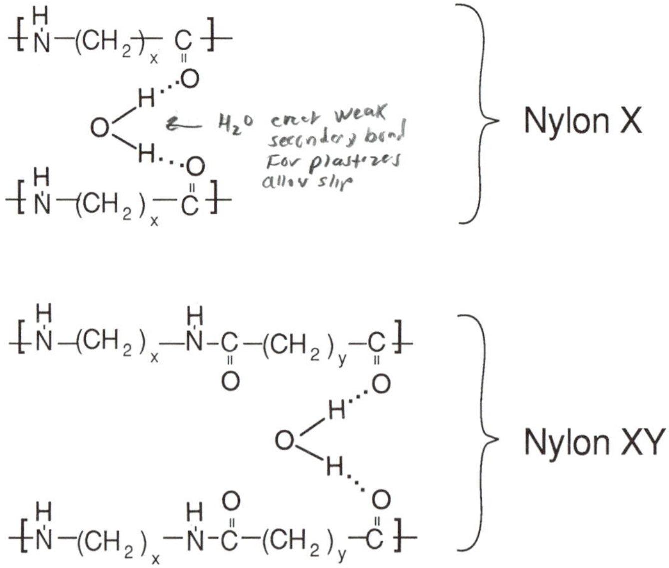

Figure 4.38.   Water, as a Plasticizer, Shown Hydrogen-Bonded to the Nylon Backbone.

Table 4.9. A partial list of small molecules that plasticize certain polymers.

Polymer	Plasticizer
Nylon	Water
High-density polyethylene	Oil
Cellulosics	Water
Polyvinyl chloride	Phthalates, such as dioctyl

densities. The space around the molecules is called "free volume". It can be devoid of matter, but frequently it is occupied by small molecules (nitrogen, oxygen, water). In some cases, the nature of the small molecule can affect the mechanical properties of the polymer. The classic example is water and polyamide (nylon), Table 4.8. The water is partially bound to the amine ($-NH-$) group on the polyamide backbone. The bonding is "hydrogen bonding" and is shown in Figure 4.38. The water molecules further separate the polymer chains. Under load, the smaller water molecules tend to lubricate the longer chains. This is called *plasticizing*. The polymer has lower modulus and strength but greater elongation as well as improved ductility and impact toughness. Other polymers that are plasticized by smaller molecules are given in Table 4.9.

Swelling

In addition to plasticizing, small molecules can swell polymers. Three-dimensional polymers, thermosets and crosslinked thermoplastics, are characteristically swelled by certain "solvents". The classic example is benzene swelling of natural rubber. Other examples are listed in Table 4.10.

Table 4.10. A partial list of small molecules that swell certain polymers..

- Crosslinked polyamides (tetralinked and octalinked poly-ε-caproamide) in *m*-cresol at 30°C.
- Polystyrene crosslinked with divinyl benzene, in 0.4N potassium hydroxide at 25°C.
- Polymethacrylic acid crosslinked with divinyl benzene in 0.1N sodium hydroxide at 30°C.
- High-density polyethylene crosslinked with dicumyl peroxide, in refluxing xylene.
- Polypropylene electron beam-crosslinked, in refluxing hexane.
- The following substances are swelling agents with natural rubber:
  Carbon tetrachloride
  Chloroform
  Carbon disulfide
  Toluene
  Petroleum ether
  n-Propyl acetate
  Ethyl acetate
  Methyl ethyl ketone
  Acetone at 25°C

Polymer–Solvent Solution

When in contact with excess solvents, uncrosslinked polymers may swell initially but are more likely to dissolve rapidly into solution. The polymer chain segment inter-molecular forces are weak when compared with polymer–solvent interaction forces. Small molecule solute–solvent interaction has been studied for nearly 200 years. If p^*_a is the vapor pressure of component A and x_a is its mole fraction, then Raoult's law states that:

$$(4.119) \qquad p_a = p^*_a\, x_a$$

where p_a is the partial pressure of component A. For a binary, solute–solvent system (A and B), the total pressure, $P = p_a + p_b$, or:

$$(4.120) \qquad P = p_a + p_b = p^*_a\, x_a + p^*_b\, x_b$$

but since $x_b = 1 - x_a$:

$$(4.121) \qquad P = p_a + p_b = p^*_a\, x_a + p^*_b\, (1 - x_a)$$

The partial pressures of an ideal solution are shown in Figure 4.39. Very few simple molecule solute–solvent pairs can be considered to be *ideal solutions*. The actual partial pressure curves for a near-ideal solution, carbon disulfide–acetone, are shown in Figure 4.40. For non-ideal solutions, the ratio, p_i/p^*_i is called the "activity". For polymeric materials, the mole fraction notation, x_i, is replaced with volume fraction, v_i. Nevertheless, for an ideal solution, a plot of activity against volume fraction should be linear, that is, $a_i = v_i$. As seen for benzene in rubber (Figure 4.41), the curve is decidedly non-linear.

Another measure of solution ideality follows from thermodynamics. For an ideal solution, the entropy of mixing is given as:

$$(4.122) \qquad \Delta S = -R \sum n_i \ln x_i$$

where n_i is the number of moles of species i, that is, $x_i = n_i/n_T$, where n_T is the total number of moles. The free energy of mixing is given as:

$$(4.123) \qquad \Delta F = \Delta H - T\,\Delta S$$

where ΔH is the enthalpy (or heat) of mixing. For an ideal solution, $\Delta H = 0$, and since ΔS exceeds zero, ΔF is less than zero, and solution results. If ΔH is large positive, ΔF can be zero or positive, and no solution results. Typically for polymer–solvent systems, if a positive molecular attraction between polymer and solvent can be achieved, ΔH is negative and so thus is ΔF and solution is achieved. For polar polymers and polar solvents, the strong molecular attraction is hydrogen bonding. On the other hand, for non-polar solvents, the solvating power is primarily due to dispersion forces

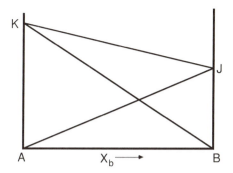

Figure 4.39.   Partial Vapor Pressures of Ideal Solution (64).

that can be relatively weak when compared with polymer–polymer attraction forces. As a result, solvation may or may not occur. One measure of the solvating potential of a polymer–solvent pair is the solubility parameter, δ:

$$(4.124) \qquad\qquad \Delta H = v_1\, v_2\, (\delta_1 - \delta_2)^2$$

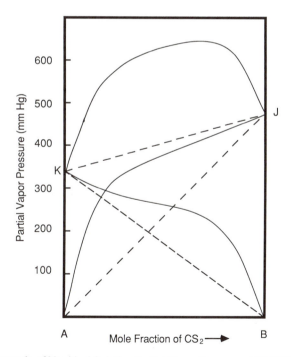

Figure 4.40.   Example of Nonideal Solution Partial Vapor Pressures at 35.2°C (65). Component A: Acetone. B: Carbon Disulfide.

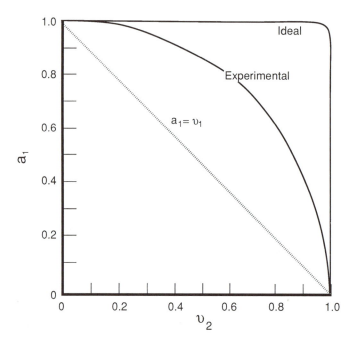

Figure 4.41.   The Activity of Benzene, a_1, in Solution with Natural Rubber (MW = 280,000) as a Function of the Volume Fraction, v_2, of Rubber (66). Dashed Line: Equating Activity with Volume Fraction of Solvent (Benzene).

where v_i is the volume fraction of the species and δ_i is its solubility parameter (Table 4.11). δ^2 is also known as the *cohesive energy density* or the *volumetric vaporization energy*.

"Good" solvents are identified as those in which polymer–solvent interactions are thermodynamically favored. This implies that individual polymer chains extend into the solvent. As dissolution continues, polymer chains become surrounded by solvent and solution occurs. Of course, polymer chains are never fully extended but usually reside as coils in random sizes. In "poor" solvents, the chains are coiled tighter than in good solvents. [For non-solvents, there is no appreciable polymer chain separation.]

Interaction Parameter

For polymers the enthalpy of mixing is given as:

(4.125) $$\Delta H = \chi_1 \, RT \, n_1 \, v_2$$

Table 4.11. Cohesive energy densities for some polymers and solvents (67).

Solvent	Cohesive energy density, δ_2 $(cal/cm^3)^{1/2}$	Polymer	Cohesive energy density, δ_2 $(cal/cm^3)^{1/2}$
n-Hexane	7.24	Polyethylene	7.9
Carbon tetrachloride	8.58	Polystyrene	8.6
2-Butanone	9.04	Polymethyl methacrylate	9.1
Benzene	9.15	Polyvinyl chloride	9.5
Chloroform	9.24	Polyethylene	
		terephthalate	10.7
Acetone	9.71	Nylon 66	13.6
Methanol	14.5	Polyacrylonitrile	15.4

where χ_1 is a polymer-specific interaction energy per solvent molecule.* The free energy of mixing can be defined in terms of χ_1 as:

$$(4.126) \qquad \Delta F = RT(n_1 \ln v_1 + n_2 \ln v_2 + \chi_1 n_1 v_2)$$

This can be written in terms of a partial molar free energy of mixing:

$$(4.127) \qquad \overline{\Delta F}_1 = RT [\ln (1 - v_2) + (1 - 1/x) v_2 + \chi_1 v_2^2]$$

where "1" is the polymer and "2" is the solvent. This expression can be used to determine osmotic pressure as a function of solvent concentration (68). From this, the *polymer* molecular weight can be obtained. The expression contains only one polymer–solvent interaction parameter, χ_1.

Historically, χ_1 was first evaluated for the binary pair, benzene–natural rubber. It was found to be independent of concentration and thus was considered to be a true material constant. As seen in Figure 4.42, χ_1 is probably linearly dependent on solvent concentration, v_2.

The interaction parameter, χ_1, is related to the cohesive energy densities of the polymer–solvent pair by:**

$$(4.129) \qquad \chi_1 = \beta_1 + (v_1/RT) (\delta_1 - \delta_2)^2$$

---

*χ_1 can be written as:

$$(4.128) \qquad \chi_1 = B v_1/RT$$

where v_1 is the molar volume of the solvent and B is an interaction energy density that is polymer–solvent pair specific, since it is a function of the molecular volume occupied by a polymer chain segment. Note then that χ_1 is inversely dependent upon solution temperature.
**This is deduced from the definitions of enthalpy of mixing and that for χ_1.

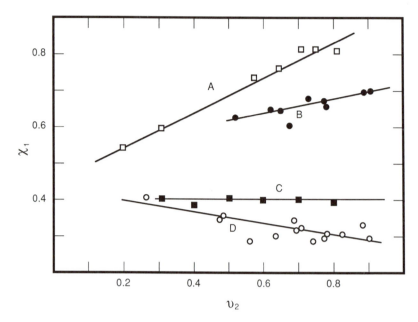

Figure 4.42. Volume-Fraction Experimental Values of χ_1 for Several Polymers (69). Curve A: (MW = 3850) Polydimethylsiloxane in Benzene. B: PS in Methyl Ethyl Ketone. C: Natural Rubber in Benzene. D: PS in Toluene.

where β_1 is a lattice constant ($\beta_1 = 0.35 \pm 0.1$) (45). The loose relationship is shown for butyl rubber in ten solvents in Figure 4.43. For polar polymer–solvent pairs, hydrogen bonding can influence solubility. In Figure 4.44, the swelling regions of cross-linked fluorocarbon rubber are compared with the solvent solubility parameter and a hydrogen bonding index (Table 4.12). The hydrogen bonding index is zero for nonpolar solvents such as benzene, hexane and other hydrocarbons, 18.7 for many aliphatic alcohols and 39 for water.

Example 4.13

Q: From the discussion above, solvation can occur when $\chi_1 \leq 0.5$. For polymethyl methacrylate (PMMA), determine the cohesive energy density range for a good solvent at 27°C.

$$R = 8.314 \text{ J/g mol } °\text{K}$$

$$\rho = 1.05 \text{ g/cm}^3$$

$$MW = 100 \text{ g/g mol}$$

$$1 \text{ cal} = 4.186 \text{ J}$$

$$\beta_1 = 0.35$$

$$\delta_1 = 8.6 \text{ (cal/g)}^{1/2} \text{ from Table 4.11}$$

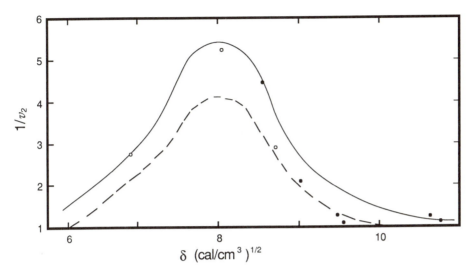

Figure 4.43.   Swelling of Butyl Rubber in Ten Solvents Having a Wide Range of Solubility Parameters (70). $1/v_2$, Specific Volume of Butyl Rubber. δ, Cohesive Energy Density. Solid Line: $v_1 = 80$. Dashed Line: $v_1 = 120$.

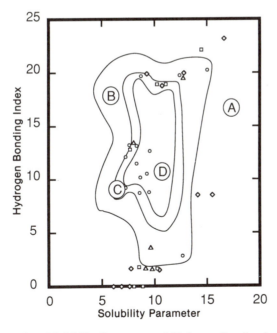

Figure 4.44.   Cross-Plot of Solubility Parameter and Hydrogen Bonding Index for Crosslinked Fluorocarbon in Various Solvents (71). Maximum Swelling Occurs When Solubility Parameter and Hydrogen Bonding Index of Solvent Closely Match Those of Polymer. Zone A: Nonswell Region. B: 25–50% Swell. C: 50–100% Swell. D: > 100% Swell.

Table 4.12. Characteristic solubility parameters for some solvents at 25°C (72).

Solvent	Solubility parameter, Hildebrands, $(cal/cm^3)^{1/2}$	Hydrogen bonding index	Molecular weight	Density (g/cm^3)
Nitrobenzene	10.0	3.2	123.11	1.200
Nitroethane	11.1	3.1	75.07	1.045
Nitromethane	12.7	3.1	61.04	1.131
1-Nitropropane	10.3	3.1	89.10	0.996
2-Nitropropane	9.9	3.1	89.10	0.983
n-Pentane	7.0	2.2	72.15	0.626
Perchlorethylene (or tetrachloroethylene)	9.3	2.2	165.83	1.631
Piperidine	8.7	10.9	85.15	0.861
Propionitrile	10.8	5.0	55.08	0.777
n-Propyl alcohol	11.9	8.9	60.10	0.800
Propylene glycol (or 1,2-propanediol)	15.0	9.4	76.10	1.036
Propylene oxide	9.2	5.8	58.08	0.829
Pyridine	10.7	8.7	79.10	0.978
Styrene	9.3	2.7	104.14	0.901
Tetrahydrofuran	9.1	5.3	72.11	0.889
Toluene	8.9	3.8	92.14	0.862
1,1,2-Trichloroethane	9.6	2.7	133.41	1.440
1,1,2-Trichloroethylene	9.3	2.5	131.39	1.462
1,1,1-Trichloro-2,2,2-trifluoroethane	7.2	2.5	187.38	1.563
VM and P naptha	7.6	2.2	–	0.750
Water	23.5	16.2*	18.02	0.997
Xylene (mixed)	8.8	3.8	106.17	1.490

A: The appropriate equation is:

$$\chi_1 = \beta_1 + (v_1/RT)(\delta_1 - \delta_2)^2$$
$$0.5 = 0.35 + (8.6 - \delta_2)^2 \times [100 \text{ (g/g mol)}/1.05 \text{ (g/cm}^3)]/[8.314 \times 300°K]$$

or:

$$(8.6 - \delta_2)^2 = 3.93$$

And the range on δ_2 is:

$$6.6 < \delta_2 < 10.6$$

From Table 4.11, potential solvents range from carbon tetrachloride through acetone.

Theta Temperature

It was noted that the positive/negative sign for free energy of mixing was key to determining thermodynamic solubility. Consider the case of $\overline{\Delta F}_1 = 0$ and $\partial\overline{\Delta F}_1/\partial\, v_2 = 0$:

$$(4.130) \qquad \frac{1}{(1 - v_2)} + \frac{1}{(1/x)} - 2\chi_1,\ v_2 = 0$$

and:

$$(4.131) \qquad \frac{1}{(1 - v_2)^2} - 2\,\chi_1 = 0$$

Eliminating χ_1:

$$(4.132) \qquad v_{2c} = \frac{1}{(1 + x^{1/2})}$$

v_{2c} is the concentration of polymer at the point where incipient solvation occurs. Thus:

$$(4.133) \qquad \chi_{1c} = \frac{(1 + x^{1/2})}{2x}$$

This can be approximated as:

$$(4.134) \qquad \chi_{1c} = \frac{1}{2} + \frac{1}{x}^{-1/2}$$

For x large (high polymer molecular weight), $\chi_{1c} = 0.5$. Since χ_{1c} is inversely dependent on temperature, a *critical solution temperature* can be defined:

$$(4.135) \qquad (0.5 - \chi_1) = \psi_1\,(1 - \theta/T)$$

where ψ_1 is an entropy parameter, given as:

$$(4.136) \qquad \overline{\Delta S}_1 = R\,\psi_1\, v_2^{2}$$

and θ is correctly defined as the "critical miscibility temperature in the limit of infinite molecular weight" (73). θ is called the "theta or θ-temperature" or the "Flory temperature". When $T = \theta$, the polymer–solvent interactions become ideal. At temperatures above θ, solution is preferred. At temperatures below θ, solution is not thermodynamically favorable and polymers in solution would tend to precipitate. Theta temperatures for several polymer–solvent binary pairs are given in Table 4.13. Keep in mind that the θ-temperature is a measure of thermodynamic suitability and many other factors may influence polymer–solvent interaction.

Table 4.13.   Theta-temperatures for several polymer-solvent combinations (74).

Polymer	Solvent	Theta temperature $\Theta(°K)$
Polystyrene	Cyclohexane (0.869)–carbon tetrachloride (0.131)	288
Polystyrene	Toluene (0.476)–*n*-heptane (0.524)	303
Polystyrene	Methyl ethyl ketone (0.889)–methanol (0.111)	303
Polystyrene	Cyclohexane	307
Polystyrene	Methylcyclohexane	343.5
Polystyrene	Ethylcyclohexane	343
Polyisobutylene	Benzene	297
Polyisobutylene	Phenetole	359
Polyisobutylene	Anisole	378.5
Polydimethyl siloxane	Methyl ethyl ketone	293
Polydimethyl siloxane	Phenetole	356

Solvent Welding and Adhesive Bonding

For certain applications, polymers of similar or identical characteristics must be joined together. One way of joining polymers is with *solvent welding*. The solvent softens or swells the polymer surfaces to a viscous solution or tacky state prior to joining. In some cases the solvent is a catalyzed monomer and is sometimes called a polymerizable cement. In other cases, the solvent contains a high concentration of dissolved polymer and can be called a dope cement. The latter is used if the surfaces to be joined are irregular or not smooth. In still other cases, the solvent contains filler in addition to the dissolved polymer. The product may have a paste or putty consistency. Solvent welding seems to work best for amorphous polymers such as PMMA and PVC. The more crystalline the polymer is, the more difficult solvent bonding becomes. It is difficult to find suitable solvents for the bonding of polyolefins, nylons (polyamides), and fluorocarbon polymers. Even though thermosets such as natural rubber are dramatically swelled by certain solvents such as benzene and acetone, they cannot be bonded in this fashion. Once the solvent has diffused away, the joint fails. Some dissimilar thermoplastics can be solvent-bonded, but the primary problem is to obtain a solvent that is suitable to both polymers. As an example, methyl ethyl ketone (MEK) can be used to solvent weld PS and PMMA.* The ideal solvent

---

*MEK is not as good a solvent for PS as it is for PMMA. As a result, room temperature MEK is applied to PS first, and the joint surface is allowed to "dry" for as much as one hour. This allows the MEK to diffuse into the PS, swelling and softening it. Care must be taken to minimize crazing when using this very fast-drying solvent. Temporarily enclosing the solvent-bond area with a cap of polyvinylidene chloride film (trademark: Saran) may help to minimize rapid solvent evaporation. Regions away from the solvent welding area must be protected from solvent attack during this step, however, to prevent unwanted solvent crazing. When bonding is desired, additional MEK is placed at the joint between PS and PMMA. The softened PS-MEK is further softened by the MEK, which also aggressively attacks the PMMA. Rapid joining under pressure is preferred, at this point. Bond strength increases very slowly with time. Drying times of a week to a month may be needed to develop high bond strength (75).

Table 4.14. Boiling points of several solvents used with polymers (76).

Solvent	Boiling point (°C)
Propylene oxide	34
Methyl dichloride	39.8
Acetone	56.2
Methyl acetate	57.2
Chloroform	61.7
Methanol	64.8
Carbon tetrachloride	76.5
Ethyl acetate	76.7
Ethanol	78.3
Methyl ethyl ketone	79.6
Benzene	80.1
Ethylene dichloride	83.5
Trichloroethylene	87.1
Isopropyl acetate	88.7
Methyl methacrylate	101
Nitromethane	101.2
Dioxane	101.3
Toluene	110.6
Nitroethane	114.0
Butanol	117.7
Acetic acid (glacial)	117.9
Perchloroethylene	121.2
Methyl cellosolve	124.6
Cellosolve (2-ethoxyethanol)	135.1
Ethyl benzene	136.2
Xylene	138–144
Methyl cellosolve acetate (2-methoxyethyl acetate)	145.1
Ethyl lactate	154.0
Cyclohexanone	155.4
Cellosolve acetate (2-ethoxyethyl acetate)	156.4
Diacetone alcohol (4-hydroxy-4-methyl-2-pentanone)	169.2
o-Dichlorobenzene	180.4
p-Diethylbenzene	183.7
Butyl lactate	188.0
Amylbenzene	202.1
Isophorone	215.2
2-Ethylnaphthalene	251

weld joint is molecularly undetectable from the unwelded polymer around it. Thus, failure at the joint will be *cohesive,* with failure occurring at polymer–polymer bonds, rather than *adhesive,* with failure occurring at the interface between the polymer and the joint material.

The key to successful solvent bonding is a good match of cohesive energy densities or solubility parameters. There are additional guidelines, however. Thermodynamically, of course, the solvent must be a good solvent. Solvation must be thorough enough to allow swelling to occur to some depth into the polymer. Then when pressure is applied, the interfaces are sufficiently soft to allow flow. This helps expel air, reduces asperities at the interfaces, and provides for excellent joint strength. Furthermore, the solvent must be sufficiently mobile to evaporate from the joint area in a short period of time. Otherwise the joint will remain rubbery after the joining force is removed, and distortion may result. Long-term residual solvents can weaken welded joints. Solvents with high vapor pressures are frequently sought, but some can cause crazing in the polymer both before and after welding. These solvent-induced crazes can lead to brittleness and loss in optical clarity. Boiling points of typical solvent cements are given in Table 4.14.

In *adhesive bonding,* the adhesive or bonding agent has a chemical composition that differs from the polymers, called *adherends.* The strength of the adhesive joint depends on the relationship between the strength in the adhesive material (cohesion) and that at the interface between the adhesive and the adherend (adhesion). Adhesives are generally selected to most nearly match the properties of the polymer, in tensile strength, flexibility, and chemical resistance. The adhesive and the adherend should exhibit similar temperature-dependent characteristics so that both are flexible or rigid in the same temperature range. Frequently adhesives are used to bond dissimilar materials, such as metal to polymer, ceramic to polymer, and polymer to paper. For these products, it must be predetermined if ultimate failure is to occur adhesively or cohesively.

4.8 Diffusion of Small Molecules into Polymers

The act of solvent bonding depends on the rate at which the small solvent molecule can move through the solid polymer chain. Whereas solution thermodynamics deals with equilibrium conditions for infinitely long times, molecular diffusion depends on time-dependent conditions. The study of time-dependent mass transfer, as with the study of solution thermodynamics, is a subject covered more thoroughly elswhere (77,78). Certain elements of diffusional mass transfer are needed here to provide understanding of the movements of small molecules through polymers. As just seen, solvent bonding is a primary application of liquid solution in polymers. This application is reconsidered briefly later.

In other applications, it is important to be able to predict the movement of small gaseous molecules through polymeric shapes. For example, the rate of gas diffusion through polymeric containers dictates the life of the product contained therein. Specifically, if CO_2 diffuses too rapidly through polyethylene terephthalate (PET) bottle walls, carbonated beverages go flat too early. If oxygen diffuses too rapidly through the same polymer, vegetable oils go rancid and beer becomes bitter. If chlorofluorocarbon gases diffuse too rapidly from polyurethane or polystyrene foams, the materials lose a substantial portion of their insulating values. Similarly, if water vapor is not allowed to diffuse through polyethylene film in building applications, it can promote early structural damage. Other examples of the influence of water in polymers are discussed below.

Concentration difference is the mass transfer driving force. The rate of mass transfer, J_a, per unit area normal to the diffusion direction is related to the concentration gradient as:

(4.137) $$J_a = D \frac{\partial c_a}{\partial x}$$

where c_a is the concentration of species a, x is the spatial coordinate and D is the *diffusion coefficient* or *diffusivity*.* This equation is sometimes called Fick's *first* law of diffusion. If J_a has the units of g mol/min^2 s and c_a has the units of g mol/m^3, D has the units of m^2/s.** The rate at which smaller molecules diffuse into the polymer can be obtained from:

(4.138) $$\partial c_a / \partial t = \nabla (D \nabla c_a)$$

If the polymer is thin in one dimension, such as a thin membrane, when compared with its in-plane dimensions, this model can be written as a transient one-dimensional form:

(4.139) $$\frac{\partial c_a}{\partial t} = \frac{\partial}{\partial x} \left[D \frac{\partial c_a}{\partial x} \right]$$

This equation is sometimes called Fick's *second* law of diffusion. There are many ways of defining and interpreting *diffusivity*. For example, for mutual diffusion of two substances, A and B, there are two diffusivities, one for substance A diffusing through substance B and one for substance B diffusing through substance A. If the substances mix ideally, the diffusivities are equal. That is, $D_{AB} \equiv D_{BA}$. On the other hand, when the substances do not mix ideally, the diffusivity of one substance through another (say, A through B) may be substantially greater than its reciprocal (say, B through A). Correctly, the diffusivities of pairs and multicomponent mixtures can be directly related to the *chemical potentials* of the substances and their self-diffusion coefficients (79). Further discussion can be found in more advanced texts on the subject.

Consider diffusion of a small molecule through a thin polymer film. At steady state, the one dimensional concentration gradient across the film is given as:

(4.140) $$(c - c_1)/(c_2 - c_1) = x/L$$

where c is the concentration at position x, c_1 and c_2 are the concentrations at $x = 0$ and $x = L$, respectively. The rate of mass transfer is given as:

(4.141) $$J = -D \, dc/dx = D (c_1 - c_2)/L$$

---

*Note that D, mass diffusivity, occupies the same relationship to mass transfer and concentration gradient as α, thermal diffusivity, does for energy flux and thermal gradient.
**The British unit system has (J_a) as *lb mol/ft^2 h*, (c_a) is *lb mol/ft^3* and (D) is *ft^2/h*.

Thus, if L, c_1 and c_2 are known and the flow rate measured, the diffusivity can be deduced. If the concentrations are known in terms of the vapor pressure of the gas, the rate of mass transfer can be written as:

$$(4.142) \qquad J = P\,(p_1 - p_2)/L$$

where p_1 and p_2 are the vapor pressures on each side of the film and P is called the permeability coefficient. The relationship between P and D can be determined if there is a relationship between the external vapor pressure and internal concentration. One such relationship is Henry's law, a linear law:

$$(4.143) \qquad c = S\,p$$

where S is called the *solubility coefficient*. If the diffusivity can be assumed to be constant:

$$(4.144) \qquad P = D\,S$$

In other words, the permeability of any small molecule species through polymers is a function not only of the diffusivity of the species but also of the solubility of the species. This point is amplified below.

4.9 Thermodynamically and Kinetically Good Solvents

Correctly there are two classes of "good" solvents. Thermodynamic solution defines a good solvent in terms of nearness of solubility parameters or favorable polymer–solvent interaction parameter, χ. Another classification is in terms of the *rate of solution*. Thermodynamically good solvents may penetrate the polymer at a velocity that does not allow swelling and establishment of a stable gel layer. Such fast-penetrating solvents crack or craze the polymer first, with the fractured polymer eventually dissolving in the solvent. Ueberreiter (80) shows thermodynamically and kinetically good and bad solvents for PS in Table 4.15. Note that toluene is a thermodynamically good solvent; its solvation power is so great, however, that it crazes rather than swells the polymer. On the other hand, although methyl ethyl ketone (MEK) is not a good solvent,

Table 4.15. A comparison of "good" and "bad" solvents for polystyrene (81).

Solvent	Thermodynamic suitability	Kinetic suitability
Toluene	+	+
Carbon tetrachloride	+	−
Ethyl acetate	−	+
Methyl ethyl ketone	−	+
Amyl acetate	−	−

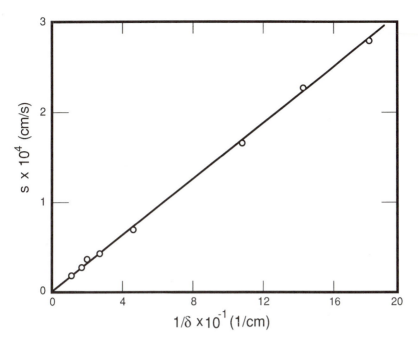

Figure 4.45. Rate of Toluene Solvent Penetration into PS as a Function of Reciprocal Surface Layer, δ (82).

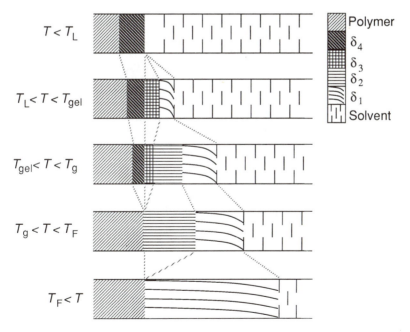

Figure 4.46. The Structure of the Solvent-to-Polymer Layers (83). T_F: Polymer Flow Temperature. T_g: Glass Transition Temperature. T_{gel}: Gel Temperature. T_L: Solution Temperature. δ_1: Liquid Layer. δ_2: Gel Layer. δ_3: Solid Swollen Layer. δ_4: Infiltration Layer.

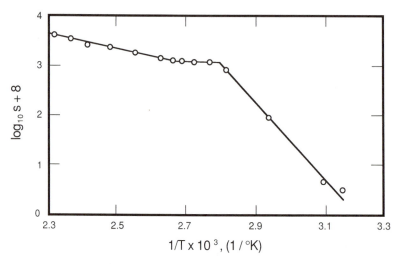

Figure 4.47. Temperature-Dependent Rate of Dimethyl Phthalate Solvent Penetration into PMMA (MW = 105,000) (84). First Transition at About 100°C is Glass Transition. Second, Sharp Transition at About 85°C is Temperature at Which Gel Formation Occurs.

it swells PS and thus can be used successfully in solvent welding (see footnote on pg 362). The velocity of penetration is related to diffusion coefficient by:

(4.145) $D = \dot{s}\,\delta$

where δ is the thickness of the swollen surface layer (cm) and $\dot{s}$ is the velocity of penetration (cm/s). As seen in Figure 4.45, the penetration velocity is inversely proportional to the swollen surface layer.

   The development of the surface or gel layer depends also on temperature. This is seen in Figure 4.46, for several temperature zones. Ueberreiter defines several solution temperatures. T_L is called the solution temperature. Below this, the solvent simply diffuses into the polymer without swelling it. Between T_L and T_{gel}, the gel temperature, a *solid swollen layer* and a polymer–solvent liquid layer are formed. This is called *elastic state dissolution*. The solid, swollen layer is the source of massive crazing and stress-cracking. As the temperature increases between T_{gel} and T_g, the glass transition temperature, the solid swollen layer decreases in thickness and a gel layer develops and increases in thickness. The gel layer is the classical swelling behavior discussed above. Between T_g and T_F, the flow temperature, the gel layer increases in thickness. Above the flow temperature, the gel layer disappears, with solution taking place only in the solvent phase. The overall rate of solution is controlled by solution in one or several of these phases. As seen in Figure 4.47 for dimethyl phthalate in PMMA, the velocity of solvent penetration decreases dramatically as the solution temperature falls below T_{gel}. This is the point at which the gel or swollen polymer layer disappears.*

*The more subtle change in slope occurs at T_g.

Solubility Coefficient

The solubility coefficient, S, is a measure of the equilibrium amount of *solvent* in a polymer. The time-dependent mass transfer rate is described by diffusivity, as the applicable material property. The solution equilibrium is described by *solubility*. This relationship is seen in Figure 4.48. For volatile vapors at low vapor concentrations, the solubility coefficient is essentially constant:

$$(4.146) \qquad S = S_0 \exp [\Delta H_s / RT]$$

where ΔH_s is the heat of solution. As noted above, for dissolved gases above their critical point:

$$(4.147) \qquad \Delta H_s = v_1 (\delta_1 - \delta_2)^2 \phi_2^2$$

where v_1 is the partial molar volume of the solvent, ϕ_2 is the volume fraction of polymer, and δ_i is the solubility parameter of species i. For ideal solutions, solubility is independent of temperature. In real cases, ΔH_s can be either positive or negative, but in most cases the value is relatively small (Table 4.16). Solubility of gases in

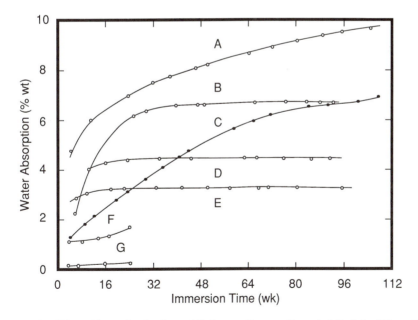

Figure 4.48.   Water Absorption by Several Polymers Over an Extended Period of Time (85). Curve A: Polyvinyl Butyral. B: Molded Phenol–Formaldehyde. C: Cast Phenol–Formaldehyde. D: Molded Urea–Formaldehyde. E: Ethyl Cellulose. F: PMMA. G: PVC–Polyvinyl Acetate Copolymer.

Table 4.16.   Heats of solution of several gases in polyvinyl acetate (86).

Gas	ΔH (kcal/mol)	
	$< T_g$	$> T_g$
Helium	−1.01	2.11
Hydrogen	−1.42	2.47
Neon	−4.62	1.05
Oxygen	−6.26	−1.10
Argon	−3.70	−1.88
Krypton	−7.14	−1.20
Methane	–	0.46

polymers is pressure dependent. Henry's law applies at relatively low pressures for rubbery polymers and simple gases:

$$(4.148) \qquad\qquad S = H' \, p$$

where H' is the solubility at one atmosphere. Solubility coefficients for volatile vapors in natural rubber are shown to be linear functions of boiling point (T_b) and critical

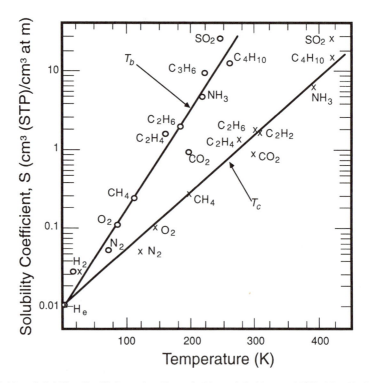

Figure 4.49.   Solubility Coefficients for Gases in Natural Rubber at 25°C (87). X: Solubility Values vs Gas Critical Temperature. ○: Solubility Values vs Gas Boiling Point.

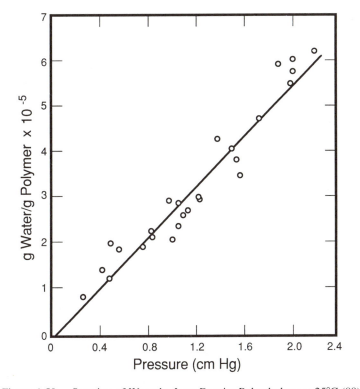

Figure 4.50.   Sorption of Water by Low-Density Polyethylene at 25°C (88).

temperature (T_c) in Figure 4.49. Henry's law works also for low solubility pairs, as seen for water vapor in polyethylene (Figure 4.50). Of course, small molecules continue to diffuse into polymers until solution equilibrium is reached.

Permeability

Permeability, that is, $P = D S$, is the equilibrium or steady-state mass transfer rate for polymer–solvent pairs. Temperature dependency of diffusivity can be written in Arrhenius form as:*

$$（4.149） \qquad D = D_0 \exp[-E_D/RT]$$

---

*Over a wide temperature range, the Arrhenius equation does not accurately describe temperature dependency of diffusivity for many polymer–volatile combinations. For a narrow temperature range, the Arrhenius plot is satisfactory. This is seen for PMMA and several gases in an ln D vs. $1/T$ plot (Figure 4.51).

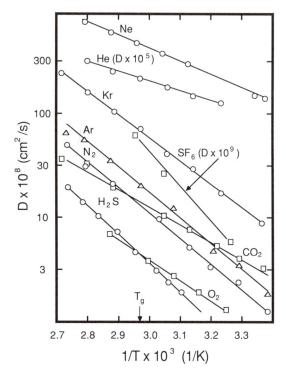

Figure 4.51.   Temperature-Dependent Diffusion Coefficients for Various Gases in PMMA (89).
Note That the Temperature Dependency Is Independent of the Glass Transition Temperature.

where E_D is the activation energy for diffusion. Thus:

$$(4.150) \qquad\qquad P = P_o \exp[-E_p/RT]$$

where $E_p = E_D + \Delta H_s$ and $P_o = D_o S_o$.

Another strong influence on diffusivity and solubility is glass transition temperature, T_g, as seen in Tables 4.17 and 4.18. As expected, increasing polymer chain mobility leads to increased diffusivity and solubility, and so to increased permeability. In semi-crystalline polymers, diffusion and solution depend not only on the glass transition temperature of the amorphous phase, but also on the role of the crystallites in hindering chain mobility.

Diffusion coefficients can be concentration dependent (Table 4.19). One proposed expression is:

$$(4.151) \qquad D = D_o \exp[\tau\, c_1] = D_o \exp[\alpha\, p_1/p{*}_1] = D_o \exp[\tau'\, \phi_1]$$

where c_1 is concentration of the solvent and τ, τ' and α are proportionality constants. As seen, c_1 can be written in terms of volatile molecule partial pressure, p, and vapor pressure, $p{*}$ or in terms of θ_1, volume fraction of sorbed vapor. As seen in Figure 4.52

Table 4.17.   Temperature dependence of diffusion above and below the polymer glass transition temperature (90).

Polymer	T_g (°C)	Gas	D_o (cm²/s)		E_D (kcal/mol)	
			$< T_g$	$> T_g$	$< T_g$	$> T_g$
Polymethyl methacrylate	90	CH_3OH	0.37	110	12.4	21.6
Polyethyl methacrylate	40	CH_3OH	0.20	6.1	9.6	14.7
Polybutyl methacrylate	10	CH_3OH	0.45	4.5	9.2	12.0
Polystyrene	88	CH_3OH	0.33	37	9.7	17.5
Polyvinyl acetate	30	CH_3OH	0.02	300	7.6	20.5
Polyvinyl acetate	30	He	0.011	0.069	4.16	5.35
Polyvinyl acetate	30	Ne	0.39	2.1	7.36	8.46
Polyvinyl acetate	30	Ar	2.7	13,000	11.4	16.5
Polyvinyl acetate	30	Kr	72.3	290,000	14.5	19.4

Table 4.18.   Concentration dependence of vapor diffusion of small molecules in certain polymers (91).

Vapor	Polymer	Temperature (°C)	α	τ (g/g)	τ'	$D_o \times 10^9$ (cm²/s)
Methyl bromide	Polyethylene (0.92 g/cm³)	30	1.3	14	–	83
	Polyethylene (0.92 g/cm³)	0	0.9	12	–	7.3
	Polyethylene (0.938 g/cm³)	0	2.9	53	–	2.9
	Polyethylene (0.954 g/cm³)	0	3.3	70	–	1.4
Methylene chloride	Polystyrene	15	–	–	39.7	0.0165
	Polystyrene	25	–	–	56.3	0.0195
	Polystyrene	35	–	–	64.2	0.0233
n-Butane	Polyisobutylene	35	–	25.5	–	3.29
	Ethyl cellulose	50	–	101	–	11.8
Isobutane	Polyisobutylene	35		24.1	–	1.46
	Ethyl cellulose	50		199	–	2.1
n-Hexane	Polyethylene (0.92 g/cm³)	30	2.5	36	–	25
	Polyethylene (0.92 g/cm³)	0	2.8	53	–	1.2
	Polyethylene (0.938 g/cm³)	0	4.6	110	–	0.15
	Polyethylene (0.954 g/cm³)	0	6.5	150	–	0.075
Benzene	Natural rubber	50	–	–	10.3	300
	Polyethylene (0.92 g/cm³)	50	–	–	28.8	132
	Ethyl cellulose	60	–	31	–	(3.0)
	Polyvinyl acetate	40	–	53	46	0.00048

Table 4.19. Permeation rates for vapor and liquid diffusants in high-density polyethylene at 80°F (92).

Compound	Permeation rate (g/24 h/100 in²/mil)	
	Vapor	Liquid
Ethylene glycol	0.37	0.87
Ethyl Alcohol	1.3	1.8
Acetic Acid	0.94	0.99
Butyric Acid	1.5	3.6
Methyl ethyl ketone	4.1	8.1
Diethyl ketone	7.7	7.7
Amyl acetate	5.5	6.8
Ethyl acetate	14	14
Butyl Iodide	52	113
Butyl Chloride	70	86
Butyl Bromide	76	113

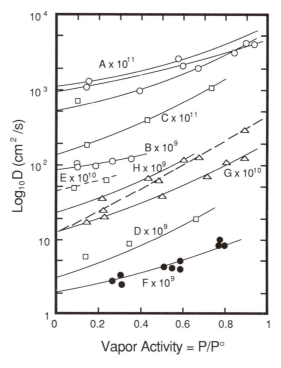

Figure 4.52. The Effect of Penetrating Vapor Activity, Shown as Ratio of Partial Pressure to Vapor Pressure, on Diffusion Coefficients in Low-Density Polyethylene (93). A: Methyl Bromide, 0°C. B: Methyl Bromide, 30°C. C: Isobutylene, −8°C. D: Isobutylene, 0°C. E: Isobutylene, 30°C. F: Benzene, 0°C. G: n-Hexane, 0°C. H: n-Hexane, 30°C.

for various vapors diffusing through low-density polyethylene at various temperatures (above T_g), the expression is satisfied in most cases. The concentration- and temperature-dependencies for methyl acetate solvent in PMMA are combined in Figure 4.53. If the solubility is considered to be concentration-independent, then permeability can be written as:

$$(4.152) \qquad\qquad P = P_o \exp [\tau\, c_1]$$

This is seen for the binary systems in Figure 4.54.

To this point, the small molecule has been variously defined as a solvent, a gas, or a volatile vapor. As seen in Table 4.20, permeabilities of small molecules as liquids and vapors are quite similar in polyethylene.

Note also that the permeability coefficient, P, is defined for a polymer–small molecule pair. From data on many polymer–small molecule pairs, contributions from individual components have been deduced. Table 4.21 lists *individual contributions* for various gases and polymers. The permeability coefficient for any polymer–small molecule pair at 30°C can be obtained by combining the F for the gas with the G for the polymer:

$$(4.153) \qquad\qquad P = F\,G$$

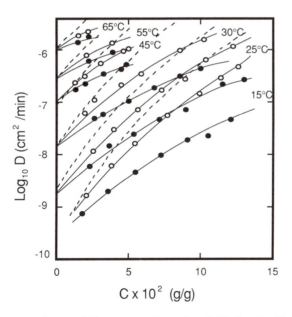

Figure 4.53. Concentration- and Temperature-Dependent Diffusion Coefficients for Methyl Acetate in PMMA (94). Diffusion Coefficients are Decidedly Nonlinear with Concentration. Three Diffusion Coefficients are Shown. Open Circles: D_a. Closed Circles: D_d. Dashed Line: Conventional Diffusion Coefficient.

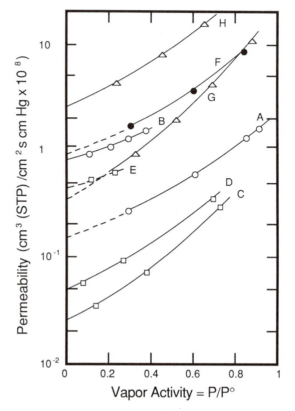

Figure 4.54.   Permeability Coefficients for Several Vapor Penetrants in Low-Density Polyethylene (95). Vapor Activity is Vapor Partial Pressure Divided by Vapor Pressure. Curve A: Methyl Bromide, 0°C. B: Methyl Bromide, 30°C. C: Isobutylene, −8°C. D: Isobutylene, 0°C. E: Isobutylene, 30°C. F: Benzene, 0°C. G: n-Hexane, 0°C. H: n-Hexane, 30°C.

4.10 Water in Polymers

As already noted, the diffusion of water vapor through thin polymer films is important in several areas. Water is a plasticizer for many condensation and polar polymers such as nylon (polyamide, PA), polyethylene terephthalate (PET) and the cellulosic family. The plasticizing effect is one of reduction in mechanical strength, stiffness and use temperature, and a corresponding increase in ductility. When the polymer is reheated, the *in situ* water vaporizes to produce cloudiness or haze. In resin pellets, absorbed water at processing temperatures can rapidly hydrolyze condensation polymers such as nylon and PET. This results in lowered molecular weight and a substantial loss in physical properties in the finished product. Polymers are frequently used as electrical isolators, as in capacitors. Sorbed water can substantially reduce the insulating ability

Table 4.20. Permeability parameters for polymer and gas pairs (*F* and *G*) (96).

Film	*F* Value*
Saran (PVDC) (tm)	0.0094
Mylar (BOPET) (tm)	0.050
Pliofilm NO (tm)	0.080
Nylon 6	0.10
Kel-F (tm)	1.3
Pliofilm FM (tm)	1.4
Hycar OR 15 (tm)	2.35
Polyvinyl butyral	2.5
Cellulose acetate	2.8
Butyl rubber	3.12
Methyl rubber	4.8
Vulcaprene (tm)	4.9
Cellulose acetate (15% wt DBP)	5.0
Hycar OR 25 (tm)	6.04
Pliofilm P4 (tm)	6.2
Perbunan (tm)	10.6
Neoprene	11.8
Polyethylene (0.92 g/cm^3)	20
Buna-S	63.5
Polybutadiene	64.5
Natural rubber	80.8
Ethyl cellulose (plasticized)	84

Gas	*G* Value*
Nitrogen	1.0
Oxygen	3.8
Hydrogen sulfide	21.9
Carbon dioxide	24.2

*The product **FG** gives permeability constant, $\times\ 10^{10}$ in cm^3 (STP)/(mm thickness)/(cm^2/s) cm Hg.

of certain polymers. Typical electrical insulating film properties are given in Table 4.22. The effects of absorbed moisture on surface resistivity of several polymers and polymer laminates are listed in Table 4.23. The water effect for polyamide and vulcanized fiber is seen in Figure 4.55 and for LDPE in Figure 4.56. Note that the drop in resistance is nearly exponential with time. In consumer products, high moisture absorption can be advantageous or disadvantageous. It is desired in carpeting, where low humidity static charge dissipation is needed or in clothing, where perspiration absorption is desired. It is unwanted in fabrics and coverings in which mold, mildew, and fungus can grow more rapidly on surfaces that retain high surface moisture. Diffusivity and permeability values for water in many polymers are given in Table 4.24.

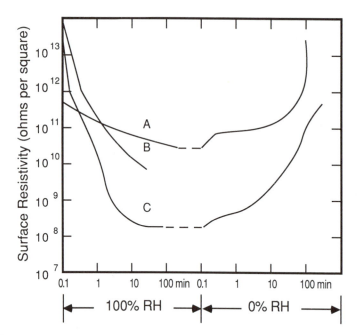

Figure 4.55. Electrical Surface Resistivity for Materials That Absorb Substantial Amounts of Water (97). Curve A: Nylon 66 (PA-66), 100% RH. B: Nylon 66 (PA-66), 61% RH. C: Vulcanized Fiber, 100% RH. Note That After 1000 min, the Samples are Transferred to 0% RH Environment and Resistivity Increases.

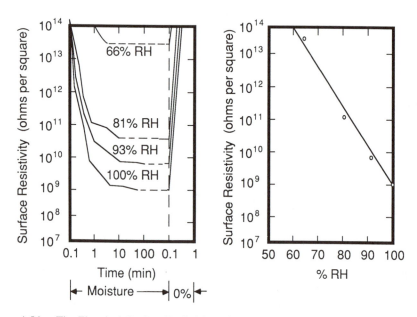

Figure 4.56. The Electrical Surface Resistivity of Polyethylene, a Low Moisture Absorbing Polymer (98). Left: the Time-Dependent Nature of Surface Resistivity. Right: the Equilibrium Value as a Function of Relative Humidity.

4.11 The Role of Plasticizers

Plasticizers are low molecular weight polymers that are added to higher molecular weight polymers to alter processibility. The classic plasticizer–polymer combinations are phthalates added to polyvinyl chloride to increase flexibility and to lower and broaden glass transition temperature. For example, the glass transition temperature of PVC homopolymer is about 80°C. This material is therefore quite brittle at room temperature. When a low-molecular weight plasticizer such as n-dioctyl phthalate having a glass transition temperature of about $-60°C$ is compounded in, the glass transition temperature of the composition decreases in proportion to its volume fraction:

(4.154)
$$T_g = V_1 T_{g,1} + V_2 T_{g,2}$$

In terms of weight fraction:

(4.155)
$$1/T_g = W_1/T_{g,1} + W_2/T_{g,2}$$

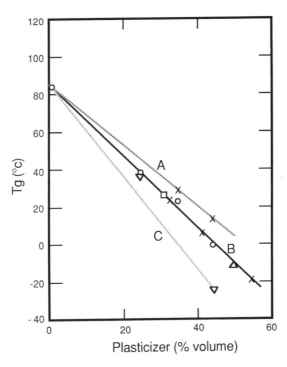

Figure 4.57.   Effect of Plasticizer on Glass Transition Temperature of PVC (99). Curve A: Phosphate Plasticizer. B: Phthalate Plasticizer. C: Sebacate Plasticizer.

where $T_{g,1}$ is in °K. For example, for *n*-dioctyl phthalate at 30% (wt), $T_g = 30°C$, but at 50% (wt), it is $-7°C$. This is seen in Figure 4.57. These relationships hold for most other plasticizer-polymer combinations as well, as seen for three different plasticizers and three polymers in Figure 4.58. In addition to shifting T_g, plasticizers broaden the flexible or ductile temperature range, as seen in Figure 4.59. The molecular nature of the plasticizer also affects the final properties of the plasticizer-polymer pair. Longer aliphatic chain plasticizers offer lower tensile strength, whereas shorter, aromatic or branched chain plasticizers yield stiffer compounds. It must be noted however, that short linear chain plasticizers also have higher diffusivities. As a result, substantial amounts of these materials can diffuse from the polymer in short times. Of course, at elevated temperature, diffusion rate is enhanced. Years ago, the "new car" smell was primarily the result of the early diffusion of low molecular weight phthalates from vinyl seat and dash covering. The oily film on interior automotive windows on hot days was also the result of high diffusion rates. While the smell is now a recognized sign of "newness" and the film is an annoyance, the loss of plasticizer signaled early brittle fracture in the PVC coverings, especially at low temperatures. Newer low diffusivity plasticizers are being used today to maintain long-term automobile value. The "new car" smell now comes in an aerosol spray.

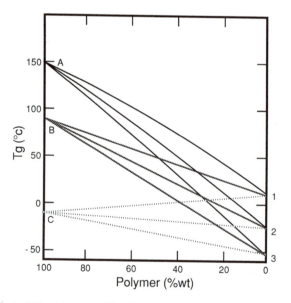

Figure 4.58. Effect of Plasticizers on Glass Transition Temperatures of Three Polymers (99). Curves A: PS-Acrylonitrile Copolymer. Curves B: Polyvinyl Acetate. Curves C: Polychloroprene. Plasticizer 1: Dimethylcyclohexyl. 2: Tricresyl Phosphate. 3: Di-*n*-Hexyl Phthalate.

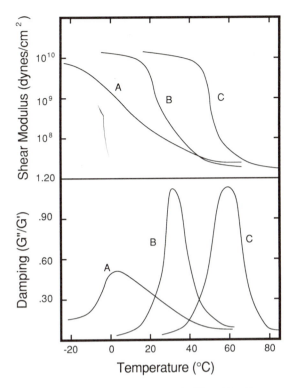

Figure 4.59. The Effect of the Type of Plasticizer on the Transition Temperature of PVC (100). Values Determined by Dynamic Mechanical Spectroscopy. Curve A: DOP, *n*-Dioctyl Phthalate, 40.2% (vol). B: DEP, Diethyl Phthalate, 25.4% (vol). C: DBP, Dibutyl Phthalate, 15.8% (vol).

4.12 Chemical Resistance

Polymers are frequently considered for application in aggressive environments, where metals, ceramics or natural materials such as wood or natural organic fibers may be unsuitable, too costly or too heavy. As seen above, certain small molecules can readily solvate certain polymers. In other polymer–small molecule pairs, diffusion and solution are negligible. Thus the polymer is considered to be impermeable to the solvent. Keep in mind that solvation is primarily a physical effect, the separation of polymer molecules by the smaller solvent molecules.

Most chemicals are also small molecules. Chemicals, on the other hand, attack the polymer structure, reducing the polymer molecular weight and thus reducing its mechanical performance. There is a blurring of definitions here, however, since some solvents are sufficiently aggressive in their solvation to cleave the polymer backbone. As was discussed above, water in condensation polymers such as nylon and PET can reduce polymer molecular weight at elevated temperature by reversing the polymerization steps. Other chemical-polymer pairs include H_2SO_4–PS, Cl_2–LDPE, HCl–PVC, enzymes–cellulosics, and aqueous NaOH–vinyl ester resin.

Oxygen is also a small molecule chemical that can produce polymer degradation in many polymers. The primary mode of oxidative degradation is sequential free radical chain scission:

$$R-O-O-H \rightarrow RO_2\bullet + \bullet H$$

(4.156)
$$\text{-}(CH_2\text{)-} + RO_2\bullet \rightarrow \text{-}(\overset{\bullet}{CH}\text{)-} + ROOH$$

$$\text{-}(\overset{\bullet}{CH}\text{)-} + O_2 \rightarrow \text{-}(\underset{\underset{O-O\bullet}{|}}{C}-H\text{)-}$$

The term on the right in the first reaction is called a *peroxy radical*. This can attack another CH bond, perpetuating the free radical unzipping. Natural and synthetic SBR and polymers containing phenyl groups (styrenics) are particularly susceptible to oxidative degradation. Ozone, a primary source of oxidative degradation, is frequently generated by electrical fields as might be present around electrical motors. Since rubber is frequently used around motors for mountings, wire coatings, couplings, electrical isolators and grommets, the opportunity for rapid deterioration is present. In Section 6.7, polymer chemical resistance tests are described in detail. In particular, environmental stress crack resistance (ESCR) is discussed.

Table 4.21. Properties of electrical insulating films (101).

Polymer	Type	Mechanical properties		Electric properties				Thermal properties	
		Tensile strength MD (N/mm^2)	Elongation at break (%)	Dielectric constant (50 Hz, 20°C)	Dissipation factor (50 Hz, 20°C)	Volume resistivity (Ω-cm)	Dielectric strength (kV/mm)	Tensile heat distortion temperature (°C)	Continuous service temperature (°C)
Polypropylene	Biaxially oriented	150	75	2.3	0.7	10^{17}	300	150	105
Polystyrene	High temperature	70	4	2.5	<0.2	10^{17}	200	110	90
Polytetrafluoro-ethylene	–	17	350	2.1	<0.3	10^{17}	100	190	<180
Polyethylene terephthalate	Biaxially oriented	210	111	3.3	2	10^{17}	300	240	130
Polycarbonate	Normal	90	110	3.1	2.5	10^{17}	240	150	130
Polycarbonate	Crystalline, oriented	250	35	2.8	1.2	10^{17}	280	230	130
Polyphenylene oxide	–	65	25	2.7	1.5	10^{16}	300	170	110
Polysulfone	–	80	65	3.1	1.2	5×10^{14}	–	185	165 to 170
Polyhydantoine	–	100	119	3.3	1.5	4×10^{16}	250	260	160
Polyamide-imide	–	180	45	4.2	9	3.5×10^{17}	200	>250	–
Polyimide	–	180	70	3.5	2	10^{17}	270	>350	>180
Cellulose triacetate	Normal	90	23	4.5	12	10^{14}	220	190	120
Cellulose triacetate	Soft	80	27	4.3	21	10^{15}	200	170	120
Cellulose acetate butyrate	Normal	80	25	4.1	11	10^{15}	230	150	120

Table 4.22.   Surface resistivities of many electrical insulating polymers (102).

Polymer	Equilibrium surface resistivity at 100% RH	Percent relative humidity to produce a 10-fold change in surface resistivity	Recovery time (min) after exposure at 100% RH for: 1 h	Recovery time (min) after exposure at 100% RH for: 16 h
Cellulose acetate butyrate	$> 2.0 \times 10^{13}$	—	0	0
Silicone rubber	1.0×10^{13}	—	0.13	—
Polytetrafluoroethylene	3.6×10^{12}	—	0.17	0.17
Polystyrene (sheet)	8.4×10^{11}	—	0.13	0.13
Polydichlorostyrene	2.9×10^{10}	7	0.17	—
Ethyl cellulose	1.3×10^{10}	9	0.33	0.5
Cellulose acetate	7.0×10^{9}	6	1.0	6
Polyvinyl chloride–acetate	5.7×10^{9}	12	6	—
Molded phenolic, mica-filled, type I	5.0×10^{9}	9	0.17	13
Nylon (polyamide), dry	3.8×10^{9}	14	200	—
Polystyrene (molded)	2.4×10^{9}	10	0.17	0.17
Polyethylene	1.3×10^{9}	9	0.17	0.17
Paper–phenolic laminate, type I	1.3×10^{9}	16	80	—
Molded phenolic, asbestos-filled	1.2×10^{9}	9	1.5	100
Paper–phenolic laminate, type II	6.6×10^{8}	15	300	—
Cotton cloth–phenolic laminate, type I	5.0×10^{8}	18	400	—
Molded phenolic, mica-filled, type II	3.2×10^{8}	8	40	—
Poly-3,4-dichlorostyrene	2.4×10^{8}	6	0.33	5.3
Molded phenolic, cellulose-filled	2.3×10^{8}	10	400	—
Glass mat–aniline formaldehyde laminate	2.4×10^{8}	9	14	1,000
Cotton clothl–phenolic laminate, type II	2.2×10^{8}	16	300	—
Vulcanized fiber	2.2×10^{8}	—	6,000	—
Glass cloth-melamine laminate	3.8×10^{7}	14	300	—
Molded phenolic, mica-filled, type III	3.0×10^{7}	11	7	—

Table 4.23. Diffusion data for water in many polymers (103)[1].

Polymer	Temperature (°C)	p/p_o	Permeability coefficient $P_{10} \times 10^9$ (cm³ (STP) cm²/s) (cm Hg/cm)	E_P (kcal/mol)	$D_o \times 10^9$ (cm²/s)	E_{Do} (kcal/mol)
Keratin (horn)	25	0.6	350	–	–	–
Keratin (horn)	25	0.9	2,900	–	–	–
Wool	35	0	–	–	0.1	–
Wool	77.7	0	–	–	0.83	>11
Regenerated cellulose	25	0.4	190	–	–	–
Regenerated cellulose	25	0.6	358	–	–	–
Regenerated cellulose	25	0.2	1,700	–	<1	–
Nylon 6	25	0.5	40	–	0.97	6.5
Nylon 6	25	0.96	140	–	–	12.5
Nylon 6	60	0.96	190	–	8	10.4
Polyethylene terephthalate	25	0 to 1.0	17.5	0.5	3.9	–
Cellulose triacetate	25	0 to 1.0	1,270	–	–	–
Cellulose triacetate	50	0 to 1.0	1,380	–	–	–
Cellulose nitrate	20	0 to 1.0	450	–	–	–
Cellulose acetate	30	0.5 to 1.0	600	–	17	5.6
Cellulose acetate	25	0 to 1.0	6,210	–	–	–
Cellulose acetate	25	0 to 1.0	15,000	–	–	–
Cellulose acetate (9% plasticizer, 56% combined acetic)	25	0 to 1.0	1,120	–	30	–
Benzyl cellulose	24	0 to 1.0	366	–	–	–
Ethyl cellulose	25	0.21	2,100	-1.5	180	6.3
Ethyl cellulose	25	0.84	2,380	–	–	–
Ethyl cellulose	80	–	1,100	-2.8	1,200	9.5
Polyvinyl butyral	25	0 to 1.0	185	-2.1	13	10.9
Polyvinyl chloride	30	–	–	–	23	42
Polyvinyl chloride	30	–	15	2.35	16	10

Table 4.23. *Continued.*

Polymer	Temperature (°C)	p/p_o	Permeability coefficient $P_{10} \times 10^9$ (cm³ (STP) cm²/s) (cm Hg/cm)	E_p (kcal/mol)	$D_o \times 10^9$ (cm²/s)	E_{Do} (kcal/mol)
Polyvinyl chloride/plasticizer (100/30)	30	0.5 to 0.9	34	–	17	–
Polyvinyl chloride/plasticizer (100/75)	25	0 to 1.0	200	–	–	–
	50	0 to 1.0	379	–	–	–
Polyvinyl chloride (87)–vinyl acetate (13) copolymer	32	0 to 0.9	–	–	60	7.7
Polyvinyl chloride–vinyl copolymer	25	0 to 0.9	28 to 32	2.35 to 4.0	–	–
Chlorinated polyvinyl chloride	25	0 to 1.0	20.7	–	–	–
Chlorinated polyvinyl chloride	50	0 to 1.0	23.5	–	–	–
Polyvinyl fluoride	39.5	0 to 1.0	41	–	–	–
Polyvinylidene chloride	30	–	0.14 to 1.0	–	–	–
Vinylidene chloride (81)–acrylonitrile (19) copolymer	25	0 to 1.0	1.6	10.3	0.32	20.2
Polyvinyl alcohol	25	0.4	1.9	–	0.051	≈14.0
Polyvinyl alcohol	25	0.6	9.6	–	1.25	–
Polyvinyl acetate	25	–	–	–	43	12.5
Polyvinyl acetate	40	0.2	600	–	150	15
Polymethyl methacrylate	50	–	250	–	130	11.6
Polyethyl methacrylate	25	0 to 1.0	≈350	0.5	105	8.7
Polyethyl methacrylate	90	0 to 1.0	≈600	4.3	3,500	15.1
Polymethacrylate	25	–	–	–	120	16
Polystyrene	25	0 to 0.8	97	–	–	–

Polystyrene	50	0 to 0.8	107	—	—	—
Polyethylene tetrasulfide	21	0 to 1.0	6	—	—	—
Bakelite	25	0 to 1.0	166	—	—	14.2
Polyethylene (0.922 g/cm^3)	25	0 to 1.0	9	8	230	—
Polyethylene (0.922 g/cm^3)	30	0 to 1.0	12.4	—	—	—
Polyethylene (0.938 g/cm^3)	25	0 to 1.0	2.5	—	—	—
Polyethylene (0.960 g/cm^3)	25	0 to 1.0	1.2	—	—	—
Polypropylene (0.907 g/cm^3)	25	0 to 1.0	5.1	10	240	16.4
Polypropylene (0.907 g/cm^3)	30	0 to 1.0	6.8	—	—	—
Rubber hydrochloride	25	—	1.4	—	0.41	14.0
Balata	25	0 to 1.0	59	—	—	—
Gutta percha	25	0 to 1.0	51	—	—	—
Soft vulcanized natural rubber	25	0 to 0.5	229	—	—	—
Vulcanized chloroprene	21	0 to 0.8	91	—	—	—
Polytrifluorochloroethylene	30	—	0.029	—	—	≈3
Polydimethylsiloxane	35	0.2	4,300	—	≈70,000	—
Polydimethylsiloxane	65	0.2	3,280	—	≈100,000	—
Polybutadiene	37.5	0 to 0.9	507	—	—	—
cis-Polyisoprene	37.5	0 to 0.9	212	—	—	—
Butadiene–styrene (6:1 Molar)	37.5	0 to 0.9	168	—	—	—
Butadiene–styrene (3:1 Molar)	37.5	0 to 0.9	101	—	—	—
Ethylene–propylene (3:2 Molar)	37.5	0 to 0.9	45	—	—	—
Chlorinated polyisobutene/isoprene	37.5	0 to 0.9	12	—	—	—
Polyisobutene	37.5	0 to 0.9	11	—	—	—
Polyisobutene	30	0 to 0.9	7.1 to 22.44	—	—	—

[1]Multiple data indicates more than one source.

Table 4.24. Power-law constants for several thermoplastic resins (104).

Polymer [mfgr identification]	Temperature (°C)	Power-law constants			Ellis Model Constants			
		Shear rate range (s^{-1})	k (Ns^n/m^2) (× 1000)	n	Shear stress range (N/m^2) (× 10^{-5})	η_0 (Ns/m^2) (× 10^4)	α	$\tau_{1/2}$ (N/m^2) (× 1000)
Polystyrene (PS) [Dylene™ 8][a]	190	100 to 4,500	44.7	0.22	1.0 to 2.7	1.4	4.51	60.4
	210	100 to 4,000	23.8	0.25	0.6 to 1.8	0.92	3.85	31.7
	225	100 to 5,000	15.6	0.28	0.5 to 1.7	0.66	3.67	22.7
High-impact polystyrene (HIPS) [LX-2400][b]	170	100 to 7,000	75.8	0.20	2.0 to 4.0	21.	4.66	50.9
	190	100 to 7,000	45.7	0.21	0.9 to 1.6	14.8	4.80	32.9
	210	100 to 7,000	23.1	0.19	0.5 to 1.9	10.5	4.74	22.2
Polystyrene-acrylonitrile (SAN) [Lustran™ 31-1000][b]	190	100 to 9,000	90.0	0.21	2.3 to 5.6	2.2	4.40	126.
	220	100 to 8,000	32.2	0.27	1.0 to 3.2	0.90	3.59	50.3
	250	100 to 8,000	11.1	0.35	0.5 to 2.0	0.42	2.76	18.0
Polyethylene vinyl acetate (EVA) [Vistaflex™ 905B][c]	200	100 to 5,000	27.5	0.27	1.3 to 2.7	3.6	3.69	24.1
	220	100 to 4,000	18.3	0.30	0.7 to 1.7	2.15	3.32	16.8
	240	100 to 4,000	19.9	0.28	0.6 to 1.6	1.35	3.36	19.9
Polyacrylonitrile-butadiene-styrene (ABS) [AM-1000][d]	170	100 to 5,500	119.	0.25	3.0 to 6.0	7.95	3.31	89.8
	190	100 to 6,000	62.9	0.25	1.8 to 5.0	4.4	3.85	47.9
	210	100 to 7,000	39.3	0.25	1.0 to 2.5	2.6	3.33	31.8
Low-density polyethylene (LDPE) [Alathon™ 1540][e]	160	100 to 4,000	9.36	0.41	0.6 to 3.0	0.63	2.56	15.2
	180	100 to 6,500	5.21	0.46	0.4 to 2.0	0.32	2.22	9.06
	200	100 to 6,000	4.31	0.47	0.3 to 1.8	0.17	2.23	12.

Polymer							
High-density polyethylene (HDPE) [Alathon™ 7040][e]	180	100 to 1,000	6.19	0.56	0.8 to 2.8	0.21	2.57	75.0
	200	100 to 1,000	4.68	0.59	0.6 to 2.5	0.152	2.51	74.9
	220	100 to 1,000	3.73	0.61	0.5 to 2.0	0.117	2.49	76.7
Polypropylene (PP) [CD 460][c]	180	100 to 4,000	67.9	0.37	0.3 to 1.4	0.421	2.72	9.57
	190	100 to 3,500	48.9	0.41	0.2 to 1.4	0.32	2.50	7.19
	200	100 to 4,000	43.5	0.41	0.2 to 1.3	0.25	2.49	7.17
Nylon 6 (PA-6) [Capron™ 8200][f]	225	100 to 2,500	2.62	0.63	0.4 to 3.0	0.16	1.64	10.6
	230	100 to 2,000	1.95	0.66	0.3 to 3.0	0.13	1.70	13.
	235	100 to 2,300	1.81	0.66	0.3 to 3.0	0.11	1.61	10.4
Polymethyl methacrylate (PMMA) [Lucite™ 147][e]	220	100 to 6,000	88.3	0.19	2.0 to 4.5	1.3	5.23	144.
	240	100 to 6,000	42.7	0.25	1.2 to 3.3	0.60	3.68	73.6
	260	100 to 7,000	26.2	0.27	0.9 to 2.4	0.29	4.06	85.6
Polycarbonate (PC) [Lexan™][g]	280	100 to 1,000	8.39	0.64	1.2 to 8.0	0.152	2.23	946.
	300	100 to 1,000	4.31	0.67	0.8 to 5.8	0.08	2.76	638.
	320	100 to 1,000	1.08	0.80	0.4 to 3.0	0.042	2.06	734.

[a] = ARCO.
[b] = Monsanto Co.
[c] = Exxon Chemical Co., USA.
[d] = Borger-Warner, Div. GE Plastics.
[e] = E.I. DuPont de Nemours and Co., Inc.
[f] = Allied Chemical Corp.
[g] = GE Plastics.

References

1. R.B. Bird, R.C. Armstrong and O. Hassager, *Dynamics of Polymer Liquids*, Vol. 1, John Wiley & Sons, New York (1977).
2. J.M. Dealy, *Rheometers for Molten Plastics*, Van Nostrand Reinhold, New York (1982).
3. L.E. Nielsen, *Polymer Rheology*, Marcel Dekker, New York (1977).
4. R.S. Lenk, *Polymer Rheology*, Applied Science Publishers, New York (1978).
5. J.A. Brydson, *Flow Properties of Polymer Melts*, Van Nostrand Reinhold, New York (1970).
6. R.G. Larson, *Constitutive Equations for Polymer Melts and Solutions*, Butterworths, Boston (1988).
7. J.A. Brydson, *Flow Properties of Polymer Melts*, Van Nostrand Reinhold, New York (1970), p. 17.
8. J.M. Dealy, *Rheometers for Molten Plastics*, Van Nostrand Reinhold, New York (1982), p. 9.
9. R.B. Bird, R.C. Armstrong and O. Hassager, *Dynamics of Polymer Liquids*, Vol. 1, John Wiley & Sons, New York (1977), p. 146.
10. R.B. Bird, R.C. Armstrong and O. Hassager, *Dynamics of Polymer Liquids*, Vol. 1, John Wiley & Sons, New York (1977), p. 146.
11. A.B. Glanvill, *The Plastics Engineer's Data Book*, Industrial Press, New York (1971), p. 165.
12. J.M. Dealy, *Rheometers for Molten Plastics*, Van Nostrand Reinhold, New York (1982), p. 18.
13. C.D. Han, *Rheology in Polymer Processing*, Academic Press, New York (1976), p. 123.
14. J.M. McKelvey, *Polymer Processing*, John Wiley & Sons, New York (1962), p. 78.
15. J.A. Brydson, *Flow Properties of Polymer Melts*, Van Nostrand Reinhold, New York (1970), p. 27.
16. R.B. Bird, W.E. Stewart, and E.N. Lightfoot, *Transport Phenomena*, John Wiley & Sons, New York (1960), p. 46.
17. K. Weissenberg, Nature, *159* (1947), p. 310.
18. A.S. Lodge, *Elastic Liquids*, Academic Press, New York (1964), p. 238.
19. N.N. Kapoor, MS Thesis in Chemical Engineering, University of Minnesota, Minneapolis (1964).
20. R.B. Bird, R.C. Armstrong and O. Hassager, *Dynamics of Polymer Liquids*, Vol. 1, John Wiley & Sons, New York (1977), p. 109.
21. E.A. Uebler, PhD Dissertation, Chemical Engineering, University of Delaware, Newark (1966).
22. A.S. Lodge, *Elastic Liquids*, Academic Press, New York (1964), p. 242.
23. R.B. Bird, R.C. Armstrong and O. Hassager, *Dynamics of Polymer Liquids*, Vol. 1, John Wiley & Sons, New York (1977), p. 101.
24. R.I. Tanner, J. Polym. Sci., A-2, *8* (1970), p. 2067.
25. R.B. Bird, R.C. Armstrong and O. Hassager, *Dynamics of Polymer Liquids*, Vol. 1, John Wiley & Sons, New York (1977), pp. 102–103.
26. P. Powell, *Engineering with Polymers*, Chapman and Hall, London (1983), p. 239.
27. R.J. Crawford, *Plastics Engineering*, Pergamon Press, New York (1981), p. 243.
28. R.J. Crawford, *Plastics Engineering*, Pergamon Press, New York (1981), p. 245.
29. R.J. Crawford, *Plastics Engineering*, Pergamon Press, New York (1981), p. 240.
30. R. J. Crawford, *Plastics Engineering*, Pergamon Press, New York (1981), p. 242.
31. P. Powell, *Engineering with Polymers*, Chapman and Hall, London (1983), p. 236.

32. P. Powell, *Engineering with Polymers,* Chapman and Hall, London (1983), p. 220.

33. J.M. Dealy, *Rheometers for Molten Plastics,* Van Nostrand Reinhold, New York (1982), p. 107.

34. R.B. Bird, R.C. Armstrong and O. Hassager, *Dynamics of Polymer Liquids,* Vol. 1, John Wiley & Sons, New York (1977), pp. 117–127.

35. J.M. McKelvey, *Polymer Processing,* John Wiley & Sons, New York (1962), p. 43.

36. J.M. McKelvey, *Polymer Processing,* John Wiley & Sons, New York (1962), p. 41.

37. J.M. McKelvey, *Polymer Processing,* John Wiley & Sons, New York (1962), p. 45.

38. R.B. Bird, SPE J., *11* (1955), pp. 35–40.

39. R.B. Bird, R.C. Armstrong and O. Hassager, *Dynamics of Polymer Liquids,* Vol. 1, John Wiley & Sons, New York (1977), pp. 247–249.

40. R.B. Bird, R.C. Armstrong and O. Hassager, *Dynamics of Polymer Liquids,* Vol. 1, John Wiley & Sons, New York (1977), p. 250.

41. J.M. Dealy, *Rheometers for Molten Plastics,* Von Nostrand Reinhold, New York (1982), p. 88.

42. S. Middleman, *The Flow of High Polymers,* Wiley Interscience, New York (1968), p. 33.

43. G. Dupp, Kunststofftechnik, *8* (1969), p. 273.

44. G. Dupp, Kunststofftechnik, *8* (1969), p. 271.

45. R. Spencer and G. Gilmore, J. Appl. Polym. Sci., *20* (1949), p. 502.

46. F. Rodriguez, *Principles of Polymer Systems,* McGraw-Hill Book Co., New York (1970), p. 21.

47. R.S. Spencer and G. Gilmore, J. Appl. Polym. Sci., *21* (1950), p. 523.

48. G. Campbell, Plastics Compounding, *11*:1 (1982), p. 77.

49. H. Saechtling, *International Plastics Handbook,* Hanser, Munich (1983), p. 354.

50. I. Klein, Plastics World, *38*:7 (1980), p. 29.

51. J.L. Throne, *Plastics Process Engineering,* Marcel Dekker, New York (1979), Chapter 14.

52. J.M. McKelvey, *Polymer Processing,* John Wiley & Sons, New York (1962), p. 121.

53. J.M. McKelvey, *Polymer Processing,* John Wiley & Sons, New York (1962), Section 4.1.

54. R.S. Spencer and R.F. Boyer, J. Appl. Phys., *17* (1946), p. 398.

55. J.M. McKelvey, *Polymer Processing,* John Wiley & Sons, New York (1962), p. 122.

56. C.D. Han, *Rheology in Polymer Processing,* Academic Press, New York (1976).

57. H.H. Hull, *An Approach to Rheology Through Multivariable Thermodynamics,* Deeds Associates, Society of Plastics Engineers, Brookfield Center, CT (1981).

58. P.J. Flory, *Principles of Polymer Chemistry,* Cornell University Press, Ithaca, NY (1953).

59. W.P. Brennan, "Characterization of Polyethylene Films in Differential Scanning Calorimetry", Thermal Analysis Application Study 24, Perkin-Elmer, Norwalk, CT (1978), p. 11.

60. O.A. Hougen and K.M. Watson, *Chemical Process Principles: Part Two, Thermodynamics,* John Wiley & Sons, New York (1947).

61. K.E. Bett, J.S. Rowlinson and G. Saville, *Thermodynamics for Chemical Engineers,* MIT Press, Cambridge (1975), Chapter 6.

62. O. Olabisi, L.M. Robeson and M.T. Shaw, *Polymer–Polymer Miscibility,* Academic Press, New York (1979).

63. R.M.N. Bonner, et al., "Properties of Molded Nylons", in M.I. Kohan, Ed., *Nylon Plastics,* Wiley-Interscience, New York (1973), p. 27.

64. K.G. Denbigh, *The Principles of Chemical Equilibrium*, Cambridge Press, London (1961), p. 221.

65. K.G. Denbigh, *The Principles of Chemical Equilibrium*, Cambridge Press, London (1961), p. 222.

66. P.J. Flory, *Principles of Polymer Chemistry*, Cornell University Press, Ithaca, NY (1953), p. 496.

67. F.W. Billmeyer, Jr., *Textbook of Polymer Science*, Wiley-Interscience, New York (1962), p. 26.

68. F.W. Billmeyer, Jr., *Textbook of Polymer Science*, Wiley-Interscience, New York (1962), p. 38.

69. P.J. Flory, *Principles of Polymer Chemistry*, Cornell University Press, Ithaca, NY (1953), p. 515.

70. F. Rodriguez, *Principles of Polymer Systems*, McGraw-Hill Book Co., New York (1970), p. 20.

71. F. Rodriguez, *Principles of Polymer Systems*, McGraw-Hill Book Co., New York (1970), p. 29.

72. F. Rodriguez, *Principles of Polymer Systems*, McGraw-Hill Book Co., New York (1970), p. 26.

73. P.J. Flory, *Principles of Polymer Chemistry*, Cornell University Press, Ithaca, NY (1953), p. 545.

74. P.J. Flory, *Principles of Polymer Chemistry*, Cornell University Press, Ithaca, NY (1953), p. 615.

75. J.L. Been, "Bonding", in N.M. Bikales, Ed., *Adhesion and Bonding*, Wiley-Interscience, New York (1971), p. 125.

76. J.L. Been, "Bonding", in N.M. Bikales, Ed., *Adhesion and Bonding*, Wiley-Interscience, New York (1971), p. 134.

77. K.G. Denbigh, *The Principles of Chemical Equilibrium*, Cambridge Press, London (1961).

78. J. Crank and G.S. Park, Eds., *Diffusion in Polymers*, Academic Press, New York (1968).

79. J. Crank and G.S. Park, "Methods of Measurement", in J. Crank and G.S. Park, Eds., *Diffusion in Polymers*, Academic Press, New York (1968).

80. K. Ueberreiter, "The Solution Process", in J. Crank and G.S. Park, Eds., *Diffusion in Polymers*, Academic Press, New York (1968).

81. K. Ueberreiter, "The Solution Process", in J. Crank and G.S. Park, Eds., *Diffusion in Polymers*, Academic Press, New York (1968), p. 247.

82. K. Ueberreiter, "The Solution Process", in J. Crank and G.S. Park, Eds., *Diffusion in Polymers*, Academic Press, New York (1968), p. 234.

83. K. Ueberreiter, "The Solution Process", in J. Crank and G.S. Park, Eds., *Diffusion in Polymers*, Academic Press, New York (1968), p. 246.

84. K. Ueberreiter, "The Solution Process", in J. Crank and G.S. Park, Eds., *Diffusion in Polymers*, Academic Press, New York (1968), p. 244.

85. C.E. Rogers, "Permeability and Chemical Resistance", in E. Baer, Ed., *Engineering Design for Plastics*, Reinhold, New York (1964), p. 643.

86. V. Stannett, "Simple Gases", in J. Crank and G.S. Park, Eds., *Diffusion in Polymers*, Academic Press, New York (1968), p. 65.

87. V. Stannett, "Simple Gases", in J. Crank and G.S. Park, Eds., *Diffusion in Polymers*, Academic Press, New York (1968), p. 65.

88. C.E. Rogers, "Permeability and Chemical Resistance", in E. Baer, Ed., *Engineering Design for Plastics*, Reinhold, New York (1964), p. 633.

89. V. Stannett, "Simple Gases", in J. Crank and G.S. Park, Eds., *Diffusion in Polymers*, Academic Press, New York (1968), p. 60.

90. C.E. Rogers, "Permeability and Chemical Resistance", in E. Baer, Ed., *Engineering Design for Plastics*, Reinhold, New York (1964), p. 647.

91. C.E. Rogers, "Permeability and Chemical Resistance", in E. Baer, Ed., *Engineering Design for Plastics*, Reinhold, New York (1964), p. 656.

92. C.E. Rogers, "Permeability and Chemical Resistance", in E. Baer, Ed., *Engineering Design for Plastics*, Reinhold, New York (1964), p. 658.

93. C.E. Rogers, "Permeability and Chemical Resistance", in E. Baer, Ed., *Engineering Design for Plastics*, Reinhold, New York (1964), p. 653.

94. H. Fujita, "Organic Vapors Above the Glass Transition Temperature", in J. Crank and G.S. Park, Eds., *Diffusion in Polymers*, Academic Press, New York (1968), p. 95.

95. C.E. Rogers, "Permeability and Chemical Resistance", in E. Baer, Ed., *Engineering Design for Plastics*, Reinhold, New York (1964), p. 655.

96. C.E. Rogers, "Permeability and Chemical Resistance", in E. Baer, Ed., *Engineering Design for Plastics*, Reinhold, New York (1964), p. 660.

97. K.N. Mathes, "Electrical Properties", in E. Baer, Ed., *Engineering Design for Plastics*, Reinhold, New York (1964), p. 519.

98. K.N. Mathes, "Electrical Properties" in E. Baer, Ed., *Engineering Design for Plastics*, Reinhold, New York (1964), p. 518.

99. R.D. Deanin, *Polymer Structures, Properties and Applications*, Cahners Books, Boston (1972), p. 95.

100. R.D. Deanin, *Polymer Structures, Properties and Applications*, Cahners Books, Boston (1972), p. 100.

101. H. Saechtling, *International Plastics Handbook*, Hanser, Munich (1983), p. 295.

102. K.N. Mathes, "Electrical Properties", in E. Baer, Ed., *Engineering Design for Plastics*, Reinhold, New York (1964), p. 520.

103. J.A. Barrie, "Water in Polymers", in J. Crank and G.S. Park, Eds., *Diffusion in Polymers*, Academic Press, New York (1968), p. 274.

104. A. Tadmor and C.G. Gogos, *Principles of Polymer Processing*, Wiley-Interscience, New York (1979), p. 694.

Glossary

Apparent Shear Rate The shear rate given by volumetric flow divided by a simple geometric factor.

Apparent Viscosity The ratio of apparent shear rate to shear stress at a wall.

Bingham Fluid A material that is Newtonian above a critical shear stress and Hookean elastic below.

Capillary Very small diameter tube used in rheological and viscometric measurements.

Cohesive Energy Density The square of the solubility parameter.

Cold Crystallization Temperature Beginning from room temperature, the temperature at which a quenched amorphous polymer begins to crystallize.

Dimensional Stability A measure of the time-dependent volume change of a polymer.

Elastic Liquid Specific class of viscoelastic liquid, defined by tensor-driven constitutive equations.

Elastoplastic Solid Solid-like at low stress, then yielding with deformation dependent on stress level.

Elastoviscous Solid   Solid-like at low stress, then yielding with rate of deformation dependent on stress level, with complete recovery after stress removal.

Energy of Activation   A measure of a material response to thermal changes.

Extrudate Swell   The ratio of extrudate diameter to tube internal diameter.

Hysteresis   The phenomenon of material response to increasing load differing from response to decreasing load.

Interaction Parameter   A thermodynamic measure of the ability of a solvent to dissolve or swell a polymer.

Inviscid Fluid   A fluid having zero viscosity. A fluid that once set in motion, remains forever in motion.

Newtonian Fluid   A fluid that has a shear- and time-independent viscosity.

Normal Stress Difference   The difference in any two stresses, given that each is applied and reacts in the same direction.

Post-Mold Shrinkage   Manifestation of time-dependent dimensional change in a polymer owing to material being in thermodynamic nonequilibrium.

Power-Law Fluid   A viscous-only fluid flow model in which the viscosity is proportional to the shear rate to a power, n.

Rheology   Study of fluid strain response to applied stress.

Solute   The polymer, when dissolved in a solvent.

Solvation   The physical migration of small molecules between polymer chains, acting to separate them, producing a solution.

Solubility Parameter   A thermodynamic measure of the polar, ionic and dispersive forces of a given substance.

Spencer–Gilmore Equation   An empirical thermodynamic P-v-T equation of state for polymers.

Steady Flow   Time-independent flow.

Steric Hindrance   Interference in molecular packing or order owing to the presence of large pendant groups or long side chains.

Thermodynamic Equation of State   An equation relating pressure, volume, and temperature of a polymer.

Theta Temperature   A temperature above which solution is preferred and below which phase separation is preferred.

True Shear Rate   Actual rate at which the fluid is being sheared.

Viscoelastic Liquid   A material that behaves like an elastoviscous solid, but does not achieve complete recovery.

Viscometric Flow   Flow in which fluid behaves as if it is in simple shear.

Viscosity   Ratio of local shear stress to local shear rate.

Viscous Dissipation   Unrecoverable work of deformation, resulting in material melt temperature increase.

Wall Shear Stress   Stress the fluid exerts on any confining solid wall.

CHAPTER 5

PROCESSING

Read

394-404
405-419
431-453
454-461

5.1 Introduction

Processing is the generic term for converting raw polymer and its adducts (additives, colorants, stabilizers, minor amounts of other polymers, fillers, reinforcements and so on) to useful products of commerce. A significant amount of the final product performance depends upon proper selection of the polymer and its adducts (1). The influence of the processing environment on the final product performance must also be considered. For commodity thermoplastics such as polyethylenes (PE) and polypropylenes (PP), the raw reactor powder is converted to resin pellets by compounding (an extrusion process whereby the polymer is heated to a plastic state, then shear-mixed at high pressures). The pellets are then converted into continuous finished goods by a second extrusion step or into discrete shapes by injection molding. The processing conditions can act to alter or even to interfere with the intrinsic properties of the chosen polymer and its adducts. For many thermosets, the processing steps convert the pre-polymer into the final product shape. Polymer-process interaction is most important to the development of the final product properties. There are two major concerns in the manufacture of any product:

- Will the finished part meet all required and specified design criteria?
- Can the part be produced at the minimum cost for the projected market size?

These appear to be very simple criteria. In reality, they are strongly interrelated and usually require complex technical and marketing analyses. A typical schematic flowchart is shown in Figure 5.1. The design engineer must realize that in many applications, several polymers, including thermoplastics and thermosets, may satisfy the design criteria. Then secondary criteria (environmental concerns, scrap disposal, reliability of machinery, labor force trained in processing a specific polymer in a specific way) may direct the selection process.

Rarely is a polymer processed into an article of commerce without applications of external heat, shear or pressure in some time-dependent fashion. There are many ways of converting polymers into finished products. Combinations of processes are employed in most commercial manufacturing operations. This chapter serves to acquaint the design engineer with some of the more common polymer processing techniques, indicate some of the important polymer properties that influence the efficiency of the processes, and demonstrate how the processes can affect the performance of the polymer in the final product. For the individual wishing to learn the fundamentals of the individual processes, many texts that deal in depth with a specific process family are listed in the *General References* section (p. 499).

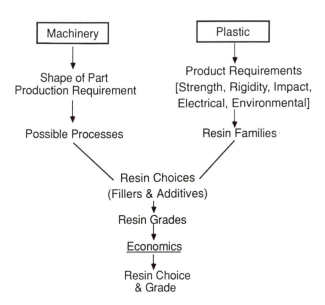

Figure 5.1.   Schematic for Choosing a Polymer and Attendant Processing Technique.

5.2 Categories of Plastics Processing Techniques

Polymer processes can be cataloged in many different ways. A common way is to consider the type of flow a polymer fluid element undergoes during shaping into the final design configuration (2,3). This is a classical approach, similar to *Unit Operations* in chemical engineering. The process engineer focuses on the rheological considerations of the polymer and the resulting residual stresses in the finished article, thus combining elements of basic fluid and solid mechanics in the study of polymers. As seen in Table 5.1, the viscosity and other rheological characteristics of polymers can dictate the suitability of a given process. Extrusion and fiber spinning are examples of continuous flow, whereas injection molding and blow molding are examples of cyclical

Table 5.1. Processing variables for major polymer processes.

The following are vertical captions
a. Thermoset
b. Thermoplastic
c. Melt Fracture
d. Extrudate Swelling
e. Melt Strength
f. Rate of Reaction
g. Rate of Crystallization

Process	a	b	Viscosity	c	d	e	f	g
Blow Molding		x	Medium			x		x
Calendering		x	Medium					
Casting	x	x	Very Low				x	x
Compression Molding	x		High				x	
Extrusion, Film		x	Medium	x	x	x		x
Extrusion, Profile		x	Medium	x	x			x
Extrusion, Sheet		x	Medium	x	x			x
Filament Winding	x		Medium				x	
Hand Layup or Sprayup	x		Medium				x	
Injection Molding, Compact		x	Low to Medium	x	x			x
Injection, Molding, Foam		x	Low to Medium	x		x		x
Injection Molding, Reactive	x		Low to Medium				x	
Machining	x	x						
Melt Flow Stamping		x	Low to Medium					x
Pultrusion	x		Medium				x	
Rotational Molding	x	x	Medium to Low				x	x
Thermoforming		x	Medium			x		x
Transfer Molding	x		Medium				x	

flow. Thermoforming is an example of near-elastic sheet forming and flowing powders are used in rotational molding.

In industrial design, the geometric shape of the final part is important. *Shape* can be used to classify the various polymer processes as seen in Table 5.2. So long as the final product is a single part, this classification is valid. However, many products are assemblies of several parts, with performance criteria and material characteristics differing from part to part. The specific interrelationships of the parts, including alignment, shrinkage, thermal expansion, color matching and the like are not considered of primary importance in this classification method.

A more acceptable engineering approach is to identify the primary design characteristics of each part insofar as:

- shape limitations,
- maximum size, and

some arbitrary design considerations such as one or more of the following:

- complex shape,
- controlled wall thickness,
- open or closed hollow shape,
- very small item, or

Table 5.2.   Part geometry as a way of classifying polymer processes.

Linear forming
   Extrusion
   Pultrusion

Formation of a solid body by injection into a cavity
   Solid (thermoplastic or thermosetting)
   Thermoplastic foam (TSF)
   Reaction injection molding (RIM)

Formation of a hollow object
   Blow molding
   Rotational molding
   Filament winding

Formation of a solid object by prefilling a cavity
   Compression molding (BMC)
   Matched die molding (SMC)
   Melt flow stamping

Sheet forming
   Thermoforming
   Stretch blow molding

- plane area $> 10 \text{ ft}^2$ (0.015 m^2),
- inserts,
- molded-in holes,
- threads.

Table 5.3 compares some of the more common processes based on these criteria. An important aspect in choosing a particular polymer and ancillary process is *prod-*

Table 5.3. Shape and size as ways of classifying polymer processes.

The following are vertical captions
a. Shape Limitation
b. Factor Limiting Maximum Size
c. Complex Shapes
d. Controlled Wall Thickness
e. Open Hollow Shapes
f. Closed Hollow Shapes
g. Very Small Items
h. Plane Area $> 10 \text{ ft}^2$
i. Inserts
j. Molded-in Holes
k. Threads

Process	a	b	c	d	e	f	g	h	i	j	k
Blow Molding	Hollow, thin Wall	Platen			x	x		x			x
Calendering	Sheet	Width of Roll		x				x			
Casting		Mold	x	x					x	x	x
Compression Molding		Platen	x	x	x				x	x	x
Extrusion, Film	Sheet or Tube	Die		x				x			
Extrusion, Profile	Linear	Die		x							
Extrusion, Sheet	Sheet	Width of Roll		x				x			
Filament Winding	Surface of Revolution			x		x		x			
Hand Layup or Sprayup	Large Thin Wall	Mold		x	x				x	x	x
Injection Molding, Compact		Platen	x	x	x		x	x	x	x	x
Injection Modling, Foam		Platen	x		x			x	x	x	x
Injection Molding, Reactive		Platen	x		x			x	x	x	x
Machining			x	x	x		x				x
Melt Flow Stamping		Platen			x				x	x	x
Pultrusion	Linear	Die		x							
Rotational Molding	Hollow					x		x	x	x	x
Thermoforming	Thin Wall	Platen				x		x			
Transfer Molding		Platen	x	x	x		x		x	x	x

uct demand. Each process has a production rate range and a capital cost associated with that range. Most processing equipment is available in a range of sizes and the capital cost related to a specific piece of processing equipment and ancillary dies and molds will vary. The following arbitrary scale allows a qualitative comparison between the unit capital costs for different processing options:

VL (very low)	<	$20,000
L (low)	<	$50,000
M (medium)	<	$100,000
H (high)	<	$250,000
VH (very high)	>	$250,000

Table 5.4 compares capital costs for small and large manufacturing units. Annual production volumes are also given. Processes can then be ranked in terms of cost effectiveness per unit cost by combining these ranges with approximate cycle times per manufacturing unit, Table 5.5. As noted earlier, polymers are either thermoplastic or thermosetting. As seen in Tables 5.6 and 5.7, this can be used as a method of classification, as well. In addition, in many practical product applications, real plastic processes can be quite economically competitive (4). Certain processes are more econom-

Table 5.4. Comparative capital costs and production rates for major polymer processes.

Process	Equipment costs	Tooling costs	Production volume/year (10^1 – 10^7)
Blow Molding	M to H	L to M	10^4 ←——————→ 10^7
Calendering	VL		10^4 ←——————→ 10^7
Casting	VL	VL	10^1 ←————→ 10^3
Compression Molding	M to H	M	10^4 ←————————→ 10^6
Extrusion, Film	H to VH	H	10^5 ←————→ 10^7
Extrusion, Profile	M to H	M	10^4 ←——————→ 10^7
Extrusion, Sheet	H to VH	H	10^6 ←——→ 10^7
Filament Winding	M		10^1 ←————→ 10^3
Hand Layup or Sprayup	L	VL to M	10^1 ←————→ 10^3
Injection Molding, Compact	M to VH	M to H	(Hand Unit) 10^1 ←→ 10^2 (Automatic Unit) 10^4 ←——————→ 10^7
Injection Molding, Foam	VH	M to H	10^4 ←————→ 10^6
Injection Molding, Reactive	L to M	L	10^4 ←————→ 10^6
Machining	L	VL	10^1 ←————→ 10^3
Melt Flow Stamping	M to H	M	10^4 ←————→ 10^6
Pultrusion	L	L	10^1 ←————→ 10^3
Rotational Molding	M to H	L	10^1 ←——→ 10^3 10^5 ←——→ 10^6
Thermoforming	M to H	L	10^1 ←——→ 10^3 10^5 ←————→ 10^7
Transfer Molding	M to H	M to H	10^4 ←————→ 10^6

Table 5.5.   Ranking of major polymer processes according to unit cost.

Process	Cost
Calendering Film Extrusion Solid Injection Molding Wire and Cable Extrusion	Very Low
Blow Molding* Foam Injection Molding Melt Flow Stamping Profile Extrusion Rotational Molding* Sheet Extrusion Thermoforming Transfer Molding Reaction Injection Molding	Low
Blow Molding* Compression Molding Powder Coating Pultrusion Rotational Molding	Medium
Casting Filament Winding Hand Layup or Sprayup Machining	High

*Varies with Part Size.

ically attractive at low product volume, others at very high volume. Thus, the design engineer must not only select the appropriate material to meet the product performance critera, he/she must determine the effect of processing on material properties and select a process that is economically viable. Unfortunately, exhaustive comparative economic analyses are beyond the scope of this book.

It should be apparent that each classification technique has its own limitations. For example, a design approach based entirely on geometric shape would not include thermoforming as a process for producing a closed liquid container. However, many fruit drinks are sold in thermoformed containers. In one case, two sections of the container are thermoformed separately, then joined by a process known as spin welding.

Then, how can rational decisions be made concerning the "correct" resin and attendant process? A logical decision-making process follows the scheme in Figure 5.1. However, when comparing equally balanced alternatives on either polymer or process or both, the design engineer must call upon prior experience. In many instances, different design engineers will solve the same design problem with different approaches. Each may be the "best" solution for the particular company, based on its present facilities, labor-force skills, and design "know-how" developed over many years with selected classes of polymers or types of process.

Table 5.6. Primary and secondary processing methods for several thermoplastics.

The following are vertical captions
a. ABS
b. Polyacetal
c. Acrylic (PMMA)
d. Fluoropolymers (melt processed)
e. Nylon (polyamide)
f. Polybutylene terephthalate (PBT)
g. Polyethylene terephthalate (PET)
h. Polycarbonate (PC)
i. Polyethylene
j. Modified polyphenylene oxide (mPPO)
k. Polypropylene (PP)
l. Polystyrene (PS)
m. Polyvinyl chloride (PVC)
n. Styrene-acrylonitrile (SAN)

Process	a	b	c	d	e	f	g	h	i	j	k	l	m	n
Blow Molding	+	+			+	+	X	+	X	+	X	+	X	+
Calendering													+	
Casting			X		+								X	
Compression Molding														
Extrusion, Film				X	X		X		X		X		X	
Extrusion, Fiber				X	X		X				X			
Extrusion, Profile	X	+	X		+			+	+		+	+	X	+
Extrusion, Sheet	+		X					+	+	+	+	X	+	+
Extrusion, Wire & Cable					X				X		+		X	
Filament Winding														
Hand Layup or Sprayup														
Injection Molding, Compact	X	X	X	+	X	X	X	X	X	X	X	X	+	X
Injection Molding, Foam	+					+		+	+	+	+	+		
Injection Molding, Reactive														
Lamination														
Melt Flow Stamping					+			+			+			
Powder Coating				+	+				+					
Pultrusion														
Rotational Molding	+				+			+	X			+		
Thermoforming	X		X				X	+	+	+	X	X	X	X
Transfer Molding														

X denotes Primary
+ denotes Secondary

Table 5.7. Primary and secondary processing methods for several thermosetting resins.

The following are vertical captions
a. Alkyd
b. Allyl (DAP)
c. Amino melamine
d. Urea
e. Epoxy
f. Phenolic
g. Unsaturated Polyester Resin (UPE)
h. Polyurethane (PUR)
i. Silicone

Process	a	b	c	d	e	f	g	h	i
Blow Molding									
Calendering									
Casting					X		X	+	X
Compression Molding	X	+	X	X	+	X	X		+
Extrusion, Film									
Extrusion, Fiber									
Extrusion, Profile									+
Extrusion, Sheet									
Extrusion, Wire & Cable									+
Filament Winding					+		X		
Hand Layup or Sprayup					+		X		
Injection Molding, Compact	+	+		+	+		+		+
Injection Molding, Foam									
Injection Molding, Reactive								X	
Lamination		+	+		+	+	+		+
Melt Flow Stamping									
Powder Coating					X		X		
Pultrusion					+		X		
Rotational Molding							+	+	
Thermoforming									
Transfer Molding	X	X		+	X	+	+		X

X denotes Primary
+ denotes Secondary

5.3 Selected Plastics Processing Techniques

It is necessary to first discuss several of the more important polymer processes and to show some interrelationships of the important processing parameters. This will help the design engineer to gain an appreciation of how processing parameters can affect the polymer molecule and how this might influence the physical characteristics of the finished part. In many cases, the thermoplastic resin process is considered first, followed by a similar thermosetting resin process. Again, the engineer must realize that the material that follows is not an in-depth treatise on polymer processes.

Extrusion

Extrusion is the most widely used way of fabricating plastic products. The process is used for thermosetting and thermoplastic resins. Many thermoplastics are produced as reactor powders. These powders are frequently mixed with additives, colorants, stabilizers, fillers, and even other polymer powders. In the process known as *compounding,* this mixture is heated to fluid state, mixed and shaped into pellets. One basic compounding device is the screw extruder.

In its simplest form, an extruder consists of a high strength constant diameter tube, called the *barrel,* in which a tapered, flighted *screw* is concentrically mounted (Figure 5.2). For thermoplastic resins, the extruder is used to melt and homogenize the resin. The resin, in the form of pellets or powders, is fed into the barrel through a hopper to a feed throat. It passes down the barrel in the space between the barrel and the channel made by the sides of the screw flights and the screw root and exits at the head end of the extruder, into a *die* of some sort. The die shapes or forms the polymer melt. Typically a *screen pack* or filter is mounted on the head end of the barrel to remove foreign substances from the melt stream before it enters the die.

One screw extruder design characteristic is inside barrel diameter (1 inch, 2 inch, 57 mm, 90 mm and so on). Screw diameters can be as large as 12 to 16 inches for reactor resin compounding and as small as 1/2 to 1 inch for laboratory use. Another characteristic is the screw length-to-inside diameter ratio or *L/D.* Short extruders have L/Ds of 12 to 15. Moderately long extruders have L/Ds of 15 to 25. Long extruders, used in compounding, film extrusion, and devolatilization, can have L/Ds of 36 to 42 or more. The screw is rotated at constant speed by an electric motor connected through a speed reducer.

A third characteristic is the compression ratio of the screw. This is the depth of the screw channel at the feed zone divided by that at the head or die end. Screw compres-

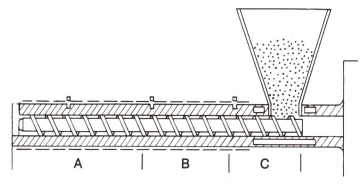

Figure 5.2. Breakaway Section of Conventional Single-Screw Extruder (5). A: Metering Section. B: Compression and Plasticating Section. C: Solids Conveying Section.

sion ratios depend on the physical form of the resin (powder or pellets) and on the nature of the polymer itself. Crystalline polymers require lower compression ratios than amorphous ones. This is because crystalline polymers have greater specific volume changes between room temperature and processing temperature than do amorphous polymers. The primary mechanism of transferring energy to the resin is by mechanical action of the screw, or "shear work." As a result, the extruder power train is usually very heavy duty. As an example, a blown film line with a 1-1/2 inch, 24:1 extruder for processing low-density polyethylene can have a 10 HP variable speed DC drive and can have screw speeds of 20 to 100 RPM.

mechanical Energy {

The screw is turned at relatively low speed (to 200 RPM) through a variable speed gearbox by a high-torque DC motor. As seen in Figure 5.3, at the motor end, the barrel is opened to allow resin to be fed continuously into the screw flights from a simple hopper. The barrel is heated, usually with electrical resistance heaters in concentric bands. Small or short L/D extruders usually have two or three heating zones, demarked by the location of the heater bands. Large or long extruders may have six to eight zones. The barrel temperature in a given zone is measured by a thermocouple embedded in the barrel wall. The barrel temperature is usually proportional-integral controlled to within 5°F (3°C) of the chosen setpoint. Typical design and operating characteristics of single screw extruders are listed in Table 5.8.

Heater Energy {

An appropriate *die* is placed on the head of the barrel at the free end of the screw. The die shapes the molten polymer into the desired shape. Generally, a cylindrical die produces a rod extrudate. A slot die produces a sheet. An annular die produces a tubular film, and so on. Typically the screw flight has the same pitch, or angle to the screw axis, throughout the screw length (Figure 5.4).

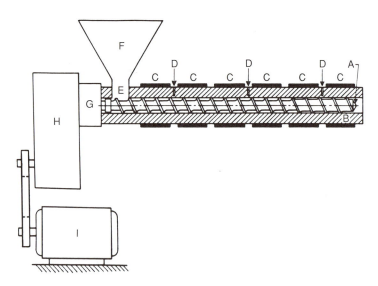

Figure 5.3. Schematic of Conventional Single-Screw Extruder (6). A: Screw. B: Barrel. C: Heater Band. D: Thermocouple Well. E: Feed Throat. F: Hopper. G: Main Thrust Bearing. H: Gear Reducing Box. I: DC Motor.

Table 5.8. Typical characteristics for single-screw extruders (7).

Extruder size		Average drive (HP)	Output		Barrel heater (kW)
(in)	(mm)		(lb/h)	(kg/h)	
1½	38.1	10 to 15	50 to 75	23 to 34	7.5
2½	63.5	20 to 30	120 to 160	54 to 73	21
3½	88.9	40 to 75	250 to 400	113 to 181	45
4½	114.3	80 to 125	400 to 700	181 to 318	75
6	152.4	150 to 225	800 to 1,200	363 to 544	140
8	203.2	300 to 500	1,500 to 2,000	680 to 907	225

The plastic is supplied to the extruder hopper as a powder, granule or pellet. As a result, it usually has a lower bulk density than either the solid or fluid polymer. As a result, the first few flights of the screw are quite deep. That is, the screw root diameter is relatively small. This section of the screw is the *solids conveying zone*. The compacted polymer is pressed against the hot barrel wall, where it melts or softens. The melt film that forms is conveyed toward the *rear* of the flight chamber (Figure 5.5), where it accumulates in a pool, called the *melt pool*. As the solid polymer melts or softens and the melt pool grows, the bulk density of the mixture decreases. To keep the polymer mixture moving toward the die, the screw root at this point is gradually increased in diameter. The narrowing of the *flow channel* also causes the pressure on the mixture to increase substantially. The melting or softening of a polymer in a shear field is *plastication*. The section of the screw where plastication usually takes place is the *compression zone*.

The last section of the screw extruder is the *metering zone*. The molten or softened polymer from the plastication zone is now homogenized and its pressure increased so that the highly viscous fluid can be pumped through the usually narrow constrictions of the die. As noted, the compression ratio of the screw is a design factor for extrusion. For certain materials such as polyamide (nylon), the compression ratio can be as low as 1.5:1. For others, such as polyethylene low-bulk-density powder, the compression

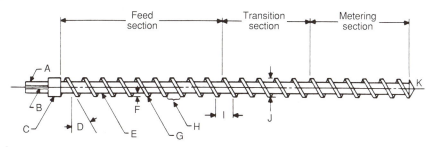

Figure 5.4. Characteristics of a Conventional Single-Flighted Extruder Screw (8). A: Shank. B: Key. C: Hub. D: Screw Flight Helix Angle. E: Screw Root. F: Channel Depth. G: Screw Flight. H: Channel Width. I: Pitch. J: Screw Diameter. K: Screw Tip.

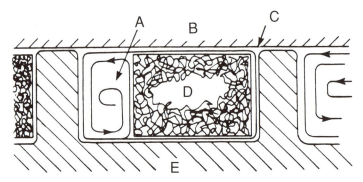

Figure 5.5.   Schematic of Melting Mechanism in a Single-Screw Extruder (9). A: Melt Pool.
B: Barrel. C: Molten Polymer Film. D: Solid Bed of Polymer Granules or Powder. E: Screw.

ratio can be as large as 4:1. For styrenics and PVC, a compression ratio of 2.5:1 or so
is used.

Example 5.1

Q: Estimate the size of the motor on an extruder that will deliver 100 lb/h (44.6 kg/h) of
polypropylene (PP) at 200°C and 2000 lb$_f$/in^2. Assume the drive train is 90% efficient and the
electric barrel heaters supply energy equal to the amount of heat loss from the barrel.

A: Since the electrical heaters supply energy sufficient to compensate for the heat loss from
the barrel, the energy supplied by the screw must equal the enthalpy change of the polymer.

$$HP = \dot{m}\ \Delta H/\eta,$$
$$= \dot{m}\ (\Delta H_1 + \Delta H_2)/\eta,$$

where $\dot{m}$ is the mass flow rate and η is the efficiency of the process. ΔH_1 is the enthalpy change
at 1 atm, $= 0.14$ KWh/kg from Figure 4.30, and ΔH_2 is the enthalpy change at constant temperature, $= \omega\ \Delta P = 0.62$ cm^3/g $(2000 - 14.7)$ lb$_f$/in^2 $= 0.0024$ KWh/kg. Thus $\Delta H = 0.142$
KWh/kg, and:

$$HP = 44.6\ (kg/hr) \times 0.14\ (KW\ h/kg)/0.9 = 6.94\ KW = 9.4\ HP$$

The extruder described above is a *single-screw extruder.* It is the most prevalent
processing device in the plastics industry. In some processes such as devolatilization
(liberating volatiles such as solvents or water from molten polymers), the screw root
diameter is decreased at some distance into the metering zone. This causes the melt
pressure to drop in that *devolatilizing zone.* If a vacuum port is provided in the barrel
at that point, volatiles can be extracted. This technique can also be used to continue
polymerization reactions of condensation polymers. After the devolatilizing zone, the

melt pressure is rebuilt with another metering zone. This screw extruder is a *two-stage* extruder. Two-stage single screw extruders are also used to produce foamed plastics. Instead of evacuating volatiles in the *pressure let-down zone,* gas or volatile liquids such as chlorofluorocarbons are injected at high pressure. Under pressure in the metering zone, these foaming agents dissolve in the melt. When the melt is passed through the die to atmospheric pressure, the foaming agent comes from solution to produce a polymeric foam.

Extruder Die

A die is mounted at the head or open end of the extruder barrel. The function of the die is to form the molten polymer into the desired shape. Extruder dies can be quite simple or very complex. If the extruder is used to *compound* polymers and adducts, the product is (usually) nearly cylindrical strands. These strands are usually water-cooled and chopped into pellets for subsequent processing. The trueness of the strand shape is not as critical as the homogeneity of the resin in the pellet. The die configuration becomes critical if the extrudate is to be a product, such as a pipe or *profile*. In addition to being hydraulic fluids, most polymers are elastic liquids. To squeeze these materials through constrictions at high speeds, the extruder must develop very high pressure in the melt, typically 2000 to 6000 lb_f/in^2 (14 to 42 MPa). The elastic nature of the liquid causes it to expand as it issues from the die. This is *extrudate swell.* This expansion is not well understood and cannot be reliably predicted from material properties. As a result, dimensional tolerances on the extrudate can be very difficult to control. Very close dimensions are obtained only by fine-tuning the die geometry. In extrusion of sheet and film, the polymer thickness is measured with a nuclear gage (beta gage) and the signal is fed back to hydraulically-controlled die lips.

The extrudates for many polymers are quite smooth. At high extrusion rates, some polymers such as high-density polyethylene yield very irregular extrudates. The rough extrudate is said to exhibit *melt fracture* due primarily to the viscoelastic nature of the polymer at the processing condition. Changes in the die geometry, processing temperature and pressure, and in the nature of the polymer are usually warranted when this effect is prevalent. Several of the more common applications are given below.

> **Compounding.** Compounding is the process in which fillers, reinforcing elements, additives and other adducts are physically mixed into a molten polymer. The adducts can be added with the resin or at some point along the extruder barrel where the polymer is fluid. The resin is then forced through many small circular holes in a single die plate, called a strand die. The strands of molten polymer are then cooled and cut into pellets, typically 1/8 inch diameter by 3/16 to 1/4 in long (3 × 5 to 6 mm).

> **Film Extrusion.** There are several ways of forming polymer films. The simplest is cast film. The film produced in this fashion is not oriented. Blown film is biaxially oriented. These are discussed briefly below.

Cast Film. Molten resin is extruded through a linear *coat hanger die* into a thin wide sheet that is *cast* onto a highly polished *chill roll,* cooled and wound onto a roll, Figure 5.6. If multiple extruders are used, several extrudates of several types of polymers can be simultaneously fed through a die-block, Figure 5.7. Typical processing rates for single component cast-film extruders are 10 to 30 lb plastic per inch of die width (0.2 to 0.6 kg/mm). This corresponds to a sheet velocity of 500 to 1500 ft/min (150 to 450 m/min) for film that is 0.0005 to 0.002 in (0.01 to 0.05 mm) thick. When the film thickness exceeds 0.002 in (0.05 mm) or when the film is embossed, the sheet velocity drops to about 500 to 700 ft/min (150 to 200 m/min).

Single Layer Blown Film. In blown film processing, the film die is an annular ring with a very small gap width. The thin ring of molten thermoplastic resin that extrudes is radially and axially stretched by maintaining a slight positive pressure on the inside closed volume or bubble of the extrudate, Figure 5.8. The thickness or gage of the film depends on the die gap width, the draw-down ratio and the degree of inflation or *blow-up ratio.* The film is cooled on its outer surface by air, forced from a ring at the die exit. In some instances, internal surface cooling is also used. The blown film bubble is retained in a metal tower. At the top is a set of *nip* rollers that maintain the inflation air within the film bubble. When multiple layers of different resins supplied from multiple extruders are combined within the die, a coextruded laminate film is produced.

Blown film lines are sized in terms of the diameter of the ring, in the typical range of 4 to 86 in (100 to 2200 mm), or in terms of the lay-flat width of the product, in the typical range of 12 to 240 in (300 to 600 mm). Blown film lines typically operate in the range of 10 to 20 lb/h per inch of annular ring circumference (0.2 to 0.4 kg/h mm). For high-density polyethylene (HDPE) film of 0.0005 to 0.001 in (0.01 to 0.02 mm) thickness, the processing rate is 200 to 400 ft/min (60 to 120 m/min). For coextrusion systems, the rates

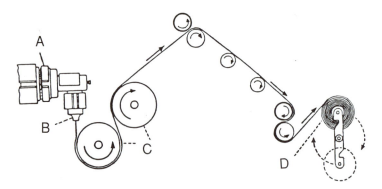

Figure 5.6. Schematic of a Cast Film Line (10). A: Extruder. B: Sheet Die. C: Chill Rolls. D: Windup Rolls.

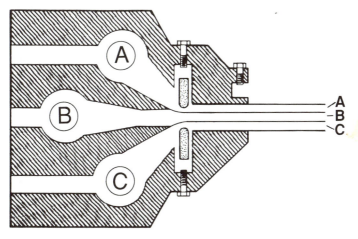

Figure 5.7.   Schematic of a Three-Layer Cast Film (A, B, C) Coextrusion Die Block (11).

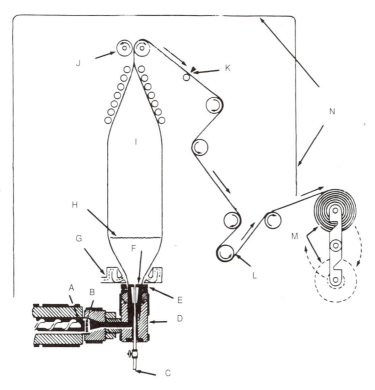

Figure 5.8.   Schematic of Blown Film Line (12). A: Screen Pack. B: Breaker Plate. C: Internal Air Line. D: Blown Film Die. E: Die Ring. F: Film Gage Mandrel. G: External Air Ring. H: Frost Line. I: Blown Film Bubble. J: Nip Rolls. K: Surface Treater Bar. L: Tension Rolls. M: Windup Rolls. N: Support Frame.

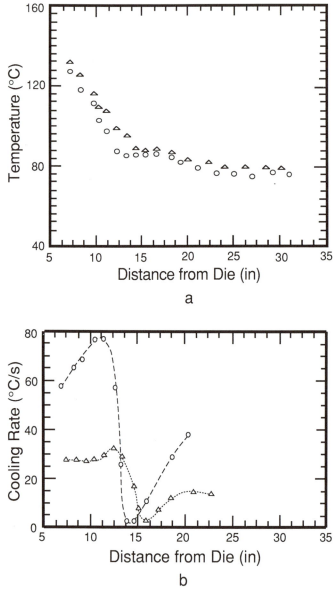

Figure 5.9. (a) Temperature and (b) Cooling Rate Profiles for Low-Density Polyethylene (LDPE) Blown Film (13). Blow Up Ratio (BUR) 3:1. 6 lb/h per in of Die Circumference. Frost Line at 17 in. Circle: 0.0015 in Thick Film. Triangle: 0.003 in Thick Film.

are 100 to 200 ft/min (30 to 60 m/min), depending on film gage and resin characteristics.

The majority of the resins used to produce biaxially oriented blown films are crystalline or semicrystalline and most of these are polyolefins. The region of maximum shear rate in the film apparently occurs just as the falling film temperature reaches the high crystallization rate region for the polymer. The time- or position-dependent film temperature can be measured with infrared means. Typical temperature profiles are seen in Figure 5.9. After dropping nearly linearly from the melt temperature, the temperature plateaus or even rises slightly for some time before again decreasing. The abrupt change marks the region of rapid crystallization. The exothermic heat of crystallization can more than offset the convective heat removal rate. As the film cools and crystallizes, it rigidifies. There is usually a sharp demarcation between the transparent extruded film and the hazy or milky translucent crystalline film. This is called the *frost line* (Figure 5.8). The film beyond the frost line is very stiff and therefore no appreciable biaxial deformation can occur beyond this point.

Characteristically, instantaneous strain rates can reach 10 to 100 s^{-1}. As seen in Figure 5.10, high-pressure low-density polyethylene (LDPE) shows rapid strain hardening whereas linear low-density polyethylene (LLDPE) does not. Rapid strain hardening can lead to brittle, splitty films. The process of nonisothermally orienting a crystallizing polymer can produce crystalline characteristics that are material-specific (14). For high-density polyethylene (HDPE), crystallites are arrayed predominantly parallel to the film surface as seen in Figure 5.11. On the other hand, when nylon 6 (PA-6) is biaxially stretched, its crystallites are predominantly arrayed perpendicularly to the film surface, Figure 5.12. The stress-strain curves of these films appear quite similar, however, as seen in Figure 5.13. Under tensile load, HDPE films pref-

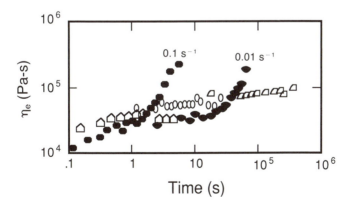

Figure 5.10.   Elongational Viscosity at 200°C for Linear Low-Density Polyethylene (LLDPE) and Low-Density Polyethylene (LDPE) (15). Open Symbols: LLDPE. Closed Symbols: LDPE. Shear rate range on open symbols: 0.01 $^{-1}$ to 0.1s^{-1}.

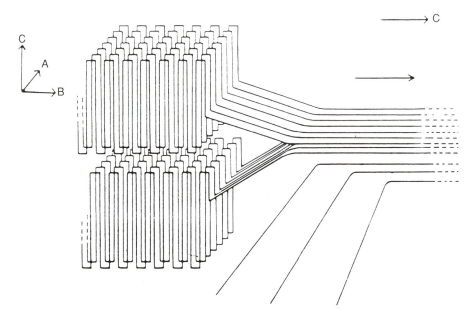

Figure 5.11.   Schematic of High-Density Polyethylene Crystallite Orientation in Biaxially Oriented Thin Film (16).

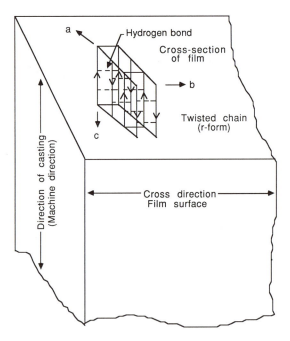

Figure 5.12.   Schematic of Nylon 6 (PA-6) Crystallite Orientation in Biaxially Oriented Thin Film (17).

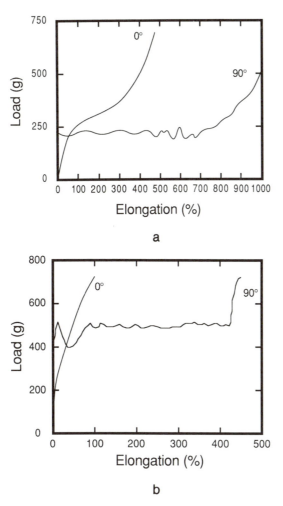

Figure 5.13. Stress-Strain Curves for Biaxially Oriented High-Density Polyethylene (a) and Nylon 6 (PA-6) (b) (18). 0° = Machine Direction, 90° = Cross-Machine Direction.

erentially neck or draw down in film *thickness*, since the force required to separate fibrils is much less than that required to additionally elongate them. When blown nylon film is oriented, the expansion in the folded chains results in decrease in the *width* of the film. The resulting effect is seen as deformation or shear bands across the film surface.

Example 5.2

Q: Estimate the blow-up ratio for a 0.05 mm equal biaxially oriented film made from a 30 cm diameter annular die with a 1 mm gap width. Consider as a first approximation that orientation is only a function of bubble inflation and neglect the effects of extrudate swell, draw down and induced stress/strain relaxation.

A: The volumetric flow rates of resin within the bubble (b) and within the die (d) are equal:

$$\pi D_b\, h_b\, L_b \equiv \pi D_d\, h_d\, L_d$$

where D, h and L are the diameter, thickness and length, respectively. The orientation in the machine direction, O_{MD}, is given by:

$$O_{MD} = L_b/L_d = h_d\, D_d/h_b\, D_b = (h_d/h_b)\,(1/B_R)$$

where B_R, the bubble blow-up ratio, is:

$$B_R = D_b/D_d$$

The orientation in the transverse or cross-machine direction, O_{TD}, is given as:

$$O_{TD} = D_b/D_d = B_R$$

And the ratio of the two orientations describes the degree of biaxial orientation, O_{BA}:

$$O_{BA} = O_{MD}/O_{TD} = h_d/h_b\, B_R^2$$

For equal biaxial orientation, $O_{BA} = 1$. Thus:

$$h_d/h_b\, B_R^2 = 1$$
$$B_R = (h_d/h_b)^{1/2} = (1/0.05)^{1/2}$$
$$B_R = 4.5.$$

Multi-Layer Blown Film.   Two or more extruders can feed polymer melts to a common die to produce multilayer film and sheet for the packaging and construction industries. The multiple polymer melt mixture can be fed to a *blown film die* to produce biaxially oriented packaging film of 0.0005 to 0.0015 in (0.01 to 0.04 mm) thickness. Frequently co-blown films are used for improved toughness at reduced weight or cost. The supertough garbage bag is a multilayer biaxially coblown lamination. One or more of the layers can be a barrier to specific diffusants such as oxygen or carbon dioxide. The contents of the package thus achieve greater shelf-life. Frequently, polyethylene and polypropylene are candidate resins for single- and multilayer laminates. To maintain toughness or clarity, the polymer crystallites must be as small as possible. Since polyolefins are rapidly crystallizing polymers, the blown film must be oriented and quenched as quickly as possible. Cold air is directed against the blown film from an air ring around the inflated polymer film bubble on the outside and from concentric collars on a tower inside the inflated polymer film bubble.

Profile Die.   In profile extrusion, the molten resin is extruded through a die in which the desired cross-sectional geometry of the linear part has been cut.

The resin that is supplied by the extruder is thus forced to assume the shape of the die. Upon exiting the die, the resin is cooled to fix the desired cross-section. Pipe, conduit and flexible tubing are the principal commercial products produced this way. Siding and window lineals are also important commercial products. Multiple resins in circular dies produce cylindrical tube preforms for barrier containers.

Example 5.3

Q: From a molecular viewpoint, describe the orientation of polymer molecules in a circular rod extruded from a long circular die (L/D > 16).

A: Assume that the velocity profile is fully developed. The shear stress distribution will vary linearly from zero at the centerline to τ_w, at the wall of the die. The degree of molecular orientation within the die, neglecting tensile strain, is directly proportional to the shear stress. Maximum molecular orientation occurs at the die wall and maximum molecular randomness is at the centerline. As the polymer emerges from the die, the highly oriented molecules near the die wall tend to recover along the cylindrical axis. This causes the surface of the extruded rod to shrink in length and swell radially. The degree of orientation in the solidified extruded rod will be highest along the rod surface, with the principal direction or molecular orientation being in the axial direction. The core will have essentially no molecular orientation. Due to the shrinkage along the rod surface, there may be a residual stress field across the cross-section of the rod. The core region will be in compression and the skin region will be in tension. The radial position of zero stress will depend upon the cooling history.

Coating. Polymer melt is extruded through a sheet die and the extrudate is laid onto a moving substrate, as seen in Figure 5.14. This technique is used extensively to form plastic/paper laminates for waterproof cardboard containers such as milk cartons and for plastic/metal laminates for flexible food packaging. The primary resins are low-density and medium-density polyethylene (LDPE and MDPE). Coating rates of up to 2000 ft/min (600 m/min)

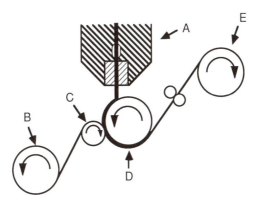

Figure 5.14.   Schematic of Extrusion Coating Line (19). A: Extrusion Die. B: Substrate Takeoff Roll. C: Pressure Roll. D: Chill Roll. E: Takeup Roll.

correspond to polyethylene extrusion rates of 1200 to 3000 lb/h (550 to 1400 kg/h).

The Extruder as a Melt Pump

Single-screw extruders are not positive displacement pumps. Rather, they pump by drag flow between the barrel and the leading edge of the rear flight in the flow channel. Because the screw does not shear-fold the polymer melt into the unmelted solid polymer plug, the single-screw extruder is a poor mixing device. Owing to these limitations, static mixers and positive displacement gear pumps can be installed between the end of the mixing zone of the single-screw extruders and the dies. The improvement in the quality of the melt dispensed to the die frequently pays for the additional investment in another device.

Twin-Screw Extruder

Recently, the twin-screw extruder has been commercialized as a compromise device (Figure 5.15). The twin-screw extruder acts as an efficient solids conveying, plastication and devolatilizing device (20). Twin-screw extruders can have screws that may or may not mesh, and the screws can rotate in the same direction (corotating) or in opposite directions (counter-rotating). The intermeshing counter-rotating twin-screw extruder can act as a positive-displacement pump and is used to extrude very viscous polymers at very high extrusion pressures. Although the intermeshing co-rotating twin-screw extruder is not a positive-displacement pump, nearly all the polymer remains in the extruder for about the same length of time. Therefore the melt is conditioned far more efficiently than a single-screw extruder. As a result, this type of extruder can be used to produce extruded products requiring quite critical tolerances, such as polyvinyl chloride (PVC) window lineals and siding. Equally important, the twin-screw extruder does not subject the polymer to the typically very high shear seen in the single-screw extruder. This is quite important in processing shear-sensitive polymers such as rigid polyvinyl chloride (RPVC) and polymers containing temperature-sensitive additives such as antioxidants, fire retardants and foaming agents.

Thermosetting Extrusion

For thermosetting resins, the single-screw extruder is used as a plasticator or mixer. The barrel is usually water-cooled to limit the temperature rise of the resin in order to retard the chemical reaction during mixing. Extruders of this type are used in transfer or compression molding processes to preheat or *plasticate* the molding compound. These processes are discussed below.

Tandem Extrusion

Multiple extrusion steps are used to produce unique products. As noted above, a two-stage single-screw extruder can be used to produce foam sheet. Foam sheet is also

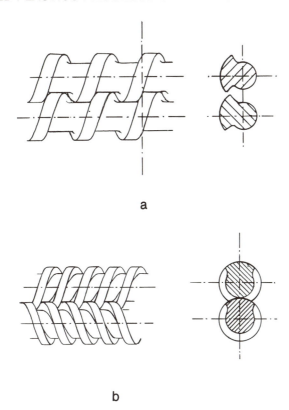

a

b

Figure 5.15.   Schematic of Twin-Screw Extruder (21). (a) Intermeshing Co-Rotating Screw Geometry. (b) Intermeshing Counter-Rotating Screw Geometry.

produced by using two extruders in tandem. The first extruder plasticates the polymer melt. A volatile foaming agent is then added and the mixture is fed to a second extruder-heat exchanger where the mixture temperature is gradually lowered. This ensures good polymer melt strength as the foamable mixture issues from the die.

The Extrusion Process—Model

As noted earlier, extrusion is the most widely used plastics process and the single-screw extruder is the most widely used extruder. The objective in extrusion is to take a continuous stream of solid discrete particles of plastic as powder, granules, or pellets, and by shearing action and heat transfer, to melt them into a homogeneous fluid. The melt is then pressure-pumped through a shaping die, after which it rigidifies into the desired continuous shape. As noted, the four geometric zones of an extruder are solids conveying, plasticating or melting, melt metering, and melt shaping by flow through a die, Figure 5.2.

Drag-Induced Solids Conveying.   Flow of material in the solids conveying portion of the screw is called *drag-induced solids conveying* (22). If pressure builds in the compacted solid powder or pellet bed, the bed is internally deformed. The solids bed is acted upon by several forces:

- Frictional force between the solid bed and the screw root,
- Frictional force between the solid bed and the trailing screw flight,
- Frictional force between the solid bed and the leading screw flight,
- Frictional force between the solid bed and the barrel surface,
- Normal forces between the solid bed and the screw flights, and
- Differential force caused by the pressure differential.

Although the force balances can easily be written, a solution to the problem is complicated by the fact that the barrel-to-plug force is applied at an angle to the down-channel plug motion. This causes the bed to be conveyed at an angle to the down-channel direction. See Figure 5.16. The standard approach to solution is to unwrap the particulate bed from the screw, then divide all

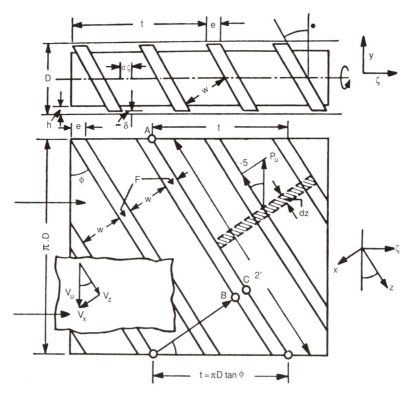

Figure 5.16.   Geometry for Modeling the Unwrapped Flights of the Screw of a Single Screw Extruder (23).

forces into vector components. This results in an expression for developed pressure in terms of θ, the solids conveying angle:

$$(5.1) \quad \ln \frac{P}{P_o} = \left\{ \frac{f_b}{H} [\cos(\theta + \phi) - f_s \sin(\theta + \phi)] - f_s \left(\frac{W + 2H}{2H} \right) \right\} z$$

where θ, the solids conveying angle, is given as:

$$(5.2) \quad \theta = \sin^{-1} \left[\frac{(1 + f_s^2 - k^2)^{1/2} - f_s k}{1 + f_s^2} \right] - \phi$$

and:

$$(5.3) \quad k = \frac{H}{f_b z} \ln \left(\frac{P}{P_o} \right) + \frac{f_s}{f_b} \left(1 + \frac{2H}{W} \right)$$

P is the down-channel pressure (at z), P_o is the initial pressure on the bed (at z = 0), H is the channel depth, W is the between-flight width, ϕ is the screw helix angle, f_b is the dynamic coefficient of friction on the barrel surface and f_s is that on the screw and flight surfaces. The solid bed moves down the channel by drag against the barrel surface. If the frictional coefficient between the barrel and solid bed is small compared with that between the screw surfaces and the solid bed, the bed velocity is small or zero. This effect is exacerbated by positive pressure on the bed.

The relationship between θ, the solids conveying angle, ϕ, the relative angle between the plates, and $\dot{m}$, the mass transfer rate, is:

$$(5.4) \quad \dot{m} = \rho\, WHV_b = \frac{\rho\, WHV_b \sin \theta}{\sin (\theta + \phi)}$$

where V_s is the velocity of the solid bed and V_b is the velocity of the moving plate. ρ is the bulk density of the polymer.

The maximum bed velocity occurs when the coefficient of friction between the bed and the screw surfaces, f_s, is zero. The solids conveying angle is given as:

$$(5.5) \quad \frac{\rho\, WHV_b}{\cos \phi} = \rho\, \pi^2 D^2 HN \tan \phi$$

and the maximum solids conveying rate is:

$$(5.6) \quad \dot{m}_{max} = \frac{\rho\, WHV_b}{\cos \phi}.$$

As the polymer slides over metal surfaces, friction generates heat. In any sliding configuration, heat generation is a product of the dynamic coefficient of friction, the normal force pressing the materials together and the sliding velocity. For a solid bed sliding against the barrel surface, the heat generation is given as:

$$(5.7) \qquad \dot{Q}'_b = Pf_b \ W \ V_b \ \frac{\sin \phi}{\sin (\theta + \phi)} \ dz$$

where $\dot{Q}'_b$ is the heat generation term (energy/time). Note that heat generation is proportional to pressure applied to the bed. The heat generated is either conducted to the barrel surface or into the powder bed. The heat transfer from the solid bed to the barrel surface is controlling. In practical cases, the barrel temperature is higher than the average bed temperature. Thus frictional and thermal energy are conducted into the solid bed.

In equation 5.1, pressure is exponentially dependent on down-channel distance. As a result, heat generation is also approximately exponentially dependent on down-channel distance. Note however that frictional coefficients strongly depend on temperature-dependent material properties such as modulus and yield strength. Contact area between the polymer and the barrel surface depends on applied pressure and temperature as well. At high pressures and barrel wall temperature, local smearing of a very thin layer of polymer on the barrel surface might be expected. Furthermore, if the material temperature at the barrel surface reaches the point at which the material undergoes extensive deformation and flow, the model is no longer valid.

Example 5.4

Q: Consider the drag flow of 30 lb/ft^3 polymer powder in a rectangular channel, W = 2.5 in by H = 0.5 in. The upper wall moves at an angle, ϕ = 15°, to the stationary wall at a velocity, V_b = 12 in/s. Determine the maximum flow rate, $\dot{m}_{max}$.

Q2: If the actual flow rate is 73.2% of the maximum, what is the solids conveying angle, θ?

A1: $\dot{m}_{max} = \rho \ WH \ V_b/\cos(\phi)$,

$$= \left[30 \times (0.5 \times 2.5/144) \times 12 \times \frac{60}{12} \right] / \cos (15)$$

$\dot{m}_{max}$ = 16.1 lb/min.

A2: $\dot{m} = \rho \ WH \ V_b \ \sin(\theta)/\sin(\theta + \phi)$.

Rearranging:

$$\sin (\theta)/\sin(\theta + 15) = (\dot{m}/\dot{m}_{max}) \cos(15) = 0.707.$$

By trial and error, θ = 30°.

The polymer properties that are most important in drag-induced solids conveying are:

- Frictional coefficients of *particulate* polymer with various types of surfaces (etched, chrome-plated, or alloyed),
- Thermal properties, including thermal diffusivity, thermal conductivity, heat capacity, and density of the particulate material, and
- Temperature-dependent modulus and yield strength of the polymer.

Plasticating or Melting.   The solid bed temperature is raised to melt or flow conditions primarily by heat transfer from the barrel wall and by friction between the barrel wall and the solid bed. For crystalline polymers, melting takes place in the thin layer between the solid bed and the barrel wall, and the melt is conveyed by virtue of the vectorial nature of the bed velocity, toward the trailing channel flight. With time, the melt is accumulated in a growing pool ahead of the trailing flight surface (Figure 5.5). The growing pool and the pressure differential force the solid bed against the rear edge of the leading flight and the barrel surface. These forces tend to keep the bed consolidated with the bed height remaining relatively constant, and the bed width diminishing with time. The general energy equation in the melt film for a Newtonian fluid can be written as:

$$(5.8) \qquad k_m \frac{d^2T}{dy^2} + \tau_{xy}\left(\frac{\partial v_x}{\partial y}\right) = 0$$

where k_m is the thermal conductivity of the polymer melt, y is the distance into the melt film, and τ_{xy} is the shear stress on the polymer melt. If H_m is the thickness of the melt film, μ is the polymer Newtonian viscosity, and v is the relative velocity between the film and the barrel wall, then:

$$(5.9) \qquad k_m \frac{d^2T}{dy^2} + \mu \left(\frac{v}{H_m}\right)^2 = 0$$

This can be rewritten in terms of a dimensionless temperature increase, θ, as a function of the dimensionless distance, $y^* = y/H$:

$$(5.10) \qquad \theta = \frac{T(y^*) - T_m}{T_b - T_m} = y^* [1 + Br(1 - y^*)]$$

The dimensionless term is the Brinkman number, Br. For Newtonian fluids, $Br = \mu \, v^2/2 \, k_m \, \Delta T_b$, where μ is the Newtonian viscosity, v is the fluid velocity, k_m is the melt thermal conductivity and ΔT_b is the temperature difference between the barrel and the bulk polymer. The Brinkman number is a ratio of the increase in temperature owing to viscous dissipation to that owning to conduction into the film. In practice, the Brinkman number is larger than unity, indicating that viscous dissipation is a primary way of melting polymers.

Example 5.5

Q: Ultra-high molecular weight polyethylene (UHMWPE) is coated on a 250°C, 1 inch diameter steel pipe by rubbing a 25°C solid resin bar against the pipe. The pipe rotates at 755 RPM. If μ = 4000 N s/m^2 and k_m = 0.2 W/m °C, determine the Brinkman number, Br. If Br >> 1, viscous dissipation is a major method of heating.

A: The tangential velocity is given as:

$$\omega = 755 \text{ RPM} \times 2\pi/60 = 79 \text{ rad/s},$$

$$v = r\omega = 0.5 \text{ in} \times 0.0254 \text{ m/in} \times 79 \text{ rad/s}$$

$$= 1 \text{ m/s}$$

$$Br = \mu v^2/(2 k_m \Delta T_b)$$

$$= 4000 \text{ (N s/m}^2) \times 1(\text{m/s})^2 / [2 \times 0.2(\text{W/m °C}) \times 225°C]$$

$$Br = 44.44$$

Since this value is much greater than unity, the primary mode of heating is by viscous dissipation.

It has been assumed that there is no clearance between the screw flight and the barrel. In practice, the clearance can be quite significant. As a result, polymer melt is forced between the flight and the barrel, reducing the melting efficiency of the plasticating section. Although this simple melting model serves to illustrate the role of viscous dissipation in melting, it is no longer used in design of extruder screws. Rauwendaal (24) has recently reviewed more sophisticated models that allow for flow of softening (amorphous) polymers and that predict bed encapsulation and break-up.

The polymer material properties that are important in the plasticating zone of an extruder are:

- Polymer melt viscosity at the local temperature and shear rate,
- Densities of melt and material in solid bed (not solid polymer density), and
- Polymer melt thermal conductivity.

In addition, the heat of fusion of a crystalline polymer or the temperature-dependent heat capacity of an amorphous polymer is important in determining the amount of heat required to melt the polymer.

Melt Conveying. The first area of the extrusion process to be studied analytically was the *melt conveying zone*. The simplest model considers the polymer melt to be Newtonian with temperature-independent properties. If the channel width is considered to be infinite when compared with thickness, a very simple equation of motion in the down-channel direction is obtained:

(5.11)
$$\left(\frac{\partial P}{\partial z}\right) = g_z = \frac{\partial \tau_{yz}}{\partial y}$$

Since pressure is a function of down-channel direction *only*, the shear stress is linear with channel thickness. For the Newtonian case, the average melt flow rate is given as:

$$(5.12) \qquad Q = \int_0^H PWv_z \, dy = \frac{PWH\, v_{bz}}{2} - \frac{PWH^3\, g_z}{12\, \mu}$$

where g_z is the constant pressure drop, $v_{bz} = \pi\, D\, N \cos(\phi)$, ϕ is the screw helix angle, and W is the channel width. The first term is the result of the *drag flow*, that is, the relative motion between the screw and barrel. The second is *pressure flow*, due entirely to the pressure gradient on the polymer melt (Figure 5.17). In many cases, the screw is building pressure in this region to provide adequate driving force to move the polymer through the die ahead of the screw. As a result, the effect of an increasing positive pressure gradient is a reduction in output. And as with the melting zone, clearance between the screw flight and barrel causes *leakage* melt flow from the region of the trailing flight of one portion of the channel to that of the leading flight of the portion immediately behind. The general effect is to decrease the magnitude of drag flow and to increase the magnitude of pressure flow. For typical extrusion processing, this results in a decrease in output, all other parameters being equal.

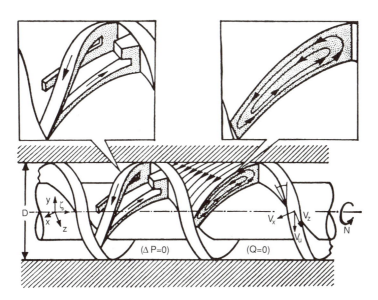

Figure 5.17.   Schematic of Fluid Flow Occurring in the Melt Conveying Section of a Single-Screw Extruder (25). Enlarged Sections: Left Demonstrates Drag Flow; Right Demonstrates Pressure Flow.

The general Newtonian melt pumping equation can be written symbolically as:

(5.13) $$Q = AN - B(\Delta P/\mu)$$

where A and B contain geometric parameters and N is the screw speed. μ is the Newtonian viscosity.

Example 5.6

Q: Develop an empirical equation for the output of an extruder, in lb/h, as a function of extruder screw diameter, using the data of Table 5.8.

A: The average throughput as a function of extruder diameter is given as:

D_b, in	1.5	2.5	3.5	4.5	6.0	8.0
$\dot{m}$, lb/h	60	140	325	550	1000	1750

The data are of the form:

$$\dot{m} = A\, D_b^{\,n}$$

Taking the logarithm of this expression:

$$\ln \dot{m} = n \ln D_b + \ln A$$
$$y = m\,x + b$$

From least-squares analysis of the data:

$$m \equiv n = 2.06 \qquad b \equiv \ln A = 3.186$$

Or:

$$\dot{m} = 24.2\, D_b^{\,2.06}$$

This correlation compares favorably with that of Van Ness, De Hoff and Bonner (107):

$$\dot{m} = 16\, D_b^{\,2.2}$$

Melt viscosity is the only major polymer property to influence extrusion output in melt conveying. The primary influences are channel geometry, screw speed, and required pressure drop. Extruder screws are designed to be drag-flow dominating, since the extruder output should not be sensitive to small changes in die opening, which of course directly affect extruder pressure drop. And if drag flow dominates melt conveying, the effect of temperature on viscosity is small.

Die Flow. As noted, an extruder delivers a melt to a shaping die. For a given polymer melt, the relationship between melt flow rate and pressure drop depends primarily on the geometry of the die and the viscosity of the polymer. The general relationship between melt flow rate and pressure drop is:

(5.14)   *melt flow rate* $Q = K \, \Delta P / \mu$

K is a constant that describes die geometry. As an example, consider melt flow in a die to be approximated by isothermal Newtonian flow in a long circular pipe. The relationship between shear stress and pressure drop is:

(5.15)
$$\frac{\partial P}{\partial z} = \frac{\mu}{r} \frac{\partial}{\partial r} \left(\frac{r \partial v}{\partial r} \right)$$

This equation is similar to that in the melt flow in an extruder channel. When this equation is integrated for flow in a pipe of radius R, the melt flow rate is given as:

(5.16)   *melt flow rate*
$$Q = \left(\frac{\pi R^4}{8L} \right) \frac{\Delta P}{\mu} = K \frac{\Delta P}{\mu}$$

It is common practice to replace the physical length of the die, L, with an equivalent length, L_c, which is equal to $L + A \, D_c$, where D_c is the diameter of the capillary die and A is a correction factor, similar to the Bagley correction factor discussed in Chapter 2. It is common practice to assume $A = 4$, unless experimental data for the particular resin are available.

Example 5.7

Q: How many die holes are needed in a compounding die to compound 150 lb/h of general-purpose polystyrene (PS) at 220°C and 500 lb_f/in^2 head pressure? The individual die hole is 3/32 inch in diameter and 1/4 inch long.

A: For an individual die hole:

$$Q = (\pi R^4 / 8L_c) \, \Delta P / \mu$$

For a power-law fluid:

$$\mu = K \, \dot{\gamma}^{n-1}$$

where:

$$\dot{\gamma} = 4Q / \pi R^3$$

Therefore:

$$\mu = K \, [4Q/\pi R^3]^{n-1}$$

Substituting into the first equation and rearranging terms:

$$Q = \frac{\pi R^3}{4} \left[\frac{R\Delta P}{2L_c K} \right]^{1/n}$$

This relationship could be obtained from:

$$\dot{\gamma}^n = \frac{\tau}{K}$$

For the general-purpose PS material, values for K and n are 2.2628 (lb_f/in^2) s^n and 0.28, respectively. Thus:

$$Q = \frac{\pi \cdot (3/64)^3}{4} \left[\frac{(3/64) \times 500}{2 \times [(1/4) + 4(3/32)] \times 2.2628} \right]^{1/0.28}$$

$$Q = 0.154 \ in^3/s = 554 \ in^3/h$$

The mass flow rate is:

$$\boxed{\dot{m} = \rho \, Q} = \rho V A$$

where the density is obtained from Figure 4.21. At 220°C and 250 lb_f/in^2:

$$\rho = 1.05 \ g/cm^3 = 0.0379 \ lb/in^3$$

Therefore:

$$\dot{m} = 554 \times 0.0379 = 21 \ lb/h$$

The number of holes, N, is given as:

$$N = \text{total mass flow rate/mass flow rate per hole,}$$

$$N = 150/21 = 7 \text{ holes.}$$

Expressions for K for other die geometries are given in standard reference texts. To determine the effect of material properties on shear heating, consider the Newtonian viscosity to depend upon temperature with the following form:

(5.17)
$$\boxed{\mu = \mu_o e^{-b(T - To)}}$$

where μ_o is the viscosity at temperature T_0. The adiabatic temperature rise is given by:

(5.18)
$$\boxed{\rho_m \, Q \, c_p \, dT = Q \, dP}$$

where ρ_m and c_p are the density and heat capacity of the polymer melt, respectively. Substitution and considerable algebraic manipulation yields:

(5.19)
$$\frac{dT}{dz} = \left(\frac{\Delta P_o}{\rho_m c_p L}\right) e^{-b(T-T_o)}$$

and:

(5.20)
$$\left(\frac{\Delta P}{\Delta P_o}\right) = \frac{\ln \beta_L}{(\beta_L - 1)}$$

where $\beta_L = 1 + b\,\Delta P_o/\rho_m\,c_p$, and $\Delta P_o = 8\,\mu_o LQ/\pi R^4$. b is a measure of the viscous energy of activation.

Again, as with thin film shear heating in the melting zone, extrusion through thin dies, as with blown film extrusion, can generate a substantial increase in temperature. The magnitude of the Brinkman number can be used to assess the relative importance of shear heating in die flow.

Example 5.8

Q1: If an extruder is melt pumping controlled, the flow rate from the extruder must match that through the die, at steady state. Show that for a Newtonian fluid, the flow rate is independent of pressure.

Q2: Then show that the maximum pressure in the screw-die system is proportional to the screw speed.

A1: From the die equation, $\Delta P/\mu = Q/K$. This is substituted into the extruder equation, eliminating $\Delta P/\mu$:

$$Q = A\,N - B\,Q/K$$

or:

$$(K + B)Q/K = A\,N$$

Rearranging:

$$Q = AKN/(K + B)$$

This equation does not contain the pressure drop term. ΔP does not appear since it is implicitly assumed that the extruder generates just enough internal pressure to overcome the flow resistance through the die.

A2: Substitute $Q = F(N)$ into the die equation:

$$\Delta P = \frac{\mu Q}{K} = \frac{\mu\,A\,N}{(K + B)}$$

The pressure drop, ΔP, is seen to be directly proportional to the screw speed, N.

Flow Effects at Die Exit.   Many polymer melts are viscoelastic. When a viscoelastic melt issues from a die, the applied stresses are set to zero, and the polymer responds elastically. Typically the elastic effect is seen as *extrudate swell,* over and above that expected from simple change in velocity profile. This subject was discussed in Section 4.4 as well. Certain polymers such as polyolefins exhibit severe extrudate distortion in certain shear rate ranges. In sheet, the effect is called *shark skin* or *orange peel.* In rod, it is called *melt fracture.* Shark skin is thought to be caused by rapid acceleration of the surface layer of the extruded sheet, leading to localized biaxial fracture. Very large irregularities, called *dragon scales,* are seen on high-speed extrusion of low-density foam. Polymers with high melt viscosity, very high viscous activation energy, and narrow molecular weight distribution (MWD) seem to be candidates for shark-skin formation. Melt fracture is thought to be caused by a combination of critical elastic deformation at the die entrance, slip-stick of the polymer in the die, and strain levels greater than an inherent critical strain level in the melt. The shear stress at which melt fracture instability begins is called the *critical shear stress.* It appears to be relatively insensitive to geometry and molecular weight distribution, but decreases with increasing number-average molecular weight and decreasing viscosity. In Chapter 4, recoverable tensile and shear deformations were related to the swell of an extruded part. Example 5.9 summarizes their application.

Example 5.9

Q1: An annular gap die with a mean radius of 50 mm with a gap of 2 mm and a land length of 40 mm is used to make pipe. A low density polyethylene (LDPE) at 170°C is extruded at a rate of 1×10^{-6} m^3/s. Estimate the dimensions of the extruded pipe.

Q2: What is the takeoff speed if the density is 730 kg/m^3 at 170°C and what is the mass flow rate?

A1: The shear rate is given as:

$$\dot{\gamma} = 6 \, Q/W \, H^2,$$
$$= 6 \times 1 \times 10^{-6}/[2 \, \pi \times 0.05 \times (0.002)^2]$$
$$= 4.8 \text{ s}^{-1}$$

To obtain the shear stress, the rheological curve, Figure 4.54, is needed.

$$\tau_{\mathrm{w}} = 5 \times 10^4 \text{ N/m}^2$$

The recoverable strain is obtained from Figure 4.58, as:

$$\gamma_{\mathrm{R}} = 2.2$$

The swelling ratios can be calculated from these data or obtained directly from Figure 4.57:

$$B_{\mathrm{ST}} = 1.13,$$
$$B_{\mathrm{SH}} = B_{\mathrm{ST}}{}^2 = 1.28.$$

Therefore the wall thickness and mean diameter are, respectively:

$$\text{Wall thickness} = 2 \times 1.28 = 2.56 \text{ mm,}$$
$$\text{Mean diameter} = 1.13 \times 50 = 56.6 \text{ mm.}$$

A2:

$$v = Q/A = 1 \times 10^{-6}/(\pi \times 0.0565 \times 0.00256),$$

$$= 2.2 \times 10^{-3} \text{ m/s} = 2.2 \text{ mm/s.}$$

$$\dot{m} = \rho Q = 730 \times 1 \times 10^{-6} \times 3600,$$

$$\dot{m} = 2.626 \text{ kg/h.}$$

Injection Molding

Injection molding is the second most widely used polymeric fabrication process. It involves the cyclic injection of a molten or liquid resin into a hollow cavity under pressure. The thermoplastic resin is fed in powder or pellet form to the drive-end of the injection molder screw and barrel assembly (or torpedo and barrel assembly in plunger injection molding machines). The resin is melted or softened by plastication in the injection unit. Typical processing conditions for polymers are listed in Table 5.9. The most common press in use today (Figure 5.18) utilizes an extruder screw to melt the resin and the screw also acts as a hydraulic ram to force the resin into the mold cavity. The melting process is similar to that in a continuous extrusion process except that as the resin is melted or softened, the screw retracts to form a melt pool in the barrel between the screw and injection nozzle (Figure 5.19). Injection molding presses are usually rated in terms of injection capacity, such as "ounces of molten polystyrene having a solid specific gravity of 1.05", and in clamp force, as "tons of clamp". The mold cavity is cut into a block of metal in such a manner that when the part has cooled or set sufficiently, it can be removed from the mold without damage to it or to the

Table 5.9. Processing temperature ranges for several extrusion-grade thermoplastics (26).

Polymer	Processing temperature range	
	(°C)	(°F)
ABS	180 to 240	356 to 464
Polyacetal (POM)	185 to 240	365 to 437
Acrylic (PMMA)	180 to 250	356 to 482
Nylon (Polyamide, PA)	260 to 290	500 to 554
Polycarbonate (PC)	280 to 310	536 to 590
Polyethylene		
Low Density (LDPE)	160 to 240	320 to 464
High Density (HDPE)	200 to 280	392 to 536
Polypropylene (PP)	200 to 300	392 to 572
Polystyrene (PS)	180 to 260	356 to 500
Polyvinyl Chloride, Rigid (RPVC)	160 to 210	320 to 410

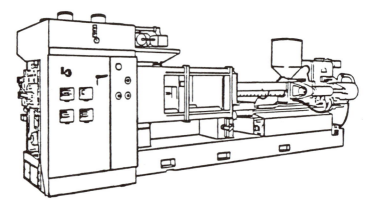

Figure 5.18.   Conventional Thermoplastic Injection Molding Press (27).

cavity. During the injection process, the various sections of the mold are held together by a clamping mechanism of one of several designs, such as mechanical, hydraulic, or hydromechanical mechanisms. See Figures 5.20 and 5.21.

Regardless of the general nature of the part being formed, the overall injection molding process is a way of *cyclically* melting, metering, shaping and cooling *thermoplastics.** In the century since its adaptation to plastics by John Wesley Hyatt and Charles Burroughs, the process has seen many modifications. Most are directed toward improvement in melt homogeneity, plasticating rate, and cooling efficiency. Plunger and screw injection molding machines both plasticate thermoplastic resins. However, screw plastication is more time- and energy-efficient. As a result, nearly all machines used today for non-specialty polymers are screw injection molding machines (Figure 5.19). A single-flighted screw turns in a high-pressure barrel. The tip of the screw has a non-return valve. The barrel is zone-heated in the fashion of screw extruders. It is also fitted with a nozzle that may have positive shut-off. The nozzle is inserted through the face of the fixed platen. This platen holds the portion of the mold containing a sprue bushing into which the nozzle is seated. The other portion of the mold is fastened to the movable platen, which rides on tie bars. The platens are held closed during mold filling by mechanical toggle or hydraulic clamps. As the screw turns either by a mechanical or hydraulic motor, polymer granules are fed from the hopper into the solids conveying portion of the screw. Plastic is then plasticated or melted in the heated barrel and the melt pumped through the non-return valve and into the barrel zone between the valve and the nozzle. The accumulated melt then applies force against the screw, forcing it backward, away from the nozzle. Hydraulic or mechanical pressure is applied to the screw to control the rate of plastication and the extent of viscous heating over

---

*Certain thermosets such as silicones, unsaturated polyester resins (UPE) and polyurethanes (PUR) can be processed in similar fashions as long as they possess a finite "fluid" time. Instead of being cooled, the thermoset is heated in the mold until the reaction rigidifies the material.

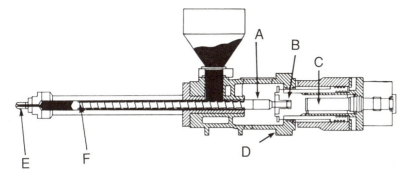

Figure 5.19.   Schematic of Modern Thermoplastic Reciprocating Screw Injection Molding Machine (28). A: Plasticating Screw. B: Hydraulic Piston. C: Electromechanical Drive. D: Frame. E: Nozzle: F: Non-Return Valve.

the flights and through the valve. A typical time-dependent pressure profile for a hydraulic injection molding system is shown in Figure 5.22.

Injection molding extruders have full-forward L/D ratios of 18:1 to 30:1. The solids conveying section is about half the screw length. The plasticating and melt pumping sections are each about one-quarter the screw length. Compression ratios for specific materials follow those recommended for extruder screw design.

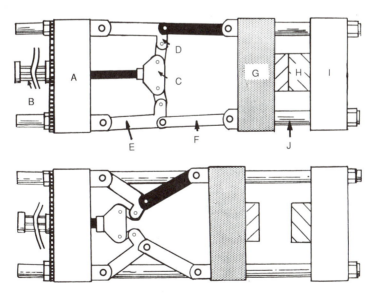

Figure 5.20.   Hydromechanical Injection Molding Machine Clamp Schematic (29). A: Tailstock Platen. B: Actuating Cylinder. C: Crosshead. D: Crosshead Link. E: Back Link. F: Front Link. G: Moving Platen. H: Mold. I: Stationary Platen. J: Tie Rod.

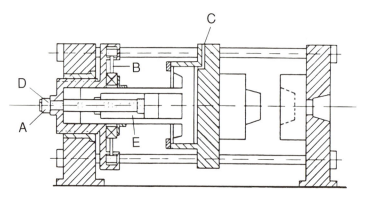

Figure 5.21. Hydraulic Injection Molding Machine Clamp Schematic (30). A: Oil Inlet for Closing Mold. B: Lock. C: Hydraulic Clamp Cylinder. D: Oil Inlet for Opening Mold. E: Rapid Traversing Cylinder.

Polymer melt is transferred at high speed from the accumulated melt pool ahead of the screw by ram-advancement of the screw, usually by hydraulic means. The typical transfer time ranges from fractions of seconds to several seconds. The non-return valve prevents accumulated melt flow back through the gap between the top of the screw flights and the barrel. The shut-off nozzle prevents pressure loss and melt drool from the nozzle. The main advantages of the injection molding process are the high rate of production, geometric flexibility in part design, a high degree of tolerance repeatability, and the very wide range of materials that can be processed. The overall cycle time is dominated by the cooling of thermoplastic resins or the curing of thermosetting resins.

Mold Geometry

Runner System. When a part is large compared with the plasticating and transfer capacity of the injection press, only one unit is molded at a time. A single-cavity mold is shown as a thin center-gated disk in Figure 5.23. The polymer flows through the machine nozzle, the *sprue* and *gate* and into the *cavity,* where it flows radially outward toward the cavity perimeter. If the part is small, many units can be formed at one time. Each mold cavity is connected to the mold sprue by a runner-gate system, as shown in Figure 5.24. If all cavities are connected to the sprue via runners of the same length and cross-sectional dimension, the runner system is said to be *balanced,* as seen in Figure 5.25a. Theoretically, with a balanced runner system, all cavities should fill with polymer at the same rate and time. The runner system in Figure 5.25b is *unbalanced.* The cavities nearest the sprue should begin to fill first. And as seen in Figure 5.26, the filling characteristic to a single cavity can also be unbalanced.

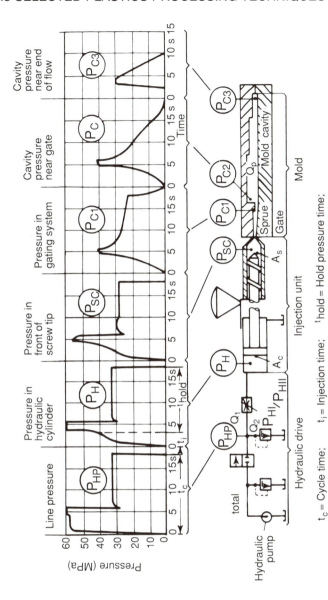

Figure 5.22. Time-Dependent Pressure Profiles at Various Points in a Standard Injection Molding Machine Cycle (31). t_c: Total Open-to-Open Cycle Time. t_i: Injection Time. t_{hold}: Mold Closed Time. P_{HP}: Hydraulic Line Pressure, First Graph. P_H: Hydraulic Cylinder Pressure, Second Graph. P_{SC}: Pressure at Screw Tip, Third Graph. P_{CI}: Pressure in the Gate, Fourth Graph. P_{C2}: Pressure in Cavity Near Gate, Fifth Graph. P_{C3}: Pressure in Cavity Near Flow End, Sixth Graph.

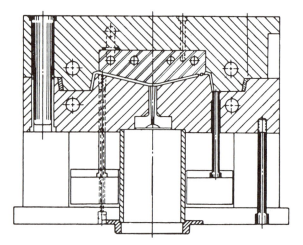

Figure 5.23. Cross-Section of Mold Showing Bottom-Center Gated Single Cavity Mold (32).

Gate. The gate is a restriction in the flow area just ahead of the cavity. The gate has two functions: flow direction and back-flow control. There are many gate styles and many rules-of-thumb regarding gate location. In many cases the gate must be placed where it is unobvious (particularly on appearance parts) or where the flow will not be seriously redirected by in-mold obstructions.

Mold Design Comments. Injection molds are usually manufactured of air- or oil-hardened tool steels (such as P-7 or H-13), and are very expensive. Thus, injection molding is used for high volume production. For thermoplastics, there are two basic types of mold construction. If the resin is allowed to

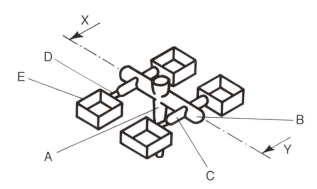

Figure 5.24. Four-Cavity Balanced Runner System. Note Symmetry Parallel and Perpendicular to the X-Y Line (33). A: Sprue. B: Primary Runner. C: Secondary Runner. D: Gate. E: Cavity.

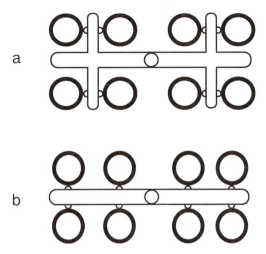

Figure 5.25. Comparison of Balanced and Unbalanced Multicavity Runner Systems (34). (a) Balanced Runner System, with Distance from Sprue to Cavity Equal for All Cavities. (b) Unbalanced Runner System, with Distance from Sprue to First Cavities Shorter than That From Sprue to Last Cavities.

cool and harden in the runner system between the injection molding machine nozzle and the part cavities, the mold is called a *cold runner mold* (Figure 5.27). For this type of molding, the cooled parts and the runner system are removed together, the parts separated from the runner, and the runner reground and recycled. If the runner material is to remain fluid, the runner is heated and the mold is a *hot runner mold* (Figure 5.28). Hot runner molds are significantly more expensive to fabricate than cold runner molds but the

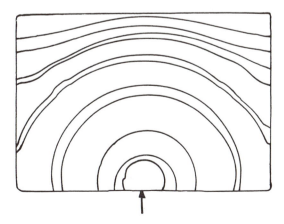

Figure 5.26. Tracing of Filling Characteristics of Edge-Gated Flat Plaque Mold.

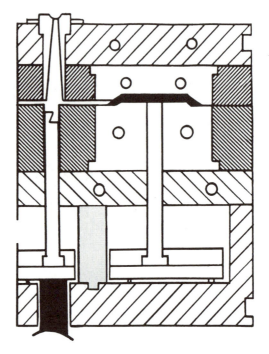

Figure 5.27.   Cold Runner Injection Mold Design (35).

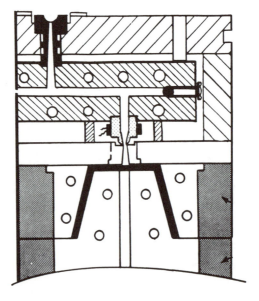

Figure 5.28.   Hot Runner Injection Mold Design (36).

minimization of regrind costs usually can offset the additional cost. When thermally sensitive resins such as rigid PVC (RPVC) are to be hot runner injection molded, care must be taken to minimize long-time exposure to high temperatures, particularly in the hot runner nozzle region.

Packing.   Once the cavity has been filled with molten polymer, additional polymer is needed to compensate for the high volume reduction (density increase) that occurs during cooling. This volume change is *shrinkage*. Additional material must be forced into the cavity under very high pressure. The processing step is called *packing*. Materials that crystallize during cooling have greater shrinkage values than amorphous polymers. As seen in Figure 5.29, shrinkage is pressure-dependent. These effects are discussed in more detail in Section 4.6 on polymer material thermodynamics.

Pressure is maintained on the softened polymer by the ram-forward action of the screw. For economic cycles, however, it is necessary to begin screw retrograde before all the material in the cavity is solidified. It can be shown that the time to cool a polymer is inversely proportional to the square of the material thickness. As a result, gates are kept relatively thin. This allows gate material to solidify long before the polymer in the cavity solidifies. This effectively pressure-seals the cavity from the accumulator at an economically early time.

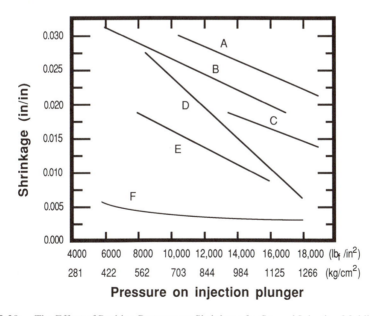

Figure 5.29.   The Effect of Packing Pressure on Shrinkage for Several Injection Molding Grade Polymers (37). A: Polyacetal (POM). B: Low-Density Polyethylene (LDPE). C: Nylon 66 (PA-66). D: Polypropylene (PP) in Flow Direction. E: Polypropylene in Cross-Flow Direction. F: Polymethyl Methacrylate (PMMA).

Example 5.10

Q: An injection molded general-purpose polystyrene part is to be molded to exact dimensions. What pressure is needed if the PS is injection molded at 170°C?

A: For the part to have the same dimensions at room conditions as it does at 170°C, the specific volume of the polymer melt must be the same as that of the room temperature solid part.

$$v_{20°C} = 0.96 \text{ g/cm}^3$$

$$(v - b)(P + \pi) = R \ T/M$$

$$(0.96 - 0.822)[P + 1840(\text{atm})] = 82.06 \times 443/104$$

Solving for P:

$$P = 693 \text{ atm} = 10,180 \text{ lb}_f/\text{in}^2$$

At any pressure less than this value, the molded part will have dimensions smaller than those of the mold cavity. This pressure is higher than that obtained under normal molding conditions. This phenomenon is known as *part shrinkage.*

Simplified Cavity Filling Models

The injection molding process can also be thought of as an unsteady, nonisothermal hydraulic process using a compressible, highly viscous polymer as the hydraulic fluid. A typical mold pressure profile is seen in Figure 5.22. Note that the injection pressure builds very rapidly to a machine characteristic constant value. The cavity sees relatively low pressures during filling. Once the cavity is full, the pressure builds rapidly to a maximum. Note that if the mold was at melt temperature, filling would be isothermal. The cavity pressure would then ultimately equal the injection pressure.

Example 5.11

Q: Estimate the clamping force for the mold design shown in Figure 5.2. The injection pressure is 5000 lb_f/in^2. The material is polycarbonate (PC). Assume the following dimensions:

> Projected cavity area: 1/2 × 1/2 (in),
> Sprue diameter: 3/8 (in),
> Main runner dimension: 1-1/2 × 1/4 (dia) (in),
> Secondary runner dimension: 1/2 × 3/16 (dia)(in).

A: Projected area = cavities + sprue + main runner + secondary runners.

$$A = 4 (1/2 \times 1/2) + (\pi/4)(3/8)^2 + (1\text{-}1/2) \times (1/4) + 4 (1/2 \times 3/16)$$

$$A = 1.86 \text{ in}^2$$

The assumed pressure is 5000 lb_f/in^2. Therefore:

$$\text{Force} = A \times 5000/2000 = 4.65 \text{ Tons}$$

Isothermal Newtonian Flow Model. An illustrative albeit very simple concept of polymer flow into an injection mold can be obtained by considering isothermal Newtonian flow into the center-gated disk cavity of Figure 5.23. If inertial terms are neglected, a momentum equation yields:

(5.21)
$$\frac{\partial P}{\partial r} = \mu \frac{\partial^2 v}{\partial z^2}$$

and continuity yields:

(5.22)
$$\frac{\partial (rv)}{\partial r} = 0$$

Now the volumetric flow rate into the cavity is given as:

(5.23)
$$Q = 4 \pi H r (dr/dt)$$

where H is the cavity half-height and r is the time-dependent radius of the leading edge. From integration of the energy and continuity equations for the case of *constant injection pressure*:

(5.24)
$$P_o = -C_1 \mu \ln \left(\frac{2r}{D} \right)$$

where D is the cavity diameter. C_1 is found from substitution into:

(5.25)
$$Q = 4 \pi \int_0^H rv \, dz$$

to obtain:

(5.26)
$$C_1 = \frac{-3Q}{4\pi H^3} = \frac{-3r}{H^2} \left(\frac{dr}{dt} \right)$$

The pressure (momentum) equation can be integrated to yield:

(5.27)
$$\left(\frac{2r}{D} \right)^2 \ln \left(\frac{2r}{D} \right) \left[\left(\frac{2r}{D} \right)^2 - 1 \right] = 2C_2 t \left(\frac{2R}{D} \right)^2$$

where:

$$(5.28) \qquad C_2 = \frac{H^2 P_0}{3\mu R^2}$$

The fill time is the time when r = R. For constant applied pressure, P_0, the fill time, t_{fill}, is given as:

$$(5.29) \qquad t_{fill} = \frac{1}{2C_2} \left[-\ln \frac{D}{2R} - \frac{1}{2} \left(1 - \left(\frac{D}{2R} \right)^2 \right) \right]$$

This is seen in Figure 5.30. The volumetric fill rate at t_{fill} is given as:

$$(5.30) \qquad Q = \frac{4\pi H^3 P_0}{3\mu \; \ln \, (2r/D)}$$

and is shown in Figure 5.31. Initially Q is infinite. This is due to the assumption of constant pressure.

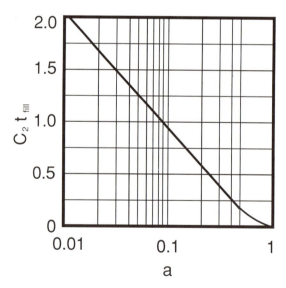

Figure 5.30.   Pressure-Dependent Fill Time for Isothermal Newtonian Fluid in a Center-Gated Disk Mold (38). $C_2 = H^2 \, P_0/3 \; \mu R^2$. a = D/2R. $t_{fill} = (1/2 \; C_2) \, [-\ln a - (1/2)(1 - a^2)]$, the Fill Time. H Is Disk Half-Thickness, P_0 Is Initial, Constant Pressure at r = 0, μ Is Newtonian Viscosity, R Is Radius of Wavefront, D Is Disk Diameter.

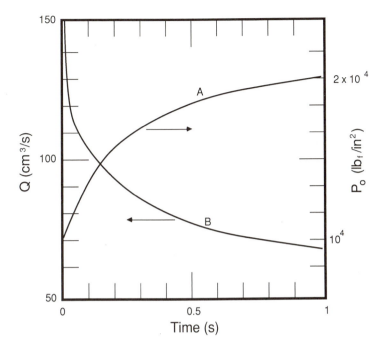

Figure 5.31.   Time-Dependent Flow Rate into Center-Gated Disk Filling With Isothermal New-tonian Polymer (39). B: $P_0(t)$ at Constant Flow Rate. B: $Q(t)$ at Constant Applied Pressure. [$L = 1$ cm, $R = 9$ cm, $D = 1$ cm, $H = 0.158$ cm, $\mu = 10^5$ poise, $\rho = 1$ g/cm^3, $t_{fill} = 1$ s.]

Example 5.12

Q: Coffee can lids are produced by injection molding low-density polyethylene (LDPE). The lids are center-gated disks and the process can be considered to be isobaric, constant pressure, $P = P_0$, $D = 1$ cm, $H = 0.0508$ cm, $R = 9$ cm, $\mu = 10^3$ Pa s, and $P_0 = 100$ MPa. Find the fill time, the average flow rate, and the flow rate at the point of maximum filling.

A:

$$C_2 = \frac{H^2 P_0}{3\mu R^2} = \frac{(0.0508)^2 \times 10^8}{3 \times 10^3 \times 9^2} = 1.062 \text{ s}^{-1}$$

$$t_{fill} = \frac{1}{2C_2}\left[-\ln\frac{D}{2R} - \frac{1}{2}\left(1 - \left(\frac{D}{2R}\right)^2\right)\right] = 1.13 \text{ s}$$

$$Q_{t,fill} = (4 \pi H^3 P_0/3 \mu)\,[\ln(2R/D)]^{-1} = 19 \text{ cm}^3/\text{s}$$

$$\text{Volume of Chamber} = 2 \pi R^2 H = 25.85 \text{ cm}^3$$

$$Q_{ave} = 25.85/1.13 = 22.88 \text{ cm}^3/\text{s}$$

A measure of viscous dissipation can be obtained by applying the arithmetic for viscous heating in an extruder die, given earlier. Usually the greatest source of viscous heating is in the gate area of a single cavity mold.

Realistic Modifications on the Model.   This simple model can be modified to better reflect real processing conditions by including inertial terms, time-dependent terms, as well as pressure-dependent viscoelasticity and an energy equation that describes material cooling and solidification at the mold walls. Inclusion of inertial and time-dependent terms greatly increases the complexity of modeling. Typically these terms are usually important only for short times and for flows very near the gate (small r). Even for very viscoelastic polymers, the elastic component can be neglected for most molding conditions and less complex, viscous-only fluid models can be used. The filling rate at constant applied pressure for power-law, viscous-only fluid flow into a runner-cavity system is given as:

$$(5.31) \quad P_0 = \left[\frac{8(1 + 3n)}{n\pi D^3} \right]^n \frac{4KLQ^n}{D} \\ + \left[\frac{2n + 1}{4n\pi H^{2 + 1/n}} \right]^n \left(\frac{K}{1 - n} \right) \left[R^{1-n} - \left(\frac{D}{2} \right)^{1-n} \right] Q^n$$

D is the runner diameter and R is the gate radius. The first term is resistance to polymer flow through the runner and the third is resistance into the cavity. The second term represents resistance to flow through the gate. Since the gate land length is usually very short, flow is never fully developed. As a result the length term can be written as:

$$(5.32) \quad \frac{L_e}{R} = \frac{L_\infty}{R} [1 - e^{-4b'Q/\pi R^3}]$$

L_e is the equivalent land length, L_∞ is an asymptotic length when Q is very large and b' is a geometric gate parameter (40,41).

The proper way to include the effect of material solidification or high viscosity non-flow is to solve the coupled set of flow and energy equations with attendant boundary conditions (42–44). The energy equation is coupled through viscous dissipation terms and the momentum equations through temperature-dependent physical properties such as viscosity. Most commercial products have very complex geometric shapes. To analyze most practical mold filling cases, computational accuracy must be compromised.

One expedient, approximate approach is to decouple the equations by assuming *a priori* that the flow channel narrows with time. For static heat transfer, the position of a plane of constant temperature moves into a cooling solid in proportion to the square root of elapsed time. For superimposed fluid flow, it appears that it is proportional to elapsed time to the $\frac{1}{3}$ power (45). This can be written as:

(5.33) $$h = C_3 \, t^n$$

where h is the position from the mold surface, n is $\frac{1}{3}$ to $\frac{1}{2}$, and C_3 is a parameter that includes thermal diffusivity, geometry and (probably) flow rate. The arithmetic leading to the flow rate and pressure drop equations can then be modified by replacing H, the constant wall thickness, with $h = h(t)$, then integrating the differential equations.

Polymer Orientation

During injection molding at high shear rates, macromolecules in the melt state can become highly oriented. This orientation can be frozen in as the polymer is cooled. Since most injection molding involves nonuniform flow during cavity filling, the extent of orientation depends on cavity geometry and distance from the injection point, the gate. Furthermore, polymers increase in density with decreasing temperature. Owing to phase change, crystalline polymers exhibit greater density increases than do amorphous polymers. Thus if a material just fills a cavity at its processing temperature, the part will be considerably smaller than the cavity at room temperature. This is known as *shrinkage* (Figure 5.29). Owing to nonuniform filling of most parts, shrinkage can be decidedly nonuniform, resulting in part warpage, optical distortion, and bowing of ribs and thin side walls. These defects were discussed in Chapter 4.

Since most polymers are compressible hydraulic fluids, additional material can be forced into the mold cavity at high pressure at the end of the injection cycle to compensate for this volume change. This is termed *packing*. Then as the part cools, the pressure falls but the part volume remains (approximately) the same as that of the mold cavity.

Packing exacerbates the effect of flow orientation. Material in the gate region experiences a combination of flow and pressure effects different from that at the part perimeter. One simple method of determining the extent of local orientation or residual stress is to section the part into squares or rectangles, and carefully measure each section dimension. Then the sections are oven-annealed and remeasured. Typically, dimensional changes in the annealed sections will be greater near the gate (Figure 5.32), at lower melt temperature, and at lower injection speed since the material is cooling more rapidly.

Birefringence is a more accurate method of determining the degree of residual stress in optically transparent polymers. A transparent material has three indices of refraction in its three primary axes. If a material is preferentially oriented, through flow or cooling conditions, the index of refraction changes in that direction. The change in any pair of indices is called birefringence in that direction. Polystyrene produces exceptionally strong birefringence under strain, primarily due to the highly anisotropic phenyl group. PMMA, on the other hand, produces weak birefringence since its pendant group is less anisotropic.

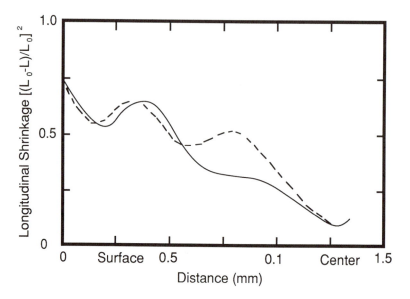

Figure 5.32.   Effect of Gate Presence on Longitudinal Shrinkage as a Function of Depth into Molded Part (46). Measured Before and After Annealing. Solid Curve: Far from Gate. Dashed Line: Near Gate.

Refractive indices are linked to the polarizability of the polymer. Certain polymers such as polyisoprene have much greater polarizability along the backbone than perpendicular to it. When an amorphous polymer is unoriented, it is not birefringent. In a strain field, in plane-polarized white light, light retardation by change in refractive index, will act to cancel out a particular wavelength from the spectrum. This leaves a spectrum of colors in the viewing field. This is termed *interference*. Black is considered as zero order, green as maximum. The classical expression for birefringence is written as:

$$(5.34) \qquad\qquad \Delta n = \lambda \, \frac{R}{D}$$

where Δn is the difference in refractive index (birefringence), λ is the wavelength of light, R is the level of interference and D is the sample thickness. R is usually determined by counting the number of green (R = 1.0) orders. Other colors are assigned fractional orders. For example, yellow = 0.3 and red = 0.7. An example of stress birefringence measured across the thickness of an injection molded PS bar is shown in Figure 5.33. The maximum stress does not occur at the bar surface, but at some distance into the bar. This is in the region in which maximum viscous dissipation would occur during injection of the viscous polymer.

For amorphous polymers, decreasing mold temperature, decreasing part wall thickness, increasing packing time (also called ram forward or hold time), and increasing

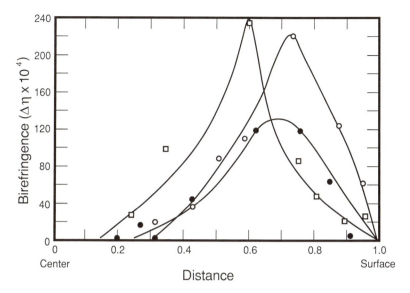

Figure 5.33. Thickness-Dependent Molded-in Stress in General-Purpose Polystyrene (GPS) as Determined by Stress Birefringence (47). Open Circles and Squares: Short Shots. Closed Circles: Full Shots, Not Flashed.

packing pressure lead to increased birefringence. Increasing gate dimensions, increasing injection speed, and for the most part, decreasing melt temperature increases birefrigence.

Typically, those process and material changes that increase birefringence in amorphous polymers also serve to decrease shrinkage (Table 5.10). Increased birefringence can lead to improved mechanical properties *in that direction* (Figures 5.34 and 5.35).

Table 5.10. Effect of process and material condition on shrinkage and stress of injection-molded amorphous polymers.

Increasing Polymer/ Process Condition	Effect on	
	Shrinkage	Stress birefringence
Melt Temperature	Decrease	Decrease*
Mold Temperature	Increase	Decrease
Injection Speed	Decrease	Increase
Packing Pressure	Decrease	Increase
Packing Hold Time	Decrease	Increase
Gate Size	Decrease	Increase
Wall Thickness	Increase	Decrease
Molecular Weight Distribution	Increase	Decrease*

*With some exceptions.

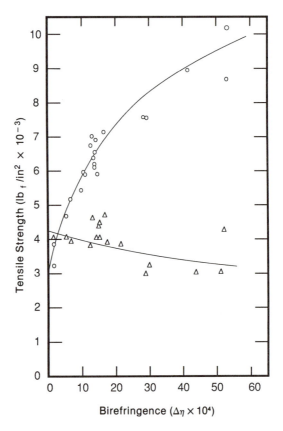

Figure 5.34. Comparison of Molding Direction Strengths with Molded-in Stresses, as Determined by Stress Birefringence for General-Purpose Polystyrene (GPS) (48). Circles: Parallel to Flow Direction. Triangles: Transverse to Flow Direction.

Excessive differential orientation leads to substantial physical property reduction in transverse directions. Impact strength is a classic example, Figure 5.36. This is discussed more fully in Chapter 6.

For semicrystalline polymers, the interpretation of birefringence requires the assumption that the polymer is indeed a system of two distinct phases, one of which is purely crystalline and the other is totally amorphous. The polarizability of crystallites can be obtained from X-ray diffraction measurements on single crystals. For oriented highly crystalline polymers, the crystalline contribution to birefringence is substantially greater than the amorphous phase contribution (Figure 5.37).

For crystalline polymers, morphology is also affected by the nonisothermal, high-shear environment of injection molding. The polymer in the vicinity of the mold is preferentially oriented in the flow direction. The thermal gradient causes *transcrystalline* lamellar growth, that is, crystal growth that occurs predominantly perpendicular to the mold surface. In the polymer at some distance from the mold surface, the shear and thermal fields are far more isotropic and so *spherulitic* crystal growth is preferred.

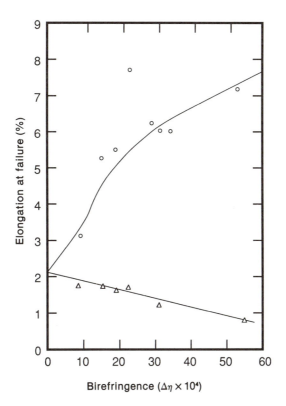

Figure 5.35. Comparison of Molding Direction Elongation at Failure with Molded-in Stresses, as Determined by Stress Birefringence on General-Purpose Polystyrene (GPS) (48). Circles: Parallel to Flow Direction. Triangles: Transverse to Flow Direction.

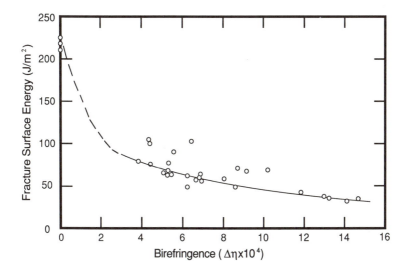

Figure 5.36. Effect of Orientation, as Measured by Stress Birefringence, on Fracture Surface Energy of Polymethyl Methacrylate (PMMA) (49).

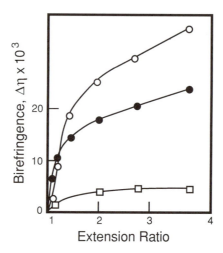

Figure 5.37. Correlation Between Stress Birefringence and Extent of Orientation for Semi-crystalline Polyethylene (PE) (50). Open Circles: Total Birefringence. Closed Circles: Birefringence Due to Crystallites. Open Squares: Birefringence Due to (Amorphous) Form Effects.

Thus two or even three distinct regions of differing crystalline nature can occur in crystalline polymers, as seen in Table 5.11 for POM (polyacetal). This effect has been observed for PP and polyphenylene sulfide (PPS), as well.

The oriented crystalline region is enhanced by increasing the temperature differential between the polymer melt and mold, increasing injection speed, decreasing part thickness, and increasing packing pressure. Since transcrystalline materials tend to be weak in the flow direction, these changes result in reduced tensile and impact properties.

Crystallization rate in injection molding is subject to a global nonisothermal environment. The Avrami equation is discussed in detail in Chapter 2. As noted, it is for isothermal crystallization and therefore correctly it should not be applied here. However, during phase change, the *local* temperature can be nearly constant. If this local temperature can be determined, the Avrami coefficient and the *local* extent of crystallinity (including $X_{1/2}$, which yields $t_{1/2}$, the *local* half-time of crystallization), can be determined (52).

Table 5.11. Crystalline characteristics of injection-molded polyacetal (POM) (51).

Nature of field	Nature of crystallite	Distance from mold surface
Thermal, Strain	Skin: Lamellae Perpendicular to Mold Surface	0 to 0.13 mm
Thermal, Viscous Heating	Transcrystalline	0.125 to 1 mm
Near-Isotropic	Spherulitic	1 mm to center

Thermoplastic Structural Foam Injection Molding

In certain injection molded part designs, thick walls are required for stiffness. To reduce part weight without substantial reduction in part stiffness, the resin is foamed. In the simplest case, a chemical foaming agent, typically a very pure, thermally sensitive chemical such as azodicarbonamide (AZ or azobisformamide, ABFA) is added as a dry fine powder at 0.25 to 1.0% (wt) to the resin at the hopper of a near-conventional injection molding machine.* The chemical decomposes in a relatively narrow temperature range, producing gas that is dissolved in the pressurized polymer melt ahead of the screw in the injection molding machine barrel.** The foamable resin is injected into the mold cavity having a volume larger than the volume of the injected melt. The drop in pressure allows the dissolved gas to come from solution, form microbubbles and expand the resin to fill the mold cavity. Since the mold surface is cold, the gas does not have time to form a foam near the mold surface. As a result, the molded product will have a dense skin and a near-uniform cellular core.

Several other techniques are commercially used to produce structural foam. A permanent gas such as nitrogen is injected directly into the pressurized polymer melt. With proper shear mixing, the permanent gas is uniformly dispersed and dissolved in the polymer. This process requires a special machine as shown in schematic in Figure 5.39. The foamable melt is then transferred to the mold cavity in a manner similar to that for conventional injection molding with the exception that the volume of the melt transferred is less than the cavity volume. In another variant, the gas can be provided as a volatile liquid such as a chlorofluorocarbon (CFC) and again added to the polymer melt under pressure. This technique is used primarily for low-density foams, however.

In *two-component thermoplastic structural foam molding,* two injection heads supply polymer melt to a single special-purpose nozzle. Only one stream of polymer

---

*To minimize gas loss from the pressurized foamable resin ahead of the screw, a shut-off nozzle between the injection molder and the mold cavity is needed. In addition, a non-return valve is needed on the tip of the injection screw to prevent polymer from flowing back along the screw during injection. For efficient structural foam processing on a conventional injection molding machine, a melt accumulator can also be used (Figure 5.38).

**The decomposition reaction for AZ is:

$$\underset{\text{H}_2\text{-N-C}}{\overset{\text{O}}{\overset{\|}{}}}\text{N=N-}\underset{}{\overset{\text{O}}{\overset{\|}{\text{C}}}}\text{-N-H}_2 \rightarrow \text{N}_2 + \text{CO} + \text{CO}_2 + \text{NH}_3 + \text{byproducts}$$

About 33% of the decomposition products of the AZ chemical is gas. The gas analysis is about 66% nitrogen, 24% CO, 5% carbon dioxide, 5% ammonia. By-products such as biurea, urazol, cyamelide and cyanuric acid comprise the remaining 67%. Despite this seemingly low amount of gas, AZ produces 220 cm^3 (STP) gas per g of blowing agent at a decomposition temperature range of 205 to 215°C (401 to 419°F). Its moderate decomposition temperature allows it to be used for most polymers being processed at a temperature range of 166 to 232°C (330 to 450°F). Since most commodity polymers are processed in this range, AZ is the most widely used chemical foaming agent.

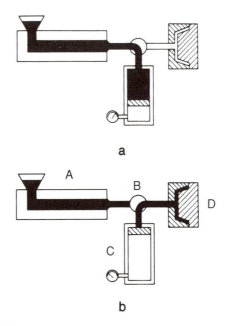

a

b

Figure 5.38.   Schematic of Low Pressure Chemical Foaming Agent Structural Foam Molding Process (53). (a) Accumulator Filling Step, Mold Empty. (b) Mold Filling Step, No Transfer From Screw Plasticizer. A: Screw Plasticizer. B: Transfer Valve. C: Accumulator. D: Mold.

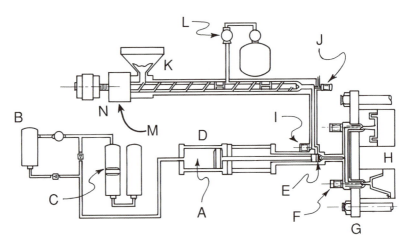

Figure 5.39.   Special-Purpose Thermoplastic Physical Foaming Agent Structural Foam Machine (54). A: Hydraulic Fluid. B: Nitrogen. C: Floating Piston. D: Injection Cylinder. E: Melt Accumulator. F: Injection Nozzle. G: Machine Platen. H: Molds. I: Shutoff Valve. J: Melt Relief Valve. K: Extruder. L: Volumetric Gas Flow Control. M: Gear Drive. N: Motor.

contains dissolved gas. The two streams flow into the mold cavity in linear flow fashion, called creeping or Hele-Shaw flow, with the outer layer being unfoamable resin and the inner core forming cells as its pressure drops in the mold cavity. While this is an expensive process, large containers having very smooth, near-injection molded quality surfaces can be formed. These containers can be FDA-approved for food and can maintain high strength and rigidity needed in materials handling. In *gas-counterpressure structural foam molding*, the mold cavity is sealed and pressurized with an inert gas prior to foamable resin injection. This back pressure prevents premature foaming and allows improved surface appearance. This technique requires specially-designed pressurized molds and careful pneumatic control. If the foamable resin is injected into a mold cavity having a volume equal to or less than the volume of resin injected, the mold pressure will approach that of conventional injection molding. When certain sections of the mold are made movable, mold expansion in those sections allow foaming to occur. This technique also requires relatively expensive molds.

Thermoplastic structural foam parts typically have overall weights of 20 to 40% lower than unfoamed parts of the same dimension. Single component foam products can be relatively large, some weighing 22 lb (10 kg) or more. Typical applications include boxes, trays, machinery covers, swimming pool filters, furniture and furniture frames. However, many smaller foamed parts such as brush handles, shoe heels, and platform shoe soles are made in molds mounted on rotating frames. Reaction injection molding (RIM) of thermoset resins, described below, is also used to make structural foam parts.

Example 5.13

Q: As a "rule of thumb", certain mechanical properties of a thermoplastic structural foam beam are functions of the square of the density ratio:

$$X/X_o = (\rho/\rho_o)^2$$

where X is the physical property and ρ is the material density, with the subscript "o" denoting the value of the unfoamed material. What is the expected apparent modulus of a unreinforced polypropylene (PP) beam, foamed to 15%?

A: From Appendix C, $E_{o,PP} = 1100$ N/m^2. The modulus of the foamed beam is:

$$E = 1100 \times (0.85)^2 = 795 \text{ N/m}^2$$

Summary on Injection Molding

Injection molding is a versatile process for mass-producing discrete plastic products with high precision. In addition to being a stand-alone process, it can be used in combination with other processes to produce unique products. As an example, when a test tube-shaped product is injection molded, cooled and ejected from the mold, it can be reheated and blow-stretched into a bottle. If it is only tempered and kept on the mold core, the core can be placed in a blow mold cavity and blown directly into a bottle.

The polymer properties that are most important in injection molding are:

- Properties that are also significant in extrusion, such as:
  - frictional coefficients,
  - thermal properties, such as thermal diffusivity, thermal conductivity, heat capacity, and density of solid particulate, and
  - polymer melt viscosity,
- Shear- and temperature-dependent viscosity over a shear rate range of 0 to 10,000 s^{-1},
- Compressibility of polymer melt (for packing), and
- Crystallization kinetics (if any).

Blow Molding

When cataloging processes as to the type of product produced, *blow molding* and rotational molding are given as the primary ways of forming a hollow object such as a bottle or drum. Rotational molding is properly considered to be a powder process and is considered below. Certain forms of blow molding such as reheat blow molding can also be considered as rubbery sheet processing and can be cataloged with thermoforming, another rubbery sheet process. In fact, as mentioned earlier, thermoforming can be used to make thin-walled hollow containers. Two halves of a shape can be thermoformed and spin, friction or ultrasonically welded together to produce a closed container.

In blow molding, a softened tube of polymer is either extruded from the melt or reheated from the solid, then clamped between two halves of a female mold. The tube is then inflated against the mold walls with air injected by a blow pin. The blown form is then cooled by conduction to the cool metal and, on occasion, by circulating cold air, possibly containing volatile fluids such as water, CO_2, or chlorofluorocarbons (CFCs), within the container. There are three commercial types of blow molding process. If the softened tube of polymer is produced by extrusion, the tube is called a **parison, and the process is called parison or extrusion blow molding.** If the softened tube is produced by injection molding and if it is not allowed to cool prior to inflation, the process is called *injection blow molding*. If the tube is reheated from a cold form, the tube is called a preform and the process is reheat, preform or *stretch blow molding*.

Extrusion Blow Molding

The polymer melt is prepared by plasticating powder or pellets in a screw extruder. If the extrudate is produced continuously, the process is called *continuous parison blow molding*. The advantages to continuous extrusion are in melt uniformity and better extrusion process control. The disadvantage is that the inflation process is a cyclical one. For high production, multiple molds are used, and parison removal must be swift and under control at all times. If the parison is extruded cyclically, the extruder must either feed a separate hydraulic (ram) accumulator or the screw must reciprocate, as in injection molding. Both of these processes are called *intermittent parison blow*

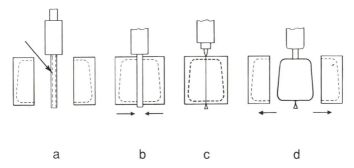

Figure 5.40.   Schematic of the Extrusion Blow Molding Process (55). (a) Parison Extrusion (Parison Denoted by Arrow). (b) Mold Halves Close onto Parison. (c) Parison Inflated Against Internal Mold Walls. (d) Mold Halves Open, Blow Molded Part Ready for Ejection.

molding. Ram accumulator blow molding is desired for fabricating large containers such as chemical drums. Intermittent parison blow molding is usually best suited for single cavity molds, and as a result is normally not economic for high-volume production runs. Extrusion blow molding is shown in schematic in Figure 5.40.

Most of the polymers that are blow molded are commodity materials (PVC, PS, PP, LDPE, HDPE). Polyethylenes dominate the dairy, laundry, and chemical markets. Certain material characteristics are critical to good parison formation. The parison is usually extruded down. As a result, gravity acts to thin the parison as it issues from the die. The longer the parison becomes, the greater is the gravity effect. On the other hand, as the material issues from the die, the elastic nature of the polymer manifests itself as extrudate swell. Higher melt temperatures and lower extrusion rates yield lower extrudate swell but greater gravity effect. Furthermore, extrudate swell is not instantaneous, since melt relaxation is highly time- and temperature-dependent.

One common processing technique used to study parison material distribution is the pillow or pinch-off mold. This mold is comprised of several longitudinal sections which pinch off and segregate measured portions of the parison. The weight of each "pillow" is related to the parison thickness at the time of pinch-off, and this to the relative effects of gravity and extrudate swell. If there was no extrudate swell or gravity effect, the parison thickness would be that of the die. The relative effects of the processing parameters on that thickness (or weight*) are seen in Figure 5.41.

To achieve nearly uniform parison wall thickness, the polymer must have good *melt strength* or *hot strength*. This implies high molecular weight, narrow molecular weight distribution, and high zero-shear viscosity at the extrusion temperature. To gain additional strength and to minimize cycle time, the parison is extruded such that inflation

---

*Weight swell is defined as W_a/W_t, where $W_a = A_aL\rho$ and $W_t = A_tL\rho$. A_a is the cross-sectional area of the parison and A_t is the cross-sectional area of the die. L is the pillow length and ρ is the resin density. Weight swell then is the ratio of local actual parison cross-sectional area to die cross-sectional area.

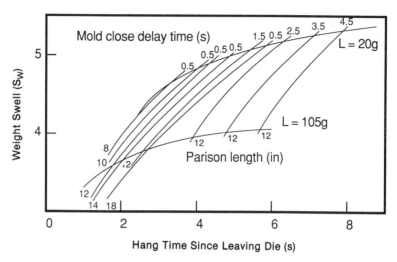

Figure 5.41. Composite Graph Showing Effect of Parison Length, Parison Weight, and Mold-Close Delay Time on Weight Swell as a Function of Time Since Material Left Die (56). Weight Swell Is Ratio of Extrudate to Die Cross-Sectional Areas. Time Is Hang Time Since Leaving Die. L Is Parison Length, in Weight Units.

occurs only a few degrees above the polymer melt temperature. Since many candidate materials are crystalline, this means that crystallization can occur throughout the inflation and cooling process. For most parison blow molding processes, the parison is captured on both ends, and so inflation initially is unsteady-state stretching of the tube in the hoop (or uniaxial tensile) direction. When any portion of the parison touches the mold wall, that portion stops stretching. The last step in inflation of the parison is biaxial stretching into three-dimensional corners of the mold. Inflation usually proceeds quite quickly (seconds), but can be considered to be isothermal only as a first approximation. The engineer should note that this process, when well-tuned for a specific resin, such as low-MI (melt index), high-density polyethylene (HDPE), can reliably produce very uniform wall thickness bottles, as seen for the one-gallon milk bottle in Figure 5.42.

Simple Parison Model. A simple parison inflation model assumes isothermal uniaxial elongation of a (viscous-only) power-law fluid (58). The hoop stress is given as:

$$(5.35) \qquad \tau = -\frac{PR(t)}{h(t)} = -\eta_e \, \dot{\varepsilon}$$

where P is the inflation pressure, R(t) is the time-dependent parison radius, h(t) is its thickness, η_e is the viscosity and $\dot{\varepsilon}$ is the strain rate. For elongational viscosity:

$$(5.36) \qquad \eta_e = \eta_{e,o} \, \dot{\varepsilon}^{n-1}$$

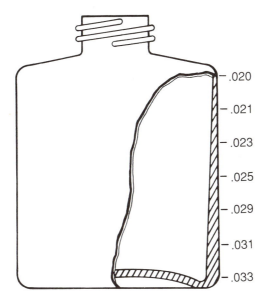

Figure 5.42.   Wall Thickness Variation for One-Gallon Milk Container Blown on a Fixed Orifice Parison Blow Molding Machine (57). Thickness in Inches.

and:

$$(5.37) \qquad \dot{\varepsilon} = \frac{1}{R(t)} \frac{dR(t)}{dt}$$

The parison volume, V, is constant:

$$(5.38) \qquad V = 2\pi R(t)\, h(t)\, L = C_1$$

where L is the parison length, also constant.

Upon substitution:

$$(5.39) \qquad C_2 PR(t)^{2+n} = \left[\frac{2\pi L}{C_1 \eta_{e,o}}\right] PR(t)^{2+n} = \left[\frac{dR(t)}{dt}\right]^n$$

where:

$$(5.40) \qquad C_2 = \left[\frac{2\pi L}{V \eta_{e,o}}\right]^{1/n}$$

If $R(t) = R_o$ when $t = 0$, then after integration, for constant pressure:

$$(5.41) \qquad t = \left(\frac{n}{2C_2}\right) [PR_o^2]^{-1/n} [(R_o/R)^{2/n} - 1]$$

The minimum inflation time, t_i, is when $R = R_m$, the mold radius:

(5.42) $$t_i = \left(\frac{n}{2C_2}\right) [PR_0^2]^{-1/n} [(R_0/R_m)^{2/n} - 1]$$

Of course this assumes no effect of crystallization during inflation and no elastic resistance to inflation. Since most blow molded materials show some degree of orientation, this model can be used only to demonstrate approximate relationships between inflation time, pressure, and geometry.

Example 5.14

Q: A blow molding die has an outside diameter of 30 mm with 1.5 mm gap. The parison is inflated with a pressure of 0.5 MPa to produce a bottle with a 60 mm outside diameter. Estimate the wall thickness of the bottle if the thickness swelling ratio, B_{SH}, is 1.8. Consider draw-down effects to be negligible.

The schematic of the system is shown in Figure 5.43. The geometry of the die is:

$$OD = 30 \text{ mm}$$

$$h_d = 1.5 \text{ mm}$$

$$ID = 27 \text{ mm}$$

$$D_d = 28.5 \text{ mm}$$

$$D_m = 60 \text{ mm}$$

A: From Chapter 4:

$$B_{SH} = B_{ST}^2$$

$$B_{ST} = (B_{SH})^{1/2} = 1.34$$

Since $B_{SH} = h_1/h_d$ and $B_{ST} = D_1/D_d$, then:

$$h_1 = h_d B_{SH}, \text{ and}$$

$$D_1 = D_d B_{ST}$$

A volume balance yields:

$$\pi D_1 h_1 = \pi D_m h$$

$$h = h_d B_{ST}^3 D_d/D_m$$

From these relationships:

$$h_1 = 2.7 \text{ mm}$$

$$D_1 = 38.2 \text{ mm, and}$$

$$h = 1.71 \text{ mm}$$

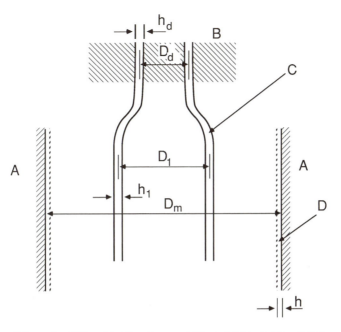

Figure 5.43.   Schematic of Blow Molding Process (59). A: Mold. B: Die. C: Polymer Parison. D: Inflated Parison.

Example 5.15

Q: Determine the approximate inflation time for a $R_o = 2.5$ cm parison of high-density polyethylene (HDPE) inflated at $P = 10$ lb$_f$/in^2 to a "Boston round" of $R_m = 10$ cm radius and $L = 20$ cm long, with an average bottle wall thickness, $h_o = 0.1$ cm, $\eta_{e.o} = 10^6$ Pa s$^{1/2}$ and $n = 0.5$.

A:

$$C_1 = V = 2\,\pi h_o R_m L = 125.7 \text{ cm}^3.$$

$$C_2 = (2\,\pi\,L/V\,\eta_{e,o})^2 = 1 \times 10^{-12} \text{ (Pa s cm}^3)^{-2}.$$

$$t_i = \left(\frac{n}{2C_2}\right) [P]^{-1/n} \left[(1/R_o)^{2/n} - (1/R_m)^{2/n}\right]$$

$$= 0.25 \times 10^{12} [10 \times 6895]^{-2} [(1/2.5)^4 - (1/10)^4]$$

$$t_i = 1.34 \text{ s, inflation time.}$$

Parison blow molding is used to produce bottles with volumes greater than about 12 ounces (350 ml). The process is used extensively to manufacture milk, detergent and bleach bottles. It is also used to produce heavy-wall containers for corrosive chemicals in 5 to 50 gallon (20 to 200 liter) sizes. Parison blow molding has been used to fabricate HDPE fuel oil tanks of 500 gallons (2000 liter) or more.

To obtain high barrier properties, multilayer parisons can be extruded. The food and beverage industries are using this process to produce barrier bottles for oxygen-sensitive ketchup and tomato juice. Although the scrap level in multilayer blow molding can be of concern, it is lower than that experienced by multilayer thermoforming. Some scrap can be recycled as an inner layer.

The polymer properties that are most important in extrusion blow molding are:

- Polymer viscosity at both high and low shear rates,
- Polymer melt strength, particularly at very low shear rates, since parison drawdown is most important,
- Polymer strain recovery, related for the most part to the molecular weight and molecular weight distribution,
- Crystallization rate (if any), and
- Melt and solid polymer thermal properties, particularly thermal diffusivity, thermal conductivity, and specific heat.

Injection Blow Molding

The tube to be blown into a container begins as an injection molded shape that remains on the *core pin,* that part of the injection mold that forms the inside of the tube. Injection molding is used to achieve very accurate wall thickness, high-quality neck finish (threads and/or closures on the top of the container), and to process polymers that cannot be extruded. The hot injection molded shape is then rotated to a blow molding station where it is inflated by relatively high internal pressure (Figure 5.44). Typically, bottles with volumes of 12 oz (350 ml) or less are formed by this process. Applications include pharmaceutical, cosmetic and single-serving liquor bottles.

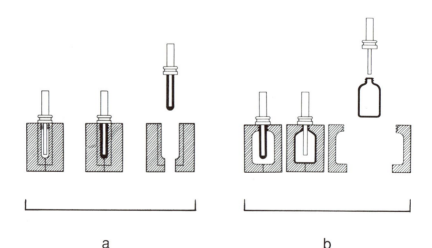

a   b

Figure 5.44.   Schematic of Injection Blow Molding Process (60). (a) Injection Cycle, Polymer Melt Supplied to Mold Halves from Injection Molding Machine, Product Formed Is Preform. (b) Heated Preform Inflated Against Cool Mold Wall Halves.

The polymer properties of interest here include those in injection molding, including shear- and temperature-dependent viscosity. In addition,

- Temperature-dependent tensile strength of the hot-rubbery polymer while on the core pin,
- Tensile elongation during inflation,
- Crystallization kinetics (if any) during cooling (tempering) on the core pin, and
- Crystallization kinetics during blowing and cooling.

Stretch-Blow Molding

Stretch blow molding begins with a *cold* extruded or injection molded polymer pre-form. The preform is heated to above the polymer glass transition temperature and for highly crystalline materials such as PP to within a few degrees of the melt temperature. The preform is usually substantially shorter than the length of the mold cavity. The preform is then simultaneously inflated and stretched with a hollow core-rod. This is shown in schematic in Figure 5.45. Inflation resembles inflation of a rubber balloon, with inflation to the mold dimension occurring first near the top of the core-rod. The deformation front then progresses toward the mold bottom. It is thought that the core-rod simply maintains axial orientation of the uninflated portion of the preform rather than substantially aiding in stretching. Stretch blow molding is reserved for difficult-to-blow crystalline and crystallizable polymers, such as PP and PET. The polymers appear to be deforming as rubbery solids (62–64).

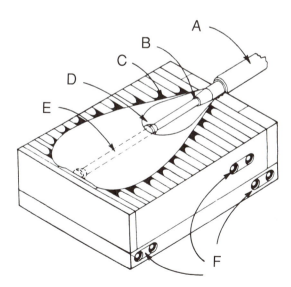

Figure 5.45.   Stretch Blow Molding Process, Breakaway Section Showing Blow Mold with Stretch-Blow Pin (61). A: Stretch-Blow Pin. B: Air Entrance. C: Mold Vents. D: Preform, E: Stretch Rod Extended. F: Cooling Channels.

Orientation in crystalline PP preforms usually requires initial partial melting of the lower temperature (145 to 150°C) imperfectly formed spherulites without melting the higher temperature (155 to 165°C) well-formed spherulites. High-speed, high-pressure deformation follows (65). Cooling then enables partial recrystallization into the desired strain shape. In PET stretching, orientation can be set by carefully reheating the formed shape above the glass transition temperature, typically to 140 to 160°F. This also acts to increase measured density and glass transition temperature. It is thought that additional fibrillar crystallites are formed (66).

The polymer properties most important in stretch blow molding deal with the tensile strengths and the effect of orientation on a polymer above its glass transition temperature. These include:

- Tensile strength and yield above T_g,
- Effect of orientation on gas permeability through the polymer, and
- Strain-oriented crystallinity levels and their effect on permeability and other properties.

Orientation and Shrinkage

The basic concepts on shrinkage and orientation given above for injection molding generally hold for blow molding with the notable exceptions that the parison freely expands (extrudate swell) *prior* to inflation and inflation results in near-*uniaxial* strain initially and near-*biaxial* strain at the end. Parison temperature, mold temperature, parison tension, parison hang time, inflation rate, rate of cooling, and the extent and rate of crystallization of the polymer affect overall shrinkage and orientation. The extent of inflation, the relationship between the geometry of the parison and that of the mold, and the extent of draw-down into two- and three-dimensional corners influence local shrinkage and orientation. Parison tension is caused by the force induced as the parison relaxes against the clamping restraints. Since the container is hollow, volumetric contraction has substantially less effect on overall shrinkage than does internal strain relaxation. Unlike injection molding, blow molding wall thickness can change quite rapidly with distance, particularly in the neck and pinch-off or weld regions. Thus local shrinkage is very difficult to predict.

As noted earlier, part density is a function of the degree of crystallinity. Part stiffness is a function of density. Since heat transfer from the material to the mold wall and to internal cooling gases control the rate at which part wall temperature decreases, thick walls imply high crystallinity and stiffer parts.

Thermoforming

In *thermoforming*, a softened sheet of polymer is forcefully deformed into or onto a single-surface mold, where it is cooled through conduction into the mold. The formed part is then trimmed from the clamped edges called *web* or trim. *Thin-gage* sheet (to 1.5 mm) is usually supplied in rolls and fed into a continuous forming machine (Figure 5.46). *Heavy-gage* sheet (to 12 mm or more) is usually supplied as cut sheet. It is fed

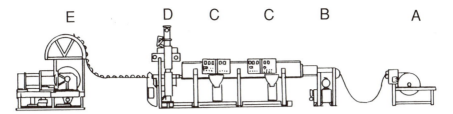

Figure 5.46.   Schematic of the Roll-Fed Thermoforming Process (67). A: Roll-Fed Sheet Take-Off Station. B: Pin Chain Engagement. C: Heating Zone. D: Forming Station. E: Trimming Station.

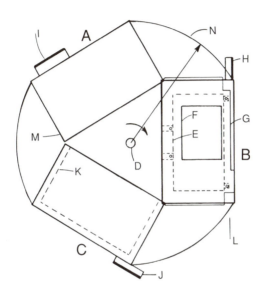

Figure 5.47.   Rotary Cut-Sheet Thermoforming (Top View) Schematic (68). A: Heating Station. B: Forming Station. C: Loading/Unloading Station. D: Drive Center. E: Platen. F: Motor Well. G: Pit. H: Platen Control Panel. I: Oven Control Panel. J: Clamp Panel. K: Clamp Frame. L: Vacuum Service. M: Air Service. N: Safety Guard.

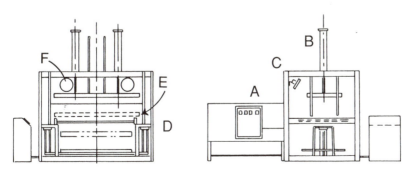

Figure 5.48.   Cut-Sheet Shuttle Thermoforming Process Schematic (69). A: Heating Zone. B: Pneumatic/Hydraulic Plug Assist. C: Auxiliary Cooling. E: Forming Table. F: Vacuum Tanks.

Table 5.12. Thermoforming temperature ranges for several thermoplastics (70,71).

Polymer	Type*	Glass Transition Temperature, T_g		Melt Temperature, T_m		Heat Distortion Temperature at 66 lb_f/in^2	
		(°C)	(°F)	(°C)	(°F)	(°C)	(°F)
ABS	A	88 to 120	190 to 248	–	–	77 to 113	170 to 235
Acrylonitrile	C	95	203	135	275	78	172
Acetate (CA)	C	70, 100	158, 212	230	445	52 to 93	125 to 200
Acrylic (PMMA)	A	100	212	–	–	74 to 113	165 to 235
Acrylic/PVC	A	105	221	–	–	81	177
Butyrate (CAB)	C	120	248	140	284	54 to 108	130 to 227
Polycarbonate (PC)	A	150	300	–	–	138	280
Polyethylene Terephthalate (PET)	C	70	258	255	490	49	120
Polyether Sulfone (PES)	A	230	445	–	–	216	420
20% Glass-Reinforced Polyether Sulfone	A	225	437	–	–	216	420
High Density Polyethylene (HDPE)	C	−110	−166	134	273	79 to 91	175 to 196
Propionate (CAP)	C	–	–	–	–	64 to 121	147 to 250
Polypropylene (PP)	C	5	41	168	334	107 to 221	225 to 250
40% Glass-Reinforced Polypropylene	C	–	–	168	334	166	330
Polysulfone (PSO$_2$)	A	190	374	–	–	181	358
Polystyrene (PS)	A	94	200	–	–	68 to 96	155 to 204
PTFE/FEP	C	–	–	275	527	70	158
Rigid PVC	A	90	194	–	–	57 to 82	135 to 180
Mod. Polyphenylene Oxide (mPPO)	A	104 to 110	219 to 230	–	–	110	230

*A = Amorphous, C = Crystalline

Table 5.12. Thermoforming temperature ranges for several thermoplastics (70, 71) (continued).

Polymer	Recommended Mold Temperature		Polymer Set Temperature		Lower Forming Temperature		Normal Forming Temperature		Upper Forming Temperature	
	(°C)	(°F)	(°C)	(°F)	(°C)	(°F)	(°C)	(°F)	(°C)	(°F)
ABS	81	180	85	185	127	260	146	295	182	360
Acrylonitrile	–	–	85	185	127	260	149	300	182	360
Acetate (CA)	–	–	71	160	127	260	154	310	182	360
Acrylic (PMMA)	88	190	85	185	149	300	177	350	193	380
Acrylic/PVC	–	–	79	175	163	325	188	370	204	400
Butyrate (CAB)	–	–	79	175	127	260	146	295	182	360
Polycarbonate (PC)	127	260	138	280	168	335	191	375	204	400
Polyethylene Terephthalate (PET)	–	–	77	170	121	260	149	300	166	330
Polyether Sulfone (PES)	–	–	204	400	274	525	316	600	371	700
20% Glass-Reinforced Polyether Sulfone	–	–	210	410	279	535	343	650	382	720
High Density Polyethylene (HDPE)	71	160	82	180	127	260	146	295	182	360
Propionate (CAP)	–	–	88	190	127	260	146	295	182	360
Polypropylene (PP)	–	–	88	190	143	290	154 to 163	310 to 325	166	330
40% Glass-Reinforced Polypropylene	–	–	91	195	132	270	204	400	232	450
Polysulfone (PSO₂)	163	325	163	325	191	375	246	475	302	575
Polystyrene (PS)	82	180	85	185	127	260	149	300	182	360
PTFE/FEP	–	–	149	300	232	450	288	550	327	620
Rigid PVC	60	140	66	150	104	220	138	280	154	210
Mod. Polyphenylene Oxide (mPPO)	–	–	99	210	163	325	188	375	204	400

one sheet at a time into a shuttle or rotary press (Figure 5.47). The simplest type of thermoforming is *vacuum forming*, shown in schematic in Figure 5.48. There are many variations of this process, including techniques to pneumatically or hydraulically pre-stretch the sheet.

Thin-gage sheet is usually radiantly heated to the forming condition. Heavy-gage sheet heat transfer is controlled by conduction of heat from the surface to the interior. As a result, energy input must be carefully controlled to minimize surface blistering or scorching. The forming temperature range is usually quite large for amorphous polymers and quite narrow for crystalline ones. As seen in Table 5.12, the minimum forming temperature is usually a few degrees above the glass transition temperature or the heat distortion temperature (DTUL at 66 lb_f/in^2 or 0.455 MPa).* The upper forming temperature is usually considered to be the temperature where sheet sag becomes unacceptable.

Thermoforming is similar to stretch or reheat blow molding in that the material can be treated as a rubbery solid. As with blow molding, when the rubbery plastic sheet locally contacts the mold surface, it stops stretching. As a result, the extent of sheet orientation is lowest in the material that touches the mold surface first and highest in the material that touches it last. For most female or negative mold parts, this means that the greatest degree of orientation occurs in two- and three-dimensional corners. Unlike preform blow molding, however, most of the materials thermoformed are amorphous. As a result, part shape is retained *only* by frozen-in stress. If a thermoformed part of amorphous resin is heated above the glass transition temperature to near its forming temperature (typically within 20°C), it will recover to a flat sheet. This supports the contention that thermoforming is a plastic solid deformation process.

Example 5.16

Q1: Obtain a scheme for estimating the effect on mechanical properties of a resin stream when regrind is added. Assume that the mechanical property *after* each processing step is X times that of the material *prior to* each processing step. Assume that for each unit of material processed, Y units are regrind. And assume that a composite mechanical property is obtainable from the properties of the virgin stream and the regrind stream via the law of mixtures.

The following definitions hold:

M_o is the mechanical property of the virgin resin,
M_r is the mechanical property of the regrind,
M_m is the mechanical property of the mixture,
M_p is the mechanical property of the processed material.

Q2: Then determine the property reduction after an infinite number of recycle steps if 50% of the resin is recycled and if recycling results in a 10% drop in physical properties.

---

*See Chapter 6 for further discussion of DTUL.

A1: The following protocol is established. For the first cycle through the process, there is no regrind:

$$M_{p1} = X\,M_{m1} = X\,M_o$$

For the second cycle, regrind is added:

$$M_{p2} = X\,M_{m2}$$

But:

$$M_{m2} = Y\,M_{r1} + (1 - Y)\,M_o$$

and:

$$M_{r1} = M_{p1}$$

Therefore:

$$M_{p2} = X\,[XY + (1 - Y)]\,M_o$$

For the third cycle:

$$M_{p3} = X\,M_{m3}$$

But:

$$M_{m3} = Y\,M_{r2} + (1 - Y)\,M_o$$

and:

$$M_{r2} = M_{p2}$$

Therefore:

$$M_{p3} = [X^3Y^2 + X^2Y(1 - Y) + X(1 - Y)]\,M_o$$

For the Nth cycle:

$$M_{PN} = \left[(X - 1) \sum_{i=0}^{N-1} (XY)^i + 1 \right] M_o$$

But:

$$\sum_{i=1}^{n} ar^i = ar + ar^2 + ar^3 + \ldots + ar^n = a\left[\frac{1 - r^n}{1 - r} \right]$$

Therefore for the Nth cycle:

$$M_{PN} = \left[1 - (1-X) \left[\frac{1 - (XY)^{N-1}}{(1-XY)} \right] \right] M_o$$

For an infinite number of cycles:

$$M_{P\infty} = \left[1 - \frac{(1-X)}{(1-XY)} \right] M_o$$

A2: For $Y = 0.5$ and $X = 0.9$, $M_{P\infty}/M_o = 0.8182$. Note that $M_{P3}/M_o = 0.835$, or the recycle properties achieve values relatively close to the final values after a very few cycles.

The material property that seems to best correlate the region of formability with the force (or pressure) needed to form a shape is the *elastic modulus* in extension or tension *at the forming temperature.** The moduli of amorphous materials show distinct rubbery region plateaus over large temperature ranges. Many crystalline materials retain high strength to within a few degrees of their melt temperature (72).

The hot creep or hot tensile test seems to yield additional useful information about the formability of a material. As discussed in Chapter 3, creep strength is determined by measuring strain on a specimen under load for an extended period of time. This is usually related then to the linear viscoelastic response of the material. In the hot creep test, the unconstrained specimen is placed in an oven. When at temperature, a load is instantaneously applied and the elongation rate determined by high-speed photographic means. Elongation rates to $10 \ s^{-1}$ can be measured this way. Usually, the slope of the stress–rate-of-strain curve is *viscosity,* a fluid property. However, since these stresses cause hot creep rupture, the test is not truly steady-state. At best, hot creep can be used as a measure of the hot polymeric material resistance to impulse loads.

The stretching process is terminated by mold contact in a very short time. Forming times of 0.5 s to 99% of final dimension have been measured. As a result, *either* a rubbery solid concept that depends on the elastic modulus at the forming temperature *or* a hot creep strength model that yields a form of extensional viscosity at very high deformation rates could be used to describe the stretching process.

Very high production rates can be achieved by continuous forming of relatively thin film and sheet. Web trimming of 40% of the sheet or more is generated. This trim must be recovered, reground, re-extruded and recycled for the process to be economical. It is difficult to reprocess certain types of multilayer sheet without degradation.

With the increased use of domestic microwave cooking, thermoformed microwave-transparent crystallized PET trays capable of withstanding conventional oven temperatures have been devloped by the food packaging industry. This process is a variant of

---

*Williams considers the correct property to be the rate of change of the strain energy function with the first principal invariant of the Cauchy strain tensor (73).

the conventional roll-fed sheet process that uses conventional thermoplastics such as PS, PETG (amorphous polyethylene terephthalate), PMMA, and PVC. The principal thermoformed products include blister packs, cups, wading pools, soaking tubs, camper tops, truck bed liners, skylights and signs. The process can also be used to make containers by forming two halves of the container and joining them together by mechanical, adhesive or solvent means.

As noted, the polymer properties that seem to be most important in thermoforming are:

- Hot strength at the forming temperature.
- Elastic modulus at the forming temperature, and/or
- Elongational viscosity of the polymer *as a melt* at the forming temperature,
- Elongation at break at the forming temperature, and
- Strain-rate hardening, which might affect the rate at which deeply drawn containers can be formed.

Reacting Systems

Thermosets are polymers that must be reacted to produce useful products. Reaction usually is induced by mixing two or more reactants or by applying heat to partially reacted polymers. Once the reaction has been initiated, the polymer must be shaped into the final product form before the polymer molecular weight reaches a level at which the bulk mixture is no longer mobile. The reacting material is then held in this shape until the reaction is, for the most part, complete.

Many of the older thermosets are based on formaldehyde reaction with such monomers as urea, melamine or phenol. These materials are usually quite brittle. As a result, fillers and reinforcements, such as mica, calcium carbonate, iron filings, dolomite, and milled glass, are added to increase stiffness. Typically these highly filled materials require the addition of heat to sustain the reaction. As a result, the polymer is frequently preheated in an oven or radio-frequency field. At their reaction temperatures, these materials are viscoelastically very stiff. As a result, these dough-like materials are typically *compression* or *transfer* molded and the compression platens are heated with steam or hot oil.

Polyurethanes, epoxies, unsaturated polyester resins and other thermosets are reacted by intensively mixing relatively low viscosity liquid prepolymers with catalysts and promoters,* then pumping these very fluid resins at low pressure into the proper reservoirs for shaping while reacting. Usually these polymers are processed without fillers. As a result, the polymers reach relatively high temperatures during reaction. The near-adiabatic shaping reservoir acts to drive the reaction to completion. Reaction casting and *reaction injection molding* are rapidly developing technologies for these polymers.

---

*Certain polyurethanes and epoxies are self-catalyzing.

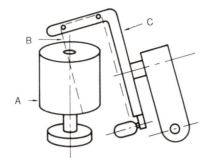

Figure 5.49.   Schematic of Planetary Filament Winding Process (74). A: Mandrel. B: Fiber Feed. C: Rotating Arm.

Hand Layup and Spray-Up

Certain thermosets can be mixed and reacted at atmospheric pressure. The primary materials are activated unsaturated polyester resin (UPE) catalyzed with a peroxide such as methyl ethyl ketone (MEK) peroxide, and polyurethane (PUR), formed by mixing an isocyanate and a polyol with appropriate catalysts. UPE is usually combined with fiberglass mat or roving and laid by hand against a mold. The mold material can be metal, wood or even fully cured UPE resin. The material reacts and cures in minutes. Boat hulls, ducts, bathtubs, shower surrounds and specialty containers are made this way. The reacting UPE can also be sprayed against the mold surface. The spray is usually mixed with chopped glass fibers in this case. After spraying, the resin/mat structure must be thoroughly rolled to express the air. Large tanks and bathtubs are sprayed this way. If a volatile liquid is added to a pressurized resin tank, and the mixture is sprayed, the energy from the exotherming reaction volatilizes the foaming agent and foaming occurs. Refrigeration systems and heated industrial tanks are insulated by spraying 2 lb/ft^3 (0.03 g/cm^3) closed-cell PUR foam against all vertical surfaces.

Filament Winding

Epoxies and UPEs can be used as reactive adhesives or matrices for continuous glass and graphite or carbon fibers to produce in-ground gasoline tanks, helicopter blades and smokestacks. The continuous fiber strands in bundles, called *roving*, are drawn through the reactive liquid resin bath, then wound in a precise fashion* on a mandrel (Figure 5.49). The entire filament-wound mass is then cured for several hours in a convection oven or a pressurized, heated tank called an autoclave. The resulting shape is that of a cylinder or body of revolution.

---

*Recent developments include placement of roving with computer-aided robots (75).

Compression Molding

Compression molding is essentially a squeezing process (Figure 5.50). A plug of polymer, usually preheated, is placed in the center of the press. The press is usually closed
at constant rate, $-dh/dt = C$, where h is the gap half-height. The applied force spreads
the plug. Pressure rapidly builds as the material reaches the mold periphery and begins
to flow into thin overflow or "flash" areas. Since the mold is hot, the resin is cured
under pressure into the shape of the cavity. When nearly cured, the part is ejected from
the mold and the process repeated. This process, and the more automated process of
transfer molding, described below, are the thermoset counterparts to thermoplastic
injection molding. Traditionally this process has been a manual operation. Robots are
now being used to insert the preform or charge and to remove the finished part. A
completely automatic operation is now available for the high production manufacture
of disposable microwave-serviceable thermoset polyester (UPE) dishes.

The time-dependent plunger force needed to close the press at a constant rate is
shown in Figure 5.51. The polymer is fluid and flowing throughout the (near-) plateau
region. For the most part, the crosslinking reaction to rigidify and cure the thermoset
occurs after the mold cavity is filled.

> **Compression Molding Model.**   The closing process involves nonisothermal
> unsteady squeezing of a highly filled, viscoelastic compressible fluid. How-
> ever, a very simple model serves to illustrate the polymer-process interaction.
> Consider radial isothermal flow of a power-law fluid. The continuity equation
> yields:

$$(5.43) \qquad \left(\frac{-dh}{dt}\right)\pi r^2 = 2\pi r \int_0^h u\, dz$$

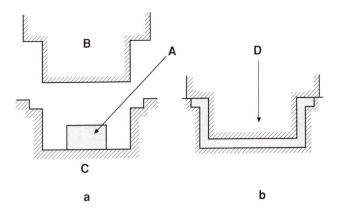

Figure 5.50.   Schematic of the Compression Molding Process. (a) Mold Open, Charge Placed.
(b) Mold Closed. A: Charge or Preform. B: Top Platen, Top Mold Half or Punch. C: Bottom
Mold Half or Cavity. D: Applied Force Direction.

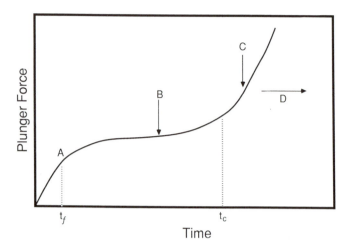

Figure 5.51. Time-Dependent Mold Closing Force in Compression Molding (76) A: Polymer Charge Heat Up Time. B: Resin Flow. C: Compression. D: Curing, t_f: Time at Which Flow Begins. t_c: Time at Which Compression Begins.

where r is the disk radius, u is the position-dependent velocity, h is the half-distance between platens, and ($-dh/dt$) is the rate of closing of the platens. The lubrication approximation relates radial pressure drop to shear rate as:

$$(5.44) \qquad \frac{dP}{dr} = \frac{-d\tau}{dz}$$

where P is the applied pressure. τ, the shear rate, for a power-law fluid is given by:

$$(5.45) \qquad \tau = -m \, (du/dz)^n$$

n and m are the power-law coefficients and z is the local thickness coordinate. The lubrication equation is then integrated for shear rate, noting that $dP/dr \neq f(z)$, and after appropriate substitutions:

$$(5.46) \qquad u = \frac{h^{1+1/n}}{m^{1/n} \, (1+1/n)} \left(\frac{dP}{dr}\right)^{1/n} \left[1 - \left(\frac{z}{h}\right)^{1+1/n}\right]$$

This is then substituted into the expression for dh/dt and integrated to yield:

$$(5.47) \qquad P = \left[\frac{m \, (2+1/n)^n}{2^n \, (n+1)} \, \frac{R^{1+n}}{h^{(1+2n)}} \left(-\frac{dh}{dt}\right)^n\right] \left[1 - \left(\frac{r}{R}\right)^{1+n}\right]$$

The closing force, F, is then the integral of pressure over the disk surface area:

$$(5.48) \qquad F = \frac{\pi m (2 + 1/n)^n}{2^n (3 + n) h^{1 + 2n}} \left(-\frac{dh}{dt} \right)^n R^{3 + n}$$

For a Newtonian polymer, $\tau = \eta \, (du/dz)$, $n = 1$, $m = \eta$, and:

$$(5.49) \qquad F_N = \frac{3\pi\eta}{8h^3} R^4 \left(-\frac{dh}{dt} \right)$$

Example 5.17

Q: Estimate the force needed to compression mold a 25 cm diameter dinner plate from a 60 mm diameter preform, 30 mm high. Assume constant applied force for a molding time of 8 sec. The molding compound Newtonian viscosity is 1000 Pa s $= 0.1$ N s/cm^2.

A: Under constant force:

$$F_N = \frac{3\pi\eta R^4}{8h^3} \left(-\frac{dh}{dt} \right)$$

For constant volume:

$$V = 2\pi R^2 h = 2\pi R_o h_o = 84.82 \text{ cm}^3$$

Thus:

$$h = 0.0864 \text{ cm.}$$

For constant force, the above equation can be rewritten in terms of V:

$$\int_0^t F dt = \frac{3\eta V^2}{32 \, \pi} \int_{h_o}^h \frac{dh}{h^5}$$

Upon integration and rearrangement:

$$Ft = \frac{3\eta V^2}{128\pi} \left[\frac{1}{h^4} - \frac{1}{h_o^4} \right]$$

or:

where h_o is the initial half-height of the preform. Since $h_o \gg h$, $1/h_o^4$ is small when compared with $1/h^4$ and the force term can be simplified as:

$$F = \frac{3 \, \eta \, V^2}{128 \, \pi \, h^4 t}$$

$$= \frac{3 \times 10^3 \times (84.82)^2}{128 \, \pi \, (0.0864)^4 \times 10^4 \times 8}$$

$$= 12 \text{ kN.}$$

Example 5.18

Q: Consider the compression molding process in Example 5.17. In this example, the closing rate of the press, $(-dh/dt)$ is held constant until the press reaches a maximum applied force of 24 kN. Determine the applied force as a function of gap, h and time. And determine the gap and time when the maximum force is reached. $(-dh/dt) = 0.25$ cm/s.

A: $(-dh/dt) \equiv C = 0.25$ cm/sec. This is integrated to give:

$$h = h_o - Ct = 1.5 - 0.25\, t \text{ (cm)}$$

The maximum closing time at constant rate is given as:

$$t_{max} = (h_o - h)/C = (1.5 - 0.0864)/20.25 = 5.65 \text{ s.}$$

Equation 5.49 is appropriate for this example:

$$F_N = \frac{3\pi\eta R^4}{8h^3}\left(\frac{-dh}{dt}\right)$$

Or:

$$F_N = \frac{3C\pi\eta R^4}{8(h_o - Ct)^3}$$

$$= \frac{3 \times 0.25 \times \pi \times 0.1 \times (12.5)^4}{8 \times (1.5 - 0.25\, t)^3} = \frac{71}{(1 - t/6)^3}$$

The easiest way to see this relationship is to tabulate F_N, h, and t:

h, cm	time, s	Force, N
1.5	0	71
1.25	1	123
1.0	2	240
0.75	3	570
0.5	4	1920
0.25	5	15,300
0.215	5.14	24,000
0.125	5.5	123,000
0.0864	5.65	372,000

Thus the maximum permitted applied force, 24 kN, occurs at 5.14 seconds or when $h = 0.215$ cm. Beyond that point, the approach used to obtain the solution to the problem in Example 5.17 applies.

For highly filled or glass-reinforced polymers, the viscous fluid assumption is replaced with a biaxial stretching model, where the stretching forces are due to friction with the mold surfaces (77). As a result, there is a flat velocity profile in the z-direction. The hydraulic frictional mechanism is given as:

(5.50) $$F' = -\mu' u$$

where u is the relative motion between the polymer and the mold, F' is the applied force, and μ' is the frictional coefficient. For a Newtonian fluid, the pressure is given as:

(5.51) $$P = 3\eta \left(-\frac{dh}{dt}\right) \left[\frac{R^2}{4h^3} + \frac{2\mu'}{\eta h}\right] \left[1 - \left(\frac{r}{R}\right)^2\right]$$

where μ' has the units of viscosity, such as Pa · s. The force is the integral of pressure over the area, as:

(5.52) $$F = \left(\frac{3\mu'\pi R^2}{h}\right) \left(-\frac{dh}{dt}\right) + \left(\frac{3\pi\eta}{8h^3} R^4\right) \left(-\frac{dh}{dt}\right)$$

The first term represents the force needed to overcome the material resistance to stretching. The second is the force needed to overcome frictional resistance.

Although most compression molded thermosets are reacting while being squeezed from the preform shape to the molded shape, reactivity is slow when compared with the squeezing process. Furthermore, since these materials are usually highly filled, the actual exothermic heat of reaction is quite small. As a result, reaction to cure is usually driven by heat transfer from the mold surfaces and the bulk of the reaction occurs *after* molding. Compressive force is required at this point, particularly if the reaction generates volatile by-products such as water. The extent of the reaction is governed by heat conduction into the polymer. Since the best simple model is one of sliding elongational flow, the resin in contact with the mold surface *always* sees the greatest level of energy. Therefore, in most cases, the surface layer achieves the highest extent of cure. The material at the centerline, in contrast, sees the smallest level of input energy and has the lowest cure rate and extent of cure.

Transfer Molding

Transfer molding is a hybrid process that combines compression molding and injection molding concepts. It is used to produce electronics "packages", or the electrically resistant polymer covering over integrated circuits. The process uses epoxy or thermoset silicone resin that has been preheated to processing temperature. The charge is placed in a cylindrical chamber in which a piston is then forced (Figure 5.52). Hydraulic pressure forces the preplasticated charge from the chamber into a heated mold. The resin flows through a runner system and gates into the individual cavities. The cavities are held under pressure until the thermoset resin is fully cured. Then the mold

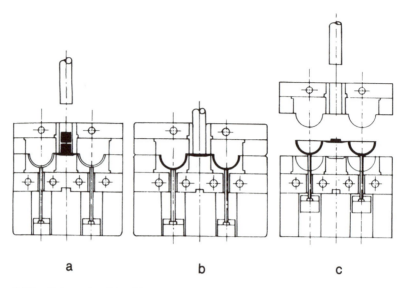

Figure 5.52. Schematic of the Thermoset Transfer Molding Process (78). Heated Resin Plug or Preform Placed Beneath Plunger. (b) Plunger Presses Resin into Cavities. (c) Ejection of Finished Parts Plus Runner and Sprue.

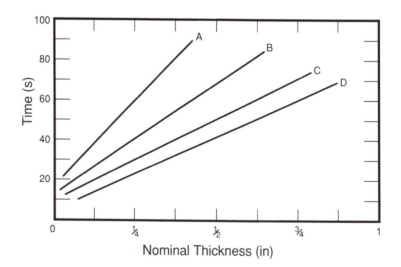

Figure 5.53. Comparative Cure Times of Compression, Transfer, and Injection-Molded General-Purpose Phenol-Formaldehyde (Phenolic) (79). A: Compression Molded, Cold Powder. B: Compression Molded, Radio-Frequency Preheated Powder. C: Transfer Molded, Radio-Frequency Preheated Powder. D: Screw Injection Molded.

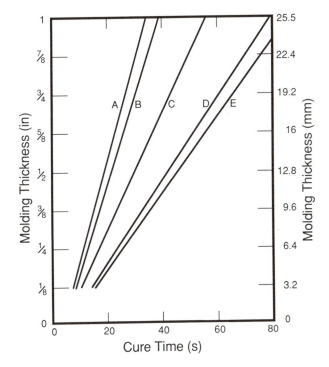

Figure 5.54.   Cure Times for Several Injection Molded Thermosetting Resins (80). A: Alkyd.
B: Urea. C: Melamine. D: Phenolic. E: DAP.

halves are separated, the parts removed and the process repeated. Transfer molding is
primarily an operator-initiated operation. Typically transfer molding cycles for thermosetting resins are substantially longer than for thermoplastics that are injection
molded. Typical mold closure times for phenolics for four types of processes are shown
in Figure 5.53. The effect of resin reactivity is given in Figure 5.54.

Example 5.19

Q: Estimate the time required to cure a radio-frequency heated phenolic transfer molded part
having a wall thickness that varies from 3/16 to 3/8 in. The mold is maintained at 160°C and
the charged is preheated to 95°C.
   A: The maximum time for curing any thermosetting resin part depends upon the maximum
thickness of the part wall, in this case, 3/8 in. From Figure 5.53, at this thickness, the estimated
cure time is 40 s.

Matched Die Molding

Matched die molding is a variant of compression molding. Here the resin charge is
usually highly filled or reinforced and is in rubbery or leathery sheet form. It is often

called *sheet molding compound* (SMC). The process uses metal-type forging or stamping presses with heated molds and extended dwell times. Very large, relatively thin-walled parts such as automotive deck lids and quarter panels and seating shells are frequently match-die molded from SMC.

BMC Molding

As noted, certain resins such as unsaturated polyester resins (UPE) and epoxies can be highly filled with calcium carbonate, talc, aluminum trihydrate and so on. The resulting resin is dough-like and is called *bulk molding compound* (BMC). BMC can be compression or transfer molded. It can also be injection molded in special injection molding machines that plunger- or coarse screw-transfer the resin from a warm barrel to a hot mold. BMC products are used extensively in electrical and electronic applications such as switch and circuit boxes and electronic cabinet bases.

Reaction Injection Molding (RIM)

Reaction injection molding (RIM) is a process of quasi-continuous high-speed impingement mixing of relatively low-viscosity polymer precursors (usually monomers) in a (typical) one-to-one ratio, and injecting the mixtures into a closed clamshell cavity under modest pressure (15 to 45 lb_f/in^2 or 0.1 to 0.3 MPa), Figure 5.55. The mold is usually moderately heated to 100 to 150°C. Materials such as epoxy and polyurethane react upon mixing. The exothermic reaction increases material temperature. This temporarily lowers resin viscosity and increases reaction rate. The filling rate must therefore be rapid enough to allow complete filling of the mold without substantial formation

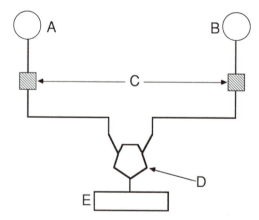

Figure 5.55.   Schematic of Reaction Injection Molding (RIM) Process (81). A: Polyol Resin Tank (Can Be Day Tank). B: Isocyanate Tank (Can Be Day Tank). C: Pump (Can Be Recirculating Pump). D: Mixing Head. E: Clamshell Mold.

of three-dimensional structure, known as gel, within the resin. (See the section on thermoset reactivity, below.)

The primary applications for RIM products are in transportation, with emphasis on automotive fascia. Most recently, RIM structural foam is competing with injection molded thermoplastic foam for equipment cabinets.

Uniformity of prepolymer mixing is considered the key to quality RIM production. Extensive investigation of high-speed impingement mixing indicates that the flow streams must be turbulent. That is, the mixing Reynolds number, $Re = \rho vD/\mu$, *for each fluid*, must be greater than about 100 and preferably greater than about 500. Here ρ is the fluid density, v is its velocity, D is the nozzle diameter, and μ is the Newtonian resin viscosity. Since polyols and epoxies have relatively high viscosities (1 to 10 Pa s or more), impingement velocities must be correspondingly high. This implies high pumping pressures. Another parameter of importance is the momentum ratio, $MR = (\rho_1 v_1^2 D_1^2)/(\rho_2 v_2^2 D_2^2)$. This is the ratio of fluid stream momenta at the moment of impingement. If MR is substantially different from unity, the flow of the fluid having the lower momentum will be effectively stopped by the impingement of the other fluid (82).

Although the reacting fluid entering the mold cavity is essentially homogeneous, flow across the mold surface is laminar, with the characterisic "volcano" flow at the leading edge. The velocity profile becomes rapidly distorted by changing viscosity owing to externally supplied heat from the mold, viscous dissipation during mixing, internal heat generation, and increasing molecular weight owing to reaction.

Reaction yields increasing resin density. RIM molds are not overpressured or packed as is done in thermoplastic injection molding. As a result, microvoids form primarily in the center or core of the part during curing. The bubbles form from dissolved air or possibly carbon dioxide or water formed during polymerization. Typical unfilled PUR RIM parts have densities of 0.95 to 1.0 g/cm^3. In comparison, fully dense PUR has a density of 1.1. The mechanical properties of foamed polymers are less than those of the unfoamed or fully dense material. The following conservative relationship can be used for tensile and flexural properties:

$$(5.53) \qquad X_{foam} = X_o (\rho_{foam}/\rho_o)^2$$

X is the property and ρ is the polymer density. The "foam" and "o" subscripts represent the foamed and unfoamed properties, respectively. Impact strength cannot be correlated in this manner, however.

Example 5.20

Q: In order to increase its stiffness, the thickness of a structural part has been increased and the part foamed to 15%. The part weight remains the same. Determine the increase in stiffness owing to foaming.

A: Since the mass, m, remains the same, the part thickness must increase. If ρ is the density and t is the thickness, then:

$$m = \rho_f t_f = \rho_o t_o$$

where "o" represents the unfoamed resin state and "f" the foamed resin state. Now:

$$t_f = (\rho_o/\rho_f)\, t_o = 1.175\, t_o$$

Stiffness, S, is the product of modulus, E, and moment of interia, I:

$$S = EI$$

Consider the part to be a beam in flexure. Then:

$$S_o = E_o\, b\, t_o^3/12$$
$$S_f = E_f\, b\, t_f^3/12$$

where b is the beam width. The modulus of a foam can be related to the unfoamed resin modulus by the square-density ratio:

$$E_f = E_o\, (\rho_f/\rho_o)^2 = 0.7225\, E_o$$

Thus the stiffness of the foamed structure, relative to that of the unfoamed structure, is:

$$S_f = 0.7225\, E_o\, b\, (1.176\, t_o)^3/12,$$
$$= 1.175\, E_o\, b\, t_o^3/12,$$
$$= 1.175\, S_o$$

Therefore the foamed structure is about 17.5% stiffer than the unfoamed structure at the same weight.

Pultrusion

Pultrusion is the process of drawing continuous inorganic reinforcing fibers such as glass or graphite/carbon through a bath of catalyzed liquid resin such as unsaturated polyester resin or epoxy. A schematic of a thermoset pultrusion line is shown in Figure 5.56. The fibers are usually in the form of a continuous fiber bundle or *rovina*, although recent advances have used woven and unwoven mat in the form of tape. The resin-impregnated filaments are then pulled through a forming die (or dies). Very high pulling forces are required to overcome the frictional resistance of the fiber bundle as it passes through the die. The die is usually heated to start the crosslinking or polymerization reaction. The shaped profile is further heated in an oven to complete the reaction. The formed section finally passes through a pulling or tractor section that provides the drawing power. Typically, a pultruded profile is 60 to 80% (wt) fiber and 20 to 40% (wt) resin. Pultrusion rates are relatively low. Applications include shaped reinforcing beams, ducting, hand rails, ladder rails and profiled treads, and truck body and indoor panel beams.

The pultrusion process is one of expressing air from a continuous fiber bundle, filling the void with moderately low viscosity resin, then holding the shaped resin-impreg-

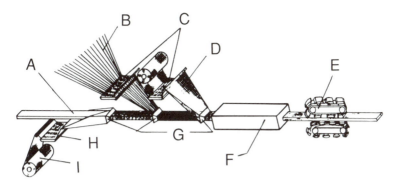

Figure 5.56. Schematic of Pultrusion Line for the Manufacture of Continuous Hollow Fiber-Reinforced Products (83). A: Steel Mandrel. B: Fiber Roving. C: Resin Bath. D: Continuous Fiber Mat. E: Puller. F: Heated Shaping Die. G: Shaping Die. H: Continuous Fiber Mat. I: Takeoff Roll.

nated bundle together until the resin reacts to rigidify the composite.* Very high pulling forces are required to achieve complete impregnation. In addition to relatively low resin viscosity, a compromise on reactivity is needed. Typically the fiber bundle is pulled through a tank or trough of reacting resin. It is apparent that the reactivity of the resin in the tank must be as low as possible. But, once the resin has coated the fiber bundle and the profile has been established, reactivity must be as high as possible in order to maintain that profile shape. Secondary concerns include adhesion between the glass or graphite/carbon fiber and the resin and air bubbles trapped in the fiber bundle. Poor adhesion leads to early delamination under bending or flexural load. Low molecular weight chemicals, called *coupling* or *sizing* agents, are applied to the fiber bundle to minimize debonding. Voids are sources of stress concentration that lead to reduction in ultimate mechanical properties such as tensile and flexural strength at break, and impact strength.

Recently, pultrusion has been extended to thermoplastic polymer impregnation of continuous glass and graphite fibers. Thermoplastics offer time-independent fluid resin properties, typically tougher, more ductile resins, and a degree of recyclability. Since thermoplastic resin viscosities are substantially greater than those of thermosetting resins, required pulling forces are substantially higher than with reactive pultrusion.

Other Reacting Schemes

There are many other commercial ways of processing reactive materials. Volatile liquids can be dissolved in one or both of the polyurethane precursors. The exothermic

---

*The engineering problem focuses on egress of the air bubbles and ingress of the resin without appreciable foaming (really frothing) of the resin. Since most relatively high viscosity resins froth easily, special chemicals known as foam breakers or anti-surfactants are added. Unfortunately these chemicals can act to reduce adhesion between the resin and the fiber bundle as well.

heat of reaction causes the liquid to vaporize. This produces a fine-celled, low-density *foam*. A flexible, open-celled version of this foam is used in shipping, seating, and mattresses. A rigid, closed-celled version is used in insulation. *Elastic reservoir molding* enjoyed much publicity when it was the primary way of reinforcing the relatively soft brushed stainless steel panels of the short-lived De Lorean automobile. The technique begins with a laminate-composite of glass mat, low-density open-celled flexible polyurethane foam, and perhaps open-mesh polymer wire. The structure is placed in a low-pressure mold containing a reactive resin such as epoxy. The foam is compressed to express the air. When the compression load is removed, the reactive resin nearly completely fills the foam, the fiber and the web structures. The material is reacted and cured in the mold. The result is a high-performance composite structure.

Some thermoplastic powders such as PMMA and PS, can be cast from a reactive syrup of monomer, catalyst, if needed, and dissolved polymer. Cast PMMA is usually made one sheet at a time. The casting machinery is called a cell. It has one floating surface to minimize microvoid formation during reaction. Continuous *cell-cast* machines need to be quite sophisticated to control heat generated during reaction and the material shrinkage as it polymerizes. Cell-cast polymers typically have very high molecular weights and are used where the need for abrasion- or solvent-resistance commands the higher processing costs. Applications for PMMA include impact-resistant glazing and furniture components.

Thermoset Reactivity

The cure of a thermoset is characterized by gelation time and vitrification. *Gelation time* is the time of formation of an immobilizing three-dimensional network. *Vitrification* is the point at which the polymer changes from a liquid or rubber to a glass, owing entirely to molecular weight increase. These time-dependent changes are strongly influenced by temperature. Time-temperature-transformation diagrams similar to those for metals have been used to illustrate these dependencies (Figure 5.57). The curves are divided horizontally by three glass transition temperatures. $T_{g,o}$ is that of the uncured resin. $T_{g,\infty}$ is that of the fully cured resin, and $T_{g,g}$ is that of the gel. Gelation usually occurs at a fixed extent of reaction *as long as* the resin is quiescent. At this point, the liquid becomes immobile. Above $T_{g,\infty}$, the polymer will gel into a rubber but will not vitrify. Above $T_{g,g}$, the polymer usually gels, then vitrifies into a gelled glass. Below $T_{g,g}$ but above $T_{g,o}$, the polymer vitrifies directly into a processable, ungelled glass. Below $T_{g,o}$, there is no reaction (85).

The degree of thermoset cure is a function of time and temperature. From an adiabatic heat balance:

(5.54) $$c_p \, dT/dt = (-\Delta H)(d\alpha/dt)$$

where c_p is the specific heat, α is the extent of reaction and $-\Delta H$ is the heat of reaction *per unit mass*. If the reaction goes to completion ($\alpha = 1$), the adiabatic temperature rise, ΔT_{ad}, is given as:

(5.55) $$\Delta T_{ad} = (-\Delta H)/c_p$$

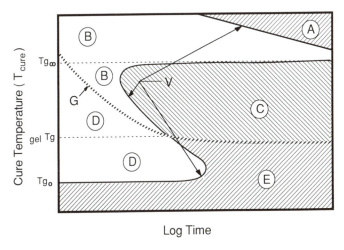

Figure 5.57. Generalized Time-Temperature-Transformation (TTT) Diagram for Thermosetting Resin (84). A: Char Region. B: Gelled Rubber Region. C: Gelled Glass Region. D: Liquid Region. E: Ungelled Glass Region. G: Gelation Curve. V: Vitrification Envelope. $T_{g,\infty}$: Glass Transition Temperature of Fully Cured Resin. gel T_g: Equal Gelation and Vitrification Temperatures. $T_{g,0}$: Glass Transition Temperature of Unreacted Resin.

The time dependency of $\alpha(t)$ is written as:

$$(5.56) \qquad \frac{d\alpha}{dt} = A\,[1 - \alpha(t)]^m\,e^{-E/RT}$$

where A is the pre-exponential coefficient, m is the order of the reaction, and E is the activation energy of the reaction. The degree of reaction at the gel point can be obtained from statistical analysis as:

$$(5.57) \qquad \alpha_{gel} = \frac{1}{[r + \rho\,(f-2)]^{1/2}}$$

where r is the ratio of A to B hydrogens, ρ is the fraction of B hydrogens that are part of the multifunctional reaction and f is the functionality of the branching units. For typical one-to-one ratio thermosets, $r = \rho = 1$, $f = 4$, and $\alpha_{gel} = 0.57$. Experimentally, α_{gel} ranges from 0.4 to 0.8 and depends strongly on the extent and nature of secondary reactions.

The weight average molecular weight of difunctional A and tetrafunctional B reactants of equal reactivity is given as:

$$(5.58) \qquad M_w = \frac{[0.5\,M_B^2\,(1 + \alpha^2) + M_A\,(1 + 3\alpha^2) + 4M_A\,M_B\,\alpha]}{(0.5\,M_B + M_A)\,(1 - 3\alpha^2)}$$

where M_A and M_B are monomer molecular weights. The molecular weight at gelation is determined by setting $\alpha = \alpha_{gel}$. And the nonisothermal bulk viscosity is given as:

$$(5.59) \quad \ln \eta = \ln \eta_\infty + \ln MW + (E_\eta/RT) - C_1(T - T_0)/[C_2 + T - T_0]$$

where η_∞ is an extrapolated viscosity, E_η is the viscous activation energy, C_1 and C_2 are the WLF constants, and $T_0 = 50 + T_g$ when $T < (50 + T_g)$.

In any real curing system, the local extent of reaction, $\alpha(T;t)$, will depend entirely on the time-dependent local temperature. For highly filled systems, as used in SMC molding or compression molding, the effect of heat of reaction can be safely ignored as a first approximation. This effectively uncouples the energy equation from the reactivity and allows for relatively easy computation of the local extent of cure. For unfilled, foamed, or very reactive systems such as those in RIM processing, the coupled model must be solved (86).

Example 5.21

Q: In Figure 5.60 are the optimum mold curing times for injection molding of several thermosetting resins. These values are substantially lower than those for compression or transfer molding. Using the data for phenolic as reference, develop curing time guidelines for the other thermosetting resins.

A: Since the curves are linear, the relative cure times for given wall thicknesses can be used to form the appropriate guidelines. For a 1/2-in wall thickness, then:

Resin	Cure Time(s)	Cure Time for Phenolic (%)
Alkyd (A)	17.5	44
Urea (B)	19.8	50
Melamine (C)	26.6	66
Phenolic (D)	40.0	100
DAP (E)	43.4	108

Powder Processing

Many thermoplastics and some solid thermosets are produced as powders by precipitation or settling from their reaction fluid. These fine powders are usually mixed with the appropriate adducts such as antioxidants, stabilizers, dyes, pigments, fillers, reinforcements, and the like, and then extruded. The extruded strands are chopped into granules or pellets. (See the discussion on extrusion given earlier in this chapter.)

In some processes, powders can be used directly. In *rotational molding,* the powder is metered into a metal clamshell-type mold. The mold is clamped closed and rotated about its equatorial and polar axes while being heated. This is shown in schematic in Figure 5.58. The overview of a commercial rotational molding machine is seen in Figure 5.59. The powder inside is heated by being tumbled against the hot mold surface. When the powder is hot enough, usually above T_m, it sticks to the metal. Ad-

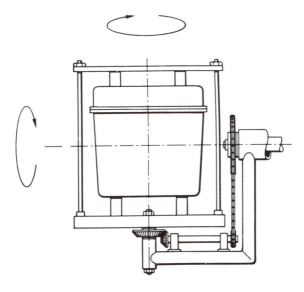

Figure 5.58.   Rotational Molding Configuration Showing Biaxial Rotation (87).

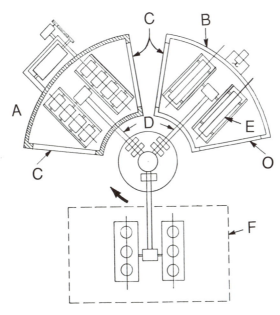

Figure 5.59.   Overview of Three-Arm Commercial Rotational Molding Machine (88). A: Forced Convection Oven. B: Water/Air Cooling Chamber. C: Door. D: Drive Spindle. E: Mold. F: Loading/Unloading Station.

ditional hot powder sticks to this powder, eventually building a porous structure. Additional heating melts the powder into a monolithic, nearly stress-free shape. The rotational mold is then carefully cooled and the hollow closed plastic shape removed. A typical molding cycle is shown in Figure 5.60. Note that the rotational molding process normally has cycle times that are much longer than cycle times of other processes. Polyethylene is the predominant polymer used in rotomolding. Typically, to be able to flow across the mold surface under gravity forces, polyethylene should have relatively low molecular weight. To gain high stress crack resistance, crosslinking agents are added to the polyethylene. One of the main advantages of this process is that parts fabricated this way are relatively stress-free. Experiments (90) show that the primary mechanism for resin consolidation is melt-sintering. Typical rotomolded parts have volumes greater than 1 ft^3 (0.03 m^3) and include fertilizer tanks, trash containers, and toys.

Polymer powder can also be blown onto a heated metal surface, then densified in an oven. *Powder coating* is used to produce corrosion-resistant surfaces for pipe and metal shelving. Stick-resistant surfaces on domestic cookware can be applied this way. In high-performance composite technology, fine powders are fluidized and blown onto heated graphite/carbon fibers to produce a *prepreg*, or uniaxial tape, one of the building blocks for laminated structures. Electrostatic coating is an alternative technology.

Recently, high-performance plastics such as polyimide and UHMWPE have been processed into functional parts directly from powder. The technology follows the tenets

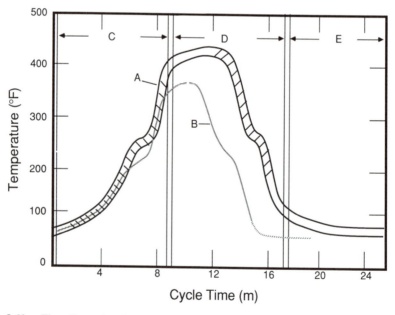

Figure 5.60. Time-Dependent Rotational Molding Temperature Profile for Polyethylene (PE) (89). Curve A: High-Density and/or Crosslinked Medium-Density Polyethylene. Curve B: Low-Density Polyethylene. Zone C: Time in Isothermal Oven. Zone D: Cooling Cycle Time. Zone E: Unloading/Refilling Time.

of powder metallurgy. Relatively coarse powder is *compacted* at very high pressure (100,000 to 200,000 lb_f/in^2 or 0.7 to 1.4 GPa) into a *near-net* shape. This is a shape that has dimensions that are only slightly larger than the final part dimensions. The part is then heated, that is, *sintered,* in an oven where particles fuse and some increase in density occurs. Usually, finished parts have densities of 75 to 85% of the compression-molded polymer density.

If a higher density or a part with more accurate dimensions is required, the hot sintered shape can be *forged* or *coined* at pressures of 10,000 to 30,000 lb_f/in^2 (70 to 140 MPa). The final part density can be as high as 95% of the compression-molded polymer density, depending on the elastoplastic nature of the polymer.

Powder can also be mixed with chopped or continuous glass fibers and oven-heated until the powder melts and fuses to the glass. The hot sheet can be consolidated by compression molding or high-pressure rolling. The resulting fiber-reinforced polymer sheet is then reheated and stamped into final shape. Shallow draw parts such as automotive seat shells and oil pans have been made this way.

The mechanical properties of sintered and densified polymer powder parts depend entirely on the degree of fusion at the powder particle interfaces. The most reliable measure of the level of fusion is part density. From limited experiments, equation (5.53), relating foam properties to unfoamed polymer properties and density ratio, seems to be appropriate here as well.

5.4 Polymer Material Properties in Computer-Aided Engineering

Computers are being used in nearly every facet of the polymer industry. Mainframe computers are used to solve extremely complex processing problems dealing with coupled heat transfer and laminar fluid flow of rheologically complex fluids with broad phase transitions in difficult-to-define process field geometries and boundary conditions. Minicomputers are used for record keeping such as material inventory flow, accounting and shipping, and in computer-aided design of polymeric parts. Microcomputers can be used for process control, quality control, as data logging and storage for material and process condition changes. These devices also interface with microprocessors, or dedicated low-level computers, that act to automatically control modern process equipment. Many smaller computers are interfaced to minicomputers to aid information flow control.

Hodgson (91) notes that the development of numerically controlled (CNC) machine tools in the 1950s was the first significant use of computers in manufacturing. Many of the first CNC products made were polymer injection molds and dies. The development of the cathode ray tube (CRT) as a display facilitated conceptualization in the 1960s. The earliest users of CAD/CAM or *Computer-Aided Design/Computer-Aided Manufacturing,* were printed circuit (PC) board and integrated circuit (IC) designers. These were followed shortly by mechanical drafters, piping and wiring diagram designers, and HVAC designers. The earliest CAD/CAM developments specifically for polymers came from computer simulation of the extrusion process. The theory for the melt pumping portion of the extrusion process had been developed in the 1920s and

1930s. In the 1950s, Dupont engineers had formulated theories for the compaction or solids conveying section of the extruder.

The plasticating section remained without mathematical modeling until Tadmor developed a succinct melt-pool concept while a summer student at Western Electric Co., Princeton NJ in the early 1960s (92). Having now a complete mathematical model for the entire extruder, Western Electric established a computer-based group to help design extruders for wire and cable extrusion. The results of these studies were published in two seminal monographs (93,94). In reality, these efforts were Computer-Aided Engineering (CAE), since they assisted the engineer in determining the interrelationships between process variables, at speeds and with accuracy superior to values determined using analog (slide rule) or mechanical (calculator) devices.

The first publications focused on simple descriptions of computer flowcharts and solution techniques for rudimentary models, based primarily on algebraic equations or exact solutions of more complex models with model boundary conditions. By 1970 (93), the computer was being used to numerically solve sets of equations for which no exact solution was known. The extruder was the obvious first device to "computerize" since it had (and has) the widest application. Furthermore, the process could be compartmentalized, at least as a first approximation, into solids conveying, plasticating, and melt pumping. And the process is steady-state continuous. The compartmentalization was an important step, since it allowed different reasonably realistic simplifying assumptions to be made in each "compartment", with the entire extruder then modeled as a series of these compartments.* Of course, it must be realized that early simulations were carried out on equally early process computers. Much early concern was on optimization of FORTRAN programs on mainframe IBM 1800 and 7070 machines. During this time, business data management programs were in an embryonic stage as well. Payroll and inventory programs were being written "in-house" by most major corporations.

---

*Compartmentalization is still an important way of visualizing and modeling polymer processes. Recently the thermoforming process has been modeled as a sequence of planar sheet heating, isothermal biaxial sheet stretching, and finally rigid sheet cooling (95).

It must be noted that the computer *must not* always be the "be all and end all" in the CAE understanding of the process. Rauwendaal expresses the following note of caution (96):

"One of the general problems of the melting theories [in a single screw extruder] is that the most realistic models are also the most complex . . . The main question is whether the more complex analysis will result in improved predictive ability . . . Therefore, one has to strike a balance between the degree of sophistication of the analysis and the practical usefulness of the analysis. This interest will depend on the particular interests of the individual . . . Many people in the extrusion industry do not have the time or inclination to work through elaborate and complex analyses of melting . . . In most cases, actual predictions of melting performance can be made with a relatively simple programmable calculator . . . At this point in time [1986], predictive computer programs to simulate the extrusion process are not fully predictive or 100 percent accurate, *despite claims to the contrary by organizations providing such programs.*" [emphasis, present authors]

Today, there are many ways of visualizing the role of computers in polymer engineering. Consider these:

CAD Computer-Aided Design. Originally CAD was considered to be mold or die design. Today CAD can include part design as well.

CAM Computer-Aided Manufacturing. This was originally part of CAD/CAM, where the "M" referred to the machining of the molds or dies. Now CAM can refer to far more than this, including computer-integrated manufacturing.

CAE Computer-Aided Engineering. As noted, this originally referred to the "electronic slide rule" that enabled the engineer to establish more realistic relationships between variables. Now it includes mold filling technology, die flow, part design and long-term performance under environmental conditions.

CIM Computer-Integrated Manufacturing. Originally this was restricted to polymer material-mold interactions and the necessary flow diagrams. Now it can include the totality of computer-based data-taking and decision-making including purchasing, part cost analysis, inventory control and scheduling.

Even though extrusion was the first process to undergo computerization, thermoplastic injection molding has seen the greatest activity recently. Some of the reasons for this are:

* injection molding is the *second* most economically important polymer process,
* there are many polymers that can be injection molded,
* although the process is cyclic, unlike extrusion, the process is relatively simple rheologically, and geometrically, since the polymer usually remains in one state during the injection process, and
* the process variables can be monitored and controlled to a large extent.

There are many areas where computers can be employed in the injection molding process. Table 5.13 (97) identifies several areas in which CIM is important in the injection molding process.* It is now well understood that no one type or size of computer can be optimum for all these functions. For example, a minicomputer is capable of handling all these functions, but unless multiple work stations are provided, critical design decisions may need to wait until payroll has been processed.

The proliferation of "software" or preprogrammed computer routines has also caused some difficulties here. Many business programs are written in *COBOL* or *Pascal*. These computer languages are designed for spreadsheet data logging but are inefficient or nonfunctional for engineering computations. *FORTRAN* and *BASIC* are scientific languages and so are optimized for complicated internal looping and conditional decision making. They are not usually used for data collection. Languages such

---

*This is a tabulation of many of Gerlitz' points (97) and not intended to be the original author's complete list or intended ranking.

Table 5.13.   Areas where computer-integrated manufacturing is important
in the injection molding process (97).

Before processing:
Material selection
Part design
  Mechanical
  Environmental
  Dimensional changes (shrinkage)
Mold design
  Mechanical (size, shape)
  Thermomechanical (waterline location)
  Hydromechanical (ejection)
  Flow (gates, runners, sprues)
  Material selection
    Chemical attack
    Erosion
    Hardiness
    Thermal conductivity
Method of mold manufacture
  Method of hobbing
  Method of hardening
Purchasing
  Bidding on molds
  Bidding on materials
Warehousing
  Bags, bins, silos
  Transportation control
    To/from site
    Within site (machine/storage)

During processing (CAE: Gerlitz):
Bulk material handling
  Conveying
  Blending
  Drying
  Coloring
The injection molding machine
  Molding process control
  Automatic mold change
  Part handling
    Robots
    Downstream automation

Temperature control
  Cooling towers
  Temperature controllers
  Chillers
  Mold cooling
Inspection systems
  Quality control/assurance
  Sorting and counting
  Cavity separation
  Incoming material control
  Melt viscosity monitoring
  Thermal properties
  Color
Material recycling
  Supervisory control

Post-processing and other:
Shipping
  Inventory control
  Warehousing
  Transportation scheduling
Maintenance
  Mold
  Injection molding machine
    Emergency
    Routine
Facilities engineering
  Building
    Maintenance
    Renovation
    HVAC
  Grounds
    Landscaping maintenance
    Driveway/parking lot maintenance
Personnel
  Management
  Labor relations
  Budgetary control
  Payroll
[And so on.]

as *"C"* appear to be designed primarily for graphical display. Complex "number crunching" for research requires rapid computation but limited graphic display. Part design requires interactive graphic display, usually in color. Many corporations have balked at purchasing computers and software in so many different forms. This view is softening however, as the computational capacity and the versatility of the microcomputer* improve. Current microcomputer capacity has surpassed the capacity of many "mainframe" computers of two decades ago. The popularity of these machines has forced the price down even with the rapid increase in memory capacity. And so it appears that computer-integrated manufacturing will succeed in using these microcomputers in several of their many versions.

The Nature of the Design Data

Regardless of the complexity of the simulation method or predictive technique and regardless of the end-use of the simulation (parts design, mold flow, etc.), the arithmetic requires material properties selected from some database. Wang and Wang (98) note that:

> "The data base of material properties is the most important [database] of all. Without reliable property data in an appropriate form, the results from the analysis would be erroneous and misleading."

In the last few years, many authors have addressed this question. A brief overview of the computer methods and the ancillary databases is given here.

Flow Problems

Polymer material properties are needed in two major types of computer-aided engineering (CAE) programs. In *flow problems,* such as polymer flow through an extruder or a die or polymer flow into an injection mold, the basic database must include rheological data [temperature- and shear-dependent viscosity values, at the very least] and thermal data [heat capacity, melting temperature, and so on]. There appear to be at least two levels of programs. The simplest is a sectioning method, in which the flow channel is divided into zero-thickness elemental shapes, such as rectangles, radial arcs and/or cylinders. Force and energy balances are struck in each of these elements, using prior time data to iterate. This might be considered a simple form of a *flow network.* The result is a marching procedure with time and pressure given in each element as

---

*The microcomputer is also called the "personal" computer. Many different types have been marketed in the past decade. Today the most common come in either of two types; the *Apple* series, with ease of operation, outstanding graphics, desktop publishing capability but only moderate "number crunching" capacity, and the *IBM* series and its clones or "IBM compatibles", with outstanding RAM and ROM capacity, many programming languages, but only moderate graphics display capability and, at times, somewhat difficult-to-learn routines.

the flow progresses along the string of elements that form the flow path. The more advanced approach uses *finite element* arithmetic to calculate these values.* Basically a triangular grid is drawn on the polymer surface, with the grid mesh or spacing small at regions of high stress. This enables higher accuracy and the modeling of geometrically thick sections. The triangular elements are connected to one another by three *nodes*, or points at which forces and deflections are transferred. If the elements are simple in geometry, the stress-strain-rate-of-strain balance within each element can be easily determined. Since the forces and deflections at each node must balance, a set of equations result. These are solved simultaneously to achieve static and/or dynamic equilibrium. For polymer fluid flow problems, the thermal equation needs to be solved at each node, as well. Many, many elements are usually needed to achieve a reasonable level of accuracy with FEA. Thus, high-speed, large-memory computers are needed. If some compromise in accuracy is made, however, the largest microcomputers can be fitted with FEA programs for polymer process design.

Material Properties

Moldflow, the simplest and most popular injection mold flow program (101), requires a minimum database of *eight* data constants: three viscosity constants, three thermal properties and two temperatures. The three viscosity constants (A, B, and C) are given in:

(5.60)   Shear viscosity = A (shear rate-B) exp (Temperature C)

This assumes a linear, Newtonian relationship between shear rate and viscosity. The three thermal properties are melt thermal conductivity, specific heat and melt density. The two temperatures are the *freeze temperature* and the *non-flow temperature*. The freeze temperature is assumed to be melt temperature for a crystallizing polymer and a temperature somewhat above the glass transition temperature for an amorphous polymer. The non-flow temperature is an artificial value, assumed to be the temperature below which a polymer does not flow when under a 100 lb_f/in^2 (0.7 MPa) pressure differential.

It is recognized that these few values cannot model a complex material like a polymer. Recall, however, that the flow network models are relatively insensitive approximations to actual flow conditions and that extensive temperature- and shear-rate-dependent data probably are unwarranted. The more sophisticated and more accurate computer programs such as FEA can utilize more complex databases. Unfortunately,

---

*Finite element methods have been used for years in structural mechanics to analyze complex structure response to complex loading schemes. In the polymer design literature, the method is called either FEM (for *Finite Element Method*) or FEA (for *Finite Element Analysis*). The foundations of this subject are clearly beyond the objective of this book. The interested reader is referred to References 99 and 100 for additional information.

these more complex databases are simply not available. Progelhof (102) notes that for one computer-based modeling system, "[m]easuring thermal conductivity directly is not easy, and there seems considerable variation from different laboratories for values (of) nominally the same material." Progelhof also notes that the earliest melt property database was collected by Westover (103). The data were presented in a series of graphs for each of dozens of *then-commercial* resins. The data included temperature-dependent shear stress-shear rate or viscosity-shear rate, temperature- and pressure-dependent density, temperature-dependent enthalpy, temperature- and pressure-dependent thermal conductivity and/or thermal diffusivity, and temperature-dependent specific heat. Many of the leading polymer suppliers provided databases of their most widely used resins. When The Society of Plastics Engineers endeavored to update this database in 1978, there was essentially no response from polymer suppliers.

Of course, this dearth of property data is not restricted to the melt phase of polymers in injection molding CAE programs. As an example, Rauwendaal notes (92) that there are very few coefficient of friction data for the interface between solid polymers and the metals used in screw extruders. He notes that the standard solids conveying computer models *require* values to *three*(!) decimals for design accuracy. Most data are available to two decimals with questionable accuracy at these values.

Part Design Programs

FEA is used primarily in static *part design programs*. In Table 5.14, Maddux (104) notes that CAE should be ideal for many types of performance characteristics. The grid structure is identical to that used in flow problems. For most designs, time-independent, isothermal stress-strain constitutive equations are used at the nodal point. Thus the part design analysis is less complex than the flow problem analysis of the same part. Nevertheless, the designer requires accurate *solid mechanical* property values for the polymers used in his design. Most important, the designer must be aware that FEA methods used in design *do not* include time-dependent material properties such as creep, stress relaxation, and brittle characteristics under impact. Currently, if these elements are required in the part design, either the FEA must be modified to include time-dependency, or the program must be run using material properties that the product may have at or beyond its design lifetime.

There is hope however. Until recently standard small-frame computer models were limited to solving FEA problems with small displacements. Now more sophisticated models are being developed for elastomeric applications, where large deformations need to be designed for. As long as the polymer part can be thought of as a thin membrane, the large deformation problem appears relatively straightforward, albeit not without difficulties. One of the more perplexing problems is how to characterize the nonlinearity of the apparent modulus with stress and the effect of local property variation with wall thickness for parts with relatively heavy walls. Thus the design engineer is cautioned to use current FEA models with great care and substantial safety factors when predicting long time performance of parts under loads in excess of 25–30% of virgin resin yield strength.

Table 5.14. Applications for computer-aided engineering in part design (104).

System
    Loads between components
    Vibration
    Displacement
    Weight and inertial distributions
    Safety
    Cost
    Fluid flow

Component
    Stress
    Fatigue life
    Deflection
    Vibration
    Noise
    Reliability
    Durability
    Mold flow
    Stability/buckling
    Heat transfer
    Energy absorption

Material Properties

The basic solid mechanical properties needed by the designer, such as tensile strength, modulus and elongation at break, are usually more accessible and more accurate than the fluid mechanical processing properties. Unfortunately, these values are not always applicable to the conditions the environment will impose on the polymer part. As is noted in Chapter 6, data obtained at one time-temperature-environmental condition, such as *2 mm/min tensile test in room temperature air,* may not translate to actual use conditions, such as *long-term creep at elevated temperature in an aggressive environment.* Thus, the questions involve the applicability of the data rather than the *availability* of the data. Some of this has been illustrated in Chapter 3 with the discussion of application of safety factors to creep modulus data.

Summary on CAE

Computers have assumed many roles in polymer manufacturing. In part design and process modeling, computers greatly aid the engineer in understanding the complex interaction of specific polymers and specific environmental conditions. By the same token, CAE *must never* be a substitute for judgment and common sense. Designers and engineers can use these elegant tools to develop a sense of *interaction sensitivity.* That is, rather than simply substituting a single value for a polymer physical property,

Table 5.15.   Sensitivity analysis of CAE program for fire-retarded high-impact polystyrene
(FR HIPS) (105). Comparison of effect of experimental and calculated
viscosities on cavity pressure drop.

Fill Time (s)	0.94	0.21	0.12
Melt Temperature (°C)	238	245	250
Mold Temperature (°C)	38	38	38
	Cavity pressure, lb_f/in^2		
Apparent Viscosity (240 to 280°C)	124	169	192
True Viscosity (240 to 280°C)	107	147	169
True Viscosity (195 to 225°C)	144	155	166
Experimental	180	148	135

the designer or engineer can run the program several times, bracketing the original value with logical upper and lower values, to determine the sensitivity of the process or design result to the accuracy of that value. Certainly, if the computed result is very sensitive to the absolute value of the property, the design cannot be completed until more accurate property values are obtained. On the other hand, if the computed result is insensitive, the design may be entirely adequate with values generated by "good guesses".

As an example of the nature of this work, Palit recently examined the sensitivity of a commercial CAE program, the CADMOULD program of Institut für Kunststoffverarbeitung (IKV), Aachen, Germany, to rheological coefficients for flame retardant high-impact polystyrene (105). The mold was a pressure- and temperature-instrumented two-cavity plaque mold. Palit chose viscosity data in the following way. Apparent viscosity data was obtained from a 30:1 L/D capillary rheometer. The non-Newtonian or Rabinowitsch corrections and the capillary end effects or Bagley corrections* were made by changing the capillary length. The temperature corrections were made using the WLF equation (see Section 3.7). The more sophisticated Carreau rheological model (106) was also used to model the data.

As seen in Table 5.15, the experimental data show a steady decrease in pressure drop with decreasing fill time. Regardless of the viscosity model chosen, the computer model predicts an increase in pressure drop with decreasing fill time. Palit notes that it is "extremely important to use the true viscosity data at the process temperature range in order to achieve more accurate CAE results". Despite this admonition, for this example, there remains an obvious disparity between the experimental and computer trends.

---

*See Chapter 6 for interpretation of these corrections.

References

1. J.L. O'Toole, "Design Guide", Modern Plastics Encyclopedia, *64*:10A (1987), p. 404.
2. R.C. Progelhof, "The Interaction of Plastic Processing Techniques and Resin Properties with Part Design", Material Supplied to PIA Seminar, Hoboken, NJ (1980).
3. S. Middleman, *Fundamentals of Polymer Processing*, McGraw-Hill Book Co., New York, (1982), Preface.
4. J.L. Throne, *Plastics Process Engineering*, Marcel Dekker, New York (1979), p. 848f.
5. Anon., "Polyolefin Injection Molding . . . An Operating Manual", U.S.I. Chemicals Corp., undated, #P2-1277, p. 15.
6. P. Richardson, *Introduction to Extrusion*, Soc. Plast. Eng., Brookfield Center, CT (1974), p. 4.
7. A.B. Glanvill, *Plastics Engineers Data Book*, Industrial Press, New York (1974), p. 37.
8. P. Richardson, *Introduction to Extrusion*, Soc. Plast. Eng., Brookfield Center, CT (1974), p. 8.
9. B. Maddock, Polym. Eng. Sci., *16* (1976), p. 284.
10. Anon., "Polyolefin Injection Molding . . . An Operating Manual", U.S.I. Chemicals Corp., undated, #P2-1277, p. 16.
11. T. Richardson, *Modern Industrial Plastics*, SAMS, Indianapolis (1974), p. 252.
12. T. Richardson, *Modern Industrial Plastics*, SAMS, Indianapolis (1974), p. 284.
13. G.A. Campbell and B. Cao, SPE Tech. Pap., *32* (1986), p. 909.
14. C. Rauwendaal, *Polymer Extrusion*, Hanser Verlag, Munich (1986), p. 20.
15. D.M. Kaylon and F.H. Moy, SPE Tech. Pap., *31* (1985), p. 684.
16. M. Horio, Int. Polym. Proc., *1* (1986), Fig. 2.
17. M. Horio, Int. Polym. Proc., *1* (1986), Fig. 7.
18. M. Horio, Int. Polym. Proc., *1* (1986), Figs. 3 and 8.
19. R.J. Baird and D.T. Baird, *Industrial Plastics*, Goodhardt Wilcox, South Holland, IL (1982), p. 105.
20. C. Rauwendaal, *Polymer Extrusion*, Hanser Verlag, Munich (1986), p. 17.
21. C. Rauwendaal, *Polymer Extrusion*, Hanser Verlag, Munich (1986), pp. 460–476.
22. C. Rauwendaal, *Polymer Extrusion*, Hanser Verlag, Munich (1986), p. 245.
23. H.R. Jacobi, *Extrusion of Plastics*, ILIFFE Books, London (1960), p. 6.
24. C. Rauwendaal, *Polymer Extrusion*, Hanser Verlag, Munich (1986), p. 27.
25. H.R. Jacobi, *Extrusion of Plastics*, ILIFFE Books, London (1960), p. 7.
26. A.B. Glanvill, *Plastics Engineers Data Book*, Industrial Press, New York (1974), p. 16.
27. HPM, Mount Gilead, OH 43338.
28. Cincinnati Milacron Co., Batavia, OH 45103.
29. J. Frados, Ed., *Plastics Engineering Handbook*, 4th Ed., Van Nostrand, New York (1976), p. 101.
30. F. Johannaber, *Injection Molding Machines*, Hanser Verlag, Munich (1979), p. 116.
31. F. Johannaber, *Injection Molding Machines*, Hanser Verlag, Munich (1979), p. 17.
32. Marland Mold Co., Pittsfield, MA.
33. J. Frados, Ed., *Plastics Engineering Handbook*, 4th Ed., Van Nostrand, New York (1976), p. 16.
34. E.I. Dupont de Nemours and Co., Inc., Wilmington, DE 19898.
35. J. Frados, Ed., *Plastics Engineering Handbook*, 4th Ed., Van Nostrand, New York (1976), p. 133.
36. J. Frados, Ed., *Plastics Engineering Handbook*, 4th Ed., Van Nostrand, New York (1976), p. 136.
37. I. Rubin, *Injection Molding*, Wiley-Interscience, New York (1972), p. 277.

38. S. Middleman, *Fundamentals of Polymer Processing*, McGraw-Hill Book Co., New York (1982), p. 269.

39. S. Middleman, *Fundamentals of Polymer Processing*, McGraw-Hill Book Co., New York (1982), p. 271.

40. J.L. Throne, *Plastics Process Engineering*, Marcel Dekker, New York (1979), p. 471.

41. W.G. Frizelle and D.C. Paulson, SPE Tech. Pap., *14* (1968), p. 405.

42. M.R. Kamal and S. Kenig, Polym. Eng. Sci., *12* (1972), pp. 294 and 302.

43. P.C. Wu, C.F. Huang, and C.G. Gogos, SPE Tech. Pap., *19* (1973), p. 197.

44. J.L. White, "The Injection Molding Process", in E.C. Bernhardt, Ed., *CAE: Computer Aided Engineering for Injection Molding*, Hanser Verlag, Munich (1983), p. 86.

45. I.T. Barrie, SPE J., *27*:8 (1971), p. 64.

46. G. Menges and G. Wubken, SPE Tech. Pap., *19* (1973), p. 512.

47. R.L. Ballman and H.L. Toor, Mod. Plast., *38*:10 (1960), p. 113.

48. G.B. Jackson and R.L. Ballman, SPE J., *16*:11 (1960), p. 1147.

49. G. Menges and G. Wubken, SPE Tech. Pap., *19* (1973), p. 513.

50. M.L. Miller, *The Structure of Polymers*, Reinhold Book Co., New York (1966), p. 572.

51. E.S. Clark, Appl. Polym. Symp., *20* (1973), p. 325.

52. K. Nakamura, T. Watanabe, K. Katayama and T. Amano, J. Appl. Polym. Sci., *16* (1972), 1077.

53. S.S. Semerdjier, *Introduction to Structured Foam*, Society of Plastics Engineers, Brookfield Center CT (1982), p. 47.

54. J.L. Throne and F.A. Shutov, "Structural Foams", in J. Kroschwitz, Ed., *Encyclopedia of Polymer Science and Engineering*, Vol. 15, John Wiley & Sons, Inc, New York (1988), p. 280.

55. R.J. Baird and D.T. Baird, *Industrial Plastics*, Goodhardt Wilcox, South Holland, IL (1982), p. 115.

56. J.L. Throne, *Plastics Process Engineering*, Marcel Dekker, New York (1979), p. 688.

57. T. Richardson, *Modern Industrial Plastics*, SAMS, Indianapolis (1974), p. 260.

58. Z. Tadmor and C.G. Gogos, *Principles of Polymer Processing*, Wiley-Interscience, New York (1979), p. 654.

59. R.J. Crawford, *Plastics Engineering*, Vol. 7, *Progress in Polymer Science*, Pergamon Press, London (1981), p. 162.

60. T. Richardson, *Modern Industrial Plastics*, SAMS, Indianapolis (1974), p. 259.

61. S.S. Schwartz and S.H. Goodman, *Plastics Materials and Processes*, Van Nostrand Reinhold, New York (1982), p. 625.

62. K. Esser and G. Menges, SPE Tech. Pap., *30* (1984), p. 924.

63. L. Erwin, H. Gonzalez, and M. Pollock, SPE Tech. Pap., *29* (1983), p. 807.

64. M. Cakmak, J.L. White and J.E. Spruiell, SPE Tech. Pap., *30* (1984), p. 920.

65. E. Winkel, H. Gross, U. Masberg and J. Wortberg, Adv. Polym. Technol., *2* (1982), 107.

66. R.J. Gartland, SPE Tech. Pap., *30* (1984), p. 753.

67. J.L. Throne, *Thermoforming*, Hanser Verlag, Munich (1986), p. 19.

68. J.L. Throne, *Thermoforming*, Hanser Verlag, Munich (1986), p. 18.

69. J.L. Throne, *Thermoforming*, Hanser Verlag, Munich (1986), p. 20.

70. J.L. Throne, *Thermoforming*, Hanser Verlag, Munich (1986), p. 53.

71. W.K. McConnell, Jr., "Thermoforming Technology for Industrial Applications", SPE Seminar, Arlington, TX (12–14 March 1985).

72. R.D. Deanin, *Polymer Structure, Properties and Applications*, Cahners Books, Boston (1972), p. 333.

73. J.G. Williams, *Stress Analysis of Polymers*, John Wiley & Sons, New York (1973), pp. 211–220.

74. Ciba-Geigy Composites Corp., Anaheim, CA 92807.

75. G. Menges, *XII Kolloquium*, Institut für Kunststoffverarbeitung, Aachen, West Germany (1986), p. 71.

76. Z. Tadmor and C.G. Gogos, *Principles of Polymer Processing*, Wiley-Interscience, New York (1979), p. 629.

77. M.R. Barone and D.A. Caulk, J. Appl. Mech., *108*:53 (1986), p. 361.

78. J. Frados, Ed., *Plastics Engineering Handbook*, 4th Ed., Van Nostrand, New York (1976), p. 224.

79. A.B. Glanvill, *Plastics Engineers Data Book*, Industrial Press, New York (1974), p. 56.

80. A.B. Glanvill, *Plastics Engineers Data Book*, Industrial Press, New York (1974), p. 57.

81. Union Carbide Corp., Danbury, CT 06705.

82. S.S. Malquarnera and N.P. Suh, SPE Tech. Pap., *22* (1976), p. 211.

83. R.W. Meyer, *Handbook of Pultrusion Technology*, Chapman and Hall, New York (1985), p. 5.

84. J. Gillham, J. Appl. Polym. Sci., *28* (1983), p. 1105.

85. J.B. Enns and J.K. Gillham, J. Appl. Polym. Sci., *28* (1983), p. 2567.

86. J.D. Domine and C.G. Gogos, SPE Tech. Pap., *22* (1975), p. 274.

87. J.M. McDonaghn, "Process Variables in Rotomolding", in P.F. Bruins, Ed. *Basic Principles of Rotational Molding*, Gordon and Breach, New York (1971), p. 34.

88. D.M. Tomo, "Rotational Molding of Polyethylene Powders", in and P.F. Bruins, Eds. *Basic Principles of Rotational Molding*, Gordon and Breach, New York (1971), p. 183.

89. D.M. Tomo, "Rotational Molding of Polyethylene Powders", in P.F. Bruins, Eds. *Basic Principles of Rotational Molding*, Gordon and Breach, New York (1971), p. 184.

90. F. Ribe, R.C. Progelhof and J.L. Throne, SPE Tech. Pap., *32* (1986), pp. 20, 23.

91. G.D. Hodgson, "CAD/CAM Technology", in E.C. Bernhardt, Ed., *CAE: Computer Aided Engineering for Injection Molding*, Hanser Verlag, Munich (1983), Chapter 3.

92. C. Rauwendaal, Seminar, Polymer Processing Hall of Fame, 10 Dec 1987, Akron, OH.

93. Z. Tadmor and I. Klein, *Engineering Principles of Plasticating Extrusion*, Van Nostrand Reinhold, New York (1970).

94. I. Klein and D.I. Marshall, Eds., *Computer Programs for Plastics Engineers*, Van Nostrand Reinhold, New York (1968).

95. J.L. Throne, *Thermoforming*, Hanser Verlag, Munich (1986), Chapters 2–4.

96. C. Rauwendaal, *Polymer Extrusion*, Hanser Verlag, Munich (1986), p. 276.

97. F. Gerlitz, SPE Tech. Pap., *33* (1987), p. 1506.

98. K.K. Wang and V.W. Wang, in A.I. Isayev, Ed., *Injection and Compression Molding Fundamentals*, Hanser Verlag, Munich (1987), p. 666.

99. K.H. Huebner, *The Finite Element Method for Engineers*, John Wiley & Sons, New York (1980).

100. R.T. Fenner, *Finite Element Methods for Engineers*, Macmillan, London (1975).

101. C. Austin, "Filling of Mold Cavities", in E.C. Bernhardt, Ed., *CAE: Computer Aided Engineering for Injection Molding*, Hanser Verlag, Munich (1983), Chapter 9.

102. R.C. Progelhof, "Data Bases", in E.C. Bernhardt, Ed., *CAE: Computer Aided Engineering for Injection Molding*, Hanser Verlag, Munich (1983), p. 406.

103. R. Westover, "Processing Properties", in E.C. Bernhardt, Ed., *Processing of Thermoplastic Materials*, Reinhold, New York (1959), Chapter X.

104. K.C. Maddux, "Part Design", in E.C. Bernhardt, Ed., *CAE: Computer Aided Engineering for Injection Molding*, Hanser Verlag, Munich (1983), p. 149.

105. K. Polit, SPE Tech. Pap., *33*(1987), p. 1483.

106. P.J. Carreau, PhD Thesis, Univ of Wisc, Madison (1968).

107. R.T. Van Ness, G.R. De Hoff, and R.M. Bonner, Mod. Plast. Encyclopedia, *45*:14A (1968), p. 672.

REFERENCES

Anon., *Rotational Molding of Plastic Powders*, Engineering Design Handbook, U.S. Army Materiel Command, AMCP 706-312 (April 1975).

W.E. Becker, *Reaction Injection Molding*, Van Nostrand Reinhold, New York (1979).

P.F. Bruins, Ed., *Basic Principles of Rotational Molding*, Gordon and Breach, New York (1971).

H.R. Jacobi, *Screw Extrusion of Plastics*, ILIFFE Books, London (1960).

L. Janssen, *Twin Screw Extrusion*, Elsevier, Holland (1978).

F. Johnnaber, *Injection Molding Machines*, Hanser Verlag, Munich (1979).

C.M. Macosko, *RIM: Fundamentals of Reaction Injection Molding*, Hanser Verlag, Munich (1989).

R.W. Meyer, *Handbook of Pultrusion Technology*, Chapman and Hall, New York (1985).

C. Rauwendaal, *Polymer Extrusion*, Hanser Verlag, Munich (1986).

D.V. Rosato and D.V. Rosato, Eds., *Blow Molding Handbook*, Hanser Verlag, Munich (1989).

I. Rubin, *Injection Molding*, Wiley-Interscience, New York (1972).

S.S. Semerdjiev, *Introduction to Structural Foam*, The Society of Plastics Engineers (1982).

F.A. Shutov, *Integral/Structural Polymer Foams*, Springer-Verlag, Berlin (1986).

J.L. Throne, *Thermoforming*, Hanser Verlag, Munich (1986).

Glossary

Blown Film   Formation of biaxially oriented extruded film, by inflating a tubular extrudate.

Blow-Up Ratio   The ratio of the diameter of the inflated tubular extrudate to the diameter of the die annulus.

BMC   Bulk molding compound, reactive polymer containing high filler loading, with a dough-like consistency. Can be injection, compression or transfer molded.

Casting   Formation of unoriented extruded film or sheet.

Cavity   The hollowed-out section in an injection mold that shapes the desired part.

Cell Cast   Reactive polymerization between coplanar surfaces, one of which is free to move toward the other during polymerization.

Chill Rolls   Horizontal, highly polished, cooled cylindrical rolls onto which polymer melt is cast.

Interaction Sensitivity   The sensitivity of a given computer-aided design to the accuracy in the values of the inputted material parameters.

Melt Fracture   Periodic irregularites on extrudate surface.

Melt Pool   In an extruder, accumulation of softened or molten polymer at the leading edge of the trailing screw flight.

Melt Strength   In blow molding, the property that allows an extruded tube of molten polymer to hang while the mold sections close around it.

Mold   A rigid shape against which softened or molten polymer is pressed and allowed to rigidify.

Non-Flow Temperature   An artificial temperature needed in Moldflow, an injection molding software program.

Non-Return Valve   A collar on the end of an injection molding screw that prevents polymer flow toward the hopper during screw advancement, in the injection step.

Orientation   Polymer molecules aligned in a preferred direction.

Packing   Putting additional polymer into the mold under pressure to minimize shrinkage.

Parison   An extruded tube of molten polymer, in blow molding.

Parting Line   A line, visible on a molded part, at which fluid streams meet.

Pinch-Off   A region or seam in a part where two softened or molten segments of polymer are squeezed together under pressure.

Pitch   In an extruder, the angle the screw flight makes with the screw and barrel axes. 17.42° pitch provides one screw flight per revolution, or a single-flighted pitch.

Plasticating Capacity   The melting capacity of a screw extruder, usually based on HDPE.

Plastication   Melting under shear and thermal gradients.

Platen   One of the faces of a press against which a mold section is clamped.

Preform   A simple solid polymer shape that is heated and stretched into a more complex shape, as in blow molding.

Pressure Flow   Differential pressure on material between two coplanar stationary surfaces, causing material to flow.

Processing   Conversion of raw polymer and adducts to useful products.

Reaction Injection Molding   Injection molding of a reactive polymer, usually a thermoset.

Regrind   Thermoplastic polymeric resin that has been processed once and has been ground for reprocessing.

Retrograde   The backward motion of a turning injection molding screw, during plastication.

Reynolds Number   Ratio of momentum transfer to viscous inertia, used to determine the degree of turbulence of a fluid stream.

RIM   See Reaction injection molding.

Roll-Fed Thermoforming   Thermoforming articles from a continuous roll of sheet.

Runner   An internal flow channel in an injection mold.

Shrinkage   Dimensional difference between molded part and mold cavity dimensions.

Sintering   Forming a fused, porous structure by heating compacted powder.

SMC   Sheet molding compound, long glass fiber reinforced reactive polymer compound, for matched die molding.

Sprue   Polymer melt directing section of a mold, between the molding machine nozzle and the runner-gate-cavity system.

Strain Hardening   Apparent increase in stiffness with increasing stress level.

Strain-Rate Hardening   Increasing resistance of a polymer to increasing rate of strain.

Strand Die   A compounding extruder die containing many like holes.

Thin-Gage Sheet   In thermoforming, sheet thickness of 1.5 mm or less.

Transcrystalline   Crystalline growth perpendicular to a cool surface.

Two-Stage Screw   An extruder screw having two compression and melt pumping zones.

Vitrification   Formation of a glassy structure.

CHAPTER 6

TESTING FOR DESIGN

6.1 Introduction

The ultimate objective of polymer processing is the fabrication of a product that may be comprised of several polymer and nonpolymer elements. This assembled product must meet certain nominal performance requirements. The materials used in the various components that make up the assembled product must have certain inherent characteristics such as stiffness, toughness, environmental resistance, or optical/electrical properties. The process of transforming the polymeric material into the product induces other characteristics such as dimensional changes or internal stresses. Assembly of the final product may impose further characteristics such as differential expansion, or friction/wear. And the characteristics of the ultimate use of the product, such as environment, time, or nature of the applied load, may further influence the material performance. As with all materials, polymers must be tested to determine their suitability in any product application. Turner (1) notes that there are several diverse reasons for testing materials:

- For quality control,
- As a basis for comparing and selecting materials,
- To build a design calculation database,
- To predict service performance,
- As a way of directing materials development,
- To devise methods of analyzing molecular behavior of materials.

Unless carried forth for purely scientific curiosity, the intent of any testing program is to provide an understanding of the nominal performance of a material in a given environment. To this end, it is important to note that molecular response to environmental changes needs to be manifested in easily measured effects (visual = birefringence, dimensional change = tensile strength, weakening = fatigue strength or creep modulus). These effects need to be quantified in some manner, with a high degree of repeatability, and the results easily and unambiguously interpreted.

Of the myriad, often overlapping, inherent material properties, only a few are critical to the performance of the final product. The final product should perhaps be considered as an assemblage of sub-products, sub-sets or "basic elements" (2). Each of these sub-sets places specific mechanical and/or chemical demands on the materials. The property set required by the final product, then, can be considered a "mean" of the properties of the basic shapes. From this, the tests required to delineate performance criteria of the final product can be ascertained. This point is amplified in a later section.

Consider the nature of the test specimen. The most obvious starting point is the part itself. Frequently, however, the size of the specimen precludes this option. Or the part may be conceptual. Appropriate materials must be selected on the basis of design and material specifications. Probably the most restrictive element is the interpretation of the test results. All test data should in some way represent material properties. There should be a unique set of relationships between material properties and parameters. Most important, there must be a genuine causal relationship. The last condition is the most difficult to achieve in direct product testing.

The test specimen can be a segment of the part. If the part is of proper configuration, test specimens can be sectioned from the part. In this manner, a measure of the effect of processing conditions on the material *in situ* can be obtained. Keep in mind, however, that many part geometries are such that suitable test specimens cannot be obtained. Furthermore, cutting or sectioning test specimens from parts can induce material changes, such as stress relief, crystallization, and warpage.

For determining material selection in general, test specimens are frequently injection or compression molded directly into the required testing shape. Specimens that are prepared specifically for property comparison, such as tensile, flexural, or impact test specimens, are usually quite simple in geometry. This allows careful analysis of the causal relationship and hence reliable determination of classical physical properties such as tensile strength and Young's modulus which are common to all materials. While these specimens can be produced with excellent control, the tester is measuring properties of a material in a state that may differ substantially from its state in the final part.

The polymer engineer must be consciously aware that unlike other materials, the properties of the final part are dominated by the nature of the polymeric molecular state. This implies that exceeding the simple geometries of these tests can lead to potentially unpredictable results. Mooney (3) calls tests that neither faithfully reproduce some standard, analyzable test routine nor model the actual in-use part performance *"hybrid" tests.* He notes that tests that are partly designed to mirror service conditions and partly to meet the speed needed in quality control rarely yield useful data. And when they do, it is usually by accident.

Of course, hybrid tests exist, particularly when tests on actual products are expensive and/or time-consuming. Typically, long-term and very-short-term tests are most prone to hybrid testing. In fatigue testing for example, a part that must perform at 1 cycle/min day in and day out for 10 years must be flex-tested under load at least 5,000,000 times. Even at 1 cycle/s (60 times design design frequency), testing will take 2 months. Internal heat generation must be considered or the accelerated test results at the much higher frequency may be meaningless.

It must further be noted that processing can impress potentially significant changes in the material characteristics. The effect of processing on data interpretation must always be considered. If the test data are to bear relation to the product in actual use, the test specimen must incorporate the appropriate influences of the process. Furthermore, since most processes have a wide range of operating conditions, the testing procedure time must be short enough to allow many of these conditions to be evaluated at a reasonable expenditure of time and money. Of course, this can seriously restrict the scope of testing. Again, the engineer must balance the need to know the causal relationship of material and environment in great detail with the pragmatic need to get the job done on time and within budget.

It has been recommended (3,4) that two *criteria of test acceptability* should be applied to every test. First, the mechanical state should be *definable in physical terms* (thickness, length, applied load, applied stress, strain, rate-of-strain, dimensional change, temperature). And second, the mechanical state should be *definable in causal mathematical terms* as well, such as stress–strain–rate-of-strain. Exceptions to these criteria must be carefully documented, particularly if the final part performance critically depends on data interpretation.

Rational tests seldom are used in isolation. Test results are usually designed to measure causal relationships between two elements, such as strain in response to stress, yielding in response to applied load, or fracture in response to impact. Interrelationships of material properties and final part performance are established through a set of tests. The weaknesses in interpreting the results of one test can to some degree be mitigated by additional, rationally chosen tests. However, the engineer is cautioned that most "practical" test results are open to multiple interpretations. It is neither economically nor pragmatically possible to fully evaluate all the material properties that might be applicable to a given application. While we wish to evaluate polymer material properties in their general molecular state, we usually try to relate properties to the polymer in some given molecular state. Furthermore, extensive repeated testing may yield information about the statistical accuracy of the test results. But more data do not necessarily contribute to bridging the gulf between test specimen properties and the final part performance (5).

In earlier chapters, polymer characterization was presented as a means of understanding the interrelationship of the material and its environment. In many ways, extensive analytical testing is impractical. The tests frequently are too expensive, too time-consuming, and not particularly relevant to the part performance criteria necessitated by the application.

When experiments become standardized, that is, when a specific procedure is used each time to measure and report the results of a specific phenomenon, the experiment is referred to as a *performance test*. Many organizations have been formed throughout the world to standardize different types of tests. In England and other regions of Europe, the International Standards Organization (ISO) is a clearinghouse.* In the United States, the most notable organization is the American Society for Testing and Materials, ASTM, founded in 1898. This scientific and technical organization was formed for:

> . . . the development of standards on characteristics and performance of materials, products, systems, and services; and the promotion of related knowledge. (6).

Data from ASTM tests are usually considered as the first test data needed when comparing and specifying polymer materials for specific applications. Typical ASTM data for several thermosetting and thermoplastic polymers are shown in Appendix C. Although many of the ASTM tests for plastics evolved from similar tests on metals, recent review procedures have been modified to be specific to polymers. It is important that the procedures peculiar to polymers be described in some detail to aid the design engineer in understanding their utilities and limitations. That is the objective of this chapter.

There are many ways of categorizing tests. For example, rheological tests are designed for fluid material evaluation, whereas tensile tests are for solid materials. As noted in Section 4.4, special tests are needed if viscoelastic properties are desired. Tests of stiffness and modulus focus on the elastic nature of the polymer. Toughness and strength deal with the plastic region of the material. Similarly, stiffness and modulus are measured by averaging the material properties. Tensile strength and toughness depend on the "weakest link" nature of the material. In this chapter, the tests are arbitrarily divided into fluid and solid tests, short-term, long-term and moderate-time mechanical tests, and environmental tests, including transparency and electrical resistance. The long-term tests are further divided into static and dynamic states. Several important mechanical properties tests for polymers are given in Table 6.1.

---

*Despite this show of unity, individual nations maintain certain national codes. For example, in West Germany, the DIN is the national standards organization. In Japan, the standards are JIS. In France, they are AFNOR. In Italy they are UNI. And in the Netherlands, they are NFT. England uses British Standards or BS code, as well as ISO. And so on.

Table 6.1. International standards for mechanical properties of rigid polymers (7).

Method	ASTM	ISO	DIN	Remarks
Preparation of specimens				
Injection and compression molding thermoplastics	D1897 D1928	293 294 2557	16770	ISO 2557, DIN 16770: Amorphous thermoplastic specimens with a designed level of shrinkage (for semi-crystallines in development)
Compression molding thermosets	D796 D956 D1896	295	53451 53457	Transfer Injection: ASTM D1896, D3881
Machining specimens		2818		
Multipurpose specimens		3168		
Stress–strain, including creep properties				
Tensile properties	D638	527	53455	Rigid plastics, for GRP:D3093, EN61
	D674	899	53444	Tensile creep
	D2991		53441	Stress relaxation test
	D882	1184	53445	Thin plastic sheeting and film
Flexural properties	D790 D2990	178 DP6602	53452	Rigid plastics, for GRP: EN63 D2990: Tensile, flexural, compression creep
Compression properties	D695	604	53454	Test specimen prism or cylinder, slenderness ratio 10 to 15
Modulus of elasticity	D639 D790 D695	527 178 604	53457	ASTM, ISO: Included in the general standards
Fatigue resistance alternating stress tests				
Flexural fatigue test	D671		53442	Flat specimens, constant amplitude of force (D671) or strain (53442)
Special tests for GRP	D3479		53398	ASTM tension–tension, DIN bending pulsation
Impact resistance				
Izod flexural impact test	D256	180		Cantilever beam. Method A, C, D swing on the same face of the notch, E reversed notch test

Method	ASTM	ISO	DIN	Remarks
Impact resistance (cont.)				
Charpy flexural impact	D256 Method B	179	53453	Horizontal beam, unnotched or notched (D256 B only notched), impact line midway, directly opposite the notch
Charpy tensile impact test	D1822		53448	Results noncomparable
Multiaxial falling weight (Tup) test	D3029 D1709 (PE Film), D2444 (Pipe)	DP6603	53443 Part 1	Failure signified by any crack or split
Impact penetration with electronic data recording device			53443 Part 2	DIN 53373 for films and thin sheet
Hardness tests				
Ball indentation hardness		2039	53456	Rigid plastics
Rockwell hardness	D785	2039		Rigid plastics
Shore durometer hardness	D2240	868	53505	Scales A and D
Barcol impression test	D2583	EN59	EN59	Portable impressor used to control the curing degree of GRP parts

GRP = Glass Reinforced Polyester, GR-UPE

6.2 Fluid Material Properties: Rheological Measurements

In Chapter 4, softened or melted polymers were considered to be highly viscous elastic liquids. Viscosity for polymeric materials is not a constant material property as it is with smaller molecules. Instead, it is shear-rate- and time-dependent. It was also pointed out that for elastic liquids, normal stress differences and elongational viscosity are important as well, particularly in such processes as blow molding and fiber spinning. In Chapter 5, it was emphasized that most plastic products are produced by melting, pumping and shaping polymers in the melt state. In addition, the pumping and shaping equipment must be robust, to overcome the intrinsically high resistance of the polymer melt. This then is the justification for determining the rheological characteristics of polymers.

There are many ways to measure rheological properties. Many of the simpler methods such as capillary rheometry are derived from techniques used to measure viscous Newtonian fluids such as asphalt or bitumen. Others are adapted from rubber techniques. For example, the enclosed bob biconical rheometer (8) is a modification of the

Mooney rubber viscometer (9). It is important to select a method of measurement that yields properties that are, in fact, independent of the measuring method. It is also important to remember that polymer viscosity is a *material parameter* and *not* a material constant. Therefore the method of measurement should produce a stress–rate-of-shear strain curve so that an isothermal shear-rate dependent viscosity curve can be established. Recall from Chapter 4 that both the power-law and the Rabinowitsch–Mooney viscosity models depend on the shape and the slope of the viscosity–shear-rate curve. To allow interpolation and, to a limited extent, extrapolation of the measured data, the constitutive equation of state must have parameters that are easily and uniquely defined in terms of the measured values. Simple geometries allow for simple analysis. Usually, the complex geometry of processing equipment does not accommodate simple analysis. As a result, special devices called *rheometers* or *viscometers** are used. As discussed in detail in Chapters 4 and 5, polymers are subjected to two general types of environmental influence:

- Steady flow processing, such as flow through constant-diameter passages,
- Highly time-dependent flow processing, such as flow through dies for fiber forming and blow molding and flow into cavities as in injection molding.

As noted in Chapter 4, the elastic nature of a polymer is manifested in short-term changes in applied forces, whereas the viscous nature is seen when forces are applied in steady fashion. Most viscometers operate in steady-state fashion. These can be grouped according to the device–fluid interface:

External continuous, where the fluid is confined within an outer surface of the device:

- Rotational viscometers, such as "cup and bob" devices used for measuring viscosities of room temperature viscous liquids (epoxies, unsaturated polyester resins)
- Translational viscometers, such as flow in a cylindrical tube or a constant cross-section slot.

Internal continuous, where the polymer surrounds the surface of the device:

- Flat plates or surfaces of revolution (cones, spheres) that are mechanically rotated or translated through an infinite reservoir of fluid,
- Freely falling objects such as darts, pins, cylinders, spheres, and cones.

---

*A "goniometer" is a device used extensively in crystallography to measure strain angles in crystals. The "rheogoniometer" is the rheological equivalent to this device, in which exacting measurements of the interaction of the complex rheological parameters for highly elastic liquids can be made. The term is usually reserved for large viscoelastic deformation measurements rather than true viscous flow measurements. The term *viscometer* is meant to include rheogoniometers.

Discontinuous steady-state, where the solid–polymer interface stops abruptly:

- Fluid jet, capillary tube and melt indexer viscometers,
- Spinneret, siphon tube or other extensional flow,
- Withdrawal of a plate or cylinder from a polymer reservoir.

Dealy (10) views rheometers in the following terms:

Rectilinear shear viscometers, such as sliding plate, capillary and slit viscometers,

Rotational viscometers, including concentric cylinder flow, and cone and plate, parallel plate, and eccentric disk viscometers,

Uniform extensional viscometers, including uniaxial and symmetrical biaxial extensiometers,

Other viscometers that involve more complex flows, including rod climbing and squeezing flow.

There are many commercial variants of these devices. Many are detailed in references 10 to 14.

This section considers four viscometers. The *melt indexer* is the most widely used commercial flow measuring device for thermoplastics. The *capillary rheometer* is an example of rectilinear viscometric shear flow. In order to interpret the data, the abrupt changes in solid–fluid interface must be understood. The *cone-and-plate rotational shear viscometer* has become the most widely used research viscometer. It can be used for both thermoplastics and reactive thermosets. The *ram-follower* is used specifically for thermosetting resins. There are many other flow measuring devices but these four are the most widely used. It is important that the engineer understand the relevance of the data from each of these devices.

The Melt Indexer: ASTM D1238

There are many traditional ways of measuring the viscosity of Newtonian fluids such as tars and bitumens. Typically, a measured amount of fluid is added to a glass vessel immersed in a constant temperature bath. The bottom of the vessel has a stoppered hole or short capillary tube that protrudes from the bath. A container is placed beneath the tube, the stopper is removed and the time to fill the container to a predetermined volume is measured. An empirical equation is used to determine the viscosity. Two popular viscometers of this type are the Engler and the Saybolt viscometers.

In the 1950s, duPont engineers adapted this concept to measurement of the viscosity of polyethylenes and called it an "extrusion plastometer". A cross-sectional view of the unit is shown as Figure 6.1. The solid polymer, preferably as powder or fine granules, is added to the 9.55 mm (0.3730 in) diameter heated plastometer chamber. The chamber has an isothermal temperature of 190°C. The polymer is prevented from flowing from the plastometer by a capillary tube. The bore diameter is 2.095 mm (0.0825 in) and its length is 8 mm (0.315 in), for an *L/D* of 3.82. The melting polymer

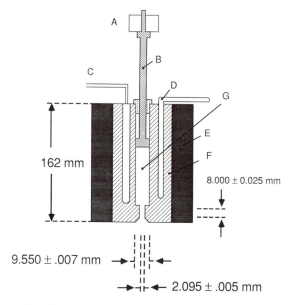

Figure 6.1. Schematic of Extrusion Plastometer, ASTM D1238. A: Weight. B: Plunger. C: Heater. D: Thermometer. E: Insulation. F: Body. G: Chamber (15).

is tamped with a close-fitting piston to remove residual air. When the polymer is molten and free of bubbles, a 2.16 kg weight is applied to the piston. The extrudate is collected over a measured (10 min) period of time and *weighed*. The weight per 10 minutes (or dg/min) is reported as "melt index" or MI, thereby giving the device the alternate name of "melt indexer".

As expected, increasing molecular weight implies increased viscosity or resistance to applied load. Therefore, the extruded weight of polyethylene per unit time decreases with increasing molecular weight. Typically a polyethylene having an MI of 20 has a substantially lower viscosity than one with an MI of 0.5. If a polymer must flow long distances under relatively low differential pressure, as in certain injection molding applications, a high MI is sought. On the other hand, if the polymer must resist unwanted flow, as in parison blow molding, a low MI is sought. The level of applied weight and the isothermal temperature can depend on the type of polymer.

Example 6.1

Q: Consider polyethylene being processed in the melt indexer.
Q1: What is the pressure applied to the melt in the reservoir during the experiment?
Q2: If 6 grams of polyethylene is extruded in a 5 minute period, what is the MI value?

Q3: What is the apparent shear rate of the polymer in the capillary? Assume $\dfrac{1}{\rho} = 1.30$ cm^3/g.

A1: The applied pressure is force per unit area, or:

$$P = F/A,$$

$$= \frac{2.16 \text{ (kg)} \times 2.2 \text{ (lb/kg)}}{[\pi \times (0.955 \text{ cm}/2.54 \text{ cm/in})^2/4]}$$

$$= 42.8 \text{ psi} = 29.5 \text{ k Pa}$$

A2:   MI, g/10 min $= (6 \text{ g}/5 \text{ min} \times 10 \text{ min}) = 12.$

A3:

$$Q = 12 \text{ (g/min)} \times (1/60) \text{ (min/s)} \times 1.3 \text{ cm}^3/\text{g},$$

$$= 0.26 \text{ cm}^3/\text{s}.$$

$$\dot{\gamma}_a = \frac{4Q}{\pi R^3} = \frac{4 \times 0.26}{[\pi (0.1047)^3]},$$

$$= 288 \text{ s}^{-1}.$$

Consider the important limitations to this test. The melt indexer applies a nearly constant pressure on the polymer sample. The pressure exerted represents the sum of the "entrance effects" and the pressure drop along the capillary. If the entrance effects are small, the shear rate and thus the viscosity of the polymer are determined by the shear stress and so a *single* viscosity value is obtained for each polymer. However, entrance effects usually cannot be ignored for polymer fluids. As a result, the "viscosity" numeric value that is determined in this fashion cannot be related to the true viscosity. Thus, data from the melt indexer should not be used in any way to calculate viscosity. Furthermore, the melt indexer gives no information about the shape of the shear stress-shear rate curve for the given material. It is also apparent that the single shear rate value changes from polymer to polymer. Thus, the MI value of a high MI polymer may be determined at a higher shear rate than that of a low MI polymer. *Even if* the MIs of two polymers are identical, there is no assurance that they have identical viscosities.

The engineer must avoid relying on melt index as a measure of "flowability" performance for any polymer, including polyethylenes, the class of polymers for which the test was developed. At best, melt indexer data allow the engineer to determine which set of polymers might be used in a given processing situation. As mentioned, extrusion blow molding grades of polyethylenes typically have fractional MIs, injection molding grades have MIs in the range of 3 to 30, and rotational molding grades have MIs greater than 30. Another concern is the use of MI for polymers other than polyethylenes and polypropylenes.* Again, the viscosity values so generated cannot be related from one polymer species to another. Mechanically, there are many problems with data from this device. For example, the *L/D* ratio does not allow for fully devel-

---

*Owing to its higher melt temperature, polypropylene must be run at 230°C. As a result, MIs from polyethylenes and from polypropylenes cannot (must not) be compared.

oped flow. Entrance and exit effects are substantial as discussed shortly for flow in capillary tubes. These limitations prevent relating MI data to *any* rheological property, including viscosity. The wide acceptance of this device by the injection molding industry is historical. In the 1950s and 1960s, there was a strong correlation between polyethylene moldability and MI value. With the development of highly tailored polyethylenes, the requirement for very precise injection molded parts, and the introduction of copolymers and blends, the plastometer no longer yields data that produce reliable correlations with moldability.

Capillary Rheometer

The capillary rheometer is an important improvement of the extrusion plastometer. It resembles the melt indexer in many regards. It features an isothermal heated cylindrical chamber stopped with a small-bore capillary tube. The solid polymer is added to the chamber and tamped during melting to eliminate bubbles. The chamber is then fitted with a close-fitting piston in a manner similar to the melt indexer. Instead of using a dead weight to apply pressure, the capillary rheometer uses a positive constant speed drive connected to a load cell. Thus, polymer flow rate through the capillary is determined as a function of the shear rate, with applied shear stress measured. A schematic of a constant velocity capillary rheometer is shown in Figure 6.2. The capillary dimensions can also be changed, with the common diameters being 0.030, 0.050 and 0.060 in and common lengths being 1, 2, 3, and 4 in. This allows a range of L/D from about 16 to 133.* The typical shear rate range for capillary viscometers is 1 to 10,000 s^{-1}.

A typical pressure drop profile is seen in Figure 6.3. The flow through the center portion of the capillary tube is fully developed. Thus the resistance to flow in this region should be given by standard pressure drop–flow rate equations, such as Poiseuille for Newtonian fluids or Rabinowitsch–Mooney for power-law fluids. In the entrance region to the capillary, however, the fluid flow is quite complex. The fluid must begin to accelerate into the constriction. A velocity profile develops. There is substantial evidence that the elastic nature of the polymer may result in secondary flow effects that persist throughout the flow time through the die. These effects can manifest as surface effects on the extrudate or surging. At the exit region of the capillary, the polymer velocity profile flattens. There is also a capillary exit effect, seen as an apparent** non-zero pressure at the exit. This is thought to be caused by the elastic liquid "anticipating" the change in flow conditions at the die exit. Traditional effects such as extrudate swell and "melt fracture" or flow irregularity are manifestations of the

---

*Note that these L/D values are substantially greater than the 4:1 value of the melt indexer. As the L/D becomes large, the polymer begins to achieve (near-) fully developed flow, and the entrance and exit effects become less significant. The smaller diameter of the capillary tube and the larger diameter chamber also aid in minimizing entrance effects.
**The value is "apparent" because it is not possible to measure pressure exactly at the exit of the capillary. The value is thus extrapolated from up-stream pressure readings.

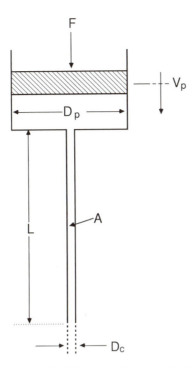

Figure 6.2.   Schematic of Capillary Die Rheometer. Compare the Capillary Length-to-Diameter with That of the Plastometer or Melt Indexer of Figure 6.1. A: Capillary Die, D_c: Capillary Diameter, D_p: Piston Diameter, F: Force, L: Capillary Length, V_p: Piston Velocity.

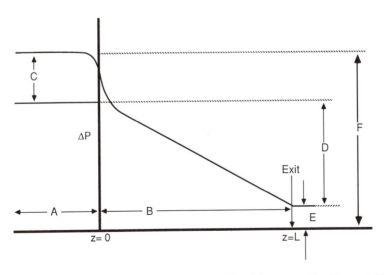

Figure 6.3.   Pressure Profile Along the Axis of a Capillary Rheometer (See Figure 6.2.) (16). A: Reservoir. B: Capillary. C: Entrance Pressure Drop. D: Effective Pressure Drop in Capillary Die. E: Exit Pressure Drop. F: Overall Pressure Drop.

elastic nature of polymers but occur *outside* the capillary pressure field and thus probably are *not* the cause of the higher exit pressure.

Bagley Correction

Bagley (17) proposed that the higher pressure drop in the entrance section of the capillary die could be represented in terms of an entrance length correction. If the reservoir pressure is P_d, the capillary radius, R, and length, L, then the entrance length, e, is related to the shear stress at the wall, τ_w, by:

$$(6.1) \qquad\qquad \tau_w = \frac{P_d}{2[L/R + e]}$$

Bagley notes that the shear stress at the wall was linear with respect to pressure, P_d. He uses a Newtonian expression for shear stress, plotting P_d against L/R with $\dot{\gamma}_{w,Newt} = (4Q/\pi R^3)$ as the parameter (Figure 6.4). The Bagley correction factor, or

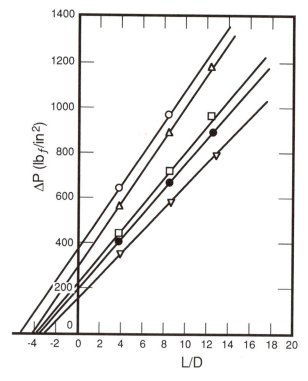

Figure 6.4. Total Pressure Drop vs. Capillary Length-to-Diameter [*L/D*] Ratio for High-Density Polyethylene [HDPE] at Various Shear Rates (20). Open Circle: 723.6 s^{-1}. Triangle: 568.9 s^{-1}. Square: 289.2 s^{-1}. Solid Circle: 234.7 s^{-1}. Inverted Triangle: 131.2 s^{-1}.

Bagley entrance correction, n_B, that represents the L/D correction factor, appears to be a sum of a viscous term and a recoverable shear term (Figure 6.5).

Exit Correction and Normal Stress Difference

Han (18) proposes that the exit pressure effect can be related to the normal stress differences. The expression is:

(6.2)
$$\left(N_1 + \frac{N_2}{2}\right)_w = P_{ec} + \frac{\tau_w}{2}\left(\frac{\partial P_{ec}}{\partial \tau_w}\right)$$

where P_{ec} is extrapolated, non-zero exit pressure, τ_w is the shear stress at the wall, and N_1 and N_2 are the first and second normal stress differences (see Chapter 4). It appears that N_1 and N_2 cannot be independently obtained from exit pressure data. For many non-elastomeric polymers, $N_1 >> N_2$, and so as a first approximation, the exit pressure can be considered to be a measure of the first normal stress difference. For additional information on this point, the reader should consult the advanced texts on rheology (12–14,19).

Hydrostatic Pressure Effect

The hydrostatic pressure can substantially affect the molten polymer density. Increasing pressure is manifested by a tighter packing of the molecules and thus, increased molecular resistance. The apparent viscosity of polycarbonate as a function of hydrostatic pressure is shown in Figure 6.6. In many molding processes, the polymer pressure can

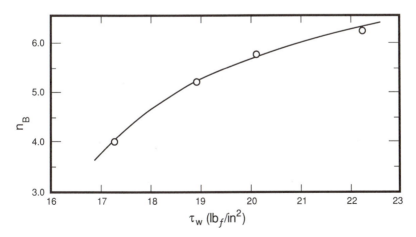

Figure 6.5. Wall Shear-Stress Dependent Bagley Entrance Correction Factor for HDPE at 180°C (21).

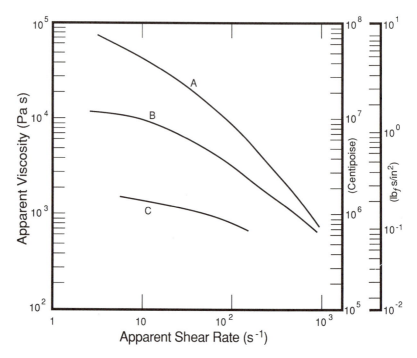

Figure 6.6.   Effect of Applied Pressure on Shear Rate-Dependent Viscosity for Polycarbonate at 180°C (22). Curve A: Pressure at 170 MPa or 25,000 lb_f/in^2. B: 69 MPa or 10,000 lb_f/in^2. C: 1.4 MPa or 2,000 lb_f/in^2.

momentarily reach 10,000 lb_f/in^2 (69 MPa). It is apparent that under these conditions, the polymer viscosity can be 5 to 10 times greater than the value measured at atmospheric pressure.

Interpretation of Capillary Rheometer Data

The shear stress-shear rate curve for nylon 6 is shown in Figure 6.7. This is a typical representation for many thermoplastic molding compounds. The apparent viscosity can be calculated for each data point from:

(6.3)
$$\eta_{apparent} = \frac{\tau}{\dot{\gamma}_{apparent}}$$

The result is shown in Figure 6.8. Polydispersity, molecular weight distribution, branching, mild crosslinking, additives packages, adducts such as fillers and fire retardants and other factors can dramatically affect the shape of the shear stress–shear rate curve. This is seen in Figure 6.9. Figure 6.10 gives a set of apparent viscosity data for a group of nylon 6 polymers having different molecular weights. Note that as the molecular weight of the polymer increases, there is a corresponding increase in the

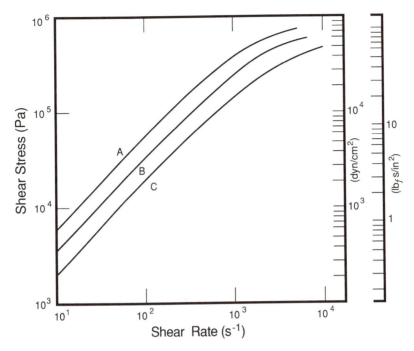

Figure 6.7. Shear Rate-Dependent Shear Stress as a Function of Temperature for Nylon 6 (PA-6) of 34,000 to 40,000 MW (23). Curve A: 232°C. B: 260°C. C: 288°C.

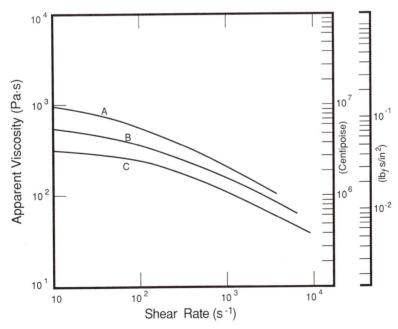

Figure 6.8. Shear-Rate Dependent Viscosity of Nylon 6 (PA-6) of 34,000 to 40,000 MW for Three Temperatures (23). Curve A: 232°C. B: 260°C. C: 288°C.

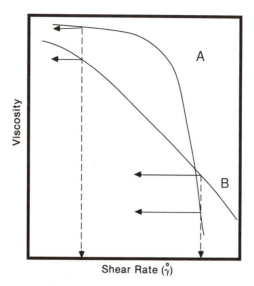

Figure 6.9. The Schematic Effect of Molecular Weight Distribution on Shear Rate-Dependent Viscosity for High-Density Polyethylenes (HDPE) of Similar Melt Indexes (24). Curve A: Narrow MWD. B: Broad MWD.

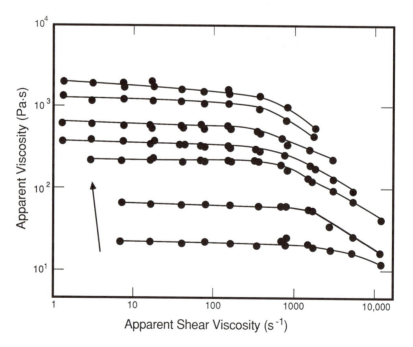

Figure 6.10. Shear Rate-Dependent Viscosity for Nylon 6 (PA-6) of 8,050 to 25,600 MW (25). Arrow Indicates Increasing MW. [Individual MW Values Not Identified in Reference.]

flow viscosity. This is expected, since in Chapter 4 it was noted that polymer viscosity is proportional to molecular weight to the 3.4 power. The viscosities of low-density polyethylenes having different MIs are compared in Figure 6.11. The shapes of the curves are similar. Thus, it can be concluded *for this example only,* that MI is related inversely to the polymer molecular weight. When it is difficult to adequately fill an injection mold with a polyethylene of a given MI, the common practice is to choose one with a higher MI grade. This, in effect, means that a polyethylene with a lower molecular weight is being used. The significance of the change in product performance that results from this substitution depends strongly on the design conditions. This is elaborated on in Section 6.5, on creep testing.

A similar effect can be seen with fillers. As the filler content is increased, the apparent viscosity also increases (Figure 6.12). One stated reason for using fillers is product cost reduction. The result of this change will undoubtedly be a change in mechanical properties. The increased polymer viscosity may also cause a reduction in the ability to successfully mold the product.

The capillary rheometer is also an excellent diagnostic tool. All thermoplastic polymers degrade at elevated temperatures. Some are also susceptible to environmental effects such as water. A classic example of the latter is thermoplastic polyester, polybutylene terephthalate (PBT). The relative stability of any polymer is related to the time-dependence of its molecular weight at the desired environmental conditions. Since

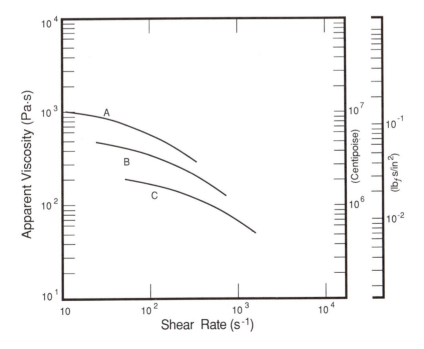

Figure 6.11.   Shear Rate-Dependent Viscosity for Three HDPEs at 210°C (26). Curve A: Melt Index (MI) = 2. B: MI = 7. C: MI = 20.

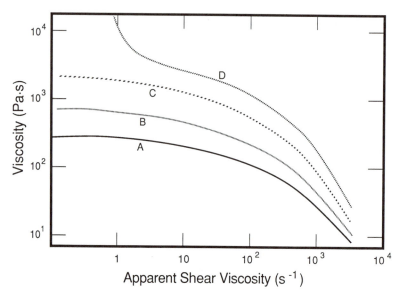

Figure 6.12.   The Shear Rate-Dependent Viscosity of Neat and Carbon-Filled Nylon. Curve A: Filler Content: 0% (wt). B: 10%. C: 20%. D: 40%.

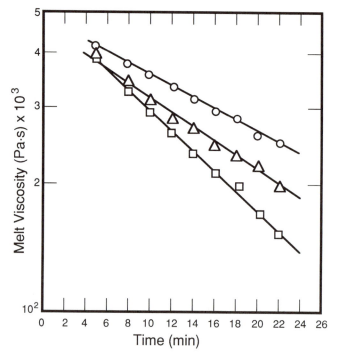

Figure 6.13.   The Effect of Moisture on the Time-Dependent Melt Viscosity of Glass-Reinforced Polybutylene Terephthalate (PBT) at 260°C and 147 s^{-1} Shear Rate (28). Circle: 0.02% (wt) Moisture. Triangle: 0.08% (wt) Moisture. Square: 0.28% (wt) Moisture.

viscosity is proportional to the molecular weight to the 3.4-power, small changes in molecular weight can appear as substantial changes in viscosity as determined by simple rheometry.

As an example, the shear viscosity of PBT as a function of residence time and moisture content is shown in Figure 6.13. It is apparent that heroic measures must be taken to dry polyesters, as well as polycarbonates, nylons and other condensation polymers, prior to molding or extrusion. Note also the apparent effect of holding the resin at processing conditions during a momentary shutdown. If the reason for the interruption cannot be resolved in a very few minutes, the molten resin should be purged from the equipment and discarded. The engineer must keep in mind that an interruption may cause abnormal results in test specimens taken from parts fabricated immediately after the interruption. Certainly effects such as color change can be quite apparent. Other effects, such as impact strength and ductility, may be masked, with the result that a small portion of the test results may be highly biased.

Example 6.2

Q: Determine the relative decrease in molecular weight after 10 minutes at 260°C for polybutylene terephthalate having 0.02% moisture content. Use Figure 6.13.

A: The interrelationship between viscosity and weight average molecular weight can be written as:

$$\frac{\eta}{\eta_o} = \left(\frac{M_w}{M_{w,o}}\right)^{3.4}$$

At t = 0 min, η_o = 4.9 Pa s.
At t = 20 min, η = 2.6 Pa s.

Therefore, the change in weight average molecular weight is given as:

$$M_w = M_{w,o} \left(\frac{2.6}{4.9}\right)^{1/3.4} = 0.83\ M_{w,o}.$$

Or the PBT resin has lost 17% of its initial molecular weight in 10 minutes at 260°C.

The rheometer can also be used to diagnose injection molding processing problems. If an injection mold cannot be filled when using a particular polymer at its recommended processing temperature, polymer temperature is usually the first processing parameter to be adjusted. This change usually meets with good success with polystyrene and PMMA, but not with polyolefins such as HDPE and PP. The reason for this can be seen in Figure 6.14, where apparent viscosity is presented at constant shear rate, 1000 s^{-1}. Note that PS and PMMA viscosities are strongly sensitive to temperature whereas that for HDPE is not. This was also pointed out in the Chapter 4 discussion of the viscous energy of activation for polymers.

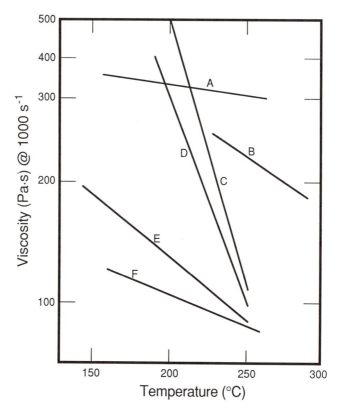

Figure 6.14.   Temperature-Dependent Viscosities for Several Polymers at 1000 s^{-1} Shear Rate (29). Curve A: HDPE. B: PA-6. C: PMMA. D: PS. E: LDPE. F: PP.

Cone-and-Plate Viscometer

The capillary die viscometer is for the most part a viscometric flow device. The cone-and-plate viscometer is also a viscometric flow device, based on a rotational principle. As seen in schematic in Figure 6.15, the polymer is held between a stationary flat plate and a rotating cone of angle θ_c. The cone-and-plate geometry is superior to parallel plate geometry in that the shear rate across the plate is uniform. As a result, viscometric material parameters are obtained directly, without data differentiation. The major operating limitation is elastic liquid distortion in the gap. At high speed, material can be physically thrown from the gap. The typical shear rate range is 0.01 to 10 s^{-1} for a cone angle of 2 to 8° or so. It can be shown that in spherical coordinates (r, radial direction along the cone, ϕ, the meridional angle, and θ, the azimuthal angle), assuming negligible inertial forces and edge effects:

(6.4)
$$\tau_{\phi\theta} = \tau_{12} = \frac{c_1}{\sin^2 \theta}$$

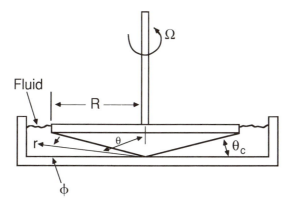

Figure 6.15.   Schematic of Cone-and-Plate Rheometer (30).

where c_1 is a constant of integration and θ is the angle in the gap between the plate and the cone. The torque, T_R, on the surface of the cone is given as:

$$(6.5) \qquad T_R = 2\pi \int_0^R (\tau_{\phi\theta})_c \, (r\cos\theta_c)^2 \, dr = 2\pi R^3 \cos^2\theta_c \, \frac{(\tau_{\phi\theta})_c}{3}$$

R is the radius of the cone. Now if $\theta_c < 2°$ or so, $\cos\theta_c \approx 1$ and $\sin\theta_c \approx \theta_c$, so:

$$(6.6) \qquad \tau_{\phi\theta} = \tau_{12} = \frac{3 \, T_R}{2 \, \pi R^3}$$

Note that, as assumed earlier, $\tau_{\phi\theta}$ is independent of angle if θ_c is small. The shear rate can be shown to be a function of angular velocity, v_ϕ, by:

$$(6.7) \qquad \dot\gamma \approx \left(\frac{1}{r}\right) \frac{\partial v_\phi}{\partial\theta},$$

for small values of ϕ. This can be integrated to give:

$$(6.8) \qquad v_\phi = r\dot\gamma \left(\theta - \frac{\pi}{2}\right)$$

If the angular velocity is ω when $\theta = \theta_c$, then:

$$(6.9) \qquad \dot\gamma = -\frac{\omega}{\theta_c}$$

Example 6.3

Q: A cone-and-plate rheometer with a 50 mm diameter cone having a 4° angle is used to measure the viscosity of a polymeric liquid. When operating at 1 RPM, a torque of 0.5 N m is measured. What is the viscosity of the polymer?

A: The relationship between shear stress and torque is given as:

$$T_R = \frac{2\,\pi R^3 \cos^2(\theta_c)\,(\tau_{\phi\theta})_c}{3}.$$

Now $\cos^2(4°) = 0.995$. Thus $\cos^2(\theta) \approx 1$. Simplifying and rearranging:

$$\tau_{\phi\theta} = \frac{3\,T_R}{2\,\pi R^3} = \frac{3 \times 0.5}{[2\,\pi(0.025)^3]},$$

$$= 15{,}000 \text{ Pa.}$$

$$-\dot{\gamma} = \frac{\omega}{\theta_c} = \frac{(2\,\pi/60)}{(4 \times 2\,\pi/360)}$$

$$= 1.5 \text{ s}^{-1}.$$

Thus, viscosity is given as:

$$\eta = \frac{\tau_{\phi\theta}}{\dot{\gamma}} = \frac{15{,}000 \text{ Pa}}{1.5 \text{ s}^{-1}} = 10{,}000 \text{ Pa} \cdot \text{s.}$$

In other words, the shear rate is uniquely a function of the angular velocity and the shear stress is uniquely a function of the torque on the rotating member. This allows the viscosity parameter, $\eta(\dot{\gamma})$, to be uniquely determined.

By employing pressure transducers in the stationary wall, normal stress differences can also be determined. It has been shown (31,32) that if F' is the net thrust measured on the stationary plate, over and above that due simply to ambient pressure, then the first normal stress difference is found from:

(6.10)
$$\tau_{\phi\phi} - \tau_{\theta\theta} = \tau_{11} - \tau_{22} = \frac{2F'}{\pi R^2}$$

The second normal stress difference is determined by measuring, $S_{\theta\theta}$, the total normal stress via pressure transducers at various positions along the stationary surface:

(6.11)
$$\frac{d\,S_{\theta\theta}}{d\ln r} = N$$

where N is a constant. It can be shown that the second normal stress difference is obtained from:

$$(6.12) \qquad \tau_{\theta\theta} - t_{rr} = \tau_{22} - \tau_{33} = \frac{[N - (\tau_{11} - \tau_{22})]}{2}$$

The second normal stress difference, $\tau_{22} - \tau_{33}$, is considered to be substantially smaller than the first normal stress difference and there is some dispute as to whether it is positive or negative. Again the reader is referred to more advanced texts for additional information.

Dealy (32) notes that cone-and-plate viscometry is only approximately controllable in that to yield unique rheological values, certain approximations are needed. Substantial error can be introduced if these approximations are invalid. Measurement errors are normally due to viscous dissipation, edge effects, mechanical effects such as bearing noise and uneven angular velocity and shear instability.

The Ram Follower for Thermosetting Resins

The viscosity functional relationship of a thermosetting resin is much more complex than a thermoplastic polymer. It is a function of temperature, shear rate and time. The elements in the initial resin system are usually low molecular weight components designed to chemically react at elevated temperatures, perhaps in the presence of accelerators and catalysts. Initially, the higher molecular weight reactants can be linear, but eventually three-dimensional structures are formed, theoretically yielding a few interpenetrating highly crosslinked, three-dimensional macromolecules. In the final state, the reaction produces a material that simply cannot flow. That is, its viscosity is infinite. Some cone-and-plate rheometers can be used to measure time-dependent thermoset viscosity, at least until slippage occurs between the hardening resin and the metal plates.

The ram follower molder is the thermosetting resin equivalent to the thermoplastic polymer melt indexer. The device is an instrumented plunger in which the position of the plunger is determined as a function of time. The plunger is mounted on a standard ASTM D3123 spiral mold. The plunger position is therefore related to the distance traveled by the resin as it fills the mold. Typical time-dependent distance curves are shown in Figure 6.16. Short distance values correspond to relatively high viscosity compression molding resins. Typical general purpose transfer molding resins have moderate distance values. The relative distance value is influenced by the filler content and type and the reactivity of the resin, either by chemical adjustment or by molding temperature. Long distance values are typical of flow characteristics of high flow, low viscosity encapsulation grade resins used primarily in large multicavity electronic semiconductor transfer molding operations. By altering the chemical initiator and fillers, these resins can be designed to have very low viscosities and significantly retarded crosslinking reactions. Prior to the development of the ram follower device, thermoset

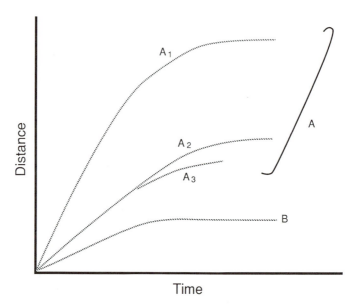

Figure 6.16. Typical Ram-Follower Thermosetting Resin Flow Curves. Curve A_1: Encapsulation Grade Transfer Molding Grade Thermosetting Resin. A_2: High Flow Transfer Molding Grade. A_3: Low Flow Transfer Molding Grade. B: Compression Molding Grade.

resin molders simply depended on the final spiral flow length, the y-axis height in Figure 6.16. However, two resins having the same final spiral flow length could exhibit significantly different cavity filling characteristics. Now the effects of modifications to such elements as chain extension mechanics and catalyst systems can be more easily seen, even though not fully quantified.

Example 6.4

Q: For the ram follower method, develop an empirical relationship to characterize the flow characteristics of a reactive resin.

A: Typical flow curves exhibit an exponential characteristic, asymptotically approaching the final spiral flow length, L_0. This can be written as:

$$\frac{L}{L_0} = 1 - e^{-\beta t}$$

The value of β can be obtained from:

$$1 - \frac{L}{L_0} = e^{-\beta t}$$

$$\ln\left(\frac{\Delta L}{L_0}\right) = -\beta t$$

or:

$$y = -mx$$

The slope of a straight line fitted through a semi-logarithmic plot of the time-dependent variable, written as $\Delta L/L_o$, yields β, the asymptote constant.

6.3 Solid Material Properties: Moderate-Time Measurements

As noted, solid polymer material response to long-term or creep loading can be substantially different from response to short-term or impact loading. Any screening procedure must, of course, include tests in the appropriate time frame. There are many more design applications where polymeric material response at moderate times is needed. Moderate-time tests generate static data such as tensile, flexural, and compressive and hardness properties, and dynamic data such as complex modulus. Not only are these data required for most polymeric applications, but they are frequently used to screen candidate polymers for short- and long-term applications. If a polymer property is unsatisfactory at moderate times, it will probably be unsatisfactory at extreme time conditions.

The *tensile test* is one of the most widely used mechanical property tests. In this test, the prepared specimen is uniaxially loaded and the uniaxial deformation as a result of this load is measured. The test procedure was originally developed for metals. The description and explanation of the technique that follows is based on an example using test data taken on a mild steel specimen, a Hookean material. The test specimen can be a section of wire or rod or a specially fabricated specimen cut or machined from sheet or barstock (Figure 6.17). Other standard geometries are specified in the ASTM test specification, ASTM E8.

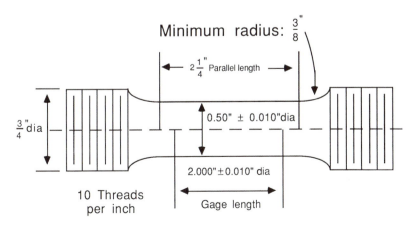

Figure 6.17.   Metal Test Specimen Geometry for ASTM E8 Tensile Test.

The standardized geometry allows quantitative comparison of data from various metals. Extreme care must be taken in test specimen preparation to minimize residual stresses and surface irregularities caused by machining. Such problems can adversely affect the test results. The test specimen is fastened into a specially designed grip that minimizes slip between the grip and the specimen during loading. The gripping system must be self-aligning and must not transmit torque to the specimen. Self-aligning grips for threaded cylindrical test specimens are shown in Figure 6.18.

The device that applies the prescribed loading characteristics to the specimen is a tensile test machine. The device can impart a controlled velocity to the movable platen containing one grip while the platen containing the other grip remains stationary. The two most common platen drive systems are the intermeshing mechanical screw and the hydraulic ram.

Between one grip and its corresponding platen is a load cell or force transducer. Thus, during loading, the applied force is transmitted to the specimen through the load cell. The load cell produces an output signal that is proportional to the applied force.

The deformation of the test specimen is measured either by an extensiometer attached to the specimen edges or by an extensiometer that measures the separation of the platens or grips. Direct contact with the test specimen is preferred for most materials. If grip separation is used as a measure of elongation, it is assumed that "grip slip" is negligible and that the sample is elongating uniformly. Under small strains, the former assumption may result in serious error. For necking-prone materials, the latter may result in substantial inaccuracies in interpretation. Typically, the sample extensiometer may be a

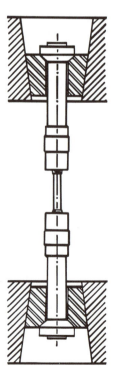

Figure 6.18.   Schematic of Self-Aligning Grip Assembly for ASTM E8 Tensile Test with Threaded Metal Test Specimen.

simple dial gage for visual measurement, an electronic transducer or LVDT that produces an output signal proportional to the displacement ends of the gage, or for very high-modulus, low elongation-to-break materials, strain gages attached directly to the test specimen. In newer systems, the signals from the load cell and the extensiometer are displaced on an XY plotter or computer CRT with the load on the Y axis and the displacement on the X axis. A typical set of tensile load–deflection data for a mild steel cylindrical test specimen is tabulated in columns 1 and 2 of Table 6.2 and graph-

Table 6.2. Experimental data for tensile test of mild steel specimen: diameter, 0.502 in (12.75 mm), gage length, 8 in (200 mm) (33).

Force (lb$_f$)	Engineering stress, F/A_o (lb$_f$/in^2)	True Stress Least dia (in)	True Stress Area (in^2)	True Stress F/A (lb$_f$/in^2)	Elongation in gage length (in)	Engineering strain, $\Delta L/L$ (in/in)	True strain, in L/L_o (in/in)
1,100	5,500	0.5020	0.1979	5,550	0.001	0.000125	0.000125
1,950	9,850	0.5015	0.1975	9,870	0.002	0.000250	0.000250
2,570	12,480	0.5012	0.1973	13,020	0.003	0.000375	0.000375
3,230	16,310	0.5012	0.1973	16,370	0.004	0.000500	0.000500
3,800	19,200	0.5012	0.1973	19,240	0.005	0.000625	0.000625
4,600	23,200	0.5012	0.1973	23,300	0.006	0.000750	0.000750
5,300	26,800	0.5012	0.1973	26,820	0.007	0.000875	0.000875
6,150	31,050	0.5012	0.1973	31,170	0.008	0.001000	0.000999
6,600	33,300	0.5012	0.1973	33,420	0.009	0.001125	0.001124
7,300	36,850	0.5012	0.1973	37,000	0.010	0.001250	0.001249
7,600	38,400	0.5012	0.1973	38,500	0.011	0.001375	0.001374
7,500	37,850	0.5012	0.1973	38,000	0.012	0.001500	0.001499
7,450	37,600	0.5012	0.1973	37,700	0.0135	0.001688	0.001686
7,560	38,200	0.5012	0.1973	38,300	0.014	0.001750	0.001748
7,600	38,400	0.5011	0.1972	38,500	0.015	0.001875	0.001873
7,780	39,300	0.5010	0.1971	39,420	0.016	0.002000	0.001997
7,700	38,900	0.4990	0.1956	39,350	0.0505	0.006310	0.006292
8,100	40,900	0.4965	0.1936	41,800	0.100	0.012500	0.012422
8,000	40,400	0.4960	0.1932	41,400	0.150	0.018750	0.018576
8,700	43,900	0.4960	0.1932	45,000	0.200	0.025000	0.024693
9,300	46,900	0.4940	0.1917	48,500	0.250	0.031250	0.030772
9,860	49,800	0.4925	0.1909	51,700	0.300	0.037500	0.036814
10,330	52,200	0.4900	0.1886	54,800	0.360	0.045000	0.044017
10,980	55,400	0.4870	0.1863	58,800	0.460	0.057500	0.055908
11,850	59,800	0.4810	0.1817	65,200	0.660	0.082500	0.079273
12,340	62,300	0.4740	0.1765	69,800	0.860	0.107500	0.102105
12,450	62,900	0.4660	0.1706	73,000	1.060	0.132500	0.124427
12,620	63,700	0.4570	0.1640	76,900	1.260	0.157500	0.146263
12,670	63,900	0.4390	0.1514	83,600	1.460	0.182500	0.167631
9,980	50,400	–	–	–	1.660	0.207500	0.188552
9,840	49,700	0.3350	0.0881	111,500	1.670	0.208750	0.189587

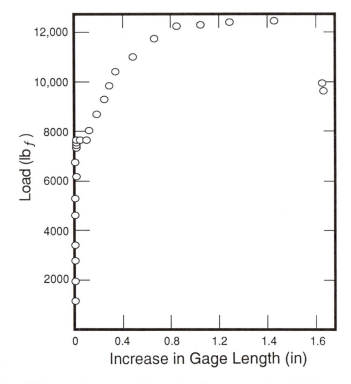

Figure 6.19. Differential Gage-Length-Dependent Tensile Load for Mild Steel According To ASTM E8. Cylindrical Test Specimen 0.502 in Diameter by 8 in Initial Gage Length. See Data in Table 6.2.

ically displayed in Figure 6.19. It must be remembered that steel below its proportional limit is a Hookean material and that, as expected, the shape of the load-deflection curve is independent of the rate of loading.

As noted above, tensile data are presented in engineering stress and strain. The engineering stress was defined as the force per unit area of unstrained specimen. The true stress is the force per minimum unit area of the specimen under load. The diameter of the test specimen decreases under increasing load in accordance with Poisson's ratio even if the specimen does not neck. As a result, the effective area under load *always* decreases with increasing load. Thus the true stress is always equal to or greater than the engineering stress. Similarly, the engineering deformation or strain of the sample is the increase in length per initial specimen length. The true strain is the natural logarithm of the ratio of the overall length under load to the initial length. The engineering and true stress-strain curves for the mild steel specimen of Table 6.2 are shown in Figure 6.20. The material property details observed from the shape of the stress-strain curve for polymers were discussed in Chapter 3 and are not repeated here. The important transitions in Figure 6.20 are:

A, the *proportional limit*, below which the material is Hookean elastic,

B, the *elastic limit*, above which the material retains a permanent set upon load removal,

C, the *yield point*, at which plastic flow or deformation begins,

D, the *ultimate stress*, or the maximum strength of the material,

E, the *rupture stress* or the strength of the material at rupture. The turning down in the curve is seen for many materials that exhibit substantial necking or local plastic flow prior to fracture.

The best measure of the toughness of a material is the total work done to break the specimen. *Modulus of toughness*, with units of *in lb$_f$* or *Pa m^3*, is obtained from the area under the stress-strain curve. The modulus of the material is obtained in one of four ways:

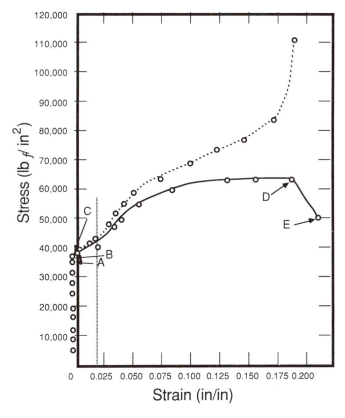

Figure 6.20. Tensile Stress–Strain for Mild Steel Specimen of Figure 6.19 and Table 6.2. Dashed Curve: True Stress–Strain Curve. Solid Curve: Engineering Stress–Strain. Vertical Line: 2% Strain Offset Line. Point A: Proportional Limit. B: Elastic Limit. C: Yield Point. D: Ultimate Strength. E: Breaking Stress.

- The initial tangent to the stress-strain curve, that is, a zero-strain tangent,
- The tangent to the stress-strain curve at some predetermined stress level,
- The secant of the stress-strain curve between the origin and some predetermined stress level,
- The chord of the stress-strain curve between any two predetermined stress levels.

These are seen in Figure 6.21. For high-modulus materials, including many polymers, the tangent modulus is most often used. For very low modulus, rubbery materials, including polymers and rubbers, the secant modulus is most often used. All tensile modulus data should state the method, however.

These mechanical property tensile tests and the reduction of the test data have been developed and perfected over the years. Similar tests for other mechanical properties such as shear modulus, bulk modulus, Poisson's ratio and impact strength have also been developed.

It is out of these ASTM metals testing programs that testing programs for mechanical properties of polymers have been developed. The significance of the results obtained for polymers depends on the criterion:

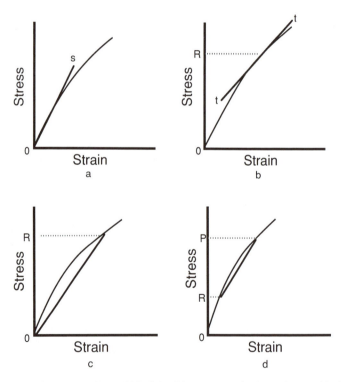

Figure 6.21.   Four Representations of Modulus Measurement in Accordance with ASTM E111 Standard. Curve a: Initial Tangent Modulus. b: Tangent Modulus at Stress R. c: Secant Modulus Between Origin and Stress R. d: Chord Modulus, Between Any Two Stresses, P and R.

> **"Did the experiment include all the variables which directly or indirectly interact with the observed phenomenon?"**

The mechanical property charts for metals were developed to measure Hookean characteristics. Polymers are viscoelastic or time-dependent materials. Many of the tests developed for metals do not yield significant design data or measure true material properties for polymers. Thus each test must be evaluated in light of the particular characteristics of polymers. For example, early ASTM tests for polymers did not include specific details about conditioning, that is, preparing the polymer test sample in the same atmospheric environment every time. Some polymers such as polyamides (nylons) are dramatically softened by atmospheric moisture. As-molded nylon is substantially stiffer and less ductile than nylon that has been conditioned for 40 hours at 50% RH and 23°C.*

As noted, various mechanical property tests for polymers can be categorized in terms of the material response to applied stress. Turner shows these responses as a series of rings about the polymer constitutive equations of state (Figure 6.22):

$$(6.13) \qquad\qquad \varepsilon = \varepsilon(\sigma, \dot{\sigma}; t, T)$$

or:

$$(6.14) \qquad\qquad \sigma = \sigma(\varepsilon, \dot{\varepsilon}; t, T)$$

where ε is the strain and $\dot{\varepsilon}$ is the rate of strain resulting from applied stress, σ, and rate of applied stress, $\dot{\sigma}$. Certainly the test time duration, t, and the environmental temperature, T, can directly affect the constitutive equation of state. Note also that the mechanical properties in the inside rings do not include non-mechanical interactions such as ultraviolet effects and chemical embrittlement. These effects are usually included as secondary interactions, even though these effects may actually dominate material performance. These effects are treated in Section 6.7, on environmental effects.

Moderate-time-duration tests are those that take minutes rather than milliseconds, as with impact tests, or weeks, as with creep tests. For a Hookean material such as low carbon steel, moderate-time duration tests can be used to measure the *design properties* of the material such as Young's modulus, shear modulus, yield strength, ultimate strength and Poisson's ratio. Regardless of the duration of the test, the Hookean material properties remain nominally time-independent. This is not so for polymers. For example, see the tensile stress-strain data for PMMA in Figure 6.23. The response of the test specimen to applied stress is a function of the speed of application of the stress. From data analysis, it is apparent that each testing speed yields a different value for

---

*At the same time, note that conditioning may in fact yield a polymer that is substantially different from that used in the end product. This, in effect, sustains the gulf between tested polymer properties and end-product performance rather than helping bridge it.

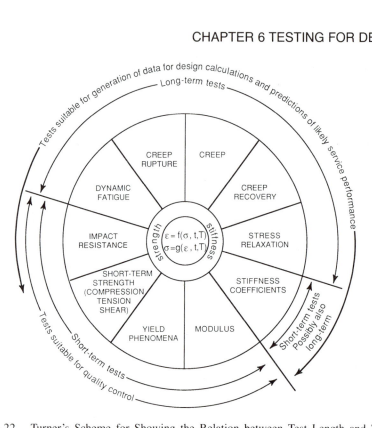

Figure 6.22. Turner's Scheme for Showing the Relation between Test Length and Typical Polymer Physical Properties (34).

Young's modulus. Since a material property should be invariant with time, it is obvious that the test is not yielding a material property. If the moderate-time duration tests do not measure the *design* or *material* properties of a polymer, of what practical use are they? Chastain (36) addresses the question this way:

> The ASTM tests for plastics were designed as quick, easy reproducible tests to provide a rough comparison among similar materials—but principally for material quality control, not for material performance. They provide benchmarks to help a user determine whether the plastic resin he buys today will have the same general processing and service properties of the batch that he bought last month or last year. Thus, these test data should be regarded only as material specifications for purchasing and quality control, not as indications of performance under long-time real-life conditions.

> Performance tests, to be valid, would have to incorporate so many variables that there would eventually be almost as many tests as there are molded parts. These variables not only involve configuration of the part and different formulations of resin compounds, but also the complex parameters of molding, such as melt temperature, mold temperature, time in the mold, pressure, flow path, gate configuration, and other factors that affect service performance. The number of possible permutations is endless.

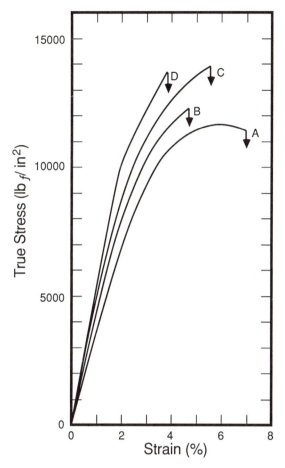

Figure 6.23. Effect of Speed of Testing on the 25°C Tensile Stress–Strain Behavior of Polymethyl Methacrylate (PMMA) (35). Curve A: Crosshead Rate 0.02 in/min. B: 0.08 in/min. C:
0.32 in/min. D: 1.28 in/min.

In other words, most ASTM tests yield *non-design* properties for polymers. To mas-
querade these as meeting the above-mentioned *criteria of test acceptability* is to label
them as Mooney hybrids. To understand the limitations and utility of several standard
moderate-time-duration ASTM tests, a brief description of each follows.

Tensile Test: ASTM D638

This test is an adaptation of the metals-based ASTM E8 test described above. For
viscoelastic polymers, additional test parameters and preconditioning of the specimen
must be specified. The test specimens must have rectangular cross-sectional geometry.

They may be molded, machined or die cut. Table 6.3 presents several specimen geometries.

Recall that the molecular state can dominate the test results. For example, the degree of crystallinity, the nature of the crystalline structure, and the degree of crystallite orientation in a crystalline polymer depend on mold and melt temperatures, shear his-

Table 6.3. ASTM D638 standard tensile test specimen geometry.

| | Specimen dimensions for thickness, T (mm) | | | | | |
| | ≤ 7 | | $7-14$ | ≤ 4 | | |
Dimensions (mm)	Type I	Type II	Type III	Type IV	Type V	Tolerances
W, width of narrow section	13	6	19	6	3.18	±0.5
L, length of narrow section	57	57	57	33	9.53	±0.5
WO, overall width, min.	19	19	29	19	–	+6.4
WO, overall width, min.	–	–	–	–	9.53	+3.18
LO, overall length, min.	165	183	246	115	63.5	No max
G, gage length	50	50	50	–	7.62	±0.25
G, gage length	–	–	–	25	–	±0.13
D, distance between grips	115	135	115	64	25.4	±5
R, radius of fillet	76	76	76	14	12.7	±1
RO, outer radius (type IV)	–	–	–	25	–	±1

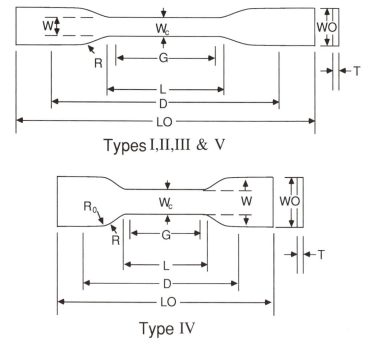

Types I, II, III & V

Type IV

tory, and other molding variables. The flow field and shear history dominate the morphological features of an amorphous polymer. Substantial test differences can therefore result between injection molded test specimens and specimens cut from extruded or compression molded sheet.

As noted above, certain polymers are very hydroscopic. Since the water molecules act as plasticizers, polymers must be preconditioned in accordance with ASTM D618 for 40 hours at 50% RH and 23 $\pm$ 2°C. This preconditioning helps minimize variation in properties due to moisture pickup.

As with the metal tensile specimen, the polymeric specimen is mounted in the grips of a universal tensile tester. The rate of movement of the moving platen depends on the geometry of the test specimen and the rigidity of the polymer under test (Table 6.4). For polymers, the speed of testing must be noted in all test results. The measured stress-strain levels are reported in engineering terms, based on the original dimension of the test specimen. As noted above, the measure of test specimen rigidity in tension can be reported *either* as zero-strain tangent modulus *or* as a secant modulus at a specific strain, usually 1% or 2%. A minimum of five specimens must be tested and the results averaged. The similar tests in Europe are BS 2782 method 301 in England and ISO R527 on the continent. Representative engineering stress-strain diagrams for several unfilled polymers are shown in Figure 6.24.

To illustrate the viscoelastic nature of a polymer, consider the effect of the rate of strain, that is, the crosshead speed, on the stress in a polymer. The effect of time-dependent strain on the stress of impact polystyrene is seen in Figure 6.25. The shapes of the curves illustrate the very strongly rate-dependent nature of the material. At lower strain rates, the material has sufficient time to plastically flow or relax the induced stress. The material thus appears as a ductile polymer. At higher extension rates, the material cannot plastically respond rapidly enough and so it appears as a brittle polymer. In Figure 6.26, the effects of strain rate on typical "design" properties are shown for a general-purpose polystyrene and a rubber-modified impact polystyrene. It is apparent that the values of the four properties—Young's modulus, breaking strength, maximum load, and breaking energy or toughness—are strongly dependent on the testing speed. Note that all impact polystyrene properties vary with rate, whereas only the general-purpose polystyrene modulus is highly rate-dependent. As a result, it is imperative that the rate of testing or crosshead speed be reported whenever "design properties" are reported. The strain rate, in *percent/unit time,* is a preferred variable, allowing specimens of different size to be compared without dimension. The tensile strength and strain at break responses to strain rate for four unfilled polymers are shown in Figure 6.27.

Temperature can also change the nature of the polymer's response to applied load. Increasing the temperature of a polymer usually increases its molecular mobility and reduces the secondary or van der Waals bonding forces that hold the molecules together. This increase in intermolecular mobility usually results in greater apparent polymer ductility and reduced ultimate stress. This phenomenon is seen in Figure 6.28 for amorphous polymethyl methacrylate. Note that the glass transition temperature of PMMA is 105°C. As the testing temperature approaches T_g, the polymer exhibits increasing rubber-like characteristics. Below about 40°C, the polymer appears brittle. Above, it has increasing ductility.

Table 6.4. ASTM D638 standard universal tensile testing machine speeds for various polymers.

Polymer	Specimen type	Grip separation speed	
		(mm/min)	(in/min)
Acrylic (PMMA)	Type I	5.0	0.2
ABS	Type I	5.0	0.2

Cellulosic:
    Acetate,
    Propionate,
    Butyrate
	Type I	5.0	0.2

Phenolic	Type I	5.0	0.2
Nylon (PA)	Type I and Type II	50.0	2.0
Polybutylene (PIB)	Type IV	500.0	20.0
Polycarbonate (PC)	Type I	50.0	2.0
Polyethylene:			
Type I	Type IV	500.0	20.0
Types II, III, III	Type IV	50.0	2.0
Polypropylene	Type I	50.0	2.0
Polystyrene	Type I	5.0	0.2
Polyvinyl chloride:			
Flexible (FPVC)	Type IV	500.0	20.0
Rigid (RPVC)	Type I	5.0	0.2

Reinforced and laminated
    thermoplastics and
    thermosets
	Type II	5.0	0.2

Note: According to ASTM D638, the following conditions should be met in all tensile testing operations:

The speed of testing shall be chosen from one of the following:
Speed A — 1.0 mm (0.05 in) per minute, ±25%
Speed B — 5.0 mm (0.2 in) per minute, ±25%
Speed C — 50 mm (2.0 in) per minute, ±10%
Speed D — 500 mm (20 in) per minute, ±10%

The speed of testing shall be determined by the specification for the material being tested, or by agreement between those concerned. when the speed is not specified, then for specimen Types I through IV, Speed B shall be used for rigid and semirigid plastics and Speed C shall be used for nonrigid plastics; for Type V specimens, Speed A shall be used for rigid and semirigid plastics . . . *Modulus determinations may be made at the speed selected for other tensile properties.*
    *as defined in ASTM D883.

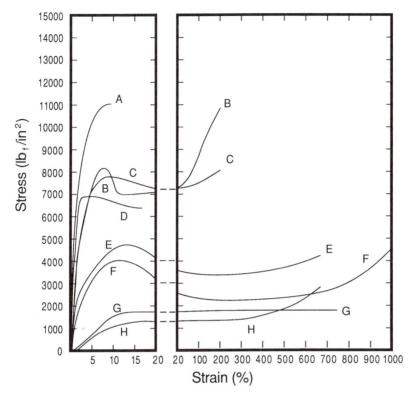

Figure 6.24.   Room-Temperature Tensile Stress–Strain Curves for Several Polymers (37).
Curve A: PMMA. B: PA-6. C: PC. D: ABS. E: PP. F: HDPE. G: PUR (Thermoplastic or Linear
Polyurethane). H: LDPE.

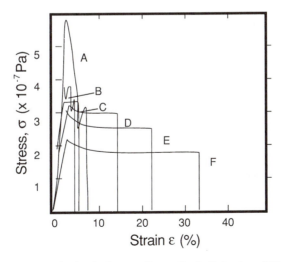

Figure 6.25.   Effect of Tensile Strain Rate on Stress–Strain Behavior of High-Impact Polystyrene (HIPS) at 21°C (38). Curve A: 16.0 m/s. B: 5.3 m/s. C: 1.3 m/s. D: 0.028 m/s. E: 0.0028
m/s. F: 0.00028 m/s.

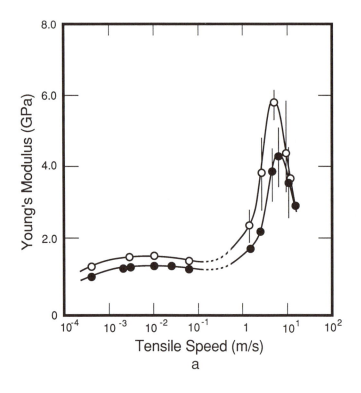

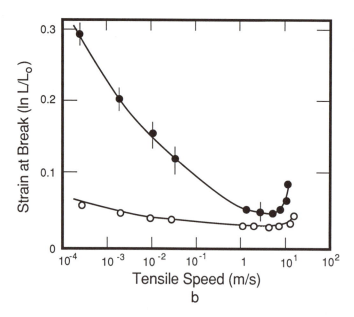

Figure 6.26. Effect of Tensile Speed on Moderate-Time Physical Properties of General Purpose PS (GPS, Open Circles) and Impact PS (HIPS, Closed Circles) at 21°C (38). a: Young's Modulus, E. b: Breaking Strain, as $\ln (L/L_o)$.

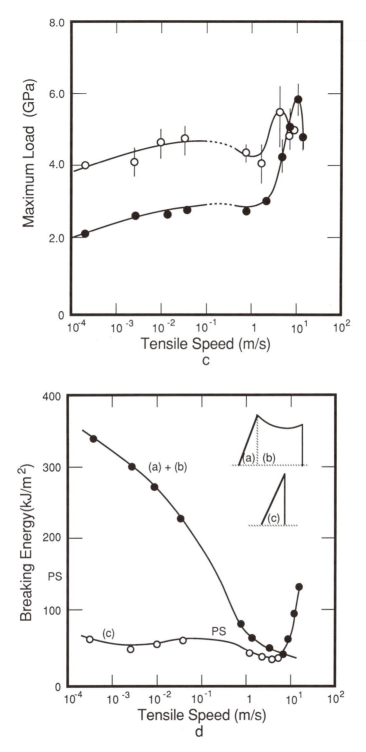

Figure 6.26 *Continued*. c: Maximum Load. d: Breaking Energy. [The Breaking Energy or Modulus of Toughness of HIPS is the Sum of the Linear Portion and the Ductile Yielding Portion of the Load–Displacement Curve.]

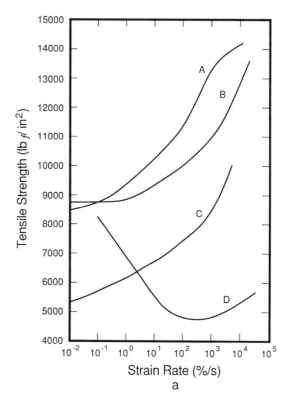

Figure 6.27.   Strain-Rate Dependent Tensile Strength (a) Curve A: PVC. B: PC. C: ABS. D: Thermoplastic Polyurethane (PUR) and Ultimate Strain;

Not all polymers that are brittle at room temperature exhibit a ductile–brittle transition. Other phenomena may substantially alter the stress–strain response. *Stress crazing* is an example of such an effect. Crazing is the initiation and growth of fine crack-like imperfections in the polymer. This effect is usually associated with amorphous polymers such as general-purpose polystyrene but it can also occur in crystalline polymers. Polyethylene is a crystalline material that stress crazes. The resistance to stress crazing can be directly related to the molecular weight of the polymer. The effect of stress crazing is seen in the unusual response of one form of general-purpose polystyrene (GPPS) to applied load (Figure 6.29). It appears that each craze-prone polymer has a temperature-dependent critical stress level. Above this stress level, the polymer crazes. Associated with this stress level is an induction time, after which the crazing is initiated. At temperatures below 80°C, the GPPS in Figure 6.29 fails by stress crazing. Both the induction time and the critical stress level decrease with increasing temperature. Thus stress crazing occurs *earlier* with increasing temperature. Once the crazes have initiated, they multiply and propagate rapidly. The part fails rapidly in a brittle nature. However, as the testing temperature approaches the lower glass transition temperature, $T_g = 100°C$, the polymer behavior changes rapidly to a ductile response.

As noted in Section 2.5, the conditions that influence transition from the glassy state to the rubbery one include molecular weight distribution and the nature of the additives.

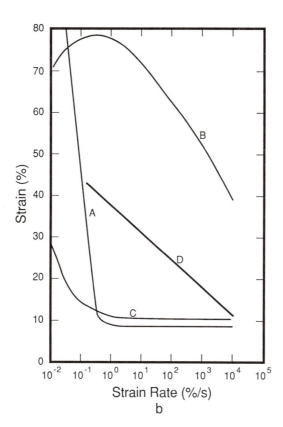

Figure 6.27 *Continued.* (b) for Several Unfilled Polymers (37). Curve A: PVC. B: PC. C: ABS. D: Thermoplastic Polyurethane (PUR)

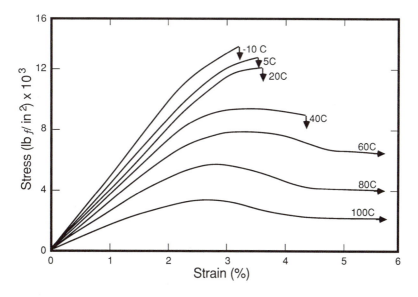

Figure 6.28. Effect of Temperature on the Stress–Strain Behavior of Polymethyl Methacrylate (PMMA) (39).

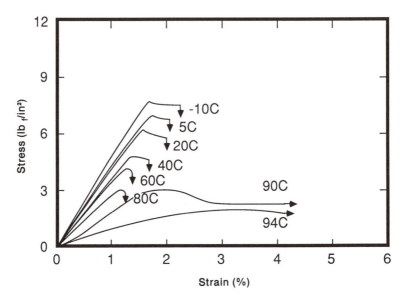

Figure 6.29.   Effect of Temperature on the Stress–Strain Behavior of General Purpose PS (GPPS) (39).

The rapid transition from brittle crazing to ductility indicates that glass transition has begun around 90°C. The rubbery phase dominates the stress crazing phenomenon, with the test specimen exhibiting distinct yielding followed by significant necking and cold drawing.

At this point, it should be apparent that ASTM D638 represents a one-point tensile test. The effects of temperature, strain rate, time and most important, localized molecular phenomena such as stress crazing cannot be adequately assessed from the data. This is an excellent illustration of the assertion that certain classical ASTM moderate-time-duration tests do not yield specific design properties for plastics in the same manner as they do for metals. Designers new to the field of plastic product design are frequently confused and alarmed by this point. This is particularly true if the designer is skilled in metal product design. Note however that the data obtained from this test or any other moderate-time duration test are *not* useless to the designer. As pointed out by Chastain (36), the tests can be and are chiefly used to provide material specifications for purchasing and quality control on incoming polymer to a manufacturing facility.

Flexural Properties of Plastics: ASTM D790

The ASTM D790 test, "Flexural Properties of Plastics and Electrical Insulating Materials", was originally based on the deflection of a Hookean beam under applied load (Figure 6.30). From classical strength of materials, the deflection of a Hookean beam can be given in terms of the radius of curvature of the beam (40):

$$(6.15) \qquad\qquad\qquad \frac{1}{\rho'} = \frac{M}{EI}$$

where ρ' is the radius of curvature of the beam, M is the applied moment, E is Young's modulus or elastic modulus and I is the moment of inertia of the beam cross-section. The flexural stress, σ, at any point within the beam at a distance y from the neutral axis, is given as:

$$(6.16) \qquad\qquad\qquad \sigma = \frac{My}{I}$$

Given the beam geometry and loading configuration along the beam, there is a direct mathematical relationship between beam deflection, δ, and the material modulus and beam dimensions. Thus, the modulus of a Hookean material can be analytically deter-mined as a function of beam deflection. Of course, this means that the mechanical test is both physically definable and mathematically definable, thus satisfying both *criteria of test acceptability*.

However, polymers are *not* Hookean materials. As with the tensile test, if ASTM D790 is used to evaluate polymers under load, it should be solely for the purposes of quality control and material specification, *not* for design purposes.

The ASTM D790 test procedure specifies two different test geometries, with either single- or two-point loading on a simply supported beam (Figure 6.31). The test spec-imen of length L* is placed on two supporting mounts separated by a distance L. The load, P, is either applied to the beam at the center of the support span [single point loading] or in two equal loads of P/2 applied to the beam at distances L/3 from each support [two-point loading]. The recommended length of the test specimen, support span, specimen thickness (d) and width (b) and applied loading rate (V) for span-to-beam thickness ratios of 16, 32, and 40 to 1 are given in tables in the ASTM standard. The most common test geometry is given in the ASTM D790 standard (Table 6.5).

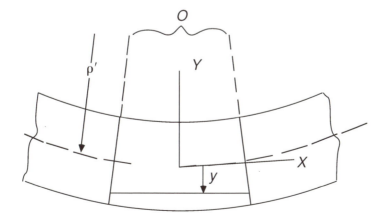

Figure 6.30.   Coordinates of a Beam in Mild Flexure About the Neutral Axis.

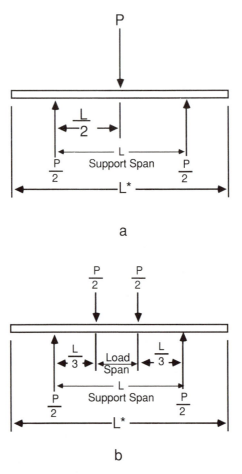

Figure 6.31.   Two Allowable Configurations for Polymeric Flexural Test Specimens According to ASTM D790. a: Three-Point Bending or Single-Point Loading. b: Four-Point Bending or Two-Point Loading.

For most ductile materials, the test specimen will not fracture as with tensile testing. Therefore the test is assumed complete when the outer fiber strain, ε, reaches 5%. The strain on the outer fiber of a single-point loaded beam is related to the beam deflection at the beam center, δ, by:

(6.17)
$$\delta = \frac{\varepsilon \, L^2}{6d}$$

The maximum fiber stress, σ_m, is related to the applied load by:

(6.18)
$$\sigma_m = \frac{3 \, PL}{2 \, bd^3}$$

Table 6.5. Most Commonly Used ASTM D790 Standard Flexural Test Conditions.

Geometry: single point loading
Span-to-depth ratio: 16:1
Specimen length: 130 mm (5 in)
Support span: 100 mm (4 in)
Beam depth: 6.4 mm (¼ in)
Beam width: 13 mm (½ in)
Cross-head moment: 2.8 mm/min (0.11 in/min)

Example 6.5

Q: Consider the plastic test specimen under Table 6.5 conditions, ASTM D790. Determine the maximum fiber *stress* when the applied load reaches 15 lb_f. Then determine the extent of beam deflection when the maximum fiber *strain* reaches 5%.

A: The beam is bent in a classic flexural mode (27). Thus:

$$\sigma = \frac{3\,PL}{2\,bd^3}$$

From the values given in Table 6.5, L = 4, b = ½, d = ¼. When P = 15:

$$\sigma = \frac{3 \times 15 \times 4}{[2 \times (\frac{1}{2}) \times (\frac{1}{4})^3]}$$

$$= 11,500 \ lb_f/in^2.$$

The maximum beam deflection is given as:

$$\delta = \frac{\varepsilon \, L^2}{6d}$$

$$= \frac{0.05 \times 4^2}{6 \times (\frac{1}{4})} = 0.53 \ \text{inch}$$

At very low strain level, the compressive and tensile stresses are symmetric about the neutral axis. The expression above can be used to obtain the maximum compressive stress just under the applied loading point. The apparent* modulus of the beam material can be measured in many ways. The simplest is the zero-strain tangent modulus, E_B:

(6.19)
$$E_B = \frac{mL^3}{4\,bd^3}$$

---

*If the material were Hookean, the measured elastic modulus would be a design property. Since polymers are viscoelastic, the measured value can only be thought of as "apparent".

where m is the slope of the tangent to the initial straight-line portion of the load–deflection curve. Note that for polymers, the tensile modulus usually does not necessarily have the same numeric value as the flexural modulus. Although this difference has been attributed to the viscoelastic nature of polymers or just measurement error, it is probably due to the variance in polymer morphology from surface to centerline in the molded bar. This can be the case for both crystalline and amorphous polymers. Recall that in tension, the measured properties of modulus, strain, and strength represent *average* values across the sample cross-section. Furthermore, yielding and failure are probably due to the "weakest link" in the cross-section. In flexure, the *skin* of the samples carries the majority of the load in compression and tension. Any variation in properties between the sample skin and the core will be amplified in the flexural test.

In interpreting tensile and flexural data, the engineer should treat the plastic test element as a molecular or microscopic composite. Undoubtedly, the difference in test data is reflected in the difference in actual performance of the polymer in a molded part.

Compressive Properties of Rigid Plastics: ASTM D695

In the compression test, the load is applied in a uniaxial compressive mode. The test specimen is either a rectangular bar (12.7 × 12.7 × 25.4 mm or ½ × ½ × 1 in) or a circular column (12.7 mm dia. or ½ in dia.). After the specimen is preconditioned it is loaded uniaxially at 1.3 (± 0.3) mm/min or 0.05 (± 0.01) in/min (Figure 6.32).

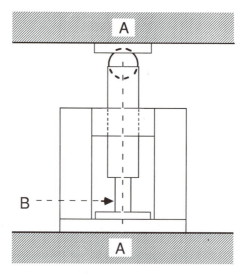

Figure 6.32.   Schematic of Compression Test in Accordance with ASTM D695. A: Testing Machine Platen. B: Test Specimen.

The apparent compressive strength and modulus of the material are then determined from the polymer deformation response to applied load. Again, these experimental results should be used only for quality control and material specification, *not* for design purposes.

Poisson's Ratio

Poisson's ratio is defined as the ratio of the increment of transverse strain, $d\varepsilon_t$, to that of longitudinal strain, $d\varepsilon_l$:

$$(6.20) \qquad \mu = \frac{d\varepsilon_t}{d\varepsilon_l}$$

The relative volume dilation during tensile elongation can be written as (41):

$$(6.21) \qquad \frac{\Delta V}{V_o} = (1 + \varepsilon_x)(1 - \mu\varepsilon_x)^2 - 1$$

where ΔV is the volume increment and V_o is the initial volume. For an isotropic material:

$$(6.22) \qquad \frac{\Delta V}{V_o} = (1 - 2\mu)\varepsilon_x$$

Poisson's ratio, μ, varies from ½ for purely elastic solids to low values for liquids to nearly zero for cellular polymers. For viscoelastic materials, Poisson's ratio depends on the testing frequency or time (42). Measurements on polymers at low frequency will yield values approaching ½. At high frequencies, the value asymptotically approaches ⅓. Poisson's ratios for several materials are given in Table 6.6. As the polymer is uniaxially strained, microvoids in unfilled polymers and interfacial debonding in filled and reinforced polymers are manifested by excessive volume dilation, that is, decreasing Poisson's ratio, with increasing strain. In Figure 6.33, the 30% GR PET exhibits microvoiding or interfacial debonding at high strain levels; 30% GR nylon 6 does not.

There are many inconsistencies in published Poisson's ratio values. The data presented for the solid materials in Table 6.6 were calculated from measured Young's and shear moduli using:

$$(6.23) \qquad \mu = \frac{E}{2G} - 1$$

The values based on measuring the lateral contraction of a simple test specimen in tension are usually lower. The reasons for this discrepancy are unclear and require closer investigation.

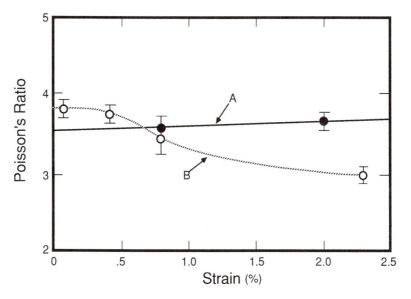

Figure 6.33. Strain-Dependent Poisson's Ratio for Two Polymers at Room Conditions (45). A: 30% (wt) Glass-Reinforced Nylon 6 (GR-PA-6). B: 30% (wt) Polyethylene Terephthalate (GR-PET). Materials Dry as Molded. Determined by Longitudinal and Transverse Measurement Techniques.

Mullen Burst Test: ASTM D774

With very thin specimens such as films, a biaxial stretching test called a Mullen burst test is commonly used. Polymer films in thicknesses of 0.01 to 0.1 mm (0.0005 to 0.005 in) are used extensively in packaging and allied industries. These films are made by casting and mechanically tenter-stretching or by extrusion-blowing and pneumatically biaxially stretching (see Chapter 5). Many tests have been developed for determining the properties of films. These include tensile and elongation tests (ASTM D882) and tearing strength tests (ASTM D1004 and D1922). The Mullen burst test, ASTM D774, gives a measure of the film resistance to short-term biaxial loading.

The Mullen burst test was developed as a way of quantifying the biaxial strength of paper. The film specimen is clamped between an upper clamping ring having a circular opening 30.48 ± 0.02 mm (1.200 ± 0.001 in) in diameter and a lower clamping or diaphragm ring having a thickness of 3.25 mm (0.128 in) and an opening 33.07 ± 0.08 mm (1.302 ± 0.003 in) in diameter. A pure gum rubber diaphragm, 0.86 mm ± 0.05 mm (0.034 ± 0.002 in) in thickness is clamped between the lower clamping ring and the pressure box. The diaphragm must be of a material that is inflated 9.5 mm (3/8 in) above the top surface of the diaphragm plate by 30 ± 5 kPa (4.3 ± 0.8 lb$_f$/in^2).

The test proceeds in this fashion. The film specimen to be tested is placed between the top and diaphragm rings and the plates tightened. Glycerin is pumped against the diaphragm at a fixed rate (95 ± 5 ml/min), inflating the diaphragm against the spec-

Table 6.6. Poisson's ratios for polymers and other materials (40, 43).

Class of materials	Material	Poisson ratio, μ
Organic and inor- ganic liquids	Benzene	0.5
	Carbon disulfide	0.5
	Water	0.5
Polymers	Low density polyethylene (LDPE)	0.49
	High density polyethylene (HDPE)	0.47
	Polypropylene (PP)	0.43
	Poly-1-butene	0.47
	Polystyrene (PS)	0.38
	Polyvinyl chloride (PVC)	0.42
	Polychlorotrifluoroethylene (FEP)	0.44
	Polytetrafluoroethylene (PTFE)	0.46
	Polymethyl methacrylate (PMMA)	0.40
	Polyphenylene oxide (PPO)	0.41
	Polyphenylene sulfide (PPS)	0.42
	Polyethylene terephthalate (PET)	0.43
	Polytetramethylene terephthalate	0.44
	Nylon 66 (PA-66)	0.46
	Nylon 6 (PA-6)	0.44
	Polycarbonate (PC)	0.42
	Polysulfone (PSO$_2$)	0.42
	Polyimide (PI)	0.42
Metals	Mercury	0.5
	Lead	0.44
	Gold	0.42
	Copper	0.34
	Steel (Mild)	0.28
	Tungsten	0.28
Inorganic solids	Granite	0.30
	Glass	0.23
	Vitreous silica	0.14

imen. The glycerin hydrostatic pressure is monitored until the specimen bursts. The Mullen burst strength equals the final value of hydrostatic pressure. This test was originally designed for paper, a nonwoven porous material. It has been adopted directly for non-woven polymer fabrics and films. For extruded and blown film polymers, vents must be applied between the polymer film and the diaphragm. Furthermore, paper is a brittle elastic substance. As a result, the biaxial stretching rate is of no importance. For viscoelastic polymers, on the other hand, the stretching rate must be considered. The Mullen burst test, as with other short-time duration tests, provides data at only one condition.

Instantaneous strain rates of 2 to 30 s^{-1} can be reached during this burst test (45). At very high rates of deformation, many polymers behave as if they are elastic liquids (46). Under constant stress or applied pressure, elastic liquid deformation rate increases without bound. It has been proposed, therefore, that applied pressure and bursting time be compared with elastic liquid response to applied stress:

(6.24)
$$\sigma = \eta_e \, \dot{\varepsilon}$$

where η_e is the biaxial extensional viscosity and $\dot{\varepsilon}$ is the principal biaxial strain rate. If σ is related to applied pressure by:

(6.25)
$$\sigma = \frac{P}{2} \left[\left(\frac{R}{2h}\right)_o + \left(\frac{R}{2h}\right)_b \right],$$

and $\dot{\varepsilon}$ is given as:

(6.26)
$$\dot{\varepsilon} = \frac{\varepsilon_b - \varepsilon_o}{\theta_b} = \frac{\Delta\varepsilon}{\theta_b},$$

where P is the applied pressure, R is the radius of the film under deformation, h is the film thickness, ε is the elongation, θ_b is the time to burst and "o" and "b" represent the initial and "at burst" conditions, then:

(6.27)
$$P \, \theta_b = \eta_e \left\{ \frac{2\Delta\varepsilon}{[(R/2h)_o + (R/2h)_b]} \right\}$$

The $(1/\theta_b)$ term has the units of strain rate, $\dot{\varepsilon}$ (s^{-1}) and (P θ_b) have the units of viscosity (Pa · s). If the term in the brackets is essentially constant, (P θ_b) is the proportional to η_e, the biaxial extensional viscosity. Figure 6.34 compares bursting time with viscosity for polyisobutylene rubber. If the biaxial extensional viscosity is inversely proportional to strain rate and is also time-independent, as with many materials such as polymethyl methacrylate and HIPS, bursting time is independent of applied pressure. On the other hand, if extensional viscosity is strongly time-dependent, as with PVC and polyethylene, increased applied pressure results in decreased burst time. Note, however, that most production films are not equally biaxially oriented. Bursting failure will be therefore manifested as splitting in a preferred direction. As with flexed-beam and flexed-plate moderate-time-duration tests, bursting tests yield single-point values that can be used only for quality control and comparative ranking of polymer films, not for design purposes.

Example 6.6

Q: Serum is stored in a bottle having a thermoplastic elastomer (TPE) septum. The septum is 1 in in diameter and 0.030 in thick. The serum is stored at a temperature at which the internal

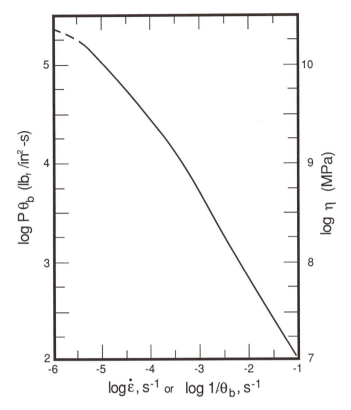

Figure 6.34. Comparison of Bursting Time with Viscosity for Polyisobutylene Rubber (47).

pressure in the bottle is 5 lb_f/in^2 (gage). The material modulus is 15,000 lb_f/in^2 and its Poisson's ratio, $\mu = 0.35$. The internal pressure causes the septum to bulge. At long times, the TPE can be considered to be a viscoelastic polymer having an elongational viscosity of 10^6 Pa · s. How long can the serum be stored at this temperature before the bulge exceeds a maximum of 0.25 in above the bottle rim?

A: This is a complex problem that requires knowledge about deflection under load. From a standard handbook on stress and strain (27), the deflection of a disk clamped at the rim and under uniform load, is given as:

$$h = \frac{3 \, pD^4 \, (1 - \mu^2)}{256 \, Et^3}$$

in which h is the extent of deflection (or height of bulge), p is the applied pressure, D is the diameter of the disk, E is the material modulus and t is its *stretched* thickness. From the numbers above:

$$h = \frac{3 \times 5 \times 1^4 \, (1 - 0.35^2)}{256 \times 15,000 \times (0.03)^3} = 0.127 \text{ in}$$

Note that this assumes that $t = 0.030$, the initial or unstretched sheet thickness.

From Ref. 27, the tensile stress, τ, can be written as a hoop stress, as:

$$\tau = \frac{P\,R_B}{2\,\delta}$$

where R_B is the radius of the expanded septum and δ is its expanded thickness:

$$R_B = \frac{h}{2} + \frac{D}{8h},$$

$$\delta = \frac{D^2\,t}{8\,h\,R_B}$$

If the initial value of $h = 0.127$ in is used:

$$R_B = \frac{0.127}{2} + \frac{8 \times 0.127}{1} = 1.048 \text{ in}$$

$$\delta = \frac{1 \times 0.03}{8 \times 0.127 \times 1.048} = 0.0282 \text{ in}$$

As is apparent, the h calculated with $t = 0.030$ is *not* correct, since $\delta = 0.0282$ is the result. To correct this, a trial-and-error procedure is used. It yields $\delta = 0.027$ in, $R_B = 0.805$ in and $h = 0.174$ in deflection. Now the initial tensile stress is given as:

$$\tau_i = \frac{5 \times 0.805}{2 \times 0.027} = 74.5 \text{ lb}_f/\text{in}^2$$

The initial elongation, ε_i, is given as:

$$\varepsilon_i = \ln\left(\frac{L}{L_o}\right) = \ln\left(\frac{2\,\alpha\,R_B}{D}\right)$$

where α, the central angle of the bulge, in radians, is:

$$\alpha_i = \cos^{-1}[1 - h/R_B] = 0.67 \text{ rad.}$$

And so:

$$\varepsilon_i = \ln\left(\frac{2 \times 0.67 \times 0.805}{1}\right) = 0.0758 \text{ or } 7.58\%$$

Now the maximum allowable value of h is 0.25 inch. From the equations above, $R_B = 0.625$ in and $\delta = 0.024$ in. Checking, the final deflection is found to be $h_f = 0.249$ in. This is close enough to the target of 0.25 in that no further iteration is needed. Now the final hoop stress is:

$$\tau_f = \frac{5 \times 0.625}{2 \times 0.024} = 65.1 \text{ lb}_f/\text{in}^2.$$

The average hoop stress, τ_{ave} is given as:

$$\tau_{ave} = \frac{\tau_0 + \tau_f}{2} = 69.8 \text{ lb}_f/\text{in}^2.$$

The final values for the central angle and elongation are:

$$\alpha_f = \cos^{-1}\left[\frac{1 - 0.25}{0.625}\right] = 0.927 \text{ rad.}$$

$$\varepsilon_f = \ln\left(\frac{2 \times 0.927 \times 0.625}{1}\right) = 0.1477 \text{ or } 14.77\%.$$

Now the elongational viscosity is defined as:

$$\eta_e = \frac{\tau}{\dot{\varepsilon}} = \frac{\tau_{ave}}{[(\varepsilon_f - \varepsilon_i)/\theta]},$$

where $\dot{\varepsilon}$ is approximated by $[\varepsilon_f - \varepsilon_i]/\theta = \Delta\varepsilon/\theta$, and θ is the time. Rearranging, the time for storage of the serum is given as:

$$\theta = \eta_e \Delta\varepsilon/\tau_{ave},$$
$$= 6.895 \times 10^9 (0.1477 - 0.0763)/69.8,$$
$$= 7.05 \times 10^6 \text{ s,}$$
$$\theta = 1960 \text{ h.}$$

Hardness

"Hardness" is generally used to describe the resistance of a material surface to indentation, scratching or marring. These phenomena are considered to be closely related. The generally accepted definition of hardness as it pertains to plastic is:

"Hardness: The resistance to penetration or indentation by another body."

Several different instruments can be used to measure indentation hardness. They differ in the shape and size of the indenter and the magnitude and duration of the applied force. The penetration and deformation of the surface is strongly dependent on the polymer viscoelastic nature. Several hardness tests are detailed below. For all tests, the polymer specimens must be conditioned according to ASTM D618 prior to testing.

Rockwell Hardness: ASTM D785

The Rockwell test was developed primarily for metals. Its European counterparts are the ball indentation hardness test, DIN 53.456 and the Vickers hardness test, ISO R81,

1967. Relationships have been developed between specimen properties and hardness (49). The Rockwell test is detailed here.

The indenting procedure is a two-step loading of a spherical indenter on the plastic surface. The specimen thickness is 6 mm or ¼ in. If necessary, this thickness can be achieved by building up thinner pieces. A steel ball is pressed into the surface of the specimen by applying an initial or minor load. The dial gage is set to zero. After the minor load has been applied for 10 s, the major load is applied. After 15 s, the major load is removed, leaving only the minor load. After an additional 15 s, with the minor load still in place, the dial gage is read. This reading corresponds to the hardness of the plastic. The size of the indenting balls and the values for the minor and major loads are given in Table 6.7. The Rockwell hardness is the value of the nonrecoverable indentation or penetration, prefixed by the letter M or R depending on the indenter. Rockwell "R" hardness values are also called α-Rockwell values, as given in ASTM D785 procedure B. Note that hardness should always be measured at least 6 mm or ¼ in from any molded edge.

Ball Indentation Hardness: DIN 53.456

The ball indentation hardness is determined in a manner similar to the Rockwell hardness. An appropriate hardness tester such as the Zwick Nr. 3106 is used. The initial or minor load is applied for 10 s, followed by the major load for 30 s. The depth of indentation is measured at 30 s while the major load is in place. In this test, the indentation depth is maintained at 0.15 mm to 0.35 mm by varying the applied major load. Sample thickness should exceed 4 mm and testing should be done 5 mm from any molded edge. Residual indentation and elastic recovery are measured immediately after removal of the major load. The elastic recovery value corresponds to the reported DIN hardness value. As seen in Figure 6.35, there is good correlation between DIN hardness and α-Rockwell hardness values for many major polymers, including glass- and mineral-filled materials.

Table 6.7. Hardness test conditions.

Hardness method	Minor load	Major load	Indenter diameter (mm)	Measure
BS hardness*	10 g$_f$	30 + 540 g$_f$	2.4	Penetration expressed on a scale of 0 to 100
Rockwell R	10 kg$_f$	60 kg$_f$	12.7	Depth of nonrecoverable penetration
Rockwell M	10 kg$_f$	100 kg$_f$	6.35	

*British Standard Test

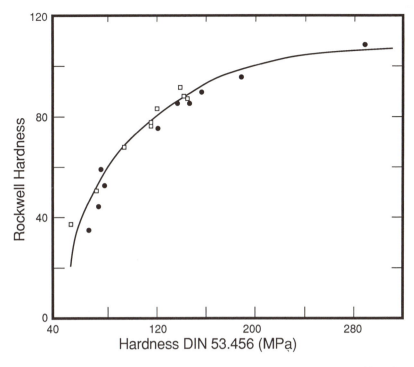

Figure 6.35.   α-Rockwell Hardness (or Rockwell "R" Hardness) Compared with Ball Indentation Hardness as Specified by DIN 53.456 (49).

Vickers Hardness: ISO R81, 1967

The indenter in the Vickers hardness test is a diamond pyramid with a quadratic base and a 136° top angle. Load is slowly applied for 8 to 10 s and a maximum load of 15 kg is applied for 20 s. After the load is removed, the diagonals of the residual impression are microscopically measured and the Vickers hardness calculated from correlations determined for metals. The Vickers test is prototypical of non-spherical indenter tests. Other indenter shapes include wedges and cones (50,51).

Shore Durometer: ASTM D2240

This test is used primarily for softer materials such as rubbers and thermoplastic elastomers. Its European equivalent is ISO R868. The Durometer device has a full or truncated conical indenter (Figure 6.36). The indenter projects 2.5 mm below the instrument base. The indenter is connected to a precalibrated linear spring and dial gage, Figure 6.37. When the base plate with the protruding indenter is pressed firmly against the surface of the plastic specimen, the dial gage registers the amount of force required to penetrate the indenter. The stiffness of the spring dictates the Durometer reading. If

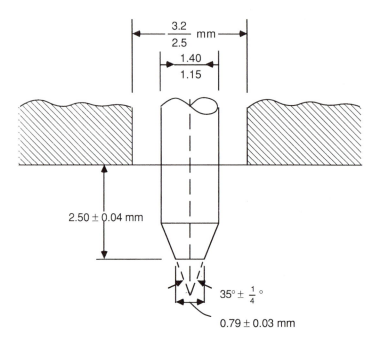

$\dfrac{3.2}{2.5}$ mm

$\dfrac{1.40}{1.15}$

2.50 ± 0.04 mm

35° ± $\dfrac{1}{4}$ °

0.79 ± 0.03 mm

a

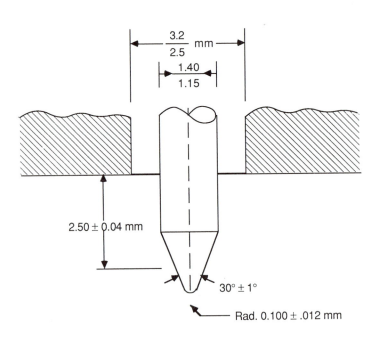

$\dfrac{3.2}{2.5}$ mm

$\dfrac{1.40}{1.15}$

2.50 ± 0.04 mm

30° ± 1°

Rad. 0.100 ± .012 mm

b

Figure 6.36. Dimensional Specifications for Shore Durometer Indenters, According to ASTM D2240. a: Shore A Durometer. b: Shore D Durometer.

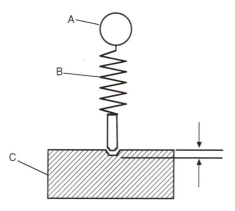

Figure 6.37.   Schematic of the Shore Durometer. The Spring for Shore A has the Range of 56 to 622 g Force. For Shore D, the Range is 0 to 10 lb$_f$. A: Indicator. B: Spring. C: Specimen.

the "A" indenter spring is used, the applied load range is 56 to 822 g$_f$. If the "D" indenter spring is used, the range is 0 to 10 lb$_f$. The Shore Durometer "A" or "D" value is read as 0 to 100. When using this device, the readings on the dial gage will change with time owing to viscoelastic relaxation of the polymer beneath the indenter. It is therefore recommended that readings be taken at about 1 s after application of the tester.

Brinell Hardness

The Brinell hardness test for soft materials such as polymers is based on the magnitude of the surface area of the impression left by a 10 mm diameter hardened steel ball pressed with a force of 500 kg$_f$ for 10 s into the surface of the plastic. The diameter of the impression is usually measured by a low-power microscope. The Brinell Hardness Number, BHN, is defined as the applied load (kg$_f$) divided by the surface area of the impression (in mm^2):

(6.28)
$$BHN = \frac{2\,W}{\pi D\,[D\,-\,\sqrt{D^2-d^2}]}$$

where W is the applied load (500 kg$_f$), D is the ball diameter, 10 mm, and d is the measured surface diameter of the crater made by the indentation.

Example 6.7

Q: What is the Brinell Hardness of a general-purpose polystyrene (GP-PS) sheet, if the diameter of the surface crater is 4.85 mm?

A: For Brinell Hardness, $W = 500 \text{ kg}_f$ and $D = 10$ mm. Therefore, the Brinell Hardness number is:

$$BHN = \frac{2W}{\pi \, D[D - \sqrt{D^2 - d^2}]},$$

$$= \frac{1000}{\pi \times 10 \times [10 - \sqrt{10^2 - 4.85^2}]},$$

$$BHN = 25.4$$

Tabor (52) shows that the BHN is not a satisfactory measure, even for metals, since the ratio of applied load, W, to indentation surface area does not yield the mean pressure, P, over the surface of the indentation.* Thus, Brinell hardness has little meaning in general.

Barcol Hardness: ASTM D2583

The Barcol hardness tester directly measures the depth of indentation of a specially designed truncated conical indenter, Figure 6.38. The indicating dial has 100 divisions, each representing a penetration unit of 0.076 mm (0.003 in). Again, the harder the material is, the higher the reading is. The indenter is spring-loaded in a hand-held unit, Figure 6.39. The spring is calibrated against a hard glassy surface and two calibrated aluminum disks having two different hardnesses. As with all other hardness testing devices used with polymers, the Barcol hardness reading will change with time. It is recommended that measurements be made within one second of applying load.

Hardness Tests: General Comments

Each measuring system has advantages and disadvantages. The actual choice of a particular test method, shape of indenter and nature of indicator depends primarily on the modulus of the plastic. This is illustrated in Table 6.8. This table is not absolute. There is considerable overlap in the different tests as well as with indenters for the

---

*Tabor notes that the mean pressure, P, is the ratio of applied load, W, to *projected* area of indentation, $\pi d^2/4$.

In 1908, Meyer proposed a hardness based mean pressure:

dD(6.29)   MHN (Meyer Hardness Number) $= \dfrac{4W}{\pi d^2}$

Experimentally, the applied load is seen to be proportional to a power of the indentation diameter:

(6.30) $W = k \, d^n$

where $2 < n < 2.5$ for metals and k is a material constant. If $n = 2$, the MHN is essentially independent of applied load and becomes simply a material constant.

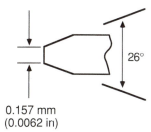

0.157 mm
(0.0062 in)

Figure 6.38. Dimensions for Barcol Hardness Indenter, in Accordance with ASTM D2583.

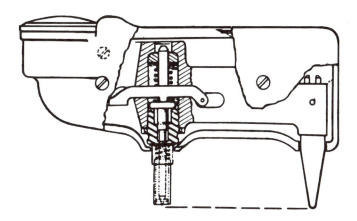

Figure 6.39. Breakaway Schematic of the Barcol Impressor, as Specified in ASTM D2583.

Table 6.8. Recommended hardness testing methods for several polymers. BS is British Standard Test.

Modulus	Polymer	Test method
Low	Rubber	BS 903 or Shore A
	Plasticized PVC	BS 2782 or Shore
	All polyethylenes	A
		Shore D
Medium	Polypropylene	Rockwell R
	Rigid PVC	Rockwell R
	High impact polystyrene	Rockwell R
	ABS	Rockwell R
	Polystyrene (PS)	Rockwell M
High	Nylon (PA)	Rockwell M
	Polyacetal (POM)	Rockwell M
	Polycarbonate (PC)	Rockwell M

same test. However, due to the lack of correlation between indenters and the viscoelastic nature of polymers, it is unadvisable to attempt correlation between testing
systems in the overlapping areas, even when the same polymer is used. Certainly
different polymers show decidedly different hardness correlations.

Owing to the simplicity of the hardness test, it is logical to seek a correlation between
material hardness and other material properties. Selden and Gustafson (49) show excellent correlations between hardness [Vickers, Ball Indentation and α-Rockwell] and
stiffness, yield strength and fracture strength for many types of polymers. These include filled and reinforced materials, amorphous and crystalline materials, and low-
and high-modulus materials. Linear correlations of Young's modulus, yield strength
and fracture strength with DIN hardness are given in Figures 6.40 to 6.42. And in
Figures 6.43 and 6.44 are given the linear correlations of yield strength and fracture
strength with Vickers hardness. Note however that even with ideally formed test specimens, free of processing-induced morphological variants, the correlations are considered only approximate. Turner notes that even if these correlations (hardness as a
function of a yield characteristic and abrasion resistance as a function of loss factor,
yield characteristics and toughness) can be shown to produce strongly linked interrelationships, this does not necessarily give them more important or significant status
than that held by less accurate correlations (53).

Darlix et al. (51) note that as with tensile and flexural tests, *hardness tests,* as
currently constructed, are *single-point tests.* They do not account for the time-dependent nature of polymers, nor do they include molding parameter effects, rate of loading,
time or temperature. Modifications of the mechanical indentation models to include an
elastoplastic zone beneath the indenter help to connect the test to the mechanical response of solid polymers under applied load. But the second criterion of test acceptability is not now met.

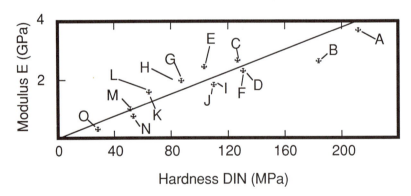

Figure 6.40.   Comparison of Young's Modulus with Ball Indentation Hardness for Several Polymers (49). Correlation Coefficient = 0.94. Polymer Names Include European Resin Trademarks. Point A: Polyether Etherketone (PEEK). B: Polyether-imide (PEI). C: Polyethersulfone
(PESO$_2$). D: Polyacetal, Polyoxymethylene (POM). E: Nylon (PA). F: Polybutylene Terephthalate (PBTP or PBT). G: Hard ABS (Varma). H: PVC. I: Polycarbonate (PC). J: Xenoy. K:
Modified Polyphenylene Oxide (mPPO). L: Soft ABS (Slag). M: Polypropylene (PP). N: High-
Density Polyethylene (HDPE). O: Olefinic Elastomer (Vistaflex).

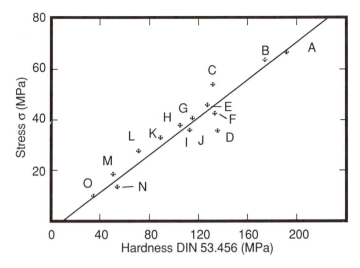

Figure 6.41.　Comparison of Yield Strength with Ball Indentation Hardness for Several Polymers (49). Correlation Coefficient = 0.95. Polymer Names Include European Resin Trademarks. Point A: Polyether Etherketone (PEEK). B: Polyether-imide (PEI). C: Polyethersulfone (PESO$_2$). D: Polyacetal, Polyoxymethylene (POM). E: Nylon (PA). F: Polybutylene Terephthalate (PBTP or PBT). G: Hard ABS (Varma). H: PVC. I: Polycarbonate (PC). J: Xenoy. K: Modified Polyphenylene Oxide (mPPO). L: Soft ABS (Slag). M: Polypropylene (PP). N: High-Density Polyethylene (HDPE). O: Olefinic Elastomer (Vistaflex).

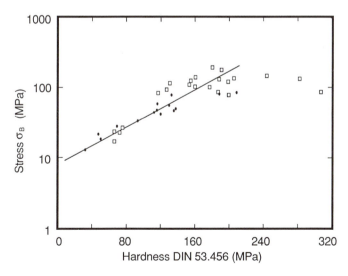

Figure 6.42.　Correlation Between Ultimate Stress (Fracture Strength) and Ball Indentation Hardness, DIN 53.456 (49). Correlation Coefficient = 0.90. Solid Circles: Filled Resins. Open Squares: Unfilled or Neat Resins.

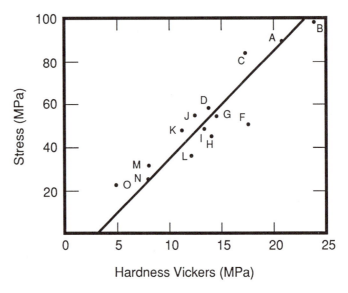

Hardness Vickers (MPa)

Figure 6.43.   Comparison of Yield Stress (Yield Strength) with Vickers Hardness for Several Polymers (49). Correlation Coefficient = 0.90. Polymer Names Include European Resin Trademarks. Point A: Polyether Etherketone (PEEK). B: Polyether-imide (PEI). C: Polyethersulfone (PESO$_2$). D: Polyacetal, Polyoxymethylene (POM). E: Nylon (PA). F: Polybutylene Terephthalate (PBTP or PBT). G: Hard ABS (Varma). H: PVC. I: Polycarbonate (PC). J: Xenoy. K: Modified Polyphenylene Oxide (mPPO). L: Soft ABS (Slag). M: Polypropylene (PP). N: High-Density Polyethylene (HDPE). O: Olefinic Elastomer (Vistaflex).

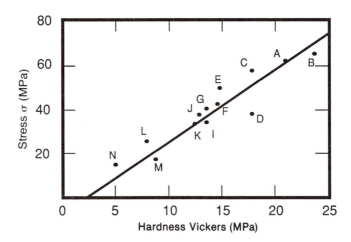

Hardness Vickers (MPa)

Figure 6.44.   Comparison of Ultimate Stress (Fracture Strength) with Vickers Hardness for Several Polymers (49). Correlation Coefficient = 0.90. Polymer Names Include European Resin Trademarks. Point A: Polyether Etherketone (PEEK). B: Polyether-imide (PEI). C: Polyethersulfone (PESO$_2$). D: Polyacetal, Polyoxymethylene (POM). E: Nylon (PA). F: Polybutylene Terephthalate (PBTP or PBT). G: Hard ABS (Varma). H: PVC. I: Polycarbonate (PC). J: Xenoy. K: Modified Polyphenylene Oxide (mPPO). L: Soft ABS (Slag). M: Polypropylene (PP). N: High-Density Polyethylene (HDPE). O: Olefinic Elastomer (Vistaflex).

Dynamic Solid Polymer
Mechanical Testing

As noted in Chapter 3, the viscoelasticity of a polymer is a function of its response to frequency-dependent loads. Pure Hookean materials exhibit in-phase response to sinusoidal loading and thus have only elastic* moduli. Purely viscous or Newtonian materials exhibit 90° ($\pi/2$) out-of-phase response and thus possess only viscous or "loss" moduli. Viscoelastic materials show both in-phase and out-of-phase characteristics. For many polymeric systems, the ratio of elastic modulus to loss modulus is strongly frequency- and temperature-dependent. These moduli can be used to identify transition temperatures. For example, glass transition temperature is identified as the region in which, at constant frequency, the elastic modulus decreases rapidly and the loss modulus increases.** Dynamic mechanical testing also yields information about the following types of polymeric systems (54):

- The degree of polymer crystallinity,
- The nature and number of polymeric transitions,
- The stress state of rubber particles in impact-modified polymers,
- The effects of annealing or heat setting,
- The extent of reactivity for a thermosetting polymer, ·
- The effect of processing aid(s) or other adducts,
- The effect of fillers, reinforcements, and other non-polymer adducts on the temperature-dependent strength or stiffness of the polymeric system,
- The extent of phase separation in multiphase polymeric systems, and
- The nature of fillers and reinforcements, the extent of dispersion and the nature of the coupling to the polymer.

Many dynamic mechanical tests have been developed. These are reviewed in Table 6.9. One of the earliest tests, the ASTM D2236, "Test Method for Dynamic Mechanical Properties of Plastics by Means of a Torsional Pendulum", uses a freely oscillating pendulum with the rectangular or circular cross-section pendulum suspended between the stationary clamp and the rotating inertial disk. This is shown in schematic in Figure 6.45. This test has also been adapted to reactive systems by replacing the specimen with a braided glass fiber strip. The strip is impregnated with the reactive resin and at various times during reaction, the pendulum is oscillated and the damping observed. It is apparent that as the resin reacts, its time-dependent viscoelastic nature changes from (near-) viscous to elastic. Thus, at a given frequency and temperature, the elastic modulus of a reactive resin increases and the loss modulus decreases in time.

Most dynamic tests are run on forced torsion testers. One end of the thermally conditioned test specimen is clamped in a torque sensing device. The other end is

---

*Many terms are used to identify the in-phase response to sinusoidal load. These include "storage modulus", "shear modulus", and "torsion modulus".
**If the temperature is held constant, the elastic modulus increases with increasing frequency in the transition region.

Table 6.9.   Dynamic Mechanical Testing Techniques, according to ASTM D4065. a: Free Torsional Pendulum. b: Dynamic Mechanical Analyzer. c: Viscoelastometer. d: Axial Mechanical Spectrometer. e: Flexural Mechanical Spectrometer. f: Torsional Mechanical Spectrometer.

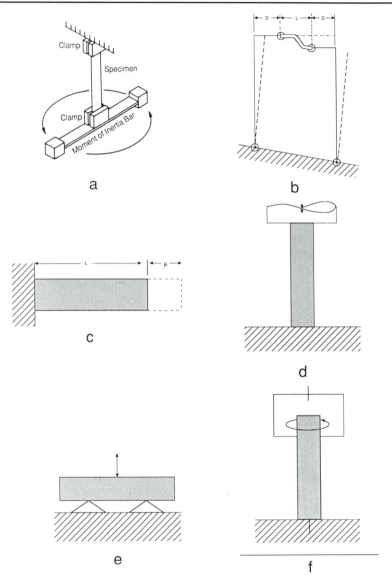

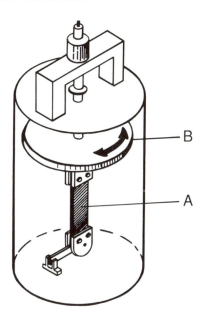

Figure 6.45.   Schematic of Torsion Pendulum According to ASTM D2236 (55). A: Test Specimen. B: Inertial Disk.

clamped in a head that can oscillate at a fixed angular displacement and frequency
selected in the range of 0.1 to 100 Hz (Figure 6.46). The in-phase portion of the
complex modulus, G^*, is the elastic modulus, G'. It is proportional to the square of
the frequency. The phase angle or dissipation factor, tan δ, is inversely proportional to
the square of the frequency. From these data, the out-of-phase portion, the loss modulus, G'', and the complex modulus, $G^* = G' + i\,G''$, can be found. The test is

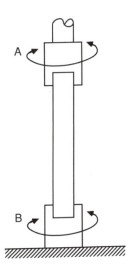

Figure 6.46.   Schematic of Driven or Forced Torsion Oscillator
(56). A: Driven Element Imparting Oscillatory Strain. B: Resulting
Torque Through Test Specimen.

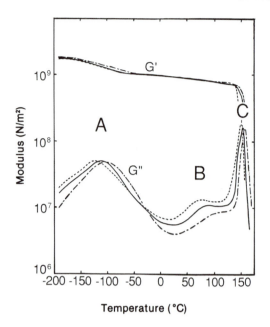

Figure 6.47.   Temperature-Dependent Loss and Storage Moduli of Polycarbonate (PC) for Three Oscillation Rates (56). G': Storage Modulus. G'': Loss Modulus. Dashed Line: 0.1 Hz. Solid Line: 1.0 Hz Dash–Dot Line: 10 Hz. Region A: Backbone Rotation. B: Crystallization. C: T_g, Glass Transition Temperature.

repeated for the instrument range of frequencies, and if equipped with a temperature chamber, at a full range of operating temperatures, as well.

A typical plot of elastic and loss moduli for polycarbonate, an amorphous polymer, is shown in Figure 6.47. Typically, the elastic modulus, G', of an amorphous polymer is independent of frequency in the range of 0.1 to 10 Hz. Note that it is relatively temperature-independent until the glass transition temperature is reached, at about 150°C. The loss modulus, G'', and the phase angle, as tan δ, are more sensitive to minor temperature-dependent molecular changes. Note that for polycarbonate, there is a molecular transition occurring at -125°C. This transition is called a secondary or β-transition. In this case, the β-transition is attributed to the rotation of the polymer backbone around the benzene ring (57). The change seen at 75°C is thought to be the annealing of a small amount of crystallinity.*

---

*Even though polycarbonate is considered to be an amorphous polymer, it may contain a small amount of crystalline polymer. Polyvinyl chloride, another polymer that is considered to be amorphous, also contains a small amount of crystallinity, at less than 20%.

Thermal Tests

The temperature-dependent elastic moduli of crystalline polymers (Figure 6.48) differ substantially from those of amorphous polymers (Figure 6.49). This is seen by comparing the PC data with that of nylon 6. Recall that polymers are never 100% crystalline. This nylon is about 50% crystalline. Below its 50°C glass transition temperature, the polymer is a molecular composite of a rigid crystalline structure and amorphous molecules in a glassy or rigid state. Thus the elastic modulus is very high (about 30×10^9 dyne/cm^2) and is essentially temperature-independent. As the specimen temperature exceeds the glass transition temperature range, the amorphous phase becomes rubbery. The result, as seen in Figure 6.48, is a drop in elastic modulus by ⅔ to about 10×10^9 dyne/cm^2. Note that as the temperature continues to increase, the elastic modulus value continues to decrease. It is thought that the increasing molecular activity in the amorphous phase tends to reduce the effective strength of the crystalline phase. Above about 180°C, the drop in elastic modulus value and the rapid increase in loss modulus value signals the beginning of melting of the crystalline phase. Note that the melting temperature for nylon 6 is usually given as 195 to 225°C, depending on the molecular weight, extent of residual monomer, and the nature of the adducts. Representative elastic modulus curves for thermosetting polymers are given in Figure 6.50. The epoxy material (curve E) and the diallyl phthalate (DAP) material (curve D) both show inflections as moduli, indicating glass transition temperatures.

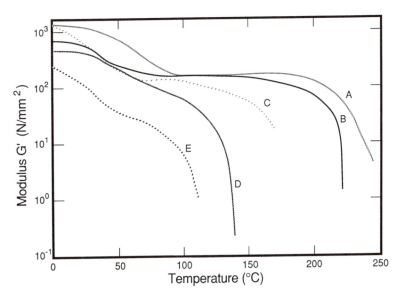

Figure 6.48.   Temperature-Dependent Storage Moduli for Several Crystalline Polymers (58). A: Nylon 66 (PA-66). B: Nylon 6 (PA-6). C: Chlorinated Polyether. D: HDPE. E: LDPE.

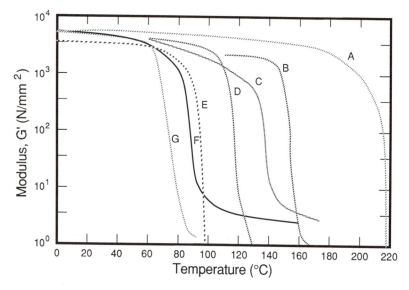

Figure 6.49.   Temperature-Dependent Storage Moduli for Several Amorphous Polymers (59). A: Polyvinyl Carbazole. B: Polycarbonate (PC). C: PMMA. D: PS. E: Polystyrene-Polybuta-diene Copolymer (Impact PS). F: Unplasticized PVC (RPVC). G: PVC/Polyvinyl Acetate Copolymer.

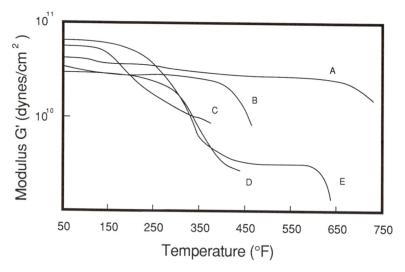

Figure 6.50.   Temperature-Dependent Storage Moduli for Several Filled Thermosetting Poly-mers (60). Curve A: Phenol–Formaldehyde. B: Melamine–Formaldehyde. C: Alkyd Resin. D: Diallyl Phthalate (DAP). E: Bisphenol A Epoxy.

Deflection Temperature: ASTM D648

The "heat resistance" of a polymer is an important design specification in many applications. The DTUL test, "Deflection Temperature of Plastics Under Flexural Load", was originally called the "Heat Distortion Test". It was proposed as a way of providing the designer with a ranking of polymers for high temperature applications. This test has some severe technical limitations, as noted below. It is included here primarily because it is used extensively to screen polymers. As with creep and fatigue tests, the test is a *one-point* test that does not describe how a material responds as a function of temperature to an applied load.

The test is relatively simple. A schematic of the DTUL apparatus is shown in Figure 6.51. The ASTM D648 test specimen is a simply supported rectangular bar 127 mm (5 in) long, 12.7 mm (0.5 in) thick, and 3.2 to 12.7 mm (0.125 to 0.5 in) wide. The specimen is simply supported to provide a 102 mm (4 in) free span. A load is applied to the center of the specimen to cause a maximum fiber stress either of 455 kPa (66 lb_f/in^2) or 1.82 MPa (264 lb_f/in^2), ± 2%. The test specimen is then lowered into an oil bath and the deflectometer reset to zero. The oil bath initially at 23°C is heated at a uniform rate of 2.0 (± 0.02) °C/min. When the deflection of the specimen under load reaches 0.254 mm (0.010 in) from its initial value, the oil temperature is recorded. This becomes the one-point deflection temperature under load (DTUL). Recently, it has been shown that, owing to simple transient heat transfer to the test specimen from the oil, the DTUL device often leads to a misconception of the actual temperature-

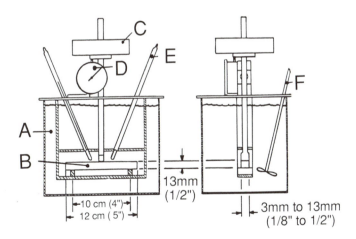

Figure 6.51.   Schematic of Apparatus for Deflection Temperature Under Load (DTUL) According to ASTM D648. A: Oil Bath. B: Test Specimen. C: Weight. D: Dial Indicator. E: Thermometer. F: Stirrer.

dependent strength of the polymer under test (61). Consider the flexing of a beam under load:

$$(6.31) \qquad \delta(t) = \frac{PL^3}{48\,S_t(T;\,t)}$$

where S_t is the temperature-dependent stiffness of the beam. Since the beam dimensions are fixed, the stiffness is directly proportional to the temperature-dependent modulus:

$$(6.32) \qquad S_t(T;\,t) = \int E_y\,(T;\,t)\,y^2\,dA$$

The temperature-dependent modulus for polyamide, a crystalline polymer, is shown in Figure 6.52. Even though the oil is heated linearly, the energy must be convected through the heat transfer film between the oil and the bar, then conducted to the bar interior. The bar does not have a uniform temperature at any time during the heating procedure, with the center colder and thus stiffer than the outer layers. The calculated temperature-dependent deflection for nylon, the crystalline polymer, and polycarbonate, an amorphous polymer, are given in Figure 6.53. The ASTM standard heating rate and an oil heat transfer coefficient of 10 Btu/ft^2 h °F are assumed. Note that the 12.7 mm (0.5 in) wide bar shows substantially higher DTUL values than the 6.35 mm (0.25 in) bar, at both loading levels. For the lower level, the DTUL values are more than 74°C (134°F) apart. As a comparison, the DTUL values for the 12.7 mm (0.5 in)

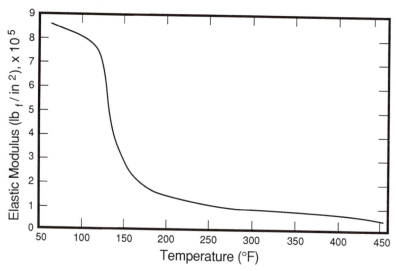

Figure 6.52. Temperature-Dependent Elastic or Young's Modulus for Monsanto Vydyne R-100 Nylon (PA)(61).

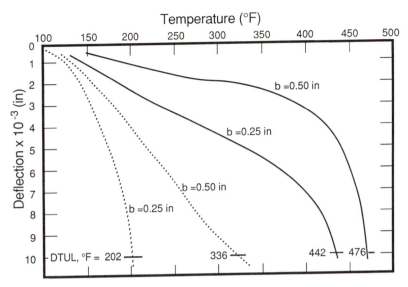

Figure 6.53.   Computed Temperature-Dependent Deflection for Nylon of Figure 6.52 (61). Heat Transfer Coefficient, $h = 10$ Btu/h ft^2°F. Dashed Line: Fiber Stress, 264 lb$_f$/in^2. Solid Line: Fiber Stress, 66 lb$_f$/in^2.

specimen of this material are listed as 238°C (460°F) at 455 kPa (66 lb$_f$/in^2) and 182°C (360°F) at 1.82 MPa (264 lb$_f$/in^2).

In Figure 6.54, a change in the heat transfer coefficient for the oil from 10 to 1 results in an increase in DTUL from 169°C (336°F) to 214°C (418°F), or an increase of 45°C (82°F). The designer must therefore realize that DTUL values, *per se,* can be very misleading, particularly if the natures of the oil bath and the sample size are not carefully defined.

This analysis also explains the results of Chastain (62) in Figure 6.55. Amorphous polymers show little sensitivity to DTUL stress level (66 or 264 lb$_f$/in^2), since the factor controlling the strength of amorphous polymers is glass transition temperature. For a semicrystalline polymer, on the other hand, an abnormally low DTUL temperature can be obtained at the higher stress level. At higher stress levels, the factor being affected is the glass transition temperature of the amorphous portion of the polymer. At lower stress levels, the dominant factor is the melting temperature of the crystalline portion of the polymer. This anomaly is shown in Figure 6.56.

It has been proposed that the DTUL values be used only to compare thermal performances of polymers within the same family, either amorphous or crystalline. Without a clear understanding of the underlying limitations to the test, this approach can also result in erroneous results. Shown in Figure 6.57 are two sets of dynamic storage modulus curves for an amorphous polymer, polycarbonate, and a crystalline one, nylon. The data show the effect of reinforcing on DTUL values. There is little effect of reinforcing on DTUL for PC, where the controlling factor is the glass transition tem-

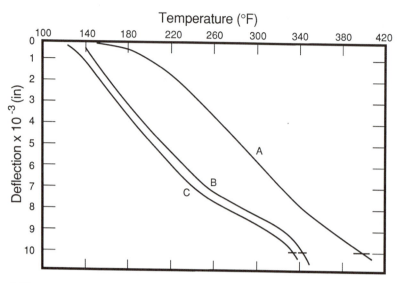

Figure 6.54. Computed Temperature-Dependent Deflection for Nylon of Figure 6.52 (61). For 264 lb$_f$/in^2 Fiber Stress, and Specimen Thickness, $b = 0.5$ in. Curve A: Heat Transfer Coefficient, $h = 1$ Btu/h ft^2°F. B: $h = 5$ Btu/h ft^2°F. C: $h = 10$ Btu/h ft^2°F.

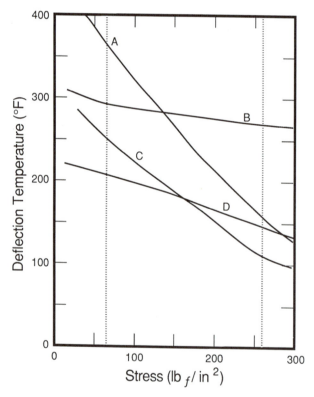

Figure 6.55. Deflection Temperature Under Load for Several Polymers (62). Deflection Temperature Measured in Accordance with ASTM D648. Dotted Vertical Lines Are at ASTM D648 Values of 66 and 264 lb$_f$/in^2. Curve A: Nylon 6 (PA-6). B: Polycarbonate (PC). C: General-Purpose PS. D: Polypropylene.

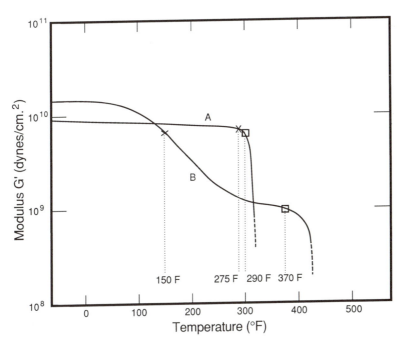

Figure 6.56. Temperature-Dependent Dynamic Elastic Modulus for Polycarbonate (Amorphous) and Nylon 6 (Crystalline) (63). Deflection Temperature Under Load (DTUL) for Each Resin Is Shown. Curve A: Polycarbonate. B: Nylon 6 (PA-6). X: Load is 264 lb_f/in^2. Square: Load is 66 lb_f/in^2.

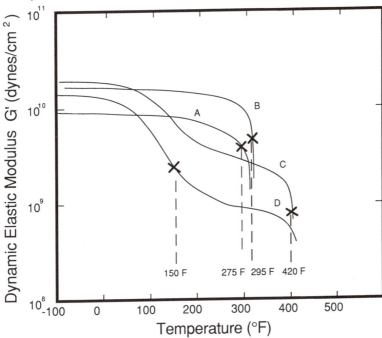

Figure 6.57. Temperature-Dependent Dynamic Elastic Modulus for Unreinforced and Glass-Reinforced Polycarbonate (Amorphous) and Nylon 6 (Crystalline) (63). Showing Deflection Temperature Under 264 lb_f/in^2 Load (X: DTUL) for Each Resin. Curve A: 20% (wt) Glass-Reinforced Polycarbonate. B: Unreinforced Polycarbonate. C: 30% (wt) Glass-Reinforced Nylon 6 (PA-6). D: Unreinforced Nylon 6 (PA-6).

perature of the resin. For the reinforced nylon, the controlling transition is the crystalline melting temperature.

In general then, the one-point value obtained from DTUL testing does not help the designer or engineer determine what phenomenon controls the heat resistance of a semicrystalline polymer. The dynamic modulus test presents a superior picture of the short-time response of a polymer to environmental temperature changes.

6.4 Solid Mechanical Properties: Short-Time Duration Tests

Most commercial polymer products must withstand some form of short-time, high-speed loading and unloading during their lifetimes. In some cases, the critical short-time duration loading occurs in shipping or even in assembly. Coffin inter-liners and automotive opera-windows are examples. The loading may be aperiodic and confined to the early period in the product life. In other cases, loading may be aperiodic but the product must be designed to resist loading throughout its lifetime. Door panels are examples. Some products see lifetime aperiodic loading over a wide temperature range. Garbage cans and refrigerator door liners are examples. In some cases, the short-time duration load is applied when the product is in full compression, as with bridge pads and floor tiles. In others, the load is applied when the product is in flexure. Equipment cabinets are examples.

As a result, most designers recognize that impact strength is usually one of the most important design properties needed when selecting an appropriate polymer for a given application. However, the designer quickly encounters a spectrum of impact tests. To achieve a functional design, the designer knows that he/she must carefully select the test or tests that yield reliable design data that correlate directly to product performance. Unfortunately, to date, no impact test or group of impact tests have shown reliable correlations to product performance. This is certainly not because ingenious individuals have not devised clever tests. Westover (64) notes:

> . . . notched and unnotched specimens have been made in various shapes and sizes and have been subjected to tensile, compressive, torsion and bending impacts. Materials have been thrown, dropped and subjected to blows from hammers, bullets, falling weights, falling balls, horizontal moving balls, pendulums and projectiles from flywheels.

The impact resistance of a plastic part depends not only on the characteristics of the base resin, but also on:

* part design,
* fillers and extenders,
* environmental conditions,
* fabricating conditions,
* nature of the shock loading, and
* type of impact tester.

Each of these parameters is known to have a profound effect on impact resistance. Turner (65) views impact failure as a short-time duration material failure. He notes that there are two elements to the failure. The first is the formal response of the material to the imposed load. It is "reversible" in the sense that the role of "excitation and response" between any two variables can be reversed. Hookean response to applied load is an example of reversibility. The second part of the response deals with "excitations" that have large amplitudes and waveforms such as ramps and pulses, as with impact. This set of phenomena does not yield reversible aspects and includes typical failure elements such as yield strength, impact strength and fatigue strength. Any constitutive equation that attempts to encompass all the important elements yields intractable mathematics. And so the second criterion of test acceptability has not been met.

Turner also points out that the very breadth of the failure "concept", including creep yielding at one extreme and brittle crack growth at the other, surely limits development of general impact testing procedures. And most assuredly failure is far more sensitive to the dominance of the polymer molecular state and the secondary aspects of processing parameters than most other material phenomena. Despite these alarming restrictions to short-time-duration impact testing, some useful insights can be extracted from standard tests.

Design plays an important role in part resistance to short-time-duration failure. Resistance to impact can be viewed as the ability of the part to absorb the energy of the applied load, the impact. Two basic energy absorbing mechanisms apparently occur simultaneously. The most obvious is the ability of the polymer to absorb the impact energy. The second, which relies on the ability of the polymer part itself to flex, bend, twist or somehow deform or distort on a gross scale, is similar to the action of a coil spring or shock absorber. The energy required to compress a spring comes directly from the initial kinetic energy supplied by the impacting projectile. Once the projectile has stopped, the energy stored by the spring acts to propel the projectile away. The spring then returns to its initial position, as Turner envisaged the Hookean response. This part deformation mechanism can be directly related to product design. Thin sections flex under load. Thick sections cannot flex as well, cannot as effectively absorb impact energy, and so will tend to fail under local impact. As a general rule for impact resistance, the energy absorbed by gross deformation is many times greater than the absorption characteristics of the bulk polymer. Thus, a part designed for an impact environment should have a substantial degree of flexibility, and part wall thickness changes should be kept to an absolute minimum.

Fillers and extenders can greatly reduce the ductility or flexibility of a polymer. This is true even if these adducts dramatically increase polymer modulus and ultimate strength. Part of this is usually the result of reduced toughness or area under the stress-strain curve. However, non-ductile adducts can also cause local stress concentration points where microcracks are initiated. This point is discussed in more detail below. Keep in mind that in energy-absorbing part design, the predominant mechanism of energy absorption is gross deformation of the plastic. Plastic with a very small elongation will therefore fail at small gross part deformation.

Recall that the plastic part must perform its designed function in a potentially aggressive environment. Probably the most important environmental parameter is temperature. As noted earlier, the glass transition temperature is the temperature above which the hard, glassy, brittle, low-impact-resistant polymer is transformed into a softer, more rubbery, ductile, relatively high-impact-resistant polymer. For example, polypropylene has very high impact resistance around room temperature, but becomes very brittle as the environmental temperature approaches 0°C, near its glass transition temperature. This will be obvious below.

Environmental aging or weathering can also be important. Aging can be the result of ultraviolet degradation, thermal degradation, or long-term diffusional loss of plasticizer. Solvent attack can also be a substantial factor. Typically, strong solvents act over the short term to dissolve the polymer. Weak solvents act over the long term to craze and embrittle the polymer. In fact, many polymers tend to craze or *whiten,* that is, develop minute surface cracks, even when the polymer is thermal- and ultraviolet-stabilized and the environment is not necessarily aggressive. Of course, crazing and embrittlement usually lead to reduction in impact resistance. It is thought that this effect is due to combined effects of long-term residual stress relief, possibly resulting in increased free volume or microvoids in the polymer, and the possible combination of oxygen or ozone degradation and slow migration of UV absorbers away from the exposed polymer surface.

Most designers do not consider the effect of fabrication on impact resistance. Two dominating effects are flow-induced orientation and molded-in or residual stresses. Flow-induced orientation can be the result of several events. Differential orientation can be caused solely by molecular orientation. It can also be caused by orientation of fillers and reinforcements and by gross flow distortions as weld and flow lines. These events yield a nonhomogeneous part having physical properties that are directionally dependent. Thus, impacting energy can be nonuniformly absorbed and the differential absorption can result in premature failure. As an example, premature impact failure results if polymer ductility transverse to molecular orientation is reduced. The continuity of molecular structure can be dramatically altered by abrupt changes in flow direction. The classic example of this is the inherent impact weakness at weld lines. Improper molding conditions can result in localized regions of high residual stress. These areas of "frozen-in stress" have a much lower resistance to shock than neighboring low stress areas. In injection molded parts, high packing pressure can cause low part impact resistance in the gate area.

Another parameter is the nature of the shock or impulse loading itself. The load can be a single sharp impulse perpendicular to the plane surface, or it can be a repeating series of low-impact blows on the corner of the part. It is not apparent at what point the designer should treat an applied load as an impact load and at what point he/she should consider it as simply a load being applied at a very high rate. Consider a similar concept in hydraulics. When a valve is closed very rapidly, a "water hammer" develops. When the valve is closed at a slower rate, there is no water hammer. The difference in these two actions is the time-rate of closure. It is apparent that when the valve is closed slowly, the system can react in an early uniform manner. When the valve is rapidly closed, a series of pressure pulses or waves is formed. The rate at which the pressure pulses propagate throughout the fluid is the *velocity of sound.* If the time-rate of closure is faster than the velocity of sound, the familiar water hammer occurs.

This analogy can be applied to impact loading in plastics. The impact loading condition can be defined as one in which the stress caused by the applied load is transferred within the polymer by stress waves of finite size initially moving at sonic velocity. The viscoelastic nature of the polymer serves to decrease the stress wave amplitude and broaden its energy peak. If the application of the load is slow, the stress waves are smaller and the assumption of instantaneous stress transference within the polymer is valid. There is certainly a broad gray area between high-rate, non-impact loading and impulse loading. Unfortunately most designers are trained in the classical approach to stress analysis. This is characterized by instantaneous stress transference. In wave phenomenon theory, the points of maximum stress depend on the nature of the applied load and upon the manner in which the stress waves are transmitted and reflected within the part. For polymers, viscoelasticity acts to confound the classical approach. For polymers, part geometry becomes critical in impact loading as well as for high-rate deformation.

The impact tests and data discussed in this section have limited utility in design. The impact values of any plastic cannot be treated as design data. The response of the polymer depends on the particular test conditions. Hence the impact ranking of plastics can have meaning for that particular test specimen configuration *only*. For material selection, then, the designer must choose a test that has loading conditions and a loading time-frame similar to what his/her part must undergo. If this is not possible, the designer can only rely on past experience, rational judgment and common sense in selecting an appropriate material. In some cases, the only choice is to fabricate prototype parts of the candidate materials and test each one under actual use conditions. From this discussion, it should be apparent to any designer that no simple test or set of tests will yield adequate information to model the complex response of an actual part to its impacting excitation.

Standard Test Methods

The standard impact test methods currently used in industry to measure polymer impact property can be divided into four broad categories:

- Pendulum-type machines,
- Falling weight to fracture,
- Constant velocity puncture, and
- Tensile impact.

Pendulum-Type Machines

On pendulum-type machines, a heavy weight or *excess-energy pendulum*, freely falls in an arc from a given height to strike the stationary specimen. After the weight has struck and fractured the specimen, it continues its motion upward along the arc. The difference in the height of the weight before and after its traverse is directly related to the energy needed to fracture the specimen. Interpretation of the data is confounded by secondary energy losses, such as the energy needed to toss the broken portion of the test specimen away from the apparatus. This factor is commonly called the "toss factor". The nature of the specimen can be either a notched or an unnotched bar, and it can be a cantilever beam (Izod) or a simply supported beam (Charpy). Schematic drawings of both systems are shown in Figure 6.58. In general, this type of test can be thought of as a pendulum-impact test against a flexed beam. The details of two of these standard tests follow.

Izod Impact: ASTM D256

The *Izod impact* test is used extensively to determine the impact resistance of brittle materials. The excess-energy pendulum falls through a vertical height of 610 mm (2.0 ft). The Izod test specimen dimensions are given in Figure 6.59. The specimen is clamped in a vertical position in a vise that is rigidly attached to the base of the test stand. The specimen can be tested without a notch. If the spectrum is notched, the notch can be placed toward or away from the striker. The center of the notch must be in a direct line with the top surface of the vise so that the point of impact on the bar is

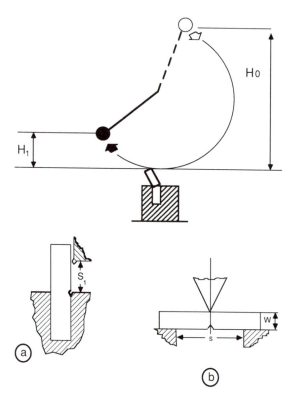

Figure 6.58.   Schematic of Excess Energy or Swinging Pendulum Impact Apparatus (66). a) Izod or Cantilever Impact. b) Charpy or Supported Beam Impact.

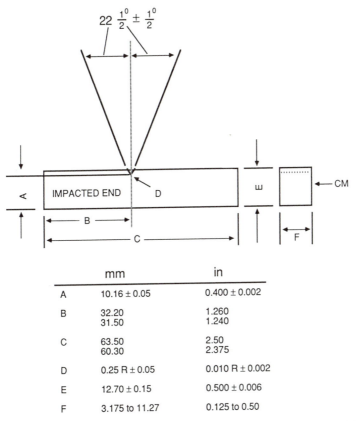

Figure 6.59.   Notched Izod Impact Test Specimen Geometry According to ASTM D256.

exactly 22 mm (0.866 in) above the top surface of the vise. At this point, the pendulum impact velocity is 3.46 m/s (11.35 ft/s). The Izod impact strength of the specimen is always reported as the energy *per unit width of notch*, J/m (ft-lb/in). Typical values of notched Izod impact strengths for several polymers are given in Table 6.10. Note the relatively wide range in values for many of the materials.

Example 6.8

Q: It is proposed to use a modified excess energy pendulum of the Izod type to measure the impact characteristics of a high-modulus composite. The pendulum arm is 48 in long and the effective mass of the striker is 16 lb$_m$. It is proposed to start the striker at a height of 3.6 ft above the impact point. It is required that the results be within the standard Izod impact guidelines, so that the results can be compared with standard impact data. The standard Izod impact test requires that the pendulum velocity at the point of impact be 3.46 m/s or 11.35 ft/s. Is this OK?

Table 6.10. Notched Izod impact strength ranges for many polymers at 23°C (67).

Polymer	Notched Izod (in-lb$_f$/in of notch)
ABS—high impact	4 to 10
ABS—medium impact	2 to 4
Polyacetal (POM), homopolymer	0.8 to 2.3
Polyacetal (POM), copolymer	0.8 to 2.3
Acrylic (PMMA), standard	0.3 to 0.5
Acrylic (PMMA), high impact	1.0 to 2.3
Cellulose acetate (CA), soft	5.0
Cellulose acetate (CA), hard	0.5
Cellulose propionate (CAP), soft	10.0
Cellulose propionate (CAP), hard	0.5
Cellulose acetate butyrate (CAB), soft	6.0
Cellulose acetate butyrate (CAB), hard	0.75
Diallyl phthalate (DAP), mineral filled	≈ 0.3
Diallyl phthalate (DAP), glass fiber filled	≈ 1.5
Epoxy, mineral filled	≈ 0.35
Epoxy, glass fiber filled	≈ 1.5
Nylon 66 (PA 66), homopolymer	0.5 to 0.7
Nylon 66 (PA 66), copolymer	12 to 22
Nylon 66 (PA 66), glass fiber filled	3.0 to 5.0
Phenol-formaldehyde (PF), wood flour filled	≈ 0.25
Phenol-formaldehyde (PF), mineral filled	≈ 1.0
Phenol-formaldehyde (PF), glass fiber filled	≈ 3.0
Polycarbonate (PC)	18.0
Polyester, thermoset (UPE)	0.2 to 3.5
Polyester, thermoplastic, glass fiber filled	1.9 to 2.5
Polyethylene, medium density (MDPE)	1.0 to 1.5
Polyethylene, high density (HDPE)	1.0 to 1.5
Modified polyphenylene oxide (mPPO)	3.0 to 7.0
Polystyrene (GPS), unmodified	3.0
Polystyrene (MIPS), rubber modified	0.7 to 3.0
Polysulfone (PSO$_2$)	1.3
Polypropylene (PP), homopolymer	0.4 to 1.0
Polypropylene (PP), copolymer	0.7 to 9.5
Polystyrene-acrylonitrile (SAN)	0.4
Urea-formaldehyde (UF), filled	0.3
Polyvinyl chloride (PVC), homopolymer	0.7
Polyvinyl chloride (PVC), rubber modified	15.0

A: The impact velocity can be calculated from:

$$V = \sqrt{2gh}$$
$$= \sqrt{2 \times 32.2 \times 3.6} = 15.2 \text{ ft/s.}$$

The test deviates from the standard by:

$$\frac{\Delta V}{V_o} = \frac{15.2 - 11.35}{11.35} = 0.34 \text{ or } 34\%.$$

Hence the test is being conducted at strain rates that differ significantly from those of the standard. The results may not be correlatable. Note that neither the length of the pendulum nor the mass of the striker is needed to come to this conclusion.

One difficulty in interpreting Izod data is the effect test parameters have on measured value. For example, the test specifications allow specimen *thickness* to range in value from 3 mm (1/8 in) to 13 mm (½ in). The effect of specimen thickness on Izod value is shown in Figure 6.60. Many polymers, such as polyphenylene oxide, nylon, cellulosics, and highly acrylonitrile-modified ABS ("hard" ABS), show very little effect

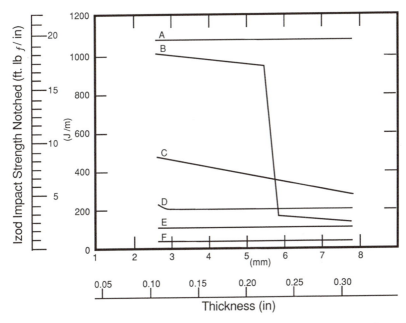

Figure 6.60. Effect of Notched Izod Impact Test Specimen Thickness on Impact Strength for Several Polymers (68). Curve A: Nylon 6 (PA-6), 8% (wt) Water. B: Polycarbonate (PC). C: Soft ABS. D: Hard ABS. E: Cellulose Acetate Butyrate (CAB). F: Modified Polyphenylene Oxide (mPPO).

of specimen thickness. Polycarbonate, on the other hand, shows a drop in Izod value with increasing thickness. The drop is so dramatic and repeatable that the drop-off point is referred to as the *critical impact thickness* of PC.

    Notch radius is also a critical parameter. The test specifications are very explicit. The notch must have a 45° ($\pi/4$ rad) apex with a tip radius of 0.25 ± 0.05 mm (0.010 ± 0.002 in). The notch depth must be approximately 10 mm (0.40 in). Care must be taken in insuring that the actual notch tip radius meets the proscribed values. Notching tools can easily be dulled by fillers, fibers, abrasive polymers, and occasional misuse. The effects of notch radius on Izod values for several polymers are seen in Figures 6.61 and 6.62. Notch sensitivity can be related to the stress concentration at the notch tip:

$$(6.33) \qquad\qquad k = 1 + 2 \left(\frac{a}{r} \right)^{1/2}$$

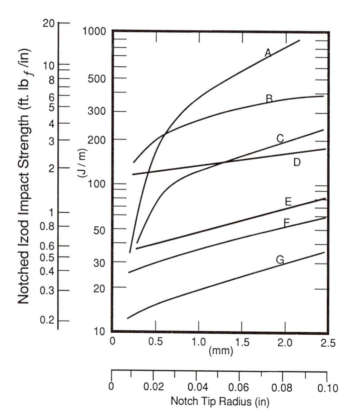

Figure 6.61.   Effect of Notch Tip Radius on Notched Izod Impact Strength at 23°C for Several Polymers (68). Curve A: PVC. B: Polyethylene (PE). C: Nylon 6 (PA-6). D: ABS. E: Polystyrene-Butadiene (HIPS). F: Melamine–Formaldehyde. G: Polymethyl Methacrylate (PMMA).

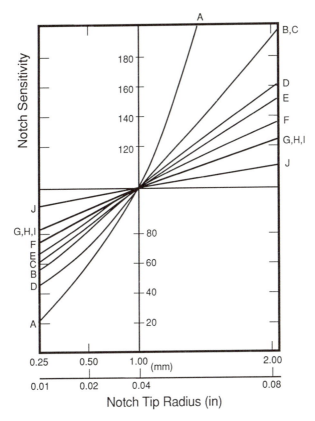

Figure 6.62.   Effect of Notch Tip Radius on Notched Izod Impact Strength Sensitivity at 23°C for Several Polymers (69). All Curves Pass Through 100 at a Notch Tip Radius of 1 mm. Curve A: Rigid PVC (RPVC). B: 3.0 MFI (Melt Flow Index) PP Homopolymer. C: 4.5 MI HDPE. D: 0.8 MFI PP Homopolymer. E: 3.0 MFI PP Copolymer. F: 3.0 MI HDPE. G: High-Impact ABS. H: 0.8 MFI PP Copolymer. I: High-Impact Polystyrene (HIPS). J: ABS.

where k is the stress concentration factor, a is the notch depth and r is the radius of the notch tip. Turner (70) suggests that a more logical approach is to consider a *critical strain energy release rate*, G:

$$(6.34) \qquad G = \frac{(\text{Impact Energy})}{b \, d \, \phi}$$

where b is a sample width, d is sample thickness before notching and ϕ is a geometrical factor related to the specific type of test (Izod, Charpy, etc.). Specimens having various notch depth-to-thickness ratios are then tested and their impact energy values plotted against bdϕ. The slope of a straight line fitted to the data should yield the critical strain energy release rate, G. This approach should yield some degree of internal consistency for a *given* polymer impacted on a *given* testing machine, but no valid extrapolation from material to material or from one impact testing technique to another should be considered.

Temperature has a profound effect on Izod impact strength (Figure 6.63). General purpose polystyrene exhibits a relatively low, nearly temperature-independent impact strength until its temperature approaches $T_g = 100°C$. Of course, at T_g, the polymer is rubbery tough and so impact resistance increases substantially. Impact-modified polystyrene contains butadiene, which has a T_g of $-90°C$. As a result, butadiene is rubbery tough at room temperature and impact energy is absorbed in the rubbery butadiene particles. This inhibits crack growth and produces an impact-resistant polymer. Below the butadiene T_g, the product is brittle. Above polystyrene T_g, it is very rubbery. The effect of the combination of specimen temperature and notch dimensions on Izod impact strength values is seen for PVC in Figure 6.64.

The sample fabrication condition can also influence the impact values. This is seen in Figure 6.65, where ABS molecular orientation at 170°C melt temperature yields dramatic differences in impact strengths along and across the injection molding flow direction. The effect is somewhat reduced by molding at 230°C melt temperature. In

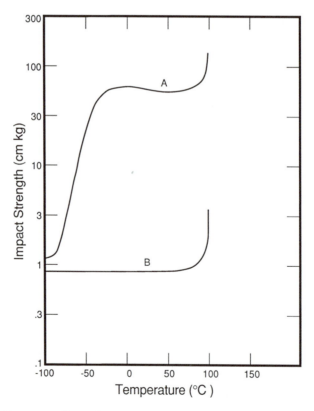

Figure 6.63. Temperature-Dependent Notched Izod Impact Strength for Two Types of Polystyrene (71). Curve A: High-Impact Polystyrene (HIPS). B: General Purpose Polystyrene (GPPS).

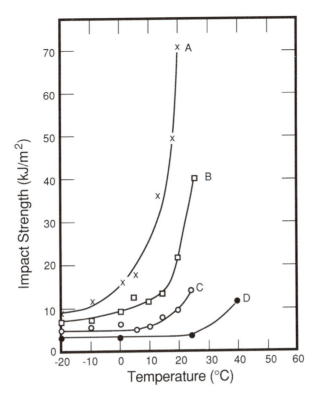

Figure 6.64.   Temperature-Dependent Notched Izod Impact Strength for PVC with Notch Radius as Parameter (72). Curve A: 2 mm Radius Notch, 2.5 mm Deep. B: 1 mm Radius. C: 0.5 mm Radius. D: 0.25 mm Radius.

Section 2.4, it was noted that, in thickness cross section, the polymer closest to the surface of the part can have greater preferred molecular orientation than that near the centerline. The molecules closest to the centerline can have longer time to relax into a more random configuration than those near the mold surface. This anisotropy overlays the more obvious anisotropic effect of lower melt temperatures. It was thought for years that mechanically cut notches would produce greater variability in values than those from molded-in notches. And it was thought that this variability was caused by uncertainty in cutting the notches through the anisotropic molecular layer at the surface. Therefore, the molded-in notch was thought to have more consistent molecular order. After much testing, it is now thought that the molecular anisotropy that occurs in the molded-in notch region simply by the change in specimen thickness at that point causes just as much variability as that caused by cutting.

Charpy Impact Test: ASTM D256

In the Charpy test, the sample is a simply supported *horizontal* beam with the notch, if any, away from but opposite to the striker (Figure 6.58). Typical Charpy impact

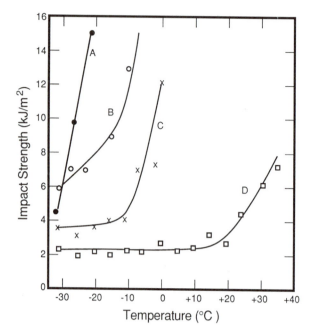

Figure 6.65.   Temperature-Dependent Notched Izod Impact Strength of Injection Molded ABS molded at two temperatures (73). Curve A: 170°C in Flow Direction. B: 230°C in Flow Direction. C: 230°C Across Flow Direction. D: 170°C Across Flow Direction.

strengths are given in Table 6.11. As one might expect, brittle materials exhibit relatively low Charpy and Izod values. Ductile materials exhibit values that are relatively high. Experiments show a direct relationship in Izod and Charpy impact energies (75). Turner (76) suggests that, in general, polymers should be classified only as:

• Tough when sharply notched (notch radius, r < 0.25 mm),
• Brittle when sharply notched but tough when bluntly notched (r > 1 mm),
• Brittle when bluntly notched but tough when unnotched,
• Brittle when unnotched.

He suggests that when testing in the flexed beam configuration, these general criteria be used rather than specific numerical rankings. The overriding reason for this, he notes, is the extensive list of potential sources of error, which includes:

• Temperature,
• Molecular orientation due to flow,
• Molecular orientation across the part thickness due to heat transfer,
• Sample thickness,
• Strain rate,
• Clamping pressure,
• Sharpness of the notching tool,

Table 6.11. Notched charpy impact strengths for several polymers at 23°C (74).

Polymer	Charpy impact strength, (ft-lb$_f$/in notch)
Rigid polyvinyl chloride (RPVC)	0.20
Rubber-modified polyvinyl chloride (FPVC)	0.54
Polystyrene-acrylonitrile (SAN)	1.38
Cellulose acetate (CA)	0.36
Polystyrene (PS), unmodified	0.07
Polystyrene (HIPS), rubber-toughened	0.25
Polymethyl methacrylate (PMMA), unmodified	0.11

- Residual or molded-in stress in specimen,
- Nature of fibers, fillers, adducts,
- Movement in the clamp during impact,
- Vibrations in the test fixture during impact,
- Energy imparted to the broken fragments, viz, "throwing energy" or "toss factor",
- Uneven fracture propagation across the notch,
- Friction in the pendulum bearing,
- [And so on].

As with notched Izod testing, notched Charpy impact strengths depend on the type and severity of the notch (Figure 6.66). In cavities G and J, impact is in the flow direction,

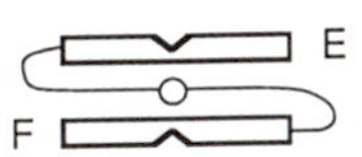

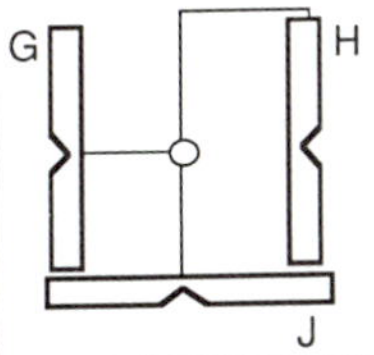

Specimen Identification*		Mean (kJ/m^2)	Standard Deviation (kJ/m^2)
	E	6.2	0.6
	F	6.2	0.4
	G	2.1	0.6
	H	3.4	0.5
	J	2.7	0.4

* Notch radius, 0.25 mm

Figure 6.66. The Effect of Flow Geometry on Notched Charpy Impact Specimens of Polyethylene-Polypropylene Copolymer at -20°C Temperature (77). Notch Radius, 0.25 mm. Testing According to British Standard (BS) 4618.

and so the impact strengths are low. When the impact strength of the cavity H specimen is compared with that for cavities E and F, gate dimension is the primary criterion. There is greater molecular orientation perpendicular to the impact direction in cavities E and F and so the impact strength is greater.

Falling Weight to Fracture: ASTM D3029

In this type of test, the specimen is a flat plate. The impacting energy is supplied by dropping a weight of known mass from a specific height. The impact strength, F_{50}, is usually defined as the amount of energy required to fracture 50% of a large number of identically formed specimens. This test is also called the "flexed-plate impact test" (78). There are two types of falling weight tests. The first seeks drop height to incipient fracture. The second measures fracture energy during penetration of the impacter. The first is described here. The second is described in the next section. The *Gardner drop-weight test* places a mandrel with a rounded tip against the horizontal specimen (Figure 6.67). The mandrel tip is 15.86 mm (5/8 in) in diameter. The early standard required the ring supporting the specimen to be 16.26 mm (0.64 in) in diameter. It was discovered that this close tolerance caused the polymer to be sheared between the mandrel tip and the support ring. To minimize shear failure, many tests now call for "ring out" data. Here the support ring is replaced with a 76.2 mm (3 in) ID × 101.6 mm (4 in) OD ring, having an inside corner radius of 1.59 mm (0.0625 in).

Standard drop-weight testers combine the mandrel and weight into a single weight having a rounded tip or "tup" that impacts directly onto the specimen plate (Figure 6.68). To cover the spectrum of material impact resistances, there are two sets of tup and ring dimensions, Table 6.12.

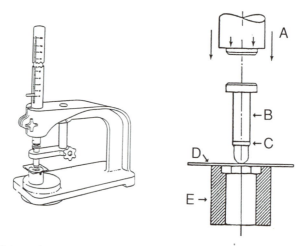

Figure 6.67.   Schematic of Gardner Drop Weight Impact Tester. The Mandrel Is Impacted by a Known Weight Dropping from a Given Height. A: Weight. B: Mandrel. C: Tup. D: Specimen. E: Pedestal.

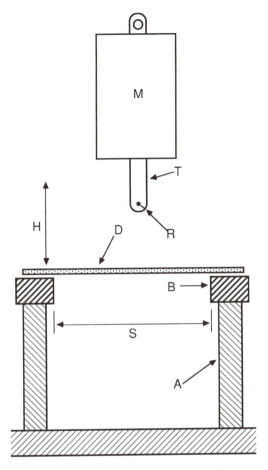

Figure 6.68.   Schematic of Simple Drop Weight Impact Tester, with Impacting Tup Attached to Weight (80). A: Support. B: Ring. D: Specimen. H: Height. M: Mass. R: Radius. S: Span. T: Tup.

Table 6.12.   ASTM D3029 standard free-falling dart tup dimensions.

Procedure	Tup diameter		Ring diameter	
	(mm)	(in)	(mm)	(in)
A	38.1	1.5	127.0	5.0
B	12.7	0.5	76.2	3.0

Many samples are needed to accurately determine the impact energy distribution for a given material. There are two methods of analysis (79). The *Probit method,* test or transformation depends on statistical testing of a large number of like specimens for a wide spectrum of impact energies. Approximately 80 specimens are needed to obtain statistical data for a single material. The cumulative failures are plotted against the impact energy (or simply drop height), as seen in Figure 6.69. If the data are plotted on probability paper as impact energy against failure percentage, a normal distribution would be a straight line, as seen in Figure 6.70. The drop-weight impact value is the point where 50% of the specimens have failed.

Note that "Probit curves" may be dramatically different for materials with the same mean impact energy. Furthermore, Probit curves are questionable for very ductile materials, materials that are alloys or interpenetrating networks, and materials that are heavily impact-modified. Furthermore, the impact energy in a typical drop-weight test is changed by changing the drop-weight *height* rather than the impact *weight.* This changes the *rate* of application of the impacting energy. The typical strain rate range for these falling weight tests is 0.1 to 1 s^{-1}. At the high strain rate, the material may undergo brittle failure at lower strain levels. In other words, the designer must be aware of a potential polymer brittle-ductile transition during changes in impacting tup height.

An alternative to the Probit method is the *Bruceton test,* also called the "up-and-down" or staircase method. The test begins at an arbitrary drop height. If the sample

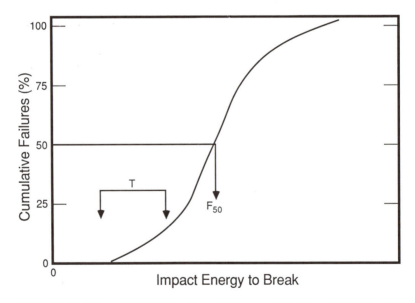

Figure 6.69.   Typical Normal Distribution Impact Energy to Fail Curve for Plastics (81). F_{50}: Cumulative Failures to 50% of the Total Number of Specimens Tested. T: Threshold Failure Energy.

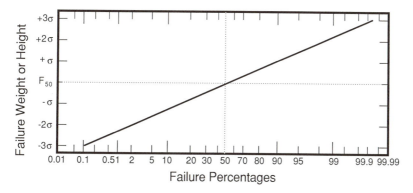

Figure 6.70. Replot of Normal Distribution Impact Energy to Fail Curve with Failure Percentage Plotted on Probability Paper (79).

fails, the next drop is at a decreased height. If it doesn't fail, the next drop is at an increased height. Each time a specimen fails, the drop height is decreased an increment. Each time it doesn't fail, the drop height is increased an increment. Statistically, then, a mean failure height can be determined and the mean failure energy, or the energy needed to produce 50% failures, can be calculated. As with the Probit test, as many as 50 specimens are needed to find the appropriate values. The breadth of the distribution is estimated by the sample standard deviation.

Example 6.9

Q: In Company A, a technician uses a 10 lb weight to determine the drop. weight impact resistance of injection molded glass reinforced polyethylene disks. He finds the F_{50} value to be 26 ft-lb$_f$. In Company B, a technican uses a 2 lb weight on an essentially identical material and finds an F_{50} value of 18.7 ft-lb$_f$. Are these materials substantially different in impact resistance? Explain your answer.

A: From the data, the following table can be constructed:

	Company A	Company B
F_{50} Drop height	2.6 ft	9.35 ft
Impact velocity ($\sqrt{2gh}$)	12.9 ft/s	24.5 ft/s

The differences in the data can be the result of several factors. For the purposes of discussion, assume that the molding conditions and the general resin characteristics such as viscosity, adduct package, and fiber loading, are essentially equivalent, and consider only the nature of the test results.

First, the polymers may be exhibiting different velocity effects. Even if they have the same relative velocity effects, they both may be strain-rate sensitive. Company B test is run at a substantially higher strain rate than that of Company A. Then, the technicians may be using different criteria for failure. One may be using a visual effect, looking for a reverse-side crack in raking light, for example. The other may be drawing a fingernail or a cotton ball over the reverse side, and looking for resistance caused by microcracks. The answer to the question, then, is that without further documentation, it is impossible to determine if these materials are substantially different in impact resistance.

Note: When conducting any test of this nature, the test conditions must be carefully documented and the *criterion of failure* clearly described.

In addition to the large number of test plaques needed for each test value, the subjective nature of "impact failure" must be considered. For brittle polymers, the nature of impact failure is clear. When the sample fails, it fractures cleanly into several pieces. For a ductile material, on the other hand, fracture may not be complete. The first sign of failure may simply be the appearance of a crack or series of cracks on the specimen surface opposite the point of tup impact. For some materials, this crack may be very difficult to see and may be detected only by running a fingernail over the surface or by using ink. Tough materials, particularly highly reinforced polymers, may show fine surface cracks at very low impact levels, but may not actually undergo brittle failure until very high impact levels are reached. The appropriate value of failure energy may then be entirely subjective. Furthermore, the bimodal energy absorption nature of many polymers may lead to substantial errors in interpretation of the data from these tests. This is seen in Figure 6.71 for the staircase method, where even after 100 test specimens, the absolute value of the failure height is unknown, and in Figure 6.72 for the

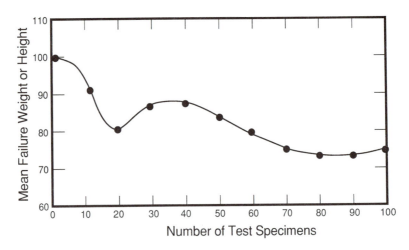

Figure 6.71.   Effect of Sample Size on Bruceton Impact Energy Stairstep Technique, Bimodal Distribution, Simulated Curve (79).

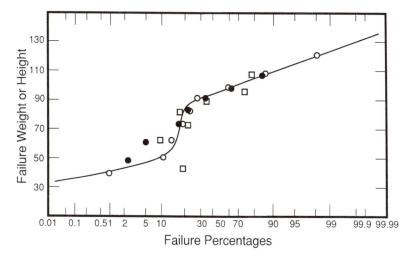

Figure 6.72.   Effect of Sample Size on Probit Impact Energy Distribution, Bimodal Distribution, Simulated Curve (79). Square: 20 Specimens. Closed Circle: 40 Specimens. Open Circle: 100 Specimens.

Probit method, where the failure height is a nonlinear function of the cumulative failure.

As with all tests, temperature *must be* considered an important variable in the interpretation of the impact properties. Recall that the Probit and Bruceton methods when only reporting a F_{50} value are *one-point* tests, that is, drop height at 50% failure level. As might be expected, the failure curves shift with temperature, as seen in Figure 6.73

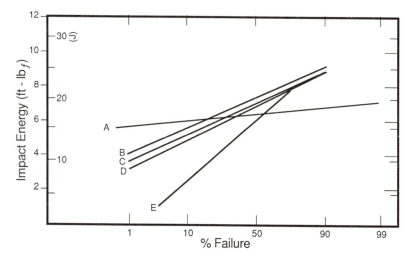

Figure 6.73.   Temperature-Dependent Probit Impact Energy Distribution for High-Impact ABS (79). Data From Ref. 63. Curve A: +20°C, Tough. B: +10°C, Tough/Bructile. C: 0°C, Tough/Bructile/Brittle Transition. D: −10°C, Brittle. E: −20°C, Brittle.

for ABS. Thus, even though the impact energy levels at $+20°C$ and $-20°C$ appear to be the same, at 6 ft-lb$_f$, the polymer at 20°C is substantially tougher to higher impact levels than at $-20°C$. Typical F$_{50}$ drop weight impact data are shown in Figure 6.74.

Note that the specimen thickness is not specified in either test. As with the flexed-beam tests, the relationship between thickness and impact energy is not simple. Turner notes that for a given polymer, thicker sections have greater load-bearing capability but can be less ductile. Typically, thicker sections can withstand higher impact energies than thinner ones. However, the increase with thickness is not the same for all materials as seen in Figure 6.75. The samples in this case were compression molded and substantially free of orientation. The impact data of PP homopolymer and medium impact polystyrene show relatively little sensitivity to thickness and so the specimens remain relatively brittle. On the other hand, the impact characteristics of rigid PVC show substantial improvement with thickness. Of course, residual-stress-free parts are ideal. Most fabricated parts are rarely free of residual orientation and frozen-in stresses. As a result, there can be a substantial susceptibility to fracture in the orientation flow direction. Certainly high melt viscosity, large thermal gradients, and long, thin flow paths exacerbate the problem. But polymer molecular weight, molecular weight distribution, and degree and rate of crystallization are also factors. The magnitude in the drop of impact strength values owing to residual molding factors differs from material to material. To compare impact strengths, then, the designer must find the variation in

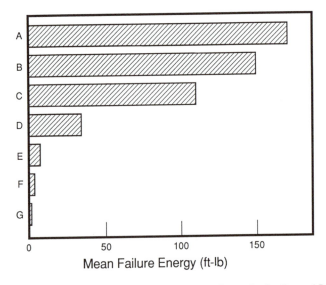

Figure 6.74.   F$_{50}$ Room Temperature Drop Weight Impact Strengths for Several Polymers (83). 1/8-inch Thick Disk. 3-inch Diameter Support Ring. A: Polycarbonate. B: Nylon 6 (PA-6) Co-polymer. C: Nylon 6 (Dry), D: Modified PPO. E: Acetal. F: Mineral-Filled Nylon 6. G: Glass Reinforced Thermoplastics.

impact strength with thickness for *both* isotropic and anisotropic specimens molded over a wide range of processing conditions.

As noted above, there are three types of impact failure. *Ductile* or *tough* failure is characterized by material yielding, perhaps just momentarily, under impact. The tup produces a depression in the specimen surface. At sufficiently high impact energies, the depression is punched or tears completely through the specimen. *Brittle* failure is characterized by complete fracture or shattering of the specimen. Microscopic examination of the fracture edge shows no yielding. If the material is very brittle, the sample shatters. For a less brittle material, the tup punches a hole through the specimen as the specimen fractures into several pieces. The third type of failure is *ductile–brittle* or "bructile" failure. Typically, some initial yielding occurs at the point of impact, followed by catastrophic failure.

An idea of the impact energy associated with each of these types of failure can be obtained by reviewing a typical load–deformation curve such as Figure 6.76. Brittle

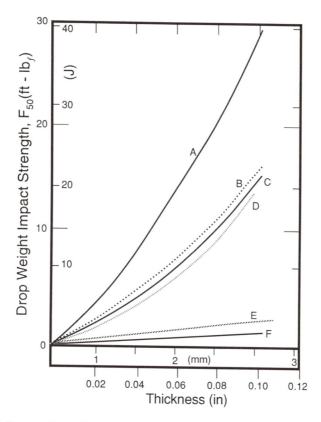

Figure 6.75. The F_{50} Room Temperature Drop Weight Impact Strengths of Several Polymers as Functions of Disk Thickness (84). Curve A: Rigid PVC. B: High-Impact ABS. C: HDPE (0.96 g/cm³, 4.0 MI). D: PP Copolymer. E: Medium-Impact PS. F: PP Homopolymer (3.0 MFI).

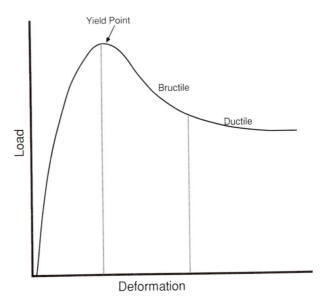

Figure 6.76. Typical Load–Deformation Curves for Polymers That Show Yielding (85). The Region Immediately Beyond the Yield Point Is Called "Bructile", That Is, Neither Brittle Nor Ductile.

failure occurs on the initial linear or near-linear part of the curve prior to the yield point. A ductile failure is associated with the high elongation under load on the final part of the load–deformation curve. A "bructile" failure probably occurs on that part of the curve beyond but near to the yield point.

Brittle failure is almost always associated with a low level of impact. In a ductile failure, deformation at failure is reasonably constant for a given striker size. The area beneath the curve up to the point of failure gives a measure of the impact energy necessary to cause failure. It is apparent that in most cases, for a given thickness, tough or ductile failure is associated with a high and reasonably constant impact energy. Impact energy in the "bructile" region is relatively high but values can be quite a bit more variable than for that in either of the other two regions. If all failures in a given test are ductile or all are brittle, a reliable and reproducible measure of the material impact strength can be obtained. If differing types of failure are possible, a much wider variation in impact values are obtained from repeated tests.

In practice, end-use performance must be judged against the *lowest* level at which impact failures are likely to occur, not against the 50% failure level obtained from Probit or Bruceton procedures. It is therefore important to have a measure of this minimum level. Some indication can be obtained by slightly modifying the Probit procedure. In this modification, sets of ten samples, are tested at each of a series of increasing energy levels, beginning at the impact level at which no samples fail. If the percentage level is then plotted against the impact energy on probability paper, the energy level for *zero* failures can be determined. The slope of the best-fit curve then gives an estimate of the variability of impact energy and the likelihood of an occasional sample failing at very low impact energy levels.

The importance of processing conditions and residual stresses on the impact level for flexed-beam impact tests has already been emphasized. These effects can also alter flexed-plate impact tests. For example, high-density polyethylene (0.96 g/cm^3, MI = 4.0), injection molded under recommended conditions, yields plaques that fail in a ductile manner. When this same material is molded under adverse conditions, such as low melt temperature, cold mold, and high injection pressure, the plaques fail in a "bructile" or even brittle manner. When these plaques are tested in a Bruceton "staircase" method, there may be little difference in the F_{50} values. But when these plaques are tested in the Probit manner, the slopes of the curves may be significantly different (Figure 6.77). This implies that the chances of a spurious failure at low energy levels are greater when the polymer is molded under less than ideal conditions.

Of course, residual stresses can be the major influence in shaping these curves. Consider F_{50} values for HDPE molded under a range of melt temperatures (Figure 6.78). The results are compared with data for essentially stress-free compression specimens. As the melt temperature increases, the residual stress level decreases. This is reflected in the increasing impact strength, ultimately approaching that of the compres-

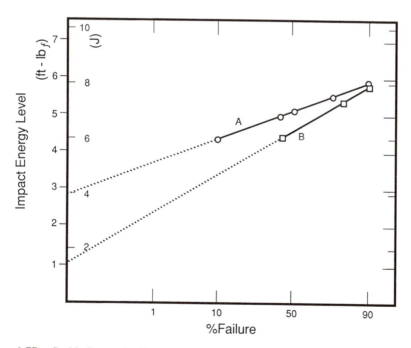

Figure 6.77. Probit Curves for HDPE Showing Ductile and Bructile Failures (86). Curve A: Ductile Failure. B: Bructile Failure.

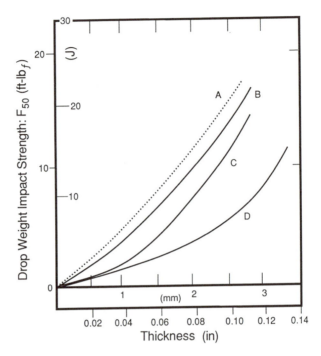

Figure 6.78.   The Effect of Specimen Thickness on F_{50} Drop Weight Impact Strength of HDPE, 0.96 g/cm³, 4.5 MI (87). Curve A: Compression Molded Specimens. B: Injection Molded Specimens, Melt Temperature 280°C. C: Same but Temperature 240°C. D: Same but Temperature 200°C.

sion molded specimens. As with flexed-beam tests, temperature can strongly influence drop-weight impact data (Figure 6.79). The reduction in impact strength with decreasing temperature is particularly marked with the otherwise higher impact-resistant polymers such as polypropylene copolymer and ABS. The superior low-temperature impact strength of HDPE is apparent. Recall that T_g for HDPE is about $-100°C$. It was noted above that ABS showed a change in the slope of the Probit energy curve with temperature (Figure 6.73). Many other materials show only a vertical shift of impact energy with temperature (Figures 6.80 and 6.81).

Constant Velocity Impact Testing

One of the major drawbacks of the *falling weight to fracture* techniques is the large number of test specimens needed to measure the fracture energy level at a given temperature and a given specimen thickness. Sample-to-sample variations within a single batch of specimens are not easily discernible. Only the spread in the data, as given by the standard deviation, for example, would indicate that a secondary effect may be

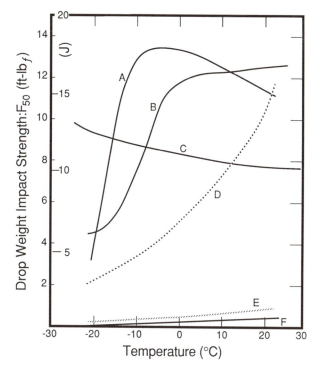

Figure 6.79.   Temperature-Dependent F_{50} Drop Weight Impact Strengths for Several Polymers Injection Molded at 260°C (88). Curve A: PP Copolymer with 0.8 MFI (Melt Flow Index). B: High-Impact ABS. C: HDPE, 0.965 g/cm^3, 4.0 MI. D: Normal Impact ABS. E: Medium Impact PS (MIPS). F: PP Homopolymer with 0.8 MFI.

present. In essence, then, these test procedures are based on the statistical probability of failure of a large group of specimens under a given type of load. The response of an individual specimen is not considered.

   Constant velocity impact systems have been developed to help quantify the response of a single sample. There are two types of constant velocity impacters used today. The *fluid actuated cylinder impacter* uses a hydraulically or pneumatically driven cylinder to advance the impacter at constant velocity. The hydraulic systems are marketed by Rheometrics, MTS and Instron. Plastechon and Pymetran market pneumatic systems. The driving system is designed so that the energy of the impacters is much greater than the energy to break the polymer specimen. As a result, there is no appreciable drop in impacter velocity as it penetrates the sample. A typical hydraulic unit is shown as Figure 6.82. The impacter velocity value is set at from 200 to 30,000 in/min (5 to 750 m/min) and is maintained with a rapidly responding closed-loop hydraulic servo valve (Figure 6.83). A high pressure hydraulic pump is used to charge the accumulator that provides power to the actuator.

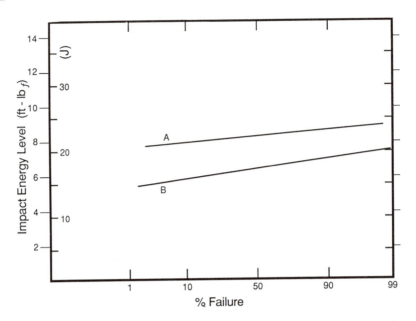

Figure 6.80.   Temperature Effect of Probit Impact Energy Level of HDPE (89). Curve A: −20°C, Tough. B: +20°C, Tough.

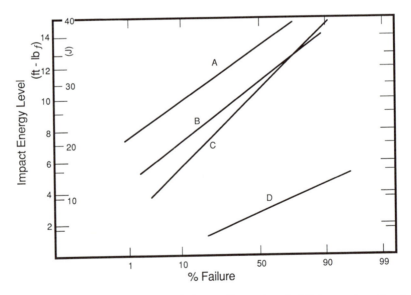

Figure 6.81.   Temperature Effect of Probit Impact Energy Level of Polypropylene Copolymer (90). Curve A: 0°C, Bructile. B: 20°C. Tough. C: −10°C, Bructile.D: −20°C, Bructile.

Figure 6.82. The Rheometrics Electrohydraulic Impact Tester (Courtesy of Rheometrics, Piscataway, NJ.).

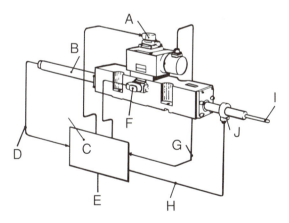

Figure 6.83. Schematic of the Servohydraulic Mechanism of the Rheometrics Impact Tester (Courtesy of Rheometrics, Piscataway, NJ.). A: High Speed Servovalve. B: Velocity Transducer. C: Control Electronics. D: Piston Velocity Feedback. E: Velocity Command. F: Low Speed Servovalve. G: Third-Stage Spool Position Feedback. H: Piston Acceleration Feedback. I: Impacting Probe. J: Accelerometer.

Figure 6.84.   The Dynatup High Mass Drop Weight Impact Tester (Courtesy of Effects Technology, Santa Barbara, CA.).

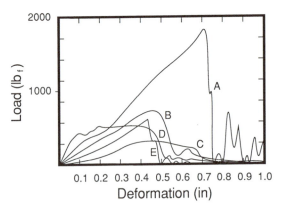

Figure 6.85.   Rheometrics Impact Tester Load–Deformation Curves for Several Polymers at 23°C (91). Tup Speed: 10,000 in/min. Sample Thickness: 0.125 in. Curve A: Polycarbonate (PC). B: ABS. C: LDPE. D: General Purpose PS. E: HDPE.

The second type of constant velocity puncture impacter is an adaptation of the *falling weight tester*. In this system, the mass of the falling weight is very large. When the impacter penetrates the sample, its momentum is not substantially decreased by the energy absorbed by the failing sample. This falling weight tester is called the Dynatup Drop Weight Tester and is commercially available from Effects Technology (Figure 6.84). Similar devices are marketed by ICI Systems, Ltd., Rheometrics, TMI and Kayeness.

In both types of impacter, a force transducer is mounted in the tup or impacting end of the probe. The position of the impacter is measured by displacement or by a velocity transducer. The data are typically displayed as the force applied by the impacter against the specimen as a function of the deflection of the specimen (really, the position of the probe relative to the original flat surface of the specimen). Force–deflection curves of several thermoplastic materials are shown in Figure 6.85. As discussed in Section 3.9, the area under the force–deflection curve is the work done by the probe during pene- tration. It is equal to the energy absorbed by the specimen during impact. As noted above, this energy is the sum of the local deformational energy and the gross defor- mational energy or displacement energy of the disk. The force–deflection curve is normally electronically integrated to provide a plot of total energy as a function of specimen deflection, as shown in Figure 6.86.

There are two types of polymer response to constant velocity puncture impact. Poly- carbonate and polystyrene exhibit *uniformly increasing deflection to break*. ABS and

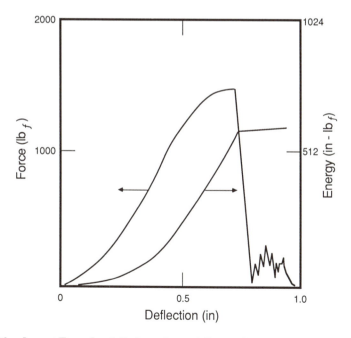

Figure 6.86.   Impact Tester Load–Deformation and Energy–Deformation Curves for ICI Vic- trex 300P Polyethersulfone ($PESO_2$) at 23°C (75). Specimen Thickness is 3.31 mm. Tup Radius is 0.5 in. Specimen Held in 3 in Diameter Ring. Tup Speed: 1000 in/min.

high-density and low-density polyethylene exhibit *increasing deflection to yield*. The yield point occurs at the point of first significant change in force–deflection curve. The point of catastrophic failure is obviously the point at which the load level drops to zero. As expected, brittle polymers exhibit a near-linear deflection to failure. However, some ductile polymers such as polycarbonate exhibit a similar behavior, with yielding and catastrophic failure occurring at essentially the same loading level. Again as expected, ductile polymers exhibit a yielding region, with the applied force decreasing thereafter until failure occurs. The energy to fracture is determined as the energy at the point of catastrophic failure. For many polymers, particularly filled or reinforced ones, the force–deflection curves do not clearly display either the yield point or the catastrophic failure point. For these cases, care must be taken in accurately detailing just how these values were obtained.

As discussed in Chapters 2 and 3, the time-dependent nature of polymers can be seen in material response to loads applied at various rates. At very high tensile rates, for example, most ductile polymers behave in a brittle fashion, failing without yielding or even appreciable deformation. The point at which a normally ductile polymer responds in a brittle fashion is called the ductile–brittle transition. This type of transition is seen in puncture impact as well. As with falling-weight tests, the size of the tup tip and the nature of the retaining ring can also affect the nature of the impact (Figure 6.87). It is apparent that to compare puncture impact data from several sources, the nature of the impacter, the type of ring support system, the thickness and condition of the test specimen, and the tup diameter must be clearly indicated. Impact energies for several unfilled thermoplastics are given in Table 6.13. Note that the test specimen thickness (1/8 in), the ring support dimension (3 in) and the tup type ($\frac{1}{2}$ in tup with spherical tip) are all specified.

Recently, Crabb (92) measured the impact resistance of a rubber-toughened PMMA by the Bruceton, Probit and constant velocity techniques. The geometries of the Bruceton and Probit systems were identical with 12.7 mm tup diameter and support ring diameter of 13 mm. The constant velocity apparatus had a 12.7 mm diameter dart and a 31.75 mm diameter support ring. The Bruceton test was conducted five times with 25 to 50 specimens used for each test. The average F_{50} value was *9.4 J*. The Probit test was conducted with 16 sets of 50 samples, as seen in Figure 6.88, the F_{50} height was 2 m. The energy of F_{50} failure was *5.69 J*.

A typical constant velocity trace for a single test specimen is shown in Figure 6.89. Note two points of failure, A and B. A is assumed to be the point at which cracking first occurs. B is the point of deflection where catastrophic fracture occurs. The area under the curve to point A was assumed to be the energy to crack the specimen. A set of 150 specimens was tested. A representative set of load–deflection curves for 50 of the samples is given in Figure 6.90. Note the significant variation in the load–deflection traces. The energy to crack each of the 150 specimens was plotted on probability coordinates, as shown in Figure 6.91. The equivalent F_{50} energy is determined to be approximately *5.2 J*. The results clearly indicate a nonlinear curve for the energy to crack. The nonlinearity of the curve confirms that this distribution did not come from a normal distribution. Crabb shows the utility of this technique in comparing the impact energy to crack of two different experimental polymers, Figure 6.92. Even though the test geometries and impact speeds used in these tests were decidedly different, the

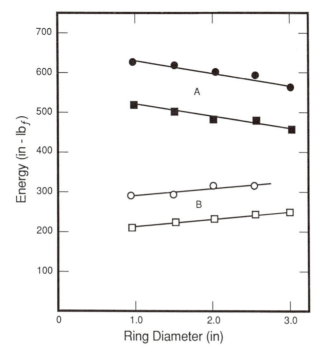

Figure 6.87.   Effect of Retaining Ring Diameter on Puncture Impact Energy for Allied Capron 8350 Nylon 6 at 23°C.Tup Diameter: 0.5 in. Curves A: Ultimate Impact Energy. B: Impact Energy at Yield. Circle: Tup Impact Velocity, 10,000 in/min. Square: Tup Impact Velocity, 500 in/min.

results were in remarkably good agreement. However, recall that PMMA is an amorphous polymer below its glass transition temperature. It is not expected that similar experiments conducted on rubbery-tough and/or semicrystalline polymers would yield the same level of agreement between testing techniques. It is apparent that the constant velocity impact data, when plotted on probability coordinates, yield a useful way of evaluating or comparing the impact resistance of many polymers.

Table 6.13.   Puncture impact energy to fracture of unfilled thermoplastics

Polymer [specific material]	Impactor speed (in/min)					
	200	1,000	3,000	10,000	20,000	30,000
Polyacetal (POM) [Celcon M90-04]	41.8[b]	65.8[d]	26.1[d]	41.6[d]	—	—
Nylon 6 (PA-6) [Capron 8250]	480[d]	460[d]	490[d]	570[d]	570[d]	580[d]
Polycarbonate (PC) [Lexan 114-12]	516[d]	530[d]	631[d]	636[d]	—	676[d]
Polycarbonate (PC) [Lexan 191 White]	420[d]	521[d]	544[d]	570[d]	—	—
Polysulfone (PSO$_2$) [Victrex 300P]	768[d]	600[d]	636[d]	673 [d]	670[d]	830[d]

Test conditions: 3-in ring, ½-in tup, spherical tip, 23°C, 1/8-in specimen thickness. Failure mode: b = brittle. d = ductile.

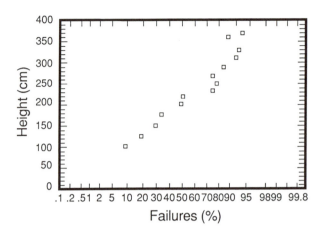

Figure 6.88.   Probit Puncture Impact Energy for Rubber-Toughened Polymethyl Methacrylate (93).

The Role of Temperature on Impact

The effect of temperature on the response of a polymer under high impact loads is very complex. Consider the four-parameter linear viscoelastic model discussed in Chapter 3. Reduction in specimen temperature is equivalent to increasing viscosity in the viscous dampers or dashpots. The spring moduli are assumed to be independent of tem-

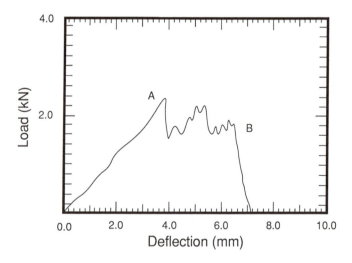

Figure 6.89.   Typical Load–Deflection Trace for Constant Velocity Puncture Impact of a Rubber-Toughened Polymethyl Methacrylate (93). Point A: Impact Yield Strength or Point of Initial Cracking. B: Ultimate Impact Strength.

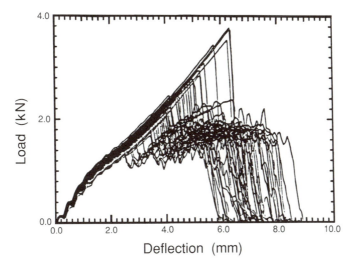

Figure 6.90.   Set of 50 Load–Deflection Traces for Constant Velocity Puncture Impact of a
Rubber-Toughened Polymethyl Methacrylate (93).

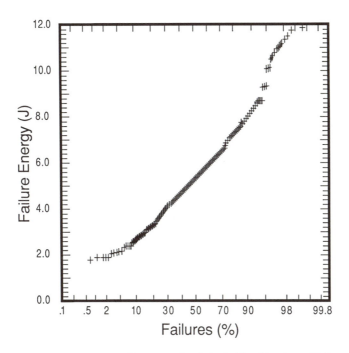

Figure 6.91.   Probit Puncture Impact Energy to Crack for Rubber-Toughened Polymethyl Methacrylate (93). 150 Test Specimens.

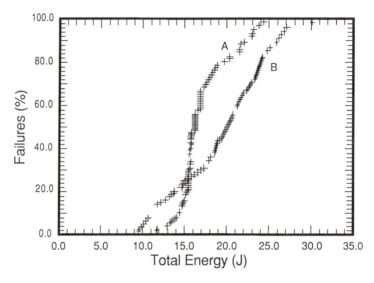

Figure 6.92.   Probit Puncture Ultimate Impact Energy for Two Rubber-Toughened Polymethyl Methacrylate Materials (93).

perature unless the polymer material undergoes some type of transition, such as T_g or T_m. With increased viscosity, the four-parameter model appears to have a higher short-term modulus. So as the temperature increases, the force–deflection curve becomes steeper and the elongation at break occurs at a lower deflection value. This is seen in Figure 6.93 for a thermoplastic rubber at three temperatures.

A similar effect is seen by Nutter (94) in Figure 6.94 at two impact velocities. For 1000 in/min (0.4 m/s), yield first occurs (D) prior to failure at about $-35°C$. Below this temperature, brittle fracture occurs. The maximum energy to fail occurs at about $-15°C$ and then slowly decreases with increasing temperature. The energy to yield continues to decrease with increasing temperature from the first indication of ductility. The energy to yield is given at the locus D–A and the energy to failure is C–B. For 10,000 in/min (4.2 m/s), no yield is observed, indicating that the material undergoes brittle failure at this impact rate at all temperatures. Note that the maximum energy at failure occurs at 20°C. This shift can be anticipated if the response of the four-parameter linear viscoelastic model to a cyclic load is recalled. As the cyclic frequency increases, or in other words, as the *impact rate* increases, the model response becomes stiffer.

Utility of Impact Data

As discussed above, the impact measurement must include the response of the polymer to the impact and the response of the tester. The value of any impact test must relate the data to real product response to real impacts. The following examples serve to illustrate the importance of this relationship.

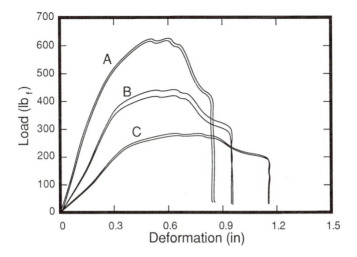

Figure 6.93.   Effect of Specimen Temperature on Puncture Impact Energy of Uniroyal Kraton GXY Thermoplastic Elastomer (TPE). (Data Courtesy of Rheometrics, Piscataway, NJ.). Curve A: $-30°C$, $E = 360$ in-lb$_f$. B: $-10°C$, $E = 290$ in-lb$_f$. C: $25°C$, $E = 210$ in-lb$_f$.

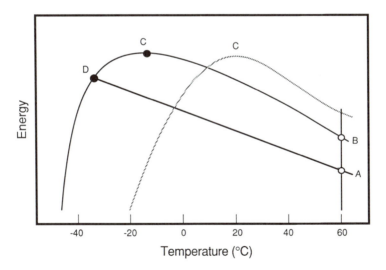

Figure 6.94.   Temperature-Dependent Puncture Impact Energy for Two Impacter Speeds (94). Solid Line: 1000 in/min. Dashed Line: 10,000 in/min. C: Maximum Energy. D: Yield.

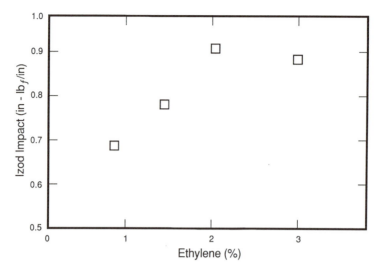

Figure 6.95.   Effect of Ethylene Comonomer Concentration on Notched Izod Impact Strength of PP Random Copolymer at 23°C (Courtesy of Rheometrics, Piscataway, NJ.).

Polypropylene is copolymerized with small amounts of ethylene to improve low-temperature impact performance. The ethylene–propylene copolymer test specimens show increased impact performance with ethylene concentration to about 2%, as seen by Izod impact (Figure 6.95). Beyond that, the Izod impact resistance drops. The designer can therefore conclude that the optimum ethylene concentration, based on these data, should not exceed 2% (or so). Consider constant velocity impact data on these same samples, however (Figure 6.96). It is apparent that the force–deflection

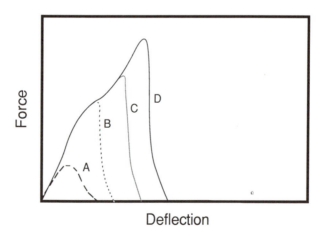

Figure 6.96.   Effect of Ethylene Comonomer Concentration on Puncture Impact Force–Deflection of PP Random Copolymer at 23°C (Courtesy of Rheometrics, Piscataway, NJ.). A: 0.7% (wt) Ethylene Comonomer. B: 1.6% (wt) Ethylene Comonomer. C: 2.1% (wt) Ethylene Comonomer. D: 3.0% (wt) Ethylene Comonomer.

curves at all ethylene concentrations have essentially the same slope. It is also apparent that the energy at failure and the ductility of the copolymer continue to increase with increasing ethylene concentration. The (near) linear relationship between energy to failure and ethylene concentration is seen in Figure 6.97. This result should also be evident from a careful study using the drop-weight test. The drop-weight test will not clearly identify the reason, increasing elongation at break with increasing ethylene concentration, for the improved toughness. Certainly this effect would be obtained from a standard tensile test (ASTM D638). Recall however that tensile tests and impact tests subject the polymer molecule to substantially different strain-rate environments. And so the experimenter might hesitate before concluding that ethylene concentration was the *real* reason for the improved impact resistance.

As discussed earlier, extrusion, thermoforming and injection molding operations produce parts that can have substantial directional molecular orientation owing to highly sheared or oriented polymers cooling rapidly under pressure. In injection molding, flow around obstructions such as inserts and cores produces "weld lines" where the fluid streams rejoin. These can also be produced in parts that have more than one filling port or gate. Weld line regions are inherently weak in impact. The polymer in the gate area is subjected to higher pressures and more uneven cooling conditions than polymer elsewhere. For crystalline polymers, the gate region morphology can be substantially different (95). These weaknesses are seen in Figure 6.98 for polypropylene. The impact-weak gate region is attributed to large spherulites, the result of slow cooling. The lower weld line ductility is attributed to greater molecular orientation and lower intermolecular penetration.

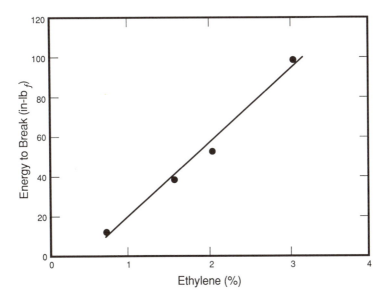

Figure 6.97.   Effect of Ethylene Comonomer Concentration on Total Impact Energy To Break of PP Random Copolymer at 23°C (Courtesy of Rheometrics, Piscataway, NJ.). (See Figure 6.96.)

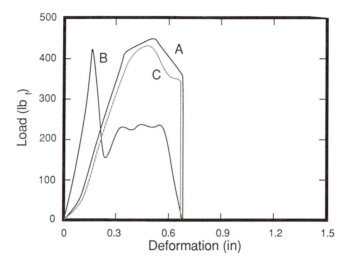

Figure 6.98. Effect of Molded-in Stresses and Molecular Orientation on Puncture Impact Load–Deflection Curves of Polypropylene at 20°C and 3000 in/min Tup Speed (Courtesy of Rheometrics, Piscataway, NJ.). Curve A: Normal Molded Segment. B: Gate Region. C: Weld Line Region.

For amorphous materials, there may be substantial anisotropy and frozen-in stresses. As a result, this region is also inherently weak in impact. Note that it would be difficult to quantify these effects with a standard drop-weight tester, since the specimens would need to be positioned identically. If this detail was not heeded, the data would show substantial scatter and probably would not allow accurate interpretation of the process effects.

Small changes in the nature of the polymer can sometimes make substantial changes in the impact resistance of polymers. The classic example is the size of the impact-modifier rubber particle in polystyrene. Recall that butadiene is block-copolymerized into polystyrene to produce impact-modified polystyrene. The polystyrene and buta-diene molecules are incompatible. As a result, they form separate but molecularly connected phases. If there is poor (or no) compatibility between the two phases, the rubber molecules form very tight, small diameter rubber particles. If there is some degree of compatibility, the rubber molecules form relatively large diameter rubber cages (containing polystyrene molecules). Since the amount of rubber is essentially constant, the (apparently) larger rubber particles act to intercept more cracks and then absorb more impact energy than do the smaller ones.

Apparently small changes in the chemical structure of polymers can yield substantial changes in its impact energy absorption. Consider polyester sheet molding compound or SMC. It is typically made from a vinyl ester resin. As seen in Figure 6.99, at 10,000 in/min (4.2 m/s), the effect of changing to isophthalic ester is a reduction in stiffness. The impact characteristics are very similar at lower impact rates, however. Of course, these SMC compounds are used for automotive and marine applications, where resis-

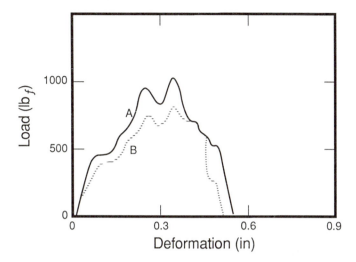

Figure 6.99.   Puncture Impact Force–Deformation Curves for Two Types of Unsaturated Poly-ester (UPE) Resin Sheet Molding Compound (SMC) at 20°C (Data Courtesy of Rheometrics, Piscataway, NJ.). Tup Speed: 10,000 in/min. A: Vinyl Ester-Based UPE. B: Isophthalic Poly-ester-Based UPE.

tance to impact at a very high rate, from sticks and stones, as examples, can be a significant design criterion. Thus this degree of testing sensitivity may be warranted. It is apparent from drop-weight and cantilever beam testing that this degree of sensi-tivity probably cannot be achieved with these test procedures.

Impact Testing: A Summary

Cantilever beam and falling weight tests are relatively simple to operate and the equip-ment needed to carry out the tests is relatively inexpensive. The engineer must keep in mind that these standard tests are *one-point* tests. That is, despite all the care in sample preparation and statistical interpretation of the data, the test produces a single value for the way in which the material responds to short-term loading. Equally im-portant, the impact loading is restricted to relatively narrow loading levels and strain rates. The recent development of reliable strain gages and ancillary electronics has allowed these measuring devices to be placed directly in the tup of a falling weight impacter. Furthermore, the impacter weight has been substantially increased to allow the tup to be driven completely through the specimen without appreciable loss in en-ergy. The strain gages allow a display of the force developed as a function of tup position as the tup penetrates the specimen. The data are commonly stored in a com-puter and later displayed on an *X–Y* plotter or computer CRT as a force–displacement or force–time curve. With a dedicated computer, the data from a set of experiments can be easily analyzed to obtain statistical fracture energy as a function of impacting energy with either drop height or drop weight, sample thickness and processing parameters. Not only is the impacting energy, the area under the force–deflection curve, more easily visualized (Figure 6.100), but the very nature of short-term ductility more apparent. However, keep in mind Turner's admonition (96) that:

> . . . energy absorbed during an impact depends on the dimensions of the specimen and is not normalized in the way that energy absorbed in a flexed-beam test is when it is expressed as the energy per unit cross-section. Thus, a datum recorded in a flexed-plate test is even less a measure of a *material property* than one recorded in a flexed beam test, and has a significance only when it is compared with data from other specimens of the same dimensions or against some standard test piece. [our emphasis supplied]

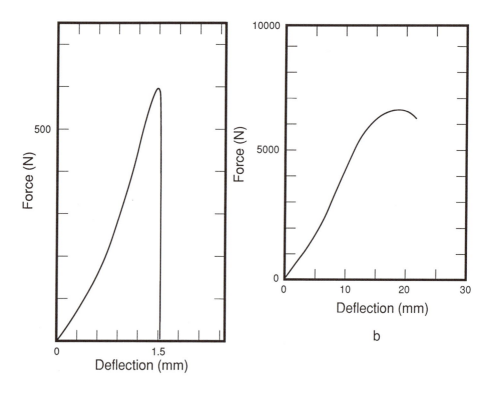

Figure 6.100.   The Nature of Ductile, Brittle and Bructile Puncture Impact Force–Deflection Curves for Amorphous and Crystalline Thermoplastic Resins (97). a) Brittle Failure, Amorphous. b) Ductile Failure, Amorphous.

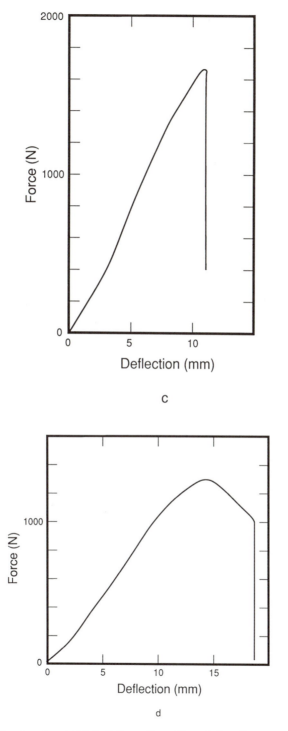

Figure 6.100 *Continued.*   c) Brittle Failure, Crystalline. d) Ductile Failure, Crystalline.

6.5 Solid Mechanical Properties: Long-Time Tests

Short-time duration tests involve material response to loads applied in micro- or milliseconds. As noted, although polymers are viscoelastic materials, most respond to short-time loads in elastic fashions. Furthermore, the typical time frames of such tests allow for many experiments to be run in a relatively short time. Creep and flexural fatigue tests are at the other end of the response spectrum and represent long-time duration testing. Many products must support continuous or cyclical loads for hundreds or thousands of hours. Applications include storage tanks, drums, fuel tanks, bridge support pads, boxes, bins, pallets, shelves, fan blades, motor housings, pneumatic and hydraulic hoses and pipes. In some cases, the long-term load is applied in a constant deformation manner, as with gaskets. Polymer materials behave differently in response to long-term *static* and long-term *dynamic* loads.

Long-Time Duration Testing: Static

In static long-term loads, a constant applied load is maintained for the life of the product. As discussed in Chapter 3, if the material in the product is elastic, the material fully recovers dimension when the load is removed. For viscoelastic polymers, there is permanent distortion. To predict the performance of a polymer part under long-term static load, a different type of test must be used. Since time is an important variable in viscoelastic materials, the property sought is *creep*. Most creep tests are conducted on simple rectangular bar test specimens loaded under uniaxial tension. Simple bending or torsion is also used. There are limited data available on compression and combined tensile and torsion tests. Polymers have been extensively tested in combined stress in pipe or annular cylindrical geometry, owing to the economic importance of hydraulic hoses and gasline piping. For preset or fixed deformation, the testing manifests as *stress relaxation* rather than creep.

Uniaxial Tensile Creep: ASTM D2990

The uniaxial tensile test is ASTM D2990. The test specimen is either a standard Type I or II bar, per ASTM D638. The test specimen is preconditioned to ASTM D618 specifications. The test apparatus is designed to insure that the applied load does not vary with time and is uniaxial to the specimen. As with other tests, the test specimen must not slip in or creep from the grips. The load must be applied to the specimen in a smooth rapid fashion in 1 to 5 s. If the test is run to specimen failure, the individual test cells must be isolated to eliminate shock loading from failure in adjacent test cells. Several tensile creep systems are shown in Figure 6.101.

When a constant uniaxial load is applied to the viscoelastic test specimen, it initially elongates in a manner similar to a Hookean material. That is, it undergoes instantaneous deformation. After this initial deformation, the specimen continues to deform, albeit

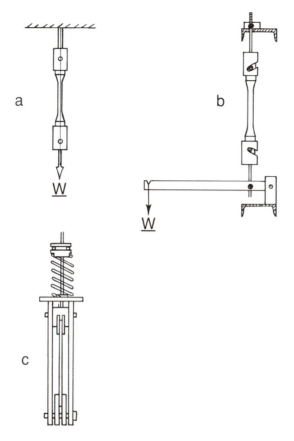

Figure 6.101.   Schematics of Three Tensile Creep Testing Fixtures (98). a) Conventional Dead Weight Tester. b) New Jersey Institute of Technology Dead Weight Tester with Mechanical Advantage. c) Allied Chemical Tester for Environmental Chamber.

at a much slower rate. Ultimately the specimen ruptures. The deformation that occurs after the Hookean response is called *creep*. The failure is called *creep rupture*. Note that creep and creep rupture are not restricted to polymers. Many soft metals, such as gold, copper and lead, creep at room temperature and many structural metals, such as steel and aluminum, creep at elevated temperature.

As discussed in Chapter 3, creep curves generally exhibit three distinct phases. *First-stage creep deformation* is characterized by rapid deformation rate that decreases slowly to a constant value. Earlier, a four-parameter model was proposed to describe long-term creep. In that model, the first-stage creep deformation was called *retarded elastic strain*. *Second-stage creep deformation* is characterized by a relatively constant, low deformation rate. In the four-parameter model, this was called *equilibrium viscous flow*. The final or third-stage creep deformation is *creep rupture*, fracture or breaking. The generalized uniaxial tensile creep behavior of polymers under constant load, isothermal temperature and given environment can be displayed either as *ductile creep behavior,* Figure 6.102 or *brittle creep behavior,* Figure 6.103. Note that at very low stress levels, both types of polymers exhibit similar first- and second-stage creep deformation. Onset of creep rupture may not occur within the lifetime of the product, let alone the test. As the stress level increases, first- and second-stage creep deformation rates remain relatively the same for these types, but of course the time of failure is considerably reduced. Note, too, that third-stage creep deformation characteristics now differ considerably. The ductile polymer exhibits typical *ductile yielding* or irreversible plastic deformation (point i) prior to fracture. The brittle polymer on the other hand exhibits no observable gross plastic deformation, only abrupt failure.

Macroscopic yielding and fracture may not always be appropriate criteria for long-time-duration material failure. For some polymers, stress crazing, stress cracking or stress whitening may signal product failure and may therefore become design limitations.

Stress Cracking.   Stress cracking implies *localized* failure that occurs when localized stresses produce excessive *localized* strain. This localized failure results in the formation of microcracks that spread rapidly throughout the local area. Brittle materials are more prone to stress crack than to stress whiten.

Stress Crazing.   Stress crazing was discussed in Section 6.3. Usually the localized failure results in a crack that is bridged by polymer microfibrils. These fibrils are oriented in the direction of applied stress. Since the fibrils are load-bearing, the microcrack usually does not open substantially before parallel microcracks form. Although fiber-forming polymers such as nylon, polyethylene and polypropylene readily stress craze, non-fiber-forming polymers such as polycarbonate and polymethyl methacrylate also stress craze in creep.

Stress Whitening.   Stress whitening is a generic term describing many different microscopic phenomena that produce a cloudy, foggy, or whitening appearance in transparent or translucent polymers in stress. The cloudy appearance is the result of a localized change in polymer refractive index. Thus, transmitted light is scattered. Microvoid clusters of dimension equal to or greater than the wavelength of light are thought to be the primary cause of stress whitening. The microvoids can be caused by delamination of fillers or fibers or can be localized failure around occlusions such as rubber particles or other impact modifiers. Although stress whitening results in visually apparent changes, the load-bearing capabilities of a specimen may not be substantially reduced during stress whitening.

The designer must keep in mind that by their very nature, polymers cannot be fabricated at their maximum packing density. Voids, known collectively as "free volume", exist around the polymeric macromolecular structure. Applied stress serves to straighten the polymer chains and perhaps redistribute the free volume such that sizable microvoids are formed. It is the nature of

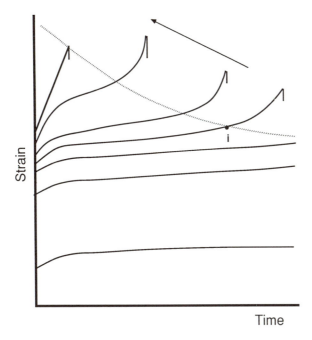

Figure 6.102.   Typical Tensile Creep or Time-Dependent Strain Curves for a Ductile Polymer (82). Dashed Line Represents Creep Rupture Region. Arrow Direction Denotes Increasing Tensile Stress. Point i: Irreversible Plastic Deformation.

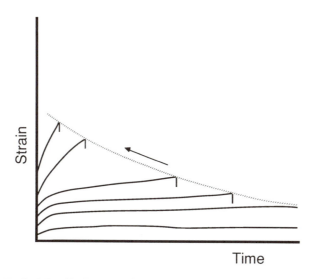

Figure 6.103.   Typical Tensile Creep or Time-Dependent Strain Curves for a Brittle Polymer (82). Dashed Line Represents Creep Rupture Region. Arrow Direction Denotes Increasing Tensile Stress.

the polymer that dictates whether these microvoids continue to grow into localized areas of stress whitening, stress crazing, stress cracking, or brittle
failure.

Analysis of Creep Data

The raw tensile creep data at constant stress are time-dependent displacement values.
Depending on the test apparatus design, these values are converted to an effective time-
dependent gage length, L(t). Traditionally, creep data are reported in engineering strain
terms (see Section 3.6):

(6.35) $$\varepsilon(t) = \frac{[L(t) - L(0)]}{L(0)} = \frac{\Delta L(t)}{L(0)}$$

where L(0) is the initial gage length. A tensile creep modulus, E(t), is defined as:

(6.36) $$E(t) = \frac{\sigma}{\varepsilon(t)}$$

where σ is the applied stress and $\varepsilon(t)$ is the time-dependent elongation. Note that E(t)
is neither a design property nor a material constant. It is a time-dependent variable that
is also a function of temperature and environment. There is no universal method of
graphically displaying tensile creep or in fact any other mode of creep. Traditionally,
the creep strain at constant load is recorded as a function of time. The data are displayed
at the top of Figure 6.104. Two sets of graphical representations can be constructed
from these data. The creep rupture envelope can be created, as shown on the two graphs
to the right of the original data presentation in Figure 6.104. The first graph is a linear
plot of creep rupture stress as a function of failure time. For most polymers that are
candidates for long-term performance, the design life can be quite long—months or
years. As a result, the dual logarithmic coordinate system has greater utility. Furthermore, creep rupture data tend to be displayed in linear form on this coordinate scheme.

Consider the second type of graphical representation, based on actual deformation
under load. As shown below the original graph, Figure 6.104, there are three methods
of analyzing these data. Each method holds one variable—stress, strain or time—
constant. For constant stress, the set of graphs to the left apply. The data can be displayed either as a set of curvilinear lines on semi-logarithmic paper or as a set of
(usually near-) linear lines on dual logarithmic paper. This second set of parallel straight
lines on dual logarithmic coordinates is called a *creep strain plot*. If the slopes of the
semi-logarithmic curves are replotted against time, a set of nearly linear lines on semi-
logarithmic paper result. This represents the time-dependent *creep modulus plot*. Most
creep design data published in the United States are reported in this manner.

If the time parameter is held constant, a set of *isochronous* or constant time stress-
strain curves result. A linear coordinate system is used to display these results. The
slopes of these *isochronous creep* curves produce the *isochronous modulus* graph below. If strain is constant, *isometric creep* curves result. The graph is usually semi-

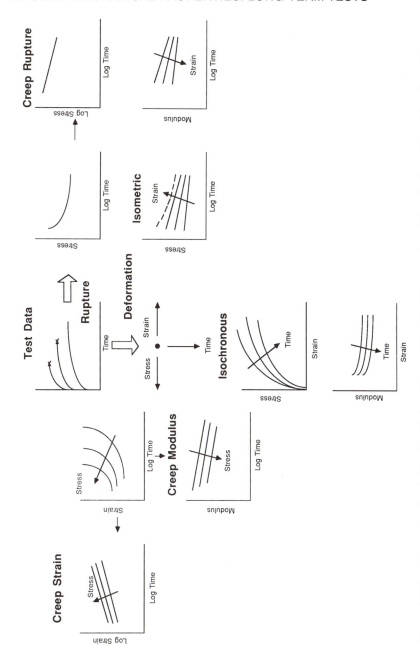

Figure 6.104.   Composite Graph Showing Various Ways of Graphical Presentation of Tensile Creep Data. Most Data Taken as Time-Dependent Strain Under Constant Stress, as Shown in Top Center Graph and in Figures 6.102 and 6.103.

logarithmic in time. Isometric creep data are used extensively in Europe. Of course, the isometric modulus data can be extracted from these curves.

It is important to note the way in which creep data are displayed. Certain sales and technical bulletins present linear creep data (Figure 6.105), to show the relative improvement in "short-time" creep owing to addition of fillers, crosslinking agents, bonding or coupling agents and fiber reinforcements. This type of presentation magnifies the short-time creep effects. Note in Figure 6.105 that the data extend 10,000 hours. Although this appears long, it is only 417 days or 1.1 years. This time may be short for many design applications. O'Toole (100) recommends that *at the same temperature*, creep curves not be extrapolated more than one decade. Therefore, at best, the data shown in Figure 6.105, collected over more than 1 year, should only be extrapolated to 100,000 hours or 11.4 years. This may be beyond the design lifetime of many products, but applications such as underground tanks, civil engineering fabrics and pipelines require lifetimes of 20 to 50 years or more.

In many respects, *creep rupture* is a more important parameter. Creep rupture represents the ultimate lifetime of a given material. As seen in Figure 6.106 for several unfilled polymers, the creep rupture curve typically is linear with time on dual logarithmic scales. The actual rupture envelope of a given polymer depends strongly on molecular weight distribution, filler type, coupling agents, and other adducts. Note however that as the lifetime of a given material increases, the critical stress level on that part decreases. In metals, the critical stress is usually independent of time. For

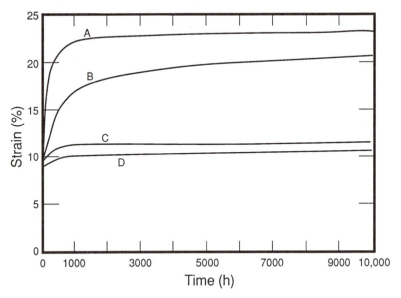

Figure 6.105.   Linear Time-Dependent Tensile Creep Strain for Poly Sulfone, (PSO$_2$), Polycarbonate (PC), ABS, and Polyacetal (POM) at 72°F Under Stress of 3000 lb$_f$/in^2 (101). Total Strain Curve A: POM. B: ABS. C: PC. D: Polysulfone (PSO$_2$).

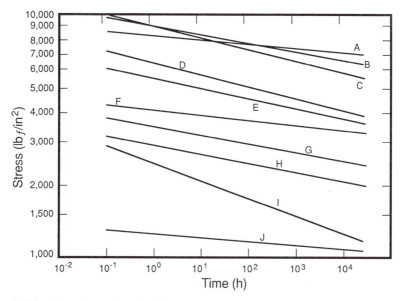

Figure 6.106.   Time-Dependent Tensile Creep Rupture Strength at 20°C for Several Polymers (102). Curve A: Polycarbonate. B: Polystyrene–Acrylonitrile (SAN). C: Polyacetal (POM). D: Polystyrene. E: ABS. F: Impact PS. G: Polypropylene. H: Cellulose Acetate Butyrate (CAB). I: HDPE. J: LDPE.

example, for gray cast-iron, the ultimate stress is 20,000 lb_f/in^2 [138 MPa] and it retains that value at 100,000 hours under load. Certainly if gray cast-iron is fabricated and loaded and does not immediately fail, it will most likely never fail due to tensile rupture.* Consider a polymer such as polystyrene in a similar tensile loading situation. Assume that it is loaded to 4600 lb_f/in^2 [32 MPa] and has the creep rupture curve shown in Figure 6.106. The part supports the load adequately for 1000 hours, at which time it fails in brittle creep rupture.

In other words, any design that requires the polymeric part to be under predetermined continuous load or stress must have a predetermined design life associated with it. It is to this design life value that a *design safety factor* is applied.** To design for long-

---

*Certainly it may fail prematurely if its environment is aggressive or it is subjected to other secondary loads such as impact.

**It is a common but entirely incorrect practice to assign a safety factor to the ultimate stress value given in supplier literature. Values given in standard data sheets are typically obtained from simple moderate-time tests such as the ASTM D638 tensile test. The polystyrene in Figure 6.106 has an ultimate tensile strength of 8500 lb_f/in^2 [5.9 MPa]. If a factor of 2 is applied to *incorrectly* obtain an actual design stress, the resulting applied load is 4250 lb_f/in^2 [2.9 MPa]. This load will cause the part to fail in about 7000 hours, or somewhat less than one year. That same safety factor, applied to the assumed load of 4600 lb_f/in^2 [3.2 MPa], will allow the part to perform satisfactorily for more than 10,000,000 hours or 100 years.

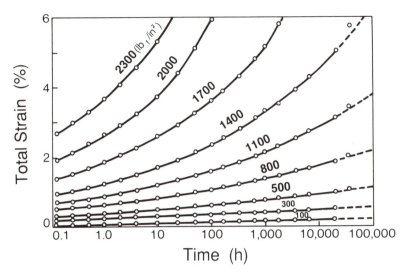

Figure 6.107.   Tensile Creep Strain Behavior of Polypropylene Copolymer at Various Stress
Levels (103).

term performance, the designer must understand the characteristics of these materials.
As a general rule of thumb, however, the designer should strive to *minimize* the effect
of long term stress on any part through intelligent redesign.

As noted in Figure 6.104, creep rupture is only one element of importance to the
designer of plastic parts. The time-dependent nature of deformation is also important.
Creep strain is usually plotted against time on semi-logarithmic plots (Figure 6.107).
Extrapolation to times beyond the data is difficult on this plot. Replotting on dual
logarithmic paper allows extrapolation, but the designer should again heed O'Toole's
words about extrapolation beyond one decade in time. The data of Figure 6.107 are
replotted in Figure 6.108. Note that this does not eliminate *all* line curvature, but for

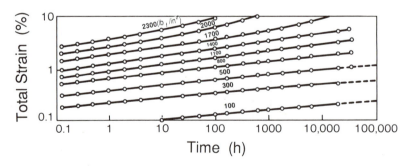

Figure 6.108.   Tensile Creep Data of Figure 6.107, Replotted on Dual Logarithmic Coordinates
to Demonstrate the Linearity of the Data (103).

small strains, the curves can be considered linear. Usually these curves are used to compare polymers at the same loading levels.

Consider deformation of polyethylene, polypropylene and polybutene at 1000 lb$_f$/in^2 [6.9 MPa] and room temperature. Supplier information on ASTM D638 tensile tests for these materials is given in Table 6.14. It appears that polypropylene is substantially stiffer than either polyethylene or polybutene. In Figure 6.109 are the creep deformation curves for these materials. It is apparent that their behaviors under long-term load are quite different. Beyond 6 min (0.1 h), the polyethylene deformation exceeds that for polybutene. Between 0.1 and 1,000,000 hours, polypropylene deformation is less than that of polybutene which is less than that for polyethylene. Beyond 1,000,000 hours, polybutene has the lowest isometric deformation. Of course, this means that for these three polymers, the ordinal ranking of stiffness beyond 10^6 hours is diametric to that in the data sheets. Certainly, one would not expect any polyolefin product to be required to support this load for 1,000,000 hours (114 years).

The point is that to determine an ordinal ranking for polymeric design candidates, an appropriate design lifetime must be included. If the part is designed to have a 10 year life, the appropriate data for comparison of polymer candidates are 10 year creep data and not short-time tensile strength data. It is also of interest to note the time-dependent slope of the creep strain curve. If part loading deformation is to be minimized, the most desirable polymeric candidate is one that has an isometric creep strain curve with a very small slope. Two plastics that have creep strain curves with characteristically small slopes are polybutene and polybutylene terephthalate (PBT).

As noted earlier, the plot of the slope of the isometric time-dependent creep strain curves is the isometric creep modulus plot. If a simple four-parameter model is used to describe viscoelasticity (Figure 3.28), the creep modulus is stress-independent. That is, doubling the applied force on the four-parameter model doubles its response, that is, its deflection. Thus, the strain is directly proportional to the applied stress at any time. Time-dependent room temperature creep moduli for 30% (wt) glass-filled nylon 6 (dry) in tension and 25% (wt) glass-filled polytetrafluoroethylene (PTFE) in compression are shown in Figure 6.110. The nylon data indicate that the simple linear viscoelastic model can be applied without substantial error. However the PTFE data show substantial viscoelastic nonlinearity. The designer must be aware that not all polymers are linear viscoelastic at all stress levels. And the designer must recall also that the concept of superposition, whereby time-dependent responses are shifted until a single master curve obtains, is valid only for linear viscoelastic materials. Before complex

Table 6.14. Short-duration ranking of three polyolefins according to modulus and strain.

Polymer	Modulus ASTM D638 (lb$_f$/in^2)	Strain, ε at 1000 lb$_f$/in^2 (in/in)
Polypropylene (PP)	170,000	0.0059
Polyethylene (PE)	100,000	0.010
Poly-1-butene	40,000	0.025

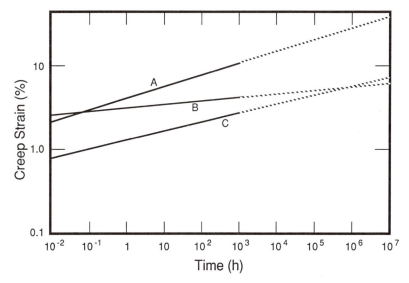

Figure 6.109. Tensile Creep Strain for Three Polyolefins, Demonstrating the Importance of Long-Term Testing (Courtesy of Allied Chemical Corporation, Morristown, NJ.). Room Temperature, Applied Tensile Stress, 1000 lb_f/in^2. Curve A: Polyethylene, Tensile Modulus, $E_o =$ 100,000 lb_f/in^2. B: Polybutene, $E_o = 40,000$ lb_f/in^2. C: Polypropylene, $E_o = 170,000$ lb_f/in^2.

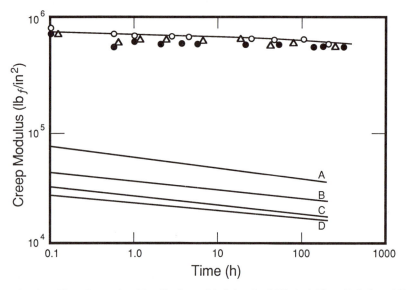

Figure 6.110.   Time-Dependent Tensile Creep Modulus for 30% (wt) Glass Reinforced Nylon 6 (PA-6) (Dry) and Time-Dependent Compressive Creep Modulus for 25% (wt) Glass Reinforced Polytetrafluoroethane (PTFE) at Various Stresses, lb_f/in^2 (104). Nylon–Open Circle: Stress at 8000 lb_f/in^2. Solid Circle: 6000 lb_f/in^2. Triangle: 4000 lb_f/in^2. PTFE–Curve A: 1000 lb_f/in^2. B: 2000 $_f/in^2$. C: 3000 lb_f/in^2. D: 4000 lb_f/in^2.

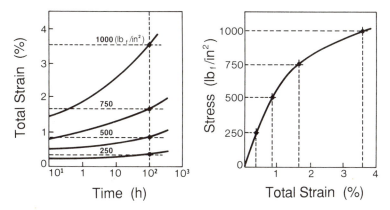

Figure 6.111.   Construction of Isochronous Creep Curves for Polypropylene Copolymer (102). Isochronous Time: 100 hours.

analyses are contemplated, the polymer must be characterized. Time-dependent iso-metric creep modulus is one of the most important tests for determining the linearity of polymer viscoelasticity.

As noted above, establishment of valid physical properties for metals is based on the stress–strain behaviors of the materials. The workhorse test to obtain these data is the simple tensile test. The data are plotted as engineering stress against engineering strain. And as noted, the early attempts to measure the physical properties of polymers followed this protocol. To present a rational stress–strain curve for viscoelastic mate-rials, however, it is necessary to fix the time. The resulting stress-strain curves are called *isochronous creep* curves. As seen in Figure 6.111, these graphs are obtained by crossplotting the linear strain-logarithmic time curves. As an example, the selected time frame is 100 hours. The strain at each stress level is measured* from the graph. For example, the strain is 0.34% at 250 lb_f/in^2 [1.72 MPa]. The appropriate stress–strain points are then plotted in linear stresss–strain coordinates. In this case, there are four data points to form the isochronous, 100 hour curve. This can be repeated at other times to form a set of isochronous stress–strain curves, as seen for polycarbonate in Figure 6.112. To illustrate how the isochronous data are used, consider the 3000 lb_f/in^2 [20.7 MPa] stress line. Polycarbonate is known to stress craze at this load. At 0.1 hour at this load, the critical strain is 0.89%. At 10,000 hours, it is 1.13%.

The isochronous creep curves for several polymers at 1000 hours are shown in Figure 6.113. The choice of 1000 hours is important, since many polymer materials do not need to withstand applied load for more than 10,000 hours (1.14 year). Furthermore, at this time essentially all retarded elastic effects have occurred and the only effect still present in the creep elongation is the viscous flow effect.

---

*With advanced computer curve-fitting and data differentiation software, this can be done nu-merically in a very short time.

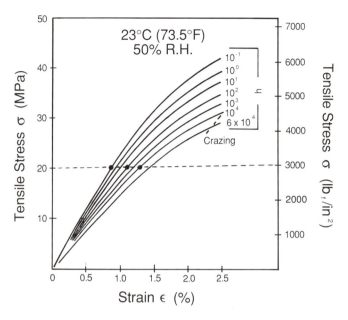

Figure 6.112.   An Isochronous Tensile Creep Stress–Strain Plot for Polycarbonate at 23°C and 50% RH (Courtesy of Mobay Chemical Co., Erie, PA.). Dashed Line: Crazing Zone.

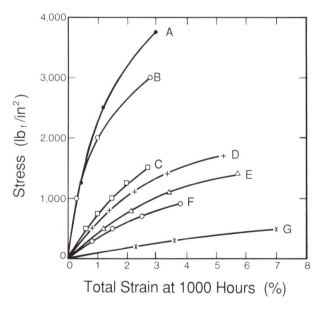

Figure 6.113.   Isochronous (1000 h) Tensile Creep Stress–Strain Diagrams for Several Polymers (103). Curve A: Unplasticized PVC Pipe Formulation. B: Normal Impact ABS. C: Polypropylene Homopolymer. D: Polypropylene Copolymer I. E: Polypropylene Copolymer II. F: HDPE, 0.958 g/cm^3. G: LDPE, 0.918 g/cm^3.

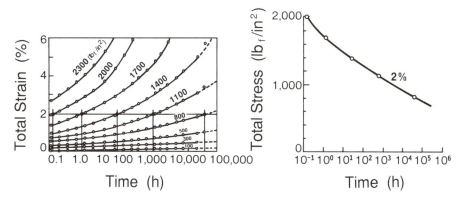

Figure 6.114. Development of Time-Dependent Isometric Total Stress for Polypropylene (PP) at 2% Strain and 23°C. Data from Ref. (103).

For isometric creep, a line of constant *strain* is plotted on a time-dependent stress plot. As seen in Figure 6.114, the time-dependent stress is shown for polypropylene at a strain level of 2%. A set of time-dependent stress curves can then be developed for other strain levels. In the isometric creep plot for unplasticized PVC, Figure 6.115, the upper line is ductile failure due to necking. The region around the dotted line denotes stress whitening failure. The dashed line represents craze initiation. The craze initiation curve has a shape similar to the ductile failure curve, but it is apparent that craze initiation occurs at stress levels substantially below those for ductile failure. Note

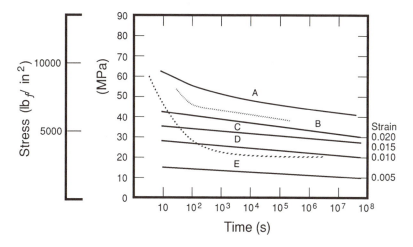

Figure 6.115. Time-Dependent Isometric Creep Rupture Stress for Unplasticized PVC (RPVC) at 23°C, 65% RH (103). Curve A: Ultimate Stress, Necking and Ductile Failure. B: 2% Strain Level. C: 1.5%. D: 1% Strain Level. E: 0.5% Strain Level. Dotted Line: Stress Whitening Strain Level. Dashed Line: Craze Initiation Strain Level.

that in order to prevent failure initiation, the strain level must be below 1%. This means that if the product must have a design lifetime of 10^8 s or 3.2 year, the applied stress cannot exceed about 2500 lb_f/in^2 [17.2 MPa].

Temperature as a Design Parameter

Temperature is considered as a design parameter when as frequently happens, it reflects the result or deliberate of accidental external conditioning of the polymer. The direct link between temperature and load bearing capability through "molecular viscosity" requires that its effect on material be presented in the same content as deformational considerations. Note that in Chapter 3, a similar discussion dealt with ways in which viscoelastic mechanical models could be altered to include temperature effects.

The effect of temperature on mechanical performance can be illustrated in a very simple fashion. Consider the drop in viscosity of a liquid with an increase in its temperature. The engine in a new car, with tight tolerances and fresh oil, is usually very difficult to turn over on a cold day. An older engine, with poor tolerances and worn-out oil, starts much more easily. On a hot day, both engines are easy to start. The difference is in the resistance of oil in the piston ring gaps. In a similar manner, for high polymer temperature, there is low resistance to flow. As the viscosity decreases, there is an increase in the slope of the deformation curve. Increasing temperature also reduces initial elastic modulus. As a result, increasing polymer temperature results in reduction of the creep modulus. This is seen in Figure 6.116. One way of displaying

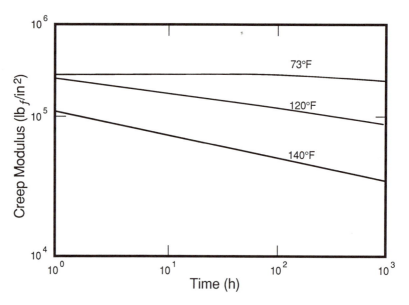

Figure 6.116.   Time-Dependent Creep Modulus of Cyclolac DFA-R(tm) ABS at 1000 lb_f/in^2 Stress Level.

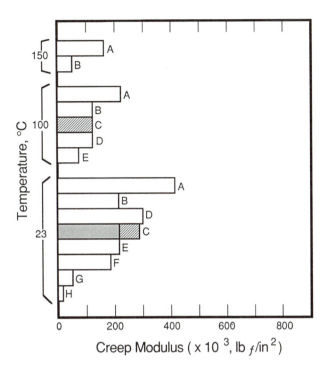

Figure 6.117. Isochronous (1000 h) Creep Modulus Values for Several Polymers at Three Temperatures (105). A: PBT. B: Nylon (PA). C: Modified Polyphenylene Oxide (mPPO). D: Polycarbonate (PC). E: Polyacetal (POM). F: Rigid PVC (RPVC). G: Polypropylene (PP). H: HDPE.

the effect of temperature on creep is to present a series of bar graphs of temperature-dependent isochronous creep moduli. Again, the 1000 hour time is commonly accepted. The creep moduli for eight unfilled polymers are shown in Figure 6.117 for three temperatures. Note that fillers, reinforcements, plasticizers and other adducts can significantly influence creep data. Consider the temperature effect on the isochronous creep plot for polycarbonate (Figure 6.118). As expected, the overall creep effect increases with increasing temperature. Then note that increasing temperature yields a greater creep effect at longer times. That is, the curves spread more at higher temperatures. The crazing boundary is also influenced by temperature. At room temperature (Figure 6.118A), polycarbonate does not craze at strain levels below about 2%. However, at 100°C (212°F), crazing is initiated at strain levels in excess of about 0.7 to 0.8% (Figure 6.118D). So, even though polycarbonate may be mechanically suitable at 100°C and 1% strain level, it may be nonuniformly optically clouded by the crazing and so may be unsatisfactory from an aesthetic viewpoint.

And finally consider the effect of temperature on creep rupture stress. As seen in Figure 6.119, increasing temperature results in decreasing ultimate strength for PMMA. Temperature-dependent isochronous creep rupture (usually at 1000 h) is frequently given for comparison purposes (Figure 6.120).

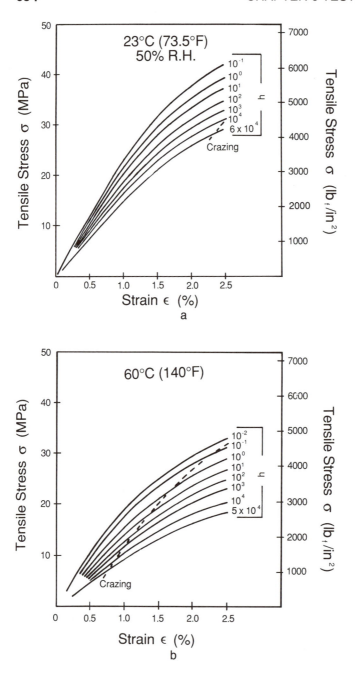

Figure 6.118.   Isochronous Stress–Strain Behavior of Polycarbonate (PC) as a Function of Temperature (Courtesy of Mobay Chemical Co., Pittsburgh, PA.). Dashed Line: Stress Crazing Line. (a): 23°C, 50% RH. (b): 40°C. (c) 80°C. (d) 100°C.

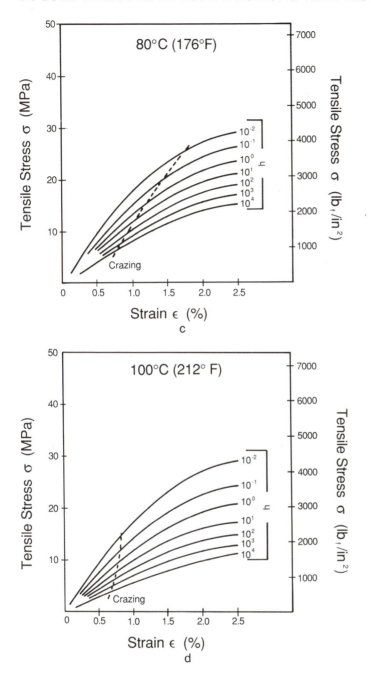

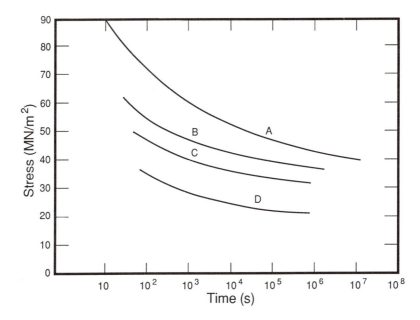

Figure 6.119. Time-Dependent Creep Rupture Stress for Polymethyl Methacrylate (PMMA) (106). Curve A: 20°C, Ductile Failure or Necking. B: 40°C. C: 60°C. D: 80°C.

Recovery After Load Removal

It is apparent that purely elastic materials recover completely when an applied load is removed. Furthermore, purely viscous materials remain fully deformed when an applied load is removed. Viscoelastic materials exhibit some recovery and some permanent deformation after load removal. Turner (107) considers the polymer deformation process to be a function of three variables:

- t, the time deformation of the prior creep process,
- ε_0 (t), the magnitude of the creep strain, and
- t_{max}, the duration of the recovery process.

Two variables can be used to correlate creep recovery data:

- FR, recovery fraction = 1 − (residual strain)/(maximum creep strain),
- t_r, reduced time = (recovery time)/(preceding creep time).

Typical creep recovery curves for polypropylene are shown in Figure 6.121. Turner notes that if the creep strain is very small, such as $\varepsilon_0 < 0.005$, *and* the time duration is less than, say, a few days, *and* the temperature is at least 40°C below the softening point, *then* the material will recover almost completely. *But* the recovery period will

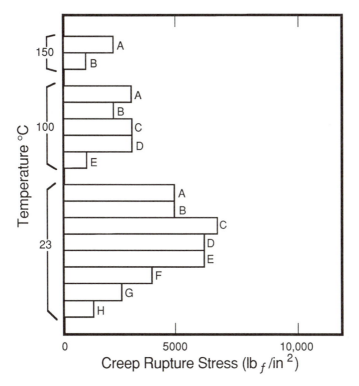

Figure 6.120.   Isochronous (1000 h) Creep Rupture Stress for Several Polymers at Three Temperatures (105). A: PBT. B: Nylon 6 (PA-6). C: Polycarbonate (PC). D: Modified Polyphenylene Oxide (mPPO). E: Polyacetal (POM). F: Rigid PVC (RPVC). G: Polypropylene (PP). H: HDPE.

be longer than the creep time duration. As the deformation and application time increase, the recovery response time increases. Certainly, if the polymer undergoes a change on the molecular level, such as short-chain segment molecular movement due to high loading or very long time load duration, complete recovery can never occur. This phenomenon is seen by a flat fractional recovery curve that asymptotically approaches a value less than 1.0. As an example, if an amorphous polymer is loaded for a very short time, a week or so, at temperatures a few degrees, say 10°, *above* T_g, it will retain a permanent set. That same polymer loaded for a very long time, a year or more, at temperatures a few degrees, say 10°, *below* T_g, will also retain permanent set. Turner notes that this type of permanent set usually does not exhibit classical signs of failure such as crazing or whitening or shear banding.

Other Forms of Creep and Creep Rupture

The discussion above was restricted to uniaxial tensile creep and creep rupture. Of course, there are many other uniaxial forms of creep. The data for most polymers undergoing uniaxial creep in shear, torsion and flexure exhibit characteristics similar

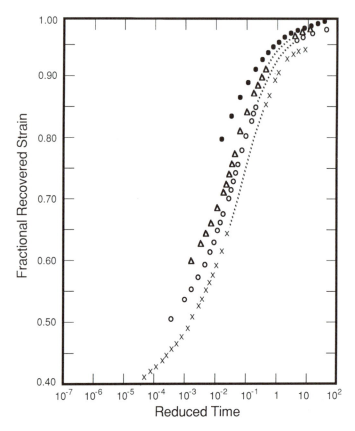

Figure 6.121.   Creep Recovery for Polypropylene at 20°C with Applied Strain of 1% (108). Reduced Time: Ratio of Recovery Time to Time of Applied Strain. Solid Circle: Creep Duration, 199 s, Creep Stress, 12.5 MN/m². Triangle: Duration, 3430 s, Stress, 10.0 MN/m². Open Circle: Duration 18600 s, Stress, 8.5 MN/m². X: Creep Duration, 1720 s, Creep Stress, 7.0 MN/m².

to those seen in tensile creep. An example of shear creep and creep recovery is the recovery of "playability" in polyvinyl chloride records. As is well known, the recorded sound which deteriorates after many consecutive plays, recovers significantly if the record is allowed to "rest" for some time. A classic example of flexural creep is in permanent deformation under long-term loading of chip- or particleboard. This material is essentially wood chips held together with a urea-based thermoset polymer. The permanent deformation can be seen in inexpensive bookcase shelving, roof and floor underlayment, and table tennis tables.

As mentioned earlier, polymers are frequently loaded in more than one axis. It is usually not possible to simply combine uniaxial loading values to determine combined stresses in complex loadings. Poisson's ratio also cannot be determined by simple combinations. The applied stresses can be flexural in one axis and torsional in another, for example. Or, they can be uniaxial in the presence of an aggressive environment. Unfortunately, as is apparent above, nearly all literature data represent the response of a particular test specimen geometrry with a specific molecular orientation. Of course, there is no assurance that the response of the specimen to its environment in any way mirrors the response of the final part to very complex stress patterns.

Stress Relaxation

As noted in Section 3.6, creep is the deformational response of the polymer to constant applied load. Stress relaxation is the polymer stress response to a constant deformation. Of course, for a Hookean elastic material, the stress under constant deformation is independent of time. In a viscoelastic material, stress decays with time. In that Section, the viscoelastic material response as a Maxwell element showed stress to decay exponentially with time. Although there are test devices for measuring tensile stress relaxation with time, the most common way of measuring stress relaxation is in the compressive mode. A simple device used to measure compressive stress relaxation is called the "gasket relaxometer", as seen in Figure 6.122. Note that the bolt is instru-

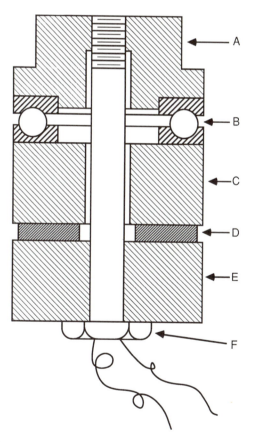

Figure 6.122.   Gasket Relaxometer Geometry for the Measurement of Stress Relaxation in Compression of Annular Flat Specimens (109). A: Upper Nut. B: Ball Bearing. C: Upper Platen. D: Gasket Test Specimen. E: Lower Platen. F: Load Cell.

mented with strain gages. A typical stress–time curve measured via the strain gages is shown in Figure 6.123. Typically, the gasket is compressed to a stress of σ_0 in a time t_1. As a result, the actual instantaneous stress is never known, and the data are usually shifted by the time t_1. Even if the data are replotted to semi-logarithmic coordinates as in Figure 6.124, the curves remain nonlinear. The geometry of the specimen can also influence stress relaxation, as seen for thickness in Figure 6.124. Certainly the change in geometry here is similar to the change in geometry of the test specimen in uniaxial tensile creep.

Prediction of Deformation Under Load

There are currently three accepted methods for analytically predicting the deformational response of a plastic structure under load. All are based on analysis of experimental data. And each has serious limitations. The three are:

- Graphical or tabular,
- Time-temperature superposition, and
- Empirical correlations.

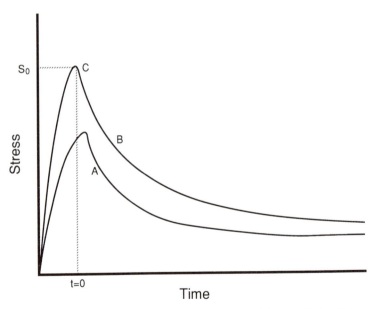

Figure 6.123.   Typical Stress Relaxation Curves as Measured by the Gasket Relaxometer of Figure 6.124 (109). S_0: Initial Applied Stress, $t = 0$, Artificially Determined at This Time. Curve A: Slow Loading Rate. B: Rapid Loading Rate. C: Peak Stress.

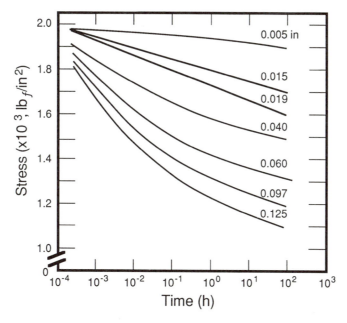

Figure 6.124.   Effect of Gasket Thickness on Compressive Stress Relaxation of Allied Halon G-700 Polytetrafluoroethylene (PTFE) at 73°F and an Initial Stress of 2000 lb_f/in^2 (110). Parameter is Gasket Thickness.

Graphical Approach.   Consider first the graphical or tabular approach. Traditionally in the United States, tensile, compressive and flexural creep data are displayed in tabular form, as creep modulus and rupture time for various creep levels. Tables 6.15 and 6.16 are reproductions of tabulated tensile creep data from a standard reference source (*Modern Plastics Encyclopedia*). These two sets of data are chosen to illustrate the way in which these data can be

Table 6.15.   Tensile creep properties at 73°F of a general-purpose polystyrene-acrylonitrile, Dow Tyril 867 (111).

Initial applied stress (lb_f/in^2)	Creep (apparent) modulus* ($\times 10^3$ lb_f/in^2) test time (h)						Time at latest test point (h)	Rupture or onset of yielding at initial applied stress in air (h)	
	1	10	30	100	1000	Latest			
4,400	500	475	455	425	385 325	250	3,500	4,300	(rupture)
4,920	500	475	455	425	385 315	290	1,705	1,410	(rupture)
5,500	490	440	410	355	285 —	285	309	284	(rupture)
6,030	475	435	400	330	— —	275	190	152	(rupture)
6,460	470	430	395	—	— —	370	47	55	(rupture)
6,870	460	—	—	—	— —	440	6	4.7	(rupture)

*Calculated from total creep strain or deflection (before rupture and onset of yielding)

642 CHAPTER 6 TESTING FOR DESIGN

used in simple mechanical design problems. Consider first the polystyrene–acrylonitrile (SAN) data of Table 6.15. The creep modulus and creep rupture data are plotted in Figures 6.125 and 6.126. The creep modulus graph indicates that SAN is nearly linearly viscoelastic. That is, the apparent modulus of SAN at 73°F (23°C) is essentially independent of strain level. Representative creep data for many polymers are presented in Appendix D.

To illustrate how these data are used, consider the following design problem:

> Find the minimum depth, h, of a simply supported rectangular SAN beam with a span, L, of 3 in and width, b, of 0.5 in that will support a load, P, of 2.5 lb$_f$ for 5 years without fracturing or having a maximum deflection, δ, of greater than 0.100 in.

From stress analysis, the fiber stress in bending, σ, and the midspan deflection, δ, of the beam are given as:

(6.37)
$$\sigma = \frac{3\,PL}{2\,bh^2}$$

(6.38)
$$\delta = \frac{PL^3}{48\,E\,bh^3}$$

Table 6.16. Tensile creep properties at several temperatures of a polyacetal (POM) homopolymer,* DuPont Delrin 570 (112).

Test temperature (°F)	Initial applied stress (lb$_f$/in^2)	Creep (apparent) modulus*,** ($\times 10^3$ lb$_f$/in^2) test time (h)							Time at latest test point (h)
		1	10	30	100	300	1000	Latest	
73	500	1,220	1,100	920	800	700	640	380	20,000
	1,000	940	830	750	670	610	540	300	20,000
	1,500	880	750	720	620	590	500	380	10,000
	2,000	870	730	700	630	560	490	320	20,000
140	500	680	590	500	410	350	300	210	20,000
	1,000	530	460	400	350	310	280	200	20,000
	2,000	420	320	290	250	220	190	110	10,000
185	500	410	340	290	260	240	190	110	10,000
	1,000	380	300	240	220	200	180	90	10,000
	1,500	330	300	280	240	220	200	160	10,000
	2,500	300	230	210	180	—	—	—	—
195	2,000	330	250	230	190	180	160	—	—

*20% (wt) glass-reinforced, general purpose, injection molding compound
**Calculated from total creep strain or deflection (before rupture and onset of yielding)

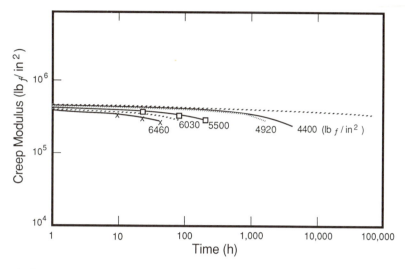

Figure 6.125.   Time-Dependent Creep Modulus of Dow Tyril 867 Polystyrene–Acrylonitrile (SAN) (113). Parameter is Applied Stress.

The data in Table 6.15 and Figure 6.126 are for tensile creep rather than flexural beam bending. As noted, most of the literature data are for tensile creep. Under beam bending, the maximum fiber stress occurs at the outer surface of the beam and decreases to zero at the neutral axis* of the beam. For a linear viscoelastic polymer, the modulus is independent of the applied stress. Thus the modulus of the beam is uniform. For a nonlinear viscoelastic beam, on the other hand, $E = f(\sigma)$. If the modulus corresponding to the maximum fiber stress, on the surface, is used, the calculated deflection will be substantially greater than that actually observed.

For the problem under discussion, the design stress is seen to be time-dependent. The stated design time is 5 years = 43,800 hours. From Figure 6.125, the creep rupture stress at this time is 3350 lbf/in² [23 MPa]. The design or working stress is the creep rupture stress at the design time divided by a safety factor. In this example, the safety factor is assumed to be 2, giving a working stress of 3350/2 = 1675 lbf/in² [11.5 MPA]. The designer should recall that when working with Hookean materials, the working stress can be assumed to be the yield stress divided by the safety factor. The yield stress for polymers is a short-time duration property that has little relevance to the creep rupture stress at, say, 5 years. For example, for SAN, the yield stress is about 8000 lbf/in² [55 MPa]. If a safety factor of 2 is used, an incorrect value of 4000 lbf/in² [27.6 MPa] would have been chosen for the working

---

*Under small deformation, the neutral axis and the centerline of the beam nearly coincide.

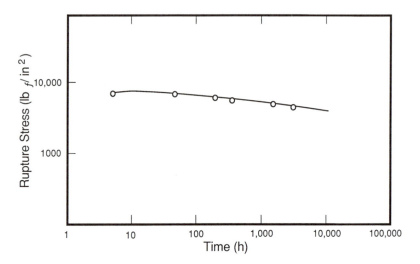

Figure 6.126.   Time-Dependent Creep Rupture Stress of Dow Tyril 867 Polystyrene–Acrylonitrile (SAN) (113).

stress. As seen in Figure 6.125, the beam would have failed in flexural creep
rupture in about 7000 hours (42 weeks). Keep in mind that this time is really
a statistical average. Certain parts will fail in shorter times and some will last
far longer. 7000 hours is the average failure time. The designer should be
aware that very few creep data include the statistical variance that would allow
a more intelligent judgment as to the actual lifetime of, say, 63, 86 or 95%*
of the molded parts.

Based on the 5 year design stress of 1675 lb_f/in^2 [11.5 MPa], the thickness
of beam, h, is given as:

$$(6.39) \qquad\qquad h = \left[\frac{3\ PL}{2\ b\sigma} \right]^{1/2} = 0.116\ in$$

To determine the maximum deflection at 5 years and an h = 0.116 in, the
creep modulus must be determined. As seen in Figure 6.125, the data do not
extend beyond 3500 hours. To get the value at 43,800 hours, a modulus-time
curve must be extrapolated from the data. There are no specific rules to allow
extrapolation, only good judgment. As noted previously, extrapolation should
not extend more than one decade beyond the last data point. The extrapolated

---

*These percentages represent variance of 1, 2 and 3 standard deviations from the mean.

creep modulus at 5 years is about 270,000 lb$_f$/in^2 [1.86 GPa]. The maximum 5 year deflection is given as:

$$(6.40) \qquad\qquad \delta = \frac{PL^3}{4 \ E \ bh^3} = 0.080 \ in$$

Thus, the calculated deflection meets the design criterion of a maximum deflection not to exceed 0.100 in. If the calculated deflection had been greater than the design criterion, the beam thickness would have needed to be increased. This would have resulted in a lower working stress and hence a greater safety factor.

Consider a second example of the application of creep data:

> Calculate the deflection of a simply supported beam of polyacetal (POM) after 1 year at 140°F under a load P, of 1 lb$_f$. The beam width, b, is 0.5 in, its thickness, h, is 0.09 in, and its span, L, is 4 in.

The polyacetal (POM) data of Table 6.16 are replotted as Figure 6.127. Note that there are no creep rupture data. Therefore it is imprudent for the designer to extrapolate any curves beyond the time frame of the experimental data. The design time of 1 year = 8760 hours is within the experimental database of Table 6.16. Note further that the creep modulus is stress-depen-

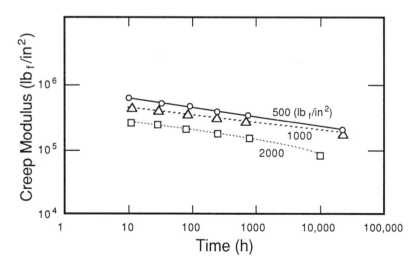

Figure 6.127.   Time-Dependent Creep Modulus of DuPont Delrin 570 Polyacetal (POM) Homopolymer (113). Parameter is Applied Stress.

dent, indicating a polymer that is nonlinearly viscoelastic. The maximum fiber stress on the beam is given as:

$$(6.41) \qquad \sigma = \frac{3 \, PL}{2 \, bh^2} = 1480 \ lb_f/in^2 \ [10 \ MPa]$$

From Figure 6.127, the tensile creep modulus at 1480 lb_f/in^2 [10 MPa] stress and 8760 hours, is about 160,000 lb_f/in^2 [1.1 GPa]. The deflection at this time is:

$$(6.42) \qquad \delta = \frac{PL^3}{4 \, E \, bh^3} = 0.274 \ in$$

Since the effective modulus at the neutral axis of the beam is less than 160,000 lb_f/in^2 [1.1 GPa], due entirely to zero stress at the neutral axis, the calculated deflection, $\delta = 0.274$ in, will be greater than the actual deflection.

The experimental data can also be plotted at constant temperature, as an isochronous plot. Consider the isochronous stress–strain plot for polycarbonate (PC) at 40°C (Figure 6.128). As an example of the way in which this graph can be used, consider this:

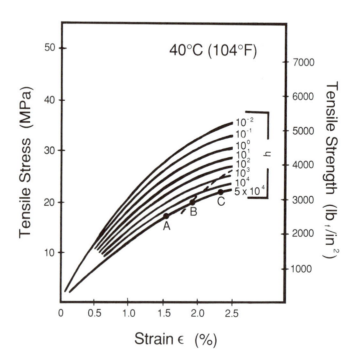

Figure 6.128.   Isochronous Stress–Strain Diagram for Polycarbonate (PC) at 40°C. Dashed Line: Crazing Stress.

A 6 in long polycarbonate (PC) square member is to hold an axial load, P, of 40 lb_f in air at 40°C. If the member cannot elongate more than 0.090 in in 4 years, what should the dimensions of the member be?

The allowable strain, $\varepsilon = 0.090/6 = 0.015$ or 1.5%. Four years is about 35,000 hours. From Figure 6.128, the design stress at 1.5% and 35,000 hours is 2600 lb_f/in^2 [18 MPa]. The side, b, of the square member is determined as:

$$(6.43) \qquad b = \left(\frac{P}{\sigma}\right)^{1/2} = 0.124 \text{ in}$$

Consider the case where the maximum deflection is 0.135 in. The strain is $\varepsilon = 0.135/6 = 2.25\%$. At this strain, the design stress is about 3200 lb_f/in^2 [22 MPa] (point b on Figure 6.128). Note that this strain is greater than the crazing strain, about 0.9%, and probably not acceptable. If the crazing strain is considered to be the upper strain limit, the stress at 35,000 hours and 40°C, cannot exceed 3000 lb_f/in^2 [20.7 MPa]. At this stress, the side of the beam is 0.115 in and the beam will deform, $\delta = 0.019 \times 6 = 0.114$ in.

Time–Temperature Superposition.   The time–temperature superposition method was described in detail in Section 3.6, where it was used to illustrate a way of analyzing the temperature effect on the time constants in mechanical models. As an example, torsional creep data for polycarbonate in Figure 6.129 were analyzed using the Williams–Landel–Ferry (WLF) shifting equation. The designer is encouraged to reread Section 3.6 to regain an appreciation of the shift factor concept, as applied to long-time duration data interpretation.

Consider Figure 6.129 for polycarbonate as the result of creep and stress relaxation experiments. The test results are obtained over a 40° temperature range. Note that increasing the test temperature from 150 to 167°C results in a ten-fold increase in modulus. As noted in Section 3.6.4, the raw data are shifted for temperature and density. Usually T_g, the glass transition temperature, is selected as the reference temperature. Here the data are shifted against a reference temperature, T_t. Each data point is therefore shifted up or down, according to:

$$(6.44) \qquad E(T_t; t) = \left[\frac{T_t}{T_i}\right] E(T_i; t)$$

The effect of the temperature on the density of the polymer, $\rho(T)$, and on the inherent modulus value must also be taken into account. Usually these corrections have opposite effects. Thus the overall correction to the raw data is usually small. The modulus change is related, then, to time as:

$$(6.45) \qquad E(T_t; t) = \left[\frac{\rho(T_t)\, T_t}{\rho(T)\, T}\right] E(T; t/a_T).$$

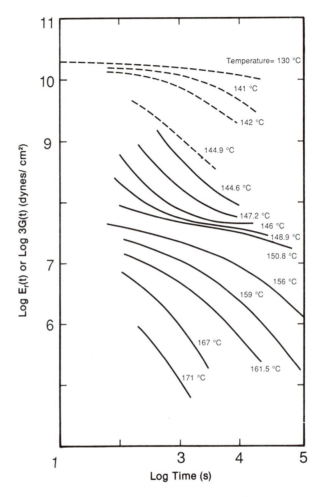

Figure 6.129. Tensile Creep and Stress Relaxation Strength of M_w = 40,000 Bisphenol A Polycarbonate (PC) (114). Dashed Line: Creep Data. Solid Line: Stress Relaxation Data.

The mathematical formula can be constructed by taking each of the corrected data curves and shifting them horizontally to the right or left to form the single "master curve" at temperature, T_t. This is shown graphically in Figure 6.130. The data of Figure 6.129 as well as a second set of data, not shown, were used to form the master curve, shown here as Figure 6.131. The shift factor, a_T, for both sets of data is shown in Figure 6.132. As noted earlier, a_T is the shift factor, the ratio of t_T to $t_{T,a}$, where t_T is the time required to reach a particular modulus value at temperature T and $t_{T,a}$ is the time to reach that same value at temperature, T_a. As noted in Chapter 3, the WLF equation

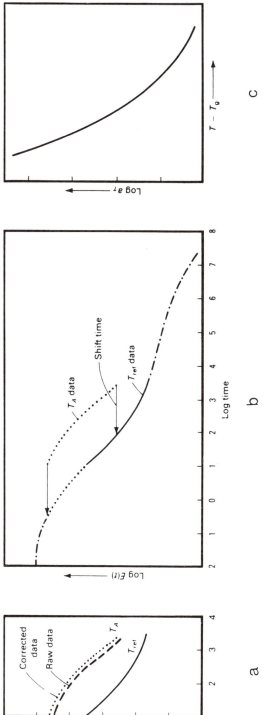

Figure 6.130.   A Schematic of the Time–Temperature Superposition Scheme. a) Raw Data. b) Data Shift. The Amount of Shift Needed to Produce a Smooth Master Curve Results in the Temperature-Dependent Shift Factor, a_T, Shown in c).

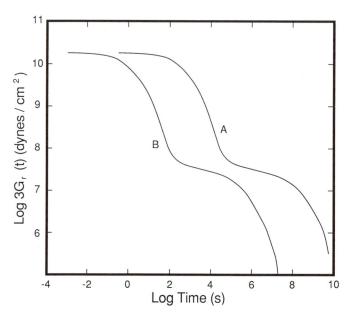

Figure 6.131.   An Example of the Master Curve for Two Molecular Weight Ranges of Bisphenol A Polycarbonate (PC) (114). Curve A: $M_w = 90,000$. B: $M_w = 40,000$.

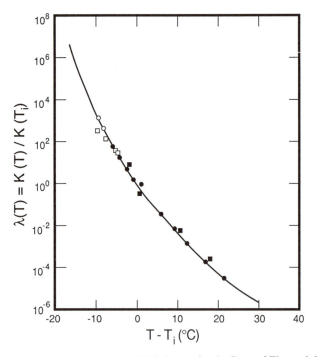

Figure 6.132.   The Time-Dependent Shift Factor for the Data of Figure 6.131 (114).

that relates a_T to temperature is a combination of experimental observations and theoretical analyses:

$$(6.46) \qquad \log a_T = \frac{-C_1 (T - T_g)}{[C_2 + (T - T_g)]}$$

For many polymers, C_1 and C_2 can be considered to be universal constants, with values of 17.44 and 51.6, respectively. For some polymers, these values are not quite correct, as seen in Table 6.17. The coefficients for the two polycarbonate resins in Figure 6.131 are $C_1 = 16.4$ and $C_2 = 56$.

Example 6.10

Q: A liquid fertilizer is packaged in a polypropylene homopolymer bottle. If the liquid has a vapor pressure of 20 lb_f/in^2 at 25°C, what is the maximum allowable stress in the resin if the container is 2 in in diameter and 0.040 in in wall thickness?

If the resin reaches this stress level at these conditions in 10,000 hours, and if the vapor pressure increases in proportion to its absolute temperature, $p/p_o = T/T_o$, at what time does the resin reach the maximum allowable stress level at 35°C? T_g for PP = -15°C.

A: The maximum stress for a thin-walled container is given as the hoop stress equation:

$$\sigma = \frac{pd}{2t}$$

in which σ is the stress [lb_f/in^2], p is internal pressure [lb_f/in^2], d is pressure vessel diameter and t is its thickness.

$$\sigma = \frac{20 \times 2}{2 \times 0.04}$$

$$= 500 \ lb_f/in^2$$

Table 6.17. Williams-Landel-Ferry (WLF) constants for several polymers; universal constants also given (115).

Polymer	C_1	C_2	$T_g(°K)$
Polyisobutylene (PIB)	16.6	104	202
Natural rubber	16.7	53.6	202
Polyurethane elastomer	15.6	32.6	238
Polystyrene (PS)	14.5	50.5	373
Polyethyl methacrylate (PEMA)	17.6	65.6	335
Polycarbonate (PC)	16.14	56	423
Universal constants	17.44	51.6	—

The pressure at 35°C is given as:

$$p = 20 \ [\mathrm{lb_f/in^2}] \times \frac{35 + 273}{25 + 273} = 20.7 \ \mathrm{lb_f/in^2}$$

$$\sigma = 500 \times \frac{20.7}{20} = 517 \ \mathrm{lb_f/in^2}$$

This is not substantially greater than the value at 25°C.

Since the C_1 and C_2 constants for polypropylene in the WLF equation are not given in Table 6.17, assume $C_1 = 17.44$ and $C_2 = 51.6$. Now:

$$\log_{10} a_{25} = \frac{-17.44 \times (25 + 15)}{(51.6 + 25 + 15)} = -7.615$$

$$a_{25} = 2.42 \times 10^{-8}$$

$$\log_{10} a_{35} = \frac{-17.44 \times (35 + 15)}{(51.6 + 35 + 15)} = -8.583$$

$$a_{35} = 2.614 \times 10^{-9}$$

$$\frac{a_{35}}{a_{25}} = \frac{t_{35}}{t_{25}} = 0.108$$

Or if the pressure were constant, the time to reach maximum stress level at 35°C would be:

$$t_{35} = 10,000 \times 0.108 = 1080 \ \mathrm{h.}$$

Since the pressure is somewhat higher, the time to reach the maximum stress level is *somewhat less than* this number. The exact number can be determined only if we know the creep stress characteristics of polypropylene at these temperatures.

Lowering the use temperature significantly increases the design lifetime of the polymer. Of course, despite its popularity, the WLF technique has some serious limitations. It works satisfactorily for amorphous polymers within 50°C of the glass transition temperature, but it can lead to substantial errors for glassy amorphous polymers and crystalline polymers. The calculated values should therefore be supported with values from one of the other techniques.

Empirical Correlations   There are many different empirical correlations for creep data. One of the most widely used is the Findlay correlation. Findlay (116) suggests that the total strain on any polymer structure under constant axial load can be predicted with a power law of the form:

(6.47) $$\varepsilon = \varepsilon'_o + \varepsilon'_T \, t^n$$

where ε'_o is the stress-dependent, time-independent initial elastic strain and ε'_T is the stress-dependent, time-independent coefficient of strain. n is a stress-independent material constant. He further shows that the stress-dependent parameters can be related to applied stress and other material parameters by:

$$(6.48) \qquad \varepsilon = \varepsilon_0 \sinh\left(\frac{\sigma}{\sigma_0}\right) + \varepsilon_T \, t^n \sinh\left(\frac{\sigma}{\sigma_T}\right),$$

Here ε_0, σ_0, ε_T, σ_T, and n are empirical material constants obtained by simple curve-fitting techniques applied to typical creep data. Representative values for these five constants for several materials are given in Table 6.18. Note that for a linear visco-elastic polymer, creep and stress relaxation effects are directly related. Thus the time for stress relaxation can be obtained directly from the $\varepsilon = \varepsilon(\sigma)$ equation:

$$(6.49) \qquad \tau = \left\{ \frac{[\varepsilon - \varepsilon_0 \sinh(\sigma/\sigma_0)]}{[\varepsilon_T \sinh(\sigma/\sigma_T]} \right\}^{1/n}.$$

Findlay's original correlation did not include temperature. Most experimental data indicate that the effect is linear over a very limited temperature range. However, it is accepted today that all temperature effects for the Findlay correlation must be obtained empirically.

The Findlay correlation is based on best fit of creep data on formaldehyde- and phenolic-based thermosets impregnated in canvas, paper, and asbestos. Several other empirical equations have been tried:

$$(6.50) \qquad \varepsilon = \varepsilon_0 + A \ln t + Bt$$

$$(6.51) \qquad \varepsilon = \varepsilon_0 + A (1 - e^{Ct}) + Bt$$

$$(6.52) \qquad \varepsilon = \varepsilon_0 + A \ln t$$

$$(6.53) \qquad \varepsilon = \varepsilon_0 + A (1 - e^{Ct})$$

$$(6.54) \qquad \varepsilon = \varepsilon_0 + m \, t^n$$

Table 6.18. Constants for the empirical Findlay equation at 77°F and 50% relative humidity (117, 118).

Polymer	n	ε_0	ε_T	σ_0	σ_T
Polyethylene (PE)	0.154	0.027	0.0021	585	230
Polychlorotrifluoroethylene (CTFE)	0.0872	0.0081	0.00099	2,600	1,475
Polyvinyl chloride (PVC)	0.305	0.00833	0.000079	4,640	1,630
Polystyrene (PS)	0.525	0.048	0.0000041	20,000	650
Melamine (glass fiber filled)	0.0186	0.00575	0.00575	35,000	35,000
Unsaturated polyester (fiber filled)	0.090	0.0034	0.000445	15,000	14,000
Unsaturated polyester (glass mat)	0.19	0.0067	0.0011	8,500	8,500

As seen in Figure 6.133, Equation (6.54) correlation fits the data best. It should be noted that the Findlay correlation assumes linear viscoelasticity of the polymer. Many polymers are nonlinear viscoelastic. As a result, the calculated values are not always in good agreement with the experimental results. The lack of a temperature effect in the correlation also limits its utility. The linear viscoelasticity assumption can be very restrictive in another way. Substantial errors can be introduced if Boltzmann's superposition principle is used to account for complex loading and unloading of a nonlinear viscoelastic material. Experimental evidence shows that the response times for real polymers after load removal can be substantially longer than values predicted from linear viscoelasticity.

It is recommended that at least two of these calculation schemes be used to obtain suitable design data. If two agree, the designer can be reasonably certain that the values are correct. If there is great disparity, the designer must either obtain experimental data for the specific load and length of time at the desired temperature or must include a substantial design safety factor.

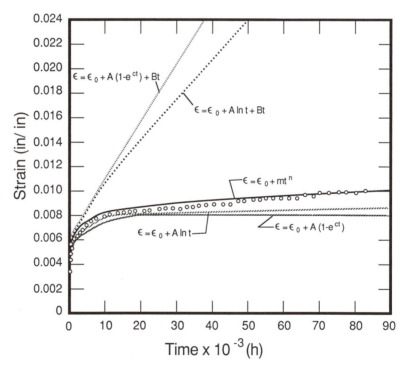

Figure 6.133.   Time-Dependent Tensile Creep Strain for a Canvas Laminate at Tensile Stress, 3000 lb_f/in^2 at 23°C. (119).

Creep: The Practical Aspects

It is important that the design engineer consider creep in practical terms. As noted, creep is the phenomenon of continuing material elongation under sustained load. The design modulus is not a material property such as Young's modulus of a Hookean material. Rather, the designer uses "apparent modulus" or "time-dependent modulus" to estimate the time-dependent deformation of an arbitrary polymer body under external forces. The designer must use this approximate approach today since there is no general constitutive equation to describe the response of an arbitrary polymer to multidimensional forces. With polymers that exhibit linear stress–strain behavior, hence time-independent moduli, the empirical approach has some basis in theory, Figure 6.125. For polymers with nonlinear stress—strain behavior or time- or stress-dependent moduli, the empirical approach is questionable, since the causal relationships between stress and strain have evolved from basic Hookean mechanics (Figure 6.127). This is particularly true at high levels of strain. Certainly improved modeling is needed to predict true time-dependent material response.

Furthermore, the design engineer should always keep in mind that on a molecular level, amorphous and semicrystalline polymers respond quite differently to applied load. Molecular weight distribution is only one parameter. The molecular crystal structure is the primary load-bearing feature of a crystalline polymer. As long as the molecular weight of the crystalline polymer is above a minimum or critical value, its elevated temperature characteristics are relatively independent of molecular weight (Figure 6.134). When the polymer is below T_g, the stiffness shows a weak dependence on molecular weight. On the other hand, the load bearing capacity of an amorphous polymer is directly related to intermolecular forces and polymer chain length. The onset of glass transition temperature is also influenced by molecular chain length, resulting in a shift in the stiffness curve, as shown in Figure 6.135. A general "rule of thumb" for amorphous polymers is that the time-dependent creep modulus curve approaches zero at the onset to glass transition temperature. Hence, the creep testing of an amor-

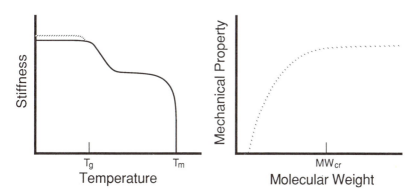

Figure 6.134.   Schematic Relationship Between Temperature-Dependent Crystalline Polymer Stiffness and the Relationship of Stiffness to Polymer Molecular Weight.

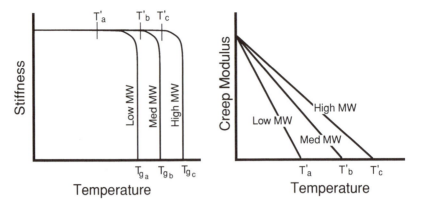

Figure 6.135.   Schematic Relationship Between Temperature-Dependent Amorphous Polymer Stiffness and Temperature-Dependent Creep Modulus with Polymer Molecular Weight.

phous polymer at room temperature will reveal no appreciable effect of molecular weight on modulus. Creep testing must therefore be conducted at elevated temperatures if the polymer is to be used near its transition temperature.

Long-Time Duration Testing: Dynamic and Periodic

In dynamic long-term loads, a load is applied, then removed, in a cyclic or possibly aperiodic fashion for the life of the product. Again, if the material is Hookean elastic, the rate of load application, the duration of loading and the duration of unloading are of no importance. The material responds only to the amount of the load. If the material is purely viscous, the extent of deformation is a function of the total time of application of the load. Once the load has been removed, the material stops moving. In viscoelastic polymers, the rate of loading, the duration of loading and unloading, and the extent of load are all factors in determining how the polymer will respond to the applied load.

As with long-term static loading, there are many ways to load dynamically. The polymer can be uniaxially loaded in tension, flexure, shear, or torsion. The loading rate can be sinusoidal, step, or ramp. The loading can be about a mean of zero, as in tension and compression to the same amplitude, or about a midpoint, by loading to a fixed value, then unloading to zero. The load can be applied in a periodic or on-off fashion, or aperiodically. The load can be complex, with more than one form of loading, such as tensile, flexural and torsional stressing in brake hoses. The complexity can be a combination of load and environmental effects, including temperature.

Most materials, even true Hookean metals, exhibit fatigue or loss in strength as a result of repeated stressing at loads far below the single-load failure point. In polymers, failure initiates in microvoids around the repeatedly stressed molecules. These microvoids connect to form microcracks. Depending on the nature of the polymer, these microcracks eventually manifest as stress whitening, stress crazing or stress cracking.

Continued repeated stressing results in *fatigue failure*. Hertzberg and Manson (120) list the following primary polymer aspects that influence fatigue behavior:

- Molecular characteristics, such as molecular weight, molecular weight distribution and the thermodynamic state,
- Chemical changes, such as bond breakage,
- Homogeneous deformation such as elastic, anelastic, or viscoelastic behavior,
- Inhomogeneous deformation such as crazing or shear banding,
- Morphological changes, such as drawing, orientation, and crystallization,
- Transition phenomena, such as glass transition and secondary transitions, and
- Thermal effects, such as hysteretic heating.

Typically, as the level of repeated stress applied to the specimen increases, the ability for the specimen to withstand the load decreases. Thus, dynamic testing results in a plot of the applied stress, also called *material fatigue strength,* as a function of the number of cycles to failure, Figure 6.136. These plots are usually called "S–N" curves, with S for stress and N for the number of cycles to failure.

As noted, for polymers, failure may not be catastrophic. Shear banding, stress whitening or other effects may signal the effective end of life for the product. Figure 6.137 shows that at a fixed repeated stress, crack initiation may occur as early as a decade before final failure.

In addition, the *nature* of the testing procedure can dramatically affect the shape of the S–N curve. Certain guidelines for testing were established early on. For example, aperiodic or random loading was excluded owing primarily to its irreproducibility. Andrews (121) recommends the following guidelines:

- The stress system must be periodically varying with a fixed stress amplitude, $\Delta\sigma$,
- The strain amplitude, $\Delta\varepsilon$, must be also periodic,
- There is to be a mean stress level, σ_m,
- And a mean strain level, ε_m,

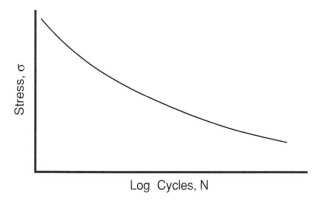

Figure 6.136.   Typical Fatigue Response of a Polymer (123).

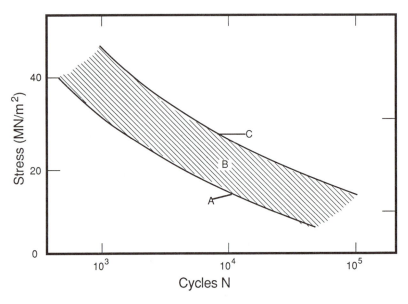

Figure 6.137.   Fatigue Response of a Polymer, Showing Lower Stress Limit of Crack Propagation (A), Crack Propagation Region (B), and Final Failure Limit (C) (124).

- And a fixed frequency,
- There must be a characteristic wave form for both stress and strain, typically sinusoidal, ramp, or square,
- The ambient and internal temperatures must be known, even if not necessarily identical, and
- The geometry must be fixed, including notches, if any.

Five tests categories have been proposed:

- Periodic loading between fixed stress limits, in tension or compression,
- Periodic loading between fixed strain limits, in tension or compression.
- Reversed bending stress, implicit in flexing a sheet in one dimension,
- Reversed bending stresses in two dimensions, such as by rotary deflection of a cylindrical specimen, and
- Reversed shear stresses obtained by torsional deformation.

Data from these five types of test cannot usually be compared. Turner notes that the two most popular tests are flexural fatigue and tension/compression. He views the flexural fatigue test as involving a relatively simple apparatus but yielding somewhat questionable data. On the other hand, the tension/compression test is quite complicated but the results are unambiguous (122). The flexural fatigue test is considered here.

Repeated Flexural Stress Tests: ASTM D671

In this test, the polymer specimen is subjected to flexural bending. The test specimen is a cantilevered beam having a triangular taper, as shown in Figure 6.138. The taper allows the beam bending moment to uniform with distance from the point of applied load. The stress is therefore uniform along the beam and fatigue failure will not occur at the gage end, as it would if the beam had no taper. This is called a *stress-controlled test*. The standard notes:

> The [flexural fatigue] results are suitable for direct application to design *only when all* design factors including magnitude and mode of stress, size and shape of part,

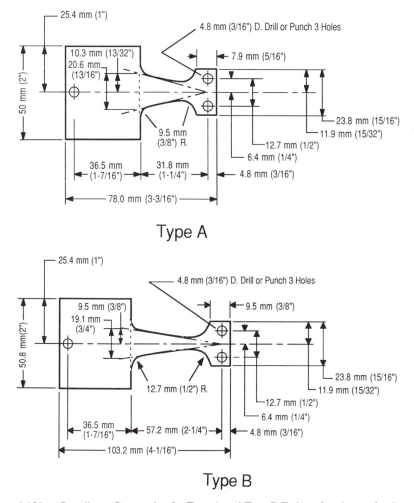

Type A

Type B

Figure 6.138.   Coordinate Geometries for Type A and Type B Fatigue Specimens for Constant Force Fatigue According to ASTM Specification D671.

ambient and part temperature, heat transfer conditions, cyclic frequency, and environmental conditions are comparable to the test conditions. [emphasis supplied by authors]

Hertzberg and Manson note that there are two primary methods of failure associated with this test. The simpler is the development and propagation of a crack across the specimen. The more complex is accumulation of hysteretic energy created during each flexing cycle. Typical stress-controlled, non-thermal fatigue failure S–N curves are shown in Figures 6.139 and 6.140. Although sinusoidally applied load is common, the square wave allows comparison of the periodically applied fixed stress with constantly applied fixed stress, as seen in Figure 6.141. It is apparent for PVC that cyclic loading results in substantially greater time-dependent damage than continuous loading at the same level.

Hysteretic heat buildup can dramatically shorten the fatigue lifetime of a polymer. It is well understood that hysteretic energy is generated volumetrically in the localized bending region and can only be liberated geometrically at the specimen surface. For polymers, there is very little conduction of energy along the specimen length to the grips. As the specimen temperature increases, the local modulus decreases. Since the load is stress-controlled, the deflections increase. This exacerbates the hysteretic energy generation. The result is an auto-accelerating temperature effect that ultimately prevents the material from supporting the applied load to the maximum deflection of the device. The standard defines thermal fatigue failure as the number of cycles in which the apparent modulus decays to about 70% of the original modulus of the spec-

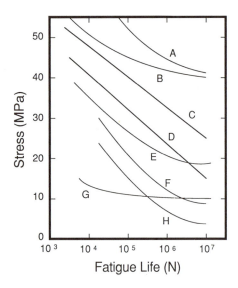

Figure 6.139.   Cyclic Fatigue Stress for Several Polymers at 23°C (125). A: Phenol–Formaldehyde. B: Epoxy. C: Polyethylene Therephthalate (PET). D: Nylon (PA), Dry. E: Modified Polyphenylene Oxide (mPPO). F: Polycarbonate (PC). G: Ethylene-Chlorinated Polytetrafluoroethylene (ECTFE). H: Polysulfone (PSO_2).

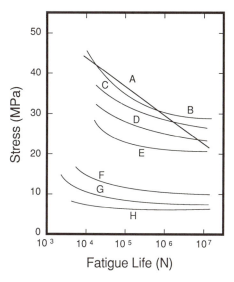

Figure 6.140.   Cyclic Fatigue Stress for Several Polymers at 23°C (125). A: Urea–Formalde-hyde. B: Diallyl Phthalate (DAP). C: Polyalkyd. D: Polyacetal (POM). E: Polymethyl Meth-acrylate (PMMA). F: Polypropylene (PP). G: Polyethylene (PE). H: Polytetrafluoroethylene (PTFE).

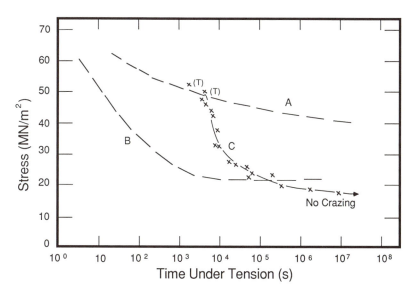

Figure 6.141.   Comparison of Continuous Load Bearing Capability of Rigid PVC (RPVC) at 20°C with Flexural Fatigue Fracture Stress (126). Curve A: Continuous Tensile Load, Necking. B: Continuous Tensile Load, Craze Initiation. C and Points X: Fracture Under Cyclic Flexural Load. ×(T): Failure by Thermal Softening (See Figures 6.143 and 6.144).

imen, as determined at the beginning of the test. The standard further calls for the tester to:

> . . . measure the temperature at failure *unless* it can be shown that the heat rise is insignificant for the specific material and test condition [emphasis supplied by authors].

The ASTM test further restricts the testing conditions to no more than 30 Hz (30 cycles/s). The effect of temperature on fatigue life is shown on a standard S–N curve for polytetrafluoroethylene (PTFE) in Figure 6.142. It is immediately apparent that thermal effects can dramatically reduce the lifetime of a part under flexural fatigue. Another view shows the cycle-dependent material temperature and the resulting J'', the loss compliance, for PTFE (Figure 6.143). Since the loss compliance increases with increasing temperature, the auto-accelerating nature of hysteretic energy buildup is assured. The rate of heat generation, H_F, is given as:

$$(6.55) \qquad H_F \; \alpha \; \frac{DF^2}{E_d}$$

where D is the damping capacity of the material, F is the applied force and E_d is the dynamic modulus. Recall that $E_d = E_d(T)$. As temperature increases, E_d decreases and H_F therefore increases. The temperature increase, ΔT, is given as:

$$(6.56) \qquad \Delta T \; \alpha \; \frac{\sigma^2 \, fd^2 D}{E_d A}$$

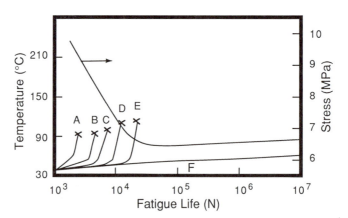

Figure 6.142. Temperature-Dependent Fatigue Fracture Curves for Polytetrafluoroethylene (PTFE) at 30 Hz Frequency (125). A: Stress = 10.3 MPa, 2000 Cycles, 100°C. B: Stress = 9.0 MPa, 4000 Cycles, 115°C. C: Stress = 8.3 MPa, 6100 Cycles, 125°C. D: Stress = 7.6 MPa, 9500 Cycles, 130°C. E: Stress = 6.9 MPa, 19,000 Cycles, 141°C. F: Stress = 6.3 MPa, 10^7 Cycles, 60°C.

where σ is the applied stress, f is the test frequency, d is the specimen diameter, and A is a heat transfer coefficient. For fixed testing conditions (σ, f, d, D), decreasing modulus results in increasing temperature rise. The energy dissipation rate, $\dot{E}$, can be written as:

$$\dot{E} = \pi \, f \, J''(f, T) \, \sigma^2 \qquad (6.57)$$

where J'' is the loss compliance. The temperature rise per unit time can then be written as:

$$\Delta \dot{T} = \frac{\pi \, f \, J''(f, T) \, \sigma^2}{\rho \, c_p} \qquad (6.58)$$

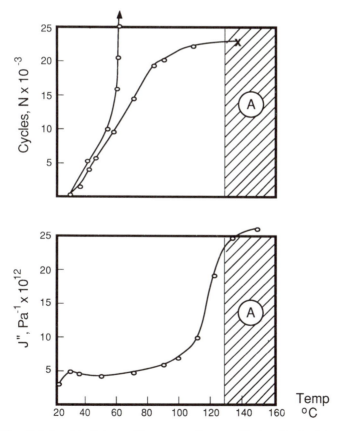

Figure 6.143.   Relationship Between Temperature Increase During Cyclic Fatigue and Temperature Dependence of Loss Compliance, J'', for Polytetrafluoroethylene (PTFE) (125). A: Fatigue Failure Region.

where ρ is the material density and c_p is its specific heat. It is apparent that thermal effects are felt most in the final stages of fatigue. Nielsen (127) showed that no unfilled polymer should show no early signs of premature failure when hysteretic effects dominate any stress-controlled testing results. Certainly, testing at lower frequency should result in increased lifetime of the test specimen. This is seen in Figure 6.144 for PTFE (165).

The relationship between mechanical (isothermal) fatigue failure and thermal fatigue failure is best seen in stress–log time plots such as Figure 6.145. Essentially, the curves labeled "T" represent failures by melting. As noted, these are highly frequency-dependent. Mechanical failures, on the other hand, appear to be essentially frequency-independent.

As noted, two other types of flexural fatigue tests are proposed to circumvent the hysteretic heating problem in the stress-controlled test. The first is a deflection-controlled test, in which the stress decay is measured as a function of the number of cycles. Failure is considered to occur when the instantaneous stress level decreases to 70% of its initial value. A typical stress–logarithmic cycle curve is shown in Figure 6.146. Note that since the stress decreases with increasing number of cycles, the hysteretic heating per cycle decreases. Therefore relatively few failures can be classified as due entirely to thermal fatigue. The most familiar strain-controlled test is the tension/compression test. As with deflection-controlled fatigue testing, there is no hysteretic heat buildup and so failure is only by mechanical fatigue. A comparison of stress- and strain-controlled tests is given in Figure 6.147 for polystyrene. Zone II is the region of stress crazing. As expected, development of crazes results in a rapid reduction in material lifetime at high strain levels.

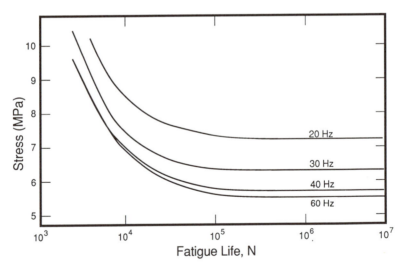

Figure 6.144. Effect of Cyclic Test Frequency on Fatigue Stress–Cycles to Fail (*S–N*) Curve for Polytetrafluoroethylene (PTFE) (125).

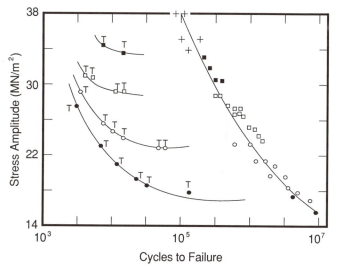

Figure 6.145.   Cyclic Fatigue Stress for Polyacetal (POM) (128). Thermal Failures Denoted with "T". +: 0.167 Hz Cyclic Frequency. Solid Square: 0.5 Hz. Open Square: 1.67 Hz. Open Circle: 5 Hz. Solid Circle: 10 Hz.

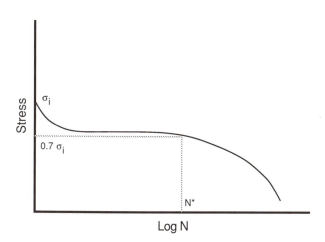

Figure 6.146.   Cyclic Stress Decay Curve. N^* Corresponds to Cycles Needed to Achieve 70% of Initial Stress Level, σ_i (129).

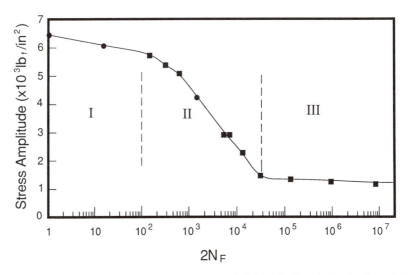

Figure 6.147.   Cyclic Stress Level for Polystyrene at 25°C (130). Zone I: Strain Control. Zone II: Transition. Zone III: Stress Control.

6.6 Friction and Wear

Frequently, polymeric materials are used as bearing surfaces in rotating devices. Applications include thrust bearings, bushings, wear pads, hinges and gears. The field of lubrication, friction and wear is known as *tribology*. The tribological characteristics of metals and some ceramics are well-documented. The understanding of friction and wear in polymers is still quite tentative, owing to the wide spectrum of polymeric materials and the more complex interaction of the macromolecular nature of the polymer surface with the friction-causing surface (131,132). It is generally accepted that the applied force, F, is a linear function of applied load, N:

$$(6.59) \qquad\qquad\qquad F = \mu\,N$$

where μ is the frictional coefficient. If a nominal contact area, A, is known, the specific frictional force, f, can be written in terms of an applied pressure, p:

$$(6.60) \qquad\qquad\qquad f = \mu\,p$$

It is common practice to consider the frictional force to be a sum of two effects, the adhesion force and the drag or deformation force. The latter is associated with the deformation of the asperities and the plowing or mechanical deformation of the softer polymer under load. Polymeric responses to frictional loading can be categorized as:

Virgin friction. The surface energy of the load is very high and no interfacial deformed layer exists. The primary mode of frictional resistance is adhesion bonding between the two surfaces. The coefficient of friction is very high.

Clean friction. The coefficient of friction is high and depends on the oxide layers on metal surfaces.

Transitional boundary friction. A very thin layer of deformed polymer forms a tertiary or interfacial layer between the bulk of the polymer and the metal load. Adhesion bonding forces decrease and deformational forces increase with time. The frictional coefficient can decrease with time as the interfacial layer builds.

Pure boundary friction. The interfacial layer thickness is stable with time. As a result, the frictional coefficient is now a function of the physical nature and mechanical properties of the metal, polymer *and* interfacial material, as well as the experimental conditions, including pressure, sliding velocity, temperature, polymer surface roughness, metal surface roughness and so on. Bartenev and Lavrentev note that this is the most complex form of friction and the most common in polymer–metal friction.

Transitional hydrodynamic friction. The interfacial boundary layer provides substantial lubrication to the sliding surface. This material can be an adduct of the polymer, an externally supplied lubricant, or even the polymer itself. In certain instances, lubricants are deliberately added to the polymer for friction-and-wear control. Under pressure and/or frictional heat, these lubricants diffuse or "bloom" to the primary polymer surface. In some cases, frictional heating can be intense enough to locally melt the polymer, thus providing the hydrodynamic layer. The classic application of this is friction welding, in which the polymer is deliberately heated to its melt state by friction. This mode of friction is very unstable and so prediction of frictional coefficients is very difficult.

Pure hydrodynamic friction. The interfacial layer is a true lubricant. The sliding performance here can be modeled with traditional lubrication theory, where the shear viscosity of the lubricant is a major property.

For metals, the relationship between frictional force and normal load can be related to the material shear strength, σ, and surface hardness, H:

$$(6.61) \qquad F = \frac{\sigma}{H} = \mu N$$

Thus, the frictional coefficient, μ, is a function only of material properties, shear strength and hardness. For polymers, the frictional force can be shown to be a function of the actual contact area, A:

$$(6.62) \qquad F = CA$$

where C, the proportionality factor, is a function of shear strength and surface hardness. Tabor (134) notes that surface hardness can also be defined as "cold-flow limit", Σ_y, or yield strength, σ_y. The linear relationship between frictional coefficient and σ_y/Σ_y is seen for steel on several polymers in Figure 6.148. As noted, the direct equality of frictional coefficient with σ_y/Σ_y is not met. The linear relationship indicates that adhesion plays a significant role in polymer response to applied loads. However, adhesion does not necessarily dominate. The bonds formed serve only to localize shear formation in the bulk of the polymer. Certainly with many polymers deformation plays as significant a role in frictional response as adhesion.

Note also that polymeric hardness increases with applied load (Figure 6.149). It is thought that amorphous polymers under moderate frictional loads act as elastically deformed structures below their glass transition temperatures and as viscoelastically-deformed structures above. Crystalline polymers are more difficult to characterize. Owing to its very low frictional coefficient, polytetrafluoroethylene (PTFE) has received great acceptance as a friction and wear polymer. As noted (133), PTFE crystallinity ranges from 65% to 95%. Although its melting point is 327°C, the glass transition temperature of the amorphous portion is only 120°C. Frictional heat-

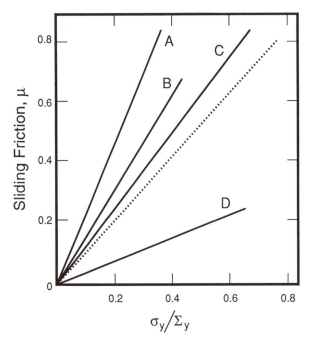

Figure 6.148.   Relationship Between Coefficient of Sliding Friction, μ, and the Ratio of Shear Strength, σ_g, to Yield Strength, σ_y, for Several Polymers Sliding Against Steel (135). Curve A: Polymethyl Methacrylate (PMMA). B: Polytrifluorochloroethylene. C: Linear Polyethylene (HDPE). D: Polytetrafluoroethylene (PTFE). Dashed Line: $\mu = \sigma_g/\sigma_y$.

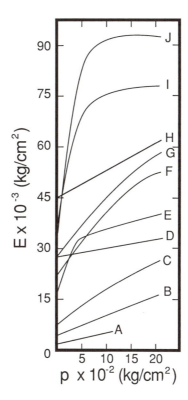

Figure 6.149. Pressure-Dependent Nature of Young's Modulus for Several Polymers at 25°C (136). Curve A: Low-Density Polyethylene (LDPE). B: Polytetrafluoroethylene (PTFE). C: High-Density Polyethylene (HDPE). D: Polycaprolactam, Nylon 6 (PA-6). E: Rigid PVC (RPVC). F: Polymethyl Methacrylate (PMMA). G: Polymer K-17-2 (Russian). H: Amino-Plastic ф-Kp-1 (Russian). I, J: Metals.

ing or continuous use temperatures in excess of this temperature will distort friction data. Furthermore, PTFE can be found in several crystalline forms, depending on the fabricating and deforming conditions.

As noted, the frictional characteristics of metals-on-polymers are functions of the nature of the metal surface, such as surface roughness, not with the type of metal used. This assumes of course that the metal has a higher modulus than the polymer. Typically, the frictional coefficient of a polymer-on-polymer pair is higher than that for a metal-on-polymer pair. For example, steel-on-nylon 6 frictional coefficient is 0.094 and nylon 6-on-nylon 6 is 0.62. The exception is PTFE, where the equivalent values are 0.049 and 0.037, respectively. The molecular reason for this disparity with PTFE is unknown. It is known that for moderate friction polymers such as PVC and polymethyl methacrylate, sliding is predominantly alternating slipping and sticking. For low friction polymers such as polyethylene and PTFE, sliding is constant once established. Some steel-on-polymer and polymer-on-polymer frictional coefficients are given in Table 6.19.

Table 6.19. Static, μ_g, and kinetic, μ_k, coefficients of friction for several polymers against steel and against themselves (137).

Polymer	Steel on polymer		Polymer on polymer	
	μ_s	μ_k	μ_s	μ_k
Polytetrafluoroethylene (PTFE)	0.10	0.05	0.04	0.04
Polytetrafluoroethylene-hexafluoropropylene copolymer (FEP)	0.25	0.18	–	–
Low density polyethylene (LDPE)	0.27	0.26	0.33	0.33
High density polyethylene (HDPE)	0.18	0.08 to 0.12	0.12	0.11
Polyacetal homopolymer (POM)	0.14	0.13	–	–
Polyvinylidene fluoride (PVDF)	0.33	0.25	–	–
Polycarbonate (PC)	0.60	0.53	–	–
Polyethylene terephthalate (PET)	0.29	0.28	0.27*	0.20*
Nylon 66 (PA-66)	0.37	0.34	0.42*	0.35*
Polychlorotrifluoroethylene (CTFE)	0.45*	0.33*	0.43*	0.32*
Polyvinyl chloride (PVC)	0.45*	0.40*	0.50*	0.40*
Polyvinylidene chloride (PVDC)	0.68*	0.45*	0.90*	0.52*

*Slip-Stick Intermittent Motion

As seen in Figure 6.150 for lubricated PMMA, there is a strong correlation between the velocity-dependent frictional force and the frequency-dependent loss tangent. For moderate-modulus elastic polymers, there is a maximum in the frictional force with velocity or rate of deformation. It is thought that this maximum corresponds with the deformation frequency of the surface asperities. Typically, the maximum occurs at about 0.01 cm/min for rigid polymers and about 0.03 cm/min for crosslinked rubbers.

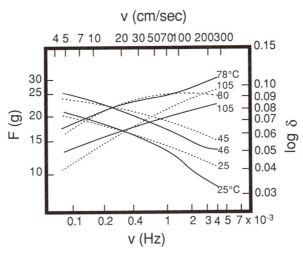

Figure 6.150. Lubricated Frictional Force of Polymethyl Methacrylate (PMMA) on Steel as a Function of Sliding Velocity. Compare with Loss Tangent of PMMA as a Function of Frequency (138). Solid Line: Sliding Force. Dotted Line: Loss Tangent.

The WLF shift equation can be used here. An increase in sliding velocity is equivalent to a decrease in material temperature. Bartenev and Lavrentev (133) propose:

$$(6.63) \qquad \frac{1}{T_m} = B - C \ln v$$

where T_m is the mechanical transition temperature, v is the cyclic-force frequency, and B and C are material constants. Temperature-dependent frictional coefficient values for several polymers are given in Figure 6.151.

The result of continuing friction of metal surfaces against polymeric ones is *wear*. Wear can be by fatigue or abrasion or by roll deformation. The primary method of wearing on high-modulus polymers is by fatigue. Most thermoplastics wear by abrasion or micro-cutting. Rubbers and most thermoplastic elastomers wear by roll deformation.

In fatigue wear, wear resistance depends on a number of factors related to the hardness of the polymer:

- Decreasing pressure decreases wear (to a power with an exponent > 1),
- Decreasing frictional coefficient decreases wear,
- Increasing modulus decreases wear,
- Increasing hardness decreases wear, so long as the extent of deformation remains fixed, and
- Increasing surface asperity radius or decreasing surface roughness decreases wear.

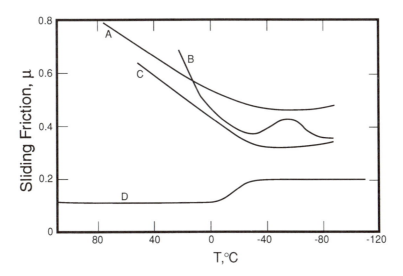

Figure 6.151.   Temperature-Dependent Coefficient of Sliding Friction for Several Polymers on Glass (139). Curve A: Polymethyl Methacrylate (PMMA). Curve B: Polyethylene (PE). Curve C: Polytrifluorochloroethylene (CTFE). Curve D: Polytetrafluoroethylene (PTFE).

By and large, the nature of the harder sliding surface determines the wear mechanism. If the polymer cannot adhere to the harder surface, the wear is primarily by fatigue. Note that polymers that begin wearing by fatigue can at some time in the wearing process begin adhering to the wearing surface. This may be due to hysteretic heating or to a change in the roughness characteristics of the polymer surface. While fatigue wear results in a very low amount of polymer removed from the surface per unit time, abrasive wear is a cutting process. Thus, much larger amounts of polymer are removed in the same time frame. Abrasive wear resembles a microscopic tearing process with the metallic surface cutting and plowing through the polymeric surface. The way in which the polymer yields to fracture is important. In abrasive wearing, there is little temperature effect on the wear rate.

As seen in Figure 6.152 for an amorphous polymer, as temperature increases toward the glass transition temperature, the elongation at break increases and the ultimate tensile strength decreases. This results in greater ductility and a resulting increase in the wear rate. In Figure 6.153, the temperature-dependent combination of increasing elongation and decreasing tensile strength at break does not result in an increase in the wear rate. As seen in Figure 6.154 for some polymers and metals, the abrasive wear resistance can be thought of as a function of the material hardness. As noted in Section 6.3, there are many ways of measuring hardness of polymers. Therefore, even if there is only a loose correlation between polymer hardness and wear resistance, polymeric materials having substantially greater surface hardness should yield surfaces with greater wear resistance.

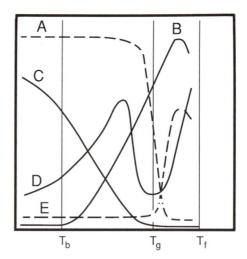

Figure 6.152.   Schematic of General Effect of Temperature on Mechanical Properties and Wear for Amorphous Polymers (140). T_b: Brittle Temperature. T_g: Glass Transition Temperature. T_f: Flow Temperature. Curve A: Hardness, HB Units. Curve B: Ultimate Elongation. Curve C: Tensile Strength. Curve D: Wear Rate. Curve E: Sliding Coefficient of Friction.

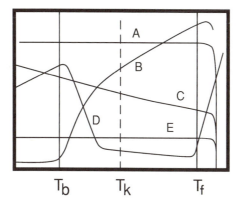

Figure 6.153.   Schematic of General Effect of Temperature on Mechanical Properties and Wear for Crystalline Polymers (141). T_b: Brittle Temperature. T_k: Room Temperature. T_f: Flow or Melt Temperature. Curve A: Hardness, HB Units. Curve B: Ultimate Elongation. Curve C: Tensile Strength. Curve D: Wear Rate. Curve E: Sliding Coefficient of Friction.

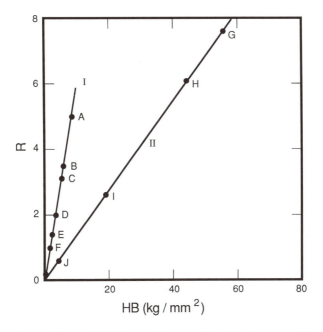

Figure 6.154.   Linear Relationship Between Abrasion Wear Resistance and Hardness for Several Polymers and Some Metals (142). Curve I: Polymers. Curve II: Metals. Point A: L54 (Russian). Point B: L68 (Russian). Point C: Nylon 6 (PA-6). Point D: HDPE. Point E: LDPE. Point F: Polytetrafluoroethylene (PTFE). Point G: Silver (Ag). Point H: Zinc (Zn). Point I: Cadmium (Cd). Point J: Lead (Pb).

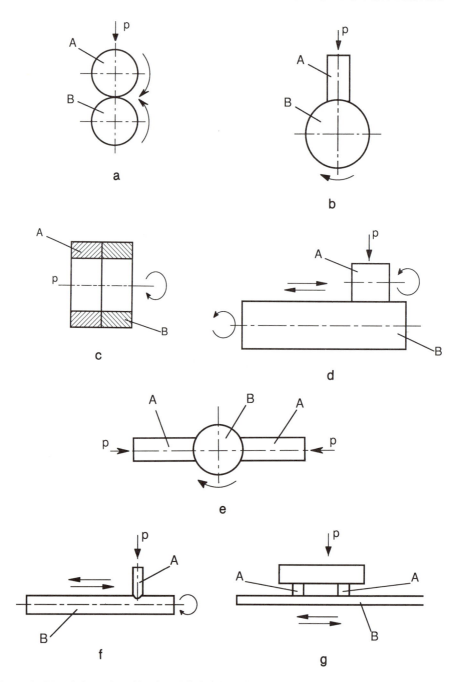

Figure 6.155. Schematics of Various Frictional Devices (144). A: Test Specimen. B: Surface Applying Force Against Specimen. a)–f): Rotating Cylinder At Interface. b), c), e), f): Specimen Stationary. a), d): Specimen Rotates. g): Oscillating Plate, Stationary Specimen.

Tabor Abrader: ASTM D1044

Friction and wear data on polymers can be obtained in many ways. Several schemes for friction pairs are shown in Figure 6.155. It has been noted that polymers can experience substantial hysteretic heating under intermittent loads. The measurement of temperature on a rotating surface is quite difficult. Of all methods proposed, static measurement with ultrafine, high-response thermocouples appears to introduce the least error. Simply put, the frictional device is stopped at predetermined times and the polymer surface temperature measured. In Europe, reciprocating-motion tribometers are extensively used (143). In England and US, rotary-motion devices are used. Most rotary devices employ the polymer as a revolving disk against which is held a stationary stylus (Figure 6.156). The stylus is mechanically indexed inward, indenting a new region of polymer with each revolution. The drag of the stylus against the polymer is determined by a strain-gage bridge circuit. These rotary devices enjoy wide ranges in applied force, material temperature and sliding velocity.

For abrasion, the ASTM method D1044, "Standard Method of Test for Resistance of Transparent Plastics to Surface Abrasion", commonly known as the *Tabor abrasion resistance test*, is recommended. The Tabor abrader, shown in Figure 6.157, is similar to the rotary friction device, except that the stylus is replaced with abrasive weighted wheels. Two abrading wheels are used. One rubs the polymer from the outside toward the center and the other rubs from the center to the outside. For plastics, two grades of "Calibrase" wheels, CS-10F and CS-17, are used. CS-10F wheels are fine-grain, medium-resilient, free-bonded abrasive wheels. These simulate light rubbing wear that

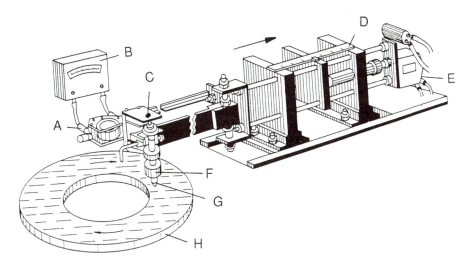

Figure 6.156. An Example of a Frictional Machine Where the Stationary Specimen, G, is Forced Against a Rotating Frictional Surface, (145). A: Dynamometer–Ring Strain-Gage Assembly. B: Force Indicator. C: Applied Static Load. D: Radial-Position Scale. E: Rider-Assembly Motor. F: Rider Holder.

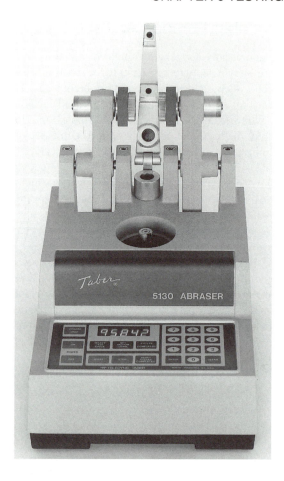

Figure 6.157.   The Tabor Abrader for Abrasion Resistance, Friction and Wear Experiments.

might be encountered in cleaning and polishing. The CS-17 wheels are coarse-grain, medium-resilient, free-bonded wheels designed to simulate abrasive scrubbing, walking or scuffing. A standard load of 500 or 1000 g is used for each of these wheels. The wheels must be redressed every 100 cycles. The material temperature is monitored only when the abrader is stopped. For highly abraded material, the loss is usually determined by very accurate weighing of the polymer disk. For lightly abraded transparent material, the extent of abrasion is determined by the level of haze produced or by the increase in light scattered in the abraded track. An example is seen in Figure 6.158.

It is not at all apparent that Tabor test data or hardness test data simulate the behavior of a polymer in a specific wear environment. Gouza (132) documents many other tests that purport to better define material response. For example, fine sand or dust is ex-

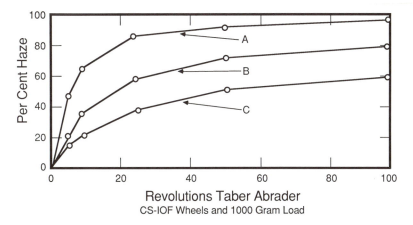

Figure 6.158. Typical Wear Characteristics for Two Polymers and Their Copolymer (147). Percent Haze is Used as a Measure of Abrasion for These Transparent Polymers. Curve A: General Purpose PS. Curve B: Copolymer. Curve C: Polymethyl Methacrylate (PMMA).

tremely abrasive to polymers used in glazing and aircraft windscreens. There seems to be only a casual relationship between data obtained from ASTM D673, dealing with resistance of polymers to marring by falling sand or carborundum, and the Tabor abrader data. The ASTM D658 test uses air-blown carborundum, perhaps simulating air-blown dust. The Marks–Conrad test (146) uses emery propelled at 200 mph, to simulate aircraft windscreens. PMMA has been used extensively in glazing for more than 40 years. The extent of haze generated by each of these tests on cell-cast PMMA is shown in Table 6.20.

The nature of scratch resistance can also be related to surface hardness, friction, and wear (132). As expected, the harder the surface is, the more scratch-resistant it is. The Mohs scale of hardness was developed in the early 1800s to rank the hardness of common minerals. Diamond was given a Mohs value of 10 and talc a value of 1. Most metals have Mohs values between about 1.5 for lead and tin and 5 to 8.5 for steels. Most plastics have Mohs values between 1 and 3. Table 6.21 gives typical Mohs values

Table 6.20. Percentage of haze on cell-cast polymethyl methacrylate as the result of four similar abrasion tests under various conditions (132).

Abrasion test	Percent haze and test conditions	
	Trial 1	Trial 2
ASTM D673 (#80 carborundum)	49.4%, 800 g	54.2%, 1,600 g
Bartoe (#40 carborundum)	34.4%, 300 g	57.5%, 600 g
Marks-Conrad (#120 Emery)	—	51.1%, 200 miles/h
ASTM D1044 (Tabor Abrader)	48%, 50 rev/min	59.9%, 100 rev/min

Table 6.21. Mohs hardness values for common materials.

Material	Hardness (Mohs)
Talc	1
Most polymers	1 to 2
Tin	1.5 to 1.8
Lead	1.5
Gypsum	2
Fingernail	2
Aluminum	2 to 2.7
Polymethyl methacrylate (PMMA)	2 to 3
Zinc	2.5
Bismuth	2.5
Copper	2.5 to 3
Gold	2.5 to 3
Silver	2.5 to 3
Calcite	3
Brass	3 to 4
Marble	3 to 4
Knife blade	4
Fluorite	4 to 5
Iron	4 to 5
Glass	4.5 to 6.5
Apatite	5
Asbestos	5
Steel	5 to 8.5
Orthoclase	6
Quartz	7
Flint	7
Diamond	8.5 to 10

for many common materials. Gouza proposes quickly determining the hardness of a new polymer by testing it against a fingernail (Mohs 2), a brass scribe (Mohs 3), a knife blade (Mohs 4), and a piece of glass (Mohs 5).

Another test is the HB or pencil hardness test. Thirteen pencils of increasing hardness from 2B (very soft) to 9H (extremely hard) are used. The procedure begins with the 2B pencil, scribed across the plastic surface. Pencils of increasing hardness are used until one leaves a visible mark on the test surface. The Mohs range in hardness for these pencils is 2 to 3. Typically, PMMA has an HB value of 9H whereas polystyrene is 2H and cellulose acetate butyrate is B. Most polymers have pencil hardness values of H to 2H or so. The disadvantage of this test is that absolute hardness does not increase uniformly with increasing pencil hardness values. The advantage is that the test is quick, relatively repeatable and easily learned.

Diamond stylus scribing can yield more accurate values but is usually restricted to laboratory testing. The Bierbaum scratch apparatus yields values of 16.4 for PMMA, 11.9 for polystyrene and 5.9 for cellulose acetate butyrate. Since the width of the scratch is measured, the test is best used for brittle or high indentation hardness polymers. Polymers that are rubbery or exhibit extensive memory or cold flow yield results that are at best approximate.

6.7 Environmental Testing

Polymers respond viscoelastically to changing environmental conditions. To this point, the conditions considered have been mechanical in nature. This section considers the effect of the nonmechanical environment, such as:

* Temperature,
* Weathering,
* Chemically aggressive environment,
* Fire, and
* Small molecule infusion (diffusion/permeation).

This section treats each of these effects as the primary influence on polymer response. The designer must always remember that many environmental effects are coupled with mechanical loads on the material. Thus, external polymer guy-wire may be under uniaxial creep load *and* exposed to atmospheric temperature extremes, ultraviolet radiation and aggressive atmospheric conditions such as salt spray and acid rain. Refrigerator doorliners are under biaxial tension *and* are exposed to vegetable oils and essences and low-temperature impact.

Environmental Stress Cracking: ASTM D1693

The chemical nature of the environment can be quite detrimental to the performance of a polymer part. The phrase "environmental stress cracking", or *ESC,* was proposed to distinguish stress cracking, hence failure, that could be attributed primarily to the environmental conditions surrounding the polymer. Howard (148) puts forth the following definition:

> Environmental stress-cracking is the failure in surface-intiated brittle fracture of a polyethylene specimen or part under polyaxial stress in contact with a medium in the absence of which fracture does not occur under the same conditions of stress. Combinations of external and/or internal stress may be involved, and the sensitizing medium may be gaseous, liquid, semi-solid, or solid.

In other words, consider a polymer that does not fail when it is stressed in a neutral environment. The same polymer, stressed in an aggressive environment, might undergo rapid brittle failure. These polymers are said to have poor resistance to environmental stress-cracking, that is, *poor ESCR*. Note that one of the important elements of ESCR is the stress field. Failure begins at a stress concentration point that can be a nick or flaw on the *surface* of the part. Failure proceeds in a brittle fashion, with conchoidal hackle and rib marks advancing from the stress concentration. Environmental stress-cracking is always distinguished from oxidative stress-cracking or solvent crazing because it is a purely *physical* effect involving an external *sensitizing agent*.

Polyethylene has been studied extensively for ESCR since PE materials are frequently used as containers for aggressive liquids, such as detergents and insecticides. Furthermore, it has been shown that lower molecular weight polyethylenes and polyethylenes with very long chain branching exhibit poor ESCR (Figure 6.159).

It is difficult to quantify ESCR. In 1950, a precursor test to ASTM D1693, "Test for Environmental Stress Cracking of Ethylene Plastics", was developed to provide a standard. The test is also known as the bent-strip test or the Bell Telephone Labs (BTL) test. A test specimen 38 ± 2.5 mm (1.5 ± 0.1 in) long, 13 ± 0.8 mm (0.5 ± 0.03 in) wide and 3.15 ± 0.15 mm (0.125 ± 0.005 in) thick is cut from a flat part or is molded to shape. The specimen is notched 0.575 ± 0.075 mm (0.0225 ± 0.025 in) length with a new razor blade. A special nicking fixture is shown in Figure 6.160. Its dimensions are shown in Table 6.22. The specimen is then bent π radians (180 deg) and inserted in a holder such that the outside specimen ends are 11.75 ± 0.05 mm (0.463 ± 0.002 in) apart. The specimen is therefore held in constant *strain*, as seen in Figure 6.161 (150). The test device dimensions are given in Table 6.23. The specimen holder is designed to hold 10 specimens. The entire holder is placed in a suitable container, the aggressive stress-cracking material added, and the entire assembly placed in a constant temperature bath. The environment must be aggressive enough to initiate

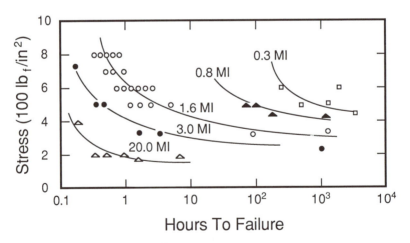

Figure 6.159.   Effect of 0.918 g/cm³ Low-Density Polyethylene Molecular Weight (Given as Melt Index, MI) on Time-Dependent Failure Stress at 60°C with Igepal Stress-Cracking Agent (149).

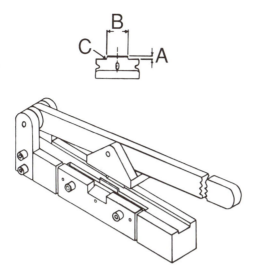

Figure 6.160.   The Environment Stress-Cracking Nicking Jig Specified by ASTM D1693.

failure in the notched region in several hours but not so aggressive that failure occurs in seconds or minutes. One accelerator is "Igepal CO–630", nonylphenoxy poly(ethyleneoxy)ethanol surfactant that is commonly used as a 1% to 10% aqueous solution at a fixed temperature, usually 50°C. Other surfactants that are used for ESCR studies are given in Table 6.24.

According to the ASTM Standard, the samples should be examined on a strict periodic timetable for failure. Failure is defined as the first appearance of cracks along the notch edges. The times are analyzed statistically. ESC times are usually reported either as "F_{50}", or the time when 50% of the samples have failed, or "F_5", or the time when 5% of the samples are failed. If the failure envelope is Gaussian (and it may not be), the F_5 data should represent the 2-standard-deviation data. Turner (151) notes that for very tough polyethylenes, first signs of cracking may be indistinguishable from fortuitous cutting nicks and surface blemishes.

Table 6.22.   ASTM D1693 standard environmental stress-cracking nicking jig dimensions.

(Figure 6.160) knicking jig location	Dimensions	
	(mm)	(in)
A	3	1/8
B	18.9 to 19.2	0.745 to 0.755
C (radius)	1.5 Max	1/16 Max

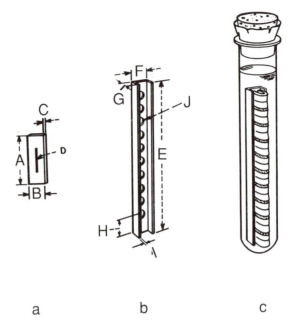

a   b   c

Figure 6.161.   The Environmental Stress-Cracking "Bent Strip" Fixture Specified by ASTM D1693. Dimensions are Given in ASTM Standard, See Table 6.23.

Table 6.23.   ASTM D1693 standard environmental stress-cracking fixture dimensions.

(Figure 6.161) fixture identification	Dimensions	
	(mm)	(in)
A	38 ± 2.5	1.5 ± 0.1
B	13 ± 0.8	0.5 ± 0.03
C	See Table 6.22	
D	See Table 6.22	
E	165	6.5
F (outside)	16	5/8
F (inside)	11.75 ± 0.05	0.463 ± 0.002
G	10	3/8
H	15	37/64
I	2	0.081 (12 B&S)
J	Ten 5 mm holes	Ten 3/16 in holes
	15 mm centers	19/32 in centers

Table 6.24. Standard environmental stress-cracking surfactants

Surfactant	Type	Concentration (% in H_2O)
Nonyl phenyl ethoxylate	Nonionic	2, 5, 10, 25, 50, 75, 100
Linear alcohol ethoxylate	Nonionic	2, 5, 10, 100
2,4,7,9-tetramethyl-5-decyn-4, 7-diol	Nonionic	100
Adducts of 2,4,7,9-tetramethyl-5-decyn-4, 7-diol, being		
40% (wt) ethylene oxide	Nonionic	100
65% (wt) ethylene oxide	Nonionic	100
85% (wt) ethylene oxide	Nonionic	100
Sodium 2-ethyl hexyl sulfate	Anionic	2
Sodium lauryl sulfate	Anionic	2
Sodium tetradecyl sulfate	Anionic	2
Sodium heptadecyl sulfate	Anionic	2

The designer should keep in mind that this ASTM test, like many others, was designed to provide a quality control for one specific product, extruded polyethylene cable sheath for under-ocean telephone lines. As seen in Figure 6.162, it is now modified to rank the ESCR performance of high molecular weight, high density polyethylene with several stress-cracking chemicals for chemical drums. Lind (152) concludes that if a liquid causes ESC in a specific HMW HDPE, then there is strong indication that the drum will fail in some fashion in service. Turner notes (153) that since accelerated flexural stress-cracking experiments are susceptible to many specific irregularities, the bent-strip test cannot be mathematically definable. As a result, it fails one of the two *criteria for test acceptability* and is therefore, by definition, a hybrid test. He further notes that:

> . . . it seems *highly unlikely* that there can be a quantitative relationship between the crack resistance of a highly annealed, compression molded specimen of polyethylene, notched, bent into a highly strained configuration and immersed in an aggressive environment at 50°C, *and* the service performance of an extruded cable insulation that may not even be in contact with its environment of sea-water mainly at 4°C. [emphasis supplied by authors]

The designer must also remember that molding conditions can dramatically change the stress-crack resistance of semicrystalline polymers such as polyethylene. In Table 6.25, the effect of cooling annealed compression molded specimens on the F_{50} values from the bent-strip test is apparent. Note also that since the bending dimension is fixed, the degree of strain applied to the polymer is also fixed. This strain may exceed the maximum recommended strain for the polymer, and this may lead to plastic yielding or fracture, even before encountering the aggressive environment. Brown (154) and Roe and Gieniewski (155) note these limitations:

- Strain level is fixed,
- Results are very sensitive to thermal history of specimens,

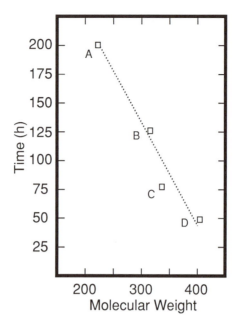

Figure 6.162. The Effect of Stress-Cracking Agent Molecular Weight on F_5 Environmental Stress-Cracking Time for 0.948 g/cm^3 HDPE of 10 MI (Nominal) in 2% Aqueous Solution at 50°C, According to ASTM D1693 (152). Point A: Sodium 2-Ethyl Hexyl Sulfate. Point B: Sodium Lauryl Sulfate. Point C: Sodium Tetradecyl Sulfate. Point D: Sodium Heptadecyl Sulfate.

- Results depend on annealing conditions of specimens,
- Precise shape of bent strip is not fixed,
- Process does not control the rate of bending or time of transfer of the strip to the fixture, so the effect of stress relaxation is unknown,
- Cracking time does *not* depend on specimen thickness, so the stress on the test piece is not proportional to thickness,
- The high strain level of the test is unsuitable for polymers other than polyethylenes.

Table 6.25. Effects on environmental stress crack resistance of annealed, compression-molded polyethylene specimens under various cooling techniques (157).

Cooling technique	F_{50}
Quenching in −25°C alcohol	> 24 h
Quenching in 23°C water	2 h
Natural cooling in 23°C air	1 h
Slow cooling in 70°C water	35 min
Very slow cooling in water (to 23°C in 24 h)	14 min

Some other tests have been proposed as more desirable alternatives to ASTM D1693. Lander (158) was the first to propose an alternate. He used tensile impact (type S) specimens that are clamped and weighted to a given stress, then placed in a heated tank containing the stress cracking liquid. When a specimen fails, its lever arm shuts off an elapse meter, thus fixing the time of failure. The general range of environmental stress rupture curves for polyethylenes of several densities at 50°C is shown in Figure 6.163. Turner notes that this constant *stress* test meets the two criteria of test acceptability, made complex only by the complex interaction of the effect of the environment on a stressed structure.

Another popular test, the Ziegler-Brown variable radius fixture, applies an increasing strain to the polymer. The point of failure is the "critical strain". Unfortunately many polymers craze or microvoid at strains lower than cracking strain. Thus, the accuracy of determining the critical strain depends on the ability of the technologist to interpret the crazing/cracking patterns from one material to another. Brown (154) reviews many other ESC tests.

Solvent and Other Forms of Stress-Cracking

Solvent stress-cracking is the aggressive action of a small mobile molecule that serves to eventually swell or even dissolve the polymer. The solvating ability of the solvent

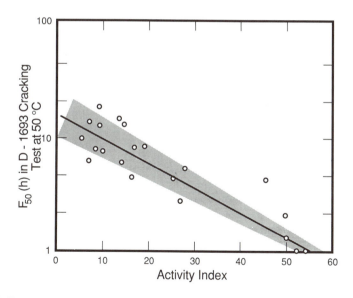

Figure 6.163. One Method of Correlating F_{50} Environmental Stress-Cracking Resistance at 50°C for Polyethylene in Igepal Against Activity Index, Being a Ratio of Pulled Film Sample Length to Surface Tension (159). In Essence, the Correlation Shows a Relationship Between the ESCR of a Material and Its Ability to Resist Wetting by Aqueous Systems.

depends on the close agreement of the free energy of the plastic and that of the solvent. All solvents are absorbed by molecular diffusion into the polymer. As the solvents are absorbed into the polymer surface, several intermediate layers are formed (Figure 6.164). In the layer nearest the solid polymer, called an infiltration layer, the solvent is molecularly diffusing, filling the void volume around the molecules. The next layer is a solid swollen layer. Here the polymer network is stretched by the diffusing solvent. A gel layer is next. Here the polymer molecules disentangle by the action of the solvent, which may have a concentration of 10% to 20% (wt). Then there is a liquid solution layer, followed by pure solvent. Aggressive solvents diffuse so rapidly into the polymer that no gel formation is apparent. The result is solvent stress cracking. The fundamental interpretation of solvation of polymers is best found in a treatise on polymer–solvent thermodynamics (161).

Polymer susceptibility to aggressive chemicals is a strong function of molecular makeup. Axioms such as "like dissolves like" give a measure of this susceptibility but cannot be considered quantitative. For example, polymers such as nylon and polyethylene terephthalate are manufactured by condensation polymerization. Water is removed as a by-product. Therefore water-based chemicals, including water itself, can have deleterious effects on the performance of the polymer. Seymour (162) notes that ionic media, viz, OH^- and H^+ groups, hydrolyze the polymer by first protonating the ester or amide. Alkaline hydrolysis, or saponification, results in polymer backbone

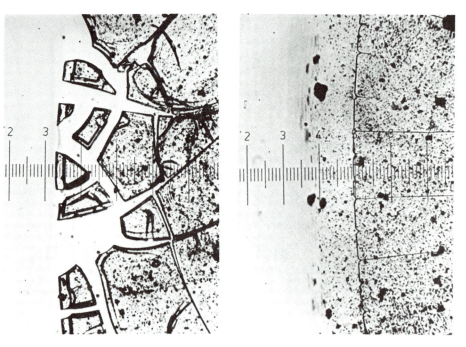

Figure 6.164.   Comparison of Solvent Cracking With (Right) and Without (Left) the Formation of a Gel Layer (160).

cleavage. Chemicals that dissolve condensation-type polymers are sometimes called "polar" chemicals.

Polyolefins are oil-based polymers. They are formed by chain addition and typically have very low affinity with water. As a result, these polymers are not normally attacked by aqueous or polar chemicals. Decomposition or backbone cleavage is normally the result of a free-radical attack. Typically, hydrocarbons serve to swell polyolefins and so are considered to be solvents rather than aggressive chemicals. Chemicals with high chlorine compositions, such as dry-cleaning chemicals, can be quite aggressive, however. Chlorination is a free-radical reaction that results in carbon–carbon bond cleavage. Certain sulfur-rich chemicals can attack polyolefins in a similar fashion. These chemicals are called "nonpolar" chemicals. And unsaturation in the polymer chain is the source of backbone cleavage when in the presence of oxygen or ozone. Thus, natural rubber and diene-based polymers rapidly crack or embrittle when used around electrical motors.

Polymers have been used as chemical barrier materials for more than a century. The ever-increasing combinations of polymeric materials and corrosive or aggressive chemicals precludes complete understanding of the interaction of a specific chemical on a specific polymer. Seymour (162) provides a tabulation of the suitability of many polymer-chemical combinations at three temperatures (25°C, 65°C, and 90°C). He ranks the suitability as "satisfactory", "unsatisfactory" or "questionable". He notes that the polymer classification is generic, such as "polyethylene", and that the suitability classification is subjective and carried out on specimens for which the residual strain level is unknown, in general. Despite these severe limitations, Seymour's selection guide is 114 pages long and contains more than 5000 entries. It is apparent that this guide can be used only to screen out certain polymer–chemical combinations. The designer must realize that a polymer classification of "satisfactory" needs to be confirmed for the polymer of choice molded into the appropriate product containing the chemical of choice under suitable use conditions for its appropriate lifetime. A very brief overview of polymer–chemical compatibility is given in Figures 6.165 and 6.166.

Weathering

Many polymers are used in outdoor products. Polyvinyl chloride is used in siding and window profiles. Polyolefins are used as outdoor chemical storage tanks and wading pools. Polymethyl methacrylate and polycarbonate are used in window glazing. Thermoset unsaturated polyesters, when combined with fiberglass, are used in portable toilets and boats and other marine applications. Typically, outdoor applications involve long exposure times (years to decades). It is known that ultraviolet (UV) radiation is the primary energy source responsible for deterioration. For PVC, for example, it is thought that UV results in free-radical generation and liberation of hydrogen chloride. This dehydrochlorination produces conjugated *chromophoric* double bonds along the PVC backbone. Heat exacerbates and accelerates the reaction. The primary effect in PVC is discoloration. Some polymers such as polystyrene already have chromophoric conjugated double bonds (in the benzyl group). As a result, UV absorption accelerates

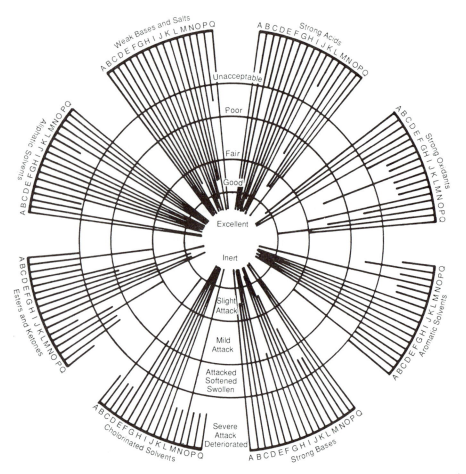

Figure 6.165.   Room-Temperature Resistance of Polymers to Various Classes of Solvents (163).
A: PVC. B: Polyvinylidene Chloride (PVDC). C: Polyethylene (PE). D: Polystyrene (PS). E:
Polytrifluorochloroethylene. F: Asbestos-Filled Phenol-Formaldehyde. G: Glass Reinforced Unsaturated Polyester Resin (GR-UPE). H: Glass-Reinforced Amine-Cured Epoxy. I: Polyvinylidene Fluoride (PVDF). J: Chlorinated Polyether. K: Asbestos-Filled Furan. L: Polymethyl Methacrylate (PMMA). M: Polycarbonate (PC). N: Nylon (PA). O: Modified Polyphenylene Oxide
(mPPO). P: Polyphenylene sulfide (PPS). Q: Polyvinyl Ester.

discoloration and ring scission. Unsaturated polyester resin and SMC–sheet molding
compound formed from fiberglass-impregnated unsaturated polyester resin–contain
styrene mers (Figure 6.167). With polyolefins, UV acts as a free-radical initiator, resulting in direct chain scission with little discoloration of the product. Typically lightly
pigmented, unstabilized white PVC will yellow in one season in Florida or Arizona.
    Since product tests are critical to the success of outdoor products and since weather
conditions are difficult to extrapolate, several techniques have been developed to measure the time-dependent UV effects in a controlled laboratory environment. The most

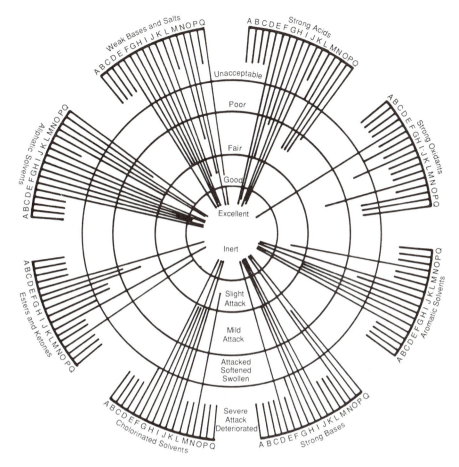

Figure 6.166. Elevated Temperature (200°F) Resistance of Polymers to Various Classes of Solvents (164).

accurate measurement of bond breakage is infrared analysis (166). The colorimeter is a device that is used primarily for product color control. Basically, color is measured and compared against a standard or in some cases, against the as-molded product. The time-dependent shift in color can be quantified and related to the product exposure time.

Special machines called "weatherometers" are used to accelerate the exposure time. Such devices expose specially prepared panels to intense UV levels from fluorescent lamps as in the QUV weathering chamber, or Xenon lamps as in the Atlas Weathero-meter. The test panels can be exposed to "sun only" conditions of fixed UV intensity or to alternating UV and water- or saltwater-spray, to simulate seaside exposure conditions. Although correlations between actual outdoor weathering and weatherometer data have been proposed, the weatherometer data must be considered to be the result of a hybrid experiment. Therefore these data should be used only to determine relative effects of adducts such as UV stabilizers and pigments, on polymer deterioration.

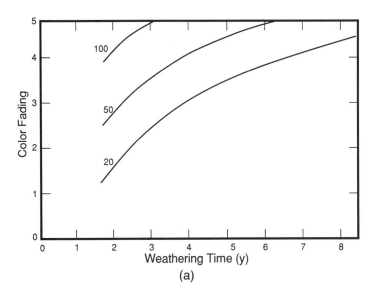

(a)

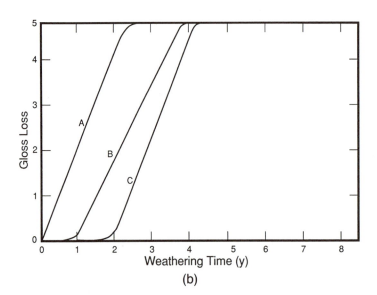

(b)

Figure 6.167. Results of Long-Term Weathering on Mechanical and Surface Qualities of Sheet Molding Compound (SMC) (165). a): Effect of $CaCO_3$ Filler Concentration (Parts Per Hundred of Resin, or pph) on Long-Term Color of Standard SMC. Color Fading, Arbitrary Units. 3 pph Iron Oxide Brown Pigment Added. b): Effect of Type of Acid in Unsaturated Polyester Resin Formulation on Loss in Gloss. Curve A: Bisphenol A or HET Acid Unsaturated Polyester Resin. Curve B: Orthophthalic Acid Unsaturated Polyester Resin. Curve C: Neopentyl Glycol/Isophthalic Acid Unsaturated Polyester Resin.

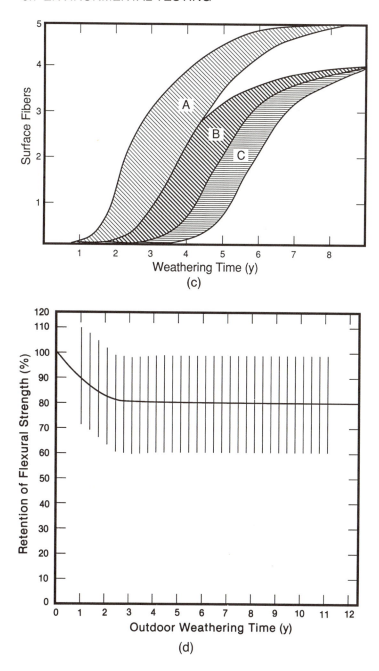

Figure 6.167 *Continued.*   c): Effect of Type of Acid in Unsaturated Polyester Resin Formulation on Surface Fiber Prominence. Curve A: Bisphenol A or HET Acid Unsaturated Polyester Resin. Curve B: Orthophthalic Acid Unsaturated Polyester Resin. Curve C: Neopentyl Glycol/Isophthalic Acid Unsaturated Polyester Resin. d): Effect of Outdoor Exposure (Weathering) on Retention of Flexural Strength for Standard SMC. Bars Indicate Statistical Variation in Data.

Again, the designer must realize that traditional materials and polymers weather in substantially different ways. Metals rust or corrode. Ceramics etch, craze or discolor. The qualitative and quantitative effects of weather are probably better understood for traditional materials than for polymers. Nevertheless, even with traditional materials, details on the molecular level are not unambiguous. Weather, as with chemicals, represents an aggressive environment to polymers. This aggression ultimately results in chain scission and loss in properties. Owing to the other unique properties of polymers (toughness, light weight, good resistance to water and atmospheric pollution, color throughout the part, and so on), a substantial effort is under way to more thoroughly understand the molecular processes. The designer working with outdoor applications can keep abreast of the technical developments by monitoring periodicals and technical treatises (167–171).

Fire Retardancy

Polymers are organic molecules. Given sufficient energy and usually a source of oxygen, polymers decompose thermally into carbonaceous products and simple molecules such as water vapor, carbon dioxide and, depending on the type of polymer, other products such as hydrogen chloride and carbon monoxide. For certain polymers, the thermal energy and oxygen concentration requirements are relatively low. For example, polyolefins such as polyethylene and polypropylene can be easily ignited and will burn with a nearly smoke-free flame. Other polymers cannot be ignited with low energy sources, are difficult to thermally decompose and may produce dense smoke while decomposing. The high chlorine content of polyvinyl chloride prevents it from burning or decomposing at low energy levels. The mechanism of burning involves at least these steps:

- The exposed surface of the polymer must increase in temperature to a point at which thermal degradation can occur rapidly,
- Thermal degradation must proceed by chain scission at a rate rapid enough to sustain itself,
- The degradation product must be smaller, more combustible molecules, and
- The smaller molecules must exothermically decompose, ignite or burn at a high enough energy level to help sustain the decomposition.

One broad flammability classification is *burning* and *non-burning*. Simply put, a flame is applied to the polymer, igniting it. If the polymer stops burning when the flame is removed, the polymer is "burning". This test has been quantified as UL (Underwriters Laboratory) Standard 94. A 127 mm (5 in) long and 12.7 mm (½ in) wide test specimen of a specific thickness (0.8, 1.6, 3.2 or 6.4 mm or 1/32, 1/16, 1/8 or ¼ in) is placed either horizontally or vertically into the flame of a laboratory bunsen burner. For the vertical bar test, the burner is adjusted to provide a 19 mm (3/4 in) blue flame against the bottom of the bar. The bar is mounted at the top with the long dimension vertical and 305 mm (12 in) above a cotton wad. The flame is applied for 10 seconds, removed and the residual flame allowed to extinguish. The application is repeated for an additional 10 seconds, as soon as the flame extinguishes. If the flame does not extinguish and the bar is consumed, the material is considered to be "flammable", and no standard rating is given. The time of flaming of each burn is recorded. In addition, observation is made if dripping materials ignite the cotton wad. The total time of flaming for *all* burns for *five* specimens is recorded.

The criteria for the "V" rating are given in Table 6.26. Note that there is a "5V" test that differs from the V-0 to V-2 test criteria. Here, the burner flame is 127 mm (5 in) high with a 38 mm (1.5 in) blue cone, and the flame is applied to one corner. The application is five times at 5 second intervals. It is apparent that the polymers that pass this test are quire fire-resistant under normal use.

There is another criterion for materials that fail the "V" test. The test specimen is tipped at a 45° angle for 30 seconds against a burner flaming with a 25 mm (1 in) blue cone. For a polymer to receive an "HB" rating, *either* the flame must extinguish before it reaches the 102 mm (4 in) mark, *or* the rate of burning must not exceed 38 mm/min (1.5 in/min) between the 25 mm (1 in) and 102 mm (4 in) marks for specimens 3.2 mm (1/8 in) thick or thicker. This test helps to establish the susceptibility of a polymer to low energy open flames and so is sometimes referred to as "the standard match" test.

For polymers used in large quantities in commercial, industrial and some residential buildings such as apartments, hotels, and boarding houses, however, the UL 94 test is insufficient. Factory Mutual, a major insurer of such buildings, has developed a full-scale fire test, called the "corner test". The test employs full-sized panels of the polymer, mounted against a fire-retarded corner structure. A fire is initiated by igniting a fixed amount of gasoline in a wooden crib containing a fixed amount of dry oak. The crib is carefully positioned near the corner. The test monitors the extent of fire propagation along the polymer surface and the amount of smoke generated, as a function

Table 6.26. Criteria for UL standard 94 vertical bar flammability ratings (172).

Criterion	5V	V-0 VTM-0	V-1 VTM-1	V-2 VTM-2
Burning up to specimen clamp	No	No	No	No
Maximum time of flaming after each burn, s	60[a]	10	30	30
Maximum total time of flaming for all burns of 5 specimens, s	—	30	250	250
Maximum time of glowing after last burn of each specimen, s	60	30	60	60
Ignition of cotton by dripping flaming particles	None	None	None	Allowed
Number of burns per specimens	5[b]	2[c]	2[c]	2[c]
Duration of each burn, s	5	10	10	10

Header spanning: Vertical flammability rating

[a]—After 5 burns on each specimen
[b]—Applied at 5 s interval
[c]—Second burn applied after initial flaming ceases

of time. The test is been used to determine fire dynamics but is too costly for routine testing.

The Steiner tunnel test (UL 723, ASTM E84), is one way of routinely evaluating polymers in larger-scale fires. The Steiner tunnel (Figure 6.168) is a 7.6 m (25 ft) long fire-brick-lined, steel duct, with a 450 mm × 305 mm (17.75 in × 12 in) opening. The polymer to be tested is laid against the open top. The polymer panels must occupy a 580 mm × 7.32 m (20 in × 24 ft) area. Thickness is not a factor. Two gas burners occupy one end and produce a 1.4 m (4.5 ft) flame that impinges on 0.65 m^2 (7 ft^2) of the test specimen. The burners release 1650 cal/s (5000 Btu/min) and create a surface temperature of 870°C (1600°F). The exhaust system at the other end produces a flow of 6.8 m^3/min (240 ft^3/s). A photometer and a thermocouple are located at the exhaust end to measure smoke obscuration from the burning plastic and temperature. Windows are provided along the chamber to allow for visual monitoring of the flame front propagation. Once the specimen has ignited, the operator calls out the location of the flame front progression to an assistant, who records location and time. The photometer continuously monitors the smoke density. The test ends when the specimen has burned to its end. A Flame Spread Index is assigned, using the time-dependent flame front propagation of red oak flooring as *100*. A Smoke Index is assigned, using the value of red oak as *100*.

The Steiner tunnel was designed to test rigid panels such as flooring, but it has been adapted to test batt and blanket insulation, wall coverings, carpeting, carpet underlayment and even drapery materials and textiles. The Flame Spread Index is used in many building codes. For example, in high-risk areas such as hallways and stairwells, the installed product must have a Flame Spread Index less than 25. Materials with values greater than 200 are generally not approved in many building codes.

Other tests have been created that allow rapid screening of candidate materials. For example, the radiant panel test, ASTM E162, employs both thermal radiation and ignition to evaluate flammability. The radiant panel test uses smaller rectangular panels, 152 mm (6 in) wide, 457 mm (18 in) long and up to 25 mm (1 in) in thickness. The flame propagation measurements are converted to an index with the range of *0* (for asbestos) to *200*. Red oak flooring has a fixed value of *100*. However, the data

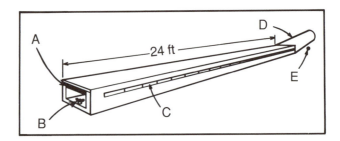

Figure 6.168.   Schematic of UL94, the Steiner Tunnel Test for Flame Spread Rate and Smoke Generation Rate Testing of Large Building Panels (173). A: Test Sample, 20-in × 24 ft. B: Gas Burner. C: Observation Window. D: Differential Manometer Tube. E: Photoelectric Cell.

from the Steiner test and this test do not correlate well. The radiant panel test gives a reliable measure of the heat of combustion of the polymer product, but the energy levels are normally quite low when compared with full-scale fire tests such as the Factory Mutual corner test.

Smoke generation can be more serious than flame. Dense smoke can hide emergency exits, for example. As a result, ASTM E662, "Optical Density of Smoke Generated by Solid Materials", was established to provide a rapid way of measuring smoke generation. The test uses a 76 mm × 76 mm (3 in × 3 in) specimen of any thickness. The specimen is mounted vertically parallel to a radiant source that produces 2.5 W/cm^2 (2.2 Btu/s ft^2) output over the center 38 mm (1.5 in) diameter of the specimen. The test can be run either in a nonflaming mode or in a flaming mode. Six gas jets are played against the bottom section of the panel. A photocell monitors the smoke obscuration. The time-dependent transmission is then monitored. The specific optical density index is determined from:

$$(6.64) \qquad D_s = \frac{V \log_{10}(100/T)}{(AL)},$$

where D_s is the Specific Optical Density, V is the chamber volume, A is the exposed area of the specimen, L is the length of the light path through the smoke and T is the percent transmission. Examples of the data are shown in Figure 6.169. A nonflaming

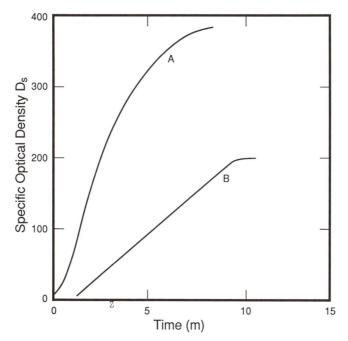

Figure 6.169.   Typical Time-Dependent Smoke Obscuration Curves for 0.045 in Thick PVC (173). Smoke Obscuration Given in Optical Density Terms. A: Flaming Sample. B: Nonflaming Sample.

specific optical density value of 450, for example, indicates a low-smoke generating polymer.

The designer must always be cognizant of the potential of combustion with plastics. There are many ways of adding fire retardants to minimize the combustion potential. For example, SMC, a form of unsaturated polyester resin and fiberglass, normally burns readily with dense black smoke and so is unusable as a building product. To achieve fire retardancy in one example, an inexpensive inorganic material, aluminum trihydrate, is added to 150 parts per hundred parts of resin (or more). As the product is heated by the fire, the inorganic material decomposes to produce water. Water requires very large amounts of energy to vaporize, and so the product temperature does not increase to an autocatalytic point where thermal decomposition sustains. In other words, SMC materials can be made quite flame-resistant by adding large amounts of aluminum trihydrate. Keep in mind, however, that some polymers such as polyimides and PVCs are naturally fire-retardant and others such as polyolefins and polyacetals are very difficult to endow with this property.

Permeability of Gases Through Polymers: ASTM D1434

There are many reasons for measuring and monitoring the transmission of gases through polymers. Food packaging is one of the major uses of polymer films and thin-gage sheet. In some cases, moisture transmission must be restricted to preserve "freshness", as with crackers and cereal. In other cases, carbon dioxide transmission is minimized to preserve carbonation, as with carbonated beverages. In the most difficult food packaging cases, oxygen transmission must be minimized to prevent spoilage. Multilayer films are being used in some cases to preserve cheese, cooked meat, beer and wine. In other areas, gas diffusion can be detrimental to the long-term function of the polymer product itself. For example, chlorofluorocarbons are used to foam polymers to produce insulation panels. When the chlorofluorocarbons diffuse from the foams and air diffuses in, the insulation value decreases. In certain applications, the film must be rapidly penetrated by one gas but nearly impermeable to other gases. New medical sterilization techniques use ethylene oxide as the sterilizing medium. The films for prepackaged medical products must be permeable to the sterilizing gas but impermeable to water vapor, for example.

The small gaseous molecules are loosely categorized as (174,175):

* Gases or permanent gases, such as oxygen, nitrogen, carbon dioxide, and
* Vapors or condensable gases, such as water vapor, certain chlorofluorocarbons and some hydrocarbons such as pentane.

As noted in Section 4.8, permeability, P, of a gas in a polymer is the product of its solubility, S, and diffusion, D:

(6.65) $P = SD$

The units on the permeability coefficient are [cm^3 (STP) cm/cm^2 s Pa]. The first part of the unit, (cm^3 (STP)), refers to the volumetric amount of gas transmitted. The second, (cm), refers to the thickness of the polymer film. The third, (cm^2), refers to the surface area of the polymer film. The fourth, (s), deals with the time of transport. The fifth, (Pa), deals with differential pressure across the film. The ASTM D1434 standard defines a *barrer* for permeability coefficient as:

$$1 \text{ barrer} = 10^{-10} \text{ cm}^3 \text{ (STP) cm/cm}^2 \text{ s mm Hg}$$

In (partial) English units:

$$1 \text{ barrer} = 167 \text{ cm}^3 \text{ mil/100 in}^2 \text{ 24 h atm}$$

Here the film thickness is measured in mils (0.001 in), pressure in atmospheres and the time as 24 hours.

Another frequently used unit is water vapor transmission rate (WVTR) or the mass transfer rate per unit area of film:

$$\text{WVTR} = \text{g/m}^2 \text{ 24 h}$$

Gas transmission rate is frequently listed in "permeability" units or *perm-cm*:

$$\text{permeability} = \text{g cm/m}^2 \text{ 24 h mm Hg} \equiv \text{perm-cm}$$

A value for the permeability coefficient can only be obtained at steady state. Lomax (174,175) chronicles many methods of measuring gas transmission through polymer films. The earliest and most popular method is the *manometric method*. It is described in standards in many countries: in the US as ASTM D1434, Method M, in England as BS 2782, in Europe as ISO 2556 and in West Germany as DIN 53380. The ASTM D1434 test is "Gas Transmission Rate of Plastic Film and Sheeting". The subject gas maintained at a fixed pressure is placed on one side of the polymer film and a vacuum is initially established on the other side. The changes in pressure and volume are measured over a predetermined length of time. A typical time-dependent pressure plot is seen in Figure 6.170. Note that most gases are soluble to some extent in most polymers. As a result, the initial portion of the pressure profile includes the time-dependent diffusion and solvation effect. Measurement of permeability depends on the establishment of a constant gas concentration gradient across the film. The pressure plot yields the steady-state gas transmission rate, the permeability coefficient, P, the diffusion coefficient, D, and the solubility coefficient, S. The diffusion coefficient can be obtained from the transient portion of the curve. As seen in Figure 6.170, by extrapolating the inflection portion of the pressure curve to zero pressure, a time constant, τ, can be found. The diffusion coefficient, D, is obtained from:

(6.66)
$$D = \frac{L^2}{6\tau},$$

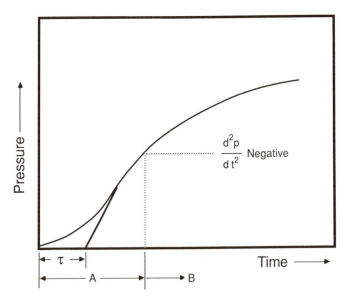

Figure 6.170.   Permeation of Gases Through Thin Plastic Films as Monitored by a Time-Dependent Pressure on One Side of a Permeation Cell (174). A: Unsteady State Region. B: Steady State Region. d^2P/dt^2: Inflection Point Between Regions A and B. τ: System Time Lag, Defined as $\tau = L^2/6D$, where L is the Total Film Thickness and D is the Diffusion Coefficient.

where L is the film thickness. The units on D are cm^2/s (ft^2/h). The solubility coefficient, S, is obtained from $S = P/D$. The solubility coefficient can also be related to applied pressure through Henry's law:

(6.67) $$H = SP$$

where H is the Henry's law coefficient, $[cm^3 (STP)/cm^3 \text{ polymer}]$, and P is pressure. Keep in mind that the manometric method yields useful data only if steady state can be achieved *and maintained* for a sufficient period of time to allow data interpretation. In highly permeable films, steady state may not be achieved. For aggressive gases, the polymer film may swell, soften or deteriorate even if an early steady state is achieved. Film thickness can be used to determine whether transient behavior dominates the data in marginally aggressive situations.

The manometer method has some serious limitations. In many cases, the subject gas is never encountered in a pure state. Oxygen, for example, is usually an element in air. Then the differential pressure required to transfer a measurable volume of gas may be so large that the film distorts or ruptures, the gas compressibility factor becomes a confounding factor, or the test cell may begin to leak. Furthermore, ASTM D1434, Method M, holds neither volume nor pressure constant. Lomax notes that many permeability tests include alternate methods. For example, ISO 1399 and BS 903: Part A17 detail constant volume techniques. ASTM D1434, BS 903: Part A30, ISO 2782,

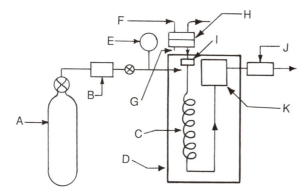

Figure 6.171.   Sweep Gas Chromatography Schematic for Measuring Permeabilities of Gases Diffusing Through Thin Plastic Membranes (174). A: Carrier Gas Supply. B: Drying Tube. C: GC Separating Column. D: Thermostatically Controlled Chamber or Oven. E: Pressure Gage. F: Test Gas. G: Carrier Gas to Cell. H: Gas Transmission Cell. I: Injector. J: Gas Flow Meter. K: Detector.

and DIN 53536 detail constant pressure techniques. For highly permeable tests, a microgravimetric method can be used (175). For very low transmission rates or very low driving pressures, infrared or gas chromatography can be used (Figure 6.171). One technique uses ultraviolent absorption bands directly to measure time-dependent gas concentration. In many cases, a carrier gas, such as nitrogen or helium, is used to maintain nearly zero differential pressure across the film.

Since the test specimens have essentially the same makeup, including thickness, as the polymer product, and since the gas makeup and conditions such as pressure and temperature, can be set to actual environmental conditions, there should be relatively little misapplication of the results. The primary problem lies in the inability to extrapolate the data from one gas to another and from one polymer to another. This is primarily due to the difficulty in understanding how small molecules move through polymer networks. And so at this point, permeability tests fail the second criterion of test acceptability.

6.8 Electrical Properties of Polymers

Polymers are considered to be *ideal dielectric insulators*. They can be easily shaped and their electrical properties readily controlled. Applications include integrated circuit "packaging", wire and cable insulation, industrial electrical switching, and high-voltage capacitor insulation. Many polymers are transparent to electromagnetic radiation. This makes them undesirable as cases for electronic equipment that might generate electromagnetic fields capable of interfering with other electronic equipment. EMI or electromagnetic interference is the subject of great current interest.

The primary electrical and electronic uses for polymers are as insulators. The electrical field on either side of the polymer film or sheet can be AC or DC and the frequency can vary from 0 Hz to 10^{10} Hz. The voltage can be microvolts as in electronic circuits or kilovolts in transformer housings. The amperage can be microamps to kiloamps. In order to select the appropriate polymer-adduct combination for a given electrical application, certain tests are desired. Mathes lists several reasons for needing to know the electrical properties of a polymer-adduct combination (176):

- Determination of the performance capability of the combination in a given application,
- Evaluation of functional properties other than electrical properties, such as thermal conductivity,
- Quality control of the manufactured product,
- Identification of the polymer and/or polymer-adduct combination,
- As an element in the research of the molecular behavior of polymers, and
- As a way of ranking various polymer and polymer-adduct combinations in terms of a cost/performance criterion.

It is important to define some terms prior to considering the details of a few of the important critical tests. *Volume resistivity*, ρ_v, is given by Ohm's Law:

$$(6.68) \qquad \rho_v = \left(\frac{E}{I_v}\right)\left(\frac{A}{t}\right) = \frac{R_v A}{t},$$

where E is the applied potential, (volts), I_v is the measured current, (amperes), R_v is the resistance of the specimen, (ohms), A is the area of the specimen in the electrical field, (cm^2), and t is the thickness of the specimen, (cm). Volume resistivity gives a measure of the electrical insulating capability of a material.

Practical electrical insulators have volume resistivities greater than 10^6 ohm-cm. Most pure polymers have volume resistivities in the range of 10^{10} to 10^{14} ohm-cm. Values as high as 10^{20} have been reported for very thin ultrapure polymer films. Volume resistivity is determined from ASTM D257. A flat disk of polymer is placed between two elecrodes, as shown in Figure 6.172. A voltage of 500 volts DC is applied across the disk. The current flow through the disk is then measured. Volume resistivity values are most often used in material selection. As expected, other environmental factors can influence the volume resistivity value.

As seen in Figure 6.173 for polyvinyl chloride, volume resistivity decreases with increasing temperature and increasing relative humidity and in Figure 6.174 for nylon 6 and 610, with increasing moisture content. In Figure 6.175, the time-dependent effect of high temperature and high humidity on volume resistivity is shown for several electrical-grade thermosetting resins. Keep in mind that only a small amount of absorbed moisture in these resins, typically less than 0.1% (wt), causes the remarkable 3- or 4-decade drop in volume resistivity. The temperature effect on volume resistivity for other polymers is shown in Figure 6.176.

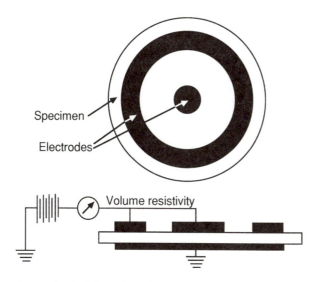

Figure 6.172. Schematic for Volume Resistivity Test for Polymers in Accordance with ASTM D257.

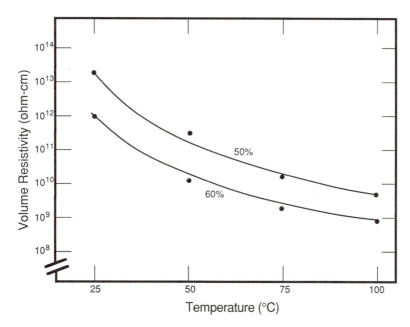

Figure 6.173. Volume Resistivity for Polyvinyl Chloride as Function of Temperature and Relative Humidity (177).

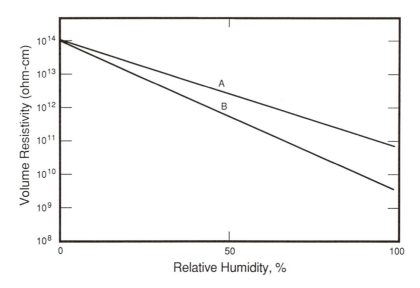

Figure 6.174.   Effect of Room Temperature (23°C) Relative Humidity on Volume Resistivity for Polyamides (177). Curve A: Nylon 610 (PA-610). Curve B: Nylon 6 (PA-6).

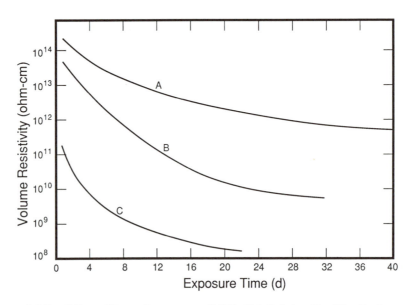

Figure 6.175.   Effect of Room Temperature (23°C) 50% Relative Humidity Environment on Volume Resistivity of Filled and Reinforced Thermosets (178). Curve A: Glass-Filled Epoxy. Curve B: Mineral-Filled Epoxy. Curve C: Glass-Filled Phenolic Resin.

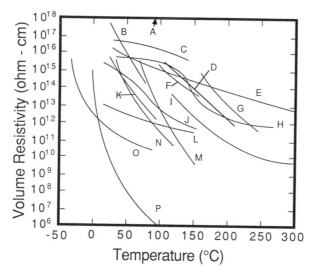

Figure 6.176. Temperature Dependence of Volume Resistivity for Several Polymers (176). Curve A: Polytetrafluoroethylene (PTFE). B: Low-Density Polyethylene (LDPE). C: Polycarbonate (PC). D: Polyester–Glass Fabric (FRP). E: Silicone Resin. F: Epoxy I. G: Polyethylene Terephthalate (PET). H: Silicone Rubber. I: Molded Silicone. J: Plasticized Cellulose Acetate Butyrate (CAB). K: Plasticized Polyvinyl Chloride (FPVC). L: Polyurethane Foam (PUR Foam). M: Epoxy II. N: Epoxy III. O: Elastomeric Polyurethane (PUR). P: Nylon (PA), Type and Conditioning Unknown.

The ultimate electrical strength of a polymer is used in determining high-voltage applications. ASTM D149, "Dielectric Breakdown Voltage and Dielectric Strength of Solid Electrical Insulating Materials at Commercial Polymer Frequencies", is used to determine electrical breakdown or the point at which voltage can no longer be maintained across the polymer specimen. There are many types of breakdown. The polymer may heat rapidly, causing the volume resistivity to drop. There may be microvoids or defects in the specimen. There may be surface conditions that exacerbate discharge. Mathes (176) notes that the time of breakdown can be used as an indicator of the nature of the failure. For surge or impulse power sources, breakdown is primarily dependent on the polymer characteristics. For example, for times less than about 10 seconds, the breakdown may be "intrinsic" to the material. If the breakdown occurs in 0.1 to 10,000 seconds, the primary mode of failure may be due to thermal heating. And if failure occurs in 100 seconds or more, the primary failure mode may be due to surface or void discharge, sometimes called corona failure. An example of time-dependent thermal failure is seen in Figure 6.177 for polyvinyl chloride cable. Dielectric strength under short-time voltage is seen to be quite thickness dependent (Figure 6.178).

Surface resistivity, ρ_s, is defined as:

$$(6.69) \qquad \rho_s = \left(\frac{E}{I_s}\right)\left(\frac{\pi D_m}{g}\right) = \frac{R_g \pi D_m}{g},$$

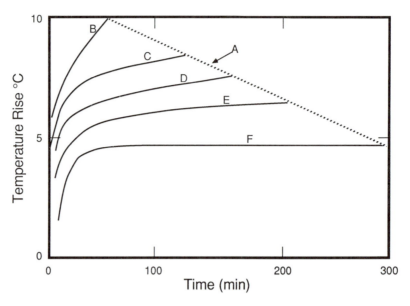

Figure 6.177. Time-Dependent Voltage Breakdown for Polyvinyl Chloride Initially at 20°C
(179). A: Thermal Instability Boundary. B: 26 kV, 121 V/mil. C: 23 kV, 108 V/mil. D: 21.9
kV, 102 V/mil. E: 20.7 kV, 96.5 V/mil. F: 19.5 kV, 91 V/mil.

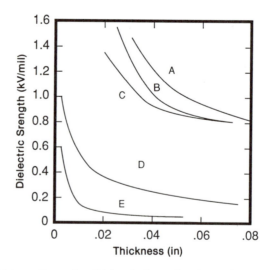

Figure 6.178. Thickness-Dependent Dielectric Strength at Room Temperature (23°C) for Sev-
eral Materials (180). A: Polystyrene (PS). B: Polyfluoroethylene–Propylene Copolymer (FEP).
C: Low-Density Polyethylene (LDPE). D: Paper–Phenolic Laminate, Dry. E: Paper–Phenolic
Laminate, Wet.

where I_s is the applied current, (amperes), D_m is the mean diameter of the gap, [cm], g is the gap width, (cm), and R_s is the resistance of the specimen *surface*, [ohms]. O'Toole (181) notes that surface resistivity is *not* a material property. It depends on the cleanliness of the surface and the volume and nature of the material in the gap.

The *dielectric constant* of any material is the ratio of its capacitance to the capacitance of vacuum or air. The dielectric constant is measured in the alternating current test circuit shown in Figure 6.179 from ASTM D150, "AC Capacitance, Dielectric Constant and Loss Characteristics of Insulating Materials". The dielectric constant is a *material property*, independent of the nature of the test. It is a measure of the ability of the polymer to act as a capacitor, that is, to store charge. On a molecular level, charge storage involves alignment of dipoles in the direction of applied field. Dipoles can be polar groups as part of the polymer chain or as pendant groups, polar adducts, pigments, fillers, plasticizers, lubricants and so on. In some instances, very weak dipoles are formed in nonpolar semicrystalline polymers due to the dielectric differences at crystallite boundaries. Dipoles do not occur in symmetric crystalline polymers such as PTFE, but replacement of a fluorine atom with a chlorine as with polytrifluorochloroethylene, produces an unsymmetric polymer that has a measurable dipole moment. The effect of temperature on the low-frequency dielectric constants for crystalline and amorphous polymers is seen in Figure 6.180.

In an AC field, the dipole oscillates from one neutral position to the other as the field reverses. The dipole motion is resisted by the nonpolar material surrounding it.

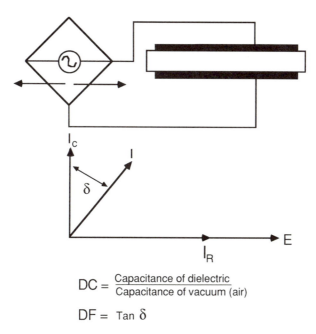

$$DC = \frac{\text{Capacitance of dielectric}}{\text{Capacitance of vacuum (air)}}$$

$$DF = \text{Tan } \delta$$

Figure 6.179. Schematic for Determination of Polymeric Dielectric Constant from ASTM D150 (48).

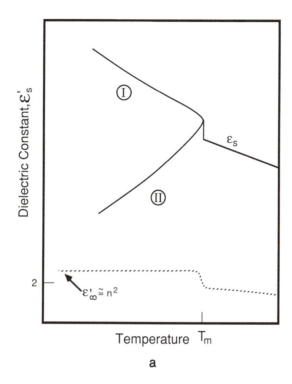

a

Figure 6.180.   Schematic Effect of Temperature on Low-Frequency Dielectric Constant for Crystalline Polymers (a) (b) and Amorphous (182) ε'_g is Dielectric Constant. Crystalline Cases I, II Represent Degrees of Unlocking of the Crystal Structure Under Oscillatory Force. T_m is Melt Temperature. Similarly, Amorphous Cases "I, II, III" Represent the Effect of the Degree of Rotation of Side Groups Under Oscillatory Force. T_g is Glass Transition Temperature. Dashed Lines Labeled [$\varepsilon'_\infty \approx n^2$] Represent High-Frequency Dielectric Constant for Comparison.

Some of the motion results in energy dissipation. At low frequencies, the dissipated heat is transferred from the polymer to its environment. The polymer part increases in temperature to a steady-state value. As the frequency increases, the dissipation rate becomes so great that the material simply heats until thermal breakdown occurs. The measure of energy dissipation is the *dissipation factor, a material property.* The dissipation factor is shown in Figure 6.180 in terms of the loss angle, δ. The complex resultant current, I, lags the capacitive current by δ. The tangent of this angle, called "tan δ", is the dissipation factor. The effect of frequency on the dielectric constant and dissipation factor for an electrical-grade epoxy is seen in Figures 6.181 and 6.182. Many polymer–adduct characteristics influence the dissipation factor, tan δ. Molecular orientation influence is seen in Figure 6.183. Frequency-dependent dissipation factors for nonpolar polymers are given in Figure 6.184. Time-dependent moisture effects are seen for both dielectric constant and tan δ for electrical-grade thermoset resins at high temperature and humidity (Figure 6.185). Electrical properties for other polymers are given in Table 6.27.

Dielectric Constant, ϵ'_s

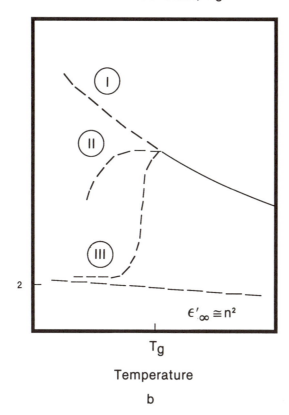

Temperature

b

Figure 6.180. Schematic Effect of Temperature on Low-Frequency Dielectric Constant for Crystalline Polymers (a) (b) and Amorphous (182) ε'_g is Dielectric Constant. Crystalline Cases I, II Represent Degrees of Unlocking of the Crystal Structure Under Oscillatory Force. T_m is Melt Temperature. Similarly, Amorphous Cases "I, II, III" Represent the Effect of the Degree of Rotation of Side Groups Under Oscillatory Force. T_g is Glass Transition Temperature. Dashed Lines Labeled [$\varepsilon'_\infty \approx n^2$] Represent High-Frequency Dielectric Constant for Comparison.

Electromagnetic Interference (EMI) and Shielding

The very rapid proliferation of electronic equipment and the subsequent development of technologies to mold polymeric equipment cases to house the electronics allowed a phenomenon known as electromagnetic interference or *EMI* to occur. EMI can cause spurious signals that can generate misinformation in delicate electronic devices such as computers and controls. In many cases, EMI can just be annoying, as with telephone cross-talk. In extreme cases, EMI can disrupt vital communications and endanger lives,

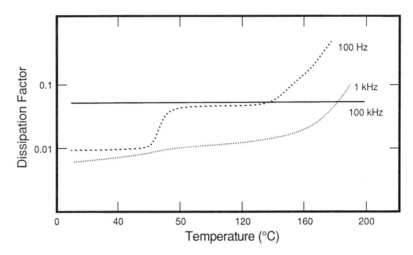

Figure 6.181.   Temperature-Dependent Dissipation Factor for Electrical-Grade Epoxy (183).

as in aircraft navigation.* The proliferation of electronic devices has prompted ASTM and other standards bodies to issue emergency standards for EMI shielding. One emergency ASTM standard is ES 7-83, "Test Method for Electromagnetic Shielding Effectiveness of Planar Materials".

O'Toole (188) notes that all AC circuits emit electromagnetic radiation, or *EMR*, with electric capacitive and magnetic inductive fields. Both can induce spurious fields

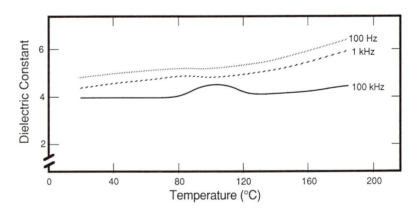

Figure 6.182.   Temperature-Dependent Dielectric Constant for Electrical-Grade Epoxy (183).

---

*It is for this reason that airlines restrict the use of cellular telephones, TVs, FM radios, powered hair dryers, razors, certain computers and handheld computerized games.

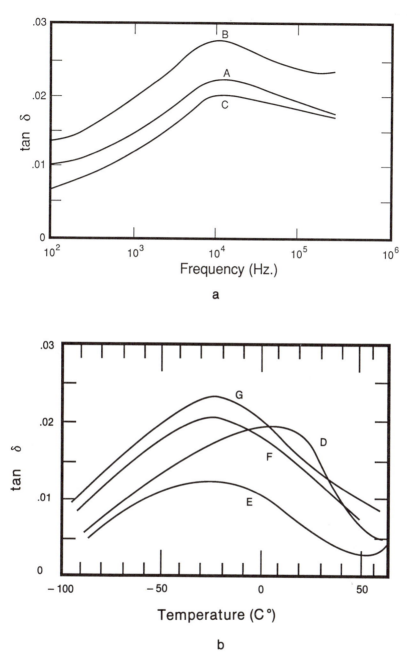

Figure 6.183.   Effect of Polymer Film Orientation on Dissipation Factor, tan δ (184). a): Nylon (PA) at 25°C as Function of Frequency, in Hz. A: Unoriented. B: Oriented, Electrical Stress Perpendicular to Orientation. C: Oriented, Stress Parallel to Orientation. b): Polyethylene Terephthalate (PET) at 1 kHz as Function of Temperature, in °C. D: Amorphous. E: Crystalline, Unoriented. F: Uniaxially Oriented. G: Biaxially Oriented.

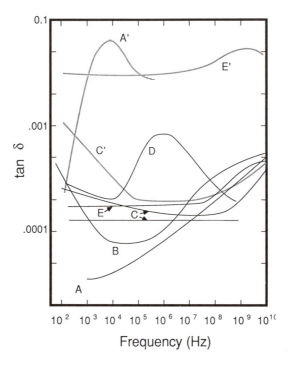

Figure 6.184.   Frequency-Dependent Dissipation Factor, tan δ, for Various Nonpolar Polymers
(185). Curve A: Polystyrene (PS) at 25°C. A': Polystyrene (PS) at 134.5° C. B: Polyisobutylene
(PIB) at 25°C. C: Polytetrafluoroethylene (PTFE) at 23°C (two sources). C': Polytetrafluoroethylene (PTFE) at 100°C. D: Polyfluoroethylene–Propylene (FEP) at 23°C. E: Polyethylene
(LDPE) at 25°C. E: Polyethylene (LDPE) at 25°C After Milling at 190°C for 30 min.

in nearby electrical circuits. The strengths of these fields depend upon the type of field,
the distance between transmitter and receiver, and the nature of the medium through
which the field is transmitted. The primary characteristic of the medium is its *impedance*. The medium provides a "shield". A highly effective shield has high electrical
conductivity and low magnetic permeability.

   In the *far field,* or the region $\lambda/2\pi$ away from the transmitter where λ is the characteristic wavelength of the interfering signal, the signal propagates as a plane wave.
The ratio of the electric field, E, to the magnetic field, H, is constant and equal to the
characteristic impedance, Z_c, of the medium. For air, the ratio is equal to 377 ohm:

(6.70) $(E/H)_{far} = Z_c$

For far field strength, only the electric field strength needs to be determined. In the
near field, where the distance is less than $\lambda/K2\pi$, the ratio, E/H, depends on the *wave*
impedance of the medium. In turn, this depends on the characteristics of the source.
If the transmission is of low current and high voltage, as with most electronic equipment, the near field will be dominated by the electric field. Magnetic field dominates

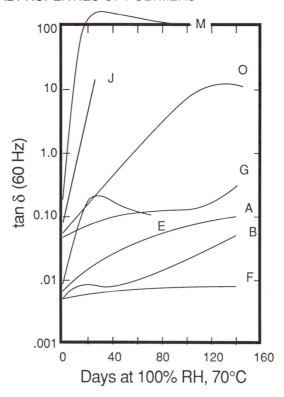

Figure 6.185.　The Effect of a Hot, Wet Environment (100% RH at 70°C) on the Dissipation Factor, tan δ, for Several Moisture-Sensitive Thermosetting Resins (186). A: Unfilled Epoxy I. B: Unfilled Epoxy II. E: Filled Epoxy I. F: Filled Epoxy II. G: Filled Epoxy III. J: Unfilled Modified Polyester (UPE). M: Filled Polyester I (UPE). O: Filled Polyester II (UPE).

when the transmission is of high current and low voltage as with most electrical equipment. This is seen in Figure 6.186.

Oberholtzer (189) notes that polymers are essentially transparent to EMR. Only metals and carbon can attenuate EMR. As a result, for shielding, carbon or metal adducts must be incorporated with the polymer or applied to the finished part as a coating. The effectiveness of the shield must be compared with a standard. The principal parameters are the distance between the EMR source and the shield, the frequency of the source, and the effective thickness of the shield. Near field, far field, electric field and magnetic field changes are achieved by changing the configuration of the transmitter, the receiver and the aperture geometry. Shielding effectiveness is given by:

$$(6.71) \qquad SE = 20 \log_{10} \left(\frac{E_t}{E_i} \right) = 20 \log_{10} \left(\frac{H_t}{H_i} \right),$$

Table 6.27. Electrical properties of several polymers (187).

Polymer	Resistivity (Ω-cm)		Dielectric constant/dissipation factor					
	Volume	Surface	100 Hz	1 kHz	1 MHz	10 MHz	100 MHz	1000 MHz
ABS	2×10^{16}	10^{14}	0.005/2.8	0.006/2.8	0.008/2.8	0.007/2.8	0.005/2.7	0.001/2.7
PMMA (acrylic)	10^{18}	10^{14}	0.062/3.6	0.058/3.2	0.04 5/3.1	0.033/2.9	—	—
Cellulose ester	3×10^{15}	10^{14}	0.006/3.8	0.011/3.6	0.024/3.3	0.022/3.2	0.020/3.0	0.014/2.1
FEP	10^{18}	10^{16}	0.0005/2.1	0.0005/2.1	0.0005/2.1	0.0005/2.1	0.0008/2.09	0.0007/2.05
Nylon 6 (PA-6)	10^{15}	10^{13}	0.031/4.2	0.024/3.8	0.031/3.8	0.020/4.0	—	—
Polycarbonate (PC)	10^{16}	10^{15}	0.001/3.1	0.0013/3.1	0.007/3.1	0.011/3.1	—	—
Polyethylene (PE)	10^{19}	10^{16}	0.0001/2.34	0.0001/2.34	0.0001/2.34	0.0001/2.34	0.0001/2.34	0.0001/2.34
Alkyd	10^{13}	10^{14}	0.02/6.0	0.02/5.8	0.015/5.4	—	—	—
DAP	10^{16}	10^{13}	0.026/3.8	0.020/3.7	0.016/3.6	—	—	—
Phenolic (thermoset)	10^{14}	10^{9}	0.013/5.4	0.013/5.3	0.0 33/4.9	—	—	—
Epoxy	10^{16}	10^{14}	0.004/3.22	0.004/3.25	0.004/3.25	—	—	—

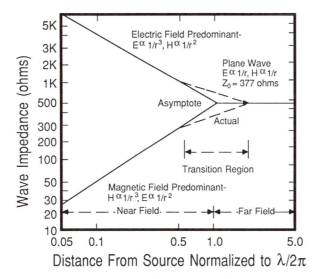

Figure 6.186.   Schematic Representation of Near-Field and Far-Field Electromagnetic Regions (156).

where the "t" and "i" subscripts refer to the transmitted and incident field strengths, respectively. The units on SE are decibels, or dB. O'Toole notes that the signal can be attenuated in three ways:

- Absorption into the shield, A,
- Reflection at the shield surfaces, R, and
- Multiple internal reflections, B.

The last is usually small for most electric fields, far fields and conditions where absorption attenuation is greater than 10 dB. Absorption and reflection losses are summed to produce shielding effectiveness:

(6.72) $SE = A + R (+ B)$

SE is found to be a function of:

- shield thickness, t,
- wave frequency, f,
- distance between transmitter and shield, r,
- magnetic permeability of shield, μ, and,
- electrical conductivity of shield, σ.

Absorption loss can be calculated from:

(6.73) $A = K_1 t \sqrt{f \mu \sigma}$

K_1 is a numerical constant and not a material constant. Note that absorption loss increases linearly with increasing thickness and is greatest at high frequencies. Reflection loss is controlled by the change in impedance at the interface between the shield and the air. The equations describing reflection loss for various field conditions are:

$$(6.74) \qquad R_p = K_2 - \log_{10}\left(\frac{\mu_f}{\sigma}\right),$$

$$(6.75) \qquad R_e = 20 \log_{10}\left[\frac{K_3}{f^{3/2}\, r\sqrt{(\mu/\sigma)}}\right],$$

$$(6.76) \qquad R_m = K_4 + 10 \log_{10}\left[\frac{fr^2\sigma}{\mu}\right],$$

The subscripts "p", "e", and "m" relate to plane wave (far field), electric and magnetic fields, respectively. The constants K_2, K_3 and K_4 are again numerical constants, not material constants. For electric fields, reflection is very large compared with absorption. Thus very thin films can successfully shield electric fields. For magnetic fields, absorption is significant. As a result, the thickness of the shield is important. The effect of frequency on shielding effectiveness for a 0.5 mm (0.020 in) copper shield is seen in Figure 6.187. A more thorough understanding of the applicable theory can be found in Ott (190).

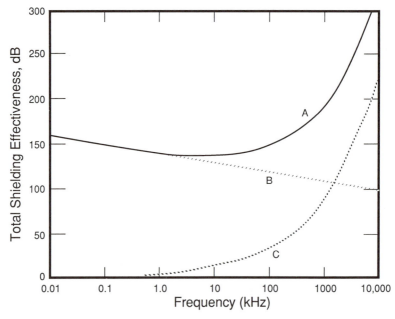

Figure 6.187. Schematic of the Relative Roles of Absorption and Reflection of Electromagnetic Plane Wave Radiation as a Function of Frequency for 0.5 mm (0.020 in) Copper Foil (156). A: Total Shielding. B: Reflection. C: Absorption.

Table 6.28. Example of data on commercial polymers available in handbooks (191)[a].

Type	Supplier	Compound or substrate trade name and grade designation	Fillers and reinforcements: identity and % by weight	Polymer coating (if any) Coating trade name, grade designation and material identity	Thickness (0.001 in)	Method of application	Composite properties at 23°C, 50% RH Density (g/cm³)	Volume resistivity (ohm-cm)	Surface resistivity (ohm-cm)
ABS	American Cyanamid	Cycolac KJB	Cycom MCG nickel-coated graphite fiber 17%	None	—	—	1.37	NA	NA

Shielding test results

Test method Test specification or apparatus used: FF = far field NF = near-field	Specimen Dimensions Type or shape	Disk or annulus: OD × ID (in)	Plate: length × (in)	Thickness (in)	Conditioning (h/°C/% RH)	Test conditions Temperature (°C)	Frequency (MHz)	Dynamic range (dB)	Shielding effectiveness Electric field, E (dB)	Magnetic field, M (dB)
NF-Dual Chamber	Plate	—	6 × 3	0.125	None	23	30	>115	56	DNM
							100	>115	50	DNM
							300	>115	51	DNM

NA - Not available
DNM - Did not measure
[a]Columns without data included to suggest range of information available; footnotes, however, have been omitted.

Since there are many ways of treating polymers to achieve a degree of shielding, resin suppliers, compounders and coatings suppliers are sources for shielding data. The data display is usually given as polymer type, type of filler or coating, test type, test temperature, frequency, specimen thickness and then shielding effectiveness in far field, near field, electric and magnetic field conditions. An example is seen in Table 6.28 for an ABS containing 17% (wt) nickel-coated graphite fiber. Note that as the frequency increases, the electric field shielding effectiveness increases.*

A good EMI shielding design insures that the generated signal remains within the enclosure. Incomplete shielding will allow the signal to transmit. Field leakage can be minimized by:

- Minimizing long narrow slots. These act as antennae. Field leakage through a small slot depends on its length rather than its area,
- Converting slots to holes, particularly if the slots are greater than 0.01 λ in dimension,
- Converting small holes to wave guides by bushing or bossing to increase the ratio of hole depth to hole diameter,
- Changing hole patterns in ventilation grids such that a large number of very small holes is used,
- Sealing all nonmetallic joints with conductive cement, adhesive or gaskets. This minimizes the slot antenna effect, and
- Positively grounding all sections of the shield. To minimize ground loops, the number and location of all grounds should be included in any circuit design.

6.9 Part Design Constraints: ASTM D955

When amorphous thermoplastics cool from the processing temperature to room temperature, density increases. For fixed weight parts, this means that the part dimensions decrease. If the part is isotropic, the decrease in dimension or shrinkage is uniform. If the strain field is nonuniform, shrinkage is nonuniform. If the stresses are severely out of balance, differential shrinkage can result in part warpage. Crystalline thermoplastics tend to have greater overall and differential shrinkage owing to greater density change and temperature-dependent crystallization rate. Thermosets exhibit similar time-dependent volume changes due to crosslinking reactions.

Shrinkage is considered to be the sum of two loosely defined components. Mold shrinkage is defined as the difference in tooling dimension and ejected part dimension. Annealing shrinkage is defined as the difference in part dimension before and after it

---

*"Dynamic range" is the attenuation of the test signal measured with an electrically opaque material such as aluminum in place of the test specimen. It is a measure of the sensitivity of the apparatus and it is the value to which test data on specimens are compared for determination of significance.

is held at an elevated temperature for a fixed period of time. Annealing is recommended for many parts to minimize part dimensional change during use. For slowly crystallizing polymers, annealing also serves to continue crystallization to some plateau value. Additional crystallization can lead to substantial dimensional change and even part distortion and warpage. As a result, parts that tend to distort are usually fixtured during annealing. The effect is seen in schematic in Figure 6.188. Since cooling and crystallizing are time-dependent phenomena, shrinkage can be thickness-dependent, as seen in Figure 6.189 for nylon 66.

Shrinkage does not give a useful measure of warpage. ASTM D955 is designed to provide shrinkage data (Figure 6.190). Rectangular bars, 125 mm × 12.7 mm × 3.2 mm (5 in × 0.5 in × 0.125 in) in dimension, are edge-gate molded. Shrinkage is determined by measuring the length (only) of the bar 48 hours after molding:

$$(6.76) \qquad \text{shrinkage} = \frac{5 - \text{measured length}}{5},$$

Data are reported as (cm/cm), (mil/in) or percentage. Since only one dimension is measured, there is no correlation to warpage. Warpage can be determined from an edge-gated disk mold, 102 mm (4 in) in diameter × 1.6 mm (0.063 in) thick. Warpage is determined by placing the disk on a flat surface, then measuring the height of the edge above the surface. Warpage is reported as:

$$(6.77) \qquad \text{warpage} = A/D$$

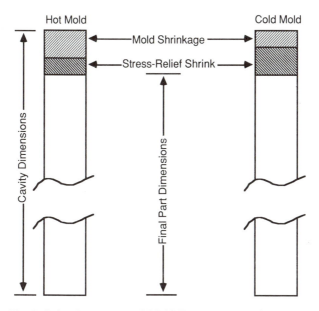

Figure 6.188.   The Relative Importance of Mold Temperature to the Nature of Shrinkage. (Courtesy of E.I. Dupont, Wilmington, DE).

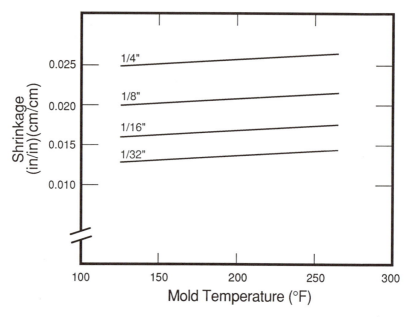

Figure 6.189.   The Effect of Injection Mold Temperature on Shrinkage of Nylon 66 (PA-66) for Various Plaque Thicknesses. 2 in × 2 in Plaques, Annealed at 325°F for 1 hour (193). Parts Tested Dry. Gate Thickness × Width = 0.5 × 1 Part Thickness.

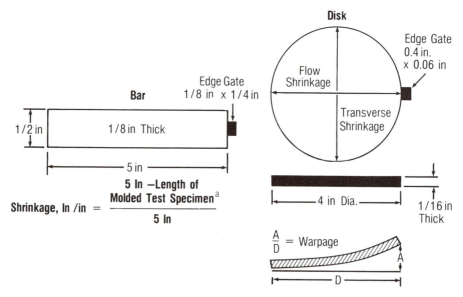

Figure 6.190.   Recommended Dimensions for Determining Shrinkage and Warpage of Injection Molded Specimens (193). Note that Shrinkage Should Be Determined 48 Hours After Molding. Moisture-Sensitive Resins Must Be Kept Dry During This Time.

where A is the maximum vertical height of the disk and D is its diameter. Units of warpage are (cm/cm), (mil/in) or percentage. Shrinkage and warpage values for several filled, reinforced and neat polymers are given in Table 6.29.

The disk can be used to measure *differential shrinkage* by measuring shrinkage in the flow and transverse directions. This is seen for various glass filler-to-fiber combinations in nylon 66 and for glass fiber loading in polyacetal (POM) (Figures 6.191 and 6.192). As seen in Table 6.30 for polyacetal, there seems to be reasonable correlation between differential shrinkage and warpage. Amorphous resins respond in roughly the same fashion, except that differential shrinkage and warpage values are substantially lower than values for crystalline polymers, Table 6.31. This is seen in Figure 6.193 for polycarbonate.

As mentioned, annealing is important for some polymers in order to minimize in-use part distortion. The effect of annealing on warpage for 30% (wt) nylon 66 is shown in Table 6.32. Annealing here is 3 hours in air at 138°C (280°F). Note that warpage increased from 190 to 270 mil/in during annealing. Table 6.32 also shows that the effects of process changes are more noticeable *before* annealing than after annealing. For example, changing mold temperature changes the unannealed warpage value from 80 to 240 mil/in but only from 270 to 280 mil/in for annealed disks. It appears that the shrinkage and differential shrinkage data from disks yield reliable data regarding warpage and so are superior to ASTM D955 for determining dimensional data.

Table 6.29.   Shrinkage and warpage data for injection-molded neat and filled thermoplastic polymers (193).

Base polymer	Modifier type	Loading level (%)	Shrinkage (in/in)[a]	Warpage (A/D)[b]
Nylon 66 (PA-66)	Unmodified	0	0.015	0.050
Nylon 66 (PA-66)	Glass fiber	10	0.006	0.060
Nylon 66 (PA-66)	Glass fiber	30	0.004	0.270
Nylon 66 (PA-66)	Glass fiber	40	0.003	0.270
Nylon 66 (PA-66)	Carbon fiber	40	0.002	0.200
Nylon 66 (PA-66)	Glass bead	40	0.010	0.008
Nylon 66 (PA-66)	Barium ferrite	80	0.008	0.002
Polyacetal (POM)	Glass fiber	30	0.003	0.300
Polypropylene (PP)	Glass fiber	30	0.004	0.380
Polypropylene (PP)	Glass fiber[c]	30	0.003	0.320
Polycarbonate (PC)	Unmodified	0	0.006	0.001
Polycarbonate (PC)	Glass fiber	10	0.003	0.001
Polycarbonate (PC)	Glass fiber	30	0.001	0.003
Polycarbonate (PC)	Carbon fiber	30	0.0005	0.002
Polystyrene-acrylonitrile (SAN)	Glass fiber	30	0.0005	0.002
Polystyrene-acrylonitrile (SAN)	Glass bead	30	0.003	0.000

[a]ASTM D955 test bar
[b]4 in diameter × 1/16 in thick disk
[c]Chemically coupled

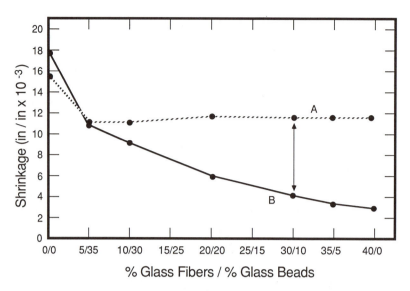

Figure 6.191. Relative Shrinkage of Filled and Reinforced Nylon 66 as Molded in 4 in Diameter $\times$ 0.062 in Thick Disk (193). A: Transverse Shrinkage. B: Flow Direction Shrinkage. Vertical Distance Considered To Be Relative Measure of Warpage.

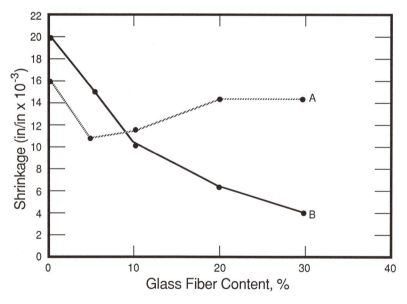

Figure 6.192. Relative Shrinkage of Reinforced Polyacetal (POM) as Molded in 4 in Diameter $\times$ 0.062 in Thick Disk (193). A: Transverse Shrinkage. B: Flow Shrinkage.

Table 6.30. Differential shrinkage and warpage data for injection-molded polyacetal (POM) disks[a] with increasing glass fiber loading (193).

Glass fiber content (%)	Flow shrinkage (in/in)	Transverse shrinkage (in/in)	Differential shrinkage (in/in $\times$ 10^{-3})	Warpage (A/D)[a]
0	0.020	0.0160	-4.0	0.075
5	0.015	0.0110	-4.0	0.060
10	0.011	0.0125	1.5	0.030
20	0.006	0.0150	9.0	0.270
30	0.004	0.0150	11.0	0.300

[a]4 in diameter $\times$ 1/16 in thick disk

Table 6.31. Comparison of plaque and disk warpage for several amorphous thermoplastics with various fiber and filler loading levels (193).

Base resin	Modifier type	Loading level (%)	Plaque warpage (in)[a]	Disk warpage (A/D)[b]
Polycarbonate (PC)	Unmodified	0	0.007	0.001
Polycarbonate (PC)	Glass fiber	10	0.007	0.001
Polycarbonate (PC)	Glass fiber	30	0.018	0.003
Polycarbonate (PC)	Carbon fiber	30	0.006	0.002
Polycarbonate (PC)	Glass bead	30	0.001	0.000
Polystyrene-acrylonitrile (SAN)	Glass fiber	30	0.001	0.002
Polystyrene-acrylonitrile (SAN)	Glass bead	30	0.001	0.000

[a]6 in $\times$ 8 in $\times$ in thick
[b]4 in diameter $\times$ 1/16 in thick

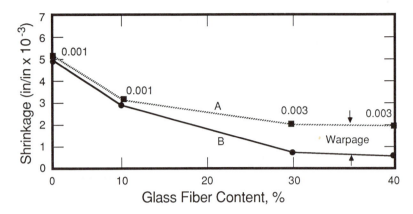

Figure 6.193.   Relative Shrinkage of Reinforced Polycarbonate (PC) as Molded in 4 in Diameter × 0.062 in Thick Disk (193). A: Transverse Shrinkage. B: Flow Shrinkage.

Table 6.32.   Effect of changing processing parameters on annealed and unannealed 30% glass fiber reinforced nylon 66 (PA-66) disks[a] (193).

	Change in warpage, A/D × 10^{-3}	
Processing parameters	Unannealed	Annealed[b]
Standard	190	270
Decreased injection rate	− 20	− 10
Increased injection pressure	+ 5	0
Decreased injection pressure	− 10	0
Increased injection hold time	0	− 10
Decreased injection hold time	+ 60	+ 20
Increased cooling time	− 20	− 5
Decreased cooling time	+ 40	+ 10
Increased melt temperature	− 10	0
Decreased melt temperature	+ 10	0
Increased mold temperature	+ 50	+ 10
Decreased mold temperature	− 110	0

[a]4 in diameter × 1/16 in thick
[b]3 hours in air at 280°F

6.10 Optical Characteristics, Including Color

Many polymers such as polystyrene, polymethyl methacrylate and polycarbonate are water-white transparent in visible light. These materials are frequently used in glazing, light displays, high-impact globes, and lenses including eyeglass, contact and camera lenses. Optical properties of some transparent polymers are given in Table 6.33. The refractive index is the ratio of the velocity of visible light in a vacuum to that in the polymer. Refractive index is always greater than one and usually depends strongly on EMR wavelength. Light scattering or haze is the result of changes in direction of light owing to occlusions such as gel particles and microvoids in the polymer. Light piping is one of the more important uses of transparent polymers. Note that many polymers have critical angles of about 40° (0.7 radian). Light that enters the polymer at angles less than the critical angle can be transmitted only along the length. With design care, the light can be "piped" around corners without substantial loss in intensity. Fiber optics has capitalized on light piping. This is shown in Figure 6.194. This concept is also used in the design of Fresnel or "fisheye" lenses, automobile taillamps and road reflectors. *Dichroism,* or the ability to selectively transmit incident light coherently, was discussed in Section 2.4 as an analytical tool for characterizing polymers. Dichroism in polymers is used to produce polarizer lenses.

Judd (194) states that:

> Color is the property of radiant energy that permits a living organism to distinguish *by eye* between two uniform, structure-free patches of identical size and shape. . . [emphasis by current authors]

The science of coloration is beyond the scope of this book. Certain elements are needed, however, to understand the ways in which color is measured and monitored today. All color is measured in the visible spectrum, that is, 380 to 780 μm. The three primary colors, red, blue, and yellow, are replaced today with three "tristimulus" values, X, Y, and Z. The secondary color field so familiar on the artist's palette is replaced with a chromaticity diagram (Figure 6.195). A given set of tristimulus values establish a vector in three-dimensional space. The length of the vector is a function of the amount of color. The direction is the quality of chromaticity of the color. Chromaticity coordinates x, y, and z are found directly from the tristimulus values:

$$(6.78) \qquad x = \frac{X}{X + Y + Z}; \quad y = \frac{Y}{X + Y + Z}; \quad z = \frac{Z}{X + Y + Z}.$$

Note that the chromaticity diagram is a two-dimensional plot in x and y. The z-value is found by difference: $z = 1 - x - y$. Pure spectrum colors, that is, red, orange, yellow, green, blue, and violet, fall on the horseshoe-shaped locus. Mixtures of these colors are always within the locus. Point C on the diagram represents the point of zero saturation. It is neutral gray and includes white. A mixture of a primary color with neutral gray produces a hue found by drawing a line between point C and the primary

Table 6.33. Some Properties of optical-grade polymers (15).

Properties	ASTM method	Units	Polymethyl Methacrylate (PMMA)	Polystyrene (PS)	Polycarbonate (PC)	Polymethyl Methacrylate–Styrene Copolymer
Refractive index (n_D)	D542	—	1.491	1.590	1.586	1.562
Abbe' value (μ)	D542	—	57.2	30.9	34.7	35
dn/dt $\times 10^{-5}$/°C	—	—	8.5	12.0	14.3	14.0
Haze	D1003	%	<2	<3	<3.0	<3
Luminous transmittance (0.125 in thickness)	D1003	%	92	88	89	90
Critical angle (i_c)	—	degree	42.2	39.0	39.1	39.6
Deflection temperature: 3.6°F/min, 264 lb_f/in^2	D648	°F	198	180	280	NA
3.6°F/min, 66 lb_f lb_f/in^2			214	230	270	212
Coefficient of thermal expansion (COE)	D696	in/in/°F $\times 10^{-5}$	3.6	3.5	3.8	3.6
Recommended maximum continuous service temperature	—	°F	198	180	255	200

Property	ASTM	Units				
Water absorption (immersed 24 h at 73°F)	D570	&	0.15	0.15	0.2	0.3
Specific gravity	D792	g/cm^3	1.09	1.20	1.06	1.19
Hardness (0.25 in) (specimen)	D785	—	M 75	M 70	M 90	M 97
Impact strength (notched Izod)	D256	ft-lb$_f$/in notch	NA	12 to 17	0.35	0.3 to 0.5
Dielectric strength	D149	V/in	450	400	500	500
Dielectric constant:	D150					
60 Hz			3.4	2.9	2.6	3.7
1 MHz			2.9	2.88	2.45	2.2
Power factor:	D150					
60 Hz			0.006	0.0007	0.0002	0.05
1 MHz			0.013	0.0075	0.0002 to 0.0004	0.03
Volume resistivity	D257	ohm-cm	10^{15}	8 × 10^{16}	> 10^{16}	10^{18}

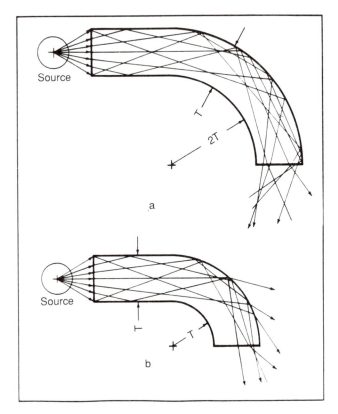

Figure 6.194. Typical Design Criteria for Light Piping (192). a): Inside Radius Equal To or Greater Than Twice the Thickness. b): Inside Radius Less Than Twice the Thickness.

color wavelength. Color matching involves establishing the X, Y, Z coordinates for two specimens, then determining the length of the vector between the two coordinate points. Color computers are now used to determine the vector length. The units are always listed as ΔE [CIELab] units.* Hemmendinger (195) notes a typical accuracy of 0.2 [CIELab] unit from colorimeter instrument to instrument. The tristimulus coordinates for a standard [CIELab] green are X = 17.33, Y = 33.87, Z = 11.22, or x = 0.278, y = 0.543, z = 0.179. This is shown on the chromaticity diagram (Figure 6.196). Other standard colors are shown on the outer envelope. The inner envelope represents "pale" colors, obtained by extrapolating the standard color to white or neutral gray. Note that on this diagram, black is shown to be shifted slightly toward blue.

As noted, some plastics such as GPPS and PMMA are water-white transparent. Others such as polyethylene are translucent. Some are milky white and nearly opaque,

---

*"CIELab" refers to Commission Internationale de l'Eclairage Laboratory, an international laboratory that established the unit system in 1931.

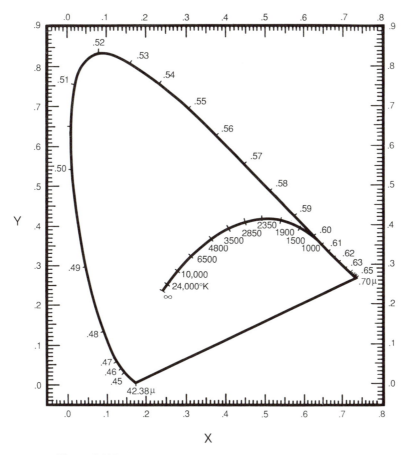

Figure 6.195.   American Standard Chromaticity Diagram (99).

like polyacetal and nylon. Some are off-white and opaque, like plasticized PVC and ABS. Some are hay- or straw-colored, such as polyimides and polyphenylene sulfide. Some, such as polysulfones, are amber. And some, such as polyamide-imide, phenolic and resorcinol are cherry red when ultra-pure. Polymers are quite easily colored and so provide products that have the color "all the way through". There are many types of colorants. Schiek (196) separates then into dyestuffs and pigments. *Dyestuffs* are usually soluble with the polymer and so are frequently used to color transparent plastics. *Pigments* are usually insoluble, with particle sizes of 0.01 to 1.0 μm. Typical pigment loading is 0.5 to 1.0% (wt). Liquid carriers and solid resin concentrates are used to produce controlled dosages of pigments. Pigments are usually added for visual effect, but in some cases, UV resistance and anti-bleed, anti-chalk features are sought. Organic pigments are more costly than inorganics, but offer brighter colors and are more transparent. Inorganic pigments are less fugitive, give greater hiding power and have greater thermal stability. Most of the organic pigments are based on azo chemistry,

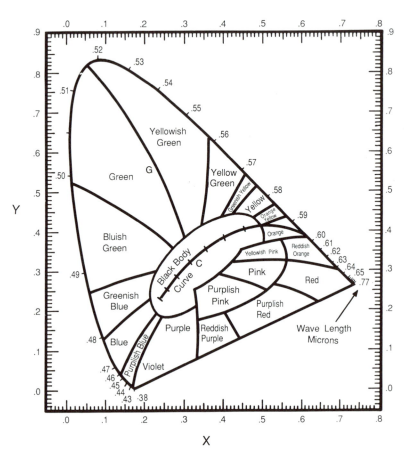

Figure 6.196.   Chromaticity Diagram Showing Single and Combination Colors (99). The *X–Y* Coordinates Are Used to Call Out Colors for Polymers.

whereas most of the inorganic pigments are oxide-based. A typical white pigment is TiO_2 and a typical black pigment is carbon black.

The designer must keep in mind that color is in "the eye" and not in the theory or analysis. Colorants must be compounded or blended with the polymer, then processed in the appropriate manner simulating the actual production process. Specimens must be taken from this step to determine color acceptability. Colorant-containing polymers can change in hue and tone when the samples are viewed first under incandescent light and then under fluorescent light. As noted above, environmental factors such as weathering, ultraviolet radiation, oxidation, and thermal degradation serve to gradually change the color with time. That is why color matching in automotive and house paint is so difficult.

References

1. S. Turner, *Mechanical Testing of Plastics*, 2nd Ed., George Godwin/PRI, London (1983), p. 1.
2. S. Turner, *Mechanical Testing of Plastics*, 2nd Ed., George Godwin/PRI, London (1983), p. 187.
3. M. Mooney, *Symposium on Consistency*, American Society for Testing and Materials, Philadelphia (29 June 1937), pp. 9–12.
4. S. Turner, *Mechanical Testing of Plastics*, 2nd Ed., George Godwin/PRI, London (1983), p. 10.
5. S. Turner, *Mechanical Testing of Plastics*, 2nd Ed., George Godwin/PRI, London (1983), p. 181.
6. R. Priemon, in R. Priemon, Ed., *Annual Book of Standards*, American Society for Testing and Materials (ASTM), Philadelphia (1987), p. iii.
7. H. Saechtling, *International Plastics Handbook*, Hanser, Munich (1983), p. 372.
8. E. Broyer and C.W. Macosko, SPE Tech. Papers, *21* (1975), p. 343.
9. M. Mooney, Ind. Eng. Chem., Anal. J., *6* (1934), p. 147.
10. J.M. Dealy, *Rheometers for Molten Plastics*, Van Nostrand Reinhold, New York (1982).
11. J.R. Van Wazer, J.W. Lyons, K.Y. Kim and R.E. Colwell, *Viscosity and Flow Measurement*, (Wiley-) Interscience, New York (1963).
12. C.D. Han, *Rheology in Polymer Processing*, Academic Press,ew York (1976).
13. S. Middleman, *The Flow of High Polymers: Continuum and Molecular Rheology*, Wiley-Interscience, New York (1968).
14. K. Walters, *Rheometry*, Chapman & Hall: Wiley, London (1975).
15. S. Levy and J.H. duBois, *Plastics Product Design Engineering Handbook*, Van Nostrand Reinhold, New York (1977), p. 299.
16. C.D. Han, *Rheology in Polymer Processing*, Academic Press, New York (1976), p. 90.
17. E.B. Bagley, J. Appl. Phys., *28* (1957), p. 624.
18. C.D. Han, *Rheology in Polymer Processing*, Academic Press, New York (1976), p. 122.
19. R.B. Bird, R.C. Armstrong, and O. Hassager, *Dynamics of Polymeric Liquids, Vol. 1: Fluid Mechanics*, John Wiley & Sons, New York (1977).
20. C.D. Han and M. Charles, Trans. Soc. Rheol., *15* (1971), p. 213.
21. C.D. Han and M. Charles, Trans. Soc. Rheol., *15* (1971), p. 371.
22. R. Westover and S. Bearder, SPE RETEC Proceedings, Newark, NJ (1977), p. 15.
23. R. Westover, in E. Bernhardt, Ed., *Processing of Thermoplastic Materials*, Reinhold, New York (1959), pp. 582–583.
24. J.L. Throne, *Plastics Process Engineering*, Marcel Dekker, New York (1979), p. 289.
25. G. Pezzin and G. Gechele, J. Appl. Polym. Sci., *89* (1964), p. 8.
26. R. Westover, "Processing Properties", in E. Bernhardt, Ed., *Processing of Thermoplastic Materials*, Reinhold, New York (1959), pp. 595, 597, and 599.
27. R.J. Roark and W.C. Young, *Formulas For Stress and Strain*, 5th Ed., McGraw-Hill Book Co., New York (1975), p. 326.
28. M. Riddell and R. Lamonte, Plast. Design Forum, *4*:3 (1976), p. 452.
29. A.B. Glanvill, *Plastics Engineers' Data Book–American Edition*, Industrial Press, New York (1974), p. 165.
30. C.D. Han, *Rheology in Polymer Processing*, Academic Press, New York (1976), p. 348.
31. C.D. Han, *Rheology in Polymer Processing*, Academic Press, New York (1976), p. 350.
32. J.M. Dealy, *Rheometers for Molten Plastics*, Van Nostrand Reinhold Co., New York (1982), p. 129.

33. D.S. Clark, *Engineering Materials and Processes,* International Textbook Co., Scranton, PA (1962), p. 76.

34. S. Turner, *Mechanical Testing of Plastics,* 2nd Ed., George Godwin/PRI, London (1983), p. 7.

35. J.K. Knowles and A.G.H. Dietz, Trans. Am. Soc. Mech. Eng., *77* (1955), p. 177.

36. C. Chastain, Machine Design, *42*:2 (23 Jan 1975), p. 108.

37. R. Kahl, *Principles of Plastic Materials Selection,* J. O'Toole, Seminar Coordinator, Center Prof. Advancement, East Brunswick, NJ (1988), p. 4B.

38. M. Yokouchi, H. Uchiyama and Y. Kobayashi, J. Appl. Polym. Sci., *25* (1980), p. 1007.

39. R. Dixon, Plast. Design Forum., *4*:6 (1979), p. 14.

40. S. Timoshenko and G.H. MacCullough, *Elements of Strength of Materials,* Van Nostrand Reinhold, New York (1954), p. 164.

41. C.S. Lee, E. Jones, and R. Kingsland, SPE Tech. Papers, *31* (1985), p. 574.

42. D.W. Van Krevelen, *Properties of Polymers,* Elsevier, New York (1976), p. 272.

43. D.W. Van Krevelen, *Properties of Polymers,* Elsevier, New York (1976), p. 303.

44. C.S. Lee, E. Jones, and R. Kingsland, SPE Tech. Papers, *31* (1985), p. 577.

45. J.L. Throne, *Thermoforming,* Hanser, Munich (1987), p. 114.

46. D.D. Joye, G.W. Poehlien, and C.D. Denson, Trans. Soc. Rheol., *17* (1973), p. 287.

47. J.L. Throne, *Thermoforming,* Hanser, Munich (1986), p. 115.

48. ASTM D257, "Test Methods for D–C Resistance or Conductance of Insulating Materials".

49. B.R. Selden and C.-G. Gustafson, SPE Tech. Papers, *30* (1984), p. 573.

50. J.-P. Chalifoux and Q.X. Nguyen, SPE Tech. Papers, *31* (1985), p. 570.

51. B. Darlix, B. Monasse and P. Montmitonnet, Polym. Testing, *6* (1986), p. 107.

52. D. Tabor, *The Hardness of Metals,* Oxford University Press, Cambridge (1951), pp. 6–8.

53. S. Turner, *Mechanical Testing of Plastics,* 2nd Ed., George Godwin/PRI, London (1983), p. 42.

54. ASTM D4065, "Practice for Determining and Reporting Dynamic Mechanical Properties of Plastics".

55. K.G. Smack and J.L. O'Toole, SPE RETEC Proceedings, New Haven, CT (1967), p. 9.

56. J. Starita, "Application News", RMS-A-4, Rheometrics, Princeton, NJ, undated.

57. E. Davis and C. Macosko, J. Macromol. Sci., Phys. B, *1* (1976), p. 1102.

58. H. Domininghaus, *Die Kunststoffe und ihre Eigenschaften,* VDI Verlag, Dusseldorf (1986), p. 697.

59. H. Domininghaus, *Die Kunststoffe und ihre Eigenschaften,* VDI Verlag, Dusseldorf (1986), p. 696.

60. K.G. Smack and J.L. O'Toole, SPE RETEC Proceedings, New Haven, CT (1967), p. 10.

61. R.C. Progelhof, A.M. Jagani, and J.L. Throne, SPE Tech. Papers, *31* (1985), p. 407.

62. C. Chastain, Machine Design, *42*:2 (23 Jan 1975), p. 110.

63. J.L. O'Toole, "Design Guide", Modern Plastics Encyclopedia, *63*:10A (1986), p. 413.

64. R. Westover, Plast. Technol., *4*:3 (1958), p. 223, and *4*:4 (1958), p. 348.

65. S. Turner, *Mechanical Testing of Plastics,* 2nd Ed., George Godwin/PRI, London (1983), p. 42.

66. A. Savadori, Polym. Testing, *5* (1985), p. 211.

67. H. Burns, in H.F. Mark, N.G. Gaylord and N.M. Bikales, Eds., *Encyclopedia of Polymer Science and Technology,* John Wiley & Sons, Inc., New York (1971), p. 607.

68. R. Kahl, *Principles of Plastic Materials Selection,* J. O'Toole, Seminar Coordinator, Center Prof. Advancement, East Brunswick, NJ (1988), p. 14B.

69. Anon., *Shell Polyolefins Engineering Design Data,* Shell International Chemical Co., Antwerp (1966), p. 50.

70. S. Turner, *Mechanical Testing of Plastics,* 2nd Ed., George Godwin/PRI, London (1983), p. 144.

71. T.T. Jones, J. Polym. Sci., *C16* (1968), p. 3845.

72. S. Turner, *Mechanical Testing of Plastics,* 2nd Ed., George Godwin/PRI, London (1983), p. 143.

73. P. Vincent, *Impact Tests and Service Performance of Thermoplastics,* Plastics Institute, London (1971), p. 29.

74. C.A. Brighton, in H.F. Mark, N.G. Gaylord and N.M. Bikales, Eds., *Encyclopedia of Polymer Science and Technology,* John Wiley & Sons, Inc., New York (1971), p. 589.

75. P. Vincent, *Impact Tests and Service Performance of Thermoplastics,* Plastics Institute, London (1971), p. 32.

76. S. Turner, *Mechanical Testing of Plastics,* 2nd Ed., George Godwin/PRI, London (1983), p. 142.

77. S. Turner, *Mechanical Testing of Plastics,* 2nd Ed., George Godwin/PRI, London (1983), p. 186.

78. S. Turner, *Mechanical Testing of Plastics,* 2nd Ed., George Godwin/PRI, London (1983), p. 146.

79. U. Patel, R.C. Progelhof, and J.L. Throne, SPE Tech. Papers, *33* (1987), p. 752.

80. J. O'Toole, "Design Guide", Modern Plastics Encyclopedia, *63*:10A (1986), p. 439.

81. J. O'Toole, "Design Guide", Modern Plastics Encyclopedia, *63*:10A (1986), p. 440.

82. J.L. O'Toole, "Design Guide", Modern Plastics Encyclopedia, *63*:10A (1986), p. 442.

83. Anon., *Shell Polyolefins Engineering Design Data,* Shell International Chemical Co., Antwerp (1966), p. 58.

84. J. O'Toole, "Design Guide", Modern Plastics Encyclopedia, *63*:10A (1986), p. 440.

85. Anon., *Shell Polyolefins Engineering Design Data,* Shell International Chemical Co., Antwerp (1966), p. 48.

86. W. Taylor, *Fundamental and Engineering Properties of Plastics,* University of Windsor, Windsor, Ontario, Canada (1968), Fig. 13.

87. Anon., *Shell Polyolefins Engineering Design Data,* Shell International Chemical Co., Antwerp (1966), p. 61.

88. Anon., *Shell Polyolefins Engineering Design Data,* Shell International Chemical Co., Antwerp (1966), p. 68.

89. Anon., *Shell Polyolefins Engineering Design Data,* Shell International Chemical Co., Antwerp (1966), p. 69.

90. Anon., *Shell Polyolefins Engineering Design Data,* Shell International Chemical Co., Antwerp (1966), p. 70.

91. Anon., Plast. Des. Forum, *5*:3 (1980), p. 13.

92. C. Crabb, SPE Tech. Papers, *34* (1988), p. 1528.

93. C. Crabb, SPE Tech. Papers, *34* (1988), p. 1530.

94. D.A. Nutter, Proc., SPI Struct. Foam Conf., *6* (1978), p. 38.

95. J.L. Throne, *Plastics Process Engineering,* Marcel Dekker, New York (1979), p. 457.

96. S. Turner, *Mechanical Testing of Plastics,* 2nd Ed., George Godwin/PRI, London (1983), p. 149.

97. S. Turner, *Mechanical Testing of Plastics,* 2nd Ed., George Godwin/PRI, London (1983), p. 148.

98. J.L. Throne and R.C. Progelhof, "Creep and Stress Relaxation", in C.A. Dostal, Sr. Ed., *Engineered Materials Handbook: Engineering Plastics,* Vol. 2, ASM International, Metals Park, OH (1989), Article 7:D.

99.   D.B. Judd, "Color Vision and Colorimetry", in E.U. Condon and H. Odishaw, Eds., *Handbook of Physics,* McGraw-Hill Book Co., New York, 1958, pp. 6–72.

100. J. O'Toole, "Design Guide", Modern Plastics Encyclopedia, *63*:10A (1986), p. 411.

101. Anon., *Udel Polysulfone: Design Engineering Data,* Union Carbide Corp., Bound Brook, NJ (1975), Brochure F-44689.

102. R. Kahl, *Principles of Plastic Materials Selection,* J. O'Toole, Seminar Coordinator, Center Prof. Advancement, East Brunswick, NJ (1988), p. 23B.

103. R. Horsley, in O. Delatycki, Ed., *Mechanical Performance and Design in Polymers,* Appl. Polym. Symp., *17* (1971), p. 117.

104. Anon., Modern Plastic Encyclopedia, *63*:10A (1986), p. 410.

105. J.L. O'Toole, "Design Guide", Modern Plastics Encyclopedia, *63*:10A (1986), p. 415.

106. K. Gotham, "Long-Term Durability", in R. Ogorkiewicz, Ed., *Thermoplastics Properties and Design,* John Wiley & Sons, London, New York (1974), p. 56.

107. S. Turner, "Deformational Behaviour", in R. Ogorkiewicz, Ed., *Thermoplastic Properties and Design,* John Wiley & Sons, London (1974), p. 60.

108. S. Turner, in R. Ogorkiewicz, Ed., *Thermoplastic Properties and Design,* John Wiley & Sons, London (1974), p. 46.

109. J.L. O'Toole, "Design Guide", Modern Plastics Encyclopedia, *63*:10A (1986), p. 428.

110. M.N. Riddell, G.A. Toelcke and J.L. O'Toole, Mod. Plast., *48*:10 (1970), p. 146.

111. Anon., "Engineering Data Bank", Modern Plastics Encyclopedia, *62*:10A (1986), Chart 4.

112. Anon., "Engineering Data Bank", Modern Plastics Encyclopedia, *62*:10A (1985), Chart 16.

113. Anon., Modern Plastic Encyclopedia, *62*:10A (1985), p. 307.

114. J. Mercier, et al., J. Appl. Polym. Sci., *9* (1965), p. 447.

115. J. Aklonis, W.J. MacKnight, and M. Shen, *Introduction to Polymer Viscoelasticity,* Wiley-Interscience, New York (1972), p. 51.

116. W.N. Findlay, SPE J., *16*:1 (1960), p. 57 and *16*:2 (1960), p. 192.

117. W.N. Findlay, SPE J., *16*:1 (1960), p. 62.

118. W.N. Findlay, Polym. Eng. Sci., *14* (1974), p. 579.

119. R.L. Thorkildsen, "Mechanical Behavior", in E. Baer, Ed., *Engineering Design for Plastics,* Reinhold, New York (1964), p. 305.

120. R.W. Hertzberg and J.A. Manson, *Fatigue of Engineering Plastics,* Academic Press, New York (1980), p. 9.

121. E.H. Andrews, in W. Brown, Ed., *Testing of Plastics,* Vol. 4, Wiley-Interscience, New York (1969), p. 128.

122. S. Turner, *Mechanical Testing of Plastics,* 2nd Ed., George Godwin/PRI, London (1983), p. 169.

123. R.W. Hertzberg and J.A. Mason, *Fatigue of Engineering Plastics,* Academic Press, New York (1980), p. 2.

124. R.W. Hertzberg and J.A. Mason, *Fatigue of Engineering Plastics,* Academic Press, New York (1980), p. 8.

125. M. Riddell, G. Koo, and J.L. O'Toole, Polym. Eng. Sci., *6* (1966), p. 182.

126. S. Turner, *Mechanical Testing of Plastics,* 2nd Ed., George Godwin/PRI, London (1983), p. 171.

127. L.E. Nielsen, *Mechanical Properties of Polymers and Composites,* Vol. 2, Marcel Dekker, New York (1974), p. 481.

128. R.J. Crawford and P.P. Benham, Polym., *16* (1975), 908.

129. R.W. Hertzberg and J.A. Manson, *Fatigue of Engineering Plastics,* Academic Press, New York (1980), p. 53.

130. P. Beardmore and S. Rabinowitz, in R.J. Arsenault, Ed., *Treatise on Materials Science and Technology,* Vol. 6, Academic Press, New York (1975), p. 267.

131. G.M. Bartenev and V.V. Lavrentev, *Friction and Wear of Polymers,* Elsevier, New York (1981).

132. J.J. Gouza, in J.V. Schmitz, Ed., *Testing of Polymers,* Vol. 2, (Wiley-) Interscience, New York (1966), Chapter 7.

133. G.M. Bartenev and V.V. Lavrentev, *Friction and Wear of Polymers,* Elsevier, New York (1981), p. 80.

134. D. Tabor, *The Hardness of Metals,* Oxford University Press, Cambridge (1951), p. 9.

135. G.M. Bartenev and V.V. Lavrentev, *Friction and Wear of Polymers,* Elsevier, New York (1981), p. 68.

136. G.M. Bartenev and V.V. Lavrentev, *Friction and Wear of Polymers,* Elsevier, New York (1981), p. 69.

137. S. Levy and J.H. DuBois, *Plastics Product Design Engineering Handbook,* Van Nostrand Reinhold, New York (1977), p. 112.

138. G.M. Bartenev and V.V. Lavrentev, *Friction and Wear of Polymers,* Elsevier, New York (1981), p. 91.

139. G.M. Bartenev and V.V. Larentev, *Friction and Wear of Polymers,* Elsevier, New York (1981), p. 95.

140. G.M. Bartenev and V.V. Larentev, *Friction and Wear of Polymers,* Elsevier, New York (1981), p. 241.

141. G.M. Bartenev and V.V. Lavrentev, *Friction and Wear of Polymers,* Elsevier, New York (1981), p. 242.

142. G.M. Bartenev and V.V. Lavrentev, *Friction and Wear of Polymers,* Elsevier, New York (1981), p. 243.

143. G.M. Bartenev and V.V. Lavrentev, *Friction and Wear of Polymers,* Elsevier, New York (1981), pp. 269–278.

144. G.M. Bartenev and V.V. Lavrentev, *Friction and Wear of Polymers,* Elsevier, New York (1981), p. 264.

145. G.M. Bartenev and V.V. Lavrentev, *Friction and Wear of Polymers,* Elsevier, New York (1981), p. 279.

146. M.E. Marks and P. Conrad, Mod. Plastics, *23*:6 (1946), p. 165.

147. J.J. Gouza, in J.V. Schmitz, Ed., *Testing of Polymers,* Vol. 2, Interscience, New York (1966), p. 271.

148. J.B. Howard, "Stress Cracking", in E. Baer, Ed., *Engineering Design for Plastics,* Reinhold, New York (1964), Chapter 11.

149. J.B. Howard, "Stress Cracking", in E. Baer, Ed., *Engineering Design for Plastics,* Reinhold, New York (1964), p. 749.

150. S.D. Kohlman, S.P. Petrie, and C.R. Desper, SPE Tech. Papers, *33* (1987), p. 472.

151. S. Turner, *Mechanical Testing of Plastics,* 2nd Ed., George Godwin/PRI, London (1983), p. 158.

152. E.V. Lind, SPE Tech. Papers, *31* (1985), p. 874.

153. S. Turner, *Mechanical Testing of Plastics,* 2nd Ed., George Godwin/PRI, London (1983), p. 164.

154. R.P. Brown, Polymer Testing, *1* (1980), p. 267.

155. R.J. Roe and C. Gieniewski, Polym. Eng. Sci., *15* (1980), p. 421.

156. J.L. O'Toole, "Design Guide", Modern Plastics Encyclopedia, *63*:10A (1986), p. 405.

157. J.B. Howard, "Stress Cracking", in E. Baer, Ed., *Engineering Design for Plastics,* Reinhold, New York (1964), p. 757.

158. L.L. Lander, SPE J., *16* (1960), p. 1329.

159. J.B. Howard, "Stress Cracking", in E. Baer, Ed., *Engineering Design for Plastics*, Reinhold, New York (1964), p. 766.
160. J.D. Uberreiter, "The Solution Process", in D. Crank and D.P. Park, Eds., *Diffusion in Polymers*, Academic Press, New York (1968), Chapter 7.
161. J.L. Throne, *Plastics Process Engineering*, Marcel Dekker, New York (1979), p. 835.
162. R.B. Seymour, *Plastics vs. Corrosives*, Wiley-Interscience, New York (1982), p. 36.
163. J.L. Throne, *Plastics Process Engineering*, Marcel Dekker, New York (1979), p. 32.
164. J.L. Throne, *Plastics Process Engineering*, Marcel Dekker, New York (1979), p. 33.
165. S. Levy and J.H. DuBois, *Plastics Product Design Engineering Handbook*, Van Nostrand Reinhold, New York (1977), pp. 247–248.
166. A. Jurriaanse and B. Zahradnik, SPE Tech. Papers, *32* (1986), p. 58.
167. F. Chevassus and R. deBroutelles, Eds., *The Stabilization of Polyvinyl Chloride*, Arnold Press, London (1963).
168. *Polymer Degradation and Stability*, journal started in 1979.
169. "Photodegradation and Photostabilization of Coatings", ACS Symposium Series 151, American Chemical Society, Washington, DC (1981).
170. A. Davis and D. Sims, *Weathering of Polymers*, Applied Science, London (1983).
171. *Durability of Building Materials*, journal started in 1982.
172. Anon., "Engineering Data Bank", Modern Plastics Encyclopedia, *63*:10A (1986), p. 446.
173. J.L. O'Toole, "Design Guide", Modern Plastics Encyclopedia, *63*:10A (1986), p. 444.
174. M. Lomax, Polym. Testing, *1* (1980), p. 105.
175. M. Lomax, Polym. Testing, *1* (1980), p. 211.
176. K.N. Mathes, "Electrical Properties", in E. Baer, Ed., *Engineering Design for Plastics*, Reinhold, New York (1964), p. 507.
177. J.L. O'Toole, "Design Guide", Modern Plastics Encyclopedia, *63*:10A (1986), p. 434.
178. J.L. O'Toole, "Design Guide", Modern Plastics Encyclopedia, *63*:10A (1986), p. 435.
179. K.N. Mathes, "Electrical Properties", in E. Baer, Ed., *Engineering Design for Plastics*, Reinhold, New York (1964), p. 463.
180. K.N. Mathes, "Electrical Properties", in E. Baer, Ed., *Engineering Design for Plastics*, Reinhold, New York (1964), p. 452.
181. J.L. O'Toole, "Design Guide", Modern Plastics Encyclopedia, *63*:10A (1986), p. 436.
182. K.N. Mathes, "Electrical Properties", in E. Baer, Ed., *Engineering Design for Plastics*, Reinhold, New York (1964), p. 540.
183. J.L. O'Toole, "Design Guide", Modern Plastics Encyclopedia, *63*:10A (1986), p. 436.
184. K.N. Mathes, "Electrical Properties", in E. Baer, Ed., *Engineering Design for Plastics*, Reinhold, New York (1964), p. 562.
185. S. Levy and J.H. DuBois, *Plastics Product Design Engineering Handbook*, Van Nostrand Reinhold, New York (1977), p. 294.
186. K.N. Mathes, "Electrical Properties", in E. Baer, Ed., *Engineering Design for Plastics*, Reinhold, New York (1964), p. 569.
187. S. Levy and J.H. DuBois, *Plastics Product Design Engineering Handbook*, Van Nostrand Reinhold, New York (1977), p. 281.
188. J.L. O'Toole, "Design Guide", Modern Plastics Encyclopedia, *63*:10A (1986), p. 441.
189. L.C. Oberholtzer, SPE Tech. Papers, *30* (1984), p. 709.
190. H.W. Ott, *Noise Reduction Techniques in Electronic Systems*, John Wiley & Sons, New York (1976).
191. Anon., "Engineering Data Bank", Modern Plastics Encyclopedia, *62*:10A (1985), pp. 556–557.
192. S. Levy and J.H. DuBois, *Plastics Product Design Engineering Handbook*, Van Nostrand Reinhold, New York (1977), p. 305.

193. P.J. Cloud and M.A. Wolverton, Plast. Technol., *24*:11 (1978), p. 107.
194. D.B. Judd, "Color Vision and Colorimetry", in E.U. Condon and H. Odishaw, Eds., *Handbook of Physics*, McGraw-Hill Book Co., New York (1958), Chapter 6-4.
195. H. Hemmendinger, SPE Tech. Papers, *29* (1983), p. 293.
196. R.C. Schiek, "Colorants", Modern Plastics Encyclopedia, *62*:10A (1985), p. 108.

Glossary

Asperity   Protrusion from a surface, source of roughness and lack of gloss.

ASTM   American Society for Testing and Materials, an American association, located in Philadelphia, for establishing standard testing and reporting procedures.

Barrer   A unit measure of permeability.

Bructile   Ductile–brittle impact failure characterized by initial yielding, then catastrophic failure.

Capillary Rheometer   An instrument for measuring rheological properties of liquids by measuring pressure drop and flow rate through a well-characterized small-diameter tube.

Causal Relationship   A direct, obvious link between cause and effect.

Charpy Impact   An excess-energy pendulum technique for measuring impact fracture energy on a doubly supported beam.

Cold Flow Limit   A measure of surface hardness.

Complex Modulus   Modulus of a material under time-dependent load application.

Conditioning   Holding a test specimen at given temperature and humidity levels for a specific time.

Cone-and-Plate Rheometer   A rotating plate instrument for measuring rheological properties of liquids.

Constitutive Relationship   A causal relationship between applied environmental condition and material response. For mechanics, the relationship is between stress and deformation and deformation rate. For thermodynamics, the relationship is between pressure and temperature and specific volume.

Core   The center section of material; it may have a morphology different from the rest of the cross-section.

Creep   The characteristic of a material to continually deform under stress.

Creep Modulus   Time-dependent material property, given as the ratio of applied stress to engineering strain.

Creep Rupture Envelope   Locus of creep failure points taken at various strain levels.

Critical Angle   The light angle incident to a planar transparent polymer surface below which the light cannot exit.

Crosshead Speed   The rate at which the grips of a tensile machine are separated.

Dielectric Constant   A material property, the ratio of its capacitance to capacitance in vacuum or air.

Dissipation Factor   A measure of dissipation of electrical energy as heat.

Ductile–Brittle Transition   The region of applied load in which a solid material can no longer absorb energy by yielding and so fails by cracking.

Embrittlement   Loss of material toughness owing to chemical, thermal, ultraviolet or solvent attack.

EMI   Electromagnetic interference.

EMR   Electromagnetic radiation.

End Correction   Correction for entrance effects in capillary rheometer.

ESCR   Environmental stress crack resistance, or the ability to resist environmental attack when highly stressed.

Far Field   In electromagnetic interference, electromagnetic radiation propagation as plane wave.

Frozen-In Stress   Flow-induced polymer orientation in molded articles.

Grip Slip   Specimen sliding from the jaws of tensile test grips.

Hardness   The resistance by one body to penetration or indentation by another.

ISO   The European international standards organization, equivalent to ASTM.

Isochronous   At constant time.

Isometric   At constant strain level.

Izod   Excess-energy pendulum test using a cantilevered test specimen.

Loss Modulus   That portion of the complex modulus that represents dissipative energy.

Material Constant   A constant relationship between two definable physical states or causal relationships.

Material Parameter   A relationship between two definable physical states or causal relationships.

Melt Indexer   A short L/D capillary tube viscometer, also called a plastometer.

Modulus   Ratio of applied stress to resulting strain.

Modulus of Rigidity   Area under the stress–strain curve.

Mooney Hybrid   A test that is neither a standardized procedure nor a model for actual in-use part performance.

Mullen Burst Test   A standardized test for films that measures a material's ability to withstand uniform biaxial loading.

Near Field   In electromagnetic interference, electromagnetic radiation depends on wave impedance and electromagnetic radiation characteristics of source.

Performance Test   A specific standardized procedure, such as an ASTM test.

Permeability   A measure of the steady-state rate of transmission of small molecules through solid materials.

Plasticizer   Small molecules that act to separate polymeric chains, changing flow properties, aiding chain disentanglement and affecting glass transition temperature and mechanical properties such as impact strength and modulus.

Poisson's Ratio   A measure of the material dimensional change in the non-load direction to dimensional change in the load direction.

Rheogoniometer   A device for measuring rheological properties.

Rheometer   A device for measuring rheological properties.

Shear Band   Coordinated dislocation in polymer under uniaxial strain.

Single-Point Test   A test that produces a single material property value, taken at a fixed set of conditions.

Skin   Surface layer on a material; it may have a morphology different from the rest of the cross-section.

Solvent   Molecules that break secondary bonds, resulting in dissolution of polymers.

Solvation   Process of dissolving or isolating polymer chains in solvent solution.

Specimen   A specific defined shape of the material to be tested.

Storage Modulus   That portion of the complex modulus that represents the elastic character of a material.

Stress Relaxation   Characteristic of a material to continue to deform on a microscopic level, thus relieving applied stress.

Swelling   Process of separating polymer chains without dissolution.

Tan δ   Phase angle or dissipation factor.

Test   A protocol by which the response of a specimen to a given environmental condition is determined.

Throwing Energy   Unmeasurable energy used to propel broken impact specimens from test fixture. Also called Toss Factor.

Toss Factor   Unmeasurable energy used to propel broken impact specimens from test fixture. Also called Throwing Energy.

Tribology   Study of lubrication, friction and wear.

Tristimulus   Three-dimensional color coordinates (X, Y, Z).

Tup   The rounded tip of an impact drop-weight.

UL   Underwriters Laboratory, an American testing laboratory for flammability and electrical standards.

Volume Resistivity   A measure of the electrical insulating capability of a material.

Warpage   Differential shrinkage on near-planar molded surface.

Young's Modulus   The slope of the stress–strain curve for a Hookean elastic material.

Zero-Strain Tangent   Modulus obtained from initial slope of stress–strain curve.

HOMEWORK

SEMESTER PROJECT 1

You are working with a group of designers who are preparing a design manual on thermosetting and thermoplastic molding compounds. In particular you have been asked/told to prepare a comprehensive technical report on the physical and environmental properties of _____

1. ABS
2. Polyacetal, POM
3. Acrylic, PMMA
4. Alkyd
5. Cellulose Acetate Butyrate, CAB
7. Polychlorotrifluoroethylene, PCTFE
8. Nylon 6, PA-6
9. Nylon 66, PA-66
10. Nylon 11, PA-11
11. Nylon 12, PA-12
12. Polyamide-imide, PAI
13. Polybutylene Terephthalate, PBT
14. Polyethersulfone, PES
15. Polyethylene Terephthalate, PET
16. Polycarbonate, PC
17. Polychlorotrifluoroethylene, PCTFE
18. Polyetheretherketone, PEEK
19. Polyether-imide, PEI
20. Polyetherketone, PEK
21. Polyethylene, PE
22. Polymethylpentene-1
23. Polyphenylene Sulfide, PPS
24. Polyphenylsulfone
25. Polypropylene, PP
26. Polystyrene, PS
27. Polysulfone, PSO_2
28. Polytetrafluoroethylene, PTFE
29. Polyurethane (linear), PUR
30. Polyvinyl Chloride, PVC

31. Styrene-Acrylonitrile, SAN
32. Modified Polyphenylene Oxide (mPPO)

Start your program by identifying and contacting the commercial resin suppliers of your particular resin. Further technical information can be obtained from texts and references, such as Modern Plastics Encyclopedia, journals, such as Polymer Engineering Science, Journal of Applied Polymer Science, and so on, and conference proceedings, such as SPE ANTECs, SPI Structural Foam Conferences, SAMPE Conference Proceedings, SPE RETECs, and so on. Your report will include a list of primary and secondary environmental concerns, description of the processes used to make the part and design data. The report cannot exceed 6 single-spaced pages. Data can be included in an Appendix.

SEMESTER PROJECT 2

You are working with a non-technical designer in a small plastics processing firm. You are responsible for picking a polymer that can be used to make _____ :

   a. Screwdriver handle
   b. Disposable ballpoint pen
   c. Space shuttle control panel
   d. New York subway seat shell (unpadded)
   e. Carbonated beverage bottle cap
   f. Window lineal
   g. Microwave oven dish
   h. F16 fighter windshield
   i. Low-friction steam iron soleplate
   j. Disposable sunglasses
   k. Sports-car steering wheel
   l. 480 V electrical junction box
   m. Hammer handle
   n. Knitting needle
   o. Disposable cold drink cup
   p. In-vitro breast augmentation sac
   q. Firefighter face-shield
   r. Ashtray
   s. Radome cover
   t. Ski boot sole
   u. Human finger joint
   v. Shipping pallet for coffins
   w. Soft seat covering
   x. Fast-food carryout container
   y. Wristwatch gear
   z. Implant capsule for 1-yr dosing of insulin
  aa. Grain silo cover
  bb. Artificial teak for boat decking

cc.   Button
dd.   Mobile home shutter

Start your work by identifying the environmental conditions the product will see. Determine possible methods for manufacturing the product. Select a set of mechanical properties that are important. Determine the time-dependency of these parameters when compared with the expected product lifetime. Then determine the best materials for the application. Your resources are similar to those of Semester Project 1.

CHAPTER 1

1.1   Determine the atomic configuration—neutrons, protons and electron shells—and atomic description of the hydrogen, nitrogen, oxygen, fluorine, silicon, sulfur and chlorine atoms.

1.2   If a polyvinyl chloride (PVC) molecule has a degree of polymerization of 948, what is its molecular weight?

1.3   If a polystyrene (PS) molecule has a molecular weight of 1,000,000, what is its degree of polymerization? How many of these molecules are in 10 grams of resin?

1.4   A polyacetal (POM) molecule has a molecular weight of 26,000. What is its degree of polymerization? How many molecules are in one gram of resin?

1.5   What is the percentage increase in length of the heptacontane molecule, Example 1.1, if the bond angles are stretched to the maximum, $\theta = 180°$?

1.6   What is the length of an unstretched polyacetal (POM) molecule with a molecular weight of 30,000?

1.7   What is the molecular weight of a fully extended linear polyethylene (PE) molecule 1 mm in length?

1.8   What are the stereoisomers of polyvinylidine chloride (PVDC)?

1.9   Both isotactic and atactic polymers of polypropylene oxide have been prepared by ring scissoring polymerization of propylene oxide:

$$\underset{\diagdown \underset{O}{} \diagup}{HC}\text{---}\underset{}{C}\text{---}CH_3$$

a.   Write the general structural formula for the polymer.
b.   Indicate how the atactic, isotactic and syndiotactic structures differ.

1.10   Find the sequence lengths for polymer A, for random linear copolymers containing 25 and 10 mole percent of monomeric material B. Use values of n of 1, 2, 3, 4, 6, 8, 10, 15, 20, 30, 40 and 50.

1.11   Draw the chemical structure for the following linear polymers and indicate which type of covalent bond would break first.

a.   Polyethylene (PE)
b.   Polyacrylonitrile (PAN)
c.   Polypropylene (PP)
d.   Polycaprolactam, Nylon 6 (PA-6)
e.   Polyacetal (POM)
f.   Polyvinyl chloride (PVC)

What property of the polymer would indicate thermal degradation?

1.12   Why are there no commercial polymers having an O-O covalent bond?

1.13   Why are some unfilled glassy polymers stiffer at room temperature than some unfilled crystalline polymers?

CHAPTER 2

2.1   Determine the number average, weight average, z average molecular weights and the polydispersity index, PI, for the following distributions:

Number of Monomer Units

Sample	10	20	30	40	50	60	70	80	90
a	1	2	3	4	5	6	7	8	9
b	9	8	7	6	5	4	3	2	1
c	10	20	30	20	10	5	4	3	2

2.2   A mixture is prepared from two essentially monodisperse fractions of linear polymers of molecular weights of 10,000 and 40,000, denoted as x and y respectively. If the mixture contains two parts by weight of x and three parts by weight of y, find the M_n, M_w, M_z and PI for the mixture?

2.3   Intrinsic viscosity was measured by a Ubbelohde viscometer for seven essentially monodisperse polystyrene polymer fractions in toluene at 30°C. Using the data below, develop a calibration curve, Figure 2.5, and fit the data with a regression analysis to obtain K and a.

M_w	$[\eta]$
76,000	0.0382
135,000	0.0592
163,000	0.0696
336,000	0.1054
440,000	0.1292
556,000	0.165
850,000	0.221

2.4   The following data were obtained for polyisobutylene in cyclohexane at 30°C.

Run	Concentration (g/dl)	Time (s)
1	0.00	87.3
2	1.00	157.0
3	0.50	120.0
4	0.25	103.2
5	0.125	95.1

Determine the intrinsic viscosity and the solution average molecular weight.

2.5   If the density of an individual crystal of polyethylene (PE) is 0.983 g/cm^3 and the density of totally amorphous polyethylene (PE) is 0.869 g/cm^3, determine the fraction of crystallinity of a polyethylene (PE) sample having a density of 0.950 g/cm^3 at the same temperature,

     a.   based on mass fraction, and

     b.   based on volume fraction

2.6   When a polymer exits an extruder die, it is completely transparent. During the cooling process the polymer exhibits a milky translucency. What type of polymer is this? Justify your answer.

2.7   Two polymers that have crystalline tendencies are copolymerized at a 70/30 ratio. The resultant copolymer does not exhibit the crystalline tendency. How do you explain this phenomenon?

2.8   Determine the sensitivity of the Avrami equation by plotting on semi-log paper the ratio ϕ/ϕ_m as a function of time for crystal growth in:

    a.   rod form,

    b.   platelet form, and

    c.   spherulitic form.

Assume the rate constant, A, is 10.

2.9   Predict the Avrami crystal growth rate of spherulites for numeric values of the rate constant of 1, 5, and 10.

2.10   Predict the time-dependent crystallinity for poly-4-methylpentene-1 at 140°C. Compare these results to the crystallization process at the temperature of the shortest induction time. Assume the maximum crystallinity in both cases is 0.97.

2.11   Estimate how long a 5 mm thick polyacetal (POM) injection molded tensile specimen should be annealed and at what temperature. Explain your procedure in detail.

2.12   Plot crystalline melting temperature versus glass transition temperature for at least ten of the polymers discussed in this text. What can you conclude from this plot?

2.13   A plasticizer in polyvinyl chloride (PVC) softens the polymer and reduces its glass transition temperature. Explain why.

2.14   The glass transition temperature is also evidenced by a change in slope of the index of refraction. Determine T_g from the following data:

Temperature (°C)	n
20	1.5913
30	1.5898
40	1.5883
50	1.5868
60	1.5853
70	1.5838
80	1.5822
90	1.5801
100	1.5766
110	1.5725
120	1.5684
130	1.5643

2.15   The glass transition temperature of a branched polymer is given by:

$$T_g = T_{g,\infty} - \frac{K\,x}{M_n}$$

where x is the number of ends per chain. A linear polymer with a number average molecular weight of 2200 has a T_g of 54°C. The T_g was increased to 78°C by increasing the number average molecular weight of 8,000. A branched version of the same polymer with a molecular weight of 4,800 has a T_g of 47°C. What is the average number of branches on this sample? List your assumptions.

2.16 Polyethylene terephthalate (PET) is cooled rapidly from 300°C (state 1) to room temperature (state 2). The resulting material is rigid and perfectly transparent. The sample is then heated to 170°C and maintained at this temperature for a short time during which the polymer contracts and becomes translucent (state 3). It is then cooled down to room temperature and is again found to be rigid, but is now translucent rather than transparent (state 4). For this polymer, $T_m = 267°C$ and $T_g = 69°C$.

    a. Sketch a general specific volume versus temperature curve for a crystallizable polymer, illustrating T and T_m.

    b. Show the locations of states 1 through 4 on the graph.

2.17 Limestone at 25% by volume is compounded into polyvinyl chloride (PVC), k = 3.9×10^{-4} cal/s cm². Estimate the thermal conductivity of the compound at 30°C. Assume the particles are "mixed sizes" with a maximum packing fraction, P, of 0.7. Does the thermal conductivity predicted by Nielsen model agree with the value predicted by the law of mixtures?

$$k = k_p V_p + k_f V_f$$

2.18 Estimate the specific heat of the polyvinyl chloride (PVC) compound in HP 2.17 and the energy necessary to heat a 0.1 cm thick sheet from room temperature, 20°C, to 150°C.

2.19 Estimate the modulus at 20°C of polyacetal (POM) compound with 30% (vol) short glass fibers. Assume the glass fiber is an E glass with an aspect ratio of 10. Use the Nielsen equation to determine the proper value.

2.20 Estimate the modulus at 20°C of nylon 6 (PA-6) compound with 20% (vol) E glass spheres.

CHAPTER 3

3.1 Give the formula for determining the shear stress in a plate, Figure HW 3.1, that is subjected to a circular rod of diameter D pressing on the plate. What is the equation for determining the stress the plate exerts on the bottom of the rod?

3.2 A uniform rod with a square cross sectional area, A_c, is hung vertically with bottom end free and the other end fixed, Figure HW 3.2. If the rod weighs w [kg/m³], determine the stress in the rod as a function of the distance from the fixed end. What are the maximum and minimum stresses and their respective locations?

3.3 Find the maximum strain, minimum strain and total elongation of the bar in problem 3.2.

3.4 Determine the elongation of a conical bar under its own weight, Figure HW 3.3, if the weight per unit volume is w lb$_m$/ft³.

3.5 Show that the shear stress in a bolt, Figure HW 3.4, is:

$$\tau = \frac{2F}{\pi D^2}$$

Find the maximum shear stress in a 0.5 cm diameter bolt carrying a load of 500 N.

3.6 A 1 mm thick Hookean sheet (E = 20 GPa, μ = 0.3) is biaxially loaded in tension with $\sigma_x = 250$ MPa and $\sigma_y = 100$ MPa. What is the new thickness of the sheet?

3.7 What would be the sheet thickness in problem 3.6 if a compressive stress of 150 MPa were applied to both surfaces of the sheet?

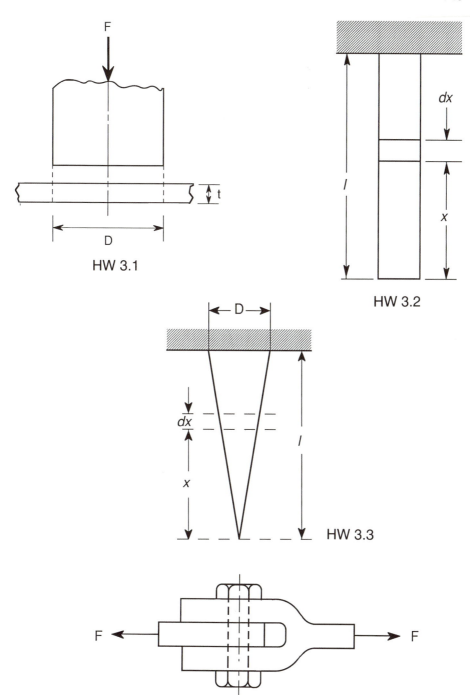

HW 3.1

HW 3.2

HW 3.3

HW 3.4

3.8 A 2 in cube of a Hookean material ($E = 200,000$ lb$_f$/in^2, $\mu = 0.35$) is submerged in the ocean. It is then compressively loaded axially with a force of 8,000 lb until the water pressure exerts a pressure of 500 lb$_f$/in^2 absolute on all surfaces. What are the axial and transverse strains?

3.9 Plot engineering strain versus true strain on two cycle log-log paper. Indicate on the graph the points where the engineering strain deviates 1%, 5%, 10% and 20% from the true strain.

3.10 A 0.5 in thick block of Hookean material, having a shear modulus of 200,000 lb$_f$/in^2, is subjected to a shear stress of 2,000 lb$_f$/in^2 on top surface.

    a. What is the shear strain in the material?

    b. What is the displacement of the top surface of the material relative to the bottom surface?

    c. What happens to the top surface when the stress is removed?

3.11 Determine the increase or decrease in diameter of a 1 in diameter steel bar if a compressive stress of 20,000 lb$_f$/in^2 is applied axially. Assume Poisson's ratio, $\mu = 0.3$, and $E = 30 \times 10^6$ lb$_f$/in^2.

3.12 Show that for very small strains, the increase per unit volume caused by strains ε_x, ε_y, and ε_z is $[\varepsilon_x + \varepsilon_y + \varepsilon_z]$.

3.13 A linear viscous fluid is continuously sheared between two infinitely long concentric cylinders with an annular gap of 0.100 cm. The outer cylinder is stationary while the inner cylinder, diameter $= 8$ cm, is rotated at 30 rev/min (RPM). The torque measured of the outer cylinder is 800 N m per meter of cylinder. What is the shear rate in the annular gap and what is the viscosity of the fluid?

3.14 A fluid subjected to a stress of 100 KPa at a given instant is being sheared at 10 s^{-1}. What is the instantaneous power in W/m^3?

3.15 A fluid is being sheared in a manner such that the power input to the fluid is given by:

$$P = 300 \, t + 80 \, t^2 \text{ [W/m}^3]$$

What is the total work done on the fluid in 20 seconds?

3.16 Derive equations 3.55 and 3.58.

3.17 The stress relaxation of a gasket can be approximated by a Maxwell element. The gasket material has a short time modulus of 2,000 lb$_f$/in^2 and a relaxation time of 300 days. If the initial strain applied to the gasket is 30% and the pressure inside the vessel is 100 lb$_f$/in^2, how long will it take before the seal leaks?

3.18 Using the 2,008 lb$_f$/in^2 tensile creep data of cellulose acetate at 25°C, Figure 3.18, find the numeric value of the springs and dashpots in the four parameter model. Assume that the initial strain at time equal to zero is 0.05%.

3.19 A four element model with:

$$E_1 = 2 \times 10^9 \text{ dynes/cm}^2$$

$$E_2 = 2.5 \times 10^9 \text{ dynes/cm}^2$$

$$\eta_1 = 5 \times 10^{11} \text{ poise}$$

$$\eta_2 = 5 \times 10^{10} \text{ poise}$$

is subjected to a uniform stress, $\sigma_a = 10^9$ dynes/cm^2 for 100 seconds after which the stress is removed. Plot the response, total deformation and apparent modulus, for 150 seconds.

3.20 Using the Boltzmann superposition principle for a material with a time dependent modulus given by:

$$E(t) = \frac{E_c \eta}{\eta + E_e t)}$$

what is the deformation as a function of time if the stress history is given by:

a) $-\infty < t < 0$ $\qquad$ $\sigma(t) = 0$
$0 < t < t_1$ $\qquad$ $\sigma(t) = K_1 t$

b) $-\infty < t < 0$ $\qquad$ $\sigma(t) = 0$
$0 < t < t'$ $\qquad$ $\sigma(t) = K_1 t$
$t' < t < t_2$ $\qquad$ $\sigma(t) = K_1 t' - K_2 (t - t')$

3.21 Graph the response of a two element Maxwell-Weichert model on log-log paper with:

$$E_1 = 3 \times 10^3 \text{ lb}_f/\text{in}^2$$

$$\tau_1 = 1 \text{ second}$$

$$E_2 = 5 \times 10^6 \text{ lb}_f/\text{in}^2$$

$$\tau_2 = 100 \text{ seconds}$$

3.22 A three element Maxwell-Weichert model can be used to describe a crosslinked polymer using the following parameters:

$$E_1 = E_2 = 10 \times 10^5 \text{ Pa}$$
$$E_3 = 1.5 \times 10^5 \text{ Pa}$$
$$\lambda_1 = 10 \text{ s}$$
$$\lambda_2 = 100 \text{ s}$$
$$\lambda_3 = \infty$$

The polymer is elongated 50% and maintained at this deformation.
    a.   What is the initial applied stress $(t = 0)$?
    b.   What is the stress after 50 seconds?
    c.   What is the stress as time approaches infinity?
3.23 A particular grade of polymethyl methacrylate (PMMA), $T_g = 105°C$, has a 6 hour modulus of 2,000 lb$_f$/in^2 at 135°C. How long will it take to reach this modulus at 150°C?
3.24 Develop the master curve for polycarbonate from the following corrected modulus data, using 150.8°C as the reference temperature:

T(°C)	Log t	Log E(t)	T(°C)	Log t	Log E(t)
130	2.00	10.25	150.8	2.00	7.97
	2.25	10.25		2.25	7.85
	2.50	10.25		2.50	7.78
	2.75	10.25		2.75	7.72
	3.00	10.22		3.00	7.69
	3.25	10.19		3.25	7.64
	3.50	10.16		3.50	7.59
	3.75	10.13		3.75	7.55
	4.00	10.06		4.00	7.50
	4.25	10.03		4.25	7.44
				4.50	7.34
141	2.00	10.19		4.75	7.25
	2.25	10.19			
	2.50	10.16	156	2.00	7.59
	2.75	10.13		2.25	7.55
	3.00	10.09		2.50	7.50
	3.25	10.03		2.75	7.41
	3.50	9.94		3.00	7.34
	3.75	9.82		3.25	7.25
	4.00	9.66		3.50	7.19
	4.25	9.50		3.75	7.06
				4.00	6.94
				4.25	6.78
				4.50	6.59
142	2.00	10.13		4.75	6.38
	2.25	10.09		5.00	6.13
	2.50	10.03			
	2.75	9.97	159	2.25	7.34
	3.00	9.88			
	3.25	9:67			
	3.50	9.63		2.50	7.25
	3.75	9.47		2.75	7.16
	4.00	9.22		3.00	7.03
				3.25	6.94
				3.50	6.78
				3.75	6.63
				4.00	6.41
144.9	2.25	9.66		4.25	6.13
	2.50	9.50		4.50	5.85
	2.75	9.28		4.75	5.53
	3.00	9.03			
	3.25	8.88	161.5	2.25	7.06
	3.50	8.66		2.50	6.97
				2.75	6.85
144.6	2.75	9.03		3.00	6.64
	3.00	8.72		3.25	6.50
	3.25	8.50		3.50	6.31
	3.50	8.25		3.75	6.06
	3.75	8.09		4.00	5.78
	4.00	7.94		4.25	5.47

T(°C)	Log t	Log E(t)	T(°C)	Log t	Log E(t)
146	2.50	8.72	167	2.25	6.72
	2.75	8.47		2.50	6.50
	3.00	8.25		2.75	6.25
	3.25	8.06		3.00	5.97
	3.50	7.94		3.25	5.63
	3.75	7.82		3.50	5.22
	4.00	7.67			
			171	2.25	6.00
				2.50	5.72
				2.75	5.41
				3.00	5.03
				3.25	4.63

CHAPTER 4

4.1 The shear rate-shear stress data for PS (T_g = 95°C) is tabulated below:

Shear rate (s^{-1})	Shear Stress (Pa) 288°C	260°C	232°C	205°C
0.1	1.2×10^1	4.1×10^1	1.4×10^2	5.0×10^2
0.316	3.8×10^1	1.3×10^2	4.4×10^2	1.5×10^3
1.0	1.1×10^2	4.0×10^2	1.4×10^3	4.0×10^3
3.16	3.8×10^2	1.2×10^3	3.7×10^3	1.1×10^4
10.0	1.1×10^3	3.2×10^3	9.0×10^3	2.5×10^4
31.6	3.2×10^3	8.0×10^3	1.8×10^4	3.8×10^4
100.	7.0×10^3	1.6×10^4	3.2×10^4	6.0×10^4
316.	1.4×10^4	2.9×10^4	5.2×10^4	8.6×10^4
1,000.	2.6×10^4	5.0×10^4	8.0×10^4	1.25×10^4
3,160.	5.0×10^4	9.5×10^4	1.3×10^5	2.0×10^4
10,000.	9.2×10^4	1.6×10^5	2.0×10^5	—

   a. Plot shear stress against shear rate on 5 cycle log-log graph paper.
   b. Calculate viscosity as a function of shear rate and temperature. Plot the results on 5 cycle log-log graph paper.
   c. Superimpose lines of constant shear stress (10^2, 10^3, 10^4, 10^5 and 10^6 Pa) on the viscosity shear rate graph.

4.2 Based upon the graphical results of problem 4.1, when can polystyrene (PS) be modeled as a Newtonian fluid?

4.3 For what range of shear rates can the PS of problem 4.1 be modeled as a power-law fluid? Using a least squares method determine K and n for each temperature.

4.4 Determine the temperature-dependent Newtonian zero shear viscosity relationship for PS of problem 4.1 using the following relationships:
   a. $\eta = A_1 e^{bT}$
   b. $\eta = A_2 e^{E/RT}$
   c. $\eta = A_3 e^{(E/R)(1/T_0 - 1/T)}$

4.5   Determine the constant stress activation energy, $E(\tau)$, at 10^3, 10^4, and 10^5 Pa.

4.6   Determine the constant shear rate activation energy, $E(\dot{\gamma})$, for shear rates of 1, 10, 100 and 1,000 s^{-1}.

4.7   A 1 cm diameter plastic sphere ($\rho = 1.2$ g/cm^3) is dropped into a graduated cylinder filled with a Newtonian fluid ($\rho_o = 1.01$ g/cm^3). According to Stokes' law, the terminal velocity of the sphere U is given by:

$$\frac{2R^2 d(\rho - \rho_o)}{9} = \eta\,U,$$

provided the Reynolds number, $Re = U\rho_o D/\eta$ is less than 0.1. If the terminal velocity of the sphere is 0.5 cm/s, what is the viscosity of the fluid?

4.8   A pipe, 3 ft long and ½ in inside diameter, isothermally transfers 750 lb/h of molten polystyrene (PS), characterized in problem 4.1, at 260°C. Estimate the pressure drop down the pipe if the PS enters the pipe at 3,000 lb$_f$/in^2. Then find the average residence time of the plastic passing through the pipe. See equations in Table 4.2.

4.9   Estimate the volume flow rate of polystyrene (PS), characterized in problem 4.1, at 288°C through the pipe described in the preceding problem if the pressure drop is 50 lb$_f$/in^2.

4.10   A 10 meter long, 1 cm internal diameter vertical pipe is filled with a liquid, $\rho = 0.980$ g/cm^3. The liquid viscosity is described by the Ellis model with $\tau_c = 400$ Pa, m = 0.68 and $\eta_o = 20$ Pa s. What is the instantaneous rate of draining of the pipe at a height of fluid is 2 meters? If the pipe was initially filled, how long will it take to drain the pipe?

4.11   What diameter pipe is neccessary to transfer 2.5 cm^3/s of polystyrene (PS), characterized in problem 4.1, at 205°C through a 25 cm long pipe if the pressure drop is 3.5 MPa?

4.12   One three constant equation used to curve fit rheological data is:

$$\dot{\gamma} = A\,\tau + B\,\tau^2 + C\,\tau^3$$

What is the zero shear viscosity, η_o?

4.13   Plot the coefficient of thermal expansion (κ) from 1 atmosphere to 4,000 lb$_f$/in^2 absolute at 150, 200 and 250°C, for each of the following polymers. Use the Spencer–Gilmore equation.

    a.   Polystyrene (PS)
    b.   Ethyl Cellulose
    c.   Polycarbonate (PC)
    d.   Cellulose Acetate Butyrate (CAB)
    e.   Polymethyl Methacrylate (PMMA)

4.14   Plot the coefficient of bulk compressibility (β) for each of the following polymers from 1 atmosphere to 4,000 lb$_f$/in^2 absolute at 100, 200 and 300°C. Use the Spencer–Gilmore equation.

    a.   Polystyrene (PS)
    b.   Ethyl Cellulose
    c.   Polycarbonate (PC)
    d.   Cellulose Acetate Butyrate (CAB)
    e.   Polymethyl Methacrylate (PMMA)

4.15   Calculate the change in enthalpy for polystyrene (PS) from 20°C, 14.7 lb$_f$/in^2 absolute to 190°C, 2,000 lb$_f$/in^2 absolute, using the Spencer–Gilmore equation. Compare your results with values obtained from Appendix E.

4.16   Find the work to compress polystyrene (PS) in a reversible isothermal compression at 140°C from 1 to 1,000 lb$_f$/in^2 absolute. Use the Spencer–Gilmore equation.

4.17   Find the temperature of polystyrene (PS) that is compressed in a reversible adiabatic process from 1 atm to 1,000 lb_f/in^2 at 140°C. Use the Spencer–Gilmore equation and compare results using data from Appendix D.

4.18 A power-law fluid flows through a slot of width, W, height, H, and length L. Derive the following:

    a. Shear stress as a function of y and $\Delta P/L$.
    b. Velocity as a function of U_{max}, y, Y and z.
    c. Shear rate as a function of volumetric flow rate per unit width of slot.

CHAPTER 5

5.1 In blown film forming, exothermic heat of crystallization and heat removal act in opposite ways to affect the film temperature. Consider the short time when the film can be considered isothermal. If the film is considered thermally thin, show that the momentary rate of crystallization is given as:

$$\frac{dX}{dt} = \frac{hA\,(T-T_a)}{\rho V \Delta H}$$

where X is the fraction of crystallinity, h is the heat transfer coefficient, A is the film unit surface area, T is the isothermal film temperature, T_a is the ambient air temperature, ρ is the film density, V is the film unit volume, $V = At'$, where t' is the film thickness, and ΔH is the heat of crystallization.

Integrate this equation and compare your result with the isothermal Avrami equation, 2.26.

5.2 Using the information in problem 5.1, consider the following isothermal conditions:

$$t' = 0.002 \text{ inch}$$
$$\rho = 0.9 \text{ g/cm}^3$$
$$\Delta H = 30 \text{ cal/g}$$
$$h = 2 \text{ Btu/ft}^2 \text{ h °F}$$
$$T = 130°C \text{ (isothermal)}$$
$$T_a = 30°C$$

If the crystallinity at the onset of the isothermal time is zero, what is it after 2 seconds?

5.3 In wire coating, a solid metal wire is pulled through a die and polymer is pressure-extruded around it, as seen in Figure HW 5.1. The annular gap between the wire and the die wall is usually approximated as a slot of height H and width, $W = \pi D_{avg}$, where D_{avg} is the geometric mean diameter of the wire and the die opening. The pressure drop down the wire in the z-direction is related to the rate of change of the shear stress across the slot in the y-direction, as:

$$\frac{\partial P}{\partial z} = \frac{\partial \tau_{yz}}{\partial y}$$

    a. Discuss why this implies that the pressure drop is linear along the wire, that is, dP/dz = constant.
    b. If the fluid is Newtonian, that is:

$$\tau_{yz} = \eta \, dU/dy$$

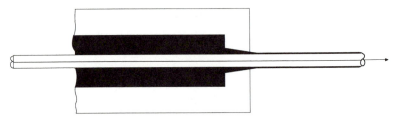

HW 5.1

show that:

$$\frac{u}{U} = \frac{y}{H} - \left(\frac{3y}{H}\right)\left(1 - \frac{y}{H}\right)\left[\frac{H^3}{6\eta U}\right]\left(\frac{dP}{dz}\right)$$

where U is the velocity of the wire.

      c.   By integrating the velocity across the slot, show that the pressure drop-volumetric flow rate expression is given as:

$$Q = \frac{UHW}{2} + \left(\frac{WH^3}{12\eta}\right)\left(\frac{\Delta P}{L}\right)$$

    5.4   In problem 5.3, what is the significance of each term on the right side of the pressure drop-flow rate equation? Specifically, what does it mean when $\Delta P = 0$? What does it mean when $U = 0$?

    5.5   Consider the wire-coating die of problem 5.3. Determine the extruder output for the following Newtonian conditions:

$$\eta = 100 \text{ Pa·s}$$
$$\Delta P = 3 \text{ MPa}$$
$$L = 5 \text{ cm}$$
$$H = 0.1 \text{ cm}$$
$$U = 10 \text{ cm/s}$$
$$D_{avg} = 1 \text{ cm}$$

    5.6   For problem 5.3, develop an expression relating the thickness of the polymer coating on the wire to the process, material and geometric variables.

    Hint: The volume of the polymer on the wire is given as:

$$Q = \pi Uh (D_w + h)$$

where D_w is the wire diameter and h is the polymer thickness. Equate this Q to that in problem 5.3.

    5.7   Given the problem of problem 5.5 and the equation in problem 5.6, determine the polymer coating thickness. What portion of the thickness is controlled by pressure flow?

    5.8   In a single screw extruder, the polymer flow due to drag of the turning screw against the fixed barrel is given as:

$$Q_d = \left(\frac{\pi DNBH}{2}\right) \cos \phi$$

where D is the screw diameter, N is the turning speed, B is the channel width, H is the channel depth, and ϕ is the angle of the flights to the axis of the screw. The flow due to pressure is given as:

$$Q_p = -\left(\frac{BH^3}{12\eta}\right)\left(\frac{\Delta P}{L}\right) \sin \phi$$

where η is the Newtonian viscosity and $\Delta P/L$ is the pressure drop.

Determine the pressure generated when the drag flow equals the pressure flow. What is the implication of this?

5.9   For the conditions of problem 5.8, what is the drag flow rate when $\Delta P = 2{,}000$ lb$_f$/in^2?

$$\eta = 100 \text{ Pa·s}$$
$$D = 5 \text{ cm}$$
$$L = 75 \text{ cm}$$
$$N = 60 \text{ RPM}$$
$$H = 0.2 \text{ cm}$$
$$\phi = 17.42°$$
$$B = 5 \text{ cm}$$

5.10   For the conditions of problems 5.8 and 5.9, what is the pressure flow rate?

5.11   In problem 5.8, what would the pressure be if the drag flow exactly balanced the pressure flow? Use the data of problem 5.9. Can this condition be achieved?

5.12   Determine the amount of heat generated in a single-screw extruder operating at maximum solids conveying angle, $\theta = 30°$.

$$L = 75 \text{ cm}$$
$$W = 5 \text{ cm}$$
$$V_b = 10 \text{ cm/s}$$
$$f_f = 1.0 \text{ (coefficient of friction)}$$
$$\Delta P = 2{,}000 \text{ lb}_f/\text{in}^2$$

5.13   What is the amount of heat generated for the conditions of problem 5.13. if the solids conveying angle, $\theta = 10°$? Explain the effect of conveying angle on heat generation.

5.14   Thermoforming is the process of stretching a heated sheet of plastic into or onto a solid form, called a mold, then holding the stretched sheet there until it rigidifies. Consider stretching a sheet into a conical mold having a diameter D, a height H, and a side having an angle α with the horizontal. Outline a method for determining the sheet thickness as a function of these geometric parameters.

5.15 The thermoforming sheet thickness ratio in the conical mold of problem 5.14 is given as:

$$\frac{h}{h_o} = \left(\frac{1 + \cos \alpha}{2}\right)\left[1 - \frac{\alpha}{H}\right]^{(1/\cos \alpha - 1)}$$

where x is the vertical distance from the top, $0 \leq x \leq H$.

What is the thickness ratio at half-height for an $\alpha = 45°$ cone?

5.16 In problem 5.15, what is the reason that $h/h_o = (1 + \cos \alpha)/2$ when $x = 0$?

5.17 Determine the value of the angle, α, if the cone of problem 5.14 is to have a thickness that decreases linearly with height.

5.18 The areal draw ratio of a thermoformed part is the ratio of the area of the formed part to that of the unformed sheet needed to produce it. What is the draw ratio of the cone in problem 5.14?

5.19 What is the areal draw ratio of a hemisphere?

5.20 One of the critical design areas in injection molding is flow through runners and gates. Consider a simple runner-gate system feeding a very large cavity, Figure HW 5.2. Show for a Newtonian fluid that the isothermal pressure drop-flow rate relationship between points 1 and 2 is:

$$Q = \left(\frac{K_G K_R}{K_{GR}}\right)(P_1 - P_2)$$

where:

$$K_G = \pi R_G 4/(8\eta L_G)$$

$$K_R = \pi R_R 4/(8\eta K_R)$$

$$K_{GR} = K_G + K_R$$

5.21 For problem 5.20, determine the pressure drop through the gate for:

$$R_G = 0.025 \text{ in}$$

$$L_G = 0.1 \text{ in}$$

$$\eta = 100 \text{ Pa·s}$$

$$R_R = 0.1 \text{ in}$$

$$L_R = 1 \text{ in}$$

$$Q = 1.0 \text{ cm}^3/\text{s}$$

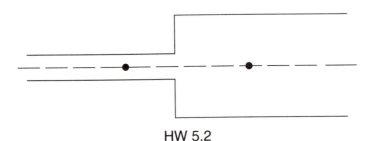

HW 5.2

5.22 For problems 5.20 and 5.21, determine the percent of combined runner-gate pressure drop that occurs through the gate.

5.23 In some texts, the power-law viscosity is related to shear stress in a three-constant equation:

$$\eta = \eta_o \, (\dot{\gamma}/\dot{\gamma}_o)^{(n-1)}$$

Determine the relationship between η_o and $\dot{\gamma}_o$ and m from:

$$\eta = m \, (\dot{\gamma})^{(n-1)}$$

5.24 Determine the pressure drop-flow rate relationship for the isothermal power-law fluid flowing through the combined runner-gate system of Figure HW 5.2.

5.25 It has been suggested that resistance to isothermal polymer flow through a gate increases with increasing shear rate. One empirical way of incorporating this in standard pressure drop-flow rate equations is to use an effective length:

$$L_e = L_\infty \, [1 - e^{-4bQ/\pi R_G^3}]$$

where b is a gate parameter and L_∞ is the effective gate length at infinite shear rate. Using this, obtain a pressure drop-flow rate relationship for the runner-gate system of problem 5.20.

5.26 Determine the pressure drop through the gate for the data of problem 5.21 if b = 0.0002 s and L_∞ = 0.2 in.

5.27 Consider replacing the round runner in problem 5.20 with a square one, d units on a side. If the pressure drop-flow rate values were to remain constant, what would the runner dimensions be, in terms of R_R?

5.28 As in problem 5.27, the round runner is replaced with a half-round runner with a radius R*. What is R*, in terms of R_R?

5.29 A mineral-filled alkyd is being compression molded into electrical box covers, 6 inches in diameter and 0.12 inch thick. If the compression press closing rate is constant at 4 in/s, determine the applied force when the charge thickness is 0.200 in. For this compound, m = 100,00 N s^n/m^2 and n = 0.5.

5.30 If the pressure force in problem 5.29 reaches a maximum at 25 kN, how long does it take to close the press from h = 0.200 in to h = 0.120 in?

5.31 What is the closing rate at the instant h = 0.120 in?

5.32 A CAE software manufacturer states that even a highly viscoelastic polymer such as a thermoplastic elastomer can be modeled as a Newtonian fluid, provided the local viscosity values are applied at each nodal value in the finite element mesh. Do you agree? Justify your answer.

CHAPTER 6

6.1 The effect of pressure on shear viscosity is typically correlated by:

$$\eta = \eta_o \, e^{\beta P}$$

Using the data in Figure 6.6, estimate the value of η_o and β at a shear rate of 10 s^{-1}.

6.2 Assuming the entrance pressure drop in a melt indexer is 40% of the overall pressure drop, estimate the wall shear stress in the capillary when testing polyethylene (PE).

6.3 Develop an empirical correlation between MI and shear viscosity for the high density polyethylene (HDPE) of Figure 6.11 at a shear rate of 100 s^{-1}.

6.4 Derive an expression for the percent change in molecular weight of polybutylene terephthalate (PBT) at 260°C and 0.08% moisture (Figure 6.13) as a function of time.

6.5 Determine the activation energies for high density polyethylene (HDPE) and polystyrene (PS) at 1000 s^{-1} using the curves in Figure 6.14.

6.6 Calculate the modulus of toughness for the steel specimen, Fig. 6.20, based upon both the true and engineering strains.

6.7 Determine the tangent modulus for polymethyl methacrylate (PMMA), Figure 6.23, and compare the results with the value tabulated in Appendix C.

6.8 Determine the 5% secant modulus for high density polyethylene (HDPE), polyurethane (PUR) and polymethyl methacrylate (PMMA) from the stress-strain data shown in Figure 6.24.

6.9 Plot temperature dependent failure strain for polymethyl methacrylate (PMMA) (Figure 6.28) and polystyrene (PS) (Figure 6.29). Develop empirical correlations for the two sets of data.

6.10 Plot the temperature (Y axis) for the amorphous homopolymers shown in Figure 6.49 where the modulus of the polymer is 1% of the modulus at 0°C versus the glass transition temperature of the polymer (X axis). What is the significance of the plot?

6.11 Replot the data of Figure 6.49 where the Y axis is G/G_0 and the X axis is T/T_g. What are your conclusions?

6.12 Plot the temperature (Y axis) for the crystalline polymers shown in Figure 6.48 where the modulus of the polymer is 50% of the modulus at 0°C against the glass transition temperature of the polymer (X axis). Repeat the process, but plot the 50% modulus ratio against the glass transition temperature. What is the significance of the two plots?

6.13 Using the data of Figure 6.55, do the following:

    a. For the amorphous polymers, polystyrene (PS) and polycarbonate (PC), replace the Y axis with DTUL T/T_g.

    b. For the crystalline tendency polymers, nylon 6 (PA-6) and polypropylene (PP), replace the Y axis with DTUL T/T_g.

    c. For the crystalline tendency polymers, nylon 6 (PA-6) and polypropylene (PP), replace the Y axis with DTUL T/T_m.

6.14 How can the DTUL test be modified to make relative comparisons between different plastics?

6.15 Replot the data of Figure 6.73 showing impact energy as a function of test temperature at 6%, 10% and 50% failure levels. What are your conclusions?

6.16 Replot the data of Figure 6.81 showing impact energy as a function of test temperature at 10% and 50% failure levels. What are your conclusions?

6.17 Develop an empirical correlation for the F_{50} impact energy (Figure 6.79) for each of the following polymers:

    a. Normal impact ABS
    b. Medium impact polystyrene (MIPS)
    c. High density polyethylene (HDPE)

What conclusions can you draw? Be specific and detailed.

6.18 From Figure 6.106, derive an empirical relationship for the time-dependent creep rupture stress for each of the following polymers:

    a. Polycarbonate (PC)
    b. Polyacetal (POM)
    c. High density polyethylene (HDPE)
    d. Low density polyethylene (LDPE)

Use the following form of equation:

$$\sigma / \sigma_0 = t^\alpha$$

What is the significance of σ_0 and α?
Hint: Start by taking the natural log of the proposed equation.

6.19   Develop an empirical correlation for the stress relaxation of polytetrafluoroethylene (PTFE), Figure 6.124.

6.20   Develop an empirical correlation for the effect of gasket thickness on stress relaxation for polytetrafluoroethylene (PTFE) gaskets less than 20 mils (0.020 in) thick, Figure 6.124.

6.21   Develop an empirical correlation for the creep modulus of polystyrene-acrylonitrile (PAN) (Dow Tyril 867), Figure 6.126.

6.22   Develop an empirical correlation for the creep modulus of polyacetal (POM) (DuPont Delrin 570), Figure 6.127.

6.23   Develop an empirical correlation for a set of isochronous creep curves. Using the data in Figure 6.128 evaluate the arbitrary constants of the proposed correlation.

6.24   A 1.00 in inside diameter thin wall polypropylene copolymer (PP) (Figures 6.107 and 6.108) pipe is subjected to an internal pressure of 150 lb_f/in^2. The service life of the pipe is 10,000 hours with a maximum strain of 2%. What should be the wall thickness, based on the criteria above?

6.25   It is proposed to fabricate a 0.9 in outside diameter tube with a 50 mil (0.050 in) wall from polycarbonate (PC). The tube will convey an inert gas at 250 lb_f/in^2 and 80°C and be used to optically observe the gas stream. The properties of the PC are given in Figure 1.118. Evaluate the design and make recommendations.

6.26   It is proposed to fabricate a polypropylene (PP) sphere 5 ft in diameter. The sphere is to hold a gas at 75 lb_f/in^2 absolute with a service life of 10,000 hours. What is the minimum wall thickness you would recommend based on a factor of safety of 2? Use the data in Figure 6.106 to estimate rupture strength. The stresses in a sphere are the same in all directions and are given by:

$$\sigma = \frac{pD}{4t}$$

What is the weight of the sphere? Repeat the design for high density polyethylene (HDPE).

6.27   A cylindrical plastic tank 3 ft in diameter and 6 ft long is to be fabricated from polyacetal (POM) (Delrin 570) (Figure 6.127). The design specifications require a design life of 20,000 hours, an internal pressure of 150 lb_f/in^2 and a maximum strain of 2.5%. Estimate the required wall thickness. The hoop and axial strains and stresses are given by:

Axial:

$$\varepsilon_x = \frac{PR(1-2\mu)}{2tE}$$

$$\sigma_x = \frac{PR}{2t}$$

Hoop:

$$\varepsilon_\theta = \frac{PR(2-\mu)}{2tE}$$

$$\sigma_\theta = \frac{PR}{t}$$

ANSWERS TO SELECTED PROBLEMS

Chapter 1
  1.2 59,254
  1.3. 9601 6.023×10^{18}
  1.4. 866 2.317×10^{19}
  1.5. 22%
  1.6. 1933 Å
  1.7. 1.24×10^7
  1.10 a) at 25%, for $n = 10$, $N_A = 0.0188$
        b) at 10%, for $n = 10$, $N_A = 0.0388$

Chapter 2
  2.1. a) $M_n = 63.33$ $M_w = 71$ $M_z = 75.7$ PI = 1.12
  2.2. $M_n = 18,200$ $M_w = 28,000$ $M_z = 35,720$ PI = 1.538
  2.3. $a = 0.71$ $K = 1.31 \times 10^{-5}$
  2.4. IV = 0.71 $M_w = 87,606$
  2.5. a) 0.735   b) 0.71
  2.8. a) For $n = 1$, $t = 0.4$ $\phi/\phi_m = 0.982$
        b) For $n = 2$, $t = 0.4$, $\phi/\phi_m = 0.798$
  2.10. $Z_n = 0.0061$ $\tau = 1.843$ $n = 2.62$
  2.11. $T = 88°C$ $t = 2.75$ h
  2.15. $T_\infty = 356k$ $K = 36,413$ $x = 5.28$
  2.17. a) k (law of mixtures) $= 16.9 \times 10^{-4}$ cal/s cm^2
        b) k (Nielson equation) $= 7.9 \times 10^{-4}$ cal/s cm^2
        Conclusion: Agreement is fair (factor of two)

Chapter 3
  3.1. a) $\tau = F/\pi Dt$ b) $\sigma = F/\pi D^2$
  3.2. $\sigma = W(L - x)$. Maximum, $x = L$. Minimum, $x = 0$
  3.3. Total Elongation, $\delta = WL/2AE$
  3.4. $\delta = wL^2/6E$
  3.5. 12.7 MPa
  3.6. $\varepsilon = -5.25 \times 10^{-3}$ $t = 0.99475$ mm
  3.7. $\varepsilon_z = 0.01275$ $t = 0.98725$ mm
  3.8. $\varepsilon_x = -0.01075$ $\varepsilon_y = \varepsilon_z = 2.75 \times 10^{-3}$
  3.10. a) 0.01 in/in   b) 0.005 in   c) zero
  3.11. $\varepsilon_y = 6.66 \times 10^{-3}$, $\Delta D = 0.002$ in
  3.13. $\dot{\gamma} = 125.7$ s^{-1}, $\eta = 603$ Pa·s
  3.14. 1×10^6 W/m^3
  3.15. 2.73×10^5 W s/m^3
  3.17. 538 days
  3.18. $E_1 = 40,000$ $E_2 = 6,400$ lb$_f$/in^2
        $\eta_1 = 4 \times 10^8$ $\eta_2 = 1.9 \times 10^6$
  3.22. a) $\sigma = 1.25 \times 10^5$ Pa
        b) $\sigma = 1.057 \times 10^5$ Pa
        c) $\sigma = 0.75 \times 10^5$ Pa
  3.23. 0.138 h

Chapter 4

4.2. When the shear stress rate, τ is less than 10^3 Pa

4.3. When the shear strain rate, $\dot{\gamma}$ is less than 5 s^{-1} or greater than 100 s^{-1}

4.4. a) b $= -0.0448$ A $= 9.81 \times 10^{12}$

4.7. $\eta = 2.071$ Pa·s

4.8. $\Delta P = 918$ lb$_f$/in^2, $\tau = 3.0$ s

4.9. Q $= 0.147$ cm^3/s

4.10. (partial) 6.42 h

4.11. D $= 1.1$ cm

4.12. $\eta = 1/A$

4.15. 180.7 Btu/lb$_m$

4.16. -2.56 Btu/lb$_m$; -16.82 Btu/lb$_m$

4.17. -44.85 Btu/lb$_m$; -42.87 Btu/lb$_m$

Chapter 5

5.2. 0.395

5.4. $\Delta P = 0$; pure drag flow.   U $= 0$; pressure flow.

5.5. 3.14 cm^3/s

5.7. 0.0917 in 45%

5.9. 47.09 cm^3/s

5.10. 1.84 cm^3/s

5.11. 51,300 lb$_f$/in^2 Unachievable pressure drop.

5.12. 1,200 Btu/min

5.13. 1,900 Btu/min

5.15. 0.64

5.17. 60°

5.19. 2

5.21. 577 lb$_f$/in^2

5.22. 96.2%

5.26. 727 lb$_f$/in^2

5.27. d $= 2R_R$

5.28. R* $= 1.637$ R$_R$

5.29. F $= 24.9$ kN

5.30. 3.55 s

5.31. 1.85 in/s

Chapter 6

6.1. $\beta = 0.019$ $\eta_0 = 1,688$

6.2. 1.68 lb$_f$/in^2

6.3. $\eta = \eta\, e^{\beta MI}$; $\beta = -0.066$ $\eta_0 = 622$ Pa·s

6.4. $\eta = \eta_0\, e^{-E/RT}$;   E $= 8,546$ cal K/g $\eta_0 = 0.0337$ Pa·s

6.7. 465,000 lb$_f$/in^2

6.8. HDPE: 70,500 lb$_f$/in^2

PUR: 24,000 lb$_f$/in^2

PMMA: 212,000 lb$_f$/in^2

6.24. t $= 0.086$ in

6.25. After 1.5 h, tube will craze. Recommend increasing wall thickness to t $= 0.113$ in to reduce stress.

6.26. (partial) PP: t $= 0.9$ in, wt $= 321$ lb

6.27. $\sigma_\theta = 1,261$ lb$_f$/in^2,   t $= 1.07$ in.

APPENDIX A

SOURCES OF POLYMERIC
MATERIALS: GENERAL STATISTICS*

*Source: This Material was Extracted in Part From Chemical Origins & Markets, SRI International and is used by permission of The Society of Plastics Industry, Inc., 1275 K Street N.W. #400, Washington D.C., 20005.

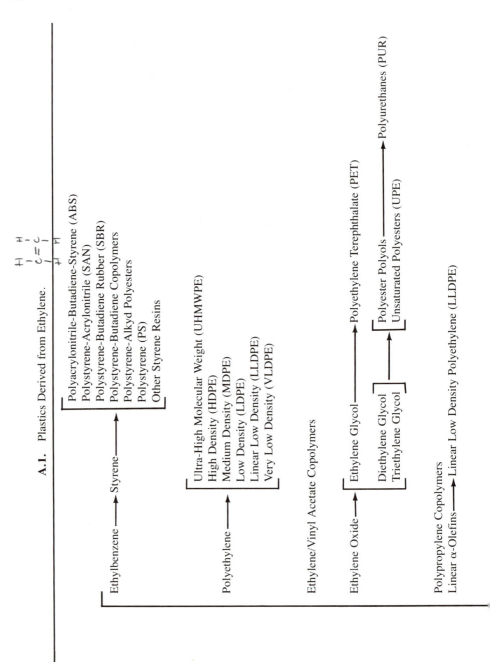

A.1. Plastics Derived from Ethylene.

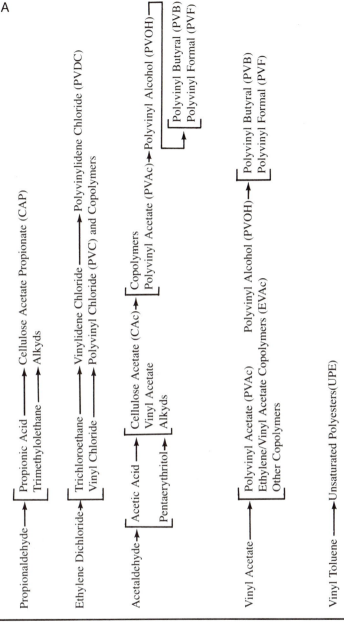

A.2. Plastics Derived from Methane.

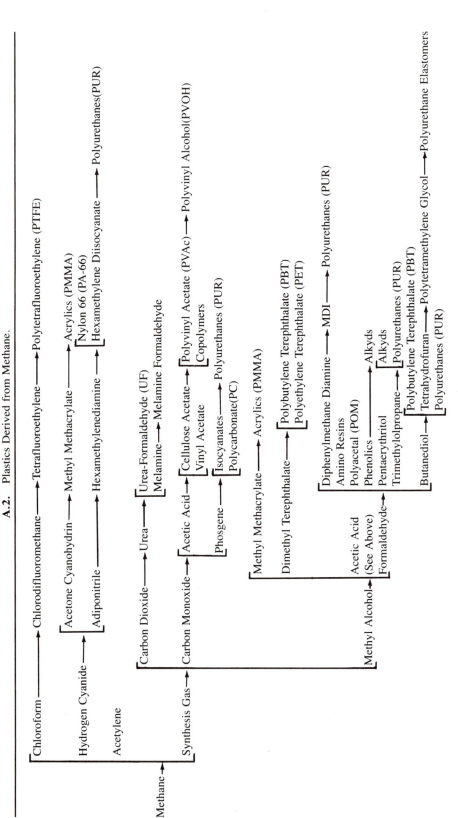

A.3. Plastics Derived from Acetylene.

Acetylene
From
Natural Gas
Methane
Olefin Coproduct

Vinyl Chloride ⟶ Polyvinyl Chloride (PVC)

Vinyl Fluoride ⟶ Polyvinyl Fluoride (PVF)

Butanediol ⟶ Polybutylene Terephthalate (PBT)
Tetrahydrofuran ⟶ Polytetramethylene Glycol ⟶ Urethane Elastomers
Polyurethanes (PUR)

A.4. Plastics Derived from Butadiene.

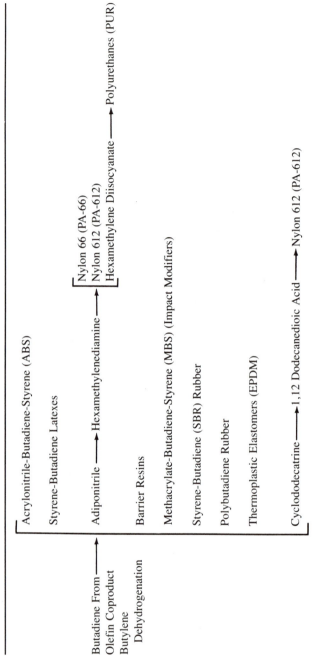

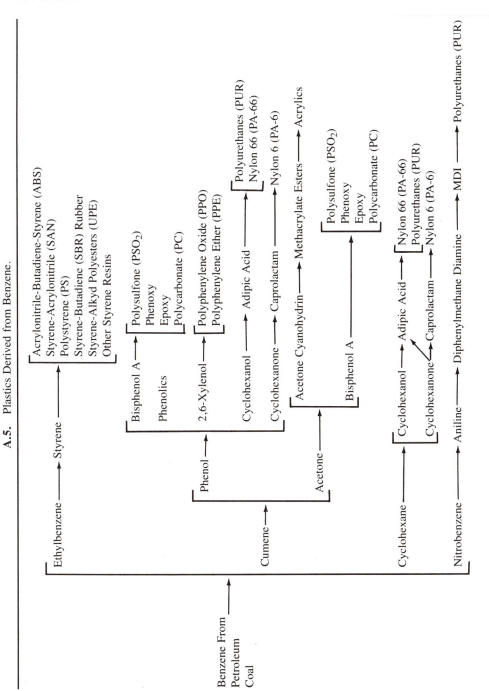

A.5. Plastics Derived from Benzene.

A.6. Plastics Derived from Propylene.

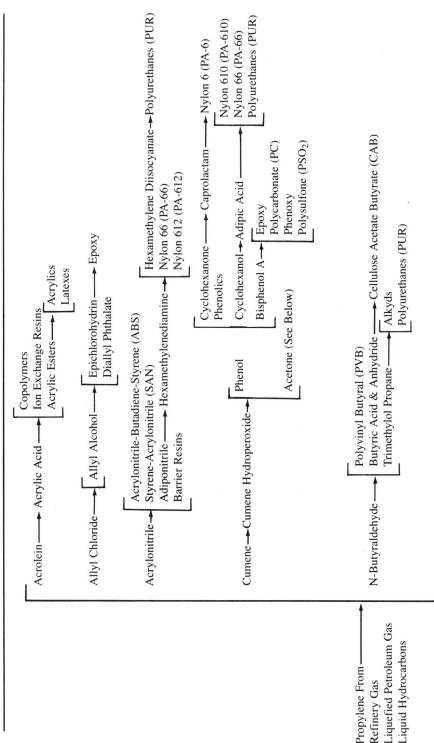

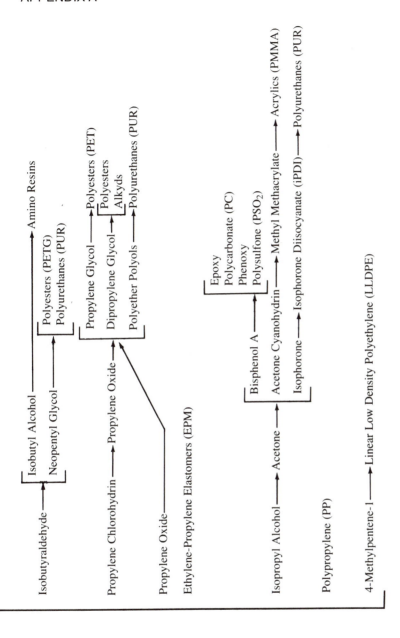

A.7. Plastics Derived from Napthalene, Toluene and Xylenes.

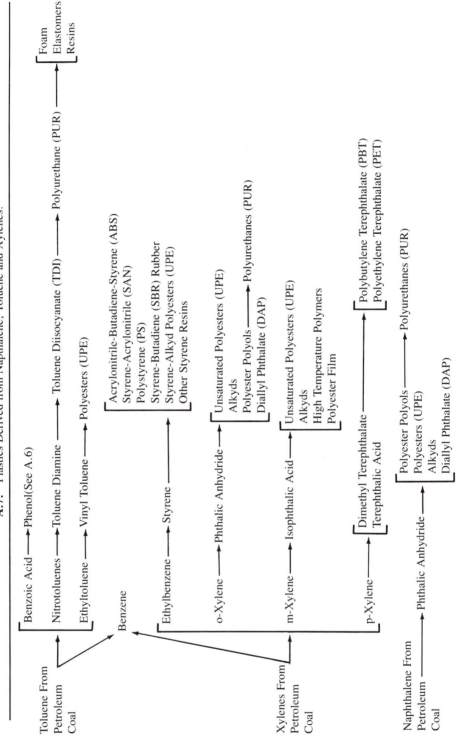

APPENDIX B

POLYMERIC FILLERS*

*Source: I. Katz and J.V. Milewski, Handbook of Fillers and Reinforcements for Plastics, (1978) Van Nostrand Reinhold Co., 135 W. 50th St., New York NY 10019. Table 2–13, p. 49. Used by permission of copyright owner.

B.1. Properties of Ground Limestone (L) and Precipitated Calcium Carbonate (C)[a].

Properties	Units	L1	L2	L3	L4	C1	C2	C3
Particle								
Mean Size	μm, esd[b]	120	14	5	3	1.0	0.07	0.07
Range Shape	μm, esd[c] / Class[d]	590 to 20 / Block	74 to 0.3 / Block	35 to 0.3 / Block	20 to 0.1 / Block	2 to 0.1 / Block	0.5 to 0.1 / Block	0.5 to 0.1 / Block
Bulk								
Specific Gravity[e]	g/cm³	2.71	2.71	2.71	2.71	2.71	2.65	2.55
Surface Area[f]	m²/g	0.3	0.9	1.2	2.2	5.3	20 to 26	11 to 17
k_E[g]		4.0	4.0	4.0	4.0	4.0	4.0	4.0
Packing Fraction, p_f[h]		0.8 to 0.83	0.8	0.68 to 0.76	0.67 to 0.7	0.46 to 0.53	0.37 to 0.54	0.39 to 0.64
Dispersibility								
Water[i]		G	G	G	G	G	G	N
Organics[j]		E	E	E	E	E	G	E
Resistance[k]		Al,W,S	Al,W,S	Al,W,S	Al,W,S	Al,W,S	Al,W,S	Al,W,S
pH		9.2 to 9.5	9.2 to 9.5	9.2 to 9.5	9.2 to 9.5	9 to 9.5	9 to 10.4	9 to 10.4
Specific Resistance[l]	ω cm $\times 10^{-3}$[l]	21 to 23	21 to 23	21 to 23	21 to 23	21 to 23	21 to 23	21 to 23
Impurities[o]		SiO_2	$MgCO_3$	$MgCO_3$	$MgCO_3$	$MgCO_3$	$CaSO_4$	$CaSO_4$

Thermal									
Conductivity	cal cm/s cm² °C × 10³	5.6	5.6	5.6	5.6	5.6	5.6	5.6	5.6
Specific Heat	cal/g °C	0.205	0.205	0.205	0.205	0.21	0.21	0.21	0.21
Coefficient of Expansion	cm/cm/°C × 10⁶	10	10	10	10	10	10	10	10
Optical									
Refractive Index									
Mean	sin i/ sin r	1.59	1.59	1.59	1.59	1.66	1.66	1.66	1.66
Range	—	1.49 to 1.66	1.49 to 1.66	1.49 to 1.66	1.49 to 1.66	1.53 to 1.69	1.53 to 1.69	1.53 to 1.69	1.53 to 1.69
Color									
Powder[m]		W	W	W	W	W	W	W	W
In DOP[n]		W	W	W	W	W	W	W	W
Physical									
Young's Modulus, E	dyn/cm² × 10⁻¹¹	2.6	2.6	2.6	2.6	2.6	2.6	2.6	2.6
Poisson's Ratio, μ[p]		0.27	0.27	0.27	0.27	0.27	0.27	0.27	0.27
Mohs Hardness[q]		2.5 to 3	2.5 to 3	2.5 to 3	2.5 to 3	2.5 to 3	2.5 to 3	2.5 to 3	2.5 to 3
Electrical									
Dielectric Constant[r]	ε = D/E	6.14	6.14	6.14	6.14	6.14	6.14	6.14	6.14

B.2. Properties of Kaolins (K)[a].

Properties	Units	K1	K2	K3	K4	K5
Particle						
Mean Size	μm, esd[b]	5	1.2	0.8	0.6	1.2
Range	μm, esd[c]	45 to 0.1	35 to 0.03	20 to 0.03	10 to 0.03	35 to 0.4
Shape	Class[d]	Block/Flake	Block/Flake	Block/Flake	Block/Flake	Block/Flake
Bulk						
Specific Gravity[e]	g/cm^3	2.58	2.58	2.58	2.58	2.63
Surface Area[f]	m^2/g	6 to 10	10 to 12	14 to 16	18 to 22	8
k_E[g]		4	4	4	4	4
Packing Fraction, p_f[h]		0.53 to 0.56	0.53 to 0.56	0.49 to 0.51	0.46 to 0.48	0.35 to 0.44
Dispersibility						
Water[i]		P	P	P	P	P
Organics[j]		G	G	G	G	G
Resistance[k]		Ac,Al,W,S	Ac,Al,W,S	Ac,Al,W,S	Ac,Al,W,S	Ac,Al,W,S
pH		3.8 to 5.2	5.3 to 6.0	3.8 to 5.4	3.8 to 7	4.2 to 6
Specific Resistance[l]	ω cm $\times$ 10^{-31}	10+	10+	10+	10+	20+
Impurities[o]		—	SiO$_2$	—	—	—

Thermal						
Conductivity	cal cm/s cm² °C $\times 10^3$	4.7	4.7	4.7	4.7	4.7
Specific Heat	cal/g °C	0.22	0.22	0.22	0.22	0.22
Coefficient of Expansion	cm/cm/°C $\times 10^6$	8	8	8	8	3
Optical						
Refractive Index						
Mean	sin i/ sin r	1.56	1.56	1.56	1.56	1.62
Range	—	1.55 to 1.57	1.55 to 1.57	1.55 to 1.57	1.55 to 1.57	
Color						
Powder[m]		W to T	W to T	W	W	W
In DOP[n]		T	T	T	T	W
Physical						
Young's Modulus, E	dyn/cm² $\times 10^{-11}$	2	2	2	2	2
Poisson's Ratio, μ^p		0.3	0.3	0.3	0.3	0.3
Mohs Hardness[q]		2	2	2	2	7
Electrical						
Dielectric Constant[r]	$\varepsilon = D/E$	2.6	2.6	2.6	2.6	1.3

B.3. Properties of Talc, Mica and Feldspar and Nepheline.

Properties	Units	T1	T2	T3	M1	M2	AS1	AS2	AS3
Particle									
Mean Size	μm, esd[b]	6 to 8	5 to 6	1.8	s	5	15	8	5
Range	μm, esd[c]	60 to 0.3	50 to 0.1	15 to 0.05	—	14 to ?	90 to 0.5	40 to 0.5	30 to 0.3
Shape	Class[d]	Block/Flake	Block/Flake	Block/Flake	Flake	Flake	Block	Block	Block
Bulk									
Specific Gravity[e]	g/cm^3	2.8	2.8	2.8	2.82	2.82	2.6	2.6	2.6
Surface Area[f]	m^2/g	6 to 8	6 to 10	17	—	30	0.8 to 0.9	1 to 1.3	1.8 to 2.3
k_E[g]		4	4	4	5	5	4	4	4
Packing Fraction, p_f[h]		0.51 to 0.56	0.52 to 0.54	0.4 to 0.42	0.45	0.38	0.68 to 0.72	0.62 to 0.67	0.56
Dispersibility									
Water[i]		P	P	P	G	G	E	E	E
Organics[j]		E	E	E	E	E	E	E	E
Resistance[k]		Ac,Al,W,S	Ac,Al,W,S	Ac,Al,W,S	Ac,Al,W,S	Ac,Al,W,S	Ac,Al,W,S	Ac,Al,W,S	Ac,Al,W,S
pH		9 to 9.5	9 to 9.5	9 to 9.5	8 to 8.5	8 to 8.5	7 to 10	7 to 10	10
Specific Resistance[l]	ω cm × 10^{-31}				—	—	4 to 15	4 to 15	3
Impurities[o]		t	t	t	—	—	u	—	u

Thermal									
Conductivity — cal cm/ s cm^2 °C $\times 10^3$	5	5	5	5	6	6	5.6	5.6	5.6
Specific Heat — cal/g °C	0.208	0.208	0.208	0.208	0.206	0.206	0.21	0.21	0.21
Coefficient of Expansion — cm/cm/°C $\times 10^6$	8	8	8	8	8	8	6.5	6.5	6.5
Optical									
Refractive Index									
Mean — sin i/ sin r	1.57	1.57	1.57	1.57	1.58	1.58	1.53	1.53	1.53
Range — —	1.54 to 1.59	1.54 to 1.59	1.54 to 1.59	1.54 to 1.59	1.56 to 1.62	1.56 to 1.62	1.528 to 1.538	1.528 to 1.538	1.528 to 1.538
Color									
Powder[m]	W	W	W	W	G	W	W	W	W
In DOP[n]	G	G	G	G	C	G	C	C	C
Physical									
Young's Modulus, E dyn/cm^2 $\times 10^{-11}$	2	2	2	2	2	2	3	3	3
Poisson's Ratio, μ[p]	0.3	0.3	0.3	0.3	0.3	0.3	0.3	0.3	0.3
Mohs Hardness[q]	2	1	1		2.5 to 3	2.5 to 3	5.5 to 6.5	5.5 to 6.5	5.5 to 6.5
Electrical									
Dielectric Constant[r] $\varepsilon = D/E$	5.5 to 7.5	5.5 to 7.5	5.5 to 7.5	5.5 to 7.5	2 to 2.6	2 to 2.6	6	6	6

[s]Available as 5 to 200 mesh products. [t]Tremolite, calcium carbonate, chlorite, etc., depending on source. [u]Feldspar - silica and mica. Nepheline - emery.

B.4. Properties of Wollastonite, Perlite, Glass and Synthetic Silicates.

Properties	Units	W	P	GB	GA	SP1	SP2
Particle							
Mean Size	μm, esd[b]	6	Large	25	50	0.02	0.08
Range	μm, esd[c]	44 to 0.1	Various	44 to 4	150 to 10	—	0.17 to 0.03
Shape	Class[d]	Block	Spheroid	Spheroid[s]	Spheroid[t]	Floc	Floc
Bulk							
Specific Gravity[e]	g/cm^3	2.9	0.15 to 0.2	2.48	0.28	2.0	1.93
Surface Area[f]	m^2/g	1.1 to 1.4	1 to 2	0.3	0.6	140 to 160	40
k_E[g]		4	3	2.5	2.5	5(?)	5(?)
Packing Fraction, p_f[h]		0.62	0.64	0.64	0.65	0.32	0.4
Dispersibility							
Water[i]		E	E	E	E	P	P
Organics[j]		G	E	E	E	G	G
Resistance[k]		Al,W,S	Ac,Al,W	Ac,W,S	Ac,W,S	Ac,W,S	Al,W,S
pH		10	7.5 to 8	9.5	9.5	6.5 to 7.3	9.5
Specific Resistance[l]	ω cm × 10^{-31}	Low	—	—	—	—	—
Impurities[o]		CaCO$_3$	—	—	—	NaCl	NaCl

		1	2	3	4	5	6
Thermal							
Conductivity	cal cm/s cm² °C × 10³	6	0.02	1.7	0.02	2	3
Specific Heat	cal/g °C	0.24	0.24	0.27	0.24	0.2	0.2
Coefficient of Expansion	cm/cm/°C × 10⁶	6.5	8.8	8.6	8.8	10	10
Optical							
Refractive Index							
Mean	sin i/ sin r	1.63	—	1.52	—	1.46	1.45
Range	—	1.63 to 1.67	—	—	—	—	—
Color							
Powderᵐ		W	W	C	W	W	W
In DOPⁿ		W	W	C	W	W	W
Physical							
Young's Modulus, E	dyn/cm² × 10⁻¹¹	3	0.01	6	0.02	2	2
Poisson's Ratio, μᵖ		0.3	0.4	0.23	0.4	0.3	0.3
Mohs Hardnessq		5 to 5.5	5.6	6	6	5 to 6	5
Electrical							
Dielectric Constantʳ	ε = D/E	6	1.5	5	1.5	9	9

s = solid. t = hollow.

B.5.   Properties of Nutshell and Wood Flours and Hydrous Alumina.

Properties	Units	O1	O2	H1	H2	H3
Particle						
Mean Size	μm, esd[b]	30	?	40	6.5 to 8.5	0.7
Range	μm, esd[c]	177 to 1	—	150 to 5	60 to 0.3	3 to 0.05
Shape	Class[d]	Block	Fiber	Floc	Floc	Plate
Bulk						
Specific Gravity[e]	g/cm^3	1.3 to 1.35	0.5 to 0.7	2.42	2.42	2.40
Surface Area[f]	m^2/g	0.4 to 0.5	—	6	6	6 to 8
k_E[g]		5(?)	10(?)	4	4	5
Packing Fraction, p_f[h]		0.56	0.4	0.64	0.56	0.5
Dispersibility						
Water[i]		G	G	G	G	G
Organics[j]		G	G	G	G	G
Resistance[k]		None	None	Ac,Al,W,S	Ac,Al,W,S	Ac,Al,W,S
pH		4 to 5	4 to 5	—	—	—
Specific Resistance[l]	ω cm $\times$ 10^{-3}[1]	Low	Low	—	—	—
Impurities[o]		H$_2$O, Lignin	H$_2$O, Lignin	—	—	—

Thermal						
Conductivity	cal cm/s cm^2 °C × 10^3	1.4	0.6	2	2	2
Specific Heat	cal/g °C	0.43	0.42	0.19	0.19	0.19
Coefficient of Expansion	cm/cm/°C × 10^6	5 to 50	5 to 50	4 to 5	4 to 5	4 to 5
Optical						
Refractive Index						
Mean	sin i/ sin r	—	—	1.57	1.57	1.57
Range	—	—	—	—	—	—
Color						
Powder[m]		T	T	W	W	W
In DOP[n]		T	T	W	W	C
Physical						
Young's Modulus, E	dyn/cm^2 × 10^{-11}	0.8	1	3	3	3
Poisson's Ratio, μ[p]		0.35	0.3	0.3	0.3	0.3
Mohs Hardness[q]		1	1	1 to 3	1	1
Electrical						
Dielectric Constant[r]	ε = D/E	5	5	7	7	7

B.6. Properties of Asbestos, Coal, Barite and Silica.

Properties	Units	AB	A	B	S1	S2	S3
Particle							
Mean Size	μm, esd[b]	0.17	14	3	10	5	2
Range	μm, esd[c]	—	100 to 2	40 to 0.1	74 to 0.5	40 to 0.5	15 to 0.1
Shape	Class[d]	Fiber	Block	Block	Block	Block	Block
Bulk							
Specific Gravity[e]	g/cm^3	2.4 to 2.6	1.47	4.4 to 4.5	2.65	2.65	2.65
Surface Area[f]	m^2/g	40	1.5	0.6	0.8	1.2	1.5
k_E[g]		10	3	3	4	4	4
Packing Fraction, p_f[h]		0.6	0.6	0.65 to 0.75	0.59 to 0.67	0.52 to 0.55	0.49 to 0.52
Dispersibility							
Water[i]		G	P	G	G	G	G
Organics[j]		G	E	G	G	G	G
Resistance[k]		Al,W,S	Al,Ac,W	Ac,Al,W,S	Ac,W,S	Ac,W,S	Ac,W,S
pH		9 to 10	6 to 7	6 to 7	6.5 to 7	7	7
Specific Resistance[l]	ω cm $\times\ 10^{-31}$	Low	Very High	Very High	25	25	25
Impurities[o]		Fe$_3$O$_4$	Mineral	—	—	—	—

Property	Units						
Thermal							
Conductivity	cal cm/s cm² °C × 10³	5.0	0.6 to 0.8	6	7	7	7
Specific Heat	cal/g °C	0.26	0.3	0.111	0.19	0.19	0.19
Coefficient of Expansion	cm/cm/°C × 10⁶	0.3	5	10	10	10	10
Optical							
Refractive Index							
Mean	sin i/ sin r	1.53	—	1.64	1.55	1.55	1.55
Range	—	1.50 to 1.55	—	1.64 to 1.65	1.54 to 1.55	1.54 to 1.55	1.54 to 1.55
Color							
Powder[m]		G	B	W	W	W	W
In DOP[n]		G	B	W	G	G	G
Physical							
Young's Modulus, E	dyn/cm² × 10⁻¹¹	14.5	2	3	3	3	3
Poisson's Ratio, μ[p]		0.2	0.3	0.3	0.3	0.3	0.3
Mohs Hardness[q]		3 to 5	3	3 to 3.5	6.5 to 7	6.5 to 7	6.5 to 7
Electrical							
Dielectric Constant[r]	ε = D/E	10	3	7.3	4.3	4.3	4.3

B.7. Producers of Polymeric Fillers.

Code	Filler		
(Appendix B.1 to Appendix B.6)	Generic Name	Sold As	Producer
L1	Casting Grade Limestone	40-200 C50 or CLF 20-200	Georgia Marble Co. Thompson, Weinman Co.
L2	Limestone	No. 1 White No. 10	Thompson, Weinman Co. Georgia Marble Co.
L3	Limestone	Calwhite Gama-Sperse 6451 Vicron 25-11 Snowflake Camel-Tex	Georgia Marble Co. C. Pfizer & Co. Thompson, Weinman Co. H.T. Campbell Co.
L4	Limestone	Atomite Gama-Sperse 140 Camel-Wite Vicron 15-15	Thompson, Weinman Co. Georgia Marble Co. H.T. Campbell Co. C. Pfizer & Co.
C1	Precipitated Calcium Carbonate	Albacar 5950	C. Pfizer & Co.
C2	Calcium Carbonate	Multifex MM Calcene NC	Diamond Shamrock PPG Industries
C3	Coated with 3.5% Fatty Acids	Calcene TM and CO Super Multifex	PPG Industries Diamond Shamrock
K1	Kaolin, Coarse, Leached and Fractionated	Hydrite Flat D ASP 400	Georgia Kaolin Co. Engelhard Ind.
K2	Air-Floated Kaolin	Huber 65A	J.M. Huber Co.
K3	Fine, Leached and Fractionated Kaolin	ASP 600 HG 80 Hydrite R Al-Sil-Ate HO2	Engelhard Ind. J.M. Huber Co. Georgia Kaolin Co. Freeport Kaolin co.
K4	Very Fine, Leached and Fractionated Kaolin	ASP 200 HG 90 Hydrite 10 Al-Sil-Ate HO	Engelhard Ind. J.M. Huber Co. Georgia Kaolin Co. Freeport Kaolin Co.
K5	Calcined Kaolin	Glomax HE Santintone Special Huber 80C Al-Sil-Ate O Whitetex	Georgia Kaolin Co. Engelhard Ind. J.M. Huber Co. Freeport Kaolin Co. Freeport Kaolin Co.
T1	Talc	Beaverwhite 3X Talcron CP 44-31	Cypress Mines Corp. International Talc Co. C. Pfizer & Co.

Code	Filler		
(Appendix B.1 to Appendix B.6)	Generic Name	Sold As	Producer
T2	Talc	C-400 325 Talcron 38-33	Cypress Mines Corp. International Talc Co. C. Pfizer & Co.
T3	Talc	Mistron Vapor 625 MP 12-50	Cypress Mines Corp. International Talc Co. C. Pfizer & Co.
M1	Dry Ground and Delaminated Flake Mica	— —	English Mica Co. Franklin Mica Co.
M2	Fine, Wet Ground and Delaminated Mica	Micro Mica C-1000 Alsibronz No. 12 No. 21 Mica	English Mica Co. Franklin Mica Co. Thompson, Weinman Co.
A	Anthracite Coal	325 325	Keystone Filler Co. Shamokin Filler Co.
B	Mineral Barium Sulfate	Baryta White and Foam A No. 22 Barytes No. 1 Barytes	NL Industries Thompson, Weinman Co. C. Pfizer & Co.
S1	Silica	Min-U-Sil 30 1240 625 Gold Bond R	PGS-Floridin Co. Illinois Mineral Co. Malvern Minerals Co. Tammsco Div./Lowe's Inc.
S2	Silica	Min-U-Sil 15 Imsil A-25 S Micron	PGS-Floridin Co. Illinois Mineral Co. Tammsco Div./Lowe's Inc.
S3	Silica	Min-U-Sil 5 A-10 S-Micron	PGS-Floridin Co. Illinois Mineral Co. Tammsco Div./Lowe's Inc.
AS1	Anhydrous Silicates	LU-390 Minex 2	Lawson-United Indusmin Ltd.
AS2	Anhydrous Silicates	LU-340 Minex 4	Lawson-United Indusmin Ltd.
AS3	Anhydrous Silicates	Minex 7	Indusmin Ltd.
W	Wollastonite	P-4	Interpace Corp.
P	Perlite	— —	Johns-Manville Corp. Silbrice Corp.
GB	Glass Beads	3000	Potters Ind.

B.7. Continued.

Code	Filler		
(Appendix B.1 to Appendix B.6)	Generic Name	Sold As	Producer
GA	Glass Microballoons	B30B Q Cel	3M Co. Philadelphia Quartz Co.
SP1	Synthetic Precipitated Silica	Hi-Sil 210, 215, 233	PPG Ind.
SP2	Synthetic Precipitated Calcium Carbonate	Silene D	PPG Ind.
O1	Nutshell Flour	WF-5 —	Agrashell Inc. Composition Materials Co.
O2	Wood Flour	— — — — —	Chicago Wood and Plastic Products Composition Materials Co. International Filler Corp. Wilner Wood Products Co. Wood Flour Co.
H1	Hydrous Alumina	C-33 RH-33	Alcoa Reynolds Chem.
H2	Hydrous Alumina	C-331 GHA-331 GHA-332 SB-331 SB-332	Alcoa Great Lakes Foundry Sand Co. Great Lakes Foundry Sand Co. Solem Ind. Solem Ind.
H3	Hydrous Alumina	Hydral 710 RH-730	Alcoa Reynolds Chem.
AB	Asbestos	7R Asbestos — — — — RG600	Asbestos Corp. Bell Asbestos Mines, Ltd. Cary-Canadian Mines, Ltd. Johns-Manville Corp. Uniroyal, Inc. Union Carbide Corp.

Footnotes to Appendices B.1 to B.6

a.   See listing of producers and trade names, Appendix B.7.
b.   Equivalent spherical diameter.
c.   Approximate, from producers' literature. Range conforms with coarseness and fineness parameters of 99.9% coarser and finer by weight, esd.
d.   Idealized class from Katz and Milewski reference.
e.   ASTM D153.
f.   Calculated when data were not available.
g.   Einstein coefficient, approximated.
h.   Approximate from oil absorption data, ASTM D281.
i.   E = Excellent. G = Good. P = Poor. N = nondispersible. Dispersibility indicated without dispersant. All fillers provide excellent dispersion with correct dispersant for a particular system.
j.   Good to excellent resistance to acids (Ac), alkalis (Al), water (W), and solvents (S).
k.   10% to 20% in water.
l.   ASTM D2448, a measure of water solubility.
m.   Powder appearance: W = white, T = tan to brown, G = Gray, B = black, C = clear.
n.   50% (wt) in DOP (di-2-ethyl hexyl phthalate). Same appearance classification as footnote^m.
o.   Predominant minor contaminants, less than about 3%.
p.   See section 4.2.1 of Katz and Milewski.
q.   See scale of section 3.3 in Katz and Milewski.
r.   See section 4.7.1 of Katz and Milewski.

APPENDIX C

PHYSICAL PROPERTIES
OF POLYMERS*

*Source: H. Domininghaus, *Die Kunststoffe und ihre Eigenschaften*, VDI Verlag, Düsseldorf, FRG (1986), Table I, after page 842. Permission granted by copyright owner.

C.1. Mechanical Properties of Polymers (I).

Polymeric Material	Abbreviation DIN 7728 B1.1	Density (g/cm³) DIN 53479	Tensile Strength (N/mm²) DIN 53455	Ultimate Elongation (%) DIN 53455	Tensile Modulus (N/mm²) DIN 53457
Low Density Polyethylene	LDPE	0.914/0.928	8/23	300/1,000	200/ 500
High Density Polyethylene	HPDE	0.94 /0.96	18/35	100/1,000	700/1,400
Ethylene-Vinyl Acetate	EVA	0.92/0.95	10/20	600/ 900	7/ 120
Polypropylene	PP	0.90/0.907	21/37	20/ 800	1,100/ 1,300
Polybutene-1	PB	0.905/0.920	30/38	250/ 280	250/ 350
Polyisobutylene	PIB	0.91/0.93	2/ 6	>1,000	—
Poly-4-methylpentene-1	PMP	0.83	25/28	13/ 22	1,100/ 1,500
Ionomer	—	0.94	21/35	250/ 300	180/ 210
Unplasticized Polyvinyl Chloride	RPVC	1.38/1.55	50/75	10/ 50	1,000/ 3,500
Plasticized Polyvinyl Chloride	FPVC	1.16/1.35	10/25	170/ 400	—
General Purpose Polystyrene	PS	1.05	45/65	3/ 4	3,200/ 3,250
Styrene-Acrylonitrile Copolymer	SAN	1.08	75	5	3,600
Styrene-Butadiene Graft Copolymer	SB	1.05	26/38	25/ 60	1,800/ 2,500
Acrylonitrile-Butadiene-Styrene Terpolymer	ABS	1.04 /1.06	32/45	15/ 30	1,900/ 2,700
Acrylonitrile-Styrene Elastomeric Copolymer	ASA	1.04	32	40	1,800

Polymethyl Methacrylate	PMMA	1.17 /1.20	50/77	2/ 10	2,700/ 3,200
Polyvinyl Carbazole	PVK	1.19	20/30	—	3,500
Polyacetal	POM	1.41 /1.42	62/70	25/ 70	2,800/ 3,200
Polytetrafluoroethylene	PTFE	2.15 /2.20	25/36	350/ 550	410
Tetrafluoro/Hexafluoropropylene Copolymer	PFEP	2.12 /2.17	22/28	250/ 330	350
Polytrifluorochloroethylene	PCTFE	2.10 /2.12	32/40	120/ 175	1,050/ 2,100
Ethylene/Tetrafluoroethylene	PETFE	1.7	35/54	400/ 500	1,100
Polyamide 6 (Nylon 6)	PA-6	1.13	70/85	200/ 300	1,400
Polyamide 66 (Nylon 66)	PA-66	1.14	77/84	150/ 300	2,000
Polyamide 11 (Nylon 11)	PA-11	1.04	56	500	1,000
Polyamide 12 (Nylon 12)	PA-12	1.02	56/65	300	1,600
Polyamide 6-3-T (Transparent Nylon)	PA-6-3-T	1.12	70/84	70/ 150	2,000
Polycarbonate	PC	1.2	56/67	100/ 130	2,100/ 2,400
Polyethylene Terephthalate	PET	1.37	47	50/ 300	3,100
Polybutylene Terephthalate	PBT	1.31	40	15	2,000
Polyphenylene Oxide, Modified	mPPO	1.06	55/68	50/ 60	2,500
Polysulfone	PSO$_2$, Ps	1.24	50/100	25/ 30	2,600/ 2,750
Polyphenylene Sulfide	PPS	1.34	75	3	3,400
Polyaryl Sulfone	PAS	1.36	90	13	2,600
Polyether Sulfone	PES	1.37	85	30/ 80	2,450
Polyaryl Ether	PAE	1.14	53	25/ 90	2,250

C.1. Continued.

Polymeric Material	Abbreviation DIN 7728 B1.1	Density (g/cm³) DIN 53479	Tensile Strength (N/mm²) DIN 53455	Ultimate Elongation (%) DIN 53455	Tensile Modulus (N/mm²) DIN 53457
Phenol-Formaldehyde (Phenolic)	PF	1.4	25	0.4/ 0.8	5,600/12,000
Urea-Formaldehyde	UF	1.5	30	0.5/ 1.0	7,000/10,500
Melamine-Formaldehyde	MF	1.5	30	0.6/ 0.9	4,900/ 9,100
Unsaturated Polyester Resin	UPE	2.0	30	0.6/ 1.2	14,000/20,000
Polydiallyl Phthalate (DAP)	DAP	1.51/1.78	40/75	—	9,800/15,500
Silicone Molding Compound	Si	1.8 /1.9	28/46	—	6,000/12,000
Polyimide	PI	1.43	75/100	4/ 9	23,000/28,000
Epoxy	EP	1.9	30/40	4	21,500
Polyurethane Thermoset	PUR	1.05	70/80	3/ 6	4,000
Polyurethane Thermoplastic Elastomer	PUR	1.20	30/40	400/ 450	700
Linear Polyurethane	PUR	1.21	30	35	1,000
Vulkan Fiber	VF	1.1 /1.45	85/100	—	—
Cellulose Acetate	CA	1.30	38	3	2,200
Cellulose Acetate Propionate	CAP	1.19 /1.23	14/55	30/ 100	420/ 1,500
Cellulose Acetate Butyrate	CAB	1.18	26	4	1,600

C.2. Mechanical Properties of Polymers (II).

Polymeric Material	Abbreviation DIN 7728 B1.1	Ball Indentation Hardness (N/mm²) DIN 53456	Impact Strength (kJ/m²) DIN 53453	Notched Impact Strength (kJ/m²) DIN 53453	(ft-lb/in Notch) ASTM D256
Low Density Polyethylene	LDPE	13/ 20	DNB	DNB	—
High Density Polyethylene	HPDE	40/ 65	DNB	DNB	—
Ethylene-Vinyl Acetate	EVA	—	DNB	DNB	DNB
Polypropylene	PP	36/ 70	DNB	3/17	0.5/20
Polybutene-1	PB	30/ 38	DNB	4/DNB	DNB
Polyisobutylene	PIB	—	DNB	DNB	DNB
Poly-4-methylpentene-1	PMP	—	—	—	0.4/0.6
Ionomer	—	—	—	—	6/15
Unplasticized Polyvinyl Chloride	RPVC	75/155	DMB/$>$20	2/50	0.4/20
Plasticized Polyvinyl Chloride	FPVC	—	DNB	DNB	—
General Purpose Polystyrene	GPPS	120/130	5/20	2/2.5	0.25/0.6
Styrene-Acrylonitrile Copolymer	SAN	130/140	8/20	2/3	0.35/0.5
Styrene-Butadiene Graft Copolymer	SB	80/130	10/80	5/13	DNB
Acrylonitrile-Butadiene-Styrene Terpolymer	ABS	80/120	70/DNB	7/20	2.5/12
Acrylonitrile-Styrene Elastomeric Copolymer	ASA	75	DNB	18	6/8
Polymethyl Methacrylate	PMMA	180/200	18	2	0.3/0.5
Polyvinyl Carbazole	PVK	200	5	2	—
Polyacetal	POM	150/170	100	8	1/2.3

C.2. Continued.

Polymeric Material	Abbreviation DIN 7728 B1.1	Ball Indentation Hardness (N/mm²) DIN 53456	Impact Strength (kJ/m²) DIN 53453	Notched Impact Strength (kJ/m²) DIN 53453	(ft-lb/in Notch) ASTM D256
Polytetrafluoroethylene	PTFE	27/ 35	DNB	13/15	3.0
Tetrafluoro/Hexafluoropropylene Copolymer	PFEP	30/ 32	—	—	DNB
Polytrifluorochloroethylene	PCTFE	65/ 70	DNB	8/10	2.5/2.8
Ethylene/Tetrafluoroethylene	PETFE	65	—	—	DNB
Polyamide 6 (Nylon 6)	PA-6	75	DNB	DNB	3.0
Polyamide 66 (Nylon 66)	PA-66	100	DNB	15/20	2.1
Polyamide 11 (Nylon 11)	PA-11	75	DNB	30/40	1.8
Polyamide 12 (Nylon 12)	PA-12	75	DNB	10/20	2/2.5
Polyamide 6-3-T (Transparent Nylon)	PA-6-3-T	160	DNB	13	—
Polycarbonate	PC	110	DNB	20/30	12/18
Polyethylene Terephthalate	PET	200	DNB	4	0.8/1.0
Polybutylene Terephthalate	PBT	180	DNB	4	0.8/1.0
Polyphenylene Oxide, Modified	mPPO	—	DNB	—	4
Polysulfone	PSO_2, Ps	—	—	—	1.3
Polyphenylene Sulfide	PPS	—	—	—	0.3
Polyaryl Sulfone	PAS	—	—	—	1/2
Polyether Sulfone	PES	—	—	—	1.6
Polyaryl Ether	PAE	—	—	—	8

Material	Abbr.			
Phenol-Formaldehyde (Phenolic)	PF	250/320	>6	>1.5	0.2/0.6
Urea-Formaldehyde	UF	260/350	>6.5	>1.5	0.4/0.5
Melamine-Formaldehyde	MF	260/410	>7	>1.5	0.2/0.3
Unsaturated Polyester Resin	UPE	240	>4.5	>3	0.5/16
Polydiallyl Phthalate (DAP)	DAP	—	—	—	0.4/15
Silicone Molding Compound	Si	—	—	—	0.3/0.8
Polyimide	PI	—	—	—	0.5/1
Epoxy	EP	—	>8	>3	2/30
Polyurethane Thermoset	PUR	—	—	—	0.4
Polyurethane Thermoplastic					
Elastomer	PUR	—	DNB	DNB	DNB
Linear Polyurethane	PUR	—	DNB	3	—
Vulkan Fiber	VF	80/140	20/120	—	—
Cellulose Acetate	CA	50	65	15	2.5
Cellulose Acetate Propionate	CAP	47/79	DNB	6/20	1.5
Cellulose Acetate Butyrate	CAB	35/45	DNB	30/35	4/5

C.3. Thermal Properties of Polymers (I).

Polymeric Material	Abbreviation DIN 7728 B1.1	Use Temperature (°C)			Heat Resistance (°C)	
		Maximum Short Time	Maximum Continuous	Minimum Continuous	VSP (Vicat 5 kg) DIN 53460	(1.86 N/mm²) (0.45 N/mm²) ASTM D648
Low Density Polyethylene	LDPE	80/ 90	60/ 75	−50	—	35
High Density Polyethylene	HPDE	90/120	70/ 80	−50	60/ 70	50
Ethylene-Vinyl Acetate	EVA	65	55	−60	—	34/ 62
Polypropylene	PP	140	100	0/ − 30	85/100	45/120
Polybutene-1	PB	130	90	0	70	60/110
Polyisobutylene	PIB	80	65	−50	—	—
Poly-4-methylpentene-1	PMP	180	120	0	—	—
Ionomer	—	120	100	−50	—	38/ 45
Unplasticized Polyvinyl Chloride	EPVC	75/100	65/ 85	− 5	75/110	60/ 82
Plasticized Polyvinyl Chloride	FPVC	55/ 65	50/ 55	0/ − 20	40	—
General Purpose Polystyrene	GPPS	60/ 80	50/ 70	− 10	78/ 99	80/110
Styrene-Acrylonitrile Copolymer	SAN	95	85	− 20	—	90/104
Styrene-Butadiene Graft Copolymer	SB	60/ 80	50/ 70	− 20	77/ 95	82/104
Acrylonitrile-Butadiene-Styrene Terpolymer	ABS	85/100	75/ 85	− 40	95/110	80/120
Acrylonitrile-Styrene Elastomeric Copolymer	ASA	85/ 90	70/ 75	− 40	92	100/110

Name	Abbrev.					
Polymethyl Methacrylate	PMMA	85/100	65/ 90	−40	70/100	60/100
Polyvinyl Carbazole	PVK	170	160	−100	180	—
Polyacetal	POM	110/140	90/100	−60	160/173	110/170
Polytetrafluoroethylene	PTFE	300	250	−200	—	−/121
Tetrafluoro/Hexafluoropropylene Copolymer	PFEP	250	205	−100	—	−/ 70
Polytrifluorochloroethylene	PCTFE	180	150	−40	—	−/126
Ethylene/Tetrafluoroethylene	PETFE	220	150	−190	—	71/104
Polyamide 6 (Nylon 6)	PA-6	140/180	80/100	−30	180	80/190
Polyamide 66 (Nylon 66)	PA-66	170/200	80/120	−30	200	105/200
Polyamide 11 (Nylon 11)	PA-11	140/150	70/ 80	−70	175	150/160
Polyamide 12 (Nylon 12)	PA-12	140/150	70/ 80	−70	165	140/150
Polyamimde 6-3-T (Transparent Nylon)	PA-6-3-T	130/140	80/100	−70	145	140/180
Polycarbonate	PC	160	135	−100	138	130/145
Polyethylene Terephthalate	PET	200	100	−20	188	—
Polybutylene Terephthalate	PBT	165	100	−30	178	50/190
Polyphenylene Oxide, Modified	mPPO	150	80	−30	148	100/140
Polysulfone	PSO$_2$, Ps	200	150	−100	—	175/180
Polyphenylene Sulfide	PPS	300	200	—	—	137/−
Polyaryl Sulfone	PAS	300	260	—	—	—
Polyether Sulfone	PES	260	200	—	—	—
Polyaryl Ether	PAE	160	120	—	—	150/160

C.3. Continued.

Polymeric Material	Abbreviation DIN 7728 B1.1	Use Temperature (°C)			Heat Resistance (°C)	
		Maximum Short Time	Maximum Continuous	Minimum Continuous	VSP (Vicat 5 kg) DIN 53460	(1.86 N/mm²) (0.45 N/mm²) ASTM D648
Phenol-Formaldehyde (Phenolic)	PF	140	110	—	—	150/190
Urea-Formaldehyde	UF	100	70	—	—	130/—
Melamine-Formaldehyde	MF	120	80	—	—	180/—
Unsaturated Polyester Resin	UPE	200	150	—	—	230/—
Polydiallyl Phthalate (DAP)	DAP	190/250	150/180	−50	—	220/—
Silicone Molding Compound	Si	250	170/180	−50	—	480/—
Polyimide	PI	400	260	−200	—	240/—
Epoxy	EP	180	130	—	—	200/—
Polyurethane Thermoset	PUR	100	80	—	—	90/—
Polyurethane Thermoplastic Elastomer	PUR	110	80	−40	—	—
Linear Polyurethane	PUR	80	60	−15	100	—
Vulkan Fiber	VF	180	105	−30	—	—
Cellulose Acetate	CA	80	70	−40	50/63	90/—
Cellulose Acetate Propionate	CAP	80/120	60/115	−40	100	73/98
Cellulose Acetate Butyrate	CAB	80/120	60/115	−40	60/75	62/71

C.4. Thermal Properties of Polymers (II).

Polymeric Material	Abbreviation DIN 7728 B1.1	Coeff. of Thermal Expansion (mm/mm K $\times$ 10^6)	Thermal Conductivity (M/m K)	Specific Heat (kJ/kg K)
Low Density Polyethylene	LDPE	250	0.32/0.40	2.1 /2.5
High Density Polyethylene	HPDE	200	0.38/0.51	2.1 /2.7
Ethylene-Vinyl Acetate	EVA	160/200	0.35	2.3
Polypropylene	PP	150	0.17/0.22	2.0
Polybutene-1	PB	150	0.20	1.8
Polyisobutylene	PIB	120	0.12/0.20	—
Poly-4-methylpentene-1	PMP	117	0.17	2.18
Ionomer	—	120	0.24	2.20
Unplasticized Polyvinyl Chloride	RPVC	70/ 80	0.14/0.17	0.85/0.9
Plasticized Polyvinyl Chloride	FPVC	150/210	0.15	0.9 /1.8
General Purpose Polystyrene	GPPS	70	0.18	1.3
Styrene-Acrylonitrile Copolymer	SAN	80	0.18	1.3
Styrene-Butadiene Graft Copolymer	SB	70	0.18	1.3
Acrylonitrile-Butadiene-Styrene Terpolymer	ABS	60/110	0.18	1.3
Acrylonitrile-Styrene Elastomeric Copolymer	ASA	80/110	0.18	1.3
Polymethyl Methacrylate	PMMA	70	0.18	1.47
Polyvinyl Carbazole	PVK	—	0.29	—
Polyacetal	POM	90/110	0.25/0.30	1.46

C.4. Continued.

Polymeric Material	Abbreviation DIN 7728 B1.1	Coeff. of Thermal Expansion (mm/mm K $\times$ 10^6)	Thermal Conductivity (M/m K)	Specific Heat (kJ/kg K)
Polytetrafluorethylene	PTFE	100	0.25	1.0
Tetrafluoro/Hexafluoropropylene Copolymer	PFEP	80	0.25	1.12
Polytrifluorochloroethylene	PCTFE	60	0.22	0.9
Ethylene/Tetrafluoroethylene	PETFE	40	0.23	0.9
Polyamide 6 (Nylon 6)	PA-6	80	0.29	1.7
Polyamide 66 (Nylon 66)	PA-66	80	0.23	1.7
Polyamide 11 (Nylon 11)	PA-11	130	0.23	1.26
Polyamide 12 (Nylon 12)	PA-12	150	0.23	1.26
Polyamide 6-3-T (Transparent Nylon)	PA-6-3-T	80	0.23	1.6
Polycarbonate	PC	60/ 70	0.21	1.17
Polyethylene Terephthalate	PET	70	0.24	1.05
Polybutylene Terephthalate	PBT	60	0.21	1.30
Polyphenylene Oxide, Modified	mPPO	60	0.23	1.40
Polysulfone	PSO$_2$, Ps	54	0.28	1.30
Polyphenylene Sulfide	PPS	55	0.25	—
Polyaryl Sulfone	PAS	47	0.16	—
Polyether Sulfone	PES	55	0.18	1.10
Polyaryl Ether	PAE	65	0.26	1.46

Material	Abbr.			
Phenol-Formaldehyde (Phenolic)	PF	30/ 50	0.35	1.30
Urea-Formaldehyde	UF	50/ 60	0.40	1.20
Melamine-Formaldehyde	MF	50/ 60	0.50	1.20
Unsaturated Polyester Resin	UPE	20/ 40	0.70	1.20
Polydiallyl Phthalate (DAP)	DAP	10/ 35	0.60	—
Silicone Molding Compound	Si	20/ 50	0.3 /0.4	0.8 /0.9
Polyimide	PI	50/ 63	0.6 /0.65	—
Epoxy	EP	11/ 35	0.88	0.8
Polyurethane Thermoset	PUR	10/ 20	0.58	1.76
Polyurethane Thermoplastic Elastomer	PUR	150	1.7	0.5
Linear Polyurethane	PUR	210	1.8	0.4
Vulkan Fiber	VF	—	—	—
Cellulose Acetate	CA	120	0.22	1.6
Cellulose Acetate Propionate	CAP	110/130	0.21	1.7
Cellulose Acetate Butyrate	CAB	120	0.21	1.6

C.5. Electrical Properties of Polymers (I).

Polymeric Material	Abbreviation DIN 7728 B1.1	Volume Resistivity (ω cm) DIN 53482	Surface Resistivity (ω) DIN 53482	Dielectric Constant DIN 53483		Dissipation (Power) Factor DIN 53483	
				50 Hz	1 MHz	50 Hz	1 MHz
Low Density Polyethylene	LDPE	$> 10^{17}$	10^{14}	2.29	2.28	1.5×10^{-4}	0.8×10^{-4}
High Density Polyethylene	HPDE	$> 10^{17}$	10^{14}	2.35	2.34	2.4×10^{-4}	2.0×10^{-4}
Ethylene-Vinyl Acetate	EVA	$< 10^{15}$	10^{13}	2.5 /3.2	2.6 /3.2	0.003/0.02	0.03/0.05
Polypropylene	PP	$> 10^{17}$	10^{13}	2.27	2.25	$< 4 \times 10^{-4}$	$< 5 \times 10^{-4}$
Polybutene-1	PB	$> 10^{17}$	10^{13}	2.5	2.2	7×10^{-4}	6×10^{-4}
Polyisobutylene	PIB	$> 10^{15}$	10^{13}	2.3	—	4×10^{-4}	—
Poly-4-methylpentene-1	PMP	$> 10^{16}$	10^{13}	2.12	2.12	7×10^{-5}	3×10^{-5}
Ionomer	—	$> 10^{16}$	10^{13}	—	—	—	—
Unplasticized Polyvinyl Chloride	RPVC	$> 10^{15}$	10^{13}	3.5	3.0	0.011	0.015
Plasticized Polyvinyl Chloride	FPVC	$> 10^{11}$	10^{11}	4 /8	4 /4.5	0.08	0.12
General Purpose Polystyrene	GPPS	$> 10^{16}$	$> 10^{13}$	2.5	2.5	$1/4 \times 10^{-4}$	$0.5/4 \times 10^{-4}$
Styrene-Acrylonitrile Copolymer	SAN	$> 10^{16}$	$> 10^{13}$	2.6 /3.4	2.6 /3.1	$6/8 \times 10^{-3}$	$7/10 \times 10^{-3}$
Styrene-Butadiene Graft Copolymer	SB	$> 10^{16}$	$> 10^{13}$	2.4 /4.7	2.4 /3.8	$4/20 \times 10^{-4}$	$4/20 \times 10^{-4}$
Acrylonitrile-Butadiene-Styrene Terpolymer	ABS	$> 10^{15}$	$> 10^{13}$	2.4 /5	2.4 /3.8	$3/8 \times 10^{-3}$	$2/15 \times 10^{-3}$
Acrylonitrile-Styrene Elastomeric Copolymer	ASA	$> 10^{15}$	$> 10^{13}$	3 /4	3 /3.5	0.02/0.05	0.02/0.03

Polymer	Abbr.						
Polymethyl Methacrylate	PMMA	> 10^{15}	10^{15}	3.3 /3.9	2.2 /3.2	0.04/0.06	0.004/0.04
Polyvinyl Carbazole	PVK	> 10^{16}	10^{14}	—	3	$6/10 \times 10^{-4}$	$6/10 \times 10^{-4}$
Polyacetal	POM	> 10^{15}	10^{13}	3.7	3.7	0.005	0.005
Polytetrafluoroethylene	PTFE	> 10^{18}	10^{17}	< 2.1	< 2.1	< 2×10^{-4}	< 2×10^{-4}
Tetrafluoro/Hexafluoropropylene Copolymer	PFEP	> 10^{18}	10^{17}	2.1	2.1	< 2×10^{-4}	< 7×10^{-4}
Polychlorotrifluoroethylene	PCTFE	> 10^{18}	10^{16}	2.3 /2.8	2.3 /2.5	1×10^{-3}	2×10^{-2}
Ethylene/Tetrafluoroethylene	PETFE	> 10^{16}	10^{13}	2.6	2.6	8×10^{-4}	5×10^{-3}
Polyamide 6 (Nylon 6)	PA-6	10^{12}	10^{10}	3.8	3.4	0.01	0.03
Polyamide 66 (Nylon 66)	PA-66	10^{12}	10^{10}	8.0	4.0	0.14	0.08
Polyamide 11 (Nylon 11)	PA-11	10^{13}	10^{11}	3.7	3.5	0.06	0.04
Polyamide 12 (Nylon 12)	PA-12	10^{13}	10^{11}	4.2	3.1	0.04	0.03
Polyamide 6-3-T (Transparent Nylon)	PA-6-3-T	10^{11}	10^{10}	4.0	3.0	0.03	0.04
Polycarbonate	PC	> 10^{17}	> 10^{15}	3.0	2.9	7×10^{-4}	1×10^{-2}
Polyethylene Terephthalate	PET	10^{16}	10^{16}	4.0	4.0	2×10^{-3}	2×10^{-2}
Polybutylene Terephthalate	PBT	10^{16}	10^{13}	3.0	3.0	2×10^{-3}	2×10^{-2}
Polyphenylene Oxide, Modified	mPPO	10^{16}	10^{14}	2.6	2.6	4×10^{-4}	9×10^{-4}
Polysulfone	PSO$_2$, Ps	> 10^{16}	—	3.1	3.0	8×10^{-4}	3×10^{-3}
Polyphenylene Sulfide	PPS	> 10^{16}	—	3.1	3.2	4×10^{-4}	7×10^{-4}
Polyaryl Sulfone	PAS	> 10^{16}	—	3.9	3.7	3×10^{-3}	13×10^{-3}
Polyether Sulfone	PES	10^{17}	—	3.5	3.5	1×10^{-3}	6×10^{-3}
Polyaryl Ether	PAE	> 10^{10}	—	3.14	3.10	6×10^{-3}	7×10^{-3}

C.5. Continued.

Polymeric Material	Abbreviation DIN 7728 B1.1	Volume Resistivity (ω cm) DIN 53482	Surface Resistivity (ω) DIN 53482	Dielectric Constant DIN 53483		Dissipation (Power) Factor DIN 53483	
				50 Hz	1 MHz	50 Hz	1 MHz
Phenol-Formaldehyde (Phenolic)	PF	10^{11}	$> 10^8$	6	4.5	0.1	0.03
Urea-Formaldehyde	UF	10^{11}	$> 10^{10}$	8	7	0.04	0.3
Melamine-Formaldehyde	MF	10^{11}	$> 10^8$	9	8	0.06	0.03
Unsaturated Polyester Resin	UPE	$> 10^{12}$	$> 10^{10}$	6	5	0.04	0.02
Polydiallyl Phthlate (DAP)	DAP	$10^{13}/10^{16}$	10^{13}	5.2	4	0.04	0.03
Silicone Molding Compound	Si	10^{14}	10^{12}	4	3.5	0.03	0.02
Polyimide	PI	$> 10^{16}$	$> 10^{15}$	3.5	3.4	2×10^{-3}	5×10^{-3}
Epoxy	EP	$> 10^{14}$	$> 10^{12}$	3.5 /5	3.5 /5	0.001	0.01
Polyurethane Thermoset	PUR	10^{16}	10^{14}	3.6	3.4	0.05	0.05
Polyurethane Thermoplastic Elastomer	PUR	10^{12}	10^{11}	6.5	5.6	0.03	0.06
Linear Polyurethane	PUR	10^{13}	10^{12}	5.8	4.0	0.12	0.07
Vulkan Fiber	VF	10^{10}	10^8	—	—	0.08	—
Cellulose Acetate	CA	10^{13}	10^{12}	5.8	4.6	0.02	0.03
Cellulose Acetate Propionate	CAP	10^{16}	10^{14}	4.2	3.7	0.01	0.03
Cellulose Acetate Butyrate	CAB	10^{16}	10^{14}	3.7	3.5	0.006	0.021

C.6. Electrical Properties of Polymers (II).

Polymeric Material	Abbreviation DIN 7728 B1.1	Dielectric Strength ASTM D149 (kV/25 μm)	Dielectric Strength DIN 53481 (kV/cm)	Tracking Resistance DIN 53480 KA	Tracking Resistance DIN 53480 KB	Tracking Resistance DIN 53480 KC
Low Density Polyethylene	LDPE	> 700	—	3[b]	> 600	> 600
High Density Polyethylene	HPDE	> 700	—	3[c]	> 600	> 600
Ethylene-Vinyl Acetate	EVA	—	620/780	3[c]	—	—
Polypropylene	PP	800	500/650	3[c]	> 600	> 600
Polybutene-1	PB	700	—	3[c]	> 600	> 600
Polyisobutylene	PIB	230	—	3[c]	> 600	> 600
Poly-4-methylpentene-1	PMP	280	700	3[c]	> 600	> 600
Ionomer	—	—	—	—	—	—
Unplasticized Polyvinyl Chloride	RPVC	200/400	350/500	2/3[b]	600	600
Plasticized Polyvinyl Chloride	FPVC	150/300	300/400	—	—	—
General Purpose Polystyrene	GPPS	500	300/700	1/2	140	150/250
Styrene-Acrylonitrile Copolymer	SAN	500	400/500	1/2	160	150/260
Styrene-Butadiene Graft Copolymer	SB	500	300/600	2	> 600	> 600
Acrylonitrile-Butadiene-Styrene Terpolymer	ABS	400	350/500	3[a]	> 600	> 600
Acrylonitrile-Styrene Elastomeric Copolymer	ASA	350	360/400	3[a]	> 600	> 600
Polymethyl Methacrylate	PMMA	300	400/500	3[c]	> 600	> 600
Polyvinyl Carbazole	PVK	500	—	3[b]	> 600	> 600
Polyacetal	POM	700	380/500	3[b]	> 600	> 600

C.6. Continued.

Polymeric Material	Abbreviation DIN 7728 B1.1	Dielectric Strength		Tracking Resistance DIN 53480		
		ASTM D149 (kV/25 µm)	DIN 53481 (kV/cm)	KA	KB	KC
Polytetrafluoroethylene	PTFE	500	480	3[c]	> 600	> 600
Tetrafluoro/Hexafluoropropylene Copolymer	PFEP	500	550	3[c]	> 600	> 600
Polychlorotrifluoroethylene	PCTFE	500	550	3[c]	> 600	> 600
Ethylene/Tetrafluoroethylene	PETFE	380	400	3[c]	> 600	> 600
Polyamide 6 (Nylon 6)	PA-6	350	400	3[b]	> 600	> 600
Polyamide 66 (Nylon 66)	PA-66	400	600	3[b]	> 600	> 600
Polyamide 11 (Nylon 11)	PA-11	300	425	3[b]	> 600	> 600
Polyamide 12 (Nylon 12)	PA-12	300	450	3[b]	> 600	> 600
Polyamide 6-3-T (Transparent Nylon)	PA-6-3-T	250	350	3[b]	> 600	> 600
Polycarbonate	PC	350	380	1	120/160	260/300
Polyethylene.Terephthalate	PET	500	420	2	—	—
Polybutylene Terephthalate	PBT	500	420	3[b]	420	380
Polyphenylene Oxide, Modified	mPPO	500	450	1	300	300
Polysulfone	PSO_2, Ps	—	425	1	175	175
Polyphenylene Sulfide	PPS	—	595	—	—	—
Polyaryl Sulfone	PAS	—	350	—	—	—
Polyether Sulfone	PES	—	400	—	—	—
Polyaryl Ether	PAE	—	430	—	—	—

Phenol-Formaldehyde (Phenolic)	PF	50/100	300/400	1	140/180	125/175
Urea-Formaldehyde	UF	80/150	300/400	3[a]	> 400	> 600
Melamine-Formaldehyde	MF	80/150	290/300	3[b]	> 500	> 600
Unsaturated Polyester Resin	UPE	120	250/530	3[c]	> 600	> 600
Polydiallyl Phthalate (DAP)	DAP	—	400	3[c]	> 600	> 600
Silicone Molding Compound	Si	—	200/400	3[c]	> 600	> 600
Polyimide	PI	—	560	1	> 300	> 380
Epoxy	EP	—	300/400	3[c]	> 300	200/600
Polyurethane Thermoset	PUR	—	240	3[b]	—	—
Polyurethane Thermoplastic Elastomer	PUR	—	300/600	3[a]	> 600	> 600
Linear Polyurethane	PUR	330	—	—	—	—
Vulkan Fiber	VF	70/180	—	—	—	—
Cellulose Acetate	CA	320	400	3[a]	> 600	> 600
Cellulose Acetate Propionate	CAP	350	400	3[a]	> 600	> 600
Cellulose Acetate Butyrate	CAB	380	400	3[a]	> 600	> 600

[a, b, c,] and KA, KB, KC - See DIN 53480 Standard for interpretation.

C.7. Optical and Absorption Properties of Polymers.

Polymeric Material	Abbreviation DIN 7728 B1.1	Refractive Index (n_D^{20}) DIN 53491	Clarity	Water Absorption (mg in 4 d) DIN 53492	(% in 24 h) ASTM D570
Low Density Polyethylene	LDPE	1.51	> Transp	< 0.01	< 0.01
High Density Polyethylene	HPDE	1.53	> Opaque	< 0.01	< 0.01
Ethylene-Vinyl Acetate	EVA	—	Transp/Opaque	—	0.05/0.13
Polypropylene	PP	1.49	Transp/Opaque	< 0.01	0.01/0.03
Polybutene-1	PB	—	> Opaque	< 0.01	< 0.02
Polyisobutylene	PIB	—	> Opaque	< 0.01	< 0.01
Poly-4-methylpentene-1	PMP	1.46	> Opaque	—	0.01
Ionomer	—	1.51	Transp	—	0.1/0.4
Unplasticized Polyvinyl Chloride	RPVC	1.52/1.55	Transp/Opaque	3/18	0.04/0.4
Plasticized Polyvinyl Chloride	FPVC	—	Transp/Opaque	6/30	0.15/0.75
General Purpose Polystyrene	GPPS	1.59	Transp	—	0.03/0.1
Styrene-Acrylonitrile Copolymer	SAN	1.57	Transp	—	0.2 /0.3
Styrene-Butadiene Graft Copolymer	SB	—	Opaque	—	0.05/0.6
Acrylonitrile-Butadiene-Styrene Terpolymer	ABS	—	Opaque	—	0.2 /0.45
Acrylonitrile-Styrene Elastomeric Copolymer	ASA	—	Transl/Opaque	—	—
Polymethyl Methacrylate	PMMA	1.49	Transp	35/45	0.1 /0.4
Polyvinyl Carbazole	PVK	—	Opaque	0.5	0.1 /0.2
Polyacetal	POM	1.48	Opaque	20/30	0.22/0.25

Material	Abbrev.				
Polytetrafluoroethylene	PTFE	1.35	Opaque	—	0
Tetrafluoro/Hexafluoropropylene Copolymer	PFEP	1.34	Transp/Transl	—	< 0.1
Polytrifluorochloroethylene	PCTFE	1.43	Transl/Opaque	—	0
Ethylene/Tetrafluoroethylene	PETFE	1.40	Transp/Opaque	—	0.03
Polyamide 6 (Nylon 6)	PA-6	1.53	Transl/Opaque	—	1.3 /1.9
Polyamide 66 (Nylon 66)	PA-66	1.53	Transl/Opaque	—	1.5
Polyamide 11 (Nylon 11)	PA-11	1.52	Transl/Opaque	—	0.3
Polyamide 12 (Nylon 12)	PA-12	—	Transl/Opaque	—	0.25
Polyamide 6-3-T (Transparent Nylon)	PA-6-3-T	1.53	Transp	—	0.4
Polycarbonate	PC	1.58	Transp	10	0.16
Polyethylene Terephthalate	PET	—	Transp/Opaque	18/20	0.30
Polybutylene Terephthalate	PBT	—	Opaque	—	0.08
Polyphenylene Oxide, Modified	mPPO	—	Opaque	—	0.06
Polysulfone	PSO_2, Ps	1.63	Transp/Opaque	—	0.02
Polyphenylene Sulfide	PPS	—	Opaque	—	0.02
Polyaryl Sulfone	PAS	1.67	Opaque	—	1.8
Polyether Sulfone	PES	1.65	Transp	—	0.43
Polyaryl Ether	PAE	—	Transl/Opaque	—	0.25
Phenol-Formaldehyde (Phenolic)	PF	—	Opaque	< 150	0.3 /1.2
Urea-Formaldehyde	UF	—	Opaque	< 300	0.4 /0.8
Melamine-Formaldehyde	MF	—	Opaque	< 250	0.1 /0.6
Unsaturated Polyester Resin	UPE	—	Opaque	< 45	0.03/0.5
Polydiallyl Phthalate (DAP)	DAP	—	Opaque	—	0.12/0.35
Silicone Molding Compound	Si	—	Opaque	—	0.02
Polyimide	PI	—	Opaque	—	0.32

C.7. Continued.

Polymeric Material	Abbreviation DIN 7728 B1.1	Refractive Index (n_D^{20}) DIN 53491	Clarity	Water Absorption (mg in 4 d) DIN 53492	Water Absorption (% in 24 h) ASTM D570
Epoxy	EP	—	Opaque	< 30	0.05/0.2
Polyurethane Thermoset	PUR	—	Transp	—	0.1 /0.2
Polyurethane Thermoplastic Elastomer	PUR	—	Transl/Opaque	—	0.7 /0.9
Linear Polyurethane	PUR	—	Transl/Opaque	130	—
Vulkan Fiber	VF	—	Opaque	—	7 /9
Cellulose Acetate	CA	1.50	Transp	130	6
Cellulose Acetate Propionate	CAP	1.47	Transp	40/ 60	1.2 /2.8
Cellulose Acetate Butyrate	CAB	1.47	Transp	40/ 60	0.9 /3.2

Transp = Transparent, Transl = Translucent, > Transp = Tendency Toward Transparency, > Opaque = Tendency Toward Opacity.

APPENDIX D

CREEP PROPERTIES
OF POLYMERS*

*Source: H. Domininghaus, *Die Kunststoffe und ihre Eigenschaften*, VDI Verlag, Düsseldorf, FRG (1986), Table II. Permission granted by copyright owner.

D.1. Creep Modulus Properties of Polyethylenes, Poly-4-Methylpentene-1 and Polyvinyl Chlorides.

Polymeric Material	Nature of Stress	Temperature (°C)	Test Stress Level, Tensile or Flexure (N/mm²)	Creep Modulus (N/mm²)			
				1 h	10 h	100 h	1000 h
High Density Polyethylene (HDPE) Injection Molding Grade	Tensile	23	1.75	—	950	760	520
			3.50	—	580	480	380
			5.25	—	480	395	295
			7.00	—	420	345	280
		40	1.75	—	320	290	265
			3.50	—	240	225	210
			5.25	—	225	205	200
			7.00	—	175	170	—
20% Glass-Reinforced HDPE	Tensile	23	14.0	2,600	2,400	2,200	—
		38	14.0	2,000	1,750	1,600	—
40% Glass-Reinforced HDPE	Flexure	23	14.0	5,900	5,200	4,800	—
		38	14.0	4,850	4,350	4,100	—
		60	14.0	4,100	3,600	3,350	—
		82	14.0	3,500	3,400	2,950	—
Low Density Polyethylene (LDPE) Injection Molding Grade	Tensile	20	2.0	127	110	98	—
			4.1	108	87	73	—
			5.1	52	46	42	39
		60	10.2	48	42	38	—
			15.0	41	37	32	—
Poly-4-Methylpentene-1	Tensile	20	5.1	1,200	850	560	—
		60	10.2	740	460	—	—
		82	2.0	220	170	145	—

Material	Test	Temp					
Polyvinyl Chloride (Impact Modified) (FPVC) Extrusion Grade	Tensile	23	3.50	1,650	1,600	1,550	—
			7.00	1,350	1,200	1,100	—
			10.50	1,100	950	775	—
		38	2.10	740	580	410	275
			5.25	740	520	350	210
		49	3.50	470	200	85	—
			7.00	415	120	—	—
Polyvinyl Chloride (Impact Modified) (FPVC) Injection Molding Grade	Tensile	23	14.0	1,850	1,750	1,650	—
			28.0	1,700	1,500	1,300	—
			35.0	1,550	1,150	—	—
25% Glass-Reinforced Polyvinyl Chloride (GR-PVC) Inj. Molding Grade	Flexure	24	35.0	—	9,500	8,800	8,500
			70.0	—	9,200	8,500	7,900
35% Glass-Reinforced Polyvinyl Chloride (GR-PVC) Inj. Molding Grade	Flexure	24	35.0	—	15,400	11,100	10,600
			70.0	—	12,000	10,800	9,900
PVC/PP Copolymer Inj. and Extrusion Grades	Tensile	23	27.1	3,900	3,200	1,700	910
			28.0	3,750	2,450	1,200	—
			30.2	3,550	1,750	1,000	—
			32.3	2,950	1,475	—	—
			34.5	2,450	1,100	—	—
			38.7	1,300	—	—	—

D.2. Creep Modulus Properties of Polyacetal (POM) Homopolymers and Copolymers.

Polymeric Material	Nature of Stress	Temperature (°C)	Test Stress Level, Tensile or Flexure (N/mm²)	Creep Modulus (N/mm²)			
				1 h	10 h	100 h	1000 h
Polyacetal (POM) (Homopolymer) Injection Molding Grade	Flexure	23	3.5	2,800	2,600	2,050	1,750
			7.0	2,800	2,550	2,050	1,750
			10.5	2,750	2,550	2,000	1,700
		46	3.5	1,850	1,500	1,200	1,050
			7.0	1,750	1,400	1,200	1,000
			10.5	1,750	1,400	1,100	900
			14.0	1,700	1,350	1,100	850
		60	3.5	1,700	1,400	1,050	850
			7.0	1,600	1,350	1,050	850
			14.0	1,350	1,000	850	700
		85	3.5	1,250	1,050	850	650
			7.0	1,200	1,000	750	550
			10.5	1,150	1,000	750	500
			14.0	1,100	900	700	—
		100	3.5	900	800	650	500
			7.0	850	700	550	400
			10.5	800	650	500	—
Polyacetal (POM) (Copolymer) Injection Molding Grade	Flexure	23	3.5	2,700	2,400	2,200	1,900
			35.0	2,700	2,400	1,900	1,550
		82	3.5	700	620	540	480
		116	3.5	51	44	40	33
			7.0	51	43	39	32

Material	Test						
25% Glass-Reinforced Polyacetal (POM) Copolymer	Flexure	82	3.5	3,750	—	2,900	2,700
			22.0	3,750	—	2,900	2,600
		116	3.5	2,300	—	1,650	1,300
25% Glass-Reinforced Polyacetal (POM) Homopolymer	Flexure	23	3.5	8,500	—	5,600	4,500
			7.0	6,500	—	4,600	3,800
			10.5	6,200	—	4,300	3,500
			14.0	6,100	5,500	4,450	3,450
		60	3.5	4,800	4,150	2,900	2,100
			7.0	3,750	3,250	2,450	2,000
			14.0	2,950	2,250	1,750	1,350
		85	3.5	2,900	2,400	1,850	1,400
			7.0	2,650	2,100	1,700	1,350
			10.5	2,300	2,100	1,550	1,250
			17.5	2,100	1,600	1,300	—
		90	14.0	2,300	1,750	1,350	1,100
30% Glass-Reinforced Polyacetal (POM) Homopolymer	Flexure	24	14	—	8,000	6,000	5,500
40% Glass-Reinforced Polyacetal (POM) Homopolyer	Flexure	23	28	8,800	7,750	6,700	—
		38	28	6,100	6,100	4,700	—
		82	28	3,150	2,400	2,100	—

D.3. Creep Modulus Properties of Polyethylenes and Halogenated Polyethylenes.

Polymeric Material	Nature of Stress	Temperature (°C)	Test Stress Level, Tensile or Flexure (N/mm²)	Creep Modulus (N/mm²)			
				1 h	10 h	100 h	1000 h
Polytetrafluoroethylene (PTFE) Molding Powder	Tensile	18	7.0	1,260	1,260	1,260	—
			14.0	670	625	515	—
			21.0	420	330	175	—
		23	3.5	420	310	225	—
			7.0	125	74	44	—
		100	1.4	115	105	90	—
			3.5	44	35	30	—
			4.1	28	22	17	—
		200	0.7	50	40	35	—
			1.4	42	32	25	—
			2.1	18	15	13	—
	Compression	23	3.5	35	30	26	—
			7.0	28	23	20	—
			12.3	15	13	—	—
		100	1.4	18	13	11	—
			3.5	10	8.5	7.1	—
			5.3	8	7	—	—
25% Glass-Reinforced Polytetrafluoroethylene (PTFE) Molding Powder	Tensile	20	1.1	140	108	84	70
			2.8	122	94	75	64
			4.2	104	85	70	60
			5.6	93	73	60	51
			7.0	84	63	51	43
			8.4	72	54	43	36
			9.8	64	48	37	30

Material	Test	Temp (°C)					
Tetrafluoro/ Hexafluoropropylene Copolymer (PFEP) Molding and Extrusion Grades	Tensile	18	2.1	610	560	510	—
			14.0	390	335	260	—
		23	3.5	470	420	370	—
			7.0	395	330	275	—
			10.5	210	130	65	—
		100	1.4	51	42	37	—
			3.5	31	20	17	—
		175	0.7	32	28	25	—
			1.4	26	21	17	—
	Compression	23	3.5	34	31	29	—
			14.0	25	22	19	—
			17.5	16	15	13	—
			21.0	12	11	—	—
		100	1.4	12	9.5	8.0	—
			3.5	9.5	7.5	6.0	—
			5.25	8.0	6.2	5.4	—
Polychlorotrifluoroethylene (PCTFE) Injection and Extrusion Grades	Tensile	23	14.0	930	830	705	585
			21.0	570	470	375	275
		66	7.0	470	325	255	—
		121	1.4	120	91	63	—
			3.5	84	56	35	—
Polyethylene Terephthalate (PET) Injection Molding Grade	Tensile	23	7.5	3,100	2,900	2,800	2,650
			15.0	3,100	2,900	2,800	2,650
			25.0	3,100	2,900	2,800	2,650
		40	7.5	2,750	2,450	2,200	1,700
			10.0	2,750	2,450	2,200	1,650
			15.0	2,750	2,450	2,100	1,600

D.3. Continued.

Polymeric Material	Nature of Stress	Temperature (°C)	Test Stress Level, Tensile or Flexure (N/mm²)	Creep Modulus (N/mm²)			
				1 h	10 h	100 h	1000 h
18% Glass-Reinforced Polyethylene Terephthalate (PET) Injection Molding Grade	Tensile	23	20	6,750	6,250	5,900	5,250
			40	6,750	6,250	5,900	5,250
			60	6,750	6,250	5,900	5,250
		40	20	6,250	5,600	5,250	4,850
			40	6,050	5,400	5,150	4,100
			60	5,750	5,000	4,100	3,350
		70	10	3,450	2,650	2,000	1,550
			20	3,100	2,400	1,850	1,350
			30	2,750	2,150	1,650	1,250
30% Glass-Reinforced Polyethylene Terephthalate (PET)	Flexure	24	14.0	—	7,750	5,600	5,050
36% Glass-Reinforced Polyethylene Terephthalate (PET)	Tensile	23	20	10,900	10,300	10,000	9,700
			40	10,800	10,200	9,900	9,600
			60	10,600	10,100	9,800	9,400
			80	10,300	10,000	9,300	8,800
			100	9,700	9,200	8,800	7,900

40	20	10,600	9,500	8,000	7,100
	40	10,500	9,400	7,800	7,050
	60	10,200	9,100	7,700	7,000
	80	9,400	8,100	7,400	6,700
	100	8,100	7,300	6,700	6,000
70	20	6,700	4,700	4,150	3,950
	40	6,200	4,650	4,100	4,000
	60	6,000	4,550	3,950	3,450
	80	5,600	4,150	3,350	—
	100	5,150	—	—	—
110	5	3,900	3,650	3,500	3,300
	10	3,800	3,300	2,900	2,700
	15	3,750	3,150	2,800	2,550
	22	3,650	3,100	2,750	2,400
	30	3,500	2,800	2,600	2,250

D.4. Creep Modulus Properties of Polysulfones and Thermoset Molding Compounds.

Polymeric Material	Nature of Stress	Temperature (°C)	Test Stress Level, Tensile or Flexure (N/mm²)	Creep Modulus (N/mm²)			
				1 h	10 h	100 h	1000 h
Polysulfone (PSO₂) Injection Molding Grade	Tensile	23 100 149	28 21 7.0	— — —	2,450 1,700 1,000	2,400 1,500 700	2,300 1,350 550
Polysulfone (VE-0 Grade) Injection Molding Grade	Flexure	60 125	29.5 14	— —	2,300 —	2,250 —	2,200 —
30% Glass-Reinforced Polysulfone Inj. Molding Grade	Flexure	100	14 21	— —	7,000 5,600	5,400 4,350	5,200 4,200
30% Glass-Reinforced Polysulfone with 15% PTFE	Flexure	24	14	—	8,450	6,400	6,050
40% Glass-Reinforced Polysulfone	Flexure	24	35 70	— —	12,000 11,000	10,600 10,200	10,400 10,200

Material	Property	Temp					
Alkyd Resin (Thermoset)	Tensile	23	21	19,000	17,600	16,200	—
		121	7	3,650	3,250	2,950	—
			14	3,050	—	—	—
Diallyl Phthalate (DAP) Molding Compound	Tensile	23	24.5	14,800	14,200	13,500	—
		121	7	4,200	3,800	3,300	—
			14	3,750	3,100	2,550	—
			21	1,850	—	—	—
Epoxy Molding Compound	Tensile	23	28	10,200	9,500	8,800	—
		121	21	6,550	5,900	5,350	4,800
			28	4,000	3,500	3,100	—
Phenolic, Glass-Filled	Tensile	23	21	30,900	30,200	26,700	—
			28	29,900	28,000	21,000	—
		121	14	6,550	6,000	5,400	—
			21	5,700	5,300	4,850	—
Phenolic, Mineral-Filled	Tensile	25	14.2	30,900	—	29,500	28,000
Unsaturated Polyester (UPE) Molding Compound, Glass-Filled	Tensile	23	14	9,900	9,200	7,750	6,550
			21	9,850	9,200	7,750	5,550

D.5. Creep Modulus Properties of Polypropylenes and Polystyrenics.

Polymeric Material	Nature of Stress	Temperature (°C)	Test Stress Level, Tensile or Flexure (N/mm²)	Creep Modulus (N/mm²)			
				1 h	10 h	100 h	1000 h
Polypropylene (PP) Injection Molding Grade	Tensile	20	3.5	1,050	810	610	465
			7.0	880	660	505	395
			10.5	730	540	410	325
		60	1.4	410	330	280	240
			2.8	380	310	265	225
			4.2	345	280	245	211
Polypropylene-Polyethylene (EPM) Copolymer	Tensile	20	20.4	820	680	535	395
			51	730	550	395	290
		60	10.2	330	265	225	200
			20.4	315	255	220	190
			35.6	240	185	140	90
20% Glass-Reinforced Polypropylene (GR-PP) Injection Molding Grade	Tensile	23	17.5	4,500	3,800	3,250	2,750
			35	3,750	3,400	3,000	2,550
		80	10.5	2,950	2,750	2,600	2,300
			21	2,650	2,400	2,000	1,700
30% Glass-Reinforced Polypropylene (GR-PP) Injection Molding Grade	Tensile	23	28	5,750	5,150	4,650	4,150
			56	5,150	4,550	3,900	3,150
		80	14	4,350	4,000	3,450	2,950
			24.5	4,000	3,450	2,950	2,400

Material	Test	Temp		Data 1	Data 2	Data 3	Data 4
40% Talc-Filled Polypropylene (PP) Injection Molding Grade	Tensile	24	7	8,100	5,600	4,100	—
40% Glass-Reinforced Polypropylene (GR-PP) Injection Molding Grade	Tensile	23	35	7,700	6,350	5,200	4,450
			56	5,900	5,000	4,100	—
			70	5,300	4,500	3,700	—
		60	17.5	5,150	4,650	4,400	3,750
			28	4,600	4,150	3,750	3,250
			32.5	4,350	4,000	3,350	—
Polystyrene (PS) Injection Molding Grade	Tensile	23	21.8	2,950	2,950	—	—
Impact-Modified Polybutadiene-Styrene (SB) Injection Molding Grade	Tensile	23	14.2	2,100	1,950	1,700	—
			16	1,850	1,500	—	—
			17	1,700	—	—	—
		38	7.4	1,800	1,750	1,450	900
			10.3	1,800	1,750	1,450	900
			14.2	1,600	—	—	—
20% Glass-Reinforced Polystyrene (GR-PS)	Flexure	23	14	6,900	6,700	6,550	—
		38	14	6,600	6,050	4,750	—
		60	14	5,550	3,650	2,900	—
35% Glass-Reinforced Polystyrene (GR-PS)	Flexure	23	56	9,500	9,000	8,750	—
		38	56	8,800	8,100	6,750	—
		60	56	6,000	3,400	3,000	—

D.6. Creep Modulus Properties of Acrylic-Based Polymers and Copolymers.

Polymeric Material	Nature of Stress	Temperature (°C)	Test Stress Level, Tensile or Flexure (N/mm²)	Creep Modulus (N/mm²)			
				1 h	10 h	100 h	1000 h
Polystyrene-Acrylonitrile (SAN) Injection Molding Grade	Tensile	23	31	3,500	3,350	3,000	2,300
			34.5	3,500	3,350	3,000	2,200
			38.7	3,450	3,100	2,500	—
			42.3	3,350	3,050	2,300	—
			45.4	3,300	3,000	—	—
			48.3	3,250	—	—	—
20% Glass-Reinforced SAN	Flexure	24	35	—	7,750	7,000	6,500
30% Glass-Reinforced SAN	Flexure	24	35	—	9,850	8,800	8,800
35% Glass-Reinforced SAN	Flexure	23	14	11,500	11,000	10,500	—
		38	14	8,000	6,250	5,050	—
		60	14	5,800	4,800	4,350	—
40% Glass-Reinforced SAN	Flexure	24	70	—	12,700	11,500	10,900

Material	Property	Temp (°C)	Rate				
Acrylonitrile-Butadiene-Styrene (ABS) Injection Molding Grade	Tensile	23	14	2,300	2,250	2,050	1,700
			21	2,300	2,150	1,850	1,500
			28	2,150	1,950	1,600	—
			35	2,000	1,650	—	—
		71	3.5	1,850	1,750	1,400	700
			7	1,850	1,650	1,000	600
			10.5	1,850	1,450	950	500
Acrylonitrile-Butadiene-Styrene (ABS) Sheet Extrusion Grade	Tensile	23	11.2	2,550	2,300	2,000	1,650
			14	2,500	2,300	1,950	1,500
			21	2,500	2,250	1,800	1,000
			28	2,500	2,050	—	—
		71	3.5	2,050	1,500	750	300
			7	2,050	1,350	600	—
			10.5	1,850	1,000	—	—
ABS/PVC Copolymer, Injection Molding Grade	Flexure	23	7	2,150	2,100	1,900	1,600
			14	2,100	2,050	1,850	1,600
ABS/PVC Copolymer, Injection and Extrusion Grades	Flexure	23	7	2,550	2,500	2,350	2,200
			14	2,550	2,500	2,350	2,200
20% Glass-Reinforced ABS	Flexure	24	14	—	5,700	5,550	5,500
			35	—	5,650	5,500	5,400

D.6. Continued.

Polymeric Material	Nature of Stress	Temperature (°C)	Test Stress Level, Tensile or Flexure (N/mm^2)	Creep Modulus (N/mm^2)			
				1 h	10 h	100 h	1000 h
40% Glass-Reinforced ABS	Flexure	24	35	—	12,000	12,000	11,600
			70	—	12,500	11,600	11,200
Polymethyl Methacrylate (PMMA) Cell-Cast	Tensile	20	10.2	2,800	2,550	2,300	—
			20.4	2,700	2,450	2,050	—
			30.6	2,350	2,000	1,650	—
		60	5.1	2,300	2,100	1,750	1,350
			15.3	2,100	1,850	1,350	—
			20.4	1,900	1,550	—	—

D.7. Creep Modulus Properties of Nylons.

Polymeric Material	Nature of Stress	Temperature (°C)	Test Stress Level, Tensile or Flexure (N/mm^2)	Creep Modulus (N/mm^2)			
				1 h	10 h	100 h	1000 h
Nylon 6 (PA-6), Dry	Tensile	23	14	2,350	2,200	1,850	—
		66	14	330	280	225	—
			21	232	197	169	—
Nylon 6 (PA-6), 50% Relative Humidity	Tensile	23	14	430	385	335	310
Nylon 66 (PA-66), Dry	Tensile	23	10.5	2,950	2,750	2,400	—
			21	2,900	2,700	2,350	—
			42	2,800	2,650	2,000	—
Nylon 66 (PA-66), 50% Relative Humidity	Tensile	23	10.5	1,100	900	800	700
			21	700	600	500	450
14% Glass-Reinforced Nylon 6, 30% Relative Humidity	Tensile	23	14	2,700	2,300	2,050	1,750
			28	2,000	1,650	1,400	1,200
30% Glass-Reinforced Nylon 6, Dry	Tensile	121	28	1,900	1,750	1,500	1,300
30% Glass-Reinforced Nylon 6, 50% Relative Humidity	Tensile	23	28	4,300	3,800	3,350	3,050

D.7. Continued.

Polymeric Material	Nature of Stress	Temperature (°C)	Test Stress Level, Tensile or Flexure (N/mm^2)	Creep Modulus (N/mm^2)			
				1 h	10 h	100 h	1000 h
30% Glass-Reinforced Nylon 6, Dry	Flexure	23	56	5,850	5,550	5,050	—
		38	56	3,600	3,400	3,250	—
		82	56	3,300	3,150	2,950	—
		116	56	2,650	2,450	2,250	—
40% Glass-Reinforced Nylon 6, Dry	Flexure	23	70	7,750	7,000	6,100	—
		38	70	6,700	6,100	5,700	—
		82	70	5,400	5,150	4,800	—
		116	70	4,650	4,350	3,900	—
40% Asbestos-Filled Nylon 6, Dry	Tensile	23	28	11,200	11,200	9,850	—
		121	14	3,150	2,550	2,100	—
			21	2,300	1,900	1,600	1,350
40% Asbestos-Filled Nylon 6, 50% Relative Humidity		23	28	3,100	2,550	2,050	—

Material	Test	Temp				
30% Glass-Reinforced Nylon 66, Dry	Tensile	23	8,800	6,700	5,850	—
		38	5,200	4,700	4,550	—
		82	4,550	4,300	4,200	—
		116	3,550	3,300	3,100	—
30% Glass-Reinforced Nylon 66, with 15% PTFE, 50% Relative Humidity	Tensile	23	—	6,550	4,200	3,950
40% Glass-Reinforced Nylon 66, Dry	Flexure	23	11,500	11,000	10,500	—
		38	9,500	9,000	8,500	—
		82	8,000	7,500	7,400	—
		116	7,500	7,000	6,700	—
30% Glass-Reinforced Nylon 610, Dry	Flexure	23	5,550	4,950	4,500	—
		38	3,250	2,950	2,900	—
		82	3,000	2,550	2,400	—
		116	2,650	2,200	1,900	—
40% Glass-Reinforced Nylon 610, Dry	Flexure	23	8,500	7,400	6,700	—
		38	6,350	6,000	5,750	—
		82	5,250	5,050	4,800	—
		116	5,150	4,950	4,550	—

D.8. Creep Modulus Properties of Polycarbonates, Modified Polyphenylene Oxide and Other Polymers.

Polymeric Material	Nature of Stress	Temperature (°C)	Test Stress Level, Tensile or Flexure (N/mm²)	Creep Modulus (N/mm²)			
				1 h	10 h	100 h	1000 h
Polycarbonate (PC) Injection Molding Grade	Flexure	23	21	2,450	2,350	2,250	2,150
		54	10.5	2,100	1,950	1,750	1,600
			14	2,100	1,850	1,600	1,400
		71	3.5	1,750	1,550	1,400	1,300
			7.0	1,700	1,500	1,350	1,250
			10.5	1,600	1,400	1,350	1,200
		121	1.7	1,050	750	490	390
			3.5	950	700	490	390
20% Glass-Reinforced Polycarbonate (GR-PC) Injection Molding Grade	Flexure	23	56	7,200	7,000	6,850	—
		38	56	6,750	6,300	5,900	—
		93	56	5,100	4,450	3,650	—
30% Glass-Reinforced Polycarbonate (GR-PC)	Flexure	24	14	—	8,400	6,300	6,050
		54	21	7,300	7,000	7,000	7,000
			28	7,050	6,100	5,300	5,300
			35	6,750	5,600	5,250	5,250
		71	21	7,600	7,050	6,900	6,700
			28	7,400	6,800	6,200	6,000
			35	6,600	6,100	5,800	5,350

Material	Test	Temp					
40% Glass-Reinforced Polycarbonate (GR-PC)	Flexure	43	28	7,700	6,600	6,100	6,050
			35	7,100	5,600	5,400	5,300
		121	21	7,500	6,250	4,700	2,950
			35	7,500	6,250	4,700	2,950
Modified Polyphenylene Oxide (mPPO), Injection Molding Grade	Flexure	23	7	2,750	2,750	2,650	2,350
			14	2,500	2,450	2,350	2,050
			21	2,400	2,300	2,200	1,850
		60	7	2,050	1,950	1,800	1,700
			14	2,050	1,750	1,500	1,250
			21	2,050	1,700	1,250	1,000
		77	5.6	1,700	1,400	1,200	1,000
			7.0	1,650	1,400	1,200	1,000
			10.5	1,650	1,250	1,050	900
			14	1,650	1,150	850	650
		100	3.5	1,750	1,450	1,100	800
			10.5	1,650	1,400	1,100	800
			14	1,650	1,350	1,000	800
			21	1,550	1,250	1,000	800
20% Glass-Reinforced Polyphenylene Oxide (mPPO) Injection Molding Grade	Flexure	77	14	5,250	4,100	3,300	3,100
			17.5	4,650	3,600	3,200	2,950
			24.5	4,400	3,700	3,000	2,250

D.8. Continued.

Polymeric Material	Nature of Stress	Temperature (°C)	Test Stress Level, Tensile or Flexure (N/mm^2)	Creep Modulus (N/mm^2)			
				1 h	10 h	100 h	1000 h
30% Glass-Reinforced Polyphenylene Oxide (mPPO) Injection Molding Grade	Flexure	23	14	8,000	8,000	8,000	7,350
			21	7,700	7,500	7,400	6,900
			28	6,800	6,800	6,700	6,600
			35	6,650	6,650	6,500	5,550
		66	14	7,800	7,300	6,850	6,300
			21	7,350	7,050	6,800	6,050
		77	14	7,050	6,150	5,900	5,000
			21	6,700	6,150	5,500	4,800
			28	6,450	6,100	5,100	4,350
			35	5,850	5,300	4,600	4,000
Polyarylether (PAE)	Tensile	23	14	2,150	2,100	1,750	1,300
			21	1,950	1,700	1,400	1,050
			28	1,900	1,650	1,300	950
			35	1,700	1,400	1,050	—
		82	3.5	2,000	1,700	1,200	720
			7	1,950	1,650	1,050	670
			10.5	1,850	1,350	1,000	570
			17.5	1,700	1,300	800	—
40% Glass-Reinforced Linear Polyurethane (PUR)	Flexure	23	3.5	—	1,100	900	875

APPENDIX E

THERMODYNAMIC TABLES FOR SEVERAL POLYMERS*

*Source: J.L. Throne, *Plastics Process Engineering*, Marcel Dekker, New York NY (1979), Tables 14.3–6 Through 14.3–14. Note: Table 14.3–7 corrected in second printing. Permission granted by copyright owner.

E.1. Thermodynamic Properties of Low-Density Polyethylene: Enthalpy, H (Btu/lb) and Entropy, S (Btu/lb) [°R datum: Enthalpy and Entropy as Zero at 14.7 lbf/in² and 32°F].

Pressure (lbf/in²)		80	120	140	160	211	248	304	322	342	360	400	440
14.7	H	22.40	42.00	52.15	62.30	92.91	118.98	166.26	181.56	200.16	216.90	254.10	278.40
	S	0.0436	0.0789	0.0962	0.1136	0.1606	0.1993	0.2687	0.2899	0.3144	0.3365	0.3855	0.4120
1,000	H	25.23	44.80	54.73	65.01	94.55	121.30	168.97	184.27	202.87	218.76	256.26	281.11
	S	0.0428	0.0780	0.0955	0.1126	0.1584	0.1940	0.2674	0.2885	0.3129	0.3341	0.3834	0.4105
2,000	H	28.10	47.28	57.31	67.71	96.19	123.61	171.68	186.98	205.58	220.63	258.62	283.93
	S	0.0421	0.0772	0.0948	0.1116	0.1562	0.1887	0.2661	0.2871	0.3115	0.3317	0.3815	0.4092
3,000	H	30.97	50.34	59.89	70.38	97.99	125.92	174.45	189.76	208.36	222.88	261.11	286.83
	S	0.0414	0.0763	0.0941	0.1106	0.1542	0.1843	0.2650	0.2859	0.3103	0.3298	0.3799	0.4080
4,000	H	33.82	53.09	62.47	73.05	99.79	128.22	177.25	192.55	211.15	235.14	263.74	289.90
	S	0.0407	0.0754	0.0935	0.1095	0.1522	0.1800	0.2639	0.2847	0.3091	0.3279	0.3784	0.4070
5,000	H	36.66	55.83	65.05	75.73	101.70	130.53	180.12	195.42	214.02	227.64	266.37	293.05
	S	0.0399	0.0745	0.0929	0.1085	0.1504	0.1764	0.2575	0.2837	0.3080	0.3264	0.3770	0.4062
6,000	H	39.53	58.55	67.63	78.40	103.63	132.88	182.99	198.29	216.89	230.14	269.01	296.25
	S	0.0391	0.0735	0.0923	0.1075	0.1486	0.1727	0.2510	0.2827	0.3070	0.3250	0.3756	0.4054
7,000	H	42.23	61.25	70.23	81.07	105.63	135.25	185.94	201.24	219.84	232.84	271.66	299.50
	S	0.0381	0.0726	0.0917	0.1064	0.1469	0.1697	0.2397	0.2818	0.3061	0.3238	0.3742	0.4047
8,000	H	44.88	63.94	72.83	83.75	107.64	137.64	188.90	204.20	222.80	235.54	274.32	302.76
	S	0.0371	0.0716	0.0911	0.1054	0.1453	0.1667	0.2284	0.2810	0.3052	0.3237	0.3728	0.4041

Temperature (°F)

9,000	H	47.49	66.63	75.45	86.45	109.71	140.04	191.91	207.21	225.81	238.37	276.99	306.01
	S	0.0359	0.0706	0.0905	0.1045	0.1438	0.1641	0.2162	0.2797	0.3044	0.3218	0.3711	0.4035
10,000	H	—	—	78.07	—	111.79	142.12	194.93	210.23	228.83	241.20	—	—
	S	—	—	0.0900	—	0.1424	0.1616	0.2040	0.2783	0.3036	0.3209	—	—
12,000	H	—	—	83.85	—	116.11	145.33	197.98	213.28	231.88	247.08	—	—
	S	—	—	0.0890	—	0.1398	0.1571	0.1786	0.2710	0.3029	0.3195	—	—
14,000	H	—	—	88.67	—	120.51	148.59	201.04	216.34	234.94	253.08	—	—
	S	—	—	0.0880	—	0.1374	0.1530	0.1735	0.2636	0.3022	0.3184	—	—
16,000	H	—	—	93.99	—	124.95	152.13	—	—	—	259.26	—	—
	S	—	—	0.0871	—	0.1352	0.1493	—	—	—	0.3175	—	—
18,000	H	—	—	99.35	—	129.44	155.87	—	—	—	265.57	—	—
	S	—	—	0.0862	—	0.1331	0.1459	—	—	—	0.3168	—	—
20,000	H	—	—	104.75	—	134.00	159.76	—	—	—	271.97	—	—
	S	—	—	0.0853	—	0.1311	0.1427	—	—	—	0.3163	—	—
22,000	H	—	—	110.19	—	138.64	163.78	—	—	—	—	—	—
	S	—	—	0.0845	—	0.1292	0.1398	—	—	—	—	—	—
24,000	H	—	—	115.63	—	143.35	167.88	—	—	—	—	—	—
	S	—	—	0.0837	—	0.1274	0.1371	—	—	—	—	—	—
26,000	H	—	—	121.11	—	148.13	172.06	—	—	—	—	—	—
	S	—	—	0.0829	—	0.1257	0.1345	—	—	—	—	—	—
28,000	H	—	—	126.63	—	152.96	176.38	—	—	—	—	—	—
	S	—	—	0.0822	—	0.1241	0.1321	—	—	—	—	—	—

E.2. Thermodynamic Properties of Polypropylene Homopolymer: Enthalpy, H (Btu/lb) and Entropy, S (Btu/lb) [°R datum: Enthalpy and Entropy as Zero at 14.7 lb$_f$/in^2 and 32°F].

Pressure (lb$_f$/in^2)		Temperature (°F)									
		80	120	160	200	240	280	320	400	440
14.7	H	22.40	42.00	62.30	85.60	112.20	146.10	179.70	254.10	278.40
	S	0.0436	0.0789	0.1136	0.1499	0.1890	0.2405	0.2875	0.3855	0.4120
1,000	H	25.23	44.80	65.01	88.24	114.52	148.41	179.91	254.26	281.11
	S	0.0428	0.0780	0.1126	0.1488	0.1874	0.2389	0.2831	0.3834	0.4105
2,000	H	28.10	47.58	67.71	90.86	116.82	150.76	180.56	258.62	283.93
	S	0.0421	0.0772	0.1116	0.1471	0.1858	0.2374	0.2793	0.3815	0.4092
3,000	H	30.97	50.34	70.38	93.47	119.12	153.13	181.45	261.11	286.83
	S	0.0414	0.0763	0.1106	0.1465	0.1842	0.2359	0.2759	0.3799	0.4086

4,000	H	33.82	53.09	73.05	96.08	121.38	155.58	182.70	263.74	289.90
	S	0.0407	0.0754	0.1095	0.1459	0.1926	0.2345	0.2729	0.3784	0.4080
5,000	H	36.66	55.83	75.73	98.72	123.65	158.05	184.01	266.37	293.05
	S	0.0399	0.0745	0.1085	0.1443	0.1810	0.2332	0.2701	0.3770	0.4062
6,000	H	39.53	58.55	78.40	101.36	125.96	160.56	185.36	269.01	296.25
	S	0.0391	0.0735	0.1075	0.1432	0.1794	0.2319	0.2672	0.3756	0.4054
7,000	H	42.23	61.25	81.07	104.01	128.28	163.11	186.73	271.66	299.50
	S	0.0381	0.0726	0.1064	0.1422	0.1779	0.2306	0.2645	0.3742	0.4047
8,000	H	44.88	63.94	83.74	106.67	130.62	165.70	188.16	274.32	302.76
	S	0.0371	0.0716	0.1054	0.1411	0.1764	0.2295	0.2618	0.3728	0.4041
9,000	H	47.49	66.63	86.45	109.34	132.97	168.32	189.65	276.99	306.01
	S	0.0359	0.0706	0.1045	0.1401	0.1750	0.2284	0.2592	0.3711	0.4035

E.3. Thermodynamic Properties of Ethylene-Propylene Copolymer: Enthalpy, H (Btu/lb) and Entropy, S (Btu/lb) [°R datum: Enthalpy and Entropy as Zero at 14.7 lb_f/in^2 and 32°F].

Pressure (lb_f/in^2)		140	180	220	Temperature (°F) 300	340	380	420	460
14.7	H	75.00	98.00	130.00	200.00	226.00	248.00	273.30	324.00
	S	0.1379	0.1738	0.2223	0.3125	0.3565	0.3795	0.4190	0.4880
1,000	H	77.36	100.11	131.88	202.87	228.90	249.97	276.05	326.71
	S	0.1364	0.1719	0.2200	0.3113	0.3553	0.3772	0.4176	0.4866
2,000	H	79.71	102.24	133.81	205.75	231.90	252.07	278.87	329.51
	S	0.1348	0.1700	0.2178	0.3101	0.3543	0.3751	0.4164	0.4853
3,000	H	82.09	104.41	135.75	208.60	235.05	254.23	281.71	332.34
	S	0.1334	0.1682	0.2157	0.3089	0.5535	0.3731	0.4152	0.4841

4,000	H	84.47	106.58	137.63	211.46	238.33	256.45	284.98	335.27
	S	0.1319	0.1664	0.2134	0.3077	0.3529	0.3712	0.4140	0.4831
5,000	H	86.85	108.72	139.52	214.31	241.58	258.70	287.29	338.25
	S	0.1305	0.1646	0.2113	0.3065	0.3522	0.3694	0.4128	0.4281
6,000	H	89.24	110.86	141.44	217.08	244.83	260.96	290.12	341.29
	S	0.1291	0.1628	0.2091	0.3053	0.3516	0.3676	0.4117	0.4812
7,000	H	91.63	113.11	143.37	219.74	248.16	263.27	292.90	344.42
	S	0.1277	0.1612	0.2071	0.3040	0.3511	0.3659	0.4106	0.4804
8,000	H	94.01	115.42	145.32	222.37	251.52	265.62	295.64	347.58
	S	0.1263	0.1597	0.2050	0.3026	0.3506	0.3642	0.4094	0.4797
9,000	H	96.37	117.70	147.30	225.02	254.84	267.98	298.38	350.81
	S	0.1249	0.1581	0.2030	0.3012	0.3502	0.3626	0.4082	0.4790

E.4. Thermodynamic Properties of Polymethyl Methacrylate: Enthalpy, H (Btu/lb) and Entropy, S (Btu/lb) [°R datum: Enthalpy and Entropy as Zero at 14.7 lb_f/in^2 and 32°F].

Pressure (lb_f/in^2)		Temperature (°F)						
		70	140	212	230.2	248	266	282
14.7	H	12.73	51.84	91.80	107.08	123.12	139.23	155.00
	S	0.0249	0.0953	0.1590	0.1828	0.2075	0.2315	0.2547
500	H	13.87	52.95	92.59	108.06	124.09	140.31	156.00
	S	0.0248	0.0951	0.1583	0.1823	0.2071	0.2313	0.2544
1,000	H	15.05	54.09	93.40	109.14	125.10	141.42	157.02
	S	0.0246	0.0949	0.1576	0.1822	0.2067	0.2310	0.2540
1,500	H	16.22	55.23	94.22	110.24	126.12	142.52	158.10
	S	0.0245	0.0947	0.1570	0.1819	0.2063	0.2307	0.2537
2,000	H	17.40	56.37	95.05	111.34	127.14	143.62	159.18
	S	0.0244	0.0945	0.1563	0.1817	0.2060	0.2305	0.2534
3,000	H	19.74	58.66	96.71	113.59	129.22	145.84	161.35
	S	0.0241	0.0941	0.1550	0.1812	0.2052	0.2300	0.2529
4,000	H	22.08	60.94	98.36	115.89	131.31	148.03	163.51
	S	0.0238	0.0937	0.1536	0.1809	0.2046	0.2294	0.2523
5,000	H	24.41	63.22	100.01	118.25	133.44	150.22	165.67
	S	0.0235	0.0933	0.1523	0.1806	0.2040	0.2289	0.2518
6,000	H	26.75	65.51	101.66	120.66	135.61	152.40	167.84
	S	0.0232	0.0930	0.1510	0.1804	0.2035	0.2284	0.2513
7,000	H	29.08	67.80	103.24	123.07	137.80	154.60	170.11
	S	0.0230	0.0926	0.1496	0.1803	0.2030	0.2279	0.2509
8,000	H	31.40	70.09	104.82	125.48	140.04	156.83	172.37
	S	0.0227	0.0923	0.1482	0.1801	0.2026	0.2275	0.2505
9,000	H	33.72	72.39	106.46	127.90	142.33	159.11	174.63
	S	0.0224	0.0920	0.1496	0.1800	0.2023	0.2272	0.2502
10,000	H	36.04	74.69	108.13	130.34	144.67	161.43	176.88
	S	0.0222	0.0917	0.1457	0.1799	0.2020	0.2269	0.2498
11,000	H	38.36	76.98	109.86	132.81	147.03	163.78	179.12
	S	0.0219	0.0914	0.1445	0.1798	0.2018	0.2267	0.2494
12,000	H	40.67	79.29	111.65	135.30	149.42	166.13	181.35
	S	0.0216	0.0911	0.1435	0.1798	0.2017	0.2265	0.2491

Pressure (lb_f/in^2)		Temperature (°F)						
		70	140	212	230.2	248	266	282
13,000	H	42.98	81.59	113.46	137.80	151.80	168.49	183.58
	S	0.0213	0.0908	0.1424	0.1798	0.2015	0.2263	0.2487
14,000	H	45.28	83.90	115.28	140.29	154.18	170.84	185.81
	S	0.0210	0.0906	0.1415	0.1798	0.2014	0.2261	0.2484
15,000	H	47.58	86.20	117.12	142.77	156.57	173.21	188.04
	S	0.0208	0.0903	0.1405	0.1798	0.2012	0.2260	0.2480
16,000	H	49.87	88.51	118.97	145.25	158.95	175.57	190.27
	S	0.0205	0.0900	0.1396	0.1798	0.2011	0.2257	0.2477
17,000	H	52.21	90.82	120.83	147.72	161.35	177.95	192.51
	S	0.0203	0.0898	0.1386	0.1798	0.2010	0.2256	0.2473
18,000	H	54.55	93.12	122.69	150.21	163.75	180.32	194.74
	S	0.0201	0.0895	0.1377	0.1798	0.2009	0.2255	0.2470
19,000	H	56.89	95.43	124.55	152.73	166.18	182.68	197.06
	S	0.0199	0.0893	0.1368	0.1798	0.2008	0.2253	0.2468
20,000	H	59.22	97.73	126.41	155.25	168.64	185.05	199.51
	S	0.0197	0.0891	0.1359	0.1798	0.2008	0.2252	0.2468
21,000	H	61.55	100.03	128.28	157.77	171.09	187.43	201.97
	S	0.0196	0.0889	0.1351	0.1798	0.2008	0.2251	0.2468
22,000	H	63.87	102.34	130.18	160.29	173.55	189.81	204.43
	S	0.0194	0.0886	0.1342	0.1798	0.2008	0.2250	0.2468
23,000	H	66.20	104.64	132.11	162.79	176.00	192.20	206.88
	S	0.0192	0.0884	0.1334	0.1798	0.2008	0.2249	0.2468
24,000	H	68.51	106.95	134.00	165.30	178.44	194.58	209.34
	S	0.0190	0.0882	0.1326	0.1798	0.2008	0.2248	0.2468
25,000	H	70.83	109.24	135.82	167.85	180.88	196.95	211.79
	S	0.0188	0.0880	0.1317	0.1798	0.2008	0.2247	0.2468
26,000	H	73.18	111.53	137.61	170.43	183.32	199.31	214.36
	S	0.0187	0.0878	0.1307	0.1798	0.2008	0.2246	0.2468
27,000	H	75.54	113.82	139.40	173.01	185.75	201.66	216.94
	S	0.0186	0.0876	0.1298	0.1798	0.2008	0.2245	0.2468
28,000	H	77.90	116.11	141.21	175.59	188.18	204.02	219.43
	S	0.0185	0.0874	0.1288	0.1798	0.2008	0.2244	0.2468

E.5. Thermodynamic Properties of Polyvinyl Chloride: Enthalpy, H (Btu/lb) and Entropy, S (Btu/lb) [°R datum: Enthalpy and Entropy as Zero at 14.7 lb$_f$/in^2 and 32°F].

Pressure (lb$_f$/in^2)		Temperature (°F)				
		70	123.8	179.6	195	206
14.7	H	9.31	23.12	56.83	69.28	74.47
	S	0.0182	0.0431	0.1010	0.1216	0.1296
500	H	10.28	24.08	57.51	70.17	75.33
	S	0.1081	0.0430	0.1004	0.1214	0.1293
1,000	H	11.29	25.07	58.25	71.07	76.21
	S	0.0180	0.0428	0.0999	0.1211	0.1290
1,500	H	12.26	26.06	59.00	71.96	77.08
	S	0.0178	0.0427	0.0994	0.1028	0.1287
2,000	H	13.25	27.04	59.77	72.84	77.96
	S	0.0177	0.0425	0.0989	0.1205	0.1284
3,000	H	15.27	29.03	61.29	74.65	79.74
	S	0.0175	0.0423	0.0979	0.1199	0.1278
4,000	H	17.30	31.01	62.80	76.52	81.68
	S	0.0173	0.0420	0.0969	0.1195	0.1274
5,000	H	19.32	32.98	64.25	78.44	83.63
	S	0.0171	0.0417	0.0958	0.1191	0.1271
6,000	H	21.34	34.94	65.65	80.42	85.58
	S	0.0169	0.0414	0.0946	0.1188	0.1268
7,000	H	23.35	36.91	67.15	82.38	87.52
	S	0.0167	0.0412	0.0936	0.1186	0.1265
8,000	H	25.37	38.90	68.75	84.31	89.45
	S	0.0166	0.0410	0.0928	0.1182	0.1262
9,000	H	27.44	40.90	70.46	86.23	91.37
	S	0.0165	0.0408	0.0921	0.1179	0.1259
10,000	H	29.44	42.91	72.23	88.17	93.29
	S	0.0163	0.0406	0.0916	0.1176	0.1256
11,000	H	31.45	44.93	74.06	90.12	95.20
	S	0.0161	0.0404	0.0911	0.1174	0.1252
12,000	H	33.45	46.95	75.92	92.11	97.11
	S	0.0159	0.0403	0.0907	0.1172	0.1249

Pressure		Temperature (°F)				
(lb$_f$/in^2)		70	123.8	179.6	195	206
13,000	H	35.45	48.97	77.80	94.10	99.01
	S	0.0158	0.0401	0.0904	0.1170	0.1246
14,000	H	37.44	51.00	79.71	96.08	100.91
•	S	0.0156	0.0400	0.0901	0.1168	0.1243
15,000	H	39.43	53.02	81.63	98.05	102.81
	S	0.0154	0.0399	0.0898	0.1166	0.1240
16,000	H	41.42	55.04	83.56	100.01	104.70
	S	0.0152	0.0397	0.0895	0.1164	0.1237
17,000	H	43.41	57.05	85.47	101.99	106.60
	S	0.0150	0.0396	0.0892	0.1162	0.1234
18,000	H	45.30	59.06	87.36	103.95	108.50
	S	0.0148	0.0395	0.0889	0.1160	0.1231
19,000	H	47.37	61.08	89.23	105.91	110.43
	S	0.0147	0.0394	0.0886	0.1158	0.1228
20,000	H	49.35	63.10	91.12	107.87	112.52
	S	0.0145	0.0393	0.0882	0.1156	0.1228
21,000	H	51.33	65.12	93.02	109.83	114.61
	S	0.0143	0.0392	0.0880	0.1154	0.1228
22,000	H	53.30	67.14	94.94	111.82	116.70
	S	0.0141	0.0391	0.0877	0.1152	0.1228
23,000	H	55.27	69.15	96.89	113.82	118.79
	S	0.0139	0.0390	0.0875	0.1151	0.1228
24,000	H	57.24	71.16	98.84	115.85	120.86
	S	0.0137	0.0389	0.0873	0.1150	0.1228
25,000	H	59.20	73.17	100.79	117.89	122.93
	S	0.0136	0.0388	0.0871	0.1150	0.1228
26,000	H	61.17	75.19	102.72	119.91	125.00
	S	0.0134	0.0387	0.0869	0.1149	0.1228
27,000	H	63.13	77.21	104.66	121.90	127.07
	S	0.0132	0.0386	0.0867	0.1148	0.1228
28,000	H	65.09	79.23	106.59	123.87	129.14
	S	0.0130	0.0385	0.0865	0.1146	0.1228

E.6. Thermodynamic Properties of General Purpose Polystyrene: Enthalpy, H (Btu/lb) and Entropy, S (Btu/lb) [°R datum: Enthalpy and Entropy as Zero at 14.7 lbf/in² and 32°F].

Pressure (lbf/in²)		Temperature (°F)							
		70	141	205	278	293	324	354	397
14.7	H	10.77	35.01	76.28	115.89	125.71	147.24	171.28	202.60
	S	0.0211	0.0643	0.1329	0.1910	0.2049	0.2349	0.2679	0.3080
500	H	11.96	36.26	76.35	117.01	126.88	148.34	172.38	203.72
	S	0.0207	0.0641	0.1308	0.1906	0.2046	0.2345	0.2674	0.3076
1,000	H	13.19	37.55	76.48	118.16	128.07	149.48	173.52	204.89
	S	0.0204	0.0638	0.1289	0.1901	0.2042	0.2340	0.2670	0.3072
1,500	H	14.55	38.84	76.67	119.31	129.27	150.62	174.66	206.07
	S	0.0203	0.0636	0.1270	0.1897	0.2038	0.2336	0.2665	0.3068
2,000	H	15.76	40.14	76.91	120.47	130.47	151.77	175.81	207.25
	S	0.0199	0.0634	0.1252	0.1893	0.2034	0.2331	0.2661	0.3064
3,000	H	18.20	42.73	77.57	122.79	132.91	154.08	178.13	209.61
	S	0.0192	0.0630	0.1218	0.1884	0.2028	0.2323	0.2653	0.3056
4,000	H	20.74	45.34	78.41	125.14	135.38	156.39	180.43	212.00
	S	0.0187	0.0626	0.1188	0.1877	0.2022	0.2315	0.2644	0.3049
5,000	H	23.28	47.95	79.44	127.48	137.87	158.72	182.72	214.41
	S	0.0182	0.0622	0.1161	0.1870	0.2016	0.2307	0.2636	0.3042

6,000	H	25.74	50.58	80.66	129.83	140.36	161.06	185.04	216.83
	S	0.0175	0.0619	0.1136	0.1861	0.2010	0.2300	0.2628	0.3036
7,000	H	28.06	53.21	82.05	132.18	142.85	163.42	187.37	219.24
	S	0.0166	0.0616	0.1114	0.1854	0.2005	0.2292	0.2621	0.3029
8,000	H	30.29	55.84	83.53	134.54	145.35	165.80	189.72	221.62
	S	0.0156	0.0612	0.1094	0.1847	0.2000	0.2286	0.2614	0.3022
9,000	H	32.70	58.47	85.06	136.90	147.86	168.17	192.07	223.98
	S	0.0148	0.0610	0.1074	0.1840	0.1995	0.2279	0.2607	0.3016
10,000	H	35.22	61.10	86.64	139.27	150.38	170.55	194.42	226.33
	S	0.0143	0.0607	0.1056	0.1834	0.1191	0.2273	0.2600	0.3009
11,000	H	37.79	63.73	88.29	141.63	152.90	172.94	196.79	228.71
	S	0.0139	0.0640	0.1038	0.1827	0.1986	0.2267	0.2593	0.3003
12,000	H	40.26	66.37	90.03	144.01	155.42	175.33	199.15	231.18
	S	0.0134	0.0601	0.1022	0.1821	0.1982	0.2261	0.2587	0.2997
13,000	H	42.75	69.01	91.82	146.39	157.94	177.72	201.52	233.70
	S	0.0128	0.0599	0.1007	0.1815	0.1978	0.2255	0.2581	0.2993
14,000	H	45.24	71.64	93.64	148.78	160.47	180.12	203.89	236.20
	S	0.0123	0.0596	0.0992	0.1809	0.1974	0.2249	0.2575	0.2989

Pressure (lbf/in²)		70	141	205	Temperature (°F) 278	293	324	354	397
15,000	H	47.84	74.27	95.47	151.18	163.00	182.52	206.27	238.65
	S	0.0120	0.0594	0.0977	0.1803	0.1970	0.2243	0.2569	0.2984
16,000	H	50.45	76.90	97.29	153.60	165.56	184.93	208.67	241.07
	S	0.0117	0.0591	0.0963	0.1798	0.1966	0.2238	0.2564	0.2978
17,000	H	53.02	79.53	99.10	156.04	168.13	187.34	211.07	243.50
	S	0.0114	0.0589	0.0948	0.1793	0.1963	0.2333	0.2558	0.2973
18,000	H	55.52	82.17	100.98	158.49	170.71	189.76	213.46	246.00
	S	0.0109	0.0586	0.0935	0.1788	0.1960	0.2228	0.2553	0.2966
19,000	H	58.08	84.80	102.87	160.97	173.29	192.17	215.85	248.52
	S	0.0106	0.0584	0.0921	0.1784	0.1958	0.2223	0.2548	0.2964
20,000	H	60.55	87.42	104.73	163.49	175.86	194.58	218.24	251.02
	S	0.0100	0.0582	0.0908	0.1781	0.1955	0.2218	0.2542	0.2962
21,000	H	62.97	90.05	106.56	166.06	178.42	197.00	220.65	253.49
	S	0.0094	0.0580	0.0894	0.1778	0.1952	0.2213	0.2537	0.2958

E.6. Continued.

22,000	H	65.38	92.67	108.38	168.66	180.98	199.43	223.06	255.94
	S	0.0088	0.0579	0.0879	0.1776	0.1949	0.2209	0.2533	0.2953
23,000	H	67.81	95.30	110.23	171.28	183.53	201.86	225.46	258.43
	S	0.0083	0.0576	0.0866	0.1774	0.1946	0.2205	0.2528	0.2950
24,000	H	70.37	97.92	112.16	173.91	186.07	204.30	227.87	260.95
	S	0.0079	0.0574	0.0853	0.1772	0.1944	0.2200	0.2523	0.2946
25,000	H	72.94	100.53	114.15	176.58	186.63	206.73	230.27	263.47
	S	0.0077	0.0572	0.0842	0.1771	0.1941	0.2196	0.2519	0.2943
26,000	H	75.48	103.15	116.14	179.28	191.23	209.17	232.68	265.98
	S	0.0073	0.0570	0.0830	0.1771	0.1939	0.2192	0.2514	0.2940
27,000	H	77.87	105.76	118.12	182.02	193.88	211.60	235.09	268.48
	S	0.0067	0.0568	0.0819	0.1771	0.1938	0.2188	0.2510	0.2937
28,000	H	80.23	108.37	120.05	184.79	196.58	214.04	237.52	270.97
	S	0.0060	0.0566	0.0807	0.1771	0.1938	0.2184	0.2506	0.2933

E.7. Thermodynamic Properties of Nylon 610 (PA-610): Enthalpy, H (Btu/lb) and Entropy, S (Btu/lb) [°R datum: Enthalpy and Entropy as Zero at 14.7 lbf/in² and 32°F].

Pressure (lbf/in²)		Temperature (°F)									
		77	151	199	250	300	351	390	399	411	421
14.7	H	17.10	59.40	86.71	122.19	152.34	187.97	241.07	305.51	419.58	563.76
	S	0.0333	0.1081	0.1517	0.2056	0.2471	0.2946	0.3681	0.4639	0.6327	0.8444
500	H	18.27	60.65	88.00	123.49	153.65	189.29	242.31	306.69	420.60	564.14
	S	0.0330	0.1080	0.1516	0.2055	0.2471	0.2945	0.3680	0.4637	0.6323	0.8433
1,000	H	19.48	61.94	89.32	124.81	154.99	190.64	243.58	307.91	421.64	564.52
	S	0.0327	0.1079	0.1515	0.2055	0.2470	0.2945	0.3679	0.4635	0.6319	0.8421
1,500	H	20.69	63.23	90.63	126.12	156.33	191.99	244.85	309.13	422.68	564.91
	S	0.0324	0.1077	0.1514	0.2053	0.2470	0.2944	0.3677	0.4633	0.6315	0.8410
2,000	H	21.89	64.52	91.93	127.41	157.65	193.34	246.13	310.37	423.72	565.29
	S	0.0321	0.1076	0.1513	0.2052	0.2469	0.2944	0.3676	0.4631	0.6311	0.8398
3,000	H	24.30	67.12	94.53	129.95	160.26	196.04	248.68	312.84	425.83	566.08
	S	0.0315	0.1074	0.1511	0.2049	0.2467	0.2943	0.3673	0.4627	0.6303	0.8375
4,000	H	26.70	69.73	97.13	132.43	162.81	198.75	251.24	315.32	427.96	566.89
	S	0.0309	0.1072	0.1509	0.2045	0.2464	0.2942	0.3670	0.4624	0.6296	0.8353
5,000	H	29.09	72.35	99.71	134.85	165.33	201.45	253.81	317.83	430.11	567.76
	S	0.0303	0.1070	0.1507	0.2041	0.2460	0.2941	0.3668	0.4620	0.6288	0.8331

6,000	H	31.48	74.97	102.38	137.22	167.81	204.15	256.38	320.35	432.28	568.73
	S	0.0297	0.1069	0.1504	0.2036	0.2456	0.2940	0.3665	0.4617	0.6281	0.8310
7,000	H	33.87	77.59	104.84	139.56	170.25	206.85	258.95	322.87	434.52	569.81
	S	0.0292	0.1067	0.1502	0.2030	0.2452	0.2939	0.3663	0.4614	0.6275	0.8291
8,000	H	36.25	80.20	107.39	141.86	172.66	209.54	261.48	325.41	436.81	570.99
	S	0.0286	0.1065	0.1499	0.2024	0.2447	0.2938	0.3660	0.4612	0.6269	0.8272
9,000	H	38.62	82.78	109.92	144.14	175.04	212.22	264.02	327.96	439.14	572.28
	S	0.0280	0.1064	0.1497	0.2018	0.2442	0.2937	0.3657	0.4609	0.6264	0.8255
10,000	H	40.98	85.34	112.46	146.40	177.37	214.87	266.59	330.53	441.50	573.64
	S	0.0274	0.1061	0.1494	0.2011	0.2437	0.2935	0.3655	0.4606	0.6259	0.8239
11,000	H	43.07	87.88	114.98	148.65	179.66	217.49	269.18	333.11	443.91	575.09
	S	0.0263	0.1059	0.1492	0.2005	0.2430	0.2934	0.3653	0.4604	0.6255	0.8224
12,000	H	45.15	90.39	117.50	150.92	181.91	220.05	271.82	335.70	446.34	576.62
	S	0.0252	0.1056	0.1489	0.1999	0.2424	0.2931	0.3651	0.4602	0.6251	0.8210
13,000	H	47.49	92.89	120.02	153.19	184.13	222.54	274.47	338.32	448.81	578.25
	S	0.0247	0.1053	0.1487	0.1993	0.2417	0.2928	0.3650	0.4600	0.6247	0.8197
14,000	H	49.83	95.37	122.54	155.49	186.31	224.98	277.16	340.93	451.32	579.97
	S	0.0241	0.1050	0.1484	0.1987	0.2409	0.2923	0.3649	0.4598	0.6244	0.8185

E.7. Continued.

Pressure (lb$_f$/in^2)		77	151	199	250	Temperature (°F) 300	351	390	399	411	421
15,000	H	52.16	97.83	125.06	157.80	188.48	227.37	279.84	343.55	453.86	581.80
	S	0.0235	0.1047	0.1482	0.1982	0.2402	0.2919	0.3648	0.4597	0.6242	0.8174
16,000	H	54.48	100.30	127.57	160.13	190.64	229.72	282.48	346.17	456.42	583.73
	S	0.0229	0.1044	0.1479	0.1977	0.2394	0.2914	0.3646	0.4595	0.6239	0.8164
17,000	H	56.81	102.76	130.07	162.47	192.79	232.04	285.13	348.77	458.98	585.76
	S	0.0223	0.1040	0.1477	0.1972	0.2387	0.2908	0.3645	0.4593	0.6237	0.8156
18,000	H	59.13	105.22	132.58	164.83	194.94	234.35	287.74	351.36	461.57	587.89
	S	0.0217	0.1037	0.1474	0.1967	0.2379	0.2903	0.3643	0.4591	0.6235	0.8149
19,000	H	61.45	107.69	135.08	167.20	197.09	236.64	290.33	353.93	464.15	590.08
	S	0.0211	0.1034	0.1471	0.1963	0.2372	0.2897	0.3641	0.4589	0.6233	0.8142
20,000	H	63.77	110.15	137.59	169.58	199.26	238.93	292.89	356.52	466.75	592.31
	S	0.0205	0.1031	0.1469	0.1959	0.2365	0.2891	0.3639	0.4587	0.6231	0.8136
21,000	H	66.09	112.62	140.09	171.97	201.44	241.21	295.44	359.11	469.35	594.58
	S	0.0199	0.1028	0.1466	0.1955	0.2358	0.2885	0.3636	0.4585	0.6229	0.8131

22,000	H	68.41	115.09	142.60	174.38	203.63	243.50	298.00	361.78	471.77	596.87
	S	0.0194	0.1025	0.1464	0.1951	0.2351	0.2880	0.3634	0.4581	0.6225	0.8125
23,000	H	69.69	117.56	145.10	176.81	205.84	245.79	300.66	363.83	473.43	600.98
	S	0.1068	0.1022	0.1461	0.1947	0.2344	0.2874	0.3633	0.4576	0.6212	0.8121
24,000	H	72.00	120.03	147.60	179.24	208.07	248.09	303.24	366.20	476.05	603.66
	S	0.0162	0.1019	0.1459	0.1944	0.2338	0.2869	0.3631	0.4572	0.6211	0.8120
25,000	H	74.32	122.49	150.10	181.69	210.32	250.40	305.83	368.73	478.67	605.97
	S	0.0156	0.1016	0.1456	0.1941	0.2332	0.2863	0.3629	0.4569	0.6210	0.8117
26,000	H	76.64	124.95	152.60	184.15	212.59	252.72	308.42	371.24	481.31	608.29
	S	0.0151	0.1013	0.1454	0.1938	0.2327	0.2858	0.3628	0.4567	0.6209	0.8115
27,000	H	78.93	127.41	155.10	186.62	214.89	255.06	311.03	373.72	483.95	610.60
	S	0.0144	0.1009	0.1451	0.1935	0.2321	0.2854	0.3626	0.4564	0.6207	0.8112
28,000	H	81.22	129.86	157.61	189.11	217.22	257.42	313.65	376.17	486.59	612.87
	S	0.0138	0.1006	0.1449	0.1932	0.2317	0.2849	0.3625	0.4560	0.6206	0.8110

E.8. Thermodynamic Properties of Nylon 66 (PA-66): Enthalpy, H (Btu/lb) and Entropy, S (Btu/lb) [°R datum: Enthalpy and Entropy as Zero at 14.7 lb_f/in^2 and 32°F].

Pressure (lb_f/in^2)		Temperature (°F)							
		77	122	167	212	257	302	347	392
14.7	H	13.75	32.60	51.25	74.20	96.88	121.25	146.25	173.20
	S	0.0157	0.0640	0.0970	0.1290	0.1620	0.1940	0.2270	0.2590
500	H	17.2	38.4	57.0	78.0	103.0	125.5	148	182
	S	—	0.0584	0.0955	0.1285	0.1615	0.1934	0.2255	0.2600
1,000	H	18.4	40.6	58.3	79.2	106	129	149.5	184
	S	—	0.0576	0.0944	0.1280	0.1600	0.1918	0.2242	0.2586
2,000	H	20.1	42.4	60.9	81.6	108.6	130.2	150.1	185.8
	S	—	0.0562	0.0938	0.1270	0.1594	0.1912	0.2240	0.2573
3,000	H	21.9	44.3	62.8	82.9	111	134	153	188
	S	—	0.0588	0.0933	0.1265	0.1590	0.1904	0.2235	0.2570
4,000	H	23.5	45.4	64.0	84.2	112	136	154	189
	S	—	0.05455	0.0930	0.1261	0.1584	0.1900	0.2228	0.2560
6,000	H	27.8	48.8	67.5	88.2	114	137	160	190
	S	—	0.0545	0.0923	0.1258	0.1580	0.1895	0.2225	0.2552
8,000	H	31.5	53.6	72.2	94.6	118	142	167	194
	S	—	0.0541	0.0920	0.1255	0.1578	0.1890	0.2224	0.2548

10,000	H	36	58	76.8	100	124	148	174	201
	S	—	0.0535	0.0918	0.1254	0.1557	0.1890	0.2224	0.2545
15,000	H	44.4	68.1	89.0	113	135.4	161	180	213
	S	—	0.0535	0.0912	0.1250	0.1555	0.1880	0.2223	0.2544
20,000	H	55.2	78.0	99.55	123.2	146.18	170.24	195.25	222.40
	S	0.0033	0.0529	0.0912	0.1243	0.1548	0.1878	0.2223	0.2543
30,000	H	74.0	98.5	124	149	173	196	224	248
	S	—	0.0500	0.0910	0.1230	0.1545	0.1876	0.2210	0.2538
40,000	H	97.4	123.29	146.95	171.60	194.68	218.65	243.55	271.10
	S	—	0.0447	0.0863	0.1213	0.1543	0.1863	0.2193	0.251
60,000	H	138	164	192	218	246	270	295	322
	S	—	0.0415	0.0845	0.1204	0.1540	0.1854	0.2186	0.2503
80,000	H	184	214.10	239.98	266.20	289.48	313.25	338.15	366.10
	S	—	0.0340	0.0795	0.1176	0.1506	0.1826	0.2156	0.2476
100,000	H	225.6	254.6	284.0	307.0	333.5	355.4	376.0	405.0
	S	—	0.0318	0.0780	0.117	0.1496	0.1808	0.2145	0.463
120,000	H	273.9	305.1	331.6	358.3	381.7	405.6	430.5	458.6
	S	—	0.0287	0.0757	0.1150	0.1479	0.1799	0.2129	0.2449

E.9. Thermodynamic Properties of Polytetrafluoroethylene (PTFE): Enthalpy, H (Btu/lb) and Entropy, S (Btu/lb) [°R datum: Enthalpy and Entropy as Zero at 14.7 lb_f/in^2 and 32°F].

Pressure (lb_f/in^2)		Temperature (°F)							
		50	75	100	125	150	175	200	225
14.7	H	4.11	12.65	19.15	25.09	31.07	37.15	43.30	49.55
	S	0.008193	0.02452	0.03641	0.04677	0.05830	0.06954	0.07905	0.08834
500	H	4.35	13.20	19.74	25.68	31.61	37.74	43.90	50.14
	S	0.007410	0.02433	0.03630	0.04666	0.05819	0.06943	0.07894	0.08823
1,000	H	4.87	13.72	20.37	26.30	32.29	38.36	44.52	50.73
	S	0.006994	0.02409	0.03619	0.04655	0.05808	0.06932	0.07883	0.08812
2,000	H	5.69	14.77	21.57	27.51	33.44	39.56	45.72	51.96
	S	0.006163	0.02353	0.03597	0.04633	0.05786	0.06910	0.07860	0.08789
3,000	H	6.65	15.83	22.81	28.75	34.73	40.80	46.96	53.16
	S	0.005334	0.02303	0.03577	0.04623	0.05765	0.06888	0.07850	0.08767

5,000	H	8.50	17.92	25.26	31.19	37.17	43.24	49.38	55.62
	S	0.003696	0.02199	0.03528	0.04573	0.05724	0.06847	0.07797	0.08724
10,000	H	13.70	23.25	31.42	37.34	43.31	49.36	55.50	61.72
	S	—	0.01997	0.03448	0.04481	0.05632	0.06752	0.07700	0.08626
25,000	H	31.19	40.12	49.10	55.16	61.63	67.78	74.03	80.37
	S	—	0.01480	0.03115	0.04166	0.05400	0.06537	0.07503	0.08444
50,000	H	60.51	69.42	80.33	86.48	87.44	94.20	104.94	111.72
	S	—	0.01051	0.03050	0.04135	0.04454	0.05663	0.07333	0.08342
100,000	H	117.69	126.59	142.11	148.04	—	—	136.46	160.68
	S	—	0.00484	—	—	0.02672	0.03959	0.06958	0.08033
150,000	H	173.20	182.09	197.61	203.53	—	—	218.96	226.19
	S	—	0.00075	—	—	0.02252	0.03540	0.06538	0.07613

CREDITS

The authors wish to acknowledge the following sources for the figures, tables, quoted text, and other referenced materials used in this text. Nearly all the sources willingly agreed to allow their copyrighted materials to be published in this volume. In those cases, the material cited is used with "expressed permission of the copyright holder in accordance with United States copyright laws. The material cannot be copied or otherwise distributed without the express consent of the copyright holder." Please note that in nearly all cases, the material cited has been redrawn and/or edited to conform to the text material. The following notation is used below. *TX.XX* is table number. *FX.XX* is figure number. *(XX)* is reference number in appropriate chapter.

PUBLISHING HOUSES

Academic Press, Inc.
115 Fifth Ave.
New York NY 10003

F2.34 (54)	T4.15 (81)	T4.16 (86)	T4.24 (104)
F4.12 (13)	F4.41 (66)	F4.39 (64)	F4.40 (65)
F4.42 (69)	F4.44 (71)	F4.46 (83)	F6.3 (16)
F6.15 (30)	F6.139 (125)	F6.140 (125)	F6.149 (136)
F6.150 (138)	F6.167 (165)		

Applied Science Publishers, Ltd.
Elsevier, Crown House, Linton Rd.
Barking Essex 1G11 8JU England

F2.45 (75)	F2.46 (75)	F3.70 (55)	F6.151 (139)
F6.152 (140)	F6.153 (141)	F6.154 (142)	F6.155 (144)
F6.156 (145)	F6.157 (145)	F6.158 (147)	F6.159 (149)

Barnes & Noble Books
10 E. 53rd St.
New York NY 10103

F5.55 (81)	F5.56 (83)	F5.57 (84)

Cahners Book Co.
Division of Cahners Publishing Company, Inc.
89 Franklin St
Boston, MA 02110

F1.11 (32)	F1.12 (33)	F1.13 (34)	F2.54 (85)
T3.7 (68)	F4.50 (88)	F4.51 (89)	F4.52 (93)

Cambridge University Press
32 East 57th St.
New York NY 10022

F4.32 (49)	F4.33 (52)

Cornell University Press
124 Roberts Place
P.O. Box 250
Ithaca NY 14851

T2.6 (8)	F2.5 (6)	T4.13 (74)	F4.34 (55)
F4.35 (55)			

Goodheart-Willcox
123 W. Taft Drive
Holland IL 60473

F5.14 (19)	F5.40 (55)

Hanser Verlag
Munich Germany

T4.22 (102)	T5.14 (104)	F5.6 (10)	F5.15 (21)
F5.21 (30)	F5.22 (31)	F5.45 (61)	F5.46 (67)
F5.47 (68)	F5.48 (69)	T6.1 (7)	F6.34 (47)

ILIFFE Books Ltd.
42 Russell Square
London W.C. 1, England

F5.16 (23)	F5.17 (25)

Industrial Press, Inc.
200 Madison Ave.
New York NY 10157

F3.17 (9)	F4.11 (11)	T5.8 (7)	T5.9 (26)
F5.59 (88)	F5.60 (89)	F6.14 (29)	

Journal of Applied Polymer Science
John Wiley & Sons, Inc.
605 Third Ave.
New York NY 10158

F2.59 (90)	F3.49 (25)	F3.53 (25)	F3.54 (29)
F5.54 (80)	F6.10 (25)	F6.133 (119)	F6.134 (119)
F6.135 (119)			

Journal of Polymer Science
John Wiley & Sons, Inc.
605 Third Ave.
New York NY 10158

F2.22 (38)	F2.30 (49)	F2.35 (55)	F2.36 (56)
F2.48 (80)	F2.55 (86)	F2.56 (87)	F2.57 (88)
F2.63 (94)	F6.64 (72)		

Marcel Dekker, Inc.
270 Madison Ave.
New York NY 10016

T2.20 (96)	F2.1 (1)	F2.70 (112)	F3.63 (41)
F5.41 (56)	F6.168 (173)	F6.169 (173)	

McGraw-Hill Book Co.
1221 Ave. of the Americas
New York NY 10020

F1.4 (18)	F2.67 (105)	F3.18 (10)	T4.12 (72)
F4.36 (59)	F4.37 (59)	F5.30 (38)	F5.31 (39)
F5.33 (47)	T6.15 (111)	T6.16 (112)	T6.26 (172)
T6.28 (191)	F6.9 (24)	F6.57 (63)	F6.58 (66)
F6.69 (81)	F6.70 (79)	F6.75 (84)	F6.103 (101)
F6.104 (102)	F6.111 (103)	F6.121 (108)	F6.123 (109)
F6.124 (110)	F6.125 (113)	F6.126 (113)	F6.127 (113)
F6.129 (114)	F6.130 (114)	F6.170 (174)	F6.171 (174)
F6.176 (178)	F6.177 (179)	F6.178 (180)	F6.182 (183)
F6.184 (184)	F6.185 (185)	F6.189 (193)	F6.190 (193)
F6.195 (99)	F6.196 (99)		

Chapman & Hall, (Routledge), Inc.
29 West 35th St.
New York NY 10001

F4.53 (94)	F4.57 (99)	F4.58 (99)

Pergamon Press, Inc.
Maxwell House, Fairview
Elmsford NY 10523

F5.58 (87)

Plastics Design Forum
P.O. Box 448
446 Southern Blvd.
Chatham NY 07928

F3.32 (15)	F6.13 (28)	F6.28 (39)	F6.29 (39)
F6.86 (75)			

Plastics Technology
Bill Communications
633 Third Ave.
New York NY 10017

F6.191 (193)	F6.192 (193)	F6.193 (193)	F6.194 (192)

Plenum Publishing Corp.
233 Spring St.
New York NY 10013

T2.9 (61)	T2.11 (66)	F2.20 (35)	F2.26 (44)
F2.27 (45)	F2.28 (46)		

Prentice-Hall Book Co.
Route 9-W
Englewood Cliffs NJ 07632

T2.12 (68)	F2.23 (40)	F2.29 (48)

Van Nostrand Reinhold Company, Inc.
115 Fifth Ave.
New York NY 10003

T2.19 (95)	T2.22 (96)	T2.23 (107)	T2.24 (108)
T2.25 (109)	T2.26 (110)	T2.27 (111)	F2.64 (100)
F2.65 (101)	F2.66 (103)	F2.71 (113)	F2.72 (115)
F2.73 (116)	F3.16 (8)	T4.17 (90)	T4.19 (92)
T4.20 (96)	T4.21 (101)	T4.23 (103)	F4.41 (66)
F4.43 (70)	F4.45 (82)	F4.47 (84)	F4.48 (85)
F4.49 (87)	F5.20 (29)	F5.27 (35)	F5.28 (36)
F5.37 (50)	F5.44 (60)	F5.51 (76)	T6.19 (137)
T6.25 (157)	T6.27 (187)	T6.33 (201)	F6.7 (23)
F6.8 (23)	F6.11 (26)	F6.136 (123)	F6.162 (152)
F6.164 (160)	F6.170 (174)	F6.179 (48)	F6.180 (182)
F6.181 (183)	F6.183 (184)	F6.186 (156)	F6.187 (156)
F6.188 (183)	F6.194 (192)		

VDI Verlag
Düsseldorf, Germany

F6.49 (59)   F6.50 (60)

John Wiley & Sons, Inc.
(Including Interscience)
605 Third Ave.
New York NY 10158

T2.4 (4)	T2.5 (7)	T2.10 (65)	F2.2 (2)
F2.7 (12)	F2.8 (22)	F2.11 (25)	F2.17 (32)
F2.19 (34)	F2.32 (52)	F2.41 (64)	F2.43 (72)
F2.68 (106)	T3.2 (14)	T3.5 (49)	F3.33 (18)
F3.46 (21)	F3.47 (22)	F3.48 (24)	F3.50 (26)
F3.51 (27)	F3.52 (28)	F3.58 (35)	F3.59 (36)
F3.60 (37)	F3.61 (38)	F3.62 (39)	F3.65 (44)
F3.66 (47)	F3.67 (50)	F3.68 (53)	F3.69 (54)
F3.75 (59)	F3.76 (60)	F3.77 (61)	F3.80 (64)
F3.81 (65)	F3.82 (66)	F3.83 (67)	F3.84 (2)
T4.3 (35)	T4.4 (36)	T4.8 (63)	T4.11 (67)
T4.14 (79)	T4.24 (104)	F4.17 (26)	F4.18 (27)
F4.19 (28)	F4.26 (41)	F4.27 (42)	F5.29 (37)
F5.39 (54)	F5.50 (74)	T6.10 (67)	T6.11 (74)
T6.17 (115)	T6.20 (132)	F6.121 (108)	F6.123 (109)
F6.159 (149)			

PROFESSIONAL SOCIETY
PUBLICATIONS

American Society of Mechanical Engineers (ASME)
345 East 47th St.
New York NY 10017

F3.20 (12)   F6.23 (35)

American Society for Testing and Materials (ASTM)
1916 Race St.
Philadelphia PA 19103

F3.57 (30)

Plastics and Rubber Institute (PRI)
(Including Godwin/PRI)
11 Hobart Place
London SW1W 0HL England

F3.72 (57)	F6.22 (34)	F6.65 (73)	F6.66 (77)
F6.64 (72)	F6.100 (97)	F6.142 (125)	

Society of Plastics Engineers, Inc. (SPE)
14 Fairfield Drive
Brookfield Center CT 06804

T2.8 (51)	F2.31 (50)	F2.62 (93)	F3.19 (11)
F3.74 (58)	T5.13 (97)	F5.3 (6)	F5.4 (8)
F5.5 (9)	F5.9 (13)	F5.10 (15)	F5.32 (46)
F5.34 (48)	F5.35 (48)	F5.36 (49)	F5.38 (53)
T6.18 (117)	F6.6 (22)	F6.33 (45)	F6.35 (49)
F6.40 (49)	F6.41 (49)	F6.42 (49)	F6.43 (49)
F6.44 (49)	F6.50 (60)	F6.52 (61)	F6.53 (61)
F6.54 (61)	F6.69 (81)	F6.70 (81)	F6.71 (79)
F6.72 (79)	F6.89 (93)	F6.90 (93)	F6.91 (93)
F6.92 (93)	F6.93 (93)	F6.142 (125)	F6.143 (125)
F6.144 (125)	F6.145 (125)	F6.163 (159)	

CORPORATIONS

Allied Chemical Corporation
Morristown NJ 07960

F6.109 (—)

Amoco Performance Products
Ridgefield CT 06877

F6.105 (101)

Boonton Plastic Molding Co.
Boonton NJ 07005

T1.7 (23)

Ciba-Geigy Composite Materials
Anaheim CA 92807

F5.49 (74)

Cincinnati Milacron
Cincinnati, OH 45103

F5.19 (28)

Dupont Chemicals Corp.
Wilmington DE 19898

F6.188 (—)

Freedonia Group Co.
Cleveland OH 43201

T1.1 (13)

General Research Corporation
Santa Barbara CA 93111

F6.84 (—)

HPM Corp.
Mt. Gilead OH 43338

F5.18 (27)

Marland Mold Co.
Address Unknown

F5.23 (32)

Mobay Chemical Co.
Pittsburgh PA 15205-9741

F6.112 (—)   F6.118 (—)

Perkin-Elmer Corporation
761 Main Ave.
Norwalk CT 06859

F2.14 (27)	F2.15 (27)	F2.18 (33)	F2.47 (79)
F2.51 (82)	F2.52 (83)	F2.53 (84)	F4.31 (48)
F4.30 (47)			

PWS Engineering
Boston MA 01002

F2.25 (43)

Rheometrics, Inc.
One Possumtown Rd.
Piscataway NJ 08854

F6.46 (56)	F6.47 (56)	F6.82 (—)	F6.83 (—)
F6.93 (—)	F6.95 (—)	F6.96 (—)	F6.97 (—)
F6.98 (—)	F6.99 (—)		

Howard W. Sams & Co.
4300 West 62nd St.
Indianapolis IN 46268

F5.7 (11)   F5.42 (57)   F5.43 (59)

Shell International Chemical Co.
Shell Centre
London SE1 7PC England

F6.63 (71)	F6.74 (83)	F6.75 (84)	F6.77 (80)
F6.78 (87)	F6.79 (88)	F6.80 (89)	

Union Carbide Corporation
Danbury CT 06817

F3.78 (62)	F5.53 (79)

OTHERWISE UNIDENTIFIED

T1.19 (26)	F1.1 (40)	F1.7 (21)	F1.14 (35)
F2.23 (40)	F2.42 (67)	F2.49 (81)	F2.58 (89)
F2.60 (92)	F3.64 (42)	F3.71 (56)	F3.79 (63)
F4.20 (29)	F4.21 (30)	F4.23 (32)	F4.25 (39)
F4.54 (95)	F4.55 (97)	F4.56 (98)	T5.11 (51)
F5.2 (5)	F5.8 (12)	F5.11 (16)	F5.12 (17)
F5.13 (18)	F5.25 (34)	T6.1 (7)	T6.29 (193)
T6.30 (193)	T6.31 (193)	T6.32 (193)	F6.4 (20)
F6.5 (21)	F6.24 (37)	F6.77 (86)	F6.94 (94)
F6.103 (82)	F6.107 (103)	F6.108 (103)	F6.110 (104)
F6.113 (103)	F6.114 (103)	F6.115 (103)	F6.117 (105)
F6.171 (174)	F6.173 (177)		

NAME INDEX

This index includes only author and editor names. Page numbers followed by a "t" denote table references. Page numbers followed by an "f" denote figure references.

POLYMER INDEX

This index includes only polymer names and acronyms. Page numbers followed by a "t" denote table references. Page numbers followed by an "f" denote figure references.

MONOMERS, ADDUCTS AND CHEMICALS INDEX

This index includes only monomer, adduct, and chemical names and acronyms. Page numbers followed by a "t" denote table references. Page numbers followed by an "f" denote figure references.

PROCESS INDEX

This index includes only polymer process terms and acronyms. Page numbers followed by a "t" denote table references. Page numbers followed by an "f" denote figure references.

SUBJECT INDEX

This index does not include specific polymers, monomers, chemicals, adducts, or processes. See individual indices for these. Page numbers followed by a "t" denote table references. Page numbers followed by an "f" denote figure references.